Index of Computer Output

Techniques	Page(s)
General	
Install Excel	10
Excel Worksheet	11
Importing Data	12
Statistical Procedures	13–14
Graphical	
Histogram	23–24
Pie chart	34
Bar chart	34
Line chart	39
Scatter diagram	43
Box plot*	76
Numerical descriptive statistics	
One-sample statistics	57
Covariance	82
Correlation	82
Probability	
Binomial	133
Poisson	140
Normal	167
Student t	176
Chi-squared	181
F	185
Random Variable	
Normal	167
Student t	176
Chi-squared	182
F	185
Inference about μ (σ known)	
Interval estimator*	238–239
Test statistic*	267–268
Probability of Type II error	286
Inference about μ (σ unknown)	
Test statistic and interval estimator	298–299
Inference about σ^2	
Test statistic and interval estimator	308
Inference about p	
Test statistic and interval estimator	314

Techniques	Page(s)
Inference about $\mu_1 - \mu_2$	
Test statistic (equal-var)	340–341
Test statistic (unequal-var)	337–338
Interval estimator (equal-var)	342
Interval estimator (unequal-var)	339
Inference about μ_D	
Test statistic	353
Interval estimator	354–355
Inference about σ_1^2/σ_2^2	
Test statistic	363
Interval estimator	366
Inference about $p_1 - p_2$	
Test statistic (Case 1)*	372
Test statistic (Case 2)*	374
Interval estimator*	372
Analysis of variance	
Independent samples	
Single-factor	416
Two-factor	442–443
Randomized blocks	429
Multiple comparison methods*	457–525
Bartlett's test*	456
Chi-squared tests	
Goodness-of-fit	469
Contingency table*	476–478
Test for normality*	487–488
Nonparametric techniques	
Wilcoxon rank sum test*	504
Sign test*	514
Wilcoxon signed rank sum test*	520
Kruskal-Wallis test*	526–527
Friedman test*	532
Linear regression	
Coefficients and tests	548–549
Multiple regression	592–593
Prediction intervals*	566 and 601
Test of correlation coefficient*	570
Spearman rank correlation*	572
Durbin-Watson statistic*	619

*Data Analysis Plus™ 3.0 add-in

Applied Statistics with Microsoft® Excel

Applied Statistics with Microsoft® Excel

Gerald Keller
Wilfrid Laurier University

DUXBURY
THOMSON LEARNING

Australia • Canada • Mexico • Singapore • Spain • United Kingdom • United States

To my parents, Hyman and Ruth Keller

Sponsoring Editor: *Carolyn Crockett*
Editorial Assistant: *Ann Day*
Marketing: *Tom Ziolkowski/Samantha Cabaluna*
Assistant Editor: *Seema Atwal*
Production Editor: *Tessa Avila*
Production Service: *Heckman & Pinette*
Manuscript Editor: *Margaret Pinette*

Permissions Editor: *Sue Ewing*
Interior Design: *Rita Naughton*
Cover Design: *Irene Morris*
Cover Photo: *Christopher Talbot Frank/Panoramic Images*
Print Buyer: *Jessica Reed*
Typesetting: *GTS Graphics*
Printing and Binding: *R. R. Donnelley/Crawfordsville*

COPYRIGHT © 2001 Wadsworth Group. Duxbury is an imprint of the Wadsworth Group, a division of Thomson Learning, Inc. Thomson Learning™ is a trademark used herein under license.

For more information about this or any other Duxbury product, contact:
DUXBURY
511 Forest Lodge Road
Pacific Grove, CA 93950 USA
www.duxbury.com
1-800-423-0563 (Thomson Learning Academic Resource Center)

ALL RIGHTS RESERVED. No part of this work covered by the copyright hereon may be reproduced or used in any form or by any means—graphic, electronic, or mechanical, including photocopying, recording, taping, Web distribution, or information storage and retrieval systems—without the prior written permission of the publisher.

For permission to use material from this work, contact us by
www.thomsonrights.com
fax: 1-800-730-2215
phone: 1-800-730-2214

Printed in the United States of America

10 9 8 7 6 5 4 3 2

Library of Congress Cataloging-in-Publication Data

Keller, Gerald
   Applied statistics with Microsoft Excel/Gerald Keller.–New ed.
     p.   cm.
   ISBN 0-534-37112-4
   1. Management—Statistical methods.   2. Economics—Statistical methods.
   3. Statistics—Computer programs.   4. Microsoft Excel for Windows   I. Title.

HD30.215.K45   2001   00-057171
658.4'033—dc21   CIP

Brief Contents

Chapter 1   What Is Statistics?   1
Chapter 2   Graphical Descriptive Techniques   15
Chapter 3   Numerical Descriptive Techniques for Interval Data   52
Chapter 4   Probability   89
Chapter 5   Random Variables and Discrete Probability Distributions   118
Chapter 6   Continuous Probability Distributions   146
Chapter 7   Sampling and Sampling Plans   188
Chapter 8   Sampling Distributions   199
Chapter 9   Introduction to Estimation   227
Chapter 10   Introduction to Hypothesis Testing   253
Chapter 11   Inference about a Single Population   293
Chapter 12   Inference about Two Populations   330
Chapter 13   Statistical Inference: Review of Chapters 11 and 12   388
Chapter 14   Analysis of Variance   405
Chapter 15   Chi-Squared Tests   464
Chapter 16   Nonparametric Statistical Techniques   496
Chapter 17   Simple Linear Regression and Correlation   542
Chapter 18   Multiple Regression   588
Chapter 19   Statistical Inference: Conclusion   636

Contents

1 WHAT IS STATISTICS? 1

    1.1   Introduction 2
    1.2   Key Statistical Concepts 6
    1.3   Statistics and the Computer 7
    1.4   World Wide Web and Learning Center 7
           APPENDIX 1.A: Introduction to Microsoft Excel 10

2 GRAPHICAL DESCRIPTIVE TECHNIQUES 15

    2.1   Introduction 16
    2.2   Types of Data 16
    2.3   Graphically Describing Interval Data: Frequency Distributions and Histograms 20
    2.4   Graphically Describing Nominal Data: Bar and Pie Charts 33
    2.5   Describing Time-Series Data: Line Charts 38
    2.6   Describing the Relationship between Two Interval Variables: Scatter Diagrams 42
    2.7   Summary 49

3 NUMERICAL DESCRIPTIVE TECHNIQUES FOR INTERVAL DATA 52

    3.1   Introduction 53
    3.2   Measures of Central Location 54
    3.3   Measures of Variability 60
    3.4   Other Measures of Shape (Optional) 70
    3.5   Measures of Relative Standing and Box Plots 71
    3.6   Measures of Linear Relationship 76
    3.7   General Guidelines for Exploring Data 84
    3.8   Summary 85

4 PROBABILITY   89

4.1 Introduction   90
4.2 Assigning Probability to Events   90
4.3 Joint, Marginal, and Conditional Probability   95
4.4 Probability Rules and Trees   103
4.5 Summary   113

CASE 4.1 Let's Make a Deal   116
CASE 4.2 To Bunt or Not to Bunt, That Is the Question   116

5 RANDOM VARIABLES AND DISCRETE PROBABILITY DISTRIBUTIONS   118

5.1 Introduction   119
5.2 Random Variables and Probability Distributions   119
5.3 Describing the Population/Probability Distribution   124
5.4 Binomial Distribution   128
5.5 Poisson Distribution   136
5.6 Summary   141

CASE 5.1 To Bunt or Not to Bunt, That Is the Question, Part II   145

6 CONTINUOUS PROBABILITY DISTRIBUTIONS   146

6.1 Introduction   147
6.2 Probability Density Functions   147
6.3 Normal Distribution   153
6.4 Other Continuous Distributions   170
6.5 Summary   187

7 SAMPLING AND SAMPLING PLANS   188

7.1 Introduction   189
7.2 Sampling   189
7.3 Sampling Plans   191
7.4 Errors Involved in Sampling   196
7.5 Summary   198

8 SAMPLING DISTRIBUTIONS 199

8.1 Introduction 200

8.2 Sampling Distribution of the Mean 200

8.3 Creating the Sampling Distribution by Computer Simulation (Optional) 212

8.4 Sampling Distribution of a Proportion 215

8.5 Sampling Distribution of the Difference between Two Means 220

8.6 From Here to Inference 223

8.7 Summary 224

9 INTRODUCTION TO ESTIMATION 227

9.1 Introduction 228

9.2 Concepts of Estimation 228

9.3 Estimating the Population Mean when the Population Standard Deviation Is Known 232

9.4 Selecting the Sample Size 245

9.5 Simulation Experiments (Optional) 247

9.6 Summary 250

10 INTRODUCTION TO HYPOTHESIS TESTING 253

10.1 Introduction 254

10.2 Concepts of Hypothesis Testing 255

10.3 Testing the Population Mean when the Population Standard Deviation Is Known 257

10.4 Calculating the Probability of a Type II Error 279

10.5 The Road Ahead 288

10.6 Summary 291

11 INFERENCE ABOUT A SINGLE POPULATION 293

11.1 Introduction 294

11.2 Inference about a Population Mean when the Standard Deviation Is Unknown 295

11.3 Inference about a Population Variance 305

11.4 Inference about a Population Proportion 311

11.5 Summary 323

    CASE 11.1 Pepsi's Exclusivity Agreement with a University 327

    CASE 11.2 Pepsi's Exclusivity Agreement with a University: The Coke Side of the Equation 328

    CASE 11.3 Number of Uninsured Motorists 328

12 INFERENCE ABOUT TWO POPULATIONS 330

12.1 Introduction 331

12.2 Inference about the Difference between Two Means: Independent Samples 332

12.3 Observational and Experimental Data 348

12.4 Inference about the Difference between Two Means: Matched Pairs Experiment 349

12.5 Inference about the Ratio of Two Variances 361

12.6 Inference about the Difference between Two Population Proportions 367

12.7 Summary 378

    CASE 12.1 Bonanza International 386

    CASE 12.2 Accounting Course Exemptions 387

13 STATISTICAL INFERENCE: REVIEW OF CHAPTERS 11 AND 12 388

13.1 Introduction 389

13.2 Guide to Identifying the Correct Technique: Chapters 11 and 12 389

    CASE 13.1 Quebec Separation: *Oui ou non*? 403

    CASE 13.2 Host Selling and Announcer Commercials 403

14 ANALYSIS OF VARIANCE 405

14.1 Introduction 406

14.2 Single-Factor (One-Way) Analysis of Variance: Independent Samples 407

14.3 Analysis of Variance Experimental Designs 423

14.4 Single-Factor Analysis of Variance: Randomized Blocks 425

14.5 Two-Factor Analysis of Variance: Independent Samples 434

14.6 Multiple Comparisons 449

14.7 Bartlett's Test   455

14.8 Summary   457

15 CHI-SQUARED TESTS   464

15.1 Introduction   465

15.2 Chi-Squared Goodness-of-Fit Test   465

15.3 Chi-Squared Test of a Contingency Table   472

15.4 Summary of Tests on Nominal Data   482

15.5 Chi-Squared Test for Normality   484

15.6 Summary   489

CASE 15.1 Predicting the Outcomes of Basketball, Baseball, Football, and Hockey Games from Intermediate Results   493

CASE 15.2 Can Exposure to a Code of Professional Ethics Help Make Managers More Ethical?   494

16 NONPARAMETRIC STATISTICAL TECHNIQUES   496

16.1 Introduction   497

16.2 Wilcoxon Rank Sum Test   499

16.3 Sign Test and Wilcoxon Signed Rank Sum Test   511

16.4 Kruskal-Wallis Test   524

16.5 Friedman Test   529

16.6 Summary   535

17 SIMPLE LINEAR REGRESSION AND CORRELATION   542

17.1 Introduction   543

17.2 Model   544

17.3 Estimating the Coefficients   546

17.4 Error Variable: Required Conditions   552

17.5 Assessing the Model   555

17.6 Using the Regression Equation   564

17.7 Coefficients of Correlation   568

17.8 Regression Diagnostics I   574

18 MULTIPLE REGRESSION   588

    17.9 Summary   580

            CASE 17.1 Predicting University Grades from High School Grades   585

            CASE 17.2 Insurance Compensation for Lost Revenues   586

18 MULTIPLE REGRESSION   588

    18.1 Introduction   589

    18.2 Model and Required Conditions   589

    18.3 Estimating the Coefficients and Assessing the Model   590

    18.4 Regression Diagnostics II   605

    18.5 Regression Diagnostics III (Time Series)   612

    18.6 Nominal Independent Variables   623

    18.7 Summary   630

            CASE 18.1 Quebec Referendum Vote: Was There Electoral Fraud?   634

            CASE 18.2 Quebec Referendum Vote: The Rebuttal   635

19 STATISTICAL INFERENCE: CONCLUSION   636

    19.1 Introduction   637

    19.2 Identifying the Correct Technique: Summary of Statistical Inference   637

            CASE 19.1 Do Banks Discriminate against Women Business Owners? I   644

            CASE 19.2 Do Banks Discriminate against Women Business Owners? II   647

    19.3 The Last Word   653

            CASE 19.3 Ambulance and Fire Department Response Interval Study   665

            CASE 19.4 *PC Magazine* Survey   666

            CASE 19.5 WLU Graduate Survey   667

            CASE 19.6 Evaluation of a New Antidepressant Drug   668

            CASE 19.7 Nutrition Education Programs   669

            CASE 19.8 Do Banks Discriminate against Women Business Owners? III   670

Appendix A Sample Statistics from Data Files in Chapters 9 and 10   A-1

Appendix B Tables   B-1

Appendix C Answers to Selected Even-Numbered Exercises   C-1

Index I-1

Preface

WHY I WROTE THIS BOOK

Until five years ago, virtually all applied statistics courses were taught with an emphasis on manual calculations. There were several reasons for this pedagogy. First, many instructors felt that students needed to be taught the mathematical principles of statistics, and the best way, it was thought, was to have students perform manual calculations. Second, although real statistics practitioners use a computer, costs of both hardware and software made it difficult or impossible to equip all students with these resources.

The emphasis on manual calculations has resulted in a variety of problems. Foremost among these is that for many students statistics is one of their least favorite courses. By focusing on the arithmetic of statistical techniques, instructors gave the false impression that this is the most important aspect of the subject. Students did not get the opportunity to see statistics as it really is.

This book was written to allow instructors to teach a course in applied statistics that demonstrates the true value of statistics.

APPROACH

The easy access to computers today allows us to change the pedagogy. Most students have their own computers or can use computers supplied by their colleges or universities. Moreover, nearly all computers have Microsoft Excel® automatically installed. As a result, college and university students know how to use the popular spreadsheet.

Changing the pedagogy involves far more than simply adding computer output to the illustrative examples. Fully taking advantage of these resources requires a major overhaul in the way we teach statistics. The void created by removing manual calculations can be enormous. A large portion of an instructor's class time must be devoted to showing how to calculate each statistic if the students are expected to solve problems the same way. However, the void can easily be filled by teaching aspects of applied statistics that have generally been ignored by instructors and by textbook authors. Simply expressed, we need to teach what statistics practitioners do before the calculations and what they do after the calculations, while letting the computer do the calculations themselves.

Before the calculations can be performed, the statistics practitioner must decide which technique or techniques should be applied. Without the ability to select the appropriate method, all other skills are irrelevant.

Once the appropriate technique has been selected, all that's needed is guidance in using the software—nothing more complicated than point-and-click instructions.

After the computer has produced the required statistics, the practitioner must interpret them. This requires an understanding of probability and statistical concepts.

FEATURES

1. **Three-step approach.** All inferential techniques are demonstrated using a three-step approach. The first step is to identify the technique. We believe that technique-identification skills are critical but seldom taught. These skills are taught throughout the book. The second step is to compute the statistics. The third step is to interpret the results, which requires an understanding of statistical concepts.

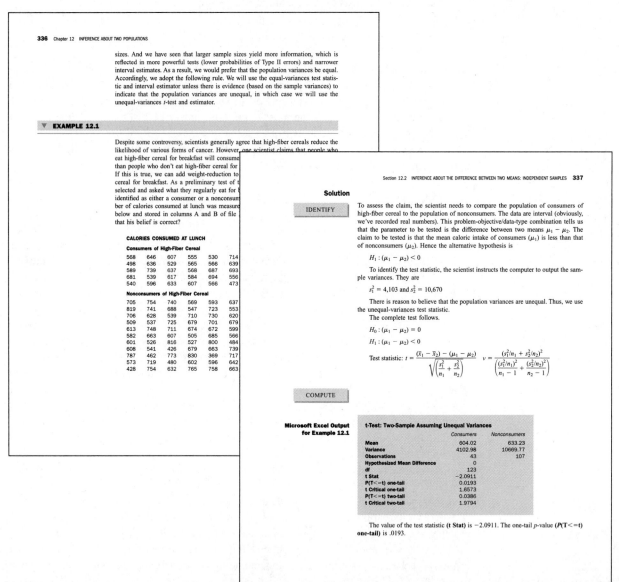

338 Chapter 12 INFERENCE ABOUT TWO POPULATIONS

Figure 12.2
Sampling distribution of the test statistic for Example 12.1.

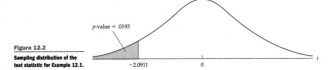

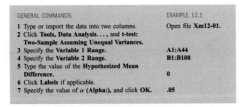

GENERAL COMMANDS	EXAMPLE 12.1
1 Type or import the data into two columns.	Open file **Xm12-01**.
2 Click **Tools, Data Analysis...**, and **t-test: Two-Sample Assuming Unequal Variances**.	
3 Specify the **Variable 1 Range**.	A1:A44
4 Specify the **Variable 2 Range**.	B1:B108
5 Type the value of the **Hypothesized Mean Difference**.	0
6 Click **Labels** if applicable.	
7 Specify the value of α (**Alpha:**), and click **OK**.	.05

Figure 12.2 depicts the Student t sampling distribution and the p-value of the test. To conduct this test from means and standard deviations or to perform a what-if analysis, activate the **t-test of 2 Means (Uneq-Var)** worksheet in **Stats-Summary.xls**.

INTERPRET

A Type I error in this example occurs when we conclude that consumers of high-fiber cereal consume fewer calories at lunch than do nonconsumers when in fact there is no difference between the two groups. A Type II error occurs when we erroneously fail to conclude that consumers of high-fiber cereal eat less at lunch than nonconsumers. It is difficult to judge which error is more costly. As a result we take a neutral position and suggest that any p-value less than 5% should be interpreted as providing enough evidence to reject the null hypothesis.

The p-value of the test is .0193. As a result, we conclude that there is sufficient evidence to infer that consumers of high-fiber [cereal consume fewer calories at lunch] than do nonconsumers. However, there are tw[o caveats to consider before con-]cluding that high-fiber cereals constitute an effe[ctive diet plan. The first caveat is that] the data were likely self-reported, which means [that the amount of high-fiber cereal a consumer ate and the num]ber of calories recorded that he or she consume[d are subjective. Addition]ally, a less subjective method of counting calori[es consumed and the conditions under] which the experiment was performed may lead t[o a different interpretation] of the data. We will discuss this important issu[e in later chapters.]

Section 12.2 INFERENCE ABOUT THE DIFFERENCE BETWEEN TWO MEANS: INDEPENDENT SAMPLES 339

In addition to testing to determine whether a difference exists, we can estimate the difference in mean caloric intake.

The interval estimator of the difference between two means with unequal population variances is

$$(\bar{x}_1 - \bar{x}_2) \pm t_{\alpha/2}\sqrt{\frac{s_1^2}{n_1} + \frac{s_2^2}{n_2}}$$

COMPUTE

The Excel output does not include the interval estimate. However, you can use the **t-Estimate of 2 Means (Uneq-Var)** worksheet of **Stats-Summary.xls**. Simply plug in the values of the sample means, sample standard deviations, and sample sizes, as well as the confidence level. The result of this action appears below.

Microsoft Excel Output of the 95% confidence Interval Estimate: Example 12.1

t-Estimate of the Difference between Two Means (Unequal-Variances)	
Sample 1	
Sample mean	604.02
Sample standard deviation	64.05
Sample size	43
Sample 2	
Sample mean	633.2
Sample standard deviation	103.3
Sample size	107
Confidence level	0.95
Degrees of freedom	122.60
Difference between means	−29.2
Bound	27.65
Lower confidence limit	−56.86
Upper confidence limit	−1.56

INTERPRET

We estimate that the average consumer of high-fiber cereal eats between 1.56 and 56.86 fewer calories at lunch than does the average nonconsumer of high-fiber cereal.

▼ **EXAMPLE 12.2**

The plant manager of a company that manufactures office equipment is attempting to determine the process that will be used to assemble a new ergonomic chair. The material, machines, and workforce have already been decided. However there are two methods under consideration. The methods differ by the order in which the separate operations are performed. To help decide which should be used, an experiment was performed. Twenty-five randomly selected workers each assembled the chair using method A, and another 25 workers each assembled the chair using method B. The assembly times in minutes were recorded and are exhibited below and stored in file Xm12-02. The plant manager would like to know whether the assembly times of the two methods differ.

2. Fewer manual calculations. All statistical procedures, ranging from graphical descriptive methods to multiple regression, will be conducted using a computer. We use a calculator to determine probability and to compute simple statistics.

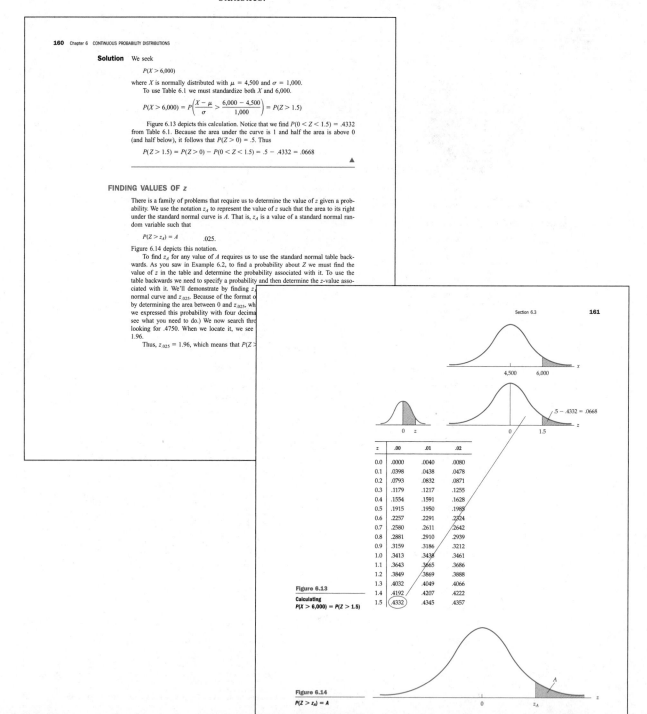

3. **Microsoft Excel.** We use the most popular spreadsheet program, showing the printouts and providing step-by-step instructions. We created macro add-ins to augment Excel's limited offering. The add-ins are stored on the CD, in Data Analysis Plus® 3.0, which, when installed, integrates the add-ins with Excel's statistical functions.

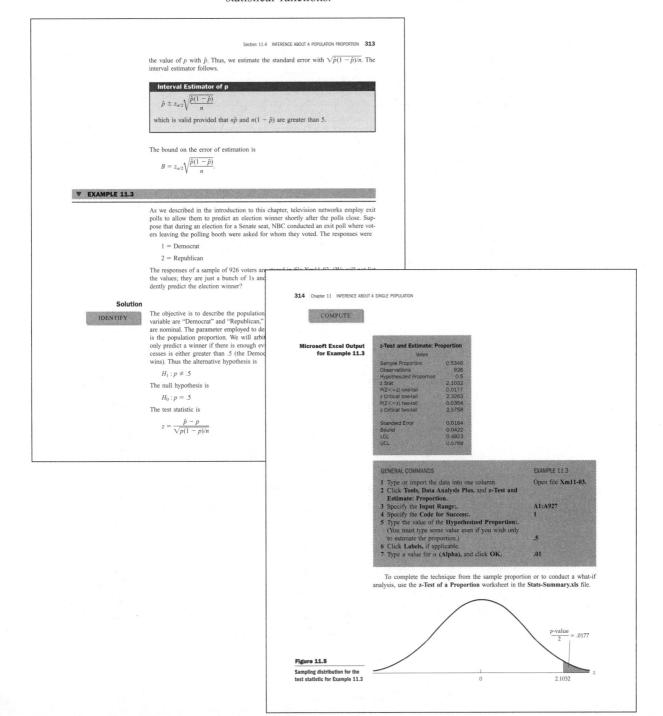

4. Emphasized applications. To motivate students and convince them of the importance of statistics, we emphasize applications. Many of the examples, exercises, and cases are adapted from real studies. Almost all feature a data set stored on the CD. There are over 600 data sets on the CD, some as large as 22,000 observations.

CASE 19.1 DO BANKS DISCRIMINATE AGAINST WOMEN BUSINESS OWNERS? I*

Increasingly, more women are becoming owners of small businesses. However, questions about how they are treated by banks and other financial institutions have been raised by women's groups. Banks are particularly important to small businesses, because studies show that bank financing represents about one-quarter of total debt and that for medium-sized businesses the proportion rises to approximately one-half. If women's requests for loans are rejected more frequently than are men's requests, or if women must pay higher interest charges than men do, women have cause for complaint. Banks might then be subject to criminal as well as civil suits. To examine this issue, a research project was launched.

The researchers surveyed a total of 1,165 business owners, of whom 115 were women. The percentage of women in the sample, 9.9%, compares favorably with other sources that indicate that women own about 10% of established small businesses. The survey asked a series of questions to men and women business owners who applied for loans during the previous month. It also determined the nature of the business, its size, and its age. Additionally, the owners were asked about their experiences in dealing with banks. The questions asked in the survey included the following:

1 What is the gender of the owner?
  1 female
  2 male

2 Was the loan approved?
  1 no
  2 yes

3 If it was approved, what interest rate (po

Of the 115 women who asked for a loan, men who asked for a loan were rejected. The rat granted were recorded. These data are stored in B (rates paid by men) in file C19-01.

What do these data disclose about possible

Solution

The problem objective is to compare two pop women and those owned by men. We can com loan applications are denied; and for loans gran

---

*Adapted from A. L. Riding and C. S. Swift, "Giving Cr and Canadian Financial Institutions," Carleton University W

19.3 The widespread use of salt on roads in Canada and the northern United States during the winter and acid precipitation throughout the year combine to cause rust on cars. Car manufacturers and other companies offer rustproofing services to help purchasers preserve the value of their cars. A consumer protection agency decides to determine whether there are any differences between the rust protection provided by automobile manufacturers and that provided by two competing types of rustproofing services. As an experiment, 60 identical new cars are selected. Of these, 20 are rustproofed by the manufacturer. Another 20 are rustproofed using a method that applies a liquid to critical areas of the car. The liquid hardens, forming a (supposedly) lifetime bond with the metal. The last 20 are treated with oil and are re-treated every 12 months. The cars are then driven under similar conditions in a Minnesota city. The number of months until the first rust appears is recorded and stored in columns A, B, and C, respectively, in file Xr19-03. Is there sufficient evidence to conclude that at least one rustproofing method is different from the others?

19.4 In the door-to-door selling of vacuum cleaners, various factors influence sales. The Birk Vacuum Cleaner Company considers its sales pitch and overall package to be extremely important. As a result, it often thinks of new ways to sell its product. Because the company's management dreams up so many new sales pitches each year, there is a two-stage testing process. In stage 1, a new plan is tested with a relatively small sample. If there is sufficient evidence that the plan increases sales, a second, considerably larger, test is undertaken. The statistical test is performed so that there is only a 1% chance of concluding that the new pitch is successful in increasing sales when it actually does not increase sales. In a stage 1 test to determine if the inclusion of a "free" ten-year service contract increases sales, 100 sales representatives were selected at random from the company's list of several thousand. The monthly sales of these representatives were recorded for one month prior to use of the new sales pitch and for one month after its introduction. The results are stored in file Xr19-04. Should the company proceed to stage 2?

19.5 Two drugs are used to treat heart attack victims. Streptokinase, which has been available since 1959, costs about $500. The second drug is t-PA, a genetically engineered product that sells for about $3,000. Both streptokinase and t-PA work by opening the arteries and dissolving blood clots, which are the cause of heart attacks. Several previous studies have failed to reveal any differences between the effects of the two drugs. Consequently, in many countries where health care is funded by governments, physicians are required to use the less expensive streptokinase. However, t-PA's maker, Genentech, Inc., contended that in the earlier studies showing no difference between the two drugs, their drug was not used in the right way. Genentech decided to sponsor a more thorough experiment. The experiment was organized in 15 countries, including the United States and Canada, and involved a total of 41,000 patients. In this study, t-PA was given to patients in 90 minutes instead of 3 hours as in previous trials. Half of the sample of 41,000 patients were treated by a rapid injection of t-PA with intravenous heparin, while the other half received streptokinase along with heparin. The number of deaths in each sample was recorded. A total of 1,497 patients treated with streptokinase died, while 1,292 patients who received t-PA died.

a Can we infer that t-PA is better than streptokinase in preventing deaths?
b Estimate with 95% confidence the cost per life saved by using t-PA.

19.6 A small but important part of a university library's budget is the amount collected in fines on overdue books. Last year, a library collected $75,652.75 in fine payments; however, the head librarian suspects that some employees are not bothering to collect the fines on overdue books. In an effort to learn more about the situation, she asked a sample of 400 students (out of a total student population of 50,000) how many books they had returned late to the library in the previous 12 months. They were also asked how many days over-due the books had been. The results indicated that the total number of days overdue ranged from 0 to 55 days. The number of days overdue was stored in file Xr19-06.

a Estimate with 95% confidence the average number of days overdue for all 50,000 students at the university.
b If the fine is 25 cents per day, estimate the amount that should be collected annually. Should the librarian conclude that not all the fines were collected?

19.7 The practice of therapeutic touch is used in hospitals all over the world and is taught in some medical and nursing schools. In this therapy, trained practitioners manipulate something that they call the "human energy field." The manipulation is carried out without actually touching the patient's body. Practitioners

5. **Review chapters with flowcharts.** To ensure that students can identify the correct technique, we provide two review chapters, Chapters 13 and 19, each with a flowchart that helps students choose the appropriate method. The exercises and cases in each review chapter require all the statistical techniques introduced up to that point.

13.1 INTRODUCTION

This chapter is more than just a review of the previous two chapters. It is a critical part of your development as a statistics practitioner. When you solved problems at the end of each section in the preceding chapters (you *have* been solving problems at the end of each section covered, haven't you?), you probably had no great difficulty identifying the correct technique to use. You used the statistical technique introduced in that section. While those exercises provided practice in setting up hypotheses, producing computer output of tests of hypothesis and interval estimators, and interpreting the results, you did not address a fundamental question faced by statistics practitioners: which technique to use. If you still do not appreciate the dimension of this problem, consider the following, which lists all the inferential methods covered thus far.

t-test and estimator of μ

χ^2-test and estimator of σ^2

z-test and estimator of p

t-test and estimator of $\mu_1 - \mu_2$ (equal variances formulas)

t-test and estimator of $\mu_1 - \mu_2$ (unequal variances formulas)

t-test and estimator of μ_D

F-test and estimator of σ_1^2/σ_2^2

z-test (cases 1 and 2) and estimator of $p_1 - p_2$

Counting tests and interval estimators of a parameter as two different techniques, a total of 17 statistical procedures have been presented thus far, and there is much left to be done. Faced with statistical problems niques (such as in real-world applications some assistance in identifying the appropri in greater detail how to make this decision. the opportunity to practice your decision skil cumulatively require all of the inferential tec Solving these problems will require you to d must analyze the problem, identify the techn ware and a computer to yield the required s

13.2 GUIDE TO IDENTIFYING THE CORRECT CHAPTERS 11 AND 12

As you've probably already discovered, the ing the correct statistical technique are the some situations, once these have been recog In other cases, however, several additional proceed. For example, when the problem ob the data are interval, three other significant measurement (central location or variability drawn, and, if so, whether the unknown po

The flowchart in Figure 13.1 represents the logical process that leads to the identification of the appropriate method. We've also included a more detailed guide (Table 13.1) to the statistical techniques that lists the formulas of the test statistics, the interval estimators, and the required conditions.

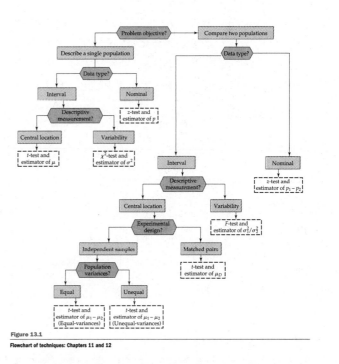

Figure 13.1

Flowchart of techniques: Chapters 11 and 12

6. **Developed concepts.** In the absence of mathematical derivations, it is necessary to emphasize the mathematical principles underlying all inferential methods. We do this in a variety of ways, including simulations. We describe how to simulate sampling distributions (Chapter 8) and interval estimators (Chapter 9). A spreadsheet—also stored on the CD—allows students to conduct "what-if" analyses (Chapters 9, 10, 11, and 12), enabling them to discover how statistical inference actually works.

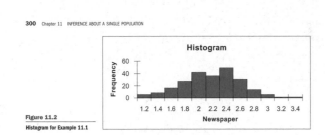

MICROSOFT EXCEL

Microsoft Excel is a spreadsheet package that performs *some* statistical analyses. Nevertheless, we chose Excel instead of a full-feature statistical package like SAS or SPSS, or a student-friendly package like Minitab, because of the problem of computer accessibility.

The high cost of SAS, SPSS, and other software requires universities to provide the statistical program on their computer systems. However, almost all universities' computer facilities are overcrowded at the best of times and heavily congested when assignments for required courses are due. This circumstance makes the frequent use of a campus computer difficult and likely to be frustrating for students. Consequently, a virtual prerequisite to using computers in a statistics course is for students to own their own computers and software.

Fortunately, an increasing proportion of students have their own computers. Moreover, many students own a version of Excel and know how to use it. Thus, Microsoft Excel was chosen as the only viable contender. Once chosen, it became one of our tasks to ensure that all the statistical techniques presented in this book could be performed with Excel. The creation of Data Analysis Plus® (a series of add-ins, now in version 3.0) fulfills this goal.

TEACHING AND LEARNING AIDS

Instructor's Solutions Manual
This manual supplies complete solutions for every exercise in the book (ISBN: 0-534-38204-5).
Student Solutions Manual
This manual, for sale to students, provides the solutions for the textbook's even-numbered exercises (ISBN: 0-534-98202-9).

ACKNOWLEDGMENTS

I would like to express my appreciation to Carolyn Crockett, the book's editor, whose advice and encouragement improved both the quality and speed of the project. Margaret Pinette and Tessa Avila did the production work, and Seema Atwal assisted with the supplementary material. I also want to thank Jeffery Keller and Brendan Paul, who produced the latest version of Data Analysis Plus®.

The following reviewers contributed helpful feedback on the manuscript: Michael Baron, The University of Texas at Dallas; John Dutton, North Carolina State University; James Guffey, Truman State University; Susan Jenkins, Southeastern Louisiana University; Robert Lee, James Madison University; and Carolyn Meitler, Concordia University Wisconsin.

Gerald Keller

KEY TO APPLICATION ICONS

 Calculations

 Definitions

 Everyday life

 Education

 Business/Economics

 Medicine/Health

 Agriculture

 Environment

 Internet

 Television

 Public opinion

 Politics

 Sports/Recreation

 Engineering/Technology

 Science

 Gambling

Chapter 1

What Is Statistics?

1.1 Introduction
1.2 Key Statistical Concepts
1.3 Statistics and the Computer
1.4 World Wide Web and Learning Center

1.1 INTRODUCTION

Statistics is a way to get information from data. That's it! Most of this textbook is devoted to describing how, when, and why statistics practitioners* conduct statistical procedures. You may ask, "If that's all there is to statistics, why is this book (and most other statistics books) so large?" The answer is that there are different kinds of information and data to which students of applied statistics should be exposed. We demonstrate some of these with three cases that are featured later in this book. (The numbering used here for these cases is the same as that used when the cases are featured later in this book. For example, Case 11.1 appears in Chapter 11.)

CASE 11.1:
PEPSI'S EXCLUSIVITY AGREEMENT WITH A UNIVERSITY

In the last few years, colleges and universities have signed exclusivity agreements with a variety of private companies. These agreements bind the university to sell that company's products exclusively on the campus. Many of the agreements involve food and beverage firms.

A large university with a total enrollment of about 50,000 students has offered Pepsi-Cola an exclusivity agreement that would give Pepsi exclusive rights to sell its products at all university facilities for the next year and an option for future years. In return the university would receive 35% of the on-campus revenues and an additional lump sum of $200,000 per year. Pepsi has been given two weeks to respond.

The management at Pepsi quickly review what they know. The market for soft drinks is measured in terms of 10-ounce cans. Pepsi currently sells an average of 22,000 cans per week (over the 40 weeks of the year that the university operates). The cans sell for an average of 75 cents each. The costs including labor amount to 20 cents per can. Pepsi is unsure of its market share but suspects it is considerably less than 50%. A quick analysis reveals that if its current market share were 25%, then, with an exclusivity agreement, Pepsi would sell 88,000 cans per week or 3,520,000 cans per year. The gross revenue would be computed as follows.

$$3,520,000 \times \$.75/can = \$2,640,000$$

---

*The term *statistician* is used to describe so many different kinds of occupations that it has ceased to have any meaning. It is used, for example, to describe both a person who calculates baseball statistics and a mathematician educated in statistical principles. We will describe the former as a *statistics practitioner* and the latter as a statistician. A statistics practitioner is a person who uses statistical techniques properly. Examples of statistics practitioners include the following:

1 A manager who converts data into information to help make decisions.
2 An economist who uses statistical models to help explain and predict.
3 A psychologist who gathers data and applies statistical methods to extract information.

Our goal in this book is to convert you into one such fortunate individual.

The term *statistician* refers to an individual who works with the mathematics of statistics. His or her work involves research that develops techniques and concepts that in the future may help the statistics practitioner. Statisticians are also statistics practitioners, frequently conducting empirical research and consulting. The author of this book is a statistician. If you're taking a statistics course, your instructor is probably a statistician.

This figure must be multiplied by 65%, because the university would rake in 35% of the gross. Thus,

65% × $2,640,000 = $1,716,000

The total cost of 20 cents per can (or $704,000) and the annual payment to the university of $200,000 is subtracted to obtain the net profit:

Net profit = $1,716,000 − $704,000 − $200,000 = $812,000

Pepsi's current annual profit is

40 weeks × 22,000 cans/week × $.55 = $484,000

If the current market share is 25%, the potential gain from the agreement is

$812,000 − $484,000 = $328,000

The only problem with this analysis is that Pepsi does not know how many soft drinks are sold weekly at the university. Coke is not likely to supply Pepsi with information about the sales of its brands, which together with Pepsi's line of products constitute virtually the entire market.

A recent university graduate volunteers that a survey of the university's students can supply the missing information. Accordingly, she organizes a survey that asks 500 students to keep track of the number of soft drinks they purchase in the next seven days. The responses are stored in a file on the disk that accompanies this book.

The information we would like to acquire is an estimate of annual profits from the exclusivity agreement. The data are the numbers of cans of soft drinks consumed by the 500 students in the sample. As a first step, we need to extract information from the sample. This is the function of **descriptive statistics.**

Descriptive statistics deals with methods of organizing, summarizing, and presenting data in a convenient and informative way. One form of descriptive statistics uses graphical techniques, which allow us to draw a picture that presents the data in such a way that we can easily see what numbers the students are reporting. Chapter 2 presents a variety of graphical methods used by statistics practitioners to present data in ways that allow the reader to extract useful information.

Another form of descriptive statistics uses numerical techniques to summarize data. One such method that you have already used frequently is the average or mean. In the same way that you calculate the average age of the employees of a company, we can compute the mean number of soft drinks consumed by the 500 students in our survey. Chapter 3 introduces several numerical statistical measures that describe different features of the data. In Case 11.1, however, we are not so much interested in what the 500 students are reporting as we are in knowing the mean number of soft drinks consumed by all 50,000 students on campus. To accomplish this goal we need another branch of statistics—**inferential statistics.**

Inferential statistics is a body of methods used to draw conclusions or inferences about characteristics of populations based on sample data. The population in question in Case 11.1 is the university's 50,000 students' soft drink consumption. The cost of interviewing each would be prohibitive and extremely time consuming. Statistical techniques make such endeavors unnecessary. Instead, we can sample a much smaller number of students (in this case the sample size is 500) and infer from the data the number of soft drinks consumed by all 50,000 students. We can then estimate annual profits for Pepsi.

CASE 13.1: QUEBEC SEPARATION—*OUI OU NON?*

Since the 1960s there has been an ongoing campaign among Quebecers to separate from Canada and form an independent nation. Should Quebec separate, the ramifications for the rest of Canada and the United States would be enormous. In 1994, the ruling party in Quebec promised to hold a referendum on separation. As with most political issues, polling plays an important role in trying to influence voters and to predict the outcome of the referendum vote. Shortly after the 1993 federal election, *The Financial Post Magazine,* in cooperation with several polling companies, conducted a survey of Quebecers.

A total of 641 adult Quebecers were interviewed. They were asked the following question. (Francophones were asked the question in French.) The pollsters recorded the answer and also the language (English or French) in which the respondent answered.

> If a referendum were held today on Quebec's sovereignty with the following question, "Do you want Quebec to separate from Canada and become an independent country?" would you vote yes or no?

The responses are stored on the data disk accompanying this book. What conclusions can you draw from these results?

This case exemplifies a very common application of statistical inference. The population we want to make inferences about is the approximately 7 million potential voters in the province of Quebec. The sample consists of the 641 Quebecers randomly selected by the polling company. The characteristic of the population that we would like to know is the proportion of the total electorate that supports separation. Specifically, we would like to know whether more than 50% of voters will vote for separation. It must be made clear that, because we will not ask every one of the 7 million potential voters how he or she will vote, we cannot predict the outcome with 100% certainty. This is a fact that statistics practitioners and even students of statistics must understand. A sample that is only a small fraction of the size of the population can lead to correct inferences only a certain percentage of the time. You will find that statistics practitioners can control that fraction and usually set it between 90 and 99%.

CASE 13.2: HOST SELLING AND ANNOUNCER COMMERCIALS[*]

A study was undertaken to compare the effects of host selling commercials and announcer commercials on children. Announcer commercials are straightforward commercials in which the announcer describes to viewers why they should buy a particular product. Host selling commercials feature a children's show personality or television character who extols the virtues of the product. In 1975, the National

---

[*]Adapted from J. H. Miller, "An Empirical Evaluation of the Host Selling Commercial and the Announcer Commercial When Used on Children." *Developments in Marketing Science* 9 (1985): 276–78.

Association of Broadcasters prohibited the use of show characters to advertise products during the same program in which the characters appear. However, this prohibition was overturned in 1982 by a judge's decree.

The objective of the study was to determine whether the two types of advertisements have different effects on children watching them. Specifically, the researchers wanted to know whether children watching host selling commercials would remember more details about the commercial and be more likely to buy the advertised product than children watching announcer commercials. The experiment consisted of two groups of children ranging in age from 6 to 10. One group of 121 children watched a program in which two host selling commercials appeared. The commercials tried to sell Canary Crunch, a breakfast cereal. A second group of 121 children watched the same program but was exposed to two announcer commercials for the same product. Immediately after the show, the children were given a questionnaire that tested their memory concerning the commercials they had watched.

Each child was rated (on a scale of 10) on his or her ability to remember details of the commercial. In addition, each child was offered a free box of cereal. The children were shown four different brands of cereal—Froot Loops (FL), Boo Berries (BE), Kangaroo Hops (KH), and Canary Crunch (CC; the advertised cereal)—and asked to pick the one they wanted. The results are stored on the data disk provided with this book. (Some of the data are shown in Table 1.1.) Are there differences in memory test scores and cereal choices between the two groups of children?

In this case, we want to compare the population of children who watch host commercials with the population of children who watch announcer commercials. The experiment consists of drawing samples of 121 children from each population. For each child, researchers recorded two observations. The first was the score out of 10 the child received on a test to measure his or her memory about the commercial. The second was the brand the child chose from among the four brands of breakfast cereals. Notice that, contrary to what you probably believed, data are not necessarily numbers. The test scores, of course, are numbers; however, the cereal choices are not. In Chapter 2, we will discuss the different types of data you will encounter in statistical applications and how to deal with them. The information sought by the researchers is whether there are differences in the test scores and the cereal selections between the two populations of children. By applying the appropriate statistical techniques, the researchers may be able to infer which type of commercial is more effective.

Table 1.1 Memory Test Scores and Cereal Choices

Children Who Watched Host Selling Commercials		Children Who Watched Announcer Commercials	
Memory Test Scores	Cereal Choices	Memory Test Scores	Cereal Choices
6	FL	8	BB
9	CC	6	FL
7	KH	10	CC
7	CC	8	FL
.	.	.	.
.	.	.	.
.	.	.	.
8	BB	9	CC

1.2 KEY STATISTICAL CONCEPTS

As the preceding cases illustrate, statistical inference problems involve three key concepts: the population, the sample, and the statistical inference. We now discuss each of these concepts in more detail.

POPULATION

A **population** is the group of all items of interest to a statistics practitioner. It is frequently very large and may, in fact, be infinitely large. In the language of statistics *population* does not refer to a group of people. It may, for example, refer to the population of diameters of ball bearings produced at a large plant. Even in situations involving a set of people, the term *population* will refer to the population of some characteristic of the people. For example, in the Pepsi-Cola case, the population of interest is the population of the number of soft drinks to be consumed by students. Thus, our population consists not of 50,000 students but of the 50,000 numbers of soft drinks to be consumed. Just imagine a giant container holding 50,000 slips of paper, each marked with the number of soft drinks to be drunk by one particular student. This container is our population of interest.

A descriptive measure of a population is called a **parameter.** The parameter of interest in Case 11.1 is the mean number of soft drinks consumed by all the students at the university. The parameter in the Quebec case (Case 13.1) is the proportion of all of Quebec's voters who will vote for separation. In the host selling commercial case (Case 13.2), the parameters are the differences in the mean scores on the memory test and the differences in the proportions of breakfast cereals selected by each population.

SAMPLE

A **sample** is a set of data drawn from the population. A descriptive measure of a sample is called a **statistic.** We use statistics to make inferences about parameters. In the Pepsi-Cola case, the statistic we would compute is the mean number of soft drinks consumed in the last week by the 500 students in the sample. We would then use the sample mean to infer the value of the population mean, which is the parameter of interest in this problem. In the Quebec separation case, we compute the proportion of the sample of 641 Quebecers who will vote for separation. The sample statistic is then used to make inferences about the population of Quebec's voters. That is, we predict the result of the referendum.

STATISTICAL INFERENCE

Statistical inference is the process of making an estimate, prediction, or decision about a population based on sample data. Because populations are almost always very large, investigating each member of the population would be impractical and expensive. It is far easier and cheaper to take a sample from the population of interest and draw conclusions or make estimates about the population on the basis of information provided by the sample. However, such conclusions and estimates are not always going to be correct. For this reason, we build into the statistical inference a measure of reliability. There are two such measures, **the confidence level** and the **significance level.** The confidence level is the proportion of times that an estimating procedure will be correct. For example, in the Pepsi-Cola case, we will produce an estimate of the average number of soft drinks to be consumed by all 50,000 students that has a confidence level of 95%. That means that, in the long run, estimates based on this form of statistical inference will be correct 95% of the time. When the purpose of the statistical inference is to draw a conclusion about a population, the significance level measures how frequently the conclusion will be wrong. For example, suppose that as a result of the analysis in the Quebec separation case we conclude that less than 50% of the electorate will vote yes on the referendum, and thus Quebec will not vote to secede. A 5% significance level means that in the long run this type of conclusion will be wrong 5% of the time.

1.3 STATISTICS AND THE COMPUTER

In virtually all applications of statistics, the statistics practitioner must deal with large amounts of data. For example, Case 11.1 involves 500 observations. To estimate annual profits, the statistics practitioner would have to perform computations on the data; although the calculations do not require any great mathematical skill, the sheer amount of arithmetic makes this aspect of the statistical method time consuming and tedious.

Fortunately, numerous commercially prepared computer programs are available to perform the arithmetic. We have chosen to use Microsoft Excel, which is a spreadsheet program. We chose Excel because we believe that it is and will continue to be the most popular spreadsheet package. One of its drawbacks is that it offers relatively few of the statistical techniques we introduce in this book. Consequently we created macros that can be loaded onto your computer to enable you to use Excel for *all* statistical procedures introduced in this book. The macros are stored on the CD that accompanies the book and, when installed, will appear as add-ins on Excel's Tools menu.

Almost all the examples, exercises, and cases will feature large data sets, also stored on the CD. We will demonstrate the solution to the statistical examples by employing Excel. Moreover, we will provide detailed instructions for all techniques. We use the Office 2000 version of Excel. A brief introduction to Microsoft Excel is provided in the appendix to this chapter.

The approach we prefer to take is to minimize the time spent on manual computations and to focus instead on selecting the appropriate method for dealing with a problem and on interpreting the output after the computer has performed the necessary computations. In this way, we hope to demonstrate that statistics can be as interesting and practical as any other subject in your curriculum.

1.4 WORLD WIDE WEB AND LEARNING CENTER

To assist students in the various aspects of using the computer to learn statistics, we have created a Web page. It offers useful information including additional exercises and cases, corrections to the different printings and supplements, and updates on the data sets and macros. Additionally, you can e-mail the author to make comments and ask questions about the installation of the files stored on the diskettes. The site can be accessed from the publisher's home page,

http://www.duxbury.com

Click **Online Book Companions,** which will take you to

http://www. duxbury.com/titles.htm

Find and click the cover of this book.

IMPORTANT TERMS

Descriptive statistics
Inferential statistics
Population
Parameter
Sample

Statistic
Statistical inference
Confidence level
Significance level

EXERCISES

1.1 In your own words, define and give an example of each of the following statistical terms.
  a Population
  b Sample
  c Parameter
  d Statistic
  e Statistical inference

1.2 Briefly describe the difference between descriptive statistics and inferential statistics.

1.3 A politician who is running for the office of mayor of a city with 25,000 registered voters commissions a survey. In the survey, 48% of the 200 registered voters interviewed say they plan to vote for her.
  a What is the population of interest?
  b What is the sample?
  c Is the value 48% a parameter or a statistic? Explain.

1.4 A manufacturer of computer chips claims that fewer than 10% of his products are defective. When 1,000 chips were drawn from a large production, 7.5% were found to be defective.
  a What is the population of interest?
  b What is the sample?
  c What is the parameter?
  d What is the statistic?
  e Does the value 10% refer to the parameter or to the statistic?
  f Is the value 7.5% a parameter or a statistic?
  g Explain briefly how the statistic can be used to make inferences about the parameter to test the claim.

1.5 Suppose you believe that, in general, graduates who have majored in *your* subject are offered higher salaries upon graduating than are graduates of other programs. Describe a statistical experiment that could help test your belief.

1.6 You are shown a coin that its owner says is fair in the sense that it will produce the same number of heads and tails when flipped a very large number of times.
  a Describe an experiment to test this claim.
  b What is the population in your experiment?
  c What is the sample?
  d What is the parameter?
  e What is the statistic?
  f Describe briefly how statistical inference can be used to test the claim.

1.7 Suppose that in Exercise 1.6 you decide to flip the coin 100 times.
  a What conclusion would you be likely to draw if you observed 95 heads?
  b What conclusion would you be likely to draw if you observed 55 heads?

c Do you believe that, if you flip a perfectly fair coin 100 times, you will always observe exactly 50 heads? If you answered no, what numbers do you think are possible? If you answered yes, how many heads would you observe if you flipped the coin twice? Try it several times, and report the results.

1.8 The owner of a large fleet of taxis is trying to estimate his costs for next year's operations. One major cost is fuel purchases. To estimate fuel purchases, the owner needs to know the total distance his taxis will travel next year, the cost of a gallon of fuel, and the fuel mileage of his taxis. The owner has been provided with the first two figures (distance estimate and cost). However, because of the high cost of gasoline, the owner has recently converted his taxis to operate on propane. He measures the propane mileage (in miles per gallon) for 50 taxis. The results are stored in file Xr01-08.

a What is the population of interest?
b What is the parameter the owner needs?
c What is the sample?
d What is the statistic?
e Describe briefly how the statistic will produce the kind of information the owner wants.

Appendix 1.A

Introduction to Microsoft® Excel

The purpose of this appendix is to introduce you to Microsoft® Excel and provide enough instruction to allow you to use Excel to produce statistical results. We suggest that you obtain an Excel instruction book to help you learn more about the software.

INSTALLING EXCEL

Installing Excel on your computer is easy. Simply follow the instructions in the booklet that accompanies your edition of Microsoft Excel. In most cases you will install some edition of Office, which includes Excel, Word, and Power Point, as well as several other programs.

After installing the software, keep the compact disk (CD) or diskettes handy; it is quite possible that you will need them again.

If the screen (called a desktop) does not show a Microsoft Excel icon, click **Start, Programs,** and **Microsoft Excel.** (Alternatively, if the desktop has an Excel icon, double-click it.) The Excel screen depicted in Figure A1.1 will appear. (Unless otherwise stated, "clicking" refers to tapping the left button on your mouse. "Double-clicking" means tapping the left button twice quickly.) The screen that appears may be slightly different, depending on which version of Excel you have.

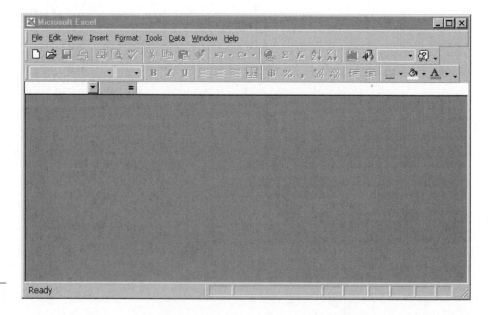

Figure A1.1

Blank Excel Screen

EXCEL SCREEN

At this point the screen is blank except for the top where five rows or bars appear. These are the locations of most of the commands that you will issue to Excel. Move the mouse, which in turn will move the pointer to different positions on the screen. To select a command, position the pointer over the command, and click once.

The first row is called the **Title bar.** At the left end of the title bar you see the name of the program, Microsoft Excel. At the right side there are three small boxes, which (moving left to right) minimize, restore, or close the Excel program. You can see what the box does before you execute it by placing the mouse pointer over it and waiting a second before moving or clicking.

The second row is the **Menu bar,** which contains a number of commands that open another list or menu of commands. These are called **drop-down menus.** The first menu item is **File.** Clicking this box results in a drop-down menu of other commands. For example, clicking the **Open** command asks Excel to open a file (which you must identify). Notice that this command can be issued by holding down the **Control (Ctrl) key** and hitting the letter **O.**

The next two rows are the **Toolbars.** The Toolbar contains icons that describe a number of functions that can also be executed from the **Menu bar.** The purpose of the icons is to make often-used commands easier to execute. For example, clicking the first icon opens a new workbook (described below).

The fifth row is the **Formula bar,** which displays the contents of the active cell (described below).

EXCEL WORKBOOK AND WORKSHEET

Excel files are called workbooks, which contain worksheets. A worksheet consists of rows and columns. The rows are numbered, and letters identify the columns. The intersection of a row and column is called a **cell,** which is a box that can store a number, word, or formula.

If you click the **New** icon (the first icon on the Toolbar) you will see Figure A1.2. Notice that the cell in row 1 column A is **active,** which means that you can type in a number, word, or formula. You can designate any cell as active by moving the mouse pointer (which now appears as a large plus sign) and clicking. Alternatively, you can use any of the four **Up, Down, Left,** or **Right** arrow keys. (These appear on your keyboard as arrows pointing up, down, left, and right, respectively.)

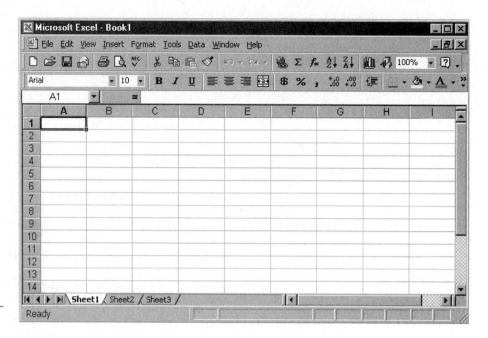

Figure A1.2

Blank Excel Screen

At the bottom of the screen you will see the word **Ready.** As you begin to type something into the active cell, the word changes to **Enter.** Above this word you will find the following tabs: **Sheet1, Sheet2,** and **Sheet3,** the three worksheets that comprise this workbook. You can operate on any of these as well as other sheets that may be created. To change the worksheet, use your mouse pointer, and click the sheet you wish to move to.

INPUTTING DATA

To input data, open a new workbook by clicking the blank icon on the **Toolbar** or click **File** and **New.** (You may have to click ⌄, which shows the commands that have been employed infrequently.) Data are usually stored in columns. Activate the cell in the first row of the column in which you plan to type the data. If you wish, you may type the name of the variable. For example, if you plan to type your assignment marks in row A you may type "Assignment Marks" in cell A1. Hit the **Enter** key, and cell A2 becomes active. Begin typing the marks, followed by **Enter.** Use the arrow key or mouse pointer to move to a new column if you wish to enter another set of numbers.

IMPORTING DATA FILES

A data file stored on the book's CD accompanies most of the examples, exercises, and cases in this book. For example, the data set accompanying Example 2.1 (see page 21) contains the times 400 people spent viewing the Barnes Exhibit. These data are stored in a file called **Xm02-01,** which is stored in a directory (or folder) called **Chapter 2.** (The **Xm** refers to files attached to e**X**amples, **Xr** is used for e**X**ercises, and **C** is used for **C**ases.)

If you follow the instructions that accompany the CD, the data files will be stored in a directory called **Datafiles.** To import a file, click the **Open** folder in the **Toolbar,** click the yellow arrow key (**Up One Level**) to locate the drive that stores the **Datafiles** directory (likely the C drive), and double-click **Datafiles,** the chapter directory, and the specific file you wish. The file will appear in the same form that it was saved. (All the files on the CD were saved by the author.) You may save in your own files (see instructions below.)

EDITING DATA

Each statistical technique requires that the data be in some specific format. If the data are not in that form, it will be necessary to edit the data. As a first step, you must highlight the data you wish to edit. To do so, place the mouse pointer over the first cell of the range, and hold the left button down as you move the mouse over the range. Alternatively, you can activate the first cell of the range, hold down the **Shift** key, and use the **Up, Down, Left,** or **Right** arrow keys to highlight the range.

To delete the range, hit the delete key.

To move the range, place the mouse pointer at the top right corner of the range, depress the left mouse button, and move the mouse until the range is where you wish it to be. Release the button. Alternatively, click **Edit** and **Cut.** Activate the cell where the top of the range will be located, and click **Edit** and **Paste.**

You can also use similar commands to copy a range of data. Instead of **Cut,** click **Copy.**

SAVING WORKBOOKS

To save a file (either one you created or one we created that you have altered) click **File** and **Save As . . . .** Use the **Up One Level** key to specify the directory, and type the **File name**. Click **Save.** If the file is already saved and you wish to use the same name, click the diskette icon on the **Menu bar,** or click **File** and **Save.**

PERFORMING STATISTICAL PROCEDURES

There are several ways to conduct a statistical analysis. These are **Data Analysis, Data Analysis Plus, Stats-summary.xls,** and the **Toolbar** function f_x.

Data Analysis/Analysis ToolPak

The **Analysis ToolPak** is a group of statistical functions that comes with Excel. The **Analysis ToolPak** can be accessed through the **Menu bar**. Click **Tools** and **Data Analysis . . . .** (Note that **Analysis ToolPak** is not the same as **Analysis ToolPak-VBA.**) If **Data Analysis . . .** does not appear, click **Add-Ins . . .** and select **Analysis ToolPak.** If **Analysis ToolPak** is not shown, you will need to install it from the original Excel or MS Office diskettes or CD-ROM. Run the setup program, and follow instructions.

There are 20 menu items in **Data Analysis . . . .** Click the one you wish to use, and follow the instructions described in this book. For example, the first technique in the menu **Anova: Single Factor** is described in Chapter 14.

Data Analysis Plus

Data Analysis Plus is the collection of macros we created to augment Excel's list of statistical procedures. **Data Analysis Plus (STATS.XLS)** is supplied on the CD that accompanies this book. The installation program that saves the data files on your computer will also save a copy of **STATS.xls** in a file called **Xlstart.** When this file is correctly saved on your computer, **Data Analysis Plus** will become a menu item in the **Tools** heading in the **Menu bar.** The instructions for **Data Analysis Plus** are also described in this book.

Stats-summary.xls

We created a workbook containing 18 worksheets that perform a variety of functions that will be described in Chapters 9, 10, 11, and 12. **Stats-summary.xls** is stored in the same directory as the data files.

Toolbar function f_x.

You find on the first **Toolbar** the f_x heading. Clicking this button produces two drop-down menus, one called **Function category** and the other called **Function name.** Selecting one from each menu allows you to perform many procedures, only a small number of which are described in this book.

CLOSING THE WORKBOOK

To close a workbook, click the **Close** button (the last item) on the **Title bar.** Alternatively, click **File** and **Close.**

STATISTICS PRACTITIONERS AND STUDENTS: BEWARE

The current versions of Excel use algorithms that are not robust. This means that under certain circumstances, Excel's calculations may be incorrect. In some extreme cases, the errors can be very large. Until Microsoft improves this aspect of its products, students are cautioned when using Excel on sets of data other than the ones included with this book. On these datasets Excel works perfectly.

Chapter 2

Graphical Descriptive Techniques

- **2.1** Introduction
- **2.2** Types of Data
- **2.3** Graphically Describing Interval Data: Frequency Distributions and Histograms
- **2.4** Graphically Describing Nominal Data: Bar and Pie Charts
- **2.5** Describing Time-Series Data: Line Charts
- **2.6** Describing the Relationship between Two Interval Variables: Scatter Diagrams
- **2.7** Summary

2.1 INTRODUCTION

Descriptive techniques are designed to summarize data so that information can be extracted. For example, if we have 250 observations of the debt incurred by university graduates in pursuing their degrees, we would like to develop various bits of information. We would be interested, for example, in knowing the size of the "typical" debt and whether the debts of different students are similar or whether they differ considerably. As we pointed out in Chapter 1, there are two ways to describe a set of data, graphically and numerically.

Graphical techniques draw a picture of the data. These methods are useful in quickly summarizing a set of data, but they are limited. Numerical techniques calculate descriptive measures that gauge various characteristics of the data. Numerical methods are more flexible but require a greater knowledge of statistics in order to interpret them. Moreover, numerical techniques are a prerequisite for statistical inference, a subject to which we devote most of this book. Statistical inference requires probability (covered in Chapters 4, 5, 6, and 8), so we will begin discussing inference in Chapter 9.

In this chapter we will introduce several graphical techniques and illustrate when and how they are employed and how we interpret the graphs. In the next chapter we will present numerical methods.

2.2 TYPES OF DATA

As we pointed out in the first paragraph of this book, the objective of statistics is to extract information from data. There are different types of data and information. To help explain this important principle, we need to define some terms.

A **variable** is some characteristic of a population or sample. For example, the mark on a statistics exam is a characteristic of statistics exams that certainly is of interest to readers of this book. Not all students achieve the same mark. The marks will vary from student to student, thus the name.

The **values** of the variable are the possible observations of the variable. The values of statistics marks are the numbers between 0 and 100 (assuming the exam is marked out of 100).

The **data** are the observed values of the variable. For example, suppose that we observe the marks of 10 students, which are:

67 74 71 83 93 55 48 82 68 62

These are the data from which we will extract the information we seek.

When most people think of the terms defined above they think of sets of numbers. However, there are three types of data. They are interval, nominal, and ordinal data.*

Interval data are real numbers, such as heights, weights, incomes, and distance. We also refer to this type of data as **quantitative** or **numerical.**

The values of **nominal** variables are categories. For example, responses to questions about marital status produce nominal data. The values of this variable are sin-

---

*There are actually four types of data, the fourth being ratio. However, for statistical purposes there is no difference between ratio and interval data. Consequently, we combine the two types.

gle, married, divorced, and widowed. Notice that the values are not numbers but instead are words describing categories. We often record nominal data by arbitrarily assigning a number to each category. For example, we could record marital status using the following codes:

| Single | 1 | Divorced | 3 |
| Married | 2 | Widowed | 4 |

However, any other numbering system is valid provided that each category has a different number assigned to it. Nominal data are also called **qualitative** or **categorical.**

The third type of data is ordinal. **Ordinal** data appear to be nominal, but their values are in order. For example, at the completion of most college and university courses, students are asked to evaluate the course. The variable is the performance rating and in a particular college the values are:

poor, fair, good, very good, and excellent

The difference between nominal and ordinal types of data is that the values of the latter are in order. Consequently, when assigning codes to the values we should maintain the order of the values. For example, we can record the students' evaluations as:

Poor	1
Fair	2
Good	3
Very good	4
Excellent	5

Because the only constraint that we impose on our choice of codes is that the order must be maintained, we can use any set of codes that are in order. For example, we can also assign the following codes:

Poor	6
Fair	18
Good	23
Very good	45
Excellent	88

As we discuss in Chapter 16, the use of any codes that preserve the order of the data will produce exactly the same result.

Students often have difficulty distinguishing between ordinal and interval data. The critical difference between them is that the intervals or differences between values of interval data are consistent and meaningful. (That's why this type of data is called interval.) For example, the difference between marks of 85 and 80 is the same five-mark difference that exists between 73 and 68. That is, we can calculate the difference and interpret the results.

Because the codes representing ordinal data are arbitrarily assigned except for the order, we cannot calculate and interpret differences. For example, using a 1-2-3-4-5 coding system to represent poor, fair, good, very good, and excellent, we note that the difference between excellent and very good is identical to the difference between good and fair. Using a 6-18-23-45-88 coding, the difference between excellent and very good is 43, and the difference between good and fair is 5. Because both coding systems are valid, we cannot use either system to compute and interpret differences.

Here is another example. Suppose that you are given the order of finish of the countries of the top five participants in the marathon race at the Olympic Games.

Finish	Country
1	Kenya
2	Italy
3	Australia
4	China
5	Japan

Does this information allow you to conclude that the difference between the times run by the Kenyan and Italian is the same as the difference in times between the Australian and Chinese? The answer is no, because we have information only about the order of the finishes, which are ordinal, and not the times, which are interval. That is, the difference between 1 and 2 is not necessarily the same as the difference between 3 and 4.

CALCULATIONS ON TYPES OF DATA

Interval Data

All calculations are permitted on interval data. We often describe a set of interval data by calculating the average. For example, the average of the 10 marks listed on page 16 is 70.3. As you will discover, there are several other important statistics that we will compute.

Nominal Data

Because the codes of nominal data are completely arbitrary, we cannot perform any calculations on these codes. To understand why, consider a survey that asks people to report their marital status. Suppose that the first 10 people surveyed responded:

single, married, married, married, widowed, single, married, married, single, divorced

Using the codes

Single	1	Divorced	3
Married	2	Widowed	4

we would record these responses as:

1, 2, 2, 2, 4, 1, 2, 2, 1, 3

The average of these numerical codes is 2.0. Does this mean that the average person is married? Now suppose four more persons were interviewed, of whom three are widowed and one is divorced. The data are:

1, 2, 2, 2, 4, 1, 2, 2, 1, 3, 4, 4, 4, 3

The average of these 14 codes is 2.5. Does this mean that the average person is married but halfway to getting divorced? The answer to both questions is an emphatic no. This example illustrates a fundamental truth about nominal data. Calculations based on the codes used to store this type of data are meaningless. All that we are permitted to do with nominal data is count the occurrences of each category. Thus, we would describe the 14 observations by counting the number of each marital status category and reporting the frequency.

Category	Code	Frequency
Single	1	3
Married	2	5
Divorced	3	2
Widowed	4	4

Ordinal Data

The most important aspect of ordinal data is the order of the values. As a result, the only permissible calculations are ones involving a ranking process. For example, we can place all the data in order and select the code that lies in the middle. As we discuss in Chapter 3, this descriptive measurement is called the *median*.

HIERARCHY OF DATA

The data types can be placed in order of their permissible calculations. At the top of the list we place the interval data type because virtually *all* computations are allowed. The nominal data type is at the bottom because *no* calculations other than determining frequencies are permitted. (We are permitted to perform calculations using the frequencies of codes. However, this differs from performing calculations on the codes themselves.) In between interval and nominal data lies the ordinal data type. Permissible calculations are ones that rank the data.

Higher-level data types may be treated as lower-level ones. For example, in universities and colleges we convert the marks in a course, which are interval, to letter grades, which are ordinal. Some graduate courses feature only a pass or fail designation. In this case, the interval data are converted to nominal. However, we cannot treat lower-level data types as higher-level types. The definitions and hierarchy are summarized in the following table.

Types of Data

- *Interval*
  Values are real numbers.
  All calculations are valid.
  Data may be treated as ordinal or nominal.
- *Ordinal*
  Values must represent the ranked order of the data.
  Calculations based on an ordering process are valid.
  Data may be treated as nominal but not as interval.
- *Nominal*
  Values are the arbitrary numbers that represent categories.
  Only calculations based on the frequencies of occurrence are valid.
  Data may not be treated as ordinal or interval.

EXERCISES

2.1 Provide two examples each of nominal, ordinal, and interval data.

2.2 For each of the following examples of data, determine the type.
  a The number of miles joggers run per week
  b The starting salaries of graduates of teaching colleges
  c The months in which children are born
  d The final letter grades received by students in a statistics course

2.3 For each of the following examples of data, determine the type.
  a The weekly closing price of the stock of Amazon.com
  b The month of highest vacancy rate at a La Quinta motel
  c The size of soft drink (small, medium, or large) ordered by a sample of McDonald's customers
  d The number of Toyotas imported monthly by the United States over the last five years
  e The marks achieved by the students in a statistics course final exam marked out of 100

2.4 The placement office at a university regularly surveys the graduates one year after graduation and asks for the following information. For each, determine the type of data.

  a What is your occupation?
  b What is your income?
  c What degree did you obtain?
  d What is the amount of your student loan?
  e How would you rate the quality of instruction?

2.5 Residents of condominiums were recently surveyed and asked a series of questions. Identify the type of data for each question.
  a What is your age?
  b On what floor is the condominium?
  c Do you own or rent?
  d How many square feet?
  e Does your condominium have a pool?

2.6 A sample of shoppers at a mall was asked the following questions. Identify the type of data each question would produce.
  a What is your age?
  b How much did you spend?
  c What is your marital status?
  d Rate the availability of parking.
  e How many stores did you enter?

2.3 GRAPHICALLY DESCRIBING INTERVAL DATA: FREQUENCY DISTRIBUTIONS AND HISTOGRAMS

Besides the different data types, there are several different types of information that we may wish to obtain from the data. We will discuss the different types of information in greater detail in Chapter 10 when we introduce *problem objectives*. However, in this and the next chapter we will use statistical techniques to describe one variable when the data are interval and when the data are nominal. The purpose or objective of conducting the statistical analyses is to describe a single set of data. In Section 2.6 the objective is to describe or analyze the relationship between two interval variables.

Consider the following illustrations.

Illustration 1

For some reason that we have been unable to determine, some college and university students seem to fear their required course in statistics. Suppose that you are such a student and that you sought reassurance by acquiring the marks in last year's final exam. There are several pieces of information that may be useful. You would like to know the approximate center of the data. Most students want to know the average mark, which is only one measure of central location, a subject that we investigate further in Chapter 3. However, there are several ways to measure central location. Moreover, there are other measures that can prove valuable. For example, you would like to know whether the marks are spread out over a wide range or are tightly grouped together. Are the marks evenly spread below and above the center, or is it more likely that many students do poorly while only a minority do well? Are there many very high marks and low marks, and very few in between?

Illustration 2

Investors always seek high rates of return on their investments. The rate of return is calculated by dividing the gain (or loss) by the value of the investment. For example, a $100 investment that is worth $106 after one year has a 6% rate of return. Suppose that you are facing a decision on where to invest that small fortune that remains after you have deducted the anticipated expenses for the next year from the earnings from your summer job. A friend has suggested a particular type of investment, and to help make the decision you acquire some rates of return from this type of investment. As in Illustration 1, you would like to know the location of the center of the data, as well as other types of information. These include whether the rates are spread out over a wide range, making the investment risky, or are grouped tightly together, indicating relatively low risk. Do the data indicate that it is possible that you can do extremely well with little likelihood of a large loss? Is it likely that you could lose money (negative rate of return)?

There are several ways to extract the information that you want from the data. In this chapter we present graphical techniques, and in Chapter 3 we introduce numerical measures. When the data are interval, as is the case with both illustrations above, we often produce a histogram. To draw a histogram we first develop the frequency distribution. Consider the following example.

▼ EXAMPLE 2.1

Several years ago the Barnes Exhibit toured major cities all over the world, with millions of people flocking to see it. Dr. Albert Barnes was a wealthy art collector who accumulated a large number of impressionist masterpieces; the total exceeds 800 paintings. When Dr. Barnes died in 1951, he stated in his will that his collection was not to be allowed to tour. However, because of the deterioration of the exhibit's home near Philadelphia, a judge ruled that the collection could go on tour to raise enough money to renovate the building. Because of the size and value of the collection, it was predicted (correctly) that in each city a large number of people would come to view the paintings. Because space was limited, most galleries had to sell tickets that were valid at one time (much like a play). In this way, they were able to control the number of visitors at any one time. To judge how many people to let in at any time, it was necessary to know the length of time people would spend at the exhibit; longer times would dictate smaller audiences, shorter times would allow for sale of more tickets. Suppose that in one city the amount of time (rounded to the nearest minute) taken to view the complete exhibit (which consisted of only 83 paintings) by each of

400 people was measured and recorded in file Xm02-01 and listed below. Graphically depict the data, and extract the information from it.

113	49	62	44	42	32	43	46	54	98	64	34	61	60	52	57	42	46	69	70
36	43	30	54	47	38	55	40	70	55	70	44	29	40	48	66	54	59	94	64
45	38	70	62	48	57	84	50	27	38	55	51	62	34	54	47	58	45	61	61
47	64	34	107	70	30	61	65	44	54	62	74	41	30	88	58	59	43	63	33
51	58	48	33	36	52	29	34	66	50	45	44	47	41	39	38	106	49	35	46
42	31	41	98	40	48	42	25	33	29	66	39	30	47	43	35	30	59	45	41
31	47	26	53	40	23	79	28	78	74	42	52	53	46	40	50	90	50	37	45
89	39	60	44	36	57	47	78	48	37	55	44	54	59	70	60	34	32	35	48
52	53	151	43	112	44	39	53	41	70	72	32	71	63	65	49	31	32	83	37
40	64	47	38	32	49	33	78	50	35	28	39	54	41	82	32	42	43	43	57
45	88	66	53	57	46	61	53	90	28	41	74	31	107	45	50	72	75	30	54
65	73	45	58	48	62	60	92	50	43	70	33	29	40	91	49	56	39	35	24
52	41	31	63	44	57	50	42	41	27	44	46	64	39	71	42	30	109	66	41
32	51	41	56	38	80	54	60	41	33	134	71	33	63	45	63	57	64	91	91
28	98	27	102	68	44	53	71	42	31	46	55	67	41	40	67	48	70	40	71
28	29	40	35	58	64	33	50	82	53	33	54	85	77	67	38	28	63	45	48
34	63	42	88	42	36	36	33	52	104	68	48	85	29	51	49	60	47	63	62
82	60	50	28	78	42	121	49	125	57	93	32	52	32	44	41	38	45	36	43
29	85	51	42	73	44	79	28	70	42	45	64	38	54	41	56	46	45	28	70
47	41	35	62	33	40	35	43	81	45	43	68	58	90	63	39	44	27	46	36

Solution If you examine the 400 observations, you acquire very little information. You may discover that the smallest number is 23 and the largest number is 151, but you will have learned very little about how the numbers are distributed between these two extremes. The most common method of obtaining information is the **histogram.** A histogram is a graphical technique that is applied to a set of interval data that describes how the data are distributed. We start by developing a **frequency distribution.**

A frequency distribution counts the number of observations that fall into each of a series of intervals, called **classes,** that cover the complete range of observations. We'll discuss how we decide the number of classes and the upper and lower limits of the intervals later. We have chosen 10 classes defined in such a way that each observation falls into one and only one class. These classes are defined below.

Classes
More than 10 but less than or equal to 25 minutes
More than 25 but less than or equal to 40 minutes
More than 40 but less than or equal to 55 minutes
More than 55 but less than or equal to 70 minutes
More than 70 but less than or equal to 85 minutes
More than 85 but less than or equal to 100 minutes
More than 100 but less than or equal to 115 minutes
More than 115 but less than or equal to 130 minutes

More than 130 but less than or equal to 145 minutes

More than 145 but less than or equal to 160 minutes

Notice that the intervals do not overlap, so that there is no uncertainty about which interval to assign to any observation. Moreover, because the smallest number is 23 and the largest is 151, every observation will be assigned to an interval. Finally, the intervals are equally wide. While this is not essential, it makes the task of reading and interpreting the graph easier.

If we were to create the frequency distribution manually, we would count the number of observations that fall into each interval. These counts are called the frequencies, hence the name. The frequency distribution would then be converted into a histogram by drawing rectangles whose bases are the intervals and whose heights are the frequencies.

While going through 400 observations and assigning each to its class interval may sound like a lot of fun, it isn't. Consequently, we now turn to the computer and Microsoft Excel to perform this monotonous function. This is the first of several dozen statistical procedures to be conducted by computer. Our usual procedure is to provide the output for the example, as well as the general Excel commands and the specific commands for the example.

The histogram consists of a series of adjacent rectangles. The bases of the rectangles are the class intervals. The numbers on the horizontal axis that are situated in the middle of the bases are the upper limits of the class intervals, not the midpoints. The heights of the rectangles represent the frequencies of the classes.

Microsoft Excel Histogram for Example 2.1

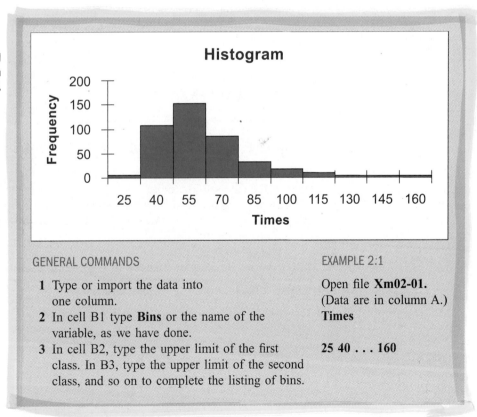

GENERAL COMMANDS

1 Type or import the data into one column.
2 In cell B1 type **Bins** or the name of the variable, as we have done.
3 In cell B2, type the upper limit of the first class. In B3, type the upper limit of the second class, and so on to complete the listing of bins.

EXAMPLE 2:1

Open file **Xm02-01.**
(Data are in column A.)
Times

25 40 . . . 160

> 4 Click **Tools, Data Analysis . . . ,** and **Histogram.**
> 5 Type the **Input Range** of the data. Include   A1:A401
>   the cell containing the name of the variable
>   (if applicable).
> 6 Type the **Input Range** of the bins.   B1:B11
> 7 If the names of the variables have been included
>   in the first row of the ranges, click **Labels.**
> 8 Click **Chart Output** and **OK.** This will create a
>   chart with gaps between the rectangles.
> 9 To remove the gaps place the cursor over one of the rectangles, and click
>   the right button of the mouse.
> 10 Click (with the left button) **Format Data Series . . . .**
> 11 Click **Options,** move the pointer to **Gap Width:,** and change the number
>   from 150 to 0. Click **OK.**
> 12 To remove the **More** class, use the left button, and click the cell
>   containing **More** and the cell containing the frequency. Click the right
>   button to produce a menu, and click **Delete.**
> 13 To improve the appearance of the histogram, click inside the box
>   containing the chart, and click **Chart** and **Chart Options.**

INTERPRETING THE RESULTS

The histogram tells us that the times are quite variable; some people stay only a short time while others linger. However, most visitors leave within 70 minutes, and very few stay longer than 100 minutes. If the gallery were to admit 400 people per hour, within a short time there would be far more people than the gallery can comfortably accommodate. There are several alternatives, one of which is to admit 150 to 200 visitors every 45 minutes.

SELECTING THE NUMBER OF CLASS INTERVALS

The number of class intervals we select depends entirely on the number of observations in the data set. The more observations we have, the larger the number of class intervals we need to use to draw the histogram. Table 2.1 provides guideline on choosing the number of classes. In Example 2.1 we had 400 observations. The table tells us to use 9 or 10 classes. We'll try 9.

Table 2.1 Approximate Number of Classes in Frequency Distributions

Number of Observations	Number of Classes
Less than 50	5–7
50–200	7–9
200–500	9–10
500–1,000	10–11
1,000–5,000	11–13
5,000–50,000	13–17
More than 50,000	17–20

We determine the width of the classes by subtracting the smallest observation from the largest and dividing the difference by the number of classes. Thus,

$$\text{Class width} = \frac{\text{Largest observation} - \text{Smallest observation}}{\text{Number of Classes}} = \frac{151 - 23}{9} = 14.2$$

We often round this number to some convenient value. We then define our class limits by arbitrarily selecting a lower limit for the first class interval and from which all other limits are calculated. The only condition we apply is that the first class interval must contain the smallest observation. In Example 2.1, we rounded the class width to 15 and set the lower limit of the first class to 10. Thus, the first class is defined as "More than or equal to 10 but less than 25 minutes." Because of the way we specified the first interval, we ended up with 10 classes.

To illustrate what happens when you use an inappropriate number of class intervals, examine the following Excel histogram constructed with only 5 classes. As you can see, the detail provided in the histogram with 10 classes is much greater, allowing us to obtain more information.

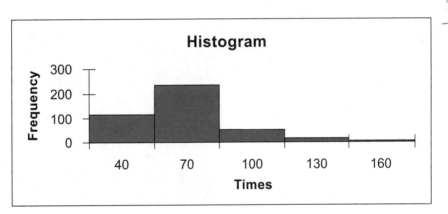

Microsoft Excel Histogram with Five Classes for Example 2.1

SHAPES OF HISTOGRAMS

The purpose of drawing histograms, like all statistical techniques, is to acquire information. Once we have the information, we frequently need to describe what we've learned to others. We describe the shape of histograms on the basis of the following characteristics.

Symmetry

A histogram is said to be **symmetric** if, when we draw a vertical line down the center of the histogram, the two sides are identical in shape and size. Figure 2.1 depicts three symmetric histograms.

Skewness

A skewed histogram is one with a long tail extending either to the right or left. The former is called **positively skewed,** and the latter is called **negatively skewed.** Figure 2.2 describes examples of both. Incomes of employees in large firms tend to be positively skewed, because there is a large number of relatively low-paid workers and a

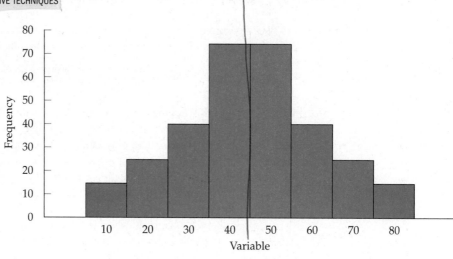
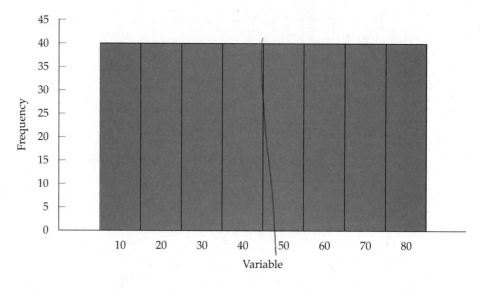
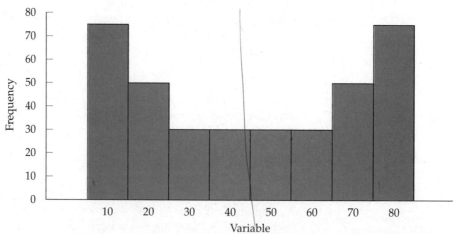

Figure 2.1

Three symmetric histograms

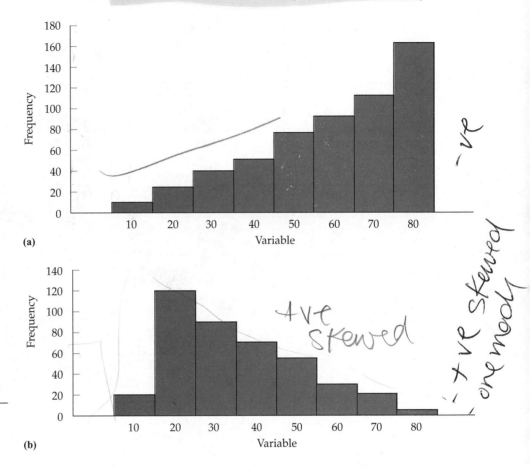

Figure 2.2

(a) A negatively skewed histogram; (b) a positively skewed histogram

small number of well-paid executives. The time taken by students to write exams is frequently negatively skewed, because few students hand in their exams early; most prefer to reread their papers and hand them in near the end of the scheduled test period.

NUMBER OF MODAL CLASSES

As we discuss in Chapter 3, a *mode* is the observation that occurs with the greatest frequency. A **modal class** is the class with the largest number of observations. A **unimodal** histogram is one with a single peak. The histogram in Figure 2.3 is unimodal.

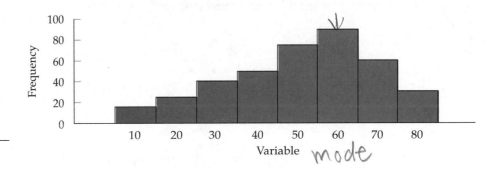

Figure 2.3

A unimodal histogram

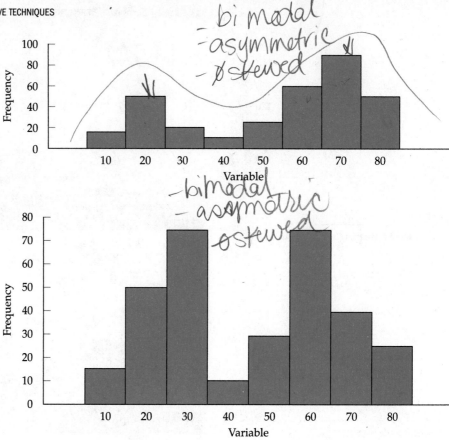

Figure 2.4

Two bimodal histograms

A **bimodal** histogram is one with two peaks, not necessarily equal in height. Figure 2.4 depicts bimodal histograms.

BELL-SHAPED HISTOGRAM

In Chapter 6 we introduce the normally distributed random variable whose histogram is bell shaped. Because of the importance of this type of variable, we frequently need to know whether the variable we're dealing with is normal. Figure 2.5 exhibits a bell-shaped histogram.

Now that we know what to look for, let's examine some examples of histograms and see what we can discover.

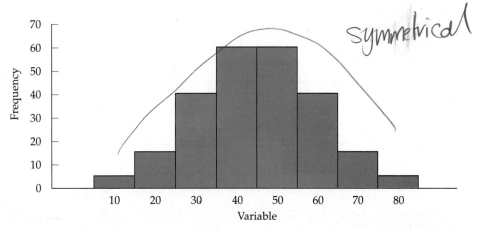

Figure 2.5

A bell-shaped histogram

▼ **EXAMPLE 2.2**

The final marks in an applied statistics course that emphasized manual calculations both during the class and on exams are stored in column A in file Xm02-02. The marks obtained by students in the same course after the emphasis was changed to using the computer and Microsoft Excel are stored in column B of the same file. Draw histograms for both groups and interpret the results.

Solution

Microsoft Excel Histogram of Marks in Course with Manual Calculations

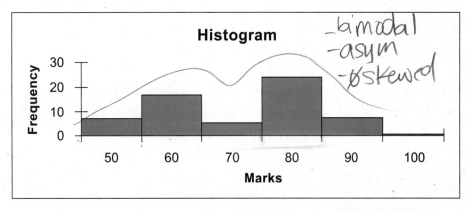

Microsoft Excel Histogram of Marks in Course with Computer

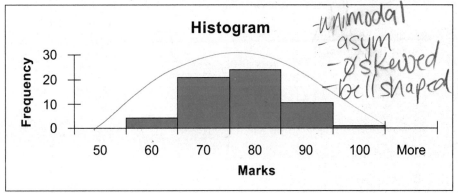

INTERPRETING THE RESULTS

The histogram of the marks in the "manual calculation" statistics course is bimodal. The larger modal class is the interval 70–80. The smaller modal class is the interval 50–60. There appear to be few marks in the 60s. This histogram suggests that there are two groups of students. Because of the emphasis on mathematical manipulation in the course, one may conclude that those who performed poorly in the course are weaker mathematically than those who performed well.

Contrast the first histogram with the second one that describes the marks on the "computer" statistics course. This histogram is unimodal and bell shaped, and it appears that the spread is less than the first histogram. One possible interpretation is that this type of course allows students who are not particularly mathematical to learn statistics as well as mathematically inclined students.

▼ **EXAMPLE 2.3**

The returns for two types of investments are stored in columns A and B of file Xm02-03. Draw histograms for each set of returns, and report on your findings. Which investment would you choose and why?

Solution

Microsoft Excel Histogram of Returns for Investment A

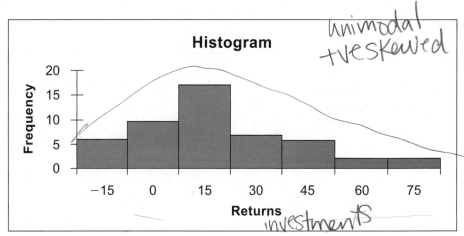

Microsoft Excel Histogram of Returns for Investment B

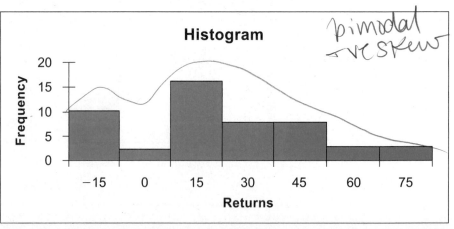

INTERPRETING THE RESULTS

Comparing the two histograms, we can extract the following information.

1. The center of the histogram of the returns of Investment A is slightly lower than that for Investment B.
2. The spread of returns for Investment A is considerably less than that for Investment B.
3. Both histograms are slightly positively skewed.

These findings suggest that Investment A is superior. Although the returns on A are slightly less than those for B, the wider spread for B makes it unappealing to most investors. Both investments allow for the possibility of a relatively large return.

The interpretation of the histograms is somewhat subjective. Other viewers may not concur with our conclusion. In such cases, numerical techniques provide the detail and precision lacking in most graphs. We will redo this example in Chapter 3 to illustrate how numerical techniques compare to graphical ones.

EXAMPLE 2.4

One hundred visitors to a Las Vegas casino were interviewed. Each person was asked how much money they had brought to gamble and what was their return on investment. (A result of −100% represents a complete loss, 0% means that the visitor broke even, and a positive figure represents winnings.) These figures are stored in file Xm02-04. Draw a histogram, and describe what information you have learned.

Solution

Microsoft Excel Histogram For Example 2.4

INTERPRETING THE RESULTS

Most of the observations are below zero. (Recall that the numbers at the bottom of the histogram represent the upper limits of the bins.) A large proportion lose between 50% and 99%, and about 10% lose all. A small fraction of gamblers make money, and a very small proportion make a lot of money.

> **Factors That Identify When to Use a Histogram**
> 1 Objective: describe a single set of data.
> 2 Data type: interval.

We complete this section with a review of when to use a histogram. Note that we use the term *objective* to identify the type of information produced by a histogram.

EXERCISES

2.7 The final exam in a third-year psychology course requires students to write several essay-style answers. The numbers of pages for a sample of 133 exams were recorded and stored in Xr02-07.
  a How many bins should a histogram of these data contain?
  b Draw a histogram using the number of bins specified in part (a).
  c Is the histogram symmetric or skewed?
  d How many modes are there?
  e Is the histogram bell shaped?

2.8 The lengths of time (in minutes) to serve 420 customers at a local restaurant are stored in file Xr02-08.
  a How many bins should a histogram of these data contain?
  b Draw a histogram using the number of bins specified in part (a).
  c Is the histogram symmetric or skewed?
  d How many modes are there?
  e Is the histogram bell shaped?

2.9 The real estate board in a wealthy suburb is investigating the prices (in $1,000s) obtained for a sample of 85 houses recently sold. The data are stored in file Xr02-09.
  a What is a suitable number of bins of a histogram for these data?
  b Draw a histogram with the number of bins suggested above.
  c Describe the histogram.
  d What does the histogram tell you about the prices?

2.10 The salaries of a sample of 480 university professors were stored in file Xr02-10.
  a What is a suitable number of bins of a histogram for these data?
  b Draw a histogram with the number of bins suggested above.
  c Describe the histogram.
  d What does the histogram tell you about the salaries?

2.11 The marks of 320 students on an economics midterm test were recorded and stored in file Xr02-11.
  a How many bins should a histogram of these data contain?
  b Draw a histogram using the number of bins specified in part (a).
  c Is the histogram symmetric or skewed?
  d How many modes are there?
  e Is the histogram bell shaped?

2.12 The lengths (in inches) of 150 newborn babies were recorded and stored in file Xr02-12. Use whichever graphical technique you judge suitable to describe these data. What have you learned from the graph?

2.13 The number of copies made by an office copier was recorded for each of the past 75 days. The data are stored in file Xr02-13.
  a Draw a histogram with an appropriate number of bins.
  b Describe the histogram.
  c What does the histogram tell you about the number of copies?

2.14 A sample of 240 tomatoes grown with a new type of fertilizer was weighed (in ounces) and the data stored in file Xr02-14. Draw a histogram, and describe your findings.

2.15 The volume of water used in a sample of 350 households was measured (in gallons) and stored in file Xr02-15.
  a Use a suitable graphical statistical method to summarize the data.
  b Describe the shape of the histogram.
  c What does the graph tell you?

2.16 The number of books shipped out daily by Amazon.com was recorded for 100 days and stored in file Xr02-16. Draw a histogram, and describe your findings.

2.17 The amounts of interest paid by 225 people over the past 12 months on their credit card accounts are stored in file Xr02-17. Draw a histogram, and describe your findings.

2.18 The amount of time (in hours) 185 employees spent on the Internet during working hours was measured and stored in file Xr02-18.

 a Use a suitable graphical statistical method to summarize the data.
 b Describe the shape of the histogram.
 c What does the graph tell you?

2.19 The number of times 250 people used an ATM (automatic teller machine) during the past year was recorded and stored in file Xr02-19. Draw a histogram, and describe your findings.

2.20 A new electronic toll road (ETR) was built north of Toronto. Users of the ETR are billed automatically, paying between 5 and 10 cents per kilometer (depending on time of day). A sample of 301 Toronto car owners was asked to report the amount of last month's ETR bill. These data were stored in file Xr02-20. Draw a histogram, and summarize your findings.

2.21 The total amount of winnings from their appearance on the popular television game show *Jeopardy* was recorded for 150 people who have won at least once and is stored in file Xr02-21. Draw a histogram of the data. What does this graph tell you?

2.22 University professors attend academic conferences to present their own research and to hear about one another's research. Most academics who attend conferences are funded by grants from outside agencies or their own universities. To acquire more information about costs, a sample of 200 professors was asked to report how much they spent on their most recent academic conference. These data are stored in file Xr02-22. Draw a suitable graph, and describe your findings.

2.4 GRAPHICALLY DESCRIBING NOMINAL DATA: BAR AND PIE CHARTS

As we discussed in Section 2.2, the only allowable calculation on nominal data is to count the frequency of each value of the variable. We can graphically display the counts in two ways, bar and pie charts. A **bar chart** is similar to a histogram. The bases of the rectangles are arbitrary intervals whose centers are the codes. The height of each rectangle represents the frequency of that category. A **pie chart** is a circle subdivided into slices whose areas are proportional to the frequencies. Pie charts emphasize the *proportion* of occurrences of each category. Bar charts focus the attention on the *frequency* of the occurrences of the categories.

To illustrate the use of both, consider the following example.

▼ EXAMPLE 2.5

The student placement office at a university conducted a survey of last year's business school graduates to determine the general areas in which the graduates found jobs. The placement office intended to use the resulting information to help decide where to concentrate its efforts in attracting companies to campus to conduct job interviews. Each graduate was asked in which area he or she found a job. The areas of employment are:

 1 Accounting.
 2 Finance.
 3 General management.
 4 Marketing/sales.
 5 Other.

The data are stored in file Xm02-05 using the codes 1, 2, 3, 4, and 5 respectively. Summarize the data by producing an appropriate chart.

Solution The data are nominal; the numbers were arbitrarily assigned. All that we are permitted to calculate from nominal data is the frequency and proportion of each category. We will instruct Excel to draw both a bar and a pie chart.

Microsoft Excel Bar Chart for Example 2.5

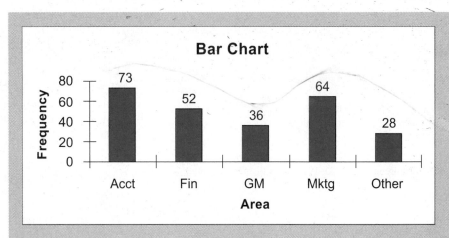

Microsoft Excel Pie Chart for Example 2.5

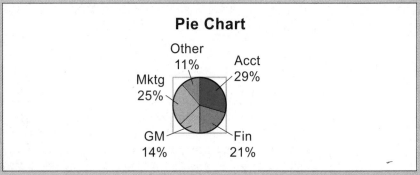

GENERAL COMMANDS

1 Proceed through the first eight steps in constructing a histogram. Use the codes as bins.
2 Click inside the boundaries of the histogram and click **Chart** and **Chart Type . . . .** Select **Column** to draw the bar chart. Select **Pie** to draw the pie chart.
3 Click **Chart** and **Chart Options . . .** to make cosmetic changes.

INTERPRETING THE RESULTS

The bar chart focuses on the frequencies. As you can see, accounting is the most popular job; 73 graduates work in this area. Least popular is the "other" category.

The pie chart focuses on the fraction of the total in each area. We can see, for example, that 29% of the graduates found jobs as accountants.

ORDINAL DATA

To describe a set of ordinal data we treat the data as nominal and draw a pie or bar chart. In Chapter 3 we introduce another graphical technique that can be used for ordinal data, the box plot.

OTHER APPLICATIONS OF PIE CHARTS AND BAR CHARTS

Pie and bar charts are used widely in newspapers, magazines, and business and government reports. One of the reasons for this appeal is that they are eye catching and can attract the reader's interest, whereas a table of numbers might not. Perhaps no one understands this better than the newspaper *USA Today*, which typically has a colored graph on the front page and others inside. Pie and bar charts are frequently used to simply present numbers associated with categories. The only reason to use a pie or bar chart in such a situation would be if the chart enhanced the reader's ability to grasp the substance of the data. It might, for example, allow the reader to more quickly recognize the relative sizes of the categories, as in the breakdown of a budget. Similarly, treasurers might use pie charts to show the breakdown of a firm's revenues by department, or university students might use pie charts to show the amount of time devoted to daily activities (e.g., eat, 10%; sleep, 30%; and study statistics, 60%).

Table 2.2 lists the daily oil production in millions of barrels for 10 countries around the world. Figure 2.6 depicts the same numbers using a bar chart. Notice that the chart merely shows the same numbers as the table but perhaps in a way that is more likely to attract readers.

Table 2.2   Daily Oil Production

Country	Oil Production (millions of barrels)
United States	8.1
Canada	2.3
Mexico	2.9
Venezuela	3.2
United Kingdom	2.9
Norway	3.3
Russia	5.8
China	3.2
Iran	3.7
Saudi Arabia	7.5

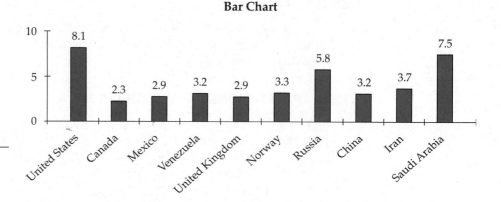

Figure 2.6

Microsoft Excel bar chart of daily oil production (millions of barrels)

We complete this section by describing when bar and pie charts are used.

> *Factors That Identify When to Use Bar and Pie Charts*
> 1 Objective: describe a single set of data.
> 2 Data type: nominal.

EXERCISES

2.23 In a taste test, 250 people were asked to taste five different brands of beer and to report which one they preferred. The results are stored in file Xr02-23 using the codes 1, 2, 3, 4, and 5 to represent the preferences.

 a Draw a bar chart.
 b Draw a pie chart.
 c What do the charts tell you about the sample of beer drinkers?

2.24 A political poll was conducted wherein a sample of 800 voters was asked which party they intended to vote for. The possible responses are:

 1. Democrat.
 2. Republican.
 3. Reform.

The responses are stored in file Xr02-24 using the codes 1, 2, and 3, respectively. The statistics practitioner wants to learn about the proportions of voters who support each party. Use an appropriate graphical method to report the data.

2.25 The vocational center at a university wanted to determine the types of jobs the graduates of the university were performing. A sample of the graduates of the university was surveyed five years after graduation and asked to report their occupations. The responses are:

 1. Unemployed.
 2. Manager.
 3. Blue-collar worker.
 4. Clerical worker.
 5. Other.

The results are stored in file Xr02-25. Summarize the data with an appropriate graphical method.

2.26 Who applies to law schools? To help determine the background of the applicants, a sample of the applicants to the university's law school was asked to report their undergraduate degree. A sample of 200 applications was drawn, and the degrees were recorded using the codes below.

1. B.A.
2. B.Sc.
3. B.B.A.
4. Other.

The codes are stored in file Xr02-26.

a Draw a bar chart.
b Draw a pie chart.
c What do the charts tell you about the sample of law school applicants?

2.27 Each day a restaurant lists on its menu five daily specials. Today's specials are:

1. Fried chicken.
2. Meat loaf.
3. Turkey pot pie.
4. Filet of sole.
5. Lasagna.

A sample of customers' choices is stored in file Xr02-27 (using the numerical codes). Draw a graph that describes the most important aspect of these data.

2.28 Most universities have several different kinds of residences on campus. To help long-term planning, one university surveyed a sample of graduate students and asked them to report their marital status. The possible responses are:

1. Single.
2. Married.
3. Divorced.
4. Widowed.

The responses for a sample of 250 students are stored in file Xr02-28 (using the numerical codes). Draw a graph that summarizes the information that you deem necessary.

2.29 After a long and distinguished career, a racehorse retired. His finishes were recorded in file Xr02-29 where 1 = first, 2 = second, and so on. Use a graphical technique to summarize the data and interpret your findings.

2.30 This statistics course requires a computer. As a result, many students buy their own. A survey asks students to identify which computer brand they have purchased. The responses are:

1. IBM.
2. Compaq.
3. Dell.
4. Other.

The results are stored in file Xr02-30.

a Draw a bar chart.
b Draw a pie chart.
c What do the charts tell you about the brands of computers used by the students?

2.31 An increasing number of statistics courses use a computer and software rather than manual calculations. A survey of statistics instructors asked each to report the software his or her course uses. The responses are:

1. Excel.
2. Minitab.
3. SAS.
4. SPSS.
5. Other.

The results are stored in file Xr02-31.

a Draw a bar chart.
b Draw a pie chart.
c What do the charts tell you about the software choices?

2.32 Opinions about the economy are important measures because they can become self-fulfilling prophecies. Annual surveys are conducted to determine the level of confidence in the future prospects of the economy. A sample of 1,000 adults was asked: Compared with last year, do you think this coming year will be:

1. Better?
2. The same?
3. Worse?

The responses are stored in file Xr02-32. Use a suitable graphical technique to summarize these data. Describe what you have learned.

2.33 The causes of aircraft crashes are a concern to a number of government agencies. To help learn more about crashes of small aircraft, the causes for the past 10 years were recorded in the following way:

When Crashes Occur	Code
Ground	1
Takeoff	2
Initial climb	3
Climb	4
Cruise	5
Descent	6
Approach	7
Landing	8

The data are stored in file Xr02-33.

a Draw a bar chart.
b Draw a pie chart.
c What do the charts tell you about small aircraft crashes?

2.34 Refer to Exercise 2.33. Suppose that a similar analysis was performed on large commercial jets. The data are stored in file Xr02-34.
  a Draw a bar chart.
  b Draw a pie chart.
  c Compare the graphs produced above with those from Exercise 2.33. What did you discover?

2.35 An analysis of female-owned small businesses was conducted. Among other questions, the owners were asked to identify the type of business using the following codes:

Business	Code
Services	1
Retail/wholesale/trade	2
Finance/insurance/real estate	3
Transportation/communication	4
Construction	5
Manufacturing	6
Agriculture	7

The responses for 260 female-owned small businesses are stored in file Xr02-35. Draw a graph to summarize these data, and describe your findings.

2.36 Refer to Exercise 2.35. A similar study involving 400 male-owned small businesses was conducted, with the results stored in file Xr02-36. Draw a graph of these data, and compare it with the graph you drew in Exercise 2.35.

2.37 Subway train riders frequently pass the time by reading a newspaper. Toronto has a subway and four newspapers. A sample of 360 subway riders who regularly read a newspaper was asked to identify that newspaper. The responses are:

1. *Globe and Mail.*
2. *National Post.*
3. *Toronto Star.*
4. *Toronto Sun.*

The responses are stored in file Xr02-37. Draw an appropriate graph to summarize the data. What does the graph tell you?

2.5 DESCRIBING TIME-SERIES DATA: LINE CHARTS

CROSS-SECTIONAL AND TIME-SERIES DATA

Besides classifying data by type, data can also be classified according to whether the observations are measured at the same time (**cross-sectional data**) or whether they represent measurements at successive points in time (**time-series data**). Marketing surveys and political opinion polls are familiar methods of collecting cross-sectional data. The data from a political survey might include, for example, the party preferences and demographic characteristics of a sample of 1,000 voters at the same point in time. Statistical techniques could be applied to the data, such as testing for differences in preferences between men and women.

To give another example, consider a real estate consultant who feels that the selling price of a house is a function of its size, age, and lot size. To estimate the specific form of the function, she samples (say) 100 homes recently sold and records the price, size, age, and lot size for each home. These data are cross-sectional, in that they all are observations at the same point in time.

The real estate consultant is also working on a separate project to forecast the monthly housing starts in the northeastern United States over the next year. To do so, she collects the monthly housing starts in this region for each of the past five years. These 60 values (housing starts) represent time-series data, because they are observations taken over time.

LINE CHART

Time-series data are often graphically depicted on a **line chart,** which is a plot of the variable over time. It is created by plotting the value of the variable on the vertical axis and the time periods on the horizontal axis. Line charts can be used for all three types of data.

If the time-series data are interval, we simply graph the values at the time periods where they were recorded. For example, suppose that a restaurant recorded the following weekly revenues for the past 10 weeks:

Week	Revenue
1	$4,854
2	$5,337
3	$5,205
4	$5,402
5	$6,001
6	$5,883
7	$6,309
8	$5,967
9	$6,040
10	$6,432

Excel is used to produce the line chart.

The chart reveals that there is upward trend in revenues.

Microsoft Excel Line Chart of Weekly Revenues

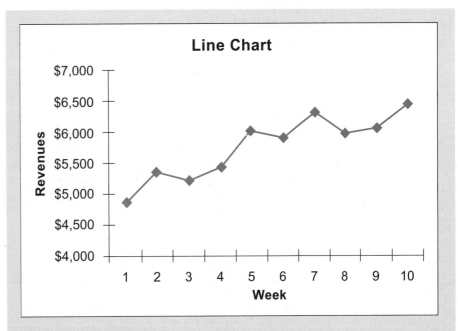

GENERAL COMMANDS

1 Type or import the data into one column.
2 Highlight the column.
3 Click **Chart Wizard, Line,** and **Finish.**
4 Click **Chart** and **Chart Options** to make whatever changes you wish.

Figure 2.7

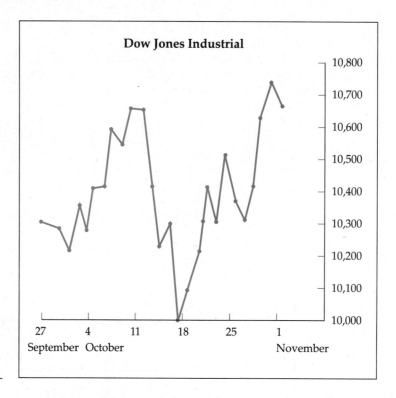

The business sections of newspapers often provide a line chart of the Dow Jones Industrial Average, an important variable that is used to measure the strength of the economy. Figure 2.7 is one such example.

When the time-series is based on nominal data, we proceed as we did in the previous section. We count the number of occurrences of a category of interest to us and graph it over time. Suppose that a series of six monthly political polls asked voters whether they support the current mayor. We identify the variable as nominal, and we count the responses. The table below shows the number of "Yes" and "No" responses. The "Don't know" responses have been omitted. (We will discuss this issue later in the book.)

Month	Responses	
	Yes	No
January	453	219
February	436	252
March	370	199
April	264	251
May	239	309
June	218	388

Because the statistics practitioner would like to gauge the level of support for the mayor and determine the trend, the line chart will plot the percentage of "Yes" responses.

We can also use a bar chart to present the same data. Both charts reveal the same information; the mayor is in trouble.

Microsoft Excel Line Chart of Percentage Support

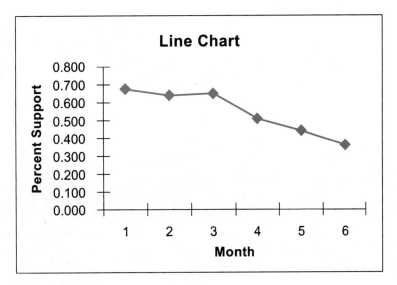

Microsoft Excel Bar Chart of Percentage Support

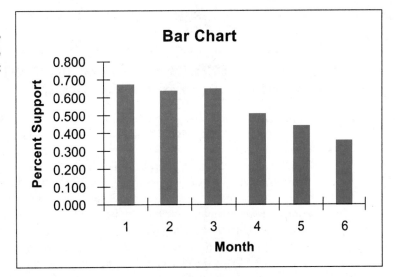

Ordinal data will be treated as nominal for the purposes of graphing a time series. For example, at the completion of most university courses, students complete a course evaluation. One question usually asks for an overall rating of the course. The question and responses may look like the following.

What is your overall rating of the course?

1 Poor   4 Very good

2 Fair   5 Excellent

3 Good

The professor may count the percentage of "Excellent" responses and graph them over his career. In that case, the data are treated as nominal.

EXERCISES

2.38 The daily closing prices of a stock on the New York Stock Exchange for the last 50 days were stored in file Xr02-38. Draw a line chart, and describe what the chart tells you.

2.39 Each semester at Wilfrid Laurier University students are asked to fill in a questionnaire that rates the courses and professors. The average ratings of a university professor for each of the 28 courses he has taught in his career are stored in sequential order in file Xr02-39.
   a Draw a line chart.
   b Briefly describe what the chart tells you about the ratings.

2.40 The retail prices of a gallon of gasoline for each of the past 20 days are stored in chronological order in Xr02-40.
   a Use a graphical technique to depict these data.
   b What have you learned about the price of gasoline?

2.41 The salary floors (minimum starting salaries) for assistant professors at a large university were recorded for the past 20 years and stored in file Xr02-41.
   a Draw a suitable chart to describe the data.
   b Briefly summarize the information you obtained.

2.42 Every month for the past two years a poll was taken to measure the perceptions of people who rate the president of the United States in his job performance. The percentages who judged that performance as satisfactory are stored in file Xr02-42. Draw a line chart, and describe what the chart tells you.

2.43 The weekly sales at a newly opened coffee shop were recorded for the first 25 weeks of operation. The data are stored in file Xr02-43.
   a Select an appropriate graphical method to present the data.
   b Describe your findings.

2.6 DESCRIBING THE RELATIONSHIP BETWEEN TWO INTERVAL VARIABLES: SCATTER DIAGRAMS

In Sections 2.3 and 2.4, we presented graphical techniques used to summarize single sets of data. For this reason, these techniques are called **univariate.** There are many situations where we wish to graphically depict the relationship between variables. Techniques such as these are called **bivariate.** The technique used to describe the relationship between two interval variables is the **scatter diagram.**

Statistics practitioners frequently need to know how two interval variables are related. For example, a criminologist wants to know how criminal behavior and poverty are related. Psychologists investigate the relationship between the intelligence of children and that of their parents. Financial analysts need to understand how the returns of individual stocks and the returns of the entire market are related.

To draw a scatter diagram, we need data for two variables. We label one variable X and the other Y. In most applications, one variable depends to some degree on the other variable. For example, the intelligence of children depends on the intelligence of their parents. In this case, we label the measure of intelligence (usually IQ) of the child Y and the intelligence measure of the parents X. We would then observe a sample of children and parents. The scatter diagram is drawn from these data. We'll illustrate when and how to draw scatter diagrams with the next example.

▼ EXAMPLE 2.6

Is the value of an education worth its cost? Education costs are easy to calculate. Costs of education include not only tuition, books, and living expenses but the cost of foregone income that a job would have provided. In an effort to learn more about the value of education a psychologist gathered data from a sample 150 30-year-old men and women. Each was asked how many years of formal education he or she had completed and his or her income for the previous 12 months. These data are stored in file Xm02-06. Draw a chart that describes the relationship between years of education and income.

Solution Using the rule outlined above we label income Y and education X.

Microsoft Excel Scatter Diagram for Example 2.6

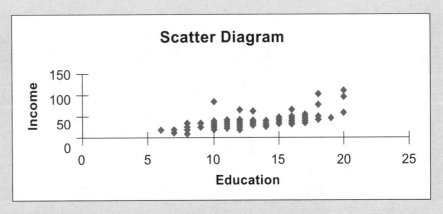

GENERAL COMMANDS

1 Type or import the data into two adjacent columns. Store variable X in the first column and variable Y in the next column.
2 Click the **Chart Wizard** icon, select **XY (Scatter)** from chart type, click on the first **Chart sub-type,** and click **Next >.**
3 Click **Data Range** (if necessary), and type the input range in the **Data range** box. Click **Next >.**
4 If you wish to label the chart and axes, click **Titles** (if necessary), and fill in the boxes.
5 Click **Gridlines,** and remove the check mark, to eliminate the horizontal lines that automatically appear. Click **Finish.**
6 If you wish to change the scale, double-click the y-axis, click **Scale,** remove the check mark under **Auto,** and change the **Minimum, Maximum,** and/or **Major** and **Minor Units.** Repeat for the x-axis.
7 To draw a straight line through the points, click **Chart and Add Trendline.** Specify **Linear,** and click **OK.**

EXAMPLE 2.6

Open file **Xm02-06.**

A1:B151

INTERPRETING THE RESULTS

The scatter diagram reveals that, in general, the greater the education the greater the income. However, there are other variables that determine income. Further analysis may reveal what these other variables are.

PATTERNS OF SCATTER DIAGRAMS

As was the case with histograms, we frequently need to describe verbally how two variables are related. The two most important characteristics are the strength and direction of the linear relationship.

Linearity

If we draw a straight line through the points of the scatter diagram in such a way that the line represents the relationship, we want to know how well the line fits. If most of the points fall close to the line, we say that there is a **linear relationship.** If most of the points appear to be scattered randomly with only a semblance of a straight line there is no, or at best a weak, linear relationship. Figure 2.8 depicts several scatter diagrams that exhibit various levels of linearity.

Note that there may well be some other type of relationship, such as a quadratic or exponential one.

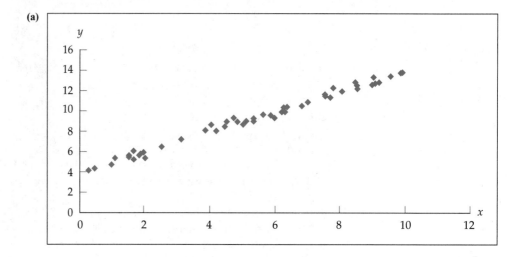

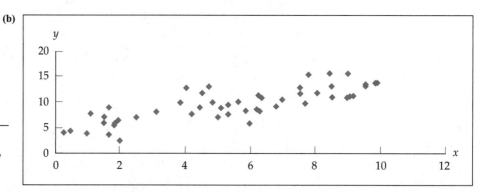

Figure 2.8

Three scatter diagrams: (a) strong linear relationship, (b) medium-strength linear relationship

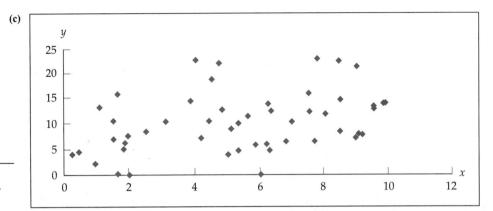

Figure 2.8

Three scatter diagrams (continued): (c) weak linear relationship

Direction

If, in general, when one variable increases, so does the other, we say that there is a **positive linear relationship.** When they tend to move in opposite directions, we describe the nature of their association as a **negative linear relationship.** (The terms *positive* and *negative* will be explained in Chapter 3.) See Figure 2.9 for examples of scatter diagrams depicting positive, negative, no relationship, and nonlinear relationship.

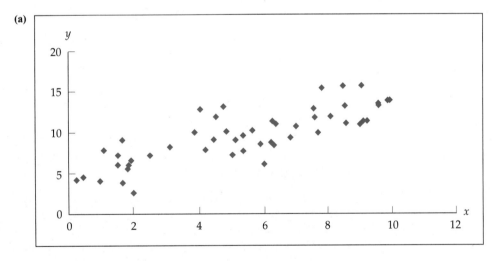

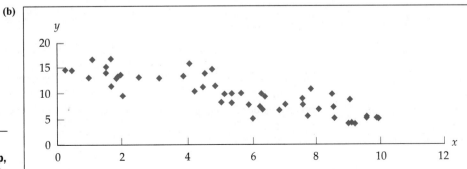

Figure 2.9

Four scatter diagrams: (a) positive linear relationship, (b) negative linear relationship

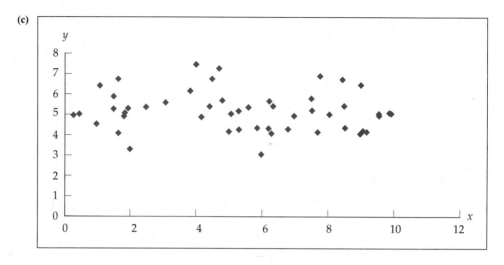

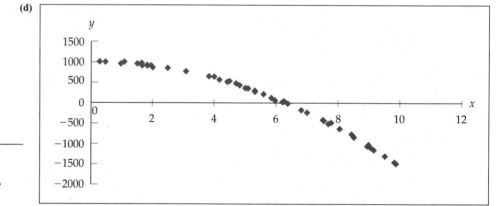

Figure 2.9

Four scatter diagrams (continued): (c) no relationship, (d) nonlinear relationship

LEAST SQUARES LINE

To determine the strength of the linear relationship, we can draw a straight line through the points. In drawing the line freehand, we would attempt to draw it so that it passes through the middle of the data. Unfortunately, different people drawing a straight line through the same set of data will produce somewhat different lines. Fortunately, statisticians have produced an objective way to draw the straight line. The method is called the least squares method. The following equation represents a straight line:

$$\hat{y}_i = b_0 + b_1 x$$

where b_0 is the y-intercept and b_1 is the slope. The term $\hat{y}$ (pronounced y-hat) is the value of y calculated from the straight line. The least squares line is determined by finding values for b_0 and b_1 that minimize the sum of squared deviations between the points and the line. That is, we minimize

$$\sum_{i=1}^{n}(y_i - \hat{y}_i)^2$$

where y_i is the actual observed value of y and $\hat{y}$ is the value of y calculated by substituting x_i into the equation of the least squares line. That is,

$$\hat{y}_i = b_0 + b_1 x_i$$

Calculus is used (we'll spare you the details) to produce equations for the coefficients. Because this chapter is devoted to graphical techniques, we will describe the least squares method in Chapter 3.

> *Factors That Identify When to Use a Scatter Diagram*
>
> 1 Objective: describe the relationship between two variables.
> 2 Data type: interval.

EXERCISES

2.44 In a study to determine the effect of a car's speed on its fuel mileage, a sample of 200 identical cars was drawn. The cars' drivers were told to drive at particular speeds ranging from 10 mph to 100 mph. The amounts of gasoline consumed over 100 miles were recorded and stored in file Xr02-44.

a Draw a scatter diagram of the data.
b On the scatter diagram, draw a straight line through the data.
c What is the direction of the line?
d Does it appear that there is a linear relationship between speed and mileage?

2.45 In a university where calculus is a prerequisite for the statistics course, a sample of students was drawn. The marks for calculus and statistics were recorded for each student. The data are stored in file Xr02-45.

a Draw a scatter diagram of the data.
b What does the graph tell you about the relationship between the marks in calculus and statistics?

2.46 Are the marks one receives in a course related to the amount of time spent studying the subject? To analyze this mysterious possibility, a student took a random sample of 60 students who had enrolled in a psychology class last semester. She asked each to report his or her mark in the course and the total number of hours spent studying psychology. These data are stored in file Xr02-46.

a Use an appropriate graphical technique to depict the data.
b What have you learned from the graph?

2.47 In Chapters 17 and 18 we introduce regression analysis, which addresses the relationships among variables. One of the first applications of regression analysis was to analyze the relationship between the heights of fathers and sons. Suppose that a sample of 80 fathers and sons was drawn. For each, the heights of the fathers and of the adult sons were measured and stored in file Xr02-47.

a Draw a scatter diagram of the data. Draw a straight line that describes the relationship.
b What is the direction of the line?
c Does it appear that there is a linear relationship between the two variables?

2.48 An increasing number of high school students hold down part-time jobs. How does this affect their performance in school? To help answer this question a sample of 300 high school students who have part-time jobs was taken. Each student reported his or her most recent average mark and the number of hours per week at his or her job. These data are stored in file Xr02-48.

a Draw a scatter diagram of the data including a straight line.
b Does it appear that there is a linear relationship between the two variables?

2.49 Who uses the Internet? To help answer this question, a random sample of 300 adults was asked to report their age and the number of hours of Internet use weekly. These data are stored in file Xr02-49.
  a Employ a suitable graph to depict the data.
  b Does it appear that there is a linear relationship between the two variables? If so, describe it.

2.50 Critics of television often refer to the detrimental effects that all the violence shown has on children. However, there may be another problem. It may be that watching television also reduces the amount of physical exercise, causing weight gains. A sample of 225 10-year-old children was taken. The number of pounds each child was overweight was recorded (a negative number indicates the child is underweight). Additionally, the number of hours of television viewing per week was also recorded. Both variables are stored in file Xr02-50.
  a Draw a scatter diagram with a straight line.
  b What is the direction of the line?
  c What does the scatter diagram tell you?

2.51 A professor of management believes that the heights of male executives affect their incomes. He takes a random sample of 90 35- to 40-year-old executives and records their height in inches and their annual incomes in thousands of dollars and stores the data in file Xr02-51.
  a Draw a scatter diagram of the data.
  b What have you learned from the scatter diagram?

2.52 A benefit of exercise is that it raises the metabolism rate. In a study to determine the effect of different durations of exercise and the metabolism rate, a sample was drawn of 100 regular exercisers. The amount of time each individual exercised and the metabolism rate one hour after the exercise were measured. The data are stored in Xr02-52.
  a Draw a scatter diagram of the data.
  b Summarize your findings.

2.53 A new antiflu vaccine has been developed, which is designed to reduce the duration of symptoms. However, the effect of the drug varies from person to person. To examine the effect of age on the effectiveness of the drug, a sample of 140 flu sufferers was drawn. Each person reported how long the symptoms of the flu persisted and his or her age. The data are stored in file Xr02-53. Draw a scatter diagram of the data, and describe your findings.

2.54 In an attempt to determine the factors that affect the amount of energy used, 200 households were analyzed. In each, the number of occupants and the amount of electricity used were measured. These data are stored in file Xr02-54.
  a Draw a scatter diagram of the data.
  b What have you learned from the scatter diagram?

2.55 The way in which most people learn to read (one word at a time) often results in slow reading. As a first step in improving the speed of reading, a group of 100 people were asked to report their number of years of education. Each person's reading speed was measured. The data are stored in file Xr02-55. Draw a graph of the data, and describe how the two variables are related.

2.56 An increasing number of consumers prefer to use debit cards in place of cash or credit cards. To analyze the relationship between the amounts of purchases made with debit cards and credit cards, 240 people were interviewed and asked to report the amount of money spent on purchases using debit and credit cards during the last month. These data are stored in file Xr02-56. Draw a scatter diagram of the data, and summarize your findings.

2.57 Refer to Exercise 2.56. The amount of money spent in cash on purchases in the last month was also recorded. These data and the debit card purchases are stored in file Xr02-57. Draw a scatter diagram, and compare this graph with the one drawn in Exercise 2.56.

2.58 The cost of repairing cars involved in accidents is one reason that insurance premiums are so high. In an experiment, 20 cars were driven into a wall. The speeds were varied between 1 and 20 mph. The costs of repair were estimated. The data are stored in file Xr02-58. Draw an appropriate graph to analyze the relationship between the two variables. What does the graph tell you?

2.7 SUMMARY

Descriptive statistical methods are used to summarize data sets so that we can extract the relevant information. In this chapter we presented graphical techniques.

Histograms are used to describe a single set of interval data. Statistics practitioners look for several aspects of the shapes of histograms. These are symmetry, number of modes, and its resemblance to a bell shape.

Bar and pie charts are employed to summarize single sets of nominal data. Because of the restrictions applied to this type of data, all that we can show is the frequency and proportion of each category.

We described the difference between time-series data and cross-sectional data. Time series are graphed by line charts. To analyze the relationship between two interval variables, we draw a scatter diagram. We look for the direction and strength of the linear relationship.

IMPORTANT TERMS

Variable	Symmetry
Values	Positively skewed
Data	Negatively skewed
Interval	Modal class
Quantitative	Unimodal
Numerical	Bimodal
Nominal	Bar chart
Qualitative	Pie chart
Categorical	Line chart
Ordinal	Univariate
Cross-sectional data	Bivariate
Time-series data	Scatter diagram
Histogram	Linear relationship
Frequency distribution	Positive linear relationship
Classes	Negative linear relationship

SYMBOLS

Symbol	Pronounced	Represents
$\hat{y}$	y-hat	Fitted or calculated value of y
b_0	b-sub-zero or b-zero	y-intercept
b_1	b-sub-one or b-one	Slope coefficient

MICROSOFT EXCEL OUTPUT AND INSTRUCTIONS

Technique	Page
Histogram	23–24
Bar chart	34
Pie chart	34
Line chart	39
Scatter diagram	43

SUPPLEMENTARY EXERCISES

2.59 The IQs of children born prematurely were measured when the children were 5 years old. The data are stored in file Xr02-59. Use whichever graphical technique you deem appropriate, and describe what the graph tells you.

2.60 Many downhill skiers eagerly look forward to the winter months and fresh snowfalls. However, winter also entails cold days. How does the temperature affect skiers' desire? To answer this question, a local ski resort recorded the temperature for 50 randomly selected days and the number of lift tickets they sold. Both variables are stored in file Xr02-60. Use a graphical technique to describe the data, and interpret your results.

2.61 There are several ways to teach applied statistics. The most popular approaches are:
1. Emphasize manual calculations.
2. Use a computer combined with manual calculations.
3. Employ a computer exclusively with no manual calculations.

A survey of 100 statistics instructors asked each to report his or her approach. The results are stored in file Xr02-61. Use a graphical method to extract the most useful information about the teaching approaches.

2.62 A sample was drawn of 220 households that recently signed on to a discount long-distance telephone supplier. The long-distance bills for the first month of service were recorded and stored in file Xr02-62. Draw a suitable graph, and briefly describe what the graph tells you.

2.63 A sample of 200 people who had purchased food at the concession stand at Yankee Stadium was asked to rate the quality of the food. The responses are:
1. Poor.
2. Fair.
3. Good.
4. Very good.
5. Excellent.

The responses (using the codes) were stored in Xr02-63. Draw a graph that describes the data. What does the graph tell you?

2.64 In another study conducted at Yankee Stadium, the concession manager wanted to know how the temperature affected beer sales. Accordingly, she took a sample of 50 games and recorded the number of beers sold and the mean temperature in the middle of the game. These data are stored in file Xr02-64. Use a suitable graphical technique to describe the data. What conclusions can you obtain from the graph?

2.65 A sample of 125 university students was asked how many books they borrowed from the library over the past 12 months. Their replies are stored in file Xr02-65. Use a suitable graphical method to depict the data. What does the graph tell you?

2.66 University and college students often play card games in their spare time. A survey of a university lounge identified the games that were played. The possibilities are:
1. Bridge.
2. Hearts.
3. Poker.
4. Other.

The results are stored in file Xr02-66. Use an appropriate graphical method to summarize the data. What does the graph tell you about the games students play in the lounge?

2.67 The following table lists the share of prime-time television viewing enjoyed by each of the major television networks in a recent year. Draw a pie chart to depict these numbers.

Network	Share
CBS	23%
ABC	20%
NBC	18%
Fox	11%

2.68 The annual budgetary expenditures for faculty salaries were recorded for the past 20 years and stored in Xr02-68. Summarize these data by creating a graph. What information does the graph produce?

2.69 One hundred students who had reported that they use their computers for at least 20 hours per week were asked to keep track of the number of crashes their computers incurred during a 12-week period. The data are stored in file Xr02-69. Using an appropriate statistical method, summarize the data. Describe your findings.

2.70 A sample of 300 Alpine (downhill) skiers was asked to report the number of days they skied the previous year. The responses are stored in file Xr02-70. Use an appropriate graphical procedure to summarize these data, and describe your findings.

2.71 The Wilfrid Laurier University bookstore conducts annual surveys of its customers. One question asks respondents to rate the prices of textbooks. The wording is "The bookstore's prices of textbooks are reasonable." The responses are:

1. Strongly disagree.
2. Disagree.
3. Neither agree nor disagree.
4. Agree.
5. Strongly agree.

The responses for a group of 115 students were stored in file Xr02-71. Graphically summarize these data, and report your findings.

2.72 A survey conducted by Colgate Palmolive (makers of toothpaste and toothbrushes) asked a random sample of adult women to report the one thing that they would like to have to improve their smiles. The responses are:

1. Whiter teeth.
2. Straighter teeth.
3. Fuller lips.

The responses are stored in file Xr02-72. Employ a suitable graphical procedure to describe the responses. Describe your findings.

2.73 Are the times of the winning runner in the New York Marathon decreasing? To answer this question, the times to complete the race for the winning male runners for the years 1978 to 1998 were recorded and stored in file Xr02-73. Draw a suitable graph, and briefly comment on the results.

2.74 The Red Lobster Restaurant chain conducts regular surveys of its customers to monitor the performance of individual restaurants. One of the questions asks customers to rate the overall quality of their visit. The listed responses are:

1. Poor.
2. Fair.
3. Good.
4. Very good.
5. Excellent.

The results of 200 customers are stored in file Xr02-74. Graphically depict these data, and describe your findings.

2.75 One way to judge children's development is to measure the size of their vocabulary. The vocabularies of 200 5-year-old children were measured and stored in file Xr02-75. Utilize a suitable graphical technique to present these data. What does the graph indicate about 5-year-old children's vocabularies?

2.76 Casino Windsor regularly conducts surveys of its customers. Among other questions, respondents were asked to give their opinion about their "overall impression of Casino Windsor." The responses are:

Excellent, Good, Average, Poor, Unacceptable.

The responses were recorded as 5, 4, 3, 2, 1, respectively, and stored in file Xr02-76. Use a graphical procedure to summarize the data, and describe your findings.

2.77 When drivers are lost, what do they do? To help answer this question, a group of drivers was asked. There were four possible responses. They are:

1. Consult a map.
2. Ask someone for directions.
3. Continue driving until location or direction is determined.
4. Other.

The responses are stored in file Xr02-77. Graphically depict the data, and describe what you have learned.

2.78 Refer to Exercise 2.73. Does temperature affect the race times? The answer to this question lies in the data stored in file Xr02-78, which stores the winning times for male runners and the temperature on the day the race was run. Use a graphical technique to present these data, and describe what you have learned.

2.79 Refer to Exercise 2.78. The winning times of female runners and temperatures were stored in file Xr02-79. Graph these data, and report your results.

2.80 It is generally believed that one of the negative effects of quitting smoking is weight gain. To examine this issue, a group of ex-smokers who had quit 12 months earlier was asked to report their weight gain. These data are stored in file Xr02-80. Draw an appropriate graph of these data, and describe what the graph tells you.

2.81 Several years ago the National Hockey League instituted a five-minute overtime period when the game was tied after the regulation 60-minute game was played. The objective was to increase the excitement of the game. One way of measuring the action is to count the number of shots during the overtime period. A sample of games where the teams played the full five minutes was taken as well as the total number of shots. These data are stored in file Xr02-81. Use a graphical technique to summarize the data. Report your findings.

2.82 Neilsen Media Research conducts surveys of television viewers. File Xr02-82 stores the results of the February 2000 (reported in *Miami Herald*, March 3, 2000) weekday prime-time viewing in South Florida. The responses are:

1. CBS.
2. ABC.
3. NBC.
4. Fox.
5. Other.
6. Television off.

Draw a graph of these data, and briefly report your findings.

Chapter 3

Numerical Descriptive Techniques for Interval Data

3.1 Introduction
3.2 Measures of Central Location
3.3 Measures of Variability
3.4 Other Measures of Shape (Optional)
3.5 Measures of Relative Standing and Box Plots
3.6 Measures of Linear Relationship
3.7 General Guidelines for Exploring Data
3.8 Summary

3.1 INTRODUCTION

In Chapter 2 we presented several graphical techniques that describe data. In this chapter, we introduce numerical descriptive techniques, which allow the statistics practitioner to be more precise in describing various characteristics of a sample or population. They also are critical to the development of statistical inference.

As we pointed out in Chapter 2, arithmetic calculations can be applied to interval data only. Consequently, the techniques introduced here may only be employed to numerically describe interval data. The exceptions are the methods presented in Section 3.5, which apply to both interval and ordinal data.

When we introduced the histogram, we commented that there are several bits of information that we look for. The first was the location of the center of the data. In Section 3.2 we will present measures of central location. Another important characteristic that we seek from a histogram is the spread of the data. The spread will be measured more precisely by measures of variability, which we'll present in Section 3.3. When verbally describing a histogram, we frequently wish to report the shape and in particular whether the histogram is symmetric or skewed and how similar it is to a bell shape. Two measures of shape will be introduced in Section 3.4. In the illustrative example about marks on a statistics exam, we noted that most students want to know where they stand relative to the rest of the class. Section 3.5 introduces measures of relative standing and another graphical technique, the box plot.

In Section 2.6 we introduced the scatter diagram, which is a graphical method that we use to analyze the relationship between two variables. The numerical counterparts to the scatter diagram are called measures of association, and they will be presented in Section 3.6.

We complete this chapter with a set of guidelines on how to explore a data set searching for information.

SAMPLE STATISTIC OR POPULATION PARAMETER

In Chapter 1 we introduced the terms *population, sample, parameter,* and *statistic*. Recall that a parameter is a descriptive measurement about a population and a statistics is a descriptive measurement about a sample. In the sections that follow, we will introduce a dozen descriptive measurements. For each, we will describe how to calculate both the population parameter and the sample statistic. However, in most realistic applications populations are very large, in fact, virtually infinite. The formulas describing the calculation of parameters are not practical and seldom used. They are provided here primarily to teach the concept and the notation. In Chapter 5 we'll introduce probability distributions, which describe populations. At that time we'll show how parameters are calculated from probability distributions. In general, small data sets of the type we feature in this book are samples.

3.2 MEASURES OF CENTRAL LOCATION

ARITHMETIC MEAN

There are three different measures that we use to describe the center of a set of data. The first is the best known, the arithmetic mean, which we'll refer to simply as the **mean.** Students may be more familiar with its other name, the *average*. The mean is computed by summing the observations and dividing by the number of observations. We label the observations in a sample $x_1, x_2, \ldots, x_n$, where x_1 is the first observation, x_2 is the second, and so on, until x_n, where n is the sample size. As a result the sample mean is denoted $\bar{x}$. The number of observations in a population is labeled N. The population mean is denoted by μ (Greek letter *mu*).

Mean

$$\text{Population mean: } \mu = \frac{\sum_{i=1}^{N} x_i}{N}$$

$$\text{Sample mean: } \bar{x} = \frac{\sum_{i=1}^{n} x_i}{n}$$

In this chapter we will use several examples to illustrate the calculation and interpretation of various statistics. Although this book is committed to calculating statistics by computer and Microsoft Excel, we believe that computing statistics from small samples is beneficial to your understanding of statistics. Accordingly, we will offer you two examples, one featuring a small sample size whose statistics we can compute manually, while a second example will describe a large sample stored on the CD that accompanies this book. It will be employed to demonstrate how Excel performs the computations.

▼ **EXAMPLE 3.1**

A sample of 10 adults was asked to report the number of hours they spent on the Internet the previous month. The results are listed below. Manually calculate the sample mean.

0, 7, 12, 5, 33, 14, 8, 0, 9, 22

Solution Using our notation we have $x_1 = 0, x_2 = 7, \ldots, x_{10} = 22$, and $n = 10$. The sample mean is

$$\bar{x} = \frac{\sum_{i=1}^{n} x_i}{n} = \frac{0 + 7 + 12 + 5 + 33 + 14 + 8 + 0 + 9 + 22}{10} = \frac{110}{10} = 11.0$$

▼ **EXAMPLE 3.2**

Refer to Example 2.1. Find the mean amount of time spent at the Barnes Exhibit by the sample of 400 visitors.

Solution Although it is possible to perform the calculations manually, there is little reason to do so. As we did in Chapter 2, we will employ a computer and Excel to do this work. If we simply want to compute the mean and no other statistics, we can proceed as follows.

> COMMANDS
>
> 1 Begin by typing or importing the data.
> 2 Activate any empty cell.
> 3 Click f_x, Function category: **Statistical,** Function name: **Average.** Specify the input range of the data.

Alternatively, type the following into any empty cell.

=**AVERAGE**([Input range])

For Example 3.2, the active cell will show 52.7.

In most applications, the statistics practitioner wants to know more than the mean of the data. To generate a longer list of statistics follow the instructions below.

> | GENERAL COMMANDS | EXAMPLE 3.2 |
> |---|---|
> | 1 Type or import the data. | Open file **Xm02-01.** |
> | 2 Click **Tools, Data Analysis...,** and **Descriptive Statistics.** | |
> | 3 Specify the **Input Range.** | A1:A401 |
> | 4 Click **Labels in First Row,** if applicable; click **Summary Statistics;** and click **OK.** | |

Excel will yield the printout below.

Microsoft Excel Output for Example 3.2

Times	
Mean	52.7
Standard Error	0.982
Median	48
Mode	41
Standard Deviation	19.64
Sample Variance	385.7
Kurtosis	2.71
Skewness	1.38
Range	128
Minimum	23
Maximum	151
Sum	21080
Count	400

We will discuss all the statistics in the printout as we proceed through this chapter. All three methods yield $\bar{x} = 52.7$. This value tells us that the average Barnes Exhibit visitor stayed for 52.7 minutes.

MEDIAN

The second most popular measure of central location is the **median.**

> **Median**
>
> The median is calculated by placing all the observations in order (ascending or descending). The observation that falls in the middle is the median. The sample and population medians are computed in the same way.

When there is an even number of observations the median is determined by averaging the two observations in the middle.

▼ EXAMPLE 3.3

Find the median for the data in Example 3.1.

Solution When placed in ascending order the data appear as follows:

0, 0, 5, 7, 8, 9, 12, 14, 22, 33

The median is the average of the fifth and sixth observations (the middle two), which are 8 and 9, respectively. Thus, the median is 8.5.

EXAMPLE 3.4

Find the median of the 400 observations in Example 2.1.

Solution Excel is induced to determine the median in the same way that it did for the mean. To calculate only the median, substitute **MEDIAN** in place of **AVERAGE.** That is, type into any empty cell =**MEDIAN**([Input Range]).

If you examine the printout from Example 3.2, you will see that the command **Descriptive Statistics** also produces the median. It is 48, which tells us that half of the times in the Barnes Exhibit are less than 48 minutes and half are greater than 48 minutes.

MODE

The third and last measure of central location that we present here is the **mode.**

> **Mode**
>
> The mode is defined as the observation (or observations) that occurs with the greatest frequency. Both the statistic and parameter are computed in the same way.

For populations and large samples, it is preferable to report the modal class, which we defined in Chapter 2.

There are several problems with using the mode as a measure of central location. First, in a small sample, it may not be a very good measure. Second, it may not be unique.

EXAMPLE 3.5

Find the mode for the data in Example 3.1.

Solution All observations except 0 occur once. There are two 0s. Thus, the mode is 0. As you can see, this is a poor measure of central location. It is nowhere near the center of the data. Compare this with the mean 11.0 and median 8.5, and you can appreciate that in this example the mean and median are superior measures.

▼ EXAMPLE 3.6

Determine the mode for Example 2.1.

Solution To compute the mode, substitute **MODE** in place of **AVERAGE** in the instructions above. The mode is also printed by **Descriptive Statistics.** However, if there is more than one mode, Excel only prints the smallest one, without indicating whether there are other modes. In this example Excel prints 41. We don't know whether other modes exist.

▲

MEAN, MEDIAN, MODE: WHICH IS BEST?

With three measures to choose, which one should we use? There are several factors to consider when making our choice of measure of central location. The mean is generally our first selection. However, there are several circumstances when the median is better. The mode is seldom the best measure of central location. One advantage the median holds is that it not as sensitive to extreme values as is the mean. To illustrate, consider the data in Example 3.1. The mean was 11.0, and the median was 8.5. Now suppose that the respondent who reported 33 hours actually reported 133 hours (obviously an Internet addict). The mean becomes

$$\bar{x} = \frac{\sum_{i=1}^{n} x_i}{n} = \frac{0 + 7 + 12 + 5 + 133 + 14 + 8 + 0 + 9 + 22}{10} = \frac{210}{10} = 21.0$$

This value is exceeded by only two of the ten observations in the sample, making this statistic a poor measure of *central* location. The median stays the same. When there is a relatively small number of extreme observations (either very small or very large but not both), the median usually produces a better measure of the center of the data.

To see another advantage of the median over the mean, suppose you and your classmates have written a statistics test, and the instructor is returning the graded tests. What piece of information is most important to you? The answer, of course, is *your* mark. What is the next important bit of information? The answer is how well you performed relative to the class. Most students ask their instructor for the class mean. This is the wrong statistic to request. You want the median. To see why, suppose that your mark was 77%. Typically you would like to know whether you are in the top half of the class. Because the median divides the class into two halves it provides this information; the mean does not. Nevertheless, the mean can also be useful in this scenario. If there are several sections of the course, the section means can be compared to determine whose class performed best (or worst).

MEASURES OF CENTRAL LOCATION FOR ORDINAL AND NOMINAL DATA

When the data are interval, we can use any of the three measures of central location. However, for ordinal and nominal data the calculation of the mean is not valid. Because the calculation of the median begins by placing the data in order, this statistic is appropriate for ordinal data. The mode, which is determined by counting the frequency of each observation, is appropriate for nominal data. However, nominal data do not have a "center," so we cannot interpret the mode of nominal data in that way. It is generally pointless to compute the mode of nominal data.

EXERCISES

3.1 A sample of 12 people was asked how much change they had in their pockets and wallets. The responses (in cents) are

52, 25, 15, 0, 104, 44, 60, 30, 33, 81, 40, 5, 71

(The data are stored in file Xr03-01.)

a Determine the mean, median, and mode for these data.
b Briefly describe what each statistic tells you.

3.2 The number of sick days due to colds and flu last year was recorded by a sample of 15 adults. The data are

5, 7, 0, 3, 15, 6, 5, 9, 3, 8, 10, 5, 2, 0, 12

(The data are stored in file Xr03-02.)

a Compute the mean, median, and mode.
b Describe what you have learned from the three statistics calculated in part (a).

3.3 A random sample of 12 joggers was asked to keep track and report the number of miles they ran last week. The responses are

5.5, 7.2, 1.6, 22.0, 8.7, 2.8, 5.3, 3.4, 12.5, 18.6, 8.3, 6.6

(The data are stored in file Xr03-03.)

a Compute the three statistics that measure central location.
b Briefly describe what each statistic tells you.

3.4 The midterm test for a statistics course has a time limit of one hour. However, like most statistics exams, this one was quite easy. To measure how easy, the professor recorded the amount of time taken by a sample of nine students to hand in their test papers. The times are (rounded to the nearest minute)

33, 29, 45 60, 42, 19, 52, 38, 36

(The data are stored in file Xr03-04.)

a Compute the mean, median, and mode.
b What you have learned from the three statistics calculated in part (a)?

3.5 The professors at Wilfrid Laurier University are required to submit their final exams to the registrar's office 10 days before the end of the semester. The exam coordinator sampled 20 professors and recorded the number of days before the final exam that each submitted his or her exam. The results are

14, 8, 3, 2, 6, 4, 9, 13, 10, 12, 7, 4, 9, 13, 15, 8, 11, 12, 4, 0

(The data are stored in file Xr03-05.)

a Compute the mean, median, and mode.
b Briefly describe what each statistic tells you.

3.6 The starting salaries of a sample of 125 recent graduates of a teachers' college who landed teaching jobs are stored in file Xr03-06.

a Determine the mean and median of these data.
b What do these two statistics tell you about the starting salaries of teachers?

3.7 To determine whether changing the color of their invoices improves the speed of payment, 200 customers were selected at random and sent their invoices on blue-colored paper. The number of days until the bills were paid was recorded and stored in file Xr03-07. Calculate the mean and median of these data. Report what you have discovered.

3.8 Refer to Exercise 2.8.

  a Calculate the mean and median of the service times of the 420 customers. (The data are in file Xr02-08.)
  b What do these two statistics tell you about service times?
  c Which statistic is more useful to you?
  d How does the information you gained from the statistics compare to what you learned from the histogram?

3.9 Refer to Exercise 2.9.
  a Calculate the mean and median of the sample of 85 house prices. (The data are stored in Xr02-09.)
  b What do these statistics tell you about house prices?
  c How does the information you gained from the statistics compare to what you learned from the histogram?

3.10 To learn more about the size of withdrawals at a banking machine, the proprietor took a sample of 75 withdrawals and stored the amounts in file Xr03-10.

  a Determine the mean and median of these data.
  b What do these two statistics tell you about the withdrawal amounts?

3.11 In an effort to slow drivers, traffic engineers painted a solid line 3 feet from the curb the entire length of a road and filled the space with diagonal lines. The lines made the road look narrower. A sample of car speeds was taken after the lines were drawn. The speeds (in miles per hour) are stored in file Xr03-11.
  a Compute the mean, median, and mode of these data.
  b Briefly describe the information you acquired from each statistic calculated in part (a).

3.12 Refer to Exercise 2.12. (The data are stored in file Xr02-12.)

  a Compute the mean and median of the lengths of newborn babies.
  b What do these statistics reveal about the data?
  c How does the information you gained from the statistics compare to what you learned from the histogram?

3.13 Increasing tuition has resulted in some students being saddled with large debts upon graduation. To examine this issue, a sample of recent graduates was asked to report whether they had student loans, and if so, how much was the debt at graduation. The amounts of debt were stored in file Xr03-13.
  a Compute all three measures of central location.
  b What do these statistics reveal about student loan debt at graduation?

3.14 Refer to Exercise 2.10. If you are a professor and want to gauge how well paid you are, which measure of location should you compute? Calculate the statistic, and describe what it tells you. (The data are stored in Xr02-10.)

3.15 For many restaurants, the amount of time customers linger over coffee and dessert negatively affect profits. To learn more about this variable, a sample of 200 restaurant groups was observed, and the amount of time (in minutes) customers spent in the restaurant was recorded and stored in file Xr03-15.
  a Calculate the mean and median of the sample of 200 groups.
  b What do these statistics tell you about the amount of time spent in this restaurant?

3.3 MEASURES OF VARIABILITY

The statistics introduced in Section 3.2 serve to provide information about the central location of the data. However, as we've already discussed in Chapter 2, there are other characteristics of data that are of interest to practitioners of statistics. One such characteristic is the spread or variability of the data. In this section we introduce three measures of variability. Let's start with the simplest.

RANGE

> **Range**
>
> Range = Largest observation − Smallest observation

EXAMPLE 3.7

Find the range of the data in Example 2.1.

Solution The largest and smallest observations are 151 and 23, respectively. Thus,

Range = 151 − 23 = 128.

The advantage of the **range** is its simplicity. The disadvantage is also its simplicity. Because the range is calculated from only two observations, it tells us nothing about the other observations. Consider the following two sets of data.

Set 1: 4, 4, 4, 4, 4, 50

Set 2: 4, 8, 15, 24, 39, 50

The range of both sets is 46. The two sets of data are completely different, and yet the range is the same. To measure variability we need other statistics that incorporate all the data and not just two observations.

VARIANCE

The **variance** and its related measure, the standard deviation, are arguably the most important statistics. They are used to measure variability, but as you will discover they play a vital role in almost all statistical inference procedures.

Variance

Population variance: $\sigma^2 = \dfrac{\sum_{i=1}^{N}(x_i - \mu)^2}{N}$

Sample variance*: $s^2 = \dfrac{\sum_{i=1}^{n}(x_i - \bar{x})^2}{n - 1}$

The population variance is represented by σ^2 (Greek letter *sigma* squared). In both the population and sample variances the numerator measures the sum of squared deviations from the mean.

---

*Technically, the sample variance is calculated by dividing the sum of squared deviations by n. The statistic computed by dividing the sum of squared deviations by $n - 1$ is called the *sample variance corrected for the mean*. Because this statistic is used extensively, we will shorten its name to sample variance.

Examine the formula for the sample variance s^2. It may seem strange that in calculating s^2 we divide the sum of squared deviations by $n - 1$ rather than by n. However, we do so for the following reason. Population parameters in practical settings are seldom known. One of the objectives of statistical inference is to estimate the parameter from the statistic. For example, we estimate the population mean μ from the sample mean $\bar{x}$. Although it is not logical, the statistic created by dividing the sum of squared deviations by $n - 1$ is a better estimator than is the one created by dividing by n. We will discuss this issue in greater detail in Chapter 9.

To compute s^2, we begin by calculating the mean. Next we compute the difference (deviation) between each observation and the mean. We square the deviations and sum. Finally, we divide the sum of squared deviations by $n - 1$.

We'll illustrate with a simple example. Suppose that we have the following observations of the numbers of hours five students spent studying statistics last week:

8, 4, 9, 11, 3

The mean is

$$\bar{x} = \frac{8 + 4 + 9 + 11 + 3}{5} = \frac{35}{5} = 7$$

For each observation we determine the deviation from the mean. The deviations are

$8 - 7 = 1$

$4 - 7 = -3$

$9 - 7 = 2$

$11 - 7 = 4$

$3 - 7 = -4$

Squaring the deviations yields

$(1)^2 = 1$

$(-3)^2 = 9$

$(2)^2 = 4$

$(4)^2 = 16$

$(-4)^2 = 16$

Summing and dividing by $n - 1$ produces

$$s^2 = \frac{1 + 9 + 4 + 16 + 16}{5 - 1} = \frac{46}{4} = 11.5$$

The calculation of this statistic raises several questions. Why do we square the deviations before averaging? If you examine the deviations, you will see that some of them are positive and some are negative. When you add them together the sum is 0. This will always be the case, because the positive and negative deviations cancel each other out, leaving a sum of 0. Consequently, we square the deviations to avoid the "canceling effect."

Is it possible to avoid the canceling effect without squaring? We could average the *absolute value* of the deviations. In fact, such a statistic has already been invented. It is called the mean absolute deviation or MAD. However, this statistic has limited utility and is seldom used.

What is the unit of measurement of the variance? Because we squared the deviations, we also squared the units. In this illustration, the units were hours (of study). Thus, the sample variance is 11.5 hours^2.

The next example illustrates how variance can be employed.

▼ EXAMPLE 3.8

Consistency is a hallmark of a good golfer. Golf equipment manufacturers are constantly seeking ways to improve their products. Suppose that a recent innovation is designed to improve a player's consistency. As a test a golfer was asked to hit 150 shots using a 7-iron, 75 of which were hit with his current club and 75 with the new innovation 7-iron. The distances were measured and recorded in column A (current club) and column B (innovation) in file Xm03-08. Which 7-iron is more consistent?

Solution We can get Excel to print the sample variance in a way similar to the measures of central location; substitute **VAR** instead of **AVERAGE**. The active cell will contain s^2. Alternatively, we can calculate all the descriptive statistics, a course of action we recommend, because we often need several statistics. The printouts for both 7-irons appear below.

Microsoft Excel Output for Example 3.8

Current		*Innovation*	
Mean	150.5	Mean	150.1
Standard Error	0.669	Standard Error	0.357
Median	151	Median	150
Mode	150	Mode	149
Standard Deviation	5.79	Standard Deviation	3.09
Sample Variance	33.55	Sample Variance	9.56
Kurtosis	0.127	Kurtosis	−0.885
Skewness	−0.430	Skewness	0.177
Range	28	Range	12
Minimum	134	Minimum	144
Maximum	162	Maximum	156
Sum	11291	Sum	11261
Count	75	Count	75

INTERPRETING THE RESULTS

The variance of the distances of the current 7-iron is 33.55 yards^2 whereas that of the innovative 7-iron is 9.56 yards^2. Based on this sample, the innovative club is more consistent. Because the mean distances are similar, it would appear that the new club is indeed superior.

▲

INTERPRETING THE VARIANCE

The variance can be used to compare two sets of data; the larger the variance, the greater the spread. However, variance cannot be interpreted by itself. One of the reasons for this is that we squared the original units. For example, the variance of the distances for the current 7-iron in Example 3.8 is 33.55 yards^2. Fortunately this problem can easily be remedied by creating another statistic, the standard deviation.

STANDARD DEVIATION

The **standard deviation** is the square root of the variance.

Standard Deviation

Population standard deviation: $\sigma = \sqrt{\sigma^2}$

Sample standard deviation: $s = \sqrt{s^2}$

In Example 3.8, the standard deviation of the distance to target for the current and the innovative 7-irons are

Current 7-iron: $s = \sqrt{s^2} = \sqrt{33.55} = 5.79$ yards

Innovative 7-iron: $s = \sqrt{s^2} = \sqrt{9.56} = 3.09$ yards

The computer also outputs these statistics.

The standard deviation has two advantages over the variance as a measure of variability. First, by taking the square root of the variance, the standard deviation's unit of measurement is the same as that of the variable we've observed. Second, we can interpret what the standard deviation means.

INTERPRETING THE STANDARD DEVIATION

Knowing the mean and standard deviation allows the statistics practitioner to extract useful bits of information. The information depends on the shape of the histogram. If the histogram is bell shaped, we can use the **Empirical Rule.**

Empirical Rule

1. Approximately 68% of all observations fall within one standard deviation of the mean.
2. Approximately 95% of all observations fall within two standard deviations of the mean.
3. Approximately 99.7% of all observations fall within three standard deviations of the mean.

Figure 3.1 depicts a histogram with the intervals and percentages described by the Empirical Rule.

A more general interpretation of the standard deviation is derived from **Chebysheff's Theorem,** which applies to all shapes of histograms.

Chebysheff's Theorem

The proportion of observations in any sample that lie with k standard deviations of the mean is at least $1 - \frac{1}{k^2}$ for $k > 1$.

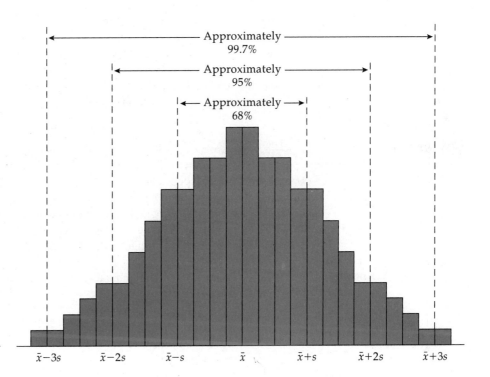

Figure 3.1

Empirical Rule

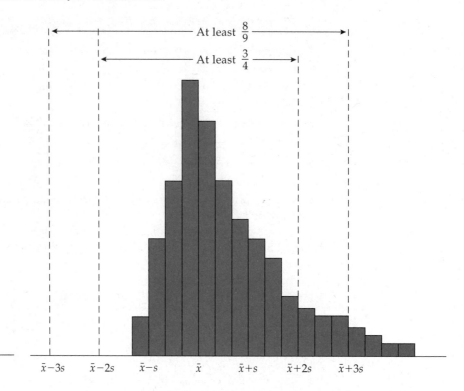

Figure 3.2

Chebysheff's Theorem

When $k = 2$, Chebysheff's Theorem states that at least three-quarters of all observation lie within two standard deviations of the mean. With $k = 3$, Chebysheff's Theorem states that at least eight-ninths of all observations lie within three standard deviations of the mean. See Figure 3.2.

Note that the Empirical Rule provides approximate proportions, while Chebysheff's Theorem provides lower bounds on the proportions contained in the intervals.

▼ EXAMPLE 3.9

At each monthly checkup, pediatricians regularly weigh their patients to determine whether they are growing too slowly. Suppose that the weights of 6-month-old babies produced a bell-shaped histogram. The mean and standard deviation were computed and found to be 18 and 1.5 pounds, respectively. Provide guidelines for pediatricians to select babies that are low weight or overweight.

Solution The intervals created by adding and subtracting one, two, and three standard deviations to and from the mean are as follows.

One standard deviation:

$\bar{x} - s = 18 - 1.5 = 16.5$

$\bar{x} + s = 18 + 1.5 = 19.5$

Two standard deviations:

$$\bar{x} - 2s = 18 - 2(1.5) = 15$$
$$\bar{x} + 2s = 18 + 2(1.5) = 21$$

Three standard deviations:

$$\bar{x} - 3s = 18 - 3(1.5) = 13.5$$
$$\bar{x} + 3s = 18 + 3(1.5) = 22.5$$

Because the histogram is bell shaped, we can employ the Empirical Rule. It states that approximately 68% of the weights of all babies will fall between 16.5 and 19.5 pounds; approximately 95% of the weights of all babies will fall between 15 and 21 pounds; and approximately 99.7% of the weights of all babies will fall between 13.5 and 22.5 pounds.

Babies whose weights fall below 13.5 pounds are certainly underweight. Those weighing more than 13.5 but less than 15 pounds are likely underweight, and those weighing less than 16.5 pounds should be monitored. A similar set of guidelines can be created to monitor overweight babies.

▲

▼ EXAMPLE 3.10

The annual salaries of the employees of a chain of computer stores produced a positively skewed histogram. The mean and standard deviation are $28,000 and $3,300, respectively. What can you say about the salaries at this chain?

Solution The intervals created by adding and subtracting two and three standard deviations to and from the mean are as follows.

Two standard deviations:

$$\bar{x} - 2s = 28{,}000 - 2(3{,}300) = 21{,}400$$
$$\bar{x} + 2s = 28{,}000 + 2(3{,}300) = 34{,}600$$

Three standard deviations:

$$\bar{x} - 3s = 28{,}000 - 3(3{,}300) = 18{,}100$$
$$\bar{x} + 3s = 28{,}000 + 3(3{,}300) = 37{,}900$$

Because the histogram is positively skewed and thus not bell shaped, we cannot use the Empirical Rule. Chebysheff's Theorem states that at least three-quarters of salaries lie between $21,400 and $34,600 and at least eight-ninths of all salaries lie between $18,100 and $37,900.

▲

COMPARING GRAPHICAL AND NUMERICAL TECHNIQUES

As we mentioned in Chapter 2, graphical techniques are useful in producing a quick picture of the data. However, numerical techniques allow for greater precision. To illustrate, recall that in Example 2.3 we drew histograms of the returns of the two investments. However, it was difficult to reach a clear-cut conclusion about which investment appeared to be better. To help judge the two investments we calculate the statistics. The Excel printouts for both sets of data appear below.

Microsoft Excel Output for Example 2.3

Return A		Return B	
Mean	10.95	Mean	12.76
Standard Error	3.10	Standard Error	3.97
Median	9.88	Median	10.755
Mode	12.89	Mode	#N/A
Standard Deviation	21.89	Standard Deviation	28.05
Sample Variance	479.3	Sample Variance	786.6
Kurtosis	−0.324	Kurtosis	−0.618
Skewness	0.545	Skewness	0.011
Range	84.95	Range	106.47
Minimum	−21.95	Minimum	−38.47
Maximum	63	Maximum	68
Sum	547.27	Sum	638.01
Count	50	Count	50

Interpreting the Results

It is now easy to see that the mean return for A is less than that for B. However, the standard deviation for A is also somewhat lower than that of B. It does appear that Investment A is better because of its lower variability.

EXERCISES

3.16 Calculate the variance of the following data.
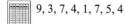
9, 3, 7, 4, 1, 7, 5, 4

3.17 Calculate the variance of the following data.
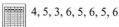
4, 5, 3, 6, 5, 6, 5, 6

3.18 Determine the variance and standard deviation of the following sample.

12, 6, 22, 31, 23, 13, 15, 17, 21

3.19 Find the variance and standard deviation of the following sample.
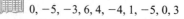
0, −5, −3, 6, 4, −4, 1, −5, 0, 3

3.20 Examine the three samples listed below. Without performing any calculations, indicate which sample has the largest amount of variation and which sample has the smallest amount of variation. Explain how you produced your answer.

a 17, 29, 12, 16, 11
b 22, 18, 23, 20, 17
c 24, 37, 6, 39, 29

3.21 Refer to Exercise 3.20. Calculate the variance for each part. Was your answer in Exercise 3.20 correct?

3.22 A friend calculates a variance and reports that it is −25.0. How do you know that he has made a serious calculation error?

3.23 Create a sample of five numbers whose mean is 10 and whose standard deviation is 0.

3.24 Refer to Exercise 2.8.
  a Calculate the range, variance, and standard deviation.
  b Which statistic is more useful to you?
  c How does the information you gained from the statistic in part (b) compare to what you learned from the histogram?

3.25 Refer to Exercise 2.9.
  a Calculate the standard deviation.
  b Which does this statistic tell you about the variability of the data?
  c How does the information you gained from the statistic in part (b) compare to what you learned from the histogram?

3.26 Three men were trying to make the football team as punters. The coach had each of them punt the ball 50 times and the distances were recorded in columns 1, 2, and 3 in file Xr03-26.
  a Compute the variance and standard deviation for each punter.
  b What do these statistics tell you about the punters?

3.27 Variance is often used to measure quality in production-line products. Suppose that a sample of steel rods that are supposed to be exactly 100 cm long is taken. The results are stored in Xr03-27. Calculate the variance and the standard deviation.

3.28 Refer to Exercise 2.12.
  a Calculate the standard deviation.
  b Which does this statistic tell you about the variability of the data?

3.29 A set of data whose histogram is bell shaped yields a sample mean and standard deviation of 50 and 4, respectively. Approximately what proportion of observations
  a are between 46 and 54?
  b are between 42 and 58?
  c are between 38 and 62?

3.30 Refer to Exercise 3.29. Approximately what proportion of observations
  a are less than 46?
  b are less than 58?
  c are greater than 54?

3.31 Refer to Exercise 2.14.
  a Compute the standard deviation of the weights of tomatoes.
  b If the histogram is bell shaped, interpret the standard deviation.

3.32 Many traffic experts argue that the most important factor in accidents is not the average speed of cars but the standard deviation. Suppose that a sample was drawn of the speeds of 200 cars over a stretch of highway that has seen numerous accidents. The data are stored in Xr03-32. Compute the variance and standard deviation of the speeds.

3.33 A set of data whose histogram is extremely skewed yields a sample mean and standard deviation of 70 and 12, respectively. What is the minimum proportion of observations that
  a are between 46 and 94?
  b are between 34 and 106?

3.34 Refer to Exercise 2.11.
  a Compute the standard deviation of the marks.
  b If the histogram is bell shaped, provide an interpretation of the standard deviation.

3.35 Everyone is familiar with waiting lines or queues. For example, people wait in line at a supermarket to go through the checkout counter. There are two factors that determine how long the queue becomes. One is the speed of service. The other is the number of arrivals at the checkout counter. The mean number of arrivals is an important number, but so is the standard deviation. Suppose that a consultant for the supermarket measures the number of arrivals per hour during a sample of 150 hours. The data are stored in Xr03-35.
  a Compute the standard deviation of the number of arrivals.
  b Assuming that the histogram is bell shaped, interpret the standard deviation.

3.4 OTHER MEASURES OF SHAPE (OPTIONAL)

We provide this section for two reasons. First, Excel prints two additional statistics using the **Tools** menu item **Descriptive Statistics,** and you should be capable of interpreting as much of the printout as possible. Second, it is helpful for you to see that there are statistics that measure different characteristics of a set of data, each producing its own form of information. When we introduced histograms in Section 2.3, we discussed the different characteristics about which we generally wish to acquire information. In this chapter we've already introduced measures of central location and measures of variability. In discussing the shapes of histograms we pointed out that we would like to know whether the histograms are symmetric and bell shaped. We employed the term *skewness* to describe the shapes of histograms that are not symmetric. Not surprisingly, statisticians have developed a statistic to measure the degree of asymmetry. (Although there are parametric counterparts to the statistics introduced below, they are rarely used.)

Sample Skewness

$$S_k = \frac{n}{(n-1)(n-2)} \sum_{i=1}^{n} \left(\frac{x_i - \bar{x}}{s}\right)^3$$

Although it is difficult to see, if the histogram is symmetric, $S_k = 0$. If the histogram displays a long tail to the right, S_k is positive. If the histogram features a long tail to the left, S_k is negative. The magnitude of S_k measures the degree of skewness.

Another characteristic of histograms that was discussed in Section 2.3 was the extent to which the histogram was bell shaped. Another measure of shape that is helpful in this regard is *kurtosis*.

Sample Kurtosis

$$K = \frac{n(n+1)}{(n-1)(n-2)(n-3)} \sum_{i=1}^{n} \left(\frac{x_i - \bar{x}}{s}\right)^4 - \frac{3(n-1)^2}{(n-2)(n-3)}$$

The kurtosis of a bell-shaped histogram is approximately 0. When K is positive, the histogram is more peaked than a bell-shaped histogram. We often describe it as having narrower shoulders and thicker tails. A histogram whose kurtosis is negative has broader shoulders and narrower tails.

These statistics provide little information about the shape of the histogram. For example, if $S_k = 0$, it doesn't necessarily mean that the histogram is symmetric. Similarly, if both S_k and K equal 0, it doesn't guarantee that the histogram is bell shaped. Hence, when we interpret the Excel output for **Descriptive Statistics,** we will use the measures of central location (particularly the mean and median), the measures of variability (particularly the standard deviation), plus several others described below. Skewness and kurtosis will seldom be used.

3.5 MEASURES OF RELATIVE STANDING AND BOX PLOTS

Measures of relative standing are designed to provide information about the position of particular values relative to the entire data set. We'll use an example we've alluded to before. Suppose that your statistics professor has just returned your midterm test with a mark of 77 on it. How did you do relative to your classmates? The median would only tell you which half of the class you fell into. The statistics we're about to introduce will give you much more detailed information.

> **Percentile**
>
> The pth percentile is the value for which p percent are less than that value and $(100 - p)\%$ are greater than that value.

For example, if you're told that 77 is the 60th percentile, it means that 60% of the other marks are below 77 and 40% are above it. You now know exactly where you stand relative to the class.

We have special names for the 25th, 50th, and 75th percentiles. Because these three statistics divide the set of data into quarters, these measures of relative standing are also called **quartiles.** The **first** or **lower quartile** is labeled Q_1. It is equal to the 25th percentile. The **second quartile,** Q_2, is equal to the 50th percentile, which is also the median. The **third** or **upper quartile,** Q_3, is equal to the 75th percentile. Incidentally, many people confuse the terms *quartile* and *quarter.* A common error is to state that someone is in the lower *quartile* of a group when they actually mean that someone is in the lower *quarter* of a group.

LOCATING PERCENTILES

The following formula allows us to approximate the location of any percentile.

> **Location of a Percentile**
>
> $$L_p = (n + 1)\frac{P}{100}$$
>
> where L_p is the location of the Pth percentile.

EXAMPLE 3.11

Calculate the 25th, 50th, and 75th percentiles (first, second, and third quartiles) of the data in Example 3.1.

Solution Placing the 10 observations in ascending order we get

0, 0, 5, 7, 8, 9, 12, 14, 22, 33

The location of the 25th percentile is

$$L_{25} = (10 + 1)\frac{25}{100} = (11)(.25) = 2.75$$

The 25th percentile is three-quarters of the distance between the second and third observations. Three-quarters of the distance is

$$(.75)(5 - 0) = 3.75$$

Because the second observation is 0, the 25th percentile is $0 + 3.75 = 3.75$.

To locate the 50th percentile, we substitute $P = 50$ into the formula and produce

$$L_{50} = (10 + 1)\frac{50}{100} = (11)(.5) = 5.5$$

which means that the 50th percentile is halfway between the fifth and sixth observations. The fifth and sixth observations are 8 and 9, respectively. The 50th percentile is 8.5.

The 75th percentile's location is

$$L_{75} = (10 + 1)\frac{75}{100} = (11)(.75) = 8.25$$

Thus, it is located one-quarter of the distance between the eighth and ninth observations, which are 14 and 22, respectively. One-quarter of the difference is

$$(.25)(22 - 14) = 2$$

which means that the 75th percentile is

$$14 + 2 = 16$$

EXCEL: PERCENTILES

Excel can be commanded to produce the percentiles we want. The dialog box for **Descriptive Statistics** included two additional entries that we have so far ignored. These are **Kth Largest** and **Kth Smallest.** Typing in values in those boxes commands Excel to include any two percentiles you specify. To illustrate, in Example 3.2 (which addresses the data in Example 2.1), we specified 100 for the **Kth Largest** and 100 for the **Kth Smallest,** meaning that we wanted the 100th largest and 100th smallest number. Because the sample size is 400 the statistics Excel produces are the 75th and 25th percentiles, respectively. The output below was produced.

Microsoft Excel Output for Example 3.2

Times	
Mean	52.7
Standard Error	0.982
Median	48
Mode	41
Standard Deviation	19.64
Sample Variance	385.7
Kurtosis	2.71
Skewness	1.38
Range	128
Minimum	23
Maximum	151
Sum	21080
Count	400
Largest(100)	63
Smallest(100)	40

The **Largest(100)** is 63, which is the number such that 300 numbers are below it and 99 numbers are above it. The **Smallest(100)** is 40, which is the number such that 100 numbers are below it and 299 numbers are above it.

We can often get an idea of the shape of the histogram from the quartiles. For example, if the first and second quartiles are closer to each other than are the second and third quartiles, the histogram is positively skewed. If the first and second quartiles are further apart than the second and third quartiles, the histogram is negatively skewed. If the difference between the first and second quartiles is approximately equal to the difference between the second and third quartiles, the histogram is approximately symmetric. The box plot described below is particularly useful in this regard.

INTERQUARTILE RANGE

The quartiles can be used to create another measure of variability, the **interquartile range**, which is defined as follows.

Interquartile Range

Interquartile range = $Q_3 - Q_1$

The interquartile range measures the spread of the middle 50% of the observations. Large values of this statistic indicate that the first and third quartiles are far apart, indicating a high level of variability.

BOX PLOTS

A **box plot** is a graphical technique that depicts the minimum and maximum observations and the first, second, and third quartiles. It also depicts other features of a set of data.

To illustrate, we have produced the box plot of the data in Example 2.1.

Microsoft Excel Box Plot for Example 2.1

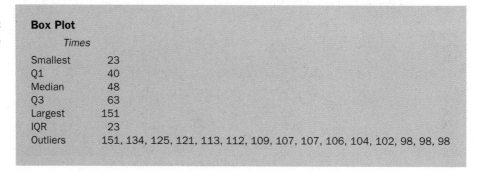

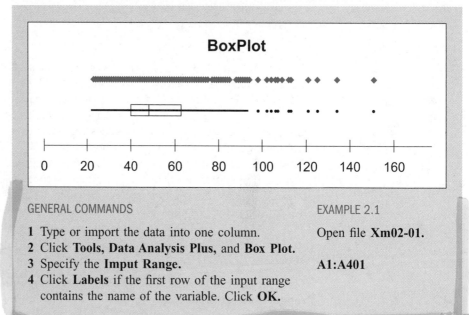

GENERAL COMMANDS	EXAMPLE 2.1
1 Type or import the data into one column.	Open file **Xm02-01.**
2 Click **Tools, Data Analysis Plus,** and **Box Plot.**	
3 Specify the **Imput Range.**	A1:A401
4 Click **Labels** if the first row of the input range contains the name of the variable. Click **OK.**	

The three vertical lines of the box are the first, second, and third quartiles. The lines extending left and right are called **whiskers.** Any points that lie outside the whiskers are called **outliers.** The whiskers extend outward to the smaller of 1.5 times the interquartile range or to the most extreme point that is not an outlier. (See Interpreting the Results, below.)

Outliers are observations that are quite different from the other values in the data set. As a result, they should be checked for accuracy and whether they actually belong to the data set.

Interpreting the Results

The smallest value is 23, and the largest is 151. The first, second, and third quartiles are 40, 48, and 63, respectively. The interquartile range is $63 - 40 = 23$. One and a half times the interquartile range is $1.5 \times 23 = 34.5$. Outliers are defined as any observations that are less than $40 - 34.5 = 5.5$ and any observations that are larger than

$63 + 34.5 = 97.5$. The whisker to the left extends only to 23, which is the smallest observation that is not an outlier. The whisker to the right extends to 94, which is the largest observation that is not an outlier.

We can see that the first and second quartiles are relatively close, indicating a positive histogram. There are several extreme observations extending to the right. The statistics practitioner should check the accuracy of all observations greater than 97.5.

Notice that the quartiles produced in the **Box Plot** are the same as those produced by **Descriptive Statistics.** This will not always be the case. The methods of determining these statistics vary slightly between the two methods. The reason and the method of calculating the quartiles in **Box Plot** are details that we will not discuss.

MEASURES OF RELATIVE STANDING AND VARIABILITY FOR ORDINAL DATA

Because the measures of relative standing are computed by ordering the data, these statistics are appropriate for ordinal as well as interval data. Furthermore, because the interquartile range is calculated by taking the difference between the upper and lower quartiles, it too can be employed to measure the variability of ordinal data.

EXERCISES

3.36 Calculate the first, second, and third quartiles of the following sample.
5, 8, 2, 9, 5, 3, 7, 4, 2, 7, 4, 10, 4, 3, 5

3.37 Find the 30th and 80th percentiles of the following data set.
26, 23, 29, 31, 24, 22, 15, 31, 30, 20

3.38 Find the 10th and 90th percentiles of the data below.
52, 61, 88, 43, 64, 71, 39, 73, 51, 60

3.39 Determine the first, second, and third quartiles of the following data.
10.5, 14.7, 15.3, 17.7, 15.9, 12.2, 10.0, 14.1, 13.9, 18.5, 13.9, 15.1, 14.7

3.40 Calculate the 30th and 60th percentiles of the data below.
7, 18, 12, 17, 29, 18, 4, 27, 30, 2, 4, 10, 21, 5, 8

3.41 A sample of Boston Marathon runners was drawn, and the times to complete the race were recorded and stored in file Xr03-41.
  a Draw the box plot.
  b What are the quartiles?
  c Identify outliers.
  d What information does the box plot deliver?

3.42 Refer to Exercise 2.8.
  a Draw the box plot.
  b What are the quartiles?
  c Are there outliers?
  d What does the box plot tell you about the service times?

3.43 A sample of condominium owners in Miami Beach was asked to report their age. The data are stored in Xr03-43.
  a Draw the box plot.
  b What does the box plot tell you about the ages?

3.44 Refer to Exercise 2.10.
  a Draw the box plot.
  b What does the box plot tell you about the salaries?

3.45 Refer to Exercise 2.11.
  a Draw the box plot.
  b What does the box plot tell you about the marks?

3.46 There appears to be a great deal of variation in gas prices. A sample of service stations in a large city was taken and the price of a gallon of gas recorded. The data are stored in Xr03-46.
  a Draw the box plot.
  b What are the quartiles?
  c Are there outliers?
  d What does the box plot tell you about the prices?

3.47 Do golfers who are members of private courses play faster than players on a public course? The amount of time taken for a sample of private-course and public-course golfers was drawn. The times (in minutes) are stored in columns A (private course) and B (public course) in Xr03-47.

a Draw box plots for each sample.
b What do the box plots tell you?

3.6 MEASURES OF LINEAR RELATIONSHIP

In Chapter 2, we introduced scatter diagrams, a graphical technique that describes the relationship between two interval variables. At that time we pointed out that we're particularly interested in the direction and strength of the linear relationship. We now present two numerical measures of linear relationship. They are the covariance and the coefficient of correlation.

COVARIANCE

As we did in Chapter 2, we label one variable X and the other Y.

Covariance

$$\text{Population covariance: COV}(X,Y) = \frac{\sum_{i=1}^{N}(x_i - \mu_x)(y_i - \mu_y)}{N}$$

$$\text{Sample covariance: cov}(x,y) = \frac{\sum_{i=1}^{n}(x_i - \bar{x})(y_i - \bar{y})}{n - 1}$$

The denominator in the calculation of the statistic is $n - 1$, not the more logical n, for the same reason we divide by $n - 1$ to calculate the sample variance. To illustrate how **covariance** measures the linear relationship, examine the following three sets of data.

Set 1

x_i	y_i	$(x_i - \bar{x})$	$(y_i - \bar{y})$	$(x_i - \bar{x})(y_i - \bar{y})$
2	13	−3	−7	21
6	20	1	0	0
7	27	2	7	14
$\bar{x} = 5$	$\bar{y} = 20$			cov(x,y) = 17.5

Set 2

x_i	y_i	$(x_i - \bar{x})$	$(y_i - \bar{y})$	$(x_i - \bar{x})(y_i - \bar{y})$
2	27	−3	7	−21
6	20	1	0	0
7	13	2	−7	−14
$\bar{x} = 5$	$\bar{y} = 20$			cov$(x,y) = -17.5$

Set 3

x_i	y_i	$(x_i - \bar{x})$	$(y_i - \bar{y})$	$(x_i - \bar{x})(y_i - \bar{y})$
2	20	−3	0	0
6	27	1	7	7
7	13	2	−7	−14
$\bar{x} = 5$	$\bar{y} = 20$			cov$(x,y) = -3.5$

In set 1, as x increases, so does y. When x is larger than its mean, y is at least as large as its mean. Thus $(x_i - \bar{x})$ and $(y_i - \bar{y})$ have the same sign or zero. Their product is also positive or zero. Consequently, the covariance is a positive number. Generally, when two variables move in the same direction (both increase or both decrease), the covariance will be a large positive number.

If you examine set 2 you will discover that as x increases, y decreases. When x is larger than its mean, y is less than or equal to its mean. As a result when $(x_i - \bar{x})$ is positive, $(y_i - \bar{y})$ is negative or zero. Their products are either negative or zero. It follows that the covariance is a negative number. In general, when two variables move in opposite directions, the covariance is a large negative number.

In set 3, as x increases, y does not exhibit any particular direction. One of the products $(x_i - \bar{x})(y_i - \bar{y})$ is zero, one is positive, and one is negative. The resulting covariance is a small number. In general, when there is no particular pattern, the covariance is a small number.

We would like to extract two pieces of information. The first is the sign of the covariance, which tells us the nature of the relationship. The second is the magnitude, which describes the strength of the association. Unfortunately, the magnitude may be difficult to judge. For example, if you're told that the covariance between two variables is 500, does this mean that there is a strong linear relationship? The answer is that it is impossible to judge without additional statistics. Fortunately, we can improve upon this statistic by creating another one.

COEFFICIENT OF CORRELATION

The **coefficient of correlation** is defined as the covariance divided by the standard deviations of the variables.

Coefficient of Correlation

Population coefficient of correlation: $\rho = \dfrac{\text{COV}(X,Y)}{\sigma_x \sigma_y}$

Sample coefficient of correlation: $r = \dfrac{\text{cov}(x,y)}{s_x s_y}$

The population parameter is denoted by the Greek letter *rho*.

The advantage that the coefficient of correlation has over the covariance is that the former has a set lower and upper limit. The limits are -1 and $+1$, respectively. That is,

$$-1 \leq r \leq +1 \quad \text{and} \quad -1 \leq \rho \leq +1$$

When $r = -1$ (or $\rho = -1$), there is a negative linear relationship, and the scatter diagram exhibits a straight line. When $r = +1$ (or $\rho = +1$), there is a perfect positive relationship. When $r = 0$ (or $\rho = 0$), there is no linear relationship. All other values of r and ρ are judged in relation to these three values.

Comparing the Scatter Diagram, Covariance, and Coefficient of Correlation

The scatter diagram depicts relationships graphically; the covariance and the coefficient of correlation describe the linear relationship numerically. Figures 3.3, 3.4, and 3.5 depict three scatter diagrams. To show how the graphical and numerical techniques compare, we calculated the covariance and the coefficient of correlation for each. (The data are stored in files Xm03-03, Xm03-04, and Xm03-05.) As you can see, Figure 3.3 depicts a strong positive relationship between the two variables. The covariance is 38.81, and the coefficient of correlation is .9641. The variables in Figure 3.4 produced a relatively strong negative linear relationship; the covariance and coefficient of correlation are -35.97 and $-.8791$, respectively. The covariance and coefficient of correlation for the data in Figure 3.5 are 2.18 and .1206, respectively. There is no apparent linear relationship in this figure.

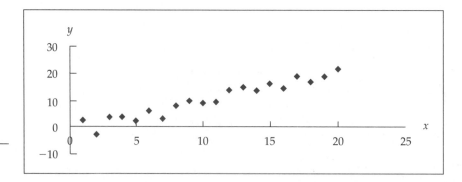

Figure 3.3

Strong positive linear relationship

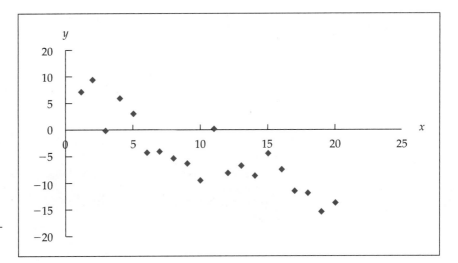

Figure 3.4

Strong negative linear relationship

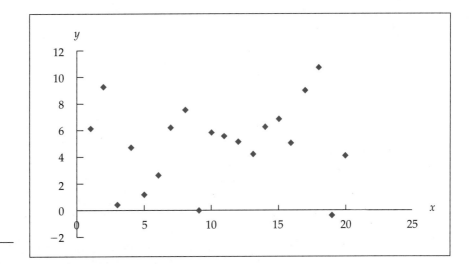

Figure 3.5

No linear relationship

▼ EXAMPLE 3.12

Determine the covariance and the coefficient of correlation between educational level and income in Example 2.6.

Solution *Covariance:* Excel produces the variance-covariance matrix, which yields the variance for both variables and the covariance. Note, however, that Office 2000* produces the population parameters COV(X,Y), σ_x^2, and σ_y^2. You can compute the sample statistics cov(x,y), s_x^2, and s_y^2 by multiplying each by ($n/n - 1$).

---

*Some versions of Excel print the sample statistics, while others output the population parameters. You will need to determine what your version of Excel outputs.

Microsoft Excel Covariance Matrix for Example 3.12

	Education	Income
Education	8.05	
Income	25.08	191.27

GENERAL COMMANDS

1 Type or import the data into two adjacent columns.
2 Click **Tools, Data Analysis,** and **Covariance.**
3 Type the **Input Range.**
4 Click **Labels in First Row,** if applicable, and click **OK.**

EXAMPLE 3.12

Open file **Xm02-06.**

A1:B151

INTERPRETING THE RESULTS

Multiplying each figure by (150/149) yields the following. The sample variances of education and income are 8.05 and 191.27, respectively. The sample covariance is 25.08. This number tells us only that education and income are positively related. As we pointed out earlier, it is difficult to judge the magnitude of the covariance.

Correlation: As is the case with covariance, Excel produces a matrix of correlation coefficients.

Microsoft Excel Correlation Matrix for Example 3.12

	Education	Income
Education	1	
Income	0.639	1

GENERAL COMMANDS

1 Click **Correlation** instead of **Covariance** in the instructions above.

INTERPRETING THE RESULTS

The correlation between education and income is .639. Thus, we see that education and income are somewhat linearly related. Incidentally, the two 1s in the matrix represent the coefficients of correlation of education with itself and of income with itself, obviously meaningless numbers. The correlation matrix will be used again later in this book with more than two variables.

LEAST SQUARES METHOD

When we presented the scatter diagram in Section 2.6, we introduced the **least squares line.** It is a straight line drawn through the points so that the sum of squared deviations between the points and the line is minimized. The line is represented by the equation

$$\hat{y} = b_0 + b_1 x$$

where $\hat{y}$ is the value of y determined by the line. The coefficients b_0 and b_1 are calculated using calculus so that we minimize the sum of squared deviations

$$\sum_{i=1}^{n}(y_i - \hat{y}_i)^2$$

The formulas are provided to show how these statistics are related to statistics already introduced in this book. We'll use the computer to perform the lengthy calculations.

Least Squares Line Coefficients

$$b_1 = \frac{\text{cov}(x,y)}{s_x^2}$$

$$b_0 = \bar{y} - b_1\bar{x}$$

To illustrate, suppose that we have the following six observations of variables x and y.

x	2	4	8	10	13	16
y	2	7	25	26	38	50

We use Excel to produce the least squares line.

Microsoft Excel Least Squares Line

	Coefficients
Intercept	−5.356
X	3.399

GENERAL COMMANDS

1. Type or import the data into two columns.
2. Click **Tools, Data Analysis ...**, and **Regression.**
3. Specify the **Input Range** of y and of x. Click **Labels,** if applicable, and click **OK.**

We only showed the part of the printout with the coefficients. The rest of the printout will be described in Chapter 17, where we'll also explain *regression*.

The intercept is −5.356, and the coefficient of X is 3.399. The least squares line is

$$\hat{y} = -5.356 + 3.399x$$

To further illustrate the principle of the least squares method, we calculate the fitted values of y, the deviations between the points and the line, and the sum of squared deviations in the table below.

Table 3.1 Calculation of Sum of Squared Deviations

i	x_i	y_i	$\hat{y}_i = -5.356 + 3.399 x_i$	$y_i - \hat{y}_i$	$(y_i - \hat{y}_i)^2$
1	2	2	1.442	0.558	0.3114
2	4	7	8.240	-1.240	1.5376
3	8	25	21.836	3.164	10.0109
4	10	13	28.634	-2.634	6.9380
5	13	38	38.831	-0.831	0.6906
6	16	50	49.028	0.972	0.9448
				$\sum(y_i - \hat{y}_i)^2 =$	20.4332

Figure 3.6 depicts the way in which the fitted values of y and the differences between the actual and fitted values of y are calculated. No other straight line will fit these data as well as does the least squares line. Try several values of the coefficients to convince yourself.

To complete this section, we provide the Excel printout of the least squares line for Example 3.12 (Example 2.6) together with the scatter diagram depicting the least squares line.

Microsoft Excel Output of the Least Squares Coefficients for Example 3.12

	Coefficients
Intercept	-2.710
Education	3.116

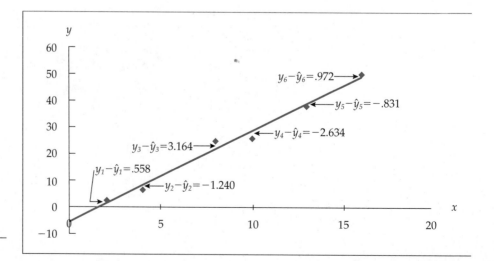

Figure 3.6

Scatter diagram

Microsoft Excel Scatter Diagram with Least Squares Line for Example 2.6

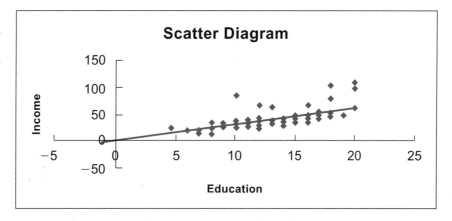

The least squares line is

$\hat{y} = -2.710 + 3.116x$

This tells us that for each additional year of education income increases on average by 3.116 thousand dollars.

EXERCISES

3.48 Calculate the covariance and coefficient of correlation for the following sample.

| X | 3 | 5 | 2 | 9 | 6 | 3 | 9 | 1 |
| Y | 8 | 4 | 9 | 2 | 4 | 7 | 3 | 7 |

3.49 Calculate the covariance and coefficient of correlation for the following sample.

| X | 45 | 22 | 16 | 50 | 44 | 31 | 48 | 27 | 30 | 28 |
| Y | 77 | 31 | 28 | 49 | 63 | 84 | 40 | 92 | 72 | 67 |

3.50 The selling price (thousands of dollars) and the size (square feet) were recorded for a sample of 10 houses that were recently sold. The data are:

Price	153	202	199	315	148
	194	250	167	305	258
Size	1,526	1,849	1,906	2,460	1,602
	1,731	2,208	2,041	2,595	2,008

a Calculate the covariance of the two variables.
b Determine the coefficient of correlation.
c What do these statistics tell you about the relationship between price and size?

3.51 Attempting to analyze the relationship between advertising and sales, the owner of a furniture store recorded the monthly advertising budget ($thousands) and the sales ($millions) for a sample of 12 months.

a Calculate the covariance of the two variables.
b Determine the coefficient of correlation.
c What do these statistics tell you about the relationship between advertising and sales?

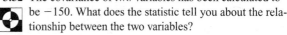

Advertising	23	46	60	54	28	33
	25	31	36	88	90	99
Sales	9.6	11.3	12.8	9.8	8.9	12.5
	12.0	11.4	12.6	13.7	14.4	15.9

3.52 The covariance of two variables has been calculated to be −150. What does the statistic tell you about the relationship between the two variables?

3.53 Refer to Exercise 3.52. You've now learned that the two sample standard deviations are 16 and 12. Calculate the coefficient of variation. What does this statistic tell you about the relationship between the two variables?

3.54 Studies of twins may reveal more about the "nature or nurture" debate. The issue being debated is whether nature or the environment has more effect on individual traits such as intelligence. Suppose that a sample of identical twins was selected and their IQs measured. These data are stored in file Xr03-54. Compute the coefficient of correlation, and describe what it tells you about the relationship between the IQs of identical twins.

3.55 Besides the known long-term effects of smoking, do cigarettes also cause short-term illnesses such as colds? To help answer this question a sample of smokers was drawn. Each person was asked to report the average number of cigarettes smoked per day and the number of sick days due to colds last year. The data are stored in Xr03-55.
   a Calculate the covariance of the two variables.
   b Determine the coefficient of correlation.
   c What do these statistics tell you about the relationship between smoking cigarettes and the incidence of colds?

3.56 Refer to Exercise 2.46.

   a Calculate the coefficient of correlation.
   b Describe what this statistic tells you about the relationship between marks and study time.

3.57 Refer to Exercise 2.49.
   a Calculate the coefficient of correlation.
   b Briefly discuss the relationship between Internet use and age.

3.58 Do better golfers play faster than poorer ones? To determine whether a relationship exists, a sample of 125 foursomes was selected. Their total scores and the amount of time taken to complete the 18 holes were recorded and stored in Xr03-58. Calculate the coefficient of correlation, and describe what this statistic tells you about the relationship between score and time.

3.59 Refer to Exercise 2.51. Calculate the coefficient of correlation, and describe what this statistic tells you about the relationship between the incomes and heights of male executives.

3.7 GENERAL GUIDELINES FOR EXPLORING DATA

The purpose of applying graphical and numerical techniques is to describe and summarize data. Statisticians usually apply graphical techniques as a first step because we need to know the shape of the distribution. The shape of the distribution helps answer the following questions.

1 Where is the approximate center of the distribution?
2 Are the observations close to one another, or are they widely dispersed?
3 Is the distribution unimodal, bimodal, or multimodal? If there is more than one mode, where are the peaks, and where are the valleys?
4 Is the distribution symmetric? If not, is it skewed? If symmetric, is it bell shaped?

Histograms and box plots provide most of the answers. We can frequently make several inferences about the nature of the data from the shape. For example, we can assess the relative risk of investments by noting their spreads. We can attempt to improve the teaching of a course by examining whether the distribution of final grades is bimodal or skewed.

The shape can also provide some guidance on which numerical techniques to use. As we noted in this chapter, the central location of highly skewed data may be more appropriately measured by the median. We may also choose to use the interquartile range instead of the standard deviation to describe the spread of skewed data.

When we have an understanding of the structure of the data, we may proceed to further analysis. For example, we often want to determine how one variable, or several variables, affects another. Scatter diagrams, covariance, and the coefficient of correlation are useful techniques for detecting relationships between variables. A number of techniques to be introduced later in this book will help uncover the nature of these associations.

3.8 SUMMARY

This chapter extended our discussion of descriptive statistics, which deals with methods of summarizing and presenting the essential information contained in a set of data. We can use numerical measures to describe the central location and variability of interval data. Three popular measures of central location are the mean, the median, and the mode. Taken by themselves, these measures provide an inadequate description of the data because they say nothing about the extent to which the data vary. Information regarding the variability of interval data is conveyed by such numerical measures as the range, variance, and standard deviation.

For the special case in which a sample of measurements has a bell-shaped histogram, the Empirical Rule provides a good approximation of the percentages of measurements that fall within one, two, or three standard deviations of the mean. Chebysheff's Theorem applies to all sets of data, no matter what the shape of the histogram.

Measures of relative standing that were presented in this chapter are percentiles and quartiles. The box plot graphically depicts these measures as well as several others. The linear relationship between two interval variables is measured by the covariance and the coefficient of correlation.

IMPORTANT TERMS

Measures of central location
Mean
Median
Mode
Range
Variance
Standard deviation
Empirical Rule
Chebysheff's Theorem
Percentiles
Quartiles

First or lower quartile
Second quartile
Third or upper quartile
Interquartile range
Box plots
Whiskers
Outlier
Covariance
Coefficient of correlation
Least squares method
Least squares (regression) line

SYMBOLS

Symbol	Pronounced	Represents
μ	mu	Population mean
σ^2	sigma-squared	Population variance
σ	sigma	Population standard deviation
ρ	rho	Population coefficient of correlation
$\sum$	sum of	Summation
$\sum_{i=1}^{n} x_i$	sum of x_i from 1 to n	Summation of n numbers
$\hat{y}$	y-hat	Fitted or calculated value of y
b_0	b-zero	y-intercept
b_1	b-one	Slope coefficient

FORMULAS

Population mean

$$\mu = \frac{\sum_{i=1}^{N} x_i}{N}$$

Sample mean

$$\bar{x} = \frac{\sum_{i=1}^{n} x_i}{n}$$

Range

Largest observation − Smallest observation

Population variance

$$\sigma^2 = \frac{\sum_{i=1}^{N}(x_i - \mu)^2}{N}$$

Sample variance

$$s^2 = \frac{\sum_{i=1}^{n}(x_i - \bar{x})^2}{n-1}$$

Population standard deviation

$$\sigma = \sqrt{\sigma^2}$$

Sample standard deviation

$$s = \sqrt{s^2}$$

Population coefficient of correlation

$$\rho = \frac{\text{COV}(X,Y)}{\sigma_x \sigma_y}$$

Sample coefficient of correlation

$$r = \frac{\text{cov}(x,y)}{s_x s_y}$$

Slope coefficient

$$b_1 = \frac{\text{cov}(x,y)}{s_x^2}$$

y-intercept

$$b_0 = \bar{y} - b_1 \bar{x}$$

MICROSOFT EXCEL OUTPUT AND INSTRUCTIONS

Technique	Page
Descriptive statistics	55
Box plot	74
Covariance	80
Correlation	80
Least squares coefficients	81

SUPPLEMENTARY EXERCISES

3.60 The temperature in December in Buffalo, New York, is often below 40 degrees Fahrenheit (4 degrees Celsius). Not surprisingly, when the National Football League Buffalo Bills play at home in December, coffee is a popular item at the concession stand. The concession manager would like to acquire more information so that he can manage inventories more efficiently. The number of cups of coffee sold during 50 games played in December in Buffalo was recorded and stored in Xr03-60.

    a Determine the mean and median.
    b Determine the variance and standard deviation.
    c Draw a box plot.
    d Briefly describe what you have learned from your statistical analysis.

3.61 Refer to Exercise 3.60. Suppose that in addition to recording the coffee sales, the manager also recorded the average temperature (measured in degrees Fahrenheit) during the game. These data together with the number of cups of coffee sold are stored in columns A and B in Xr03-61.

    a Compute the covariance and coefficient of correlation.
    b Determine the coefficients of the least squares line.
    c What have you learned about the relationship between the number of cups of coffee sold and the temperature from the statistics calculated above?
    d Discuss the information obtained here and in Exercise 3.60. Which is more useful to the manager?

3.62 Chris Golfnut loves the game of golf. Chris also loves statistics. Combining both passions, Chris records a sample of 100 scores and stores the data in Xr03-62.

    a What statistics should Chris compute to describe the scores?
    b Calculate the mean and standard deviation of the scores.
    c Briefly describe what the statistics computed in part (b) divulge.

3.63 Refer to Exercise 3.62. For each score Chris recorded the number of putts and stored these data in Xr03-63. Conduct an analysis of both sets of data. What conclusions can be achieved from the statistics?

3.64 The Internet is growing rapidly with an increasing number of regular users. However, among people older than 50, Internet use is still relatively low. To learn more about this issue, a sample of 250 men and women older than 50 who had used the Internet at least once was selected. The number of hours on the Internet during the past month was recorded and stored in Xr03-64.

    a Calculate the mean and median.
    b Calculate the variance and standard deviation.
    c Draw a box plot.
    d Briefly describe what you have learned from the statistics you calculated.

3.65 Refer to Exercise 3.64. In addition to Internet use, suppose that we have also recorded the number of years of education. These two variables are stored in Xr03-65.
  a Compute the covariance and coefficient of correlation.
  b Determine the coefficients of the least squares line.
  c Describe what these statistics tell you about the relationship between Internet use and education.
  d Discuss the information obtained here and in Exercise 3.64.

3.66 A sample was drawn of one-acre plots of land planted with corn. The crop yields were recorded and stored in Xr03-66. Calculate the descriptive statistics you judge to be useful. Interpret these statistics.

3.67 Refer to Exercise 3.66. For each plot we recorded the amount rainfall and stored both variables in Xr03-67.
  a Compute the covariance and coefficient of correlation.
  b Determine the coefficients of the least squares line.
  c Describe what these statistics tell you about the relationship between crop yield and rainfall.
  d Discuss the information obtained here and in Exercise 3.66.

3.68 Refer to Exercise 3.66. For each plot we recorded the amount of fertilizer and stored both variables in Xr03-68.
  a Compute the covariance and coefficient of correlation.
  b Determine the coefficients of the least squares line.
  c Describe what these statistics tell you about the relationship between crop yield and rainfall.
  d Discuss the information obtained here and in Exercise 3.66.

3.69 Refer to Exercise 2.75.
  a Calculate the mean and median. What do these statistics tell you about the vocabularies of 5-year-old children?
  b Calculate the variance and standard deviations. What information can be obtained from these statistics?
  c Determine the three quartiles. What do they tell you about the size of the vocabularies?

3.70 Refer to Exercise 2.78.
  a Calculate the coefficient of correlation.
  b What does this statistic tell you about temperature and winning times among male runners of the New York Marathon?
  c Compare the information you obtained from the scatter diagram in Exercise 2.78 and the statistic above.

3.71 Refer to Exercise 2.79.
  a Calculate the coefficient of correlation.
  b What does this statistic tell you about temperature and winning times among female runners of the New York Marathon?
  c Compare the information you obtained from the scatter diagram in Exercise 2.79 and the statistic above.

3.72 Refer to Exercise 2.80. Calculate a number of descriptive statistics, and report your findings.

Chapter 4

Probability

4.1 Introduction
4.2 Assigning Probability to Events
4.3 Joint, Marginal, and Conditional Probability
4.4 Probability Rules and Trees
4.5 Summary

4.1 INTRODUCTION

In Chapters 2 and 3 we introduced graphical and numerical descriptive methods. While the methods are useful on their own, we are particularly interested in developing statistical inference. As we pointed out in Chapter 1, statistical inference is the process by which we acquire information about populations from samples. A critical component of inference is probability, because it provides the link between the population and the sample.

Our primary objective in this and the following two chapters is to develop the probability-based tools that are at the basis of statistical inference. However, probability is also a useful subject by itself. Here are several illustrations of probability's almost universal applicability

1 Weather forecasters provide probabilities when they predict the next day's weather. For example, suppose that a forecaster states that there is a 30% chance of rain tomorrow. What does this number tell you?

2 In recent years we have seen an increase in the use of DNA evidence in criminal and civil court cases. Experts may report that there is one chance in 4 million that two people have the same DNA traits as those seen in the evidence. How can the jury in such a trial properly interpret this number?

3 A friend has just told you that a diagnostic test reveals that he has a possibly fatal disease. He further informs you that the probability that the diagnostic test is correct is at least 90%. What advice can you provide regarding the interpretation of the diagnostic test?

In this chapter we'll provide answers by presenting the fundamental concepts and techniques of probability. In Section 4.2 we define probability and describe how we assign probability to events. Section 4.3 introduces combinations of events, and in Section 4.4 we describe how to calculate the probability of these combinations.

4.2 ASSIGNING PROBABILITY TO EVENTS

To introduce probability we need to define a random experiment.

> **Random Experiment**
>
> A random experiment is an action or process that leads to one of several possible outcomes. The actual outcome cannot be determined in advance.

Here are six illustrations of random experiments and their outcomes.

1 Experiment: Flip a coin
   Outcomes: Heads and tails

2 Experiment: Record marks on a statistics test (out of 100)
Outcomes: Numbers between 0 and 100

3 Experiment: Record grade on a statistics test
Outcomes: A, B, C, D, and F

4 Experiment: Record student evaluations of a course
Outcomes: Poor, fair, good, very good, and excellent

5 Experiment: Measure the time to assemble a computer
Outcomes: Numbers whose smallest possible value is 0 seconds with no predefined upper limit

6 Experiment: Record the party that a voter will vote for in an upcoming election
Outcomes: Party A, Party B, . . .

The first step in assigning probabilities is to produce a list of the outcomes. The listed outcomes must be **exhaustive,** which means that all possible outcomes must be included. Additionally, the outcomes must be **mutually exclusive,** which means that no two outcomes can occur at the same time.

To illustrate the concept of exhaustive outcomes consider this list of the outcomes of the toss of a die:

1, 2, 3, 4, and 5

This list is not exhaustive, because we have omitted 6.

The concept of mutual exclusiveness can be seen by listing the following outcomes in illustration 2 above:

0–50, 50–60, 60–70, 70–80, and 80–100

These outcomes are not mutually exclusive because two outcomes can occur for any student. For example, if a student receives a mark of 70, both the third and fourth outcomes occur.

It should be noted that we could produce more than one list of exhaustive and mutually exclusive events. For example, here is another list of outcomes for illustration 3 above:

Pass and fail

A list of exhaustive and mutually exclusive outcomes is called a *sample space,* denoted by S. The outcomes are denoted by $O_1, O_2, \ldots, O_k$.

Sample Space

A **sample space** of a random experiment is a list of all possible outcomes of the experiment. The outcomes must be exhaustive and mutually exclusive.

Using set notation we represent the sample space and its outcomes as

$S = \{O_1, O_2, \ldots, O_k\}$

Once a sample space has been prepared, we begin the task of asigning probabilites to the outcomes. There are three ways to assign probability to outcomes. However it is done, there are two rules governing probabilites.

> **Requirements of Probabilities**
>
> Given a sample space $S = \{O_1, O_2, \ldots, O_k\}$, the probabilities assigned to the outcomes must satisfy two requirements:
>
> 1 The probability of any outcome must lie between zero and one. That is,
>
> $0 \leq P(O_i) \leq 1 \quad$ for each i
>
> [Note: $P(O_i)$ is the notation we employ to represent the probability of outcome i.]
>
> 2 The sum of the probabilities of all the outcomes in a sample space must be one. That is,
>
> $$\sum_{i=1}^{k} P(O_i) = 1$$

THREE APPROACHES TO ASSIGNING PROBABILITIES

The **classical approach** is used by mathematicians to help determine probability associated with games of chance. For example, the classical approach specifies that the probabilities of heads and tails in the flip of a balanced coin are equal to each other. Because the sum of the probabilities must be one, the probability of heads and the probability of tails are both 50%. Similarly, the six possible outcomes of the toss of a balanced die have the same probability; each is assigned a probability of 1/6. In some experiments it is necessary to develop mathematical ways to count the number of outcomes.

The **relative frequency approach** defines probability as the long-run relative frequency with which an outcome occurs. For example, suppose that we know that of the last 1,000 students who took the statistics course you're now taking, 200 received a grade of A. The relative frequency of A's is then 200/1,000, or 20%. This figure represents an estimate of the probability of obtaining a grade of A in the course. It is only an estimate because the relative frequency approach defines probability as the "long-run" relative frequency. One thousand students do not constitute the long run. The larger the number of students whose grades we have observed, the better the estimate becomes. In theory we would have to observe an infinite number of grades to determine the exact probability.

When it is not reasonable to use the classical approach and there is no history of the outcomes, we have no alternative but to employ the **subjective approach.** In the subjective approach, we define probability as the degree of belief that we hold in the occurrence of an event. An excellent example is derived from the field of investment. Here an investor would like to know the probability that a particular stock will increase in value. Using the subjective approach, the investor would analyze a number of factors associated with the stock and the stock market in general and, using his or her judgment, assign a probability to the outcomes of interest.

DEFINING EVENTS

An individual outcome of a sample space is called a **simple event.** All other events are composed of the simple events in a sample space.

> ### Event
> An **event** is a collection or set of one or more simple events in a sample space.

In illustration 2 we can define the event, achieve a grade of A, as the set of numbers that lie between 80 and 100, inclusive. Using set notation we have

$A = \{80, 81, 82 \ldots, 99, 100\}$

Similarly,

$F = \{0, 1, 2 \ldots, 48, 49\}$

PROBABILITY OF EVENTS

We can now define the probability of any event.

> ### Probability of an Event
> The probability of an event is the sum of the probabilities of the simple events that constitute the event.

For example, suppose that in illustration 3 we employed the relative frequency approach to assign probabilities to the simple events, as follows:

$P(A) = .20$

$P(B) = .30$

$P(C) = .25$

$P(D) = .15$

$P(F) = .10$

The probability of the event, pass the course, is

$$P(\text{pass the course}) = P(A) + P(B) + P(C) + P(D)$$
$$= .20 + .30 + .25 + .15 = .90$$

INTERPRETING PROBABILITY

No matter what method was used to assign probability, we interpret it using the relative frequency approach for an infinite number of experiments. For example, an investor may have used the subjective approach to determining that there is a 65% probability that a particular stock's price will increase over the next month. However, we interpret the 65% figure to mean that if we had an infinite number of stocks with exactly the same economic and market characteristics as the one the investor will buy, 65% of them will increase in price over the next month. Similarly, we can determine that the probability of throwing a 5 with a balanced die is 1/6. We may have used the classical approach to determine this probability. However, we interpret the number as the proportion of times that a 5 is observed on a balanced die thrown an infinite number of times.

This relative frequency approach is useful to interpret probability statements such as those heard from weather forecasters or scientists. You will also discover that this is the way we link the population and the sample in statistical inference.

EXERCISES

4.1 The weather forecaster reports that the probability of rain tomorrow is 10%.
 a Which approach was used to arrive at this number?
 b How do you interpret the probability?

4.2 A sportscaster states that he believes that the probability that the New York Yankees will win the World Series this year is 25%.
 a Which method was used to assign that probability?
 b How would you interpret the probability?

4.3 A quiz contains a multiple-choice question with five possible answers, only one of which is correct. A student plans to guess the answer because the student knows nothing about the subject.
 a Produce the sample space for this experiment.
 b Assign probabilities to the simple events in the sample space you produced.
 c Which approach did you use to answer part (b)?
 d Interpret the probabilities you assigned in part (b).

4.4 An investor tells you that in her estimation there is a 60% probability that the Dow-Jones Industrial Index will increase tomorrow.
 a Which approach was used to produce this figure?
 b Interpret the 60% probability.

4.5 The sample space of the toss of a die is

$$S = \{1, 2, 3, 4, 5, 6\}$$

If the die is balanced, each simple event has the same probability. Find the probability of the following events:
 a An even number.
 b A number less than or equal to 4.
 c A number greater than or equal to 5.

4.6 Four candidates are running for mayor. The four candidates are Adams, Brown, Collins, and Dalton. Determine the sample space of the results of the election.

4.7 Refer to Exercise 4.6. Employing the subjective approach a political scientist has assigned the following probabilities:

$P(\text{Adams wins}) = .42$

$P(\text{Brown wins}) = .09$

$P(\text{Collins wins}) = .27$

$P(\text{Dalton wins}) = .22$

Determine the probabilities of the following events:
 a Adams loses.
 b Either Brown or Dalton wins.
 c Either Adams, Brown, or Collins wins.

4.8 The manager of a computer store has kept track of the number of computers sold per day. On the basis of this information, the manager produced the following list of the number of daily sales.

Number of computers sold	Probability
0	.08
1	.17
2	.26
3	.21
4	.18
5	.10

a If we define the experiment as observing the number of computers sold tomorrow, determine the sample space.
b Use set notation to define the event (sell more than three computers).
c What is the probability of selling five computers?
d What is the probability of selling two, three, or four computers?
e What is the probability of selling six computers?

4.9 Three contractors (call them Contractors 1, 2, and 3) bid on a project to build a new bridge. What is the sample space?

4.10 Refer to Exercise 4.9. Suppose that you believe that Contractor 1 is twice as likely to win as Contractor 3 and that Contractor 2 is three times as likely to win as Contactor 3. What are the probabilities of winning for each contractor?

4.11 Shoppers can pay for their purchases with cash, a credit card, or a debit card. Suppose that the proprietor of a shop determines that 60% of her customers use a credit card, 30% pay with cash, and the rest use a debit card.

a Determine the sample space for this experiment.
b Assign probabilities to the simple events.
c Which method did you use in part (b)?

4.12 Refer to Exercise 4.11.
a What is the probability that a customer does not use a credit card?
b What is the probability that a customer pays in cash or with a credit card?

4.13 A survey asks adults to report their marital status. The sample space is $S = \{$single, married, divorced, widowed$\}$. Use set notation to represent the event (the adult is not married).

4.14 Refer to Exercise 4.13. Suppose that in the city in which the survey is conducted, 50% of adults are married, 15% are single, 25% are divorced, and 10% are widowed.

a Assign probabilities to each simple event in the sample space.
b Which approach did you use in part (a)?

4.15 Refer to Exercises 4.13 and 4.14. Find the probability of each of the following events:
a The adult is single.
b The adult is not divorced.
c The adult is either widowed or divorced.

4.3 JOINT, MARGINAL, AND CONDITIONAL PROBABILITY

In the previous section, we described how to produce a sample space and assign probabilities to the simple events in the sample space. While this method of determining probability is useful, we need to develop more sophisticated methods. In this section we discuss how to calculate the probability of more complicated events from the probability of related events. Here is an illustration of the process.

The sample space for the toss of a die is

$$S = \{1, 2, 3, 4, 5, 6\}$$

If the die is balanced, the probability of each simple event is 1/6. In most parlor games and casinos, players toss two dice. In order to determine playing and wagering strategies, players need to compute the probabilities of various totals of the two dice. For example, the probability of tossing a total of 3 with two dice is 2/36. This probability was derived by creating combinations of the simple events. There are several different types of combinations. One of the most important types is the intersection of two events.

INTERSECTION OF TWO EVENTS

> **Intersection of Events *A* and *B***
>
> The **intersection** of events A and B is the event that occurs when both A and B occur. It is denoted as
>
> A and B

For example, one way to toss a 3 with two dice is to toss a 1 on the first die *and* a 2 on the second die, which is the intersection of two simple events. Incidentally, to compute the probability of a total of 3 we need to combine this intersection with another intersection, namely, a 2 on the first die and a 1 on the second die. This type of combination is called a *union* of two events and it will be to be described later in this section.

Here is another illustration. It is known that smoking and lung disease are related events. Suppose that an experiment consists of selecting one 60-year-old man. Let A represent the event that he smokes and B the event that he has lung disease. The event A and B is defined as the event that he is a smoker *and* that he has lung disease.

The probability of the intersection is called the **joint probability.** There are several ways to compute the joint probability. We can estimate the probability by observing a number of 60-year-old men and computing the relative frequency as an estimate of the joint probability. Alternatively we can use the probabilities of related events and the multiplication law (described in the next section) to find the joint probability.

Suppose that we have produced the following table of joint probabilities of smoking and lung disease.

	He is a smoker	He is a nonsmoker
He has lung disease	.12	.03
He does not have lung disease	.19	.66

This table tells us that the joint probability that the 60-year-old man is a smoker and has lung disease is .12. That is, 12% of all 60-year-old men smoke and have lung disease. The other three joint probabilities are defined similarly. That is,

P(he is a nonsmoker and has lung disease) = .03

P(he is a smoker and does not have lung disease) = .19

P(he is a nonsmoker and does not have lung disease) = .66

MARGINAL PROBABILITY

The joint probabilities in the table above allow us to compute various probabilities. **Marginal probabilities,** computed by adding across rows and down columns, are so named because they are calculated in the margins of the table.

Adding across the first row produces

P(he has lung disease) = P(he is a smoker and has lung disease)
+ P(he is a nonsmoker and has lung disease) = .12 + .03 = .15

Notice that both intersections state that the 60-year-old man has lung disease. Thus, adding the two joint probabilities yields the probability that he has lung disease. Expressed as relative frequency, 15% of all 60-year-old men have lung disease.

Adding across the second row:

P(he does not have lung disease) =
P(he is a smoker and does not have lung disease)
+ P(he is a nonsmoker and does not have lung disease) = .19 + .66 = .85

This probability tells us that 85% of all 60-year-old men do not have lung disease. Notice that the probabilities of having lung disease and not having lung disease add to 1.

Adding down the columns produces the following marginal probabilities.

Column 1: P(he is a smoker) = P(he is a smoker and has lung disease)
+ P(he is a smoker and does not have lung disease) = .12 + .19 = .31

Column 2: P(he is a nonsmoker) = P(he is a nonsmoker and has lung disease)
+ P(he is a nonsmoker and does not have lung disease) = .03 + .66 = .69

These marginal probabilities tell us that 31% of all 60-year-old men smoke and 69% do not smoke.

The following table lists all the joint and marginal probabilities.

	He is a smoker	He is a nonsmoker	Totals
He has lung disease	.12	.03	.15
He does not have lung disease	.19	.66	.85
Totals	.31	.69	1

CONDITIONAL PROBABILITY

We frequently need to know how two events are related. In particular, we would like to know the probability of one event given the occurrence of another related event. Consider the smoking/lung disease illustration above. We would certainly like to know the probability that a 60-year-old male smoker has lung disease. This probability is called **conditional probability** because we want to know the probability that a 60-year-old man has lung disease *given* the condition that he smokes. The conditional probability that we seek is represented by

P(he has lung disease | he is a smoker)

where the "|" represents the word "given." Here is how we compute this conditional probability.

The marginal probability of finding a 60-year-old man who smokes is .31, which is made up of the two intersections (he is a smoker and has lung disease) and (he is a smoker and does not have lung disease), which have joint probabilities .12 and .19, respectively. We can interpret these numbers in the following way. On average for every 100 60-year-old men, 31 smoke. Of these 31 men, 12 have lung disease. Thus, the conditional probability is 12/31 = .387. Notice that this ratio is the same as the ratio of the joint probability to the marginal probability .12/.31. All conditional probabilities can be computed this way.

> **Conditional Probability**
>
> The probability of event B given event A is
>
> $$P(B \mid A) = \frac{P(A \text{ and } B)}{P(A)}$$
>
> The probability of event A given event B is
>
> $$P(A \mid B) = \frac{P(A \text{ and } B)}{P(B)}$$

▼ **EXAMPLE 4.1**

Find the probability that a 60-year-old male nonsmoker has lung disease.

Solution We wish to determine the conditional probability

$P(\text{he has lung disease} \mid \text{he is a nonsmoker})$

Using the conditional probability formula we find

$P(\text{he has lung disease} \mid \text{he is a nonsmoker})$

$$= \frac{P(\text{he is a nonsmoker and has lung disease})}{P(\text{he is a nonsmoker})} = \frac{.03}{.69} = .0435$$

Notice that the probability that a nonsmoker has lung disease is considerably smaller than the probability that a smoker has lung disease. The comparison of the two conditional probabilities raises the question of whether the two events, smoking and lung disease, are related, a subject we tackle next.

▲

INDEPENDENCE

One of the objectives of calculating conditional probability is to determine whether two events are related. In particular, we would like to know whether they are **independent**.

> **Independent Events**
>
> Two events A and B are said to be independent if
>
> $P(A \mid B) = P(A)$
>
> or
>
> $P(B \mid A) = P(B)$

Put another way, two events are independent if the probability of one event is not affected by the occurrence of another event.

▼ EXAMPLE 4.2

Determine whether smoking and lung disease among 60-year-old men are independent events.

Solution From the table we determined the marginal probability

P(he has lung disease) = .15

We further calculated that

P(he has lung disease | he is a smoker) = .387

Because the two probabilities are not equal, we conclude that the two events are dependent. Note that we would draw the same conclusion by using the other conditional probability

P(he has lung disease | he is a nonsmoker) = .0435

and observing that it is not equal to P(he has lung disease) = .15. ▲

UNION OF TWO EVENTS

Another event that is the combination of other events is the **union**.

> **Union of Events A and B**
>
> The union of events A and B is the event that occurs when either A or B or both occur. It is denoted as
>
> A or B

EXAMPLE 4.3

Determine the probability that a 60-year-old man smokes or has lung disease.

Solution The union consists of three events. They are (he is a smoker and has lung disease), (he is a nonsmoker and has lung disease), and (he is a smoker and does not have lung disease). Their probabilities are .12, .03, and .19, respectively. Thus the probability of the union of the events he is a smoker and he has lung disease is the sum of the three probabilities. That is,

P(he is a smoker or has lung disease) = P(he is a smoker and has lung disease)
+ P(he is a nonsmoker and has lung disease)
+ P(he is a smoker and does not have lung disease) = .12 + .03 + .19 = .34

EXERCISES

4.16 Given the following table of joint probabilities, calculate the marginal probabilities.

	A_1	A_2	A_3
B_1	.1	.3	.2
B_2	.2	.1	.1

4.17 Given the following table of joint probabilities, calculate the marginal probabilities.

	A_1	A_2
B_1	.4	.3
B_2	.2	.1

4.18 Refer to Exercise 4.17.
  a Determine $P(A_1 | B_1)$.
  b Determine $P(A_2 | B_1)$.
  c Did your answers to parts a and b sum to 1? Is this a coincidence? Explain.

4.19 Refer to Exercise 4.17. Calculate the following probabilities.
  a $P(A_1 | B_2)$.
  b $P(B_2 | A_1)$.
  c Did you expect the answers to parts a and b to be reciprocals? That is,

$$P(A_1 | B_2) = \frac{1}{P(B_2 | A_1)}?$$

  Why is this impossible (unless both probabilities are 1)?

4.20 Are the events in Exercise 4.17 independent? Explain.

4.21 Refer to Exercise 4.17. Compute the following.
  a $P(A_1 \text{ or } B_1)$
  b $P(A_1 \text{ or } B_2)$
  c $P(A_1 \text{ or } A_2)$

4.22 Suppose that you have been given the following joint probabilities. Are the events independent? Explain.

	A_1	A_2
B_1	.20	.60
B_2	.05	.15

4.23 Suppose that you have been given the following joint probabilities. Are the events independent? Explain.

	A_1	A_2
B_1	.20	.15
B_2	.60	.05

4.24 Suppose we have the following joint probabilities.

	A_1	A_2	A_3
B_1	.15	.20	.10
B_2	.25	.25	.05

Compute the marginal probabilities.

4.25 Refer to Exercise 4.24.
  a Compute $P(A_2 | B_2)$.
  b Compute $P(B_2 | A_2)$.
  c Compute $P(B_1 | A_2)$.

4.26 Refer to Exercise 4.24.
  a Compute $P(A_1 \text{ or } A_2)$.
  b Compute $P(A_2 \text{ or } B_2)$.
  c Compute $P(A_3 \text{ or } B_1)$.

4.27 The female instructors at a large university recently lodged a complaint about the most recent round of promotions from assistant professor to associate professor. An analysis of the relationship between gender and promotion was undertaken with the joint probabilities in the following table being produced.

	Promoted	Not promoted
Female	.03	.12
Male	.17	.68

a What is the rate of promotion among female assistant professors?
b What is the rate of promotion among male assistant professors?
c Is it reasonable to accuse the university of gender bias?

4.28 A department store analyzed the most recent sales and determined the relationship between the way the customer paid for the item and the price category of the item. The joint probabilities in the following table were calculated.

	Cash	Credit card	Debit card
Under $20	.09	.03	.04
$20–$100	.05	.21	.18
Over $100	.03	.23	.14

a What proportion of purchases are paid by debit card?
b Find the probability that a credit card purchase is over $100.
c Determine the proportion of purchases made by credit card or by debit card.

4.29 An analysis of the relationship between gender and whether a vote was cast in last mayoral election in Miami Beach produced the following probabilities.

	Female	Male
Voted in last mayoral election	.25	.18
Did not vote in last mayoral election	.33	.24

a What proportion of the electorate voted in the last election?
b Are gender and whether a vote was cast in last mayoral election in Miami Beach independent? Explain how you arrived at your conclusion.

4.30 Why are some mutual fund managers more successful than others? One possible factor is where the manager earned his or her MBA. The following table of joint probabilities describes the relationship between whether the mutual fund outperformed the market and whether the fund manager's MBA was completed at one of the top-20-rated MBA programs.

	Mutual fund outperforms market	Mutual fund does not outperform market
Top-20 MBA program	.11	.29
Not top-20 MBA program	.13	.47

a What proportion of mutual fund managers graduated with an MBA from a top-20 MBA program?
b What proportion of mutual fund managers outperform the market?
c Find the proportion of mutual funds that outperform the market when the fund manager graduated from a top-20 MBA program.
d Is the performance of the mutual fund dependent on whether the fund manager graduated from a top-20 MBA program?

4.31 The method of instruction in college and university applied statistics courses is changing. Historically, most courses were taught with an emphasis on manual calculation. The alternative is to employ a computer and a software package to perform the calculations. An analysis of applied statistics courses investigated whether the instructor's educational background is primarily mathematics (or statistics) or some other field. The result of this analysis is the table of joint probabilities below.

	Statistics course emphasizes manual calculations	Statistics course employs computer and software
Mathematics or statistics education	.23	.36
Other education	.11	.40

a What is the probability that a randomly selected applied statistics course instructor whose education was in statistics emphasizes manual calculations?
b What proportion of applied statistics courses employ a computer and software?
c Are the educational background of the instructor and the way his or her course is taught independent?

4.32 A restaurant chain routinely surveys customers and, among other questions, asks each customer whether he or she would return and to rate the quality of food. Summarizing hundreds of thousands of questionnaires produced the table of joint probabilities below.

Rating	Customer will return	Customer will not return
Poor	.02	.10
Fair	.08	.09
Good	.35	.14
Excellent	.20	.02

a What proportion of customers say that they will return and rate the restaurant's food as good?
b What proportion of customers who say that they will return rate the restaurant's food as good?
c What proportion of customers who rate the restaurant's food as good say that will return?
d Discuss the differences in your answers to parts a, b, and c.

4.33 To determine whether drinking alcoholic beverages has an effect on the bacteria that cause ulcers, researchers developed the following table of joint probabilities.

Number of alcoholic drinks per day	Ulcer	No ulcer
None	.01	.22
One	.03	.19
Two	.03	.32
More than two	.04	.16

a What proportion of people have ulcers?
b What is the probability that a teetotaler (no alcoholic beverages) develops an ulcer?
c What is the probability that someone who has an ulcer does not drink alcohol?
d Are ulcers and the drinking of alcohol independent? Explain.

4.34 When male and female drivers are lost, what do they do? After a thorough analysis the following joint probabilities were developed.

Action	Male	Female
Consult a map	.25	.14
Ask someone for directions	.12	.28
Continue driving until location is determined	.13	.08

a What is the probability that a man would ask for directions?
b What proportion of drivers consult a map?
c Are gender and consulting a map independent? Explain.

4.35 Many critics of television claim that there is too much violence and that it has a negative impact on society. However, there may also be a negative effect on advertisers. To examine this researchers developed two versions of a cops-and-robbers made-for-television movie. One version depicted several violent crimes, and the other removed these scenes. In the middle of the movie one 60-second commercial was shown advertising a new product and brand name. At the end of the movie viewers were asked to name the brand. After observing the results the researchers produced the following table of joint probabilities.

	Watch violent movie	Watch nonviolent movie
Remember the brand name	.15	.18
Do not remember the brand name	.35	.32

a What proportion of viewers remember the brand name?
b What proportion of viewers who watch the violent movie remember the brand name?
c Does watching a violent movie affect whether the viewer will remember the brand name? Explain.

4.36 Is there a relationship between the male hormone testosterone and criminal behavior? To answer this question medical researchers measured the testosterone level of penitentiary inmates and recorded whether they were convicted of murder. After analyzing the results they produced the following table of joint probabilities.

	Murderer	Commit other felony
Above average testosterone level	.27	.24
Below average testosterone level	.21	.28

a What proportion of murderers have above average testosterone levels?
b Are levels of testosterone and the crime committed independent? Explain.

4.4 PROBABILITY RULES AND TREES

In Section 4.3, we introduced intersection and union and described how to determine the probability of the intersection and the union of two events. In this section we present other methods of determining these probabilities. We introduce three rules, which enable us to calculate the probability of more complex events from the probability of simpler events.

COMPLEMENT RULE

The complement of event A is the event that occurs when event A does not occur. The complement of event A is denoted by A^C. The complement rule defined below derives from the fact that the probability of an event and the probability of the event's complement must sum to one.

> **Complement Rule**
> $$P(A^C) = 1 - P(A)$$
> for any event A.

Although the derivation of this rule is quite simple, it is used frequently. We will demonstrate its use in several examples after we introduce the next rule.

MULTIPLICATION RULE

This rule is used to calculate the joint probability of two events. It is based on the formula for conditional probability supplied in the previous section. That is, from the following formula

$$P(B \mid A) = \frac{P(A \text{ and } B)}{P(A)}$$

we derive the multiplication rule simply by multiplying both sides by $P(A)$.

> **Multiplication Rule**
> The joint probability of any two events A and B is
> $$P(A \text{ and } B) = P(B \mid A)P(A)$$
> or
> $$P(A \text{ and } B) = P(A \mid B)P(B)$$

If A and B are independent events, $P(B \mid A) = P(B)$ and $P(A \mid B) = P(A)$. It follows that the joint probability of two independent events is simply the product of the probabilities of the two events. We can express this as a special form of the multiplication rule.

> **Multiplication Rule for Independent Events**
>
> The joint probability of any two independent events A and B is
>
> $P(A \text{ and } B) = P(A)P(B)$

▼ EXAMPLE 4.4

A graduate statistics course has seven male and three female students. The professor wants to select two students at random to help her conduct a research project. What is the probability that the two students chosen are female?

Solution Let A represent the event that the first student chosen is female and B represent the event that the second student chosen is also female. We want the joint probability $P(A \text{ and } B)$. Consequently, we apply the multiplication rule.

$$P(A \text{ and } B) = P(B \mid A)P(A)$$

Because there are three female students in a class of ten, the probability that the first student chosen is female is

$$P(A) = \frac{3}{10}$$

After the first student is chosen, there are only nine students left. Given that the first student chosen was female there are only two female students left. It follows that

$$P(B \mid A) = \frac{2}{9}$$

Thus the joint probability is

$$P(A \text{ and } B) = P(B \mid A)P(A) = \left(\frac{2}{9}\right)\left(\frac{3}{10}\right) = \left(\frac{6}{90}\right) = .067$$

▲

▼ EXAMPLE 4.5

Refer to Example 4.4. The professor who teaches the course is suffering from the flu and will be unavailable for two classes. The professor's replacement will teach the next two classes. His style is to select one student at random and pick on him or her to answer questions during that class. What is the probability that the two students chosen are female?

Solution The form of the question is the same as in Example 4.4; we wish to compute the probability of choosing two female students. However, the experiment is slightly different. It is now possible to choose the *same* student in each of the two classes the replacement teaches. Thus A and B are independent events. Thus, we apply the special multiplication rule for independent events:

$$P(A \text{ and } B) = P(A)P(B)$$

The probability of choosing a female student in each of the two classes is the same. That is,

$$P(A) = \frac{3}{10} \quad \text{and} \quad P(B) = \frac{3}{10}$$

Hence,

$$P(A \text{ and } B) = P(A)P(B) = \left(\frac{3}{10}\right)\left(\frac{3}{10}\right) = \frac{9}{100} = .09$$

▲

ADDITION RULE

The addition rule enables us to calculate the probability of the union of two events.

> **Addition Rule**
>
> The probability that event A or event B, or both, occur is
>
> $$P(A \text{ or } B) = P(A) + P(B) - P(A \text{ and } B)$$

If you're like most students, you're wondering why we subtract the joint probability from the probabilities of A and B. To understand why this is necessary, return to Example 4.3. Recall that we had calculated the probability that he is a smoker or he has lung disease as .34. If we now attempt to calculate the probability of the union by summing the probability that he is a smoker and the probability that he has lung disease, we find

$P(\text{he is a smoker}) + P(\text{he has lung cancer}) = .31 + .15 = .46$

As you can see, the sum of the probabilities is not equal to the probability of the union. The reason why can be seen by reviewing the way the marginal probabilities were computed.

$P(\text{he is a smoker}) = P(\text{he is a smoker and has lung disease})$
$+ P(\text{he is a smoker and does not have lung disease})$

$P(\text{he has lung disease}) = P(\text{he is a smoker and has lung disease})$
$+ P(\text{he is a nonsmoker and has lung disease})$

When we add the probabilities that he is a smoker and that he has lung disease, we add the joint probability $P(\text{he is a smoker and has lung disease})$ twice. Thus, to compute the probability of the union of two events we add the probabilities of the two events, but we then subtract the joint probability.

Thus, applying the addition rule we perform the following calculation:

P(he is a smoker or he has lung disease)
$= P$(he is a smoker) $+ P$(he has lung disease)
$- P$(he is a smoker and has lung disease) $= .31 + .15 - .12 = .34$

which is the probability we obtained in Example 4.3.

As was the case with the multiplication rule, there is a special form of the addition rule. When two events are mutually exclusive, their joint probability is 0.

Addition Rule for Mutually Exclusive Events

The probability of the union of two mutually exclusive events A and B is

$P(A \text{ or } B) = P(A) + P(B)$

▼ EXAMPLE 4.6

In a large city, two newspapers are published, the *Sun* and the *Post*. The circulation departments report that 22% of the city's households have a subscription to the *Sun* and 35% subscribe to the *Post*. A survey reveals that 6% of all households subscribe to both newspapers. What proportion of the city's households subscribe to either newspaper?

Solution We can express this question as: What is the probability of selecting a household at random that subscribes to the *Sun*, the *Post*, or both? It is now clear that we seek the probability of the union, and we must apply the addition rule. Let $A =$ event the household subscribes to the *Sun* and $B =$ event the household subscribes to the *Post*. We perform the following calculation.

$P(A \text{ or } B) = P(A) + P(B) - P(A \text{ and } B) = .22 + .35 - .06 = .51$

The probability that a randomly selected household subscribes to either newspaper is .51. Expressed as relative frequency, 51% of the city's households subscribe to either newspaper.

▲

PROBABILITY TREES

An effective and simpler method of applying the probability rules is the probability tree, wherein the events in an experiment are represented by lines. The resulting figure resembles a tree, hence the name. We will illustrate the probability tree with several examples, including two that we addressed using the probability rules alone.

In Example 4.4 we wanted to find the probability of choosing two female students, where the two choices had to be different. The tree diagram below describes this experiment. Notice that the first two branches represent the two possibilities, female and male students, on the first choice. The second set of branches represents the two possibilities on the second choice. The probabilities of female and male student chosen first are 3/10 and 7/10, respectively. The probabilities for the second set of branches are conditional probabilities based on the gender of the first student selected.

We calculate the joint probabilities by multiplying the probabilities on the linked branches. Thus, the probability of choosing two female students is $P(F \text{ and } F) = (3/10)(2/9) = 6/90$. The remaining joint probabilities are computed similarly.

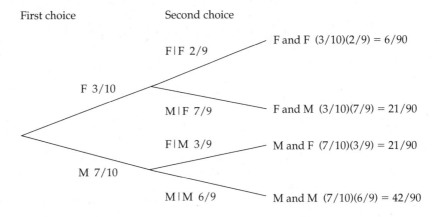

In Example 4.5, the experiment was similar to that of Example 4.4. However, the student selected on the first choice was returned to the pool of students and was eligible to be chosen again. Thus, the probabilities on the second set of branches remains the same as the probabilities on the first set, and the probability tree is drawn with these changes.

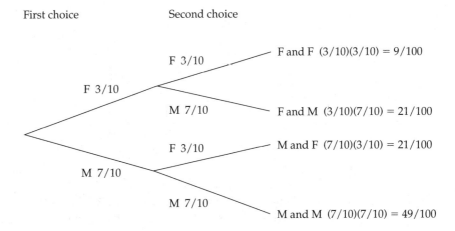

The advantage of a probability tree on this type of problem is that it restrains its users from the wrong calculation. Once the tree is drawn and the probabilities of the branches inserted, virtually the only allowable calculation is the multiplication of the probabilities of linked branches. An easy check on those calculations is available. The joint probabilities at the ends of the branches must sum to 1, because all possible events are listed. Notice in both figures that the joint probabilities do indeed sum to 1.

The special form of the addition rule for mutually exclusive events can be applied to the joint probabilities. In both probability trees we can compute the probability that one student chosen is female and one is male simply by adding the joint probabilities. For the tree in Example 4.4, we have

$$P(\text{female and male}) + P(\text{male and female}) = \frac{21}{90} + \frac{21}{90} = \frac{42}{90}$$

In the probability tree in Example 4.5, we find

$$P(\text{female and male}) + P(\text{male and female}) = \frac{21}{100} + \frac{21}{100} = \frac{42}{100}$$

▼ EXAMPLE 4.7

Students who graduate from law schools must still pass a bar exam before becoming lawyers. Suppose that in a particular jurisdiction the pass rate for first-time test takers is 72%. Candidates who fail the first exam may take it again several months later. Of those who fail their first test, 88% pass their second attempt. Find the probability that a randomly selected law school graduate becomes a lawyer. Assume that candidates cannot take the exam more than twice.

Solution The following probability tree is employed to describe the experiment. Note that we use the complement rule to determine the probability of failing each exam.

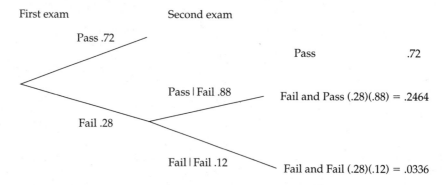

We apply the multiplication rule to calculate $P(\text{fail and pass})$, which we find to be .2464. We then apply the addition rule for mutually exclusive events to find the probability of passing the first or second exam:

$P(\text{pass[on first exam]}) + P(\text{fail [on first exam] and pass [on second exam]})$
$= .72 + .2464 = .9664$

Thus, 96.64% of applicants become lawyers by passing the first or second exam.

▲

BAYES' LAW

A useful application of probability rules and trees is demonstrated with the following example.

▼ EXAMPLE 4.8

Physicians routinely perform medical tests on their patients when they suspect various diseases. However, few tests are 100% accurate. Most can produce false-positive or false-negative results. (A false-positive result is one in which the patient does not have the disease, but the test shows positive. False-negative results are ones where the patient does have the disease, but the test produces a negative result.) Many people misinterpret medical test results. When the disease is often fatal, such misconceptions are themselves serious and need to be corrected.

Suppose that a particular test correctly identifies those with a certain serious disease 94% of the time and correctly diagnoses those without the disease 98% of the time. A friend has just informed you that he has received a positive result and asks for your advice about how to interpret these probabilities. He knows nothing about probability, but he feels that because the test is quite accurate, the probability that he does have the disease is quite high, likely in the 95% range. Before attempting to address your friend's concerns, you research the illness and discover that 4% of men have this disease. There is now enough information to draw a probability tree. Let

D = Has the disease

D^C = Does not have the disease

PT = Positive test result

NT = Negative test result

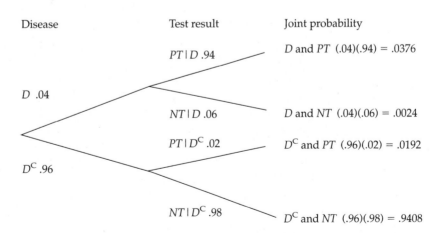

The tree allows you to determine the probability of obtaining a positive test result. It is

$$P(PT) = P(D \text{ and } PT) + P(D^C \text{ and } PT) = .0376 + .0192 = .0568$$

Your friend has been given a positive test result. Thus, the probability that you now calculate is the conditional probability that your friend has the disease *given* that he has a positive test result:

$$P(D \mid PT) = \frac{P(D \text{ and } PT)}{P(PT)} = \frac{.0376}{.0568} = .6620$$

There is a 66.2% probability that he has the disease. The probability is high, but considerably lower than your friend feared.

▲

Thomas Bayes first employed the calculation of conditional probability performed above in the eighteenth century. Accordingly, it is called **Bayes' Law.** Although there is a formula defining Bayes' Law, we will not offer it here, preferring instead to use a probability tree to conduct all such calculations.

The probabilities $P(D)$ and $P(D^C)$ are called **prior probabilities** because they are determined *prior* to the test results. The conditional probabilities are called **likelihood probabilities** for reasons that are beyond the mathematics in this book. Finally, the conditional probability $P(D \mid PT)$ and similar conditional probabilities $P(D^C \mid PT)$, $P(D \mid NT)$, and $P(D^C \mid NT)$ are called **posterior probabilities,** because these probabilities are determined *after* the test is conducted.

DEVELOPING AN UNDERSTANDING OF PROBABILITY CONCEPTS

If you review the computations made above you'll realize that the prior probabilities are as important as the probabilities associated with the test results (the likelihood probabilities) in determining the posterior probabilities. To understand this point, suppose that additional research indicates that age is a factor in the disease and that only 1% of men your friend's age contract the disease. Recalculating the probabilities from the tree produces the following:

$$P(PT) = P(D \text{ and } PT) + P(D^C \text{ and } PT) = .0094 + .0198 = .0292$$

$$P(D \mid PT) = \frac{P(D \text{ and } PT)}{P(PT)} = \frac{.0094}{.0292} = .3219$$

As you can see, the probability that your friend has the disease has decreased by about half.

Now suppose that more research tells us that 40% of men of the same ethnic group and age have the disease. What is the posterior probability? Repeating the calculations with the new prior probabilities produces these probabilities:

$$P(PT) = P(D \text{ and } PT) = P(D^C \text{ and } PT) = .3760 + .0120 = .3880$$

$$P(D \mid PT) = \frac{P(D \text{ and } PT)}{P(PT)} = \frac{.3760}{.3880} = .9691$$

Notice that as the prior probability increases, so does the posterior probability. There is a wide range of applications of Bayes' Law. We have featured just one. In the exercise set that accompanies this section, we offer several more.

IDENTIFYING THE CORRECT METHOD

As we've previously pointed out, the emphasis in this book will be on identifying the correct statistical technique to use. In Chapters 2 and 3 we showed how to summarize data by first identifying the appropriate method to use. While it is difficult to offer strict rules on which probability method to use, nevertheless we can provide some general guidelines.

In the examples and exercises in this text (and most other introductory statistics books), the key issue is whether joint probabilities are provided or are required.

Joint Probabilities Are Given

In Section 4.3 we addressed problems where the joint probabilities were given. In these problems, we can compute marginal probabilities by adding across rows and down columns. We can use the joint and marginal probabilities to compute conditional probabilities, for which a formula is available. This allows us to determine whether the events described by the table are independent or dependent.

We can also apply the addition rule to compute the probability that either of two events occur.

Joint Probabilities Are Required

This section introduced three probability rules and probability trees. We need to apply some or all of these rules in circumstances where one or more joint probabilities are required. We apply the multiplication rule (either by formula or through a probability tree) to calculate the probability of intersections. In some problems, we're interested in adding these joint probabilities. We're actually applying the addition rule for mutually exclusive events here. We also frequently use the complement rule. We can also calculate new conditional probabilities using Bayes' Law.

EXERCISES

4.37 Given the following probabilities, draw a probability tree to compute the joint probabilities.
$P(A) = .9 \qquad P(A^C) = .1$
$P(B \mid A) = .4 \qquad P(B \mid A^C) = .7$

4.38 Given the following probabilities, draw a probability tree to compute the joint probabilities.
$P(A) = .8 \qquad P(A^C) = .2$
$P(B \mid A) = .4 \qquad P(B \mid A^C) = .7$

4.39 Draw a probability tree to compute the joint probabilities from the following probabilities.
$P(A) = .5 \qquad P(A^C) = .5$
$P(B \mid A) = .4 \qquad P(B \mid A^C) = .7$

4.40 Given the following probabilities, draw a probability tree to compute the joint probabilities.
$P(A) = .8 \qquad P(A^C) = .2$
$P(B \mid A) = .3 \qquad P(B \mid A^C) = .3$

4.41 Given the following probabilities, find the joint probability $P(A$ and $B)$:
$P(A) = .7 \qquad P(B \mid A) = .3$

4.42 Refer to Exercise 4.41. Suppose further that $P(B \mid A^C) = .6$. Find $P(A \mid B)$. (*Hint:* Use a probability tree.)

4.43 The Chartered Financial Analyst (CFA) is a designation earned after taking three annual exams (CFA I, II, and III). The exams are taken in early June. Candidates who pass an exam are eligible to take the exam for the next level in the following year. The pass rates for levels I, II, and III are .57, .73, and .85, respectively. Suppose that 3,000 candidates take the level I exam, 2,500 take the level II exam, and 2,000 take the level III exam. Suppose that one student is selected at random. What is the probability that he or she has passed the exam?

4.44 Approximately 10% of people are left handed. If two people are selected at random, what is the probability of the following events?

a Both are right handed.
b Both are left handed.
c One is right handed and the other is left handed.
d At least one is right handed.

4.45 Refer to Exercise 4.44. Suppose that three people are selected at random.

a Draw a probability tree to depict the experiment.
b If we use the notation *RRR* to describe the selection of three right-handed people, similarly describe the remaining seven events. (Use *L* for left-hander.)
c How many of the events yield no right-handers, one right-hander, two right-handers, three right-handers?
d Find the probability of no right-handers, one right-hander, two right-handers, three right-handers.

4.46 Suppose that there are 100 students in your psychology class, of whom 10 are left handed. Two students are selected at random.

a Draw a probability tree, and insert the probabilities for each branch. What is the probability of the following events?
b Both are right handed.
c Both are left handed.
d One is right handed and the other is left handed.
e At least one is right handed.

4.47 Refer to Exercise 4.46. Suppose that three people are selected at random.

a Draw a probability tree, and insert the probabilities of each branch.
b What is the probability of no right-handers, one right-hander, two right-handers, three right-handers?

4.48 Refer to Exercise 4.30. The following probabilities were provided instead of the joint probabilities:

P(mutual fund manager graduated from top-20 MBA program) = .40

P(outperform market | graduated from top-20 MBA program) = .305

P(outperform market | did not graduate from top-20 MBA program) = .217

a Find the probability that a mutual fund outperformed the market and that the manager graduated from a top-20 MBA program.
b Find the probability that a manager graduated from a top-20 MBA program, given that the mutual fund outperformed the market.

4.49 Refer to Exercise 4.29. Suppose that, instead of providing the joint probabilities, the analysis yielded the following information:

1 Proportion of males in the electorate = 42%
2 Gender and voting record are independent
3 Proportion of eligible voters who voted in the last election = 43%

Use a probability tree or the probability rules to compute the following probabilities:

a P(voted in last election and male)
b P(voted in last election and female)
c P(female | voted in the last election)

4.50 Refer to Exercise 4.43. A randomly selected candidate who took a CFA exam tells you that he has passed the exam. What is the probability that he took the CFA I exam?

4.51 A foreman for an injection-molding firm admits that on 10% of his shifts, he forgets to shut off the injection machine on his line. This causes the machine to overheat, increasing the probability that a defective molding will be produced during the early morning run from 2% to 20%. What proportion of moldings from the early morning run is defective?

4.52 Refer to Exercise 4.51. The plant manager randomly selects a molding from the early morning run and discovers it is defective. What is the probability that the foreman forgot to shut off the machine the previous night?

4.53 A telemarketer calls people and tries to sell them a subscription to a daily newspaper. On 20% of her calls, there is no answer or the line is busy. She sells subscriptions to 5% of the remaining calls. For what proportion of calls does she make a sale?

4.54 Government tax auditors regularly audit the tax returns. There are several indicators that tell the auditors that a return may be fraudulent. One of them is a particular type of deduction. A total of 4% of all tax returns feature this type of deduction. Only 3% of returns without this deduction make fraudulent claims. However, 14% of returns with the deduction make fraudulent claims. What proportion of returns make fraudulent claims?

4.55 Refer to Exercise 4.54. A randomly selected return has no fraudulent claim. What is the probability that the deduction was claimed?

4.56 A survey of middle-aged men reveals that 28% of them are balding at the crown of their heads. Moreover it is known that such men have an 18%

probability of suffering a heart attack in the next ten years. Men who are not balding in this way have an 11% probability of a heart attack.

a Find the probability that a middle-aged man will suffer a heart attack sometime in the next ten years.

b Suppose that a middle-aged man has had a heart attack. What is the probability that he is balding at the crown?

4.57 Researchers at the University of Pennsylvania School of Medicine have determined that children under two years old who sleep with the lights on have a 36% chance of becoming myopic before they are 16. Children who sleep in darkness have a 21% probability of becoming myopic. A survey indicates that 28% of children under two sleep with some light on.

a Find the probability that a child under 16 is myopic.

b Find the probability that a myopic child under 16 slept with lights on before he or she was two years old.

4.58 The Nickels restaurant chain regularly conducts surveys of its customers. Respondents are asked to assess food quality, service, and price. The responses are:

Excellent Good Fair

They are also asked whether they would come back.

After analyzing the responses, an expert in probability determined that 87% of customers say that they will return. Of those who so indicate, 57% rate the restaurant as excellent, 36% rate it as good, and the remainder rate it as fair. Of those who say that they won't return, the probabilities are 14%, 32%, and 54%, respectively.

a What proportion of customers rate the restaurant as good?

b What is the probability that someone who rates the restaurant as excellent says that he or she won't return?

4.59 Refer to the medical test and disease illustration on pages 109 and 110.

a Find the probability that your friend has the disease when the false-positive and false-negative rates are .01 and .02, respectively.

b Find the probability that your friend has the disease when the false-positive and false-negative rates are .15 and .30, respectively.

4.5 SUMMARY

The first step in assigning probability is to create an exhaustive and mutually exclusive list of outcomes. The second step is to use the classical, relative frequency, or subjective approach to assign probability to the outcomes. There are a variety of methods available to compute the probability of other events. These methods include probability rules and trees.

An important application of these rules is Bayes' Law, which allows us to compute conditional probabilities from other forms of probability.

IMPORTANT TERMS

Exhaustive
Mutually exclusive
Sample space
Classical approach
Relative frequency approach
Subjective approach
Simple event
Event
Intersection

Joint probability
Marginal probability
Conditional probability
Independent events
Union
Bayes' Law
Prior probability
Likelihood probability
Posterior probability

FORMULAS

Conditional probability

$$P(A \mid B) = \frac{P(A \text{ and } B)}{P(B)}$$

Complement rule

$$P(A^C) = 1 - P(A)$$

Multiplication rule

$$P(A \text{ and } B) = P(A \mid B)P(B)$$

Addition rule

$$P(A \text{ or } B) = P(A) + P(B) - P(A \text{ and } B)$$

SUPPLEMENTARY EXERCISES

4.60 The following table lists the joint probabilities of achieving grades of A and not in two courses.

	Achieve a grade of A in psychology	Does not achieve a grade of A in psychology
Achieve a grade of A in statistics	.06	.13
Does not achieve a grade of A in statistics	.23	.58

a What is the probability that a student achieves a grade of A in psychology?
b What is the probability that a student achieves a grade of A in psychology, given that he or she does not achieve a grade of A in statistics?
c Are achieving grades of A in psychology and statistics independent events? Explain.

4.61 A construction company has bid on two contracts. The probability of winning contract A is .3. If the company wins contract A the probability of winning contract B is .4. Find the probability of the following events:

a Winning both contracts
b Winning exactly one contract
c Winning at least one contract

4.62 Laser surgery to fix short-sightedness is becoming more popular. However, for some people, a second procedure is necessary. The following table lists the joint probability of needing a second procedure and whether the patient has a corrective lens with a factor (diopter) of minus 8 or less.

	Vision corrective factor of more than minus 8	Vision corrective factor of less than minus 8
Procedure is successful	.66	.15
Second procedure is required	.05	.14

a Find the probability that a second procedure is required.
b Determine the probability that someone whose corrective lens factor is minus 8 or less does not require a second procedure.
c Are the events independent? Explain your answer.

4.63 The effect of an antidepressant drug varies from person to person. Suppose that the drug is effective on 80% of women and 65% of men. It is known that 66% of the people who take the drug are women. What is the probability that the drug is effective?

4.64 Refer to Exercise 4.63. Suppose that you are told that the drug is effective. What is the probability that the drug-taker is a man?

4.65 In a four-cylinder engine, there are four spark plugs. If any one of them malfunctions, the car will idle roughly, and power will be lost. Suppose that for a certain brand of spark plugs the probability that a spark plug will function properly after 5,000 miles is .90. Assuming that the spark plugs operate independently, what is the probability that the car will idle roughly after 5,000 miles?

4.66 A telemarketer sells magazine subscriptions over the telephone. The probability of a busy signal or no answer is 65%. If the telemarketer does make contact, the probability of 0, 1, 2, or 3 magazine subscriptions is .5, .25, .20, and .05, respectively. Find the probability that in one call she sells no magazines.

4.67 Recent studies appear to conclude that television viewers are more likely to remember the brand name of the product being advertised when the show is nonviolent. The following table of joint probabilities describes this phenomenon.

	Show is violent	Show is nonviolent
Brand name is remembered	.08	.53
Brand name is not remembered	.11	.28

a Find the probability that the brand name is remembered.
b Determine the probability that a viewer remembers the brand name when the show is violent.
c Determine the probability that a viewer remembers the brand name when the show is nonviolent.
d Are the events independent? Explain your answer.

4.68 A study by the Correction Service of Canada has found that a new law designed to keep dangerous offenders behind bars longer has failed. The law created a special category of inmates based on whether they had committed crimes involving violence or drugs. Such criminals are subject to detention if the Correction Services judges them highly likely to reoffend. Prisoners who are not so detained are automatically paroled after serving two-thirds of their sentence. Those detained under this new law serve an average of 415 additional days in prison. Recent statistics reveal that 37% of those not detained reoffend within two years. However, only 16% of those detained reoffend within two years of their release.

a Suppose that 50% of all prisoners are detained under the new law. What is the probability that a criminal who has reoffended within two years of his or her release from prison was paroled after serving two-thirds of the sentence?
b Repeat part (a) assuming that 90% of all prisoners are released after serving two-thirds of their sentence.

4.69 Casino Windsor conducts surveys to determine the opinions of its customers. Among other questions respondents were asked to give their opinion about "your overall impression of Casino Windsor." The responses are:

Excellent Good Average Poor

Additionally, the gender of the respondent is noted. After analyzing the results the following table of joint probabilities was produced.

Rating	Women	Men
Excellent	.27	.22
Good	.14	.10
Average	.06	.12
Poor	.03	.06

a What proportion of customers rate Casino Windsor as excellent?
b Determine the probability that a male customer rates Casino Windsor as excellent.
c Find the probability that a customer who rates Casino Windsor as excellent is a man.
d Are gender and rating independent? Explain your answer.

4.70 How does level of affluence affect health care? To address one dimension of the problem a group of heart attack victims was drawn. Each was categorized as a low-, medium-, or high-income earner. Each was also categorized as having survived or died. A demographer notes that in our society 21% fall into the low-income group, 49% are in the medium-income group, and 30% are in the high-income group. Furthermore, an analysis of heart attack victims reveals that 12% of low-income people, 9% of medium-income people, and 7% of high-income people die of heart attacks. Find the probability that a survivor of a heart attack is in the low-income group.

Case 4.1   Let's Make a Deal

A number of years ago, there was a popular television game show called *Let's Make a Deal*. The host, Monty Hall, would randomly select contestants from the audience and, as the title suggests, he would make deals for prizes. Contestants would be given relatively modest prizes and then would be offered the opportunity to risk them to win better ones.

Suppose that you are a contestant on this show. Monty has just given you a free trip touring toxic waste sites around the country. He now offers you a trade: Give up the trip in exchange for a gamble. On the stage are three curtains, A, B, and C. Behind one of them is a brand new car worth $20,000. Behind the other two curtains the stage is empty. You decide to gamble and select Curtain A. In an attempt to make things more interesting, Monty then exposes an empty stage by opening Curtain C (he knows there is nothing behind Curtain C). He then offers you the free trip again if you quit or, if you like, a chance to change your choice of curtains (i.e., you can keep your choice of Curtain A or switch to Curtain B). What do you do?

To help you answer that question, try first answering these questions.

1. Before Monty shows you what's behind Curtain C, what is the probability that the car is behind Curtain A? What is the probability that the car is behind Curtain B?

2. After Monty shows you what's behind Curtain C, what is the probability that the car is behind Curtain A? What is the probability that the car is behind Curtain B?

Case 4.2   To Bunt or Not to Bunt, That Is the Question

No sport generates as many statistics as baseball. Reporters, managers, and fans argue and discuss strategies on the basis of these statistics. An article in *Chance* ("A Statistician Reads the Sports Page," Hal S. Stern, Vol. 1, Winter 1997) offers baseball lovers another opportunity to analyze numbers associated with the game. Table 1 below lists the probabilities of scoring at least one run in situations that are defined by the number of outs and the bases occupied. For example, the probability of scoring at least one run when there are no outs and a man is on first base is .39. If the bases are loaded with one out, the probability of scoring any runs is .67.

Table 1   Probability of Scoring Any Runs

Bases Occupied	0 Out	1 Out	2 Outs
Bases empty	.26	.16	.07
First base	.39	.26	.13
Second base	.57	.42	.24
Third base	.72	.55	.28
First base and second base	.59	.45	.24
First base and third base	.76	.61	.37
Second base and third base	.83	.74	.37
Bases loaded	.81	.67	.43

Probabilities are based on results from the American League during the 1989 season. The results for the National League are also shown in the article and are similar. This table allows us to determine the best strategy in a variety of circumstances. This case will concentrate on the strategy of the sacrifice bunt. The purpose of the sacrifice bunt is to sacrifice the batter to move base runners to the next base. It can be employed when there are fewer than two outs and men on base. Ignoring the suicide squeeze, four outcomes can occur:

1. The bunt is successful. The runner (or runners) advances one base, and the batter is out.
2. The batter is out but fails to advance the runner.
3. The batter bunts into a double play.
4. The batter is safe (hit or error), and the runner advances.

Suppose that you are an American League manager. The game is tied in the middle innings of a game, and there is a runner on first base with no one out. Given the following probabilities of the four outcomes of a bunt for the batter at the plate, should you signal the batter to sacrifice bunt?

$P(\text{outcome 1}) = .75$

$P(\text{outcome 2}) = .10$

$P(\text{outcome 3}) = .10$

$P(\text{outcome 4}) = .05$

Assume for simplicity that after the hit or error in outcome 4 there will be men on first and second base and no one out.

Chapter 5

Random Variables and Discrete Probability Distributions

5.1 Introduction

5.2 Random Variables and Probability Distributions

5.3 Describing the Population/Probability Distribution

5.4 Binomial Distribution

5.5 Poisson Distribution

5.6 Summary

5.1 INTRODUCTION

In this chapter, we extend the concepts and techniques of probability introduced in Chapter 4. We present random variables and probability distributions, which are essential in the development of statistical inference.

Here is a brief glimpse into the wonderful world of statistical inference. Suppose that you flip a coin 100 times and count the number of heads. The objective is to determine whether we can infer from the count that the coin is not balanced. It is reasonable to believe that observing a large number of heads (say, 90) or a small number (say, 15) would be a statistical indication of an unbalanced coin. However, where do we draw the line? At 75 or 65 or 55? Without knowing the probability of the frequency of the number of heads from a balanced coin, we cannot draw any conclusions from the sample of 100 coin flips.

The concepts and techniques of probability introduced in Chapter 4 will allow us to calculate the probability we seek. Our objective is to develop the *sampling distribution*. However, to produce the sampling distribution, we first must present random variables and probability distributions, subjects that are introduced in this and the next chapter.

5.2 RANDOM VARIABLES AND PROBABILITY DISTRIBUTIONS

Consider an experiment where we flip two balanced coins and observe the results. We can represent the events as

Heads on the first coin and heads on the second coin

Heads on the first coin and tails on the second coin

Tails on the first coin and heads on the second coin

Tails on the first coin and tails on the second coin

However, we can list the events in a different way. Instead of defining the events by describing the outcome of each coin, we can count the number of heads (or, if we wish, the number of tails). Thus, the events are now

2 heads

1 heads

1 heads

0 heads

The number of heads is called the **random variable.** We often label the random variable X, and we're interested in the probability of each value of X. Thus, in this illustration the values of X are 0, 1, and 2.

Here is another example. In many parlor games as well as in the game of craps played in casinos, the player tosses two dice. One way of listing the events is to describe the number on the first die and the number on the second die as follows.

1,1	1,2	1,3	1,4	1,5	1,6
2,1	2,2	2,3	2,4	2,5	2,6
3,1	3,2	3,3	3,4	3,5	3,6
4,1	4,2	4,3	4,4	4,5	4,6
5,1	5,2	5,3	5,4	5,5	5,6
6,1	6,2	6,3	6,4	6,5	6,6

However, in almost all games the player is primarily interested in the total. Accordingly, we can list the totals of the two dice instead of the individual numbers.

2	3	4	5	6	7
3	4	5	6	7	8
4	5	6	7	8	9
5	6	7	8	9	10
6	7	8	9	10	11
7	8	9	10	11	12

If we define the random X as the total of the two dice, X can equal 2, 3, 4, 5, 6, 7, 8, 9, 10, 11, and 12.

> **Random Variable**
>
> A random variable is a function or rule that assigns a number to each outcome of an experiment.

In some experiments the outcomes are numbers. For example, when we record the mark out of 100 on a statistics test or measure the amount of time to assemble a computer, the experiment produces events that are numbers. Put simply, the value of a random variable is a numerical event.

There are two types of random variables, discrete and continuous. A **discrete random variable** is one that can take on a countable number of values. For example, if we define X as the number of heads observed in an experiment that flips a coin 10 times, the values of X are 0, 1, 2, ... 10. There is a total of 11 values that X can assume. Obviously, we counted the number of values; hence X is discrete.

A **continuous random variable** is one whose values are uncountable. An excellent example of a continuous random variable is the amount of time to complete a task. For example, let $X =$ time to write a statistics exam in a university where the time limit is 3 hours and students cannot leave before 30 minutes. The smallest value of X is 30 minutes. If we attempt to count the number of values that X can take on we need to identify the next value. Is it 30.1 minutes? 30.01 minutes? 30.001 minutes? None of these is the second possible value of X, because there exist numbers larger than 30 and smaller than 30.001. It becomes clear that we cannot identify the second, or third, or any other values of X (except for the largest value, 180 minutes). Thus, we cannot count the number of values, and X is continuous.

A **probability distribution** is a table, formula, or graph that describes the values of a random variable and the probability associated with these values. We will address discrete probability distributions in the rest of this chapter and cover continuous distributions in Chapter 6.

Incidentally, we will use the following notation. An upper-case letter will represent the *name* of the random variable, usually X. The *value* of the random variable will be represented by its lower-case counterpart. Thus, we represent the probability that the random variable X will equal x as

$P(X = x)$

We will also refer to this probability more simply as $p(x)$.

DISCRETE PROBABILITY DISTRIBUTIONS

The probability of the values of a discrete random variable may be derived by means of probability tools such as tree diagrams or by applying one of the definitions of probability. However, two fundamental requirements apply. They are:

Requirements for a Distribution of a Discrete Random Variable

If a discrete random variable X can assume values x_i

1. $0 \leq p(x_i) \leq 1$ for all x_i
2. $\sum_{\text{all } x_i} p(x_i) = 1$

These requirements are equivalent to the rules of probability provided in Chapter 4. To illustrate, consider the following example.

▼ EXAMPLE 5.1

A statistics student is about to take the first class quiz and is concerned about her mark. The professor has provided the frequency distribution of the performance of last year's class, which is listed below. Develop the probability distribution of the random variable defined as the mark out of 10 on the statistics quiz.

Mark	Frequency
0	0
1	2
2	4
3	6
4	7
5	13
6	13
7	21
8	17
9	12
10	5

Solution Because we have information about the past performance of students on the quiz, the logical choice of which definition of probability to use is relative frequency. Thus, we estimate the probability of each value of X by the relative frequency from last year. To estimate each probability we divide the frequency by the total, which is 100, producing the following probability distribution.

Mark	Probability
0	0
1	.02
2	.04
3	.06
4	.07
5	.13
6	.13
7	.21
8	.17
9	.12
10	.05

Notice that the requirements are satisfied. Each probability lies between 0 and 1 and the sum is 1.

▲

One of the reasons to study probability distributions is the relationship between a probability distribution and a population. No matter how the distribution is derived, we interpret it in the following way. The distribution represents an infinitely large population whose relative frequencies are provided by the distribution. In Example 5.1 we have an infinitely large population of statistics marks. The probability that a student achieves a mark of, for example, 10 is 5%. The probability, for example, that he or she receives a mark of 6 is 13%.

It should be understood that even though the probabilities were estimated from only 100 students we interpret probability on the basis of the long run. For example, we interpret $P(X = 10) = .05$ to mean that 5% of *all* students in the future will achieve a mark of 10 on the exam.

We can also create new events and determine their probabilities. For example, the probability of achieving a mark of 5 or more is the sum of the probabilities of $X = 5, 6, 7, 8, 9,$ and 10. Thus,

$$P(X \geq 5) = \sum_{x=5}^{10} p(x) = .13 + .13 + .21 + .17 + .12 + .05 = .81$$

There is a 81% probability of passing this test, which means that in the long run 81% of all students will pass this test.

Incidentally, because the sum of the probabilities must be 1, we also know that 19% will receive a mark less than 5. That is, using the complement rule,

$$P(X < 5) = 1 - P(X \geq 5) = 1 - .81 = .19$$

Here is another peek at statistical inference.

Suppose that you are skeptical about the probability distribution produced in Example 5.1. Accordingly, you take a sample of 10 students and ask each to report his or her mark. They are:

3, 6, 2, 3, 5, 3, 9, 7, 4, 2

These marks do not appear to be consistent with the distribution in Example 5.1. For one thing, 60% of the sample (6 marks out of 10) is below 5, whereas according to the probability distribution the probability that a student fails the quiz is 19%. Is this difference large enough to conclude that the distribution produced above is incorrect? To answer this question, we need a variety of additional tools. In the next five chapters we will develop the necessary tools for the job.

EXERCISES

5.1 The number of pizzas delivered to university students each month is a random variable with the following probability distribution.

x	0	1	2	3
p(x)	.1	.3	.4	.2

Find the probability that a student has received delivery of two or more pizzas this month.

5.2 The probability that a university graduate will be offered no jobs within one month of graduation is estimated to be 5%. The probability of receiving one, two, and three job offers has similarly been estimated to be 43%, 31%, and 21%, respectively. Determine the following probabilities.

  a A graduate is offered fewer than two jobs.
  b A graduate is offered more than one job.

5.3 Use a probability tree to compute the probability of the following events when flipping two fair coins.

  a Heads on the first coin and heads on the second coin
  b Heads on the first coin and tails on the second coin
  c Tails on the first coin and heads on the second coin
  d Tails on the first coin and tails on the second coin

5.4 Refer to Exercise 5.3. Find the following probabilities.

  a No heads
  b One heads
  c Two heads
  d At least one heads

5.5 Draw a probability tree to describe the flipping of three fair coins.

5.6 Refer to Exercise 5.5. Find the following probabilities.

  a Two heads
  b One heads
  c At least one heads
  d At least two heads

5.7 After watching a number of children playing games at a video arcade, a statistics practitioner estimated the following probability distribution of the number of games per visit X.

x	1	2	3	4	5	6	7
p(x)	.05	.15	.15	.25	.20	.10	.10

  a What is the probability that a child will play more than four games?
  b What is the probability that a child will play at least two games?

5.8 A survey of Amazon.com shoppers reveals the following probability distribution of the number of books purchased per hit.

x	0	1	2	3	4	5	6	7
p(x)	.35	.25	.20	.08	.06	.03	.02	.01

  a What is the probability that an Amazon.com visitor will buy four books?
  b What is the probability that an Amazon.com visitor will buy eight books?
  c What is the probability that an Amazon.com visitor will not buy any books?
  d What is the probability that an Amazon.com visitor will buy at least one book?

5.9 A university librarian produced the following probability distribution of the number of times a student walks into the library.

x	0	5	10	15	20	25	30	40	50	75	100
p(x)	.22	.29	.12	.09	.08	.05	.04	.04	.03	.03	.01

Find the following probabilities.

a $P(X \geq 20)$
b $P(X = 60)$
c $P(X > 50)$
d $P(X > 100)$

5.10 After analyzing the frequency with which cross-country skiers participate in their sport, the following probability distribution was created. (X = number of times per year cross-country skiers ski.)

x	0	1	2	3	4	5	6	7	8
p(x)	.04	.09	.19	.21	.16	.12	.08	.06	.05

Find the following.

a $P(X = 3)$
b $P(X \geq 5)$
c $P(5 \leq X \leq 7)$

5.11 The natural remedy echinacea is reputed to boost the immune system, which will reduce flu and colds. A six-month study was undertaken to determine whether the remedy works. From this study, the following probability distribution of the number of respiratory infections per year (X) for echinacea users was produced.

x	0	1	2	3	4
p(x)	.45	.31	.17	.06	.01

Find the following probabilities.

a An echinacea user has more than one infection per year.
b An echinacea user has no infections per year.
c An echinacea user has between one and three (inclusive) infections per year.

5.3 DESCRIBING THE POPULATION/ PROBABILITY DISTRIBUTION

In Chapter 3, we showed how to calculate the mean, variance, and standard deviation of a population. The formulas we provided were based on knowing each of the observations in the population. For example, if we want to know the mean annual income of all North American blue-collar workers we would record each of their incomes and use the formula

$$\mu = \frac{\sum_{i=1}^{N} x_i}{N}$$

where x_1 is the income of the first blue-collar worker, x_2 is the second worker's income, and so on. It is likely that N equals several million. As you can appreciate, this formula is seldom used in practical applications because populations are so large. It is unlikely that we would be able to record all the incomes in the population of North American blue-collar workers. As we noted above, probability distributions represent populations. Rather than record each of the many observations in a population, we list the values and their associated probabilities. These are then used to compute the mean as well as other parameters.

The population mean is the weighted average of all of its values. The weights are the probabilities. This parameter is also called the **expected value** of X and is represented by $E(X)$.

Population Mean

$$E(X) = \mu = \sum_{\text{all } x} xp(x)$$

The population variance $V(X)$ is the weighted average of the squared deviations from the mean.

Population Variance

$$V(X) = \sigma^2 = \sum_{\text{all } x} (x - \mu)^2 p(x)$$

Population Standard Deviation

$$\sigma = \sqrt{\sigma^2}$$

▼ EXAMPLE 5.2

Find the mean, variance, and standard deviation of the population of marks in Example 5.1.

Solution The mean of X is

$$E(X) = \mu = \sum_{\text{all } x} xp(x) = 0p(0) + 1p(1) + 2p(2) + \cdots + 10p(10)$$
$$= 0(0) + 1(.02) + 2(.04) + 3(.06) + 4(.07) + 5(.13) + 6(.13)$$
$$\quad + 7(.21) + 8(.17) + 9(.12) + 10(.05)$$
$$= 6.4$$

Notice that the marks assume integer values only, yet the mean is 6.4.

The variance of X is

$$V(X) = \sigma^2 = \sum_{\text{all } x} (x - \mu)^2 p(x)$$
$$= (0 - 6.4)^2 p(0) + (1 - 6.4)^2 p(1) + \cdots + (10 - 6.4)^2 p(10)$$
$$= (0 - 6.4)^2 (0) + (1 - 6.4)^2 (.02) + (2 - 6.4)^2 (.04)$$
$$\quad + (3 - 6.4)^2 (.06) + (4 - 6.4)^2 (.07) + (5 - 6.4)^2 (.13)$$
$$\quad + (6 - 6.4)^2 (.13) + (7 - 6.4)^2 (.21) + (8 - 6.4)^2 (.17)$$
$$\quad + (9 - 6.4)^2 (.12) + (10 - 6.4)^2 (.05)$$
$$= 4.70$$

The standard deviation is

$$\sigma = \sqrt{\sigma^2} = \sqrt{4.70} = 2.17$$

These parameters tell us that the mean and standard deviation of the marks of all the students in the course (past, present, and future) are 6.4 and 2.17, respectively.

LAWS OF EXPECTED VALUE AND VARIANCE

As you will discover, we often create new variables that are functions of other random variables. The formulas below allow us to quickly determine the expected value and variance of these new variables. In the notation below X and Y are random variables and c is a constant.

Laws of Expected Value

1. $E(c) = c$
2. $E(X + c) = E(X) + c$
3. $E(cX) = cE(X)$
4. $E(X + Y) = E(X) + E(Y)$
5. $E(X - Y) = E(X) - E(Y)$

Laws of Variance

1. $V(c) = 0$
2. $V(X + c) = V(X)$
3. $V(cX) = c^2 V(X)$

The following laws are valid if X and Y are independent. Independence means that the value that one variable assumes does not affect the value that the other variable assumes.

4. $V(X + Y) = V(X) + V(Y)$ if X and Y are independent
5. $V(X - Y) = V(X) + V(Y)$ if X and Y are independent

▼ EXAMPLE 5.3

The monthly sales at a computer store have a mean of $25,000 and a standard deviation of $4,000. Profits are calculated by multiplying sales by 30% and subtracting fixed costs of $6,000. Find the mean and standard deviation of monthly profits.

Solution We can describe the relationship between profits and sales by the following equation:

$$\text{Profit} = .30(\text{Sales}) - 6{,}000$$

The expected or mean profit is

$$E(\text{Profit}) = E[.30(\text{Sales}) - 6{,}000]$$

Applying the second Law of Expected Value, we produce

$$E(\text{Profit}) = E[.30(\text{Sales})] - E[6{,}000]$$

Applying Laws 1 and 3 yields

$$E(\text{Profit}) = .30 E(\text{Sales}) - 6{,}000 = .30(25{,}000) - 6{,}000 = 1{,}500$$

Thus, the mean monthly profit is $1,500.

The variance is

$$V(\text{Profit}) = V[.30(\text{Sales}) - 6{,}000]$$

The second Law of Variance states

$$V(\text{Profit}) = V[.30(\text{Sales})]$$

and Law 3 yields

$$V(\text{Profit}) = (.30)^2 V(\text{Sales}) = .09(4{,}000)^2 = 1{,}440{,}000$$

Thus, the standard deviation of monthly profits is

$$\sigma_{\text{Profit}} = \sqrt{1{,}440{,}000} = \$1{,}200$$

EXERCISES

5.12 Refer to Exercise 5.1
  a Calculate the mean number of pizzas delivered to each student.
  b Find the standard deviation of the number of pizzas delivered to each student.

5.13 Given the following probability distribution calculate the mean, variance, and standard deviation.

x	0	1	2	3
P(x)	.4	.3	.2	.1

5.14 Refer to Exercise 5.13. Suppose that $Y = 3X + 2$. For each value of X, determine the value of Y. What is the probability distribution of Y?

5.15 Refer to Exercise 5.14. Calculate the mean, variance, and standard deviation from the probability distribution of Y.

5.16 Refer to Exercises 5.13 and 5.14. Use the laws of expected value and variance to calculate the mean, variance, and standard deviation of Y from the mean, variance, and standard deviation of X. Compare these answers to those obtained in Exercise 5.15. Are they the same (except for rounding)?

5.17 Refer to Exercise 5.7.
  a Determine the expected value of the number of games played.
  b Determine the standard deviation of the number of games.

5.18 Refer to Exercise 5.7. Suppose that each game costs the player 25 cents. Determine the probability distribution of the amount of money the arcade takes in per child.

5.19 Refer to Exercise 5.18. Use the definitions of expected value and variance to calculate the mean and standard deviation of the amount of money the arcade makes per child.

5.20 Refer to Exercise 5.18. Use the laws of expected value and variance to calculate the mean and standard deviation of the amount of money the arcade makes per child.

5.21 Refer to Exercise 5.1. If the pizzeria makes a profit of $3 per pizza, determine the mean and standard deviation of the profits per student.

5.22 A shopping mall estimates the probability distribution of the number of stores mall customers actually enter. It is

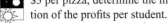

x	0	1	2	3	4	5	6
p(x)	.04	.19	.22	.28	.12	.09	.06

Find the mean and standard deviation of the number of stores entered.

5.23 Refer to Exercise 5.22. Suppose that, on average, customers spend 10 minutes in each store they enter. Find the mean and standard deviation of the amount of time customers spend in stores.

5.24 Refer to Exercise 5.8. Find the expected value and standard deviation of the number of books sold per hit.

5.25 When parking a car in a downtown parking lot, drivers pay according to the number of hours or parts thereof. The probability distribution of the number of hours cars are parked has been estimated as follows.

x	1	2	3	4	5	6	7	8
P(x)	.24	.18	.13	.10	.07	.04	.04	.20

Find the mean and standard deviation of the number of hours that cars are parked in the lot.

5.26 Refer to Exercise 5.25. The cost of parking is $2.50 per hour. Calculate the mean and standard deviation of the amount of revenue each car generates.

5.4 BINOMIAL DISTRIBUTION

In Chapter 2, we introduced the three types of data. We pointed out that when the data are nominal all we can do is calculate the frequency with which each value occurs. In this section we provide a link between a population and a sample of nominal data by introducing the binomial distribution. This distribution allows us to calculate the probability of the frequencies of nominal variables.

The binomial distribution is the result of a binomial experiment.

Binomial Experiment

1. The **binomial experiment** consists of a fixed number of trials. We represent the number of trials by *n*.
2. On each trial there are two possible outcomes. We label one outcome a *success,* and the other is called a *failure.*
3. The probability of success is *p.* The probability of failure is $1 - p$.
4. The trials are independent, which means that the outcome of one trial does not affect the outcomes of any other trials.

The random variable is defined as the number of successes in the *n* trials. It is called the **binomial random variable.**

Here are several examples of binomial experiments.

1 Flip a coin 10 times. The two outcomes per trial are heads and tails. The terms "success" and "failure" are arbitrary. We can label either outcome success. However, generally, we call success anything we're looking for. For example, if we were betting on heads, we would label heads a success. If the coin is fair, the probability of heads is 50%. Thus, $p = .5$. Finally, we can see that the trials are independent, because the outcome of one coin flip cannot possibly affect the outcomes of other flips.

2 Draw five cards out of a shuffled deck. We can label as success whatever card we seek. For example, if we wish to know the probability of receiving five clubs, clubs is labeled a success. On the first draw, the probability of a club is $13/52 = 25\%$. However, if we draw a second card without replacing the first card and shuffling, the trials are not independent. To see why, suppose that the first draw is a club. If we draw again without replacement the probability of drawing a second club is $12/51$, which is not 25%. In this experiment, the trials are *not* independent. Hence, this is not a binomial experiment. However, if we replace the card and shuffle before drawing again the experiment is binomial. Note that in most card games we do not replace the card, and as a result the experiment is not binomial.

3 A political survey asks 1,500 voters whom they intend to vote for in an approaching election. In most elections in the United States there are only two candidates, the Republican and Democratic nominees. Thus, we have two outcomes per trial. The trials are independent, because the choice of one voter does not affect the choice of other voters. In Canada, and in other countries with a parliamentary system of government, there are usually several candidates in the race. However, we can label a vote for our favored candidate (or the party that is paying us to do the survey) a success and regard all the others as failures.

As you will discover, the third example is a common application of statistical inference. The actual value of p is unknown, and the job of the statistics practitioner is to estimate its value. By understanding the probability distribution that uses p, we will be able to develop the statistical tools to estimate p.

BINOMIAL RANDOM VARIABLE

The binomial random variable is the number of successes in the experiment's n trials. It can take on values $0, 1, 2, \ldots n$. Thus, the random variable is discrete. In order to proceed, we need to be capable of calculating the probability associated with each value.

Using a probability tree we draw a series of branches as depicted in Figure 5.1. The stages represent the outcomes for each of n trials. At each stage there are two branches representing success and failure. To calculate the probability that there are x successes in n trials, we note that for each success in the sequence we must multiply by p. And, if there are x successes, there must be $n - x$ failures. For each failure in the sequence, we multiply by $1 - p$. Thus, the probability for each sequence of branches that represent x successes and $n - x$ failures has probability

$$p^x(1-p)^{n-x}$$

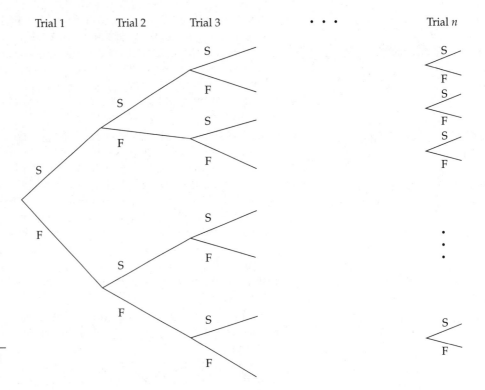

Figure 5.1

Probability tree for a binomial experiment

There are a number of branches that yield x successes and $n - x$ failures. For example, there are two ways to produce exactly one success and one failure in two trials—SF and FS. To count the number of branch sequences that produce x successes and $n - x$ failures, we use the combinatorial formula

$$C_x^n = \frac{n!}{x!(n-x)!}$$

where $n! = n(n-1)(n-2)\ldots(2)(1)$.

For example, $3! = 3(2)(1) = 6$. Incidentally, although it may not appear to be logical, $0! = 1$.

Pulling together the two components of the probability distribution yields the following.

> **Binomial Probability Distribution**
>
> The probability of x successes in a binomial experiment with n trials and probability of success $= p$ is
>
> $$P(X = x) = p(x) = \frac{n!}{x!(n-x)!} p^x (1-p)^{n-x} \quad \text{for } x = 0, 1, 2, \ldots, n$$

EXAMPLE 5.4

Pat Statsdud is a student taking a statistics course. Unfortunately, Pat is not a good student. Pat does not read the textbook before class, does not do homework, and regularly misses class. Pat intends to rely on luck to pass the next quiz. The quiz consists of ten multiple-choice questions. Each question has five possible answers, only one of which is correct. Pat plans to guess the answer to each question.

1 What is the probability that Pat gets no answers correct?
2 What is the probability that Pat gets two answers correct?

Solution The experiment consists of ten identical trials, each with two possible outcomes and where success is defined as a correct answer. Because Pat intends to guess, the probability of success is 1/5 or .2. Finally, the trials are independent because the outcome of any of the questions does not affect the outcomes of any other questions. These four properties tell us that the experiment is binomial with $n = 10$ and $p = .2$.

1 The probability of zero successes is

$$P(X = 0) = \frac{n!}{x!(n-x)!}p^x(1-p)^{n-x}$$

where $n = 10$, $p = .2$, and $x = 0$. Hence,

$$P(X = 0) = \frac{10!}{0!(10-0)!}(.2)^0(1-.2)^{10-0}$$

The combinatorial part of the formula is 10!/0!10!, which is 1. This is the number of ways to get zero correct and ten incorrect. Obviously, there is only one way to produce $X = 0$. And because $(.2)^0 = 1$,

$$P(X = 0) = 1(1)(.8)^{10} = .1074$$

2 The probability of two correct answers is computed similarly by substituting $n = 10$, $p = .2$, and $x = 2$.

$$P(X = 2) = \frac{n!}{x!(n-x)!}p^x(1-p)^{n-x}$$

$$= \frac{10!}{2!(10-2)!}(.2)^2(1-.2)^{10-2}$$

$$= \frac{(10)(9)(8)(7)(6)(5)(4)(3)(2)(1)}{(2)(1)(8)(7)(6)(5)(4)(3)(2)(1)}(.04)(.1678)$$

$$= 45(.006712)$$

$$= .3020$$

In this calculation, we discovered that there are 45 ways to get exactly two correct answers and eight incorrect, and that each such outcome has probability .006712. Multiplying the two numbers produces a probability of .3020.

CUMULATIVE PROBABILITY

The formula of the binomial distribution allows us to determine the probability that X equals individual values. In Example 5.4, the values were 0 and 2. There are many circumstances where we wish to find the probability that a random variable is less than or equal to a value. That is, we want to determine $P(X \leq x)$. Such a probability is called a **cumulative probability.**

▼ EXAMPLE 5.5

Find the probability that Pat fails the quiz. A mark is considered a failure if it is less than 50%.

Solution In this quiz, a mark of less than 5 is a failure. Because the marks must be integers, a mark of 4 or less is a failure. We wish to determine $P(X \leq 4)$.

$$P(X \leq 4) = p(0) + p(1) + p(2) + p(3) + p(4)$$

From Example 5.4, we know $p(0) = .1074$ and $p(2) = .3020$. Using the binomial formula, we find $p(1) = .2684$, $p(3) = .2013$, and $p(4) = .0881$.

$$P(X \leq 4) = .1074 + .2684 + .3020 + .2013 + .0881 = .9672$$

There is a 96.72% probability that Pat will fail the quiz by guessing the answer for each question.

▲

BINOMIAL TABLE

Table 1 in Appendix B provides cumulative binomial probabilities for several values of n and p. We can use this table to answer the question in Example 5.5 where we need $P(X \leq 4)$. Refer to Table 1, find $n = 10$, and in that table find $p = .20$. The values in that column are $P(X \leq k)$ for $k = 0, 1, 2, \ldots, 10$, which are shown in Table 5.1.

The first cumulative probability is $P(X \leq 0)$, which is $p(0) = .1074$. The probability we need for Example 5.5 is $P(X \leq 4) = .9672$, which is the same value we obtained manually.

We can use the table and the complement rule to determine probabilities of the type $P(X \geq x)$. For example, to find the probability that Pat will pass the quiz, we note that

$$P(X \leq 4) + P(X \geq 5) = 1.$$

Thus,

$$P(X \geq 5) = 1 - P(X \leq 4) = 1 - .9672 = .0328$$

Using Table 1 (Appendix B) to Find the Binomial Probability $P(X \geq x)$

$$P(X \geq x) = 1 - P(X \leq [x - 1]).$$

Table 5.1 **Cumulative Binomial Probabilities with $n = 10$ and $p = .2$**

k	$P(X \leq k)$
0	.1074
1	.3758
2	.6778
3	.8791
4	.9672
5	.9936
6	.9991
7	.9999
8	1.000
9	1.000
10	1.000

The table is also useful in determining the probability of one individual value of X. For example, to find the probability that Pat will get exactly two right answers we note that

$$P(X \leq 2) = p(0) + p(1) + p(2)$$

and

$$P(X \leq 1) = p(0) + p(1)$$

The difference between these two cumulative probabilities is $p(2)$. Thus

$$p(2) = P(X \leq 2) - P(X \leq 1) = .6778 - .3758 = .3020$$

Using Table 1 (Appendix B) to Find the Binomial Probability $P(X = x)$

$$p(x) = P(X \leq x) - P(X \leq [x - 1])$$

We can employ Excel to compute cumulative probabilities and the probability of individual values of a binomial random variable.

Using Microsoft Excel

COMMANDS

1 Click f_x, **Function category: Statistical** and **Function name: BINOMDIST**. Click **OK**.
2 Type the value of x (**Number_s**), the number of trials, n (**Trials**), the probability of success, p (**Probability_s**), and **true** (**Cumulative**) to yield a cumulative probability or **false** for an individual probability. The probability appears on the right side of the dialog box. Clicking **OK** will print the probability in the active cell.

MEAN AND VARIANCE OF A BINOMIAL DISTRIBUTION

Statisticians have developed general formulas for the mean, variance, and standard deviation of a binomial random variable. They are

$$\mu = np$$
$$\sigma^2 = np(1-p)$$
$$\sigma = \sqrt{np(1-p)}$$

▼ EXAMPLE 5.6

Suppose that a professor has a class full of students like Pat (a nightmare!). What is the mean mark? What is the standard deviation?

Solution The mean mark for a class of Pat Statsduds is

$$\mu = np = 10(.2) = 2.$$

The standard deviation is

$$\sigma = \sqrt{np(1-p)} = \sqrt{10(.2)(1-.2)} = 1.26$$

EXERCISES

5.27 Given a binomial random variable with $n = 10$ and $p = .3$, use the formula to find the following probabilities.
   a $P(X = 3)$
   b $P(X = 5)$
   c $P(X = 8)$

5.28 Repeat Exercise 5.27 using Table 1 in Appendix A.

5.29 Repeat Exercise 5.27 using Excel.

5.30 X is a binomial random variable with $n = 5$ and $p = .4$. Use the formula to find the following probabilities.
   a $P(X = 0)$
   b $P(X = 2)$
   c $P(X \leq 3)$
   d $P(X \geq 2)$

5.31 Repeat Exercise 5.30 using Table 1 in Appendix A.

5.32 Repeat Exercise 5.30 using Excel.

5.33 X is a binomial random variable with $n = 25$ and $p = .7$. Use Table 1 to find the following.
   a $P(X = 18)$
   b $P(X = 15)$
   c $P(X \leq 20)$
   d $P(X \geq 16)$

5.34 Repeat Exercise 5.33 using Excel.

5.35 X is a binomial random variable with $n = 100$ and $p = .22$. Use Excel to find the following.
   a $P(X = 24)$
   b $P(X \leq 25)$
   c $P(X \geq 20)$

5.36 In a recent election the mayor received 60% of the vote. Last week a survey was undertaken that asked 100 people whether they would vote for the mayor. Assuming that her popularity has not changed, what is the probability that more than 50 people in the sample would vote for the mayor?

5.37 The probability of winning a game of craps (a dice-throwing game played in casinos) is 244/495.

    a What is the probability of winning more than 5 times in 10 games?
    b What is the expected number of wins in 100 games?

5.38 In the game of blackjack as played in casinos in Las Vegas, Atlantic City, Niagara Falls, as well as many other cities, the dealer has the advantage. Most players do not play very well. As a result the probability that the average player wins a hand is about 45%. One such player is randomly selected. Find the probability that the player wins

    a twice in 5 hands.
    b more than 10 times in 25 hands.

5.39 There are several books that teach blackjack players the "basic strategy," which increases the probability of winning any hand to 50%. Repeat Exercise 5.38, assuming the player plays the basic strategy.

5.40 The best way of winning at blackjack is to "case the deck," which involves counting tens, non-tens, and aces. For card counters the probability may increase to 52%. Repeat Exercise 5.38 for a card counter.

5.41 The leading brand of dishwasher detergent has a 30% market share. A sample of 25 dishwasher customers was taken. What is the probability that fewer than ten customers chose the leading brand?

5.42 A certain type of tomato seed germinates 90% of the time. A backyard farmer planted 25 seeds.

    a What is the probability that exactly 20 germinate?
    b What is the probability that at least 20 germinate?
    c What is the probability that no more than 24 germinate?
    d What is the expected number of seeds that germinate?

5.43 A production line regularly produces a defective rate of 3%. A production run of 100 units was run. Find the probability that no defectives were produced.

5.44 The defective rate of new computers is 5%. Find the probability that in a batch of 10 computers, none are defective.

5.45 Repeat Exercise 5.44 for a batch of 25 computers.

5.46 Repeat Exercise 5.44 for a batch of 100 computers.

5.47 In the game of roulette, a steel ball is rolled onto a wheel that contains 18 red, 18 black, and 2 green slots. If the ball is rolled 25 times, find the probabilities of the following events.

    a The ball falls into the green slots more than twice.
    b The ball does not fall into any green slots.
    c The ball falls into black slots more than 15 times.
    d The ball falls into red slots fewer than 10 times.

5.48 In the United States voters who are neither Democrat nor Republican are called Independent. It is believed that 10% of all voters are Independent. A survey asked 25 people to identify themselves as Democrat, Republican, or Independent.

    a What is the probability that none of the people are Independent?
    b What is the probability that fewer than five people are Independent?
    c What is the probability that more than two people are Independent?

5.49 A coin that is believed to be balanced is flipped 20 times. What is the probability of observing more than 15 heads?

5.50 Repeat Exercise 5.49 assuming that the probability of heads is 60%.

5.51 Repeat Exercise 5.49 assuming that the probability of heads is 75%.

5.52 Most Internet service providers attempt to provide a large enough service so that customers seldom encounter a busy signal. Suppose that the customers of one ISP encounter busy signals 8% of the time. During the week a customer of this ISP called 25 times. What is the probability that she did not encounter any busy signals?

5.53 Major software manufacturers offer a help line that allows customers to call and receive assistance in solving their problems. However, because of the volume of calls, customers frequently are put on hold. One software manufacturer claims that only 20% of callers are put on hold. Suppose that 100 customers call. What is the probability that more than 40 of them are put on hold?

5.54 A study of drivers reveals the fact that, when lost, 45% will stop and ask for directions, 30% will consult a map, and 25% will continue driving until the location has been determined. Suppose that sample of 200 drivers was asked to report what they do when lost. Find the following probabilities.

a At least 100 stop and ask directions.
b At most 55 continue driving.
c Between 50 and 75 (inclusive) consult a map.

5.5 POISSON DISTRIBUTION

Another useful discrete probability distribution is the Poisson distribution, named after its French creator. Like the binomial random variable, the Poisson random variable is the number of occurrences of events, which we'll continue to call successes. The difference between the two random variables is that a binomial random variable is the number of successes in a set number of trials and a Poisson random variable is the number of successes in an interval of time or specific region of space. Here are several examples of Poisson random variables

1. The number of car arrivals at a service station in one hour. (The interval of time is one hour.)
2. The number of flaws in a bolt of cloth. (The specific region is a bolt of cloth.)
3. The number of accidents in one day on a particular stretch of highway. (The interval is defined by both time, one day, and space, the particular stretch of highway.)

The Poisson experiment is described below.

Poisson Experiment

A **Poisson experiment** is characterized by the following properties:

1. The number of successes that occur in any interval is independent of the number of successes that occur in any other interval.
2. The probability of a success in an interval is the same from all equal-sized intervals.
3. The probability of a success is proportional to the size of the interval.
4. The probability of more than one success in an interval approaches zero as the interval becomes smaller.

As a general rule, a Poisson random variable is the number of occurrences of a *relatively rare* event that occurs *randomly* and *independently*. The number of hits on an active Web site is not a Poisson random variable, because the hits are not rare. The number of people arriving at a restaurant is not Poisson, because restaurant patrons usually arrive in groups, which violates the independence condition.

Poisson Random Variable

The **Poisson random variable** is the number of successes that occur in a period of time or an interval of space in a Poisson experiment.

There are several ways to derive the probability distribution of a Poisson random variable. However, all are beyond the mathematical level of this book. We simply provide the formula and illustrate how it is used.

Poisson Probability Distribution

If X is a Poisson random variable, the probability that it assumes a value of x is

$$P(X = x) = p(x) = \frac{e^{-\mu}\mu^x}{x!} \quad \text{for } x = 0, 1, 2, \ldots$$

where μ is the mean number of successes in the interval or region and e is the base of the natural logarithm (approximately 2.71828).

EXAMPLE 5.7

A statistics instructor has observed that the number of typographical errors in new editions of textbooks varies considerably from book to book. After some analysis he concludes that the number of errors is Poisson distributed with a mean of 1.5 per 100 pages. The instructor randomly selects 100 pages of a new book. What is the probability that there are no typos?

Solution We want to determine the probability that a Poisson random variable with a mean of 1.5 is equal to 0. Thus, we substitute $x = 0$ and $\mu = 1.5$ into the formula for the Poisson distribution.

$$P(X = 0) = \frac{e^{-\mu}\mu^x}{x!} = \frac{e^{-1.5}1.5^0}{0!} = \frac{(2.71828)^{-1.5}(1)}{1} = .2231$$

The probability that in the 100 pages selected there are no errors is .2231.

Notice that in Example 5.7 we wanted to find the probability of 0 typos in *100 pages* given a mean of 1.5 typos in *100 pages*. The next example illustrates how we calculate the probability of events where the intervals or regions do not match.

EXAMPLE 5.8

Refer to Example 5.7. Suppose that the instructor has just received a copy of a new statistics book. He notices that there are 400 pages.

1 What is the probability that there are no typos?
2 What is the probability that there are five or fewer typos?

Solution The specific region that we're interested in is 400 pages. To calculate Poisson probabilities associated with this region, we need to determine the mean number of typos per 400 pages. Because the mean is specified as 1.5 per 100 pages, we multiply this figure by 4 to convert to 400 pages. Thus $\mu = 6$ typos per 400 pages.

1 The probability of no typos is

$$P(X=0) = \frac{e^{-\mu}\mu^x}{x!} = \frac{e^{-6}6^0}{0!} = \frac{(2.71828)^{-6}(1)}{1} = .002479$$

2 We want to determine the probability that a Poisson random variable with a mean of 6 is 5 or less. That is, we want to calculate

$$P(X \leq 5) = p(0) + p(1) + p(2) + p(3) + p(4) + p(5)$$

To produce this probability we need to compute the six probabilities in the summation.

$$p(0) = .002479$$

$$p(1) = \frac{e^{-\mu}\mu^x}{x!} = \frac{e^{-6}6^1}{1!} = \frac{(2.71828)^{-6}(6)}{1} = .01487$$

$$p(2) = \frac{e^{-\mu}\mu^x}{x!} = \frac{e^{-6}6^2}{2!} = \frac{(2.71828)^{-6}(36)}{2} = .04462$$

$$p(3) = \frac{e^{-\mu}\mu^x}{x!} = \frac{e^{-6}6^3}{3!} = \frac{(2.71828)^{-6}(216)}{6} = .08924$$

$$p(4) = \frac{e^{-\mu}\mu^x}{x!} = \frac{e^{-6}6^4}{4!} = \frac{(2.71828)^{-6}(1296)}{24} = .1339$$

$$p(5) = \frac{e^{-\mu}\mu^x}{x!} = \frac{e^{-6}6^5}{5!} = \frac{(2.71828)^{-6}(7776)}{120} = .1606$$

Thus,

$$P(X \leq 5) = .002479 + .01487 + .04462 + .08924 + .1339 + .1606 = .4457$$

The probability of observing 5 or fewer typos in this book is .4457.

POISSON TABLE

As was the case with the binomial distribution, a table is available that makes it easier to compute Poisson probabilities of individual values of X as well as cumulative and related probabilities.

Table 2 in Appendix B provides cumulative Poisson probabilities for several values of μ. This table makes it easy to find cumulative probabilities like Example 5.8, part 2, where we found $P(X \leq 5)$.

To do so, find $\mu = 6$ in Table 2. The values in that column are $P(X \leq k)$ for $k = 0, 1, 2, \ldots$, which are shown in Table 5.2.

Table 5.2 Cumulative Poisson Probabilities for $\mu = 6$

k	$P(X \leq k)$
0	.002
1	.017
2	.062
3	.151
4	.285
5	.446
6	.606
7	.744
8	.847
9	.916
10	.957
11	.980
12	.991
13	.996
14	.999
15	1.000

Theoretically, a Poisson random variable has no upper limit. The table provides cumulative probabilities until the sum is 1.000 (using three decimal places).

The first cumulative probability is $P(X \leq 0)$, which is $p(0) = .002$. The probability we need for Example 5.8, part 2, is $P(X \leq 5) = .446$, which is the same value (using three decimal places) we obtained manually.

Like Table 1 for binomial probabilities, Table 2 can be used to determine probabilities of the type $P(X \geq x)$. For example, to find the probability that in Example 5.8 there are more than five typos, we note that $P(X \leq 5) + P(X \geq 6) = 1$. Thus,

$$P(X \geq 6) = 1 - P(X \leq 5) = 1 - .446 = .554$$

> **Using Table 2 (Appendix B) to Find the Poisson Probability $P(X \geq x)$**
>
> $$P(X \geq x) = 1 - P(X \leq [x - 1])$$

We can also use the table to determine the probability of one individual value of X. For example, to find the probability that the book contains exactly 10 typos we note that

$$P(X \le 10) = p(0) + p(1) + \cdots + p(9) + p(10)$$

and

$$P(X \le 9) = p(0) + p(1) + \cdots + p(9)$$

The difference between these two cumulative probabilities is $p(10)$. Thus

$$p(10) = P(X \le 10) - P(X \le 9) = .957 - .916 = .041$$

> **Using Table 2 (Appendix B) to Find the Poisson Probability $P(X = x)$**
>
> $$p(x) = P(X \le x) - P(X \le [x-1])$$

We can employ Excel to compute cumulative probabilities and the probability of individual values of a Poisson random variable.

Using Microsoft Excel

COMMANDS

1 Click f_x, **Function category: Statistical,** and **Function name: POISSON.** Click **OK.**
2 Type the value of x **(X)**, the mean **(Mean)**, and **true (Cumulative)** to yield a cumulative probability or **false** for an individual probability. The probability appears on the right side of the dialog box. Clicking **OK** will print the probability in the active cell.

EXERCISES

5.55 The number of accidents that occur at a busy intersection is Poisson distributed with a mean of 3.5 per week. Find the probability of the following events.

a No accidents in one week
b More than five accidents in a week
c One accident today

5.56 Snowfalls occur randomly and independently over the course of winter in a Minnesota city. The average is one every three days.

a What is the probability of five snowfalls in two weeks?
b Find the probability of a snowfall today.

5.57 The number of students who seek assistance with their statistics assignments is Poisson distributed with a mean of two per day.

a What is the probability that no students seek assistance tomorrow?
b Find the probability that ten students seek assistance in a week.

5.58 Hits on personal Web site occur quite infrequently. They occur randomly and independently with an average of five per week.

a Find the probability that the site gets more than ten hits in a week.
b Determine the probability that the site gets more than 20 hits in two weeks.

5.59 The author's computer has a variety of software including Office 2000 as well as several other packages. Additionally, there are numerous Word and Excel files. As a result, the computer crashes approximately once every day. If the number of crashes is Poisson distributed, find the probability that the computer will not crash at all over the next two days.

5.60 The number of bank robberies that occur in a large North American city are Poisson distributed with a mean of 1.8 per day. Find the probabilities of the following events.
   a More than 3 bank robberies in a day
   b Between 10 and 15 (inclusive) robberies during a five-day period

5.61 Flaws in a carpet tend to occur randomly and independently at a rate of one every 200 square feet. What is the probability that a carpet that is 8 feet by 10 feet contains no flaws?

5.62 Complaints about an Internet brokerage firm occur at a rate of five per day. The number of complaints appears to be Poisson distributed.
   a Find the probability that the firm receives more than 10 complaints in a day.
   b Find the probability that the firm receives more than 25 complaints in a five-day period.

5.6 SUMMARY

There are two types of random variable. A discrete random variable is one whose values are countable. A continuous random variable can assume an uncountable number of values.

In this chapter we discussed discrete random variables and their probability distributions. Probability distributions represent populations. We defined the expected value, variance, and standard deviation of a population represented by a discrete probability distribution.

The two most important discrete distributions, the binomial and the Poisson, were introduced.

IMPORTANT TERMS

Random variable
Discrete random variable
Continuous random variable
Probability distribution
Expected value
Binomial experiment
Binomial random variable
Binomial probability distribution
Cumulative probability
Poisson experiment
Poisson random variable
Poisson probability distribution

SYMBOLS

Symbol	Pronounced	Represents
$\sum_{\text{all } x} x$	Sum of x for all values of x	Summation
C_x^n	n-choose x	Number of combinations
$n!$	n-factorial	$n(n-1)(n-2)\ldots(3)(2)(1)$
e		$2.71828\ldots$

FORMULAS

Expected value (mean)

$$E(X) = \mu = \sum_{\text{all } x} xp(x)$$

Variance

$$V(x) = \sigma^2 = \sum_{\text{all } x} (x - \mu)^2 p(x)$$

Standard deviation

$$\sigma = \sqrt{\sigma^2}$$

Laws of expected value

1. $E(c) = c$
2. $E(X + c) = E(X) + c$
3. $E(cX) = cE(X)$
4. $E(X + Y) = E(X) + E(Y)$
5. $E(X - Y) = E(X) - E(Y)$

Laws of variance

1. $V(c) = 0$
2. $V(X + c) = V(X)$
3. $V(cX) = c^2 V(X)$
4. $V(X + Y) = V(X) + V(Y)$ if X and Y are independent
5. $V(X - Y) = V(X) + V(Y)$ if X and Y are independent

Binomial probability

$$P(X = x) = \frac{n!}{x!(n-x)!} p^x (1-p)^{n-x}$$

Poisson probability

$$P(X = x) = \frac{e^{-\mu} \mu^x}{x!}$$

MICROSOFT EXCEL INSTRUCTIONS

Probability distribution	Page
Binomial	133
Poisson	140

SUPPLEMENTARY EXERCISES

5.63 The final exam in a one-term statistics course is taken in the December exam period. Students who are sick or have other legitimate reasons for missing the exam are allowed to write a deferred exam scheduled for the first week in January. A statistics professor has observed that only 2% of all students legitimately miss the December final exam. Suppose that the professor has 40 students registered this term.

  a How many students can the professor expect to miss the December exam?
  b What is the probability that the professor will not have to create a deferred exam?

5.64 The number of magazine subscriptions per household is represented by the following probability distribution.

Magazine subscriptions per household	0	1	2	3	4
	.48	.35	.08	.05	.04

  a Calculate the mean number of magazine subscriptions per household.
  b Find the standard deviation.

5.65 The number of arrivals at a gas station is Poisson distributed with a mean of eight per hour.

  a What is the probability that ten cars will arrive in the next hour?
  b What is the probability that more than five cars will arrive in the next hour?
  c What is the probability that fewer than 12 cars will arrive in the next hour?

5.66 The percentage of customers who enter a restaurant and ask for a smoking section is 15%. Suppose that 100 people enter the restaurant.

  a What is the expected number of people who request a smoking table?
  b What is the standard deviation of the number of request a smoking table?
  c What is the probability that 20 or more people request a smoking table?

5.67 Lotto 6/49 is a government-run lottery wherein players pick six numbers between 1 and 49. The lottery then selects six numbers and tickets that match three, four, five, or all six are considered winners. One player believes that the way the lottery selects numbers tends to favor two-digit numbers.

  a If all numbers are equally likely, what is the probability of selecting no one-digit numbers when the lottery picks its six numbers?
  b What is the probability of picking exactly one one-digit number?

5.68 The number of 60-minute cassettes that can be played on a Walkman before the battery expires is a variable. The distribution is shown below.

Number of 60-minutes cassettes on a Walkman before battery expires	5	6	7	8	9	10
	.05	.16	.41	.27	.07	.04

  a Calculate the mean number of cassettes.
  b Find the standard deviation.

5.69 Shutouts in the National Hockey League occur randomly and independently at a rate of 1 every 20 games. Calculate the probability of the following events.

  a 2 shutouts in the next 10 games
  b 25 shutouts in 400 games
  c A shutout in tonight's game

5.70 Most Miami Beach restaurants offer "early-bird" specials. These are lower-priced meals that are available only from 4:00 to 6:00. However, not all customers who arrive between 4:00 and 6:00 order the special. In fact, only 70% do.

  a Find the probability that of 80 customers between 4:00 and 6:00, more than 65 order the special.
  b What is the expected number of customers who order the special?
  c What is the standard deviation?

5.71 Psychologists generally believe that women who work and have children lead stressful lives. To investigate, a research project was undertaken. A random sample of women was selected. The women were asked whether they were working or stay-at-home mothers. Respondents were asked how many times in the past week they experienced depression. After examining the results the researchers produced the probability distribution below. The random variable X is the number of bouts of depression suffered by working mothers per week.

x	0	1	2	3	4
p(x)	.35	.25	.18	.13	.09

Find the mean and standard deviation of this distribution.

5.72 Researchers at the University of Pennsylvania School of Medicine theorized that children under 2 years old who sleep in rooms with the light on have a 40% probability of becoming myopic by age 16. Suppose that researchers found 25 children who slept with the light on before they were 2.

a What is the probability that ten of them will become myopic before the age of 16?
b What is the probability that fewer than five of them will become myopic before the age of 16?
c What is the probability that more than 15 of them will become myopic before the age of 16?

5.73 Refer to Exercise 5.72. Suppose that in a sample of 25 children who slept with the light on before the age of 2, 4 become myopic. What does the result tell you about the researchers' theory?

5.74 A pharmaceutical researcher working on a cure for baldness noticed that middle-aged men who are balding at the crown of their heads have a 45% probability of suffering a heart attack over the next decade. In a sample of 100 middle-aged balding men, what are the following probabilities?

a More than 50 will suffer a heart attack in the next decade.
b Fewer than 44 will suffer a heart attack in the next decade.
c Exactly 45 will suffer a heart attack in the next decade.

5.75 Advertising researchers have developed a theory that states that commercials that appear in violent television shows are less likely to be remembered and thus will be less effective. After examining samples of viewers who watch violent and nonviolent programs and asking them a series of five questions about the commercials, the researchers produced the following probability distributions of the number of correct answers.

Viewers of violent shows

x	0	1	2	3	4	5
p(x)	.36	.22	.20	.09	.08	.05

Viewers of nonviolent shows

x	0	1	2	3	4	5
p(x)	.15	.18	.23	.26	.10	.08

a Calculate the mean and standard deviation of the number of correct answers among viewers of violent television programs.
b Calculate the mean and standard deviation of the number of correct answers among viewers of nonviolent television programs.

5.76 It is recommended that women over 40 have a mammogram annually. A recent report indicated that if a woman has annual mammograms over a 10-year period, there is a 60% probability that there will be at least one false positive result. (A false-positive mammogram test result is one that indicates the presence of cancer when in fact there is no cancer.) If the annual test results are independent, what is the probability that in any one year a mammogram will produce a false-positive result? (*Hint:* Find the value of p such that the probability that a binomial random variable with $n = 10$ is greater than or equal to 1 is .60.)

Case 5.1 To Bunt or Not to Bunt, That Is the Question (Part II)

In Case 4.2 we presented the probabilities of scoring at least one run and asked you to determine whether the manager should signal for the batter to sacrifice bunt. The decision was made on the basis of comparing the probability of scoring at least one run when the manager signaled for the bunt and when he signaled the batter to swing away. Another factor that should be incorporated into the decision is the *number* of runs the manager expects his team to score. In the same article referred to in Case 4.2, the author also computed the expected number of runs scored for each situation. Table 1 below lists the expected number of runs in situations that are defined by the number of outs and the bases occupied.

Table 1 Expected Number of Runs Scored

Bases Occupied	0 Out	1 Out	2 Outs
Bases empty	.49	.27	.10
First base	.85	.52	.23
Second base	1.06	.69	.34
Third base	1.21	.82	.38
First base and second base	1.46	1.00	.48
First base and third base	1.65	1.10	.51
Second base and third base	1.94	1.50	.62
Bases loaded	2.31	1.62	.82

Assume that the manager wishes to score as many runs as possible. Using the same probabilities of the four outcomes of a bunt, determine whether the manager should signal the batter to sacrifice bunt.

Chapter 6

Continuous Probability Distributions

6.1 Introduction

6.2 Probability Density Functions

6.3 Normal Distribution

6.4 Other Continuous Distributions

6.5 Summary

6.1 INTRODUCTION

This chapter completes our presentation of probability by introducing continuous random variables and their distributions. Recall from Section 5.4 that the binomial distribution is employed to calculate the probability associated with a nominal variable. This distribution allows us to determine the probability of the number of times one particular value of a nominal variable (which we call a success) occurs. In this way, we connect the population represented by the probability distribution with a sample of nominal data. In this chapter, we introduce continuous probability distributions, which are used to calculate the probability associated with an interval variable. By doing so, we develop the link between a population and a sample of interval data.

Section 6.2 introduces probability density functions and demonstrates, with the uniform density function, how probability is calculated. In Section 6.3, we focus on the normal distribution, one of the most important distributions because of its role in the development of statistical inference. In Section 6.4, we introduce three additional continuous distributions, which will be used in statistical inference throughout the book.

6.2 PROBABILITY DENSITY FUNCTIONS

A continuous random variable is one that can assume an uncountable number of values. Because this type of random variable is so different from a discrete variable, we need to treat it completely differently. First, we cannot list the possible values, because there are an infinite number of them. Second, because there are an infinite number of values, the probability of each individual value is virtually zero. Consequently, we can only determine the probability of a range of values. To illustrate how this is done, consider the histogram we created for the amounts of time Barnes Exhibit viewers remain (Example 2.1), which is depicted in Figure 6.1.

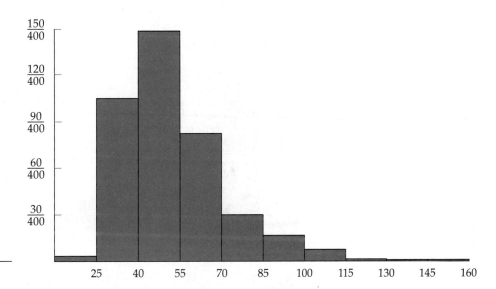

Figure 6.1

Histogram for Example 2.1

We found, for example, that the relative frequency of the interval 40 to 55 was 150/400. Using the relative frequency approach, we estimate that the probability that a Barnes Exhibit viewer visits for a period between 40 and 55 minutes is 150/400 = 37.5%. We can easily estimate the probabilities of the other intervals in the histogram.

Interval	Relative frequency
$0 < X \leq 25$	3/400
$25 < X \leq 40$	106/400
$40 < X \leq 55$	150/400
$55 < X \leq 70$	83/400
$70 < X \leq 85$	30/400
$85 < X \leq 100$	16/400
$100 < X \leq 115$	8/400
$115 < X \leq 130$	2/400
$130 < X \leq 145$	1/400
$145 < X \leq 160$	1/400

Notice that the sum of the probabilities equals one. To proceed, we set the values along the vertical axis so that the *area* in all the rectangles together adds to 1. We accomplish this by dividing the relative frequencies by the width of the interval, which is 15. In effect what we are doing is creating rectangles whose *areas* equal the probability that the random variable will fall into that interval. This is done to allow us to determine the probability of other intervals.

To determine probabilities of ranges other than the ones created when we drew the histogram, we apply the same approach. For example, the probability that a Barnes Exhibit visitor will stay for a period between 60 and 90 minutes is equal to the area between 60 and 90, as shown in Figure 6.2.

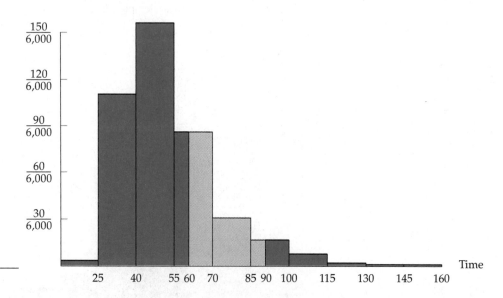

Figure 6.2

Probability that *X* falls between 60 and 90

Based on Figure 6.2, the following probability was determined.

Interval	Height of rectangle	Base Multiplied by Height
$60 < X \leq 70$	$83/(400 \times 15) = .01383$	$10 \times .01383 = .1383$
$70 < X \leq 85$	$30/(400 \times 15) = .00500$	$15 \times .00500 = .0750$
$85 < X \leq 90$	$16/(400 \times 15) = .00267$	$5 \times .00267 = .0133$
		Total = .2266

We estimate that the probability that a visitor stays in the museum between 60 and 90 minutes is .2266.

If the histogram is drawn with a large number of small intervals, we can smooth the edges of the rectangles, producing a curve as shown in Figure 6.3. In many cases, it is possible to determine a function $f(x)$ that approximates the curve. The function is called a **probability density function.**

Integral calculus can often be used to calculate the area under a curve. Fortunately, the continuous probability distributions that we deal with do not require this mathematical tool to compute probability. The distributions will be either simple or too complex for calculus. Let's start with the simplest continuous distribution.

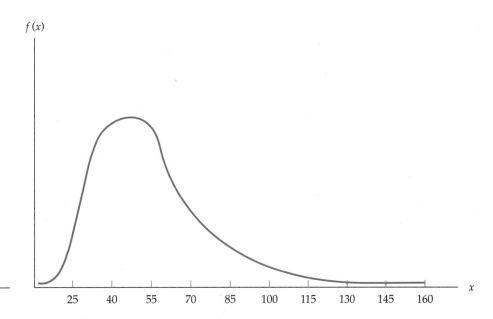

Figure 6.3

Density function

UNIFORM DISTRIBUTION

To illustrate how we find the area under the curve that describes a probability density function, consider the **uniform probability distribution,** also called the **rectangular probability distribution.**

Figure 6.4

Uniform distribution

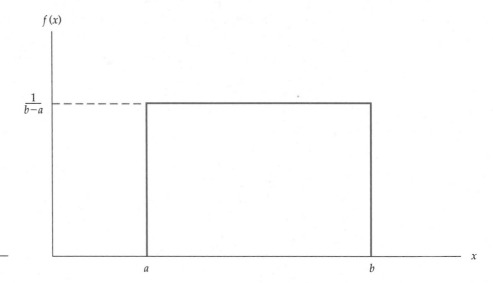

Uniform Probability Density Function

The uniform distribution is described by the function

$$f(x) = \frac{1}{b-a} \quad \text{where } a \leq x \leq b$$

When graphed, the function produces Figure 6.4. You can see why the distribution is called rectangular.

To calculate the probability of any interval, simply find the area under the curve. For example, to find the probability that X falls between x_1 and x_2, determine the area in the rectangle whose base is $x_2 - x_1$ and whose height is $1/(b - a)$. Figure 6.5 depicts the area we wish to find. As you can see, it is a rectangle, and the area in a rectangle is found by multiplying the base times the height.

Thus,

$$P(x_1 < X < x_2) = \text{Base} \times \text{Height} = (x_2 - x_1) \times \frac{1}{b-a}$$

▼ EXAMPLE 6.1

The daily sale of gasoline at a service station is uniformly distributed with a minimum of 2,000 gallons and a maximum of 5,000 gallons.

1. Find the probability that daily sales will fall between 2,500 and 3,000 gallons.
2. What is the probability that the service station will sell more than 4,000 gallons?
3. What is the probability that the station will sell exactly 2,500 gallons?

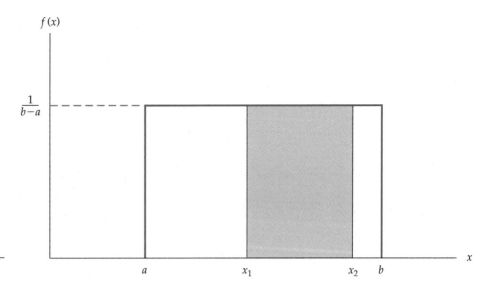

Figure 6.5

$P(x_1 < X < x_2)$

Solution The probability density function is

$$f(x) = \frac{1}{5{,}000 - 2{,}000} = \frac{1}{3{,}000} \quad 2{,}000 \leq x \leq 5{,}000$$

1 The probability that X falls between 2,500 and 3,000 is the area under the curve between 2,500 and 3,000, as depicted in Figure 6.6a. The area in a rectangle is the base times the height. Thus,

$$P(2{,}500 \leq X \leq 3{,}000) = (3{,}000 - 2{,}500) \times \left(\frac{1}{3{,}000}\right) = .1667$$

2 $P(X \geq 4{,}000) = (5{,}000 - 4{,}000) \times \left(\dfrac{1}{3{,}000}\right) = .3333 \quad$ (See Figure 6.6b.)

3 $P(X = 2{,}500) = 0$

Because there are an uncountably infinite number of values of X, the probability of each individual value is zero. Moreover, as you can see from Figure 6.6c, the area in a line is zero.

USING A CONTINUOUS DISTRIBUTION TO APPROXIMATE A DISCRETE DISTRIBUTION

In our definition of discrete and continuous random variables, we distinguish between them by noting whether the variables are countable or uncountable. However, in practice, we frequently use a continuous distribution to approximate a discrete one when the number of values the variable can assume is countable but large. For example, weekly income is countable. The values of weekly income expressed in dollars are

152 Chapter 6 CONTINUOUS PROBABILITY DISTRIBUTIONS

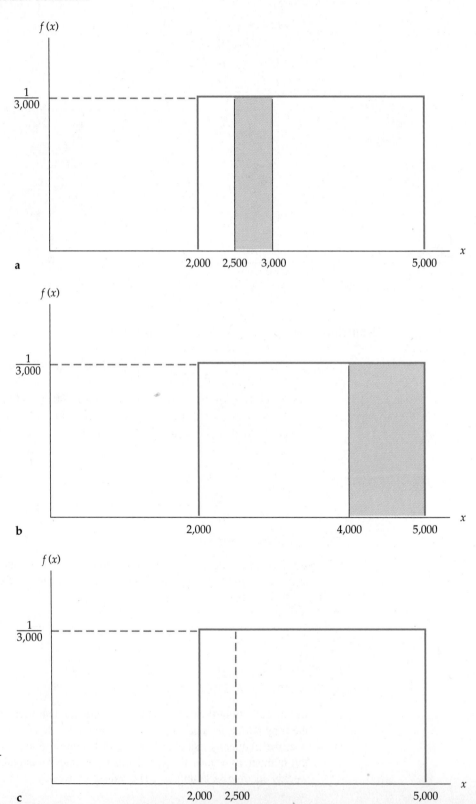

Figure 6.6

(a) $P(2{,}500 < X < 3{,}500)$;
(b) $P(4{,}000 < X < 5{,}000)$;
(c) $P(X = 2{,}500)$

0, 0.01, 0.02, ... Although there is no set upper limit we can easily identify (and thus, count) all the possibilities. Consequently, weekly income is a discrete random variable. However, because it can assume such a large number of values we prefer to employ a continuous probability distribution to determine the probability associated with such variables. In the next section we introduce the normal distribution, which is often used to describe discrete random variables that can assume a large number of values.

EXERCISES

6.1 A random variable has the following density function.

$$f(x) = 1 - .5x \quad 0 < x < 2$$

 a Graph the density function.
 b Find $P(X > 1)$.
 c Find $P(X < .5)$.
 d Find $P(X = 1.5)$.

6.2 The following function is the density function for the random variable X.

$$f(x) = \frac{x - 1}{8} \quad 1 < x < 5$$

 a Graph the density function.
 b Find the probability that X lies between 2.0 and 4.0.
 c What is the probability that X is less than 3.0?

6.3 A random variable is uniformly distributed between 5 and 25.

 a Draw the density function.
 b Find $P(X > 25)$.
 c Find $P(10 < X < 15)$.
 d Find $P(5.0 < X < 5.1)$.

6.4 The following is a graph of a density function.

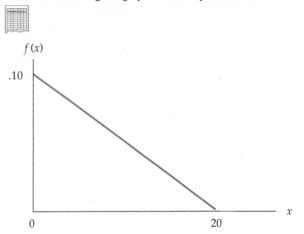

 a Determine the density function.
 b Find the probability that X is greater than 10.
 c Find the probability that X lies between 6 and 12.

6.3 NORMAL DISTRIBUTION

The **normal distribution** is the most important of all probability distributions because there are many real-life variables that are normally distributed.

> **Normal Density Function**
>
> The probability density function of a **normal random variable** is
>
> $$f(x) = \frac{1}{\sigma\sqrt{2\pi}} e^{-\frac{1}{2}\left(\frac{x-\mu}{\sigma}\right)^2} \quad -\infty < x < \infty$$
>
> where $e = 2.71828\ldots$ and $\pi = 3.14159\ldots$

Figure 6.7 depicts a normal distribution. Notice that the curve is symmetric about its mean, and the random variable ranges between $-\infty$ and ∞.

The normal distribution is described by two parameters, the mean μ and the standard deviation σ. In Figure 6.8, we describe the effect of changing the value of μ. Obviously, increasing μ shifts the curve to the right, and decreasing μ shifts it to the left.

Figure 6.9 describes the effect of σ. Larger values of σ widen the curve, and smaller ones narrow it.

To calculate the probability that a normal random variable falls into any interval, we need to compute the area in the interval under the curve. Unfortunately, the function is not as simple as the uniform. In fact, the function is so complicated that, even using integral calculus, we cannot determine the area. There is only one method to use, probability tables. At first glance it appears that we will need a table for every combination of values of μ and σ. However, because there is an infinite number of such combinations, this approach is not feasible. Instead, we reduce the number of tables needed to one by standardizing the random variable. We standardize a random variable by subtracting its mean and dividing by its standard deviation. When the

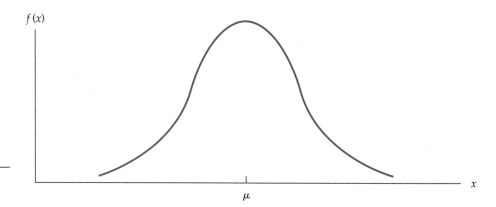

Figure 6.7

Symmetrical, bell-shaped normal distribution

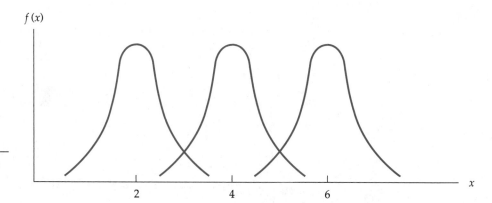

Figure 6.8

Normal distributions with the same variance but different means

variable is normal, the transformed variable is called a **standard normal random variable** and denoted by Z. That is,

$$Z = \frac{X - \mu}{\sigma}$$

The probability statement about X is transformed by this formula into Z. To illustrate how we proceed, consider the following problem.

Suppose that the amount of time to assemble a computer is normally distributed with a mean of 50 minutes and a standard deviation of 10 minutes. We would like to know the probability that a computer is assembled in a time between 45 and 60 minutes.

We want to find the probability

$P(45 < X < 60)$

Figure 6.10 describes a normal curve with mean 50, standard deviation 10, and the area we want to find.

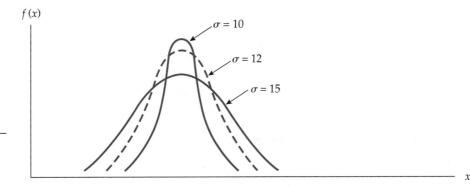

Figure 6.9

Normal distributions with the same mean but different standard deviations

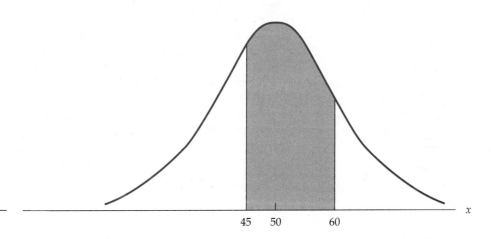

Figure 6.10

$P(45 < X < 60)$

Figure 6.11

$P(-.5 < Z < 1)$

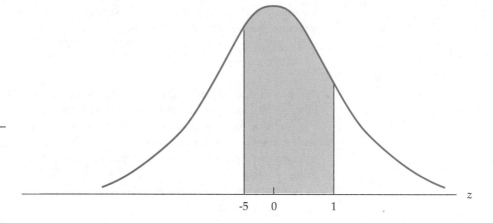

The first step is to standardize X. However, if we perform any operations on X we must perform the same operations on 45 and 60. Thus,

$$P(45 < X < 60) = P\left(\frac{45 - 50}{10} < \frac{X - \mu}{\sigma} < \frac{60 - 50}{10}\right) = P(-.5 < Z < 1)$$

Figure 6.11 describes the transformation that has taken place. Notice that the variable X was transformed into Z, 45 was transformed into $-.5$, and 60 was transformed into 1. However, the area has not changed. That is, the probability that we wish to compute, $P(45 < X < 60)$, is identical to $P(-.5 < Z < 1)$.

The values of Z specify the location of the corresponding value of X. A value $Z = -.5$ corresponds to a value of X that is one-half a standard deviation *below* the mean. That is, $X = 45$ is 5 minutes below the mean, and the standard deviation is 10. A value of $Z = 1$ corresponds to a value of X that is 1 standard deviation above the mean. Notice as well that the mean of Z, which is zero, corresponds to the mean of X.

If we know the mean and standard deviation of a normally distributed random variable, we can always transform the probability statement about X into a probability statement about Z. Consequently, we need only one table, Table 3 in Appendix B, the standard normal probability table, which is reproduced here as Table 6.1.

The table provides the probability that a standard normal random variable falls between 0 and values of z. For example, the probability $P(0 < Z < 2.00)$ is found by finding 2.0 in the left margin and, under the heading .00, finding .4772. The probability $P(0 < Z < 2.01)$ is found in the same row but under the heading .01. It is .4778. Table 6.2 depicts both values.

The numbers in the left column describe the values of z to one decimal place, and the column headings specify the second decimal place. Thus, to use this table we must always round to two decimal places.

Returning to our example, we note that the probability that we seek is actually the sum of two probabilities.

$$P(-.5 < Z < 1) = P(0 < Z < 1) + P(-.5 < Z < 0)$$

Table 6.1 Normal Curve Areas (from Appendix B, Table 3)

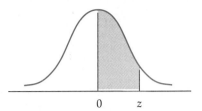

z	.00	.01	.02	.03	.04	.05	.06	.07	.08	.09
0.0	.0000	.0040	.0080	.0120	.0160	.0199	.0239	.0279	.0319	.0359
0.1	.0398	.0438	.0478	.0517	.0557	.0596	.0636	.0675	.0714	.0753
0.2	.0793	.0832	.0871	.0910	.0948	.0987	.1026	.1064	.1103	.1141
0.3	.1179	.1217	.1255	.1293	.1331	.1368	.1406	.1443	.1480	.1517
0.4	.1554	.1591	.1628	.1664	.1700	.1736	.1772	.1808	.1844	.1879
0.5	.1915	.1950	.1985	.2019	.2054	.2088	.2123	.2157	.2190	.2224
0.6	.2257	.2291	.2324	.2357	.2389	.2422	.2454	.2486	.2517	.2549
0.7	.2580	.2611	.2642	.2673	.2704	.2734	.2764	.2794	.2823	.2852
0.8	.2881	.2910	.2939	.2967	.2995	.3023	.3051	.3078	.3106	.3133
0.9	.3159	.3186	.3212	.3238	.3264	.3289	.3315	.3340	.3365	.3389
1.0	.3413	.3438	.3461	.3485	.3508	.3531	.3554	.3577	.3599	.3621
1.1	.3643	.3665	.3686	.3708	.3729	.3749	.3770	.3790	.3810	.3830
1.2	.3849	.3869	.3888	.3907	.3925	.3944	.3962	.3980	.3997	.4015
1.3	.4032	.4049	.4066	.4082	.4099	.4115	.4131	.4147	.4162	.4177
1.4	.4192	.4207	.4222	.4236	.4251	.4265	.4279	.4292	.4306	.4319
1.5	.4332	.4345	.4357	.4370	.4382	.4394	.4406	.4418	.4429	.4441
1.6	.4452	.4463	.4474	.4484	.4495	.4505	.4515	.4525	.4535	.4545
1.7	.4554	.4564	.4573	.4582	.4591	.4599	.4608	.4616	.4625	.4633
1.8	.4641	.4649	.4656	.4664	.4671	.4678	.4686	.4693	.4699	.4706
1.9	.4713	.4719	.4726	.4732	.4738	.4744	.4750	.4756	.4761	.4767
2.0	.4772	.4778	.4783	.4788	.4793	.4798	.4803	.4808	.4812	.4817
2.1	.4821	.4826	.4830	.4834	.4838	.4842	.4846	.4850	.4854	.4857
2.2	.4861	.4864	.4868	.4871	.4875	.4878	.4881	.4884	.4887	.4890
2.3	.4893	.4896	.4898	.4901	.4904	.4906	.4909	.4911	.4913	.4916
2.4	.4918	.4920	.4922	.4925	.4927	.4929	.4931	.4932	.4934	.4936
2.5	.4938	.4940	.4941	.4943	.4945	.4946	.4948	.4949	.4951	.4952
2.6	.4953	.4955	.4956	.4957	.4959	.4960	.4961	.4962	.4963	.4964
2.7	.4965	.4966	.4967	.4968	.4969	.4970	.4971	.4972	.4973	.4974
2.8	.4974	.4975	.4976	.4977	.4977	.4978	.4979	.4979	.4980	.4981
2.9	.4981	.4982	.4982	.4983	.4984	.4984	.4985	.4985	.4986	.4986
3.0	.4987	.4987	.4987	.4988	.4988	.4989	.4989	.4989	.4990	.4990

Source: Abridged from Table 1 of A. Hald, *Statistical Tables and Formulas* (New York: Wiley & Sons, Inc.), 1952. Reproduced by permission of A. Hald and the publisher, John Wiley & Sons, Inc.

The first probability on the right is easily determined from the table. It is $P(0 < Z < 1.00) = .3413$. The table also provides the second probability in an indirect way. Recall that the entire area under the curve is 1 and that the normal distribution is symmetric about its mean. As a result, $P(-.5 < Z < 0)$ is equal to $P(0 < Z < .5)$, which is found in the table. It is .1915. Thus,

$$P(-.5 < Z < 1) = .3413 + .1915 = .5328$$

Figure 6.12 shows how this calculation is performed.

Table 6.2 Normal Curve Areas (from Appendix B, Table 3)

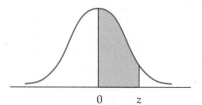

z	.00	.01	.02	.03	.04	.05	.06	.07	.08	.09
0.0	.0000	.0040	.0080	.0120	.0160	.0199	.0239	.0279	.0319	.0359
0.1	.0398	.0438	.0478	.0517	.0557	.0596	.0636	.0675	.0714	.0753
0.2	.0793	.0832	.0871	.0910	.0948	.0987	.1026	.1064	.1103	.1141
0.3	.1179	.1217	.1255	.1293	.1331	.1368	.1406	.1443	.1480	.1517
0.4	.1554	.1591	.1628	.1664	.1700	.1736	.1772	.1808	.1844	.1879
0.5	.1915	.1950	.1985	.2019	.2054	.2088	.2123	.2157	.2190	.2224
0.6	.2257	.2291	.2324	.2357	.2389	.2422	.2454	.2486	.2517	.2549
0.7	.2580	.2611	.2642	.2673	.2704	.2734	.2764	.2794	.2823	.2852
0.8	.2881	.2910	.2939	.2967	.2995	.3023	.3051	.3078	.3106	.3133
0.9	.3159	.3186	.3212	.3238	.3264	.3289	.3315	.3340	.3365	.3389
1.0	.3413	.3438	.3461	.3485	.3508	.3531	.3554	.3577	.3599	.3621
1.1	.3643	.3665	.3686	.3708	.3729	.3749	.3770	.3790	.3810	.3830
1.2	.3849	.3869	.3888	.3907	.3925	.3944	.3962	.3980	.3997	.4015
1.3	.4032	.4049	.4066	.4082	.4099	.4115	.4131	.4147	.4162	.4177
1.4	.4192	.4207	.4222	.4236	.4251	.4265	.4279	.4292	.4306	.4319
1.5	.4332	.4345	.4357	.4370	.4382	.4394	.4406	.4418	.4429	.4441
1.6	.4452	.4463	.4474	.4484	.4495	.4505	.4515	.4525	.4535	.4545
1.7	.4554	.4564	.4573	.4582	.4591	.4599	.4608	.4616	.4625	.4633
1.8	.4641	.4649	.4656	.4664	.4671	.4678	.4686	.4693	.4699	.4706
1.9	.4713	.4719	.4726	.4732	.4738	.4744	.4750	.4756	.4761	.4767
2.0	.4772	.4778	.4783	.4788	.4793	.4798	.4803	.4808	.4812	.4817
2.1	.4821	.4826	.4830	.4834	.4838	.4842	.4846	.4850	.4854	.4857
2.2	.4861	.4864	.4868	.4871	.4875	.4878	.4881	.4884	.4887	.4890
2.3	.4893	.4896	.4898	.4901	.4904	.4906	.4909	.4911	.4913	.4916
2.4	.4918	.4920	.4922	.4925	.4927	.4929	.4931	.4932	.4934	.4936
2.5	.4938	.4940	.4941	.4943	.4945	.4946	.4948	.4949	.4951	.4952
2.6	.4953	.4955	.4956	.4957	.4959	.4960	.4961	.4962	.4963	.4964
2.7	.4965	.4966	.4967	.4968	.4969	.4970	.4971	.4972	.4973	.4974
2.8	.4974	.4975	.4976	.4977	.4977	.4978	.4979	.4979	.4980	.4981
2.9	.4981	.4982	.4982	.4983	.4984	.4984	.4985	.4985	.4986	.4986
3.0	.4987	.4987	.4987	.4988	.4988	.4989	.4989	.4989	.4990	.4990

Source: Abridged from Table 1 of A. Hald, *Statistical Tables and Formulas* (New York: Wiley & Sons, Inc.), 1952. Reproduced by permission of A. Hald and the publisher, John Wiley & Sons, Inc.

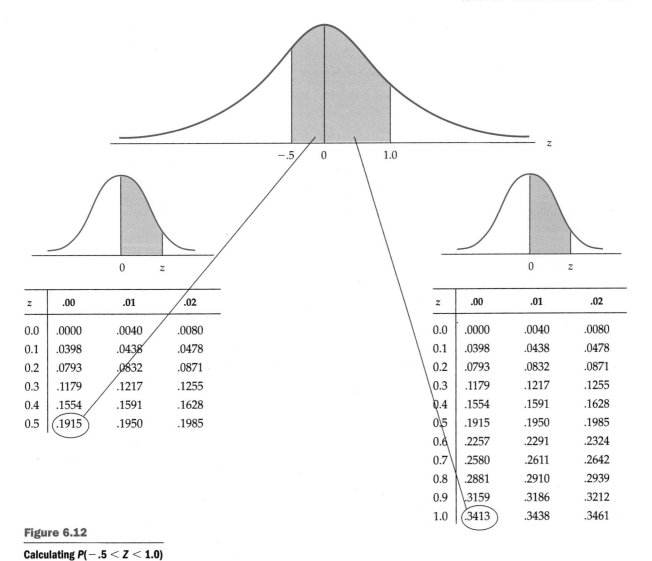

Figure 6.12

Calculating $P(-.5 < Z < 1.0)$

Therefore, the probability that a randomly selected computer takes between 45 and 60 minutes to assemble is .5328.

▼ EXAMPLE 6.2

The amount of money that college and university students earn in the summer is a normally distributed random variable with a mean of $4,500 and a standard deviation of $1,000. Find the probability that a student earns more than $6,000.

Solution We seek

$$P(X > 6,000)$$

where X is normally distributed with $\mu = 4,500$ and $\sigma = 1,000$.
To use Table 6.1 we must standardize both X and 6,000.

$$P(X > 6,000) = P\left(\frac{X - \mu}{\sigma} > \frac{6,000 - 4,500}{1,000}\right) = P(Z > 1.5)$$

Figure 6.13 depicts this calculation. Notice that we find $P(0 < Z < 1.5) = .4332$ from Table 6.1. Because the area under the curve is 1 and half the area is above 0 (and half below), it follows that $P(Z > 0) = .5$. Thus

$$P(Z > 1.5) = P(Z > 0) - P(0 < Z < 1.5) = .5 - .4332 = .0668$$

FINDING VALUES OF z

There is a family of problems that require us to determine the value of z given a probability. We use the notation z_A to represent the value of z such that the area to its right under the standard normal curve is A. That is, z_A is a value of a standard normal random variable such that

$$P(Z > z_A) = A$$

Figure 6.14 depicts this notation.

To find z_A for any value of A requires us to use the standard normal table backwards. As you saw in Example 6.2, to find a probability about Z we must find the value of z in the table and determine the probability associated with it. To use the table backwards we need to specify a probability and then determine the z-value associated with it. We'll demonstrate by finding $z_{.025}$. Figure 6.15 depicts the standard normal curve and $z_{.025}$. Because of the format of the standard normal table, we begin by determining the area between 0 and $z_{.025}$, which is $.5 - .025 = .4750$. (Notice that we expressed this probability with four decimal places to make it easier for you to see what you need to do.) We now search through the probability part of the table looking for .4750. When we locate it, we see that the z-value associated with it is 1.96.

Thus, $z_{.025} = 1.96$, which means that $P(Z > 1.96) = .025$.

Section 6.3 NORMAL DISTRIBUTION **161**

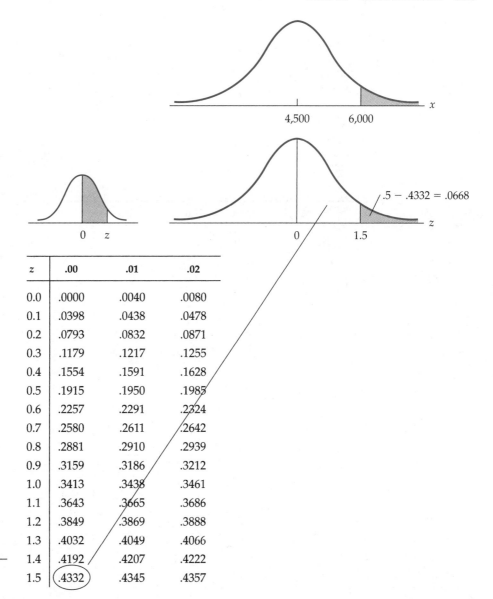

z	.00	.01	.02
0.0	.0000	.0040	.0080
0.1	.0398	.0438	.0478
0.2	.0793	.0832	.0871
0.3	.1179	.1217	.1255
0.4	.1554	.1591	.1628
0.5	.1915	.1950	.1985
0.6	.2257	.2291	.2324
0.7	.2580	.2611	.2642
0.8	.2881	.2910	.2939
0.9	.3159	.3186	.3212
1.0	.3413	.3438	.3461
1.1	.3643	.3665	.3686
1.2	.3849	.3869	.3888
1.3	.4032	.4049	.4066
1.4	.4192	.4207	.4222
1.5	.4332	.4345	.4357

Figure 6.13

Calculating $P(X > 6{,}000) = P(Z > 1.5)$

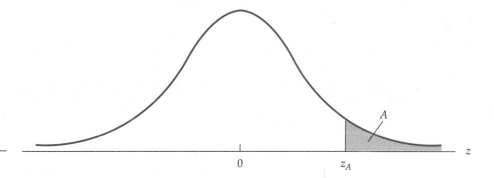

Figure 6.14

$P(Z > z_A) = A$

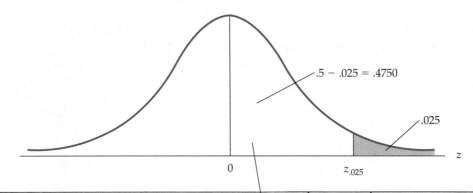

z	.00	.01	.02	.03	.04	.05	.06	.07	.08	.09
0.0	.0000	.0040	.0080	.0120	.0160	.0199	.0239	.0279	.0319	.0359
0.1	.0398	.0438	.0478	.0517	.0557	.0596	.0636	.0675	.0714	.0753
0.2	.0793	.0832	.0871	.0910	.0948	.0987	.1026	.1064	.1103	.1141
0.3	.1179	.1217	.1255	.1293	.1331	.1368	.1406	.1443	.1480	.1517
0.4	.1554	.1591	.1628	.1664	.1700	.1736	.1772	.1808	.1844	.1879
0.5	.1915	.1950	.1985	.2019	.2054	.2088	.2123	.2157	.2190	.2224
0.6	.2257	.2291	.2324	.2357	.2389	.2422	.2454	.2486	.2517	.2549
0.7	.2580	.2611	.2642	.2673	.2704	.2734	.2764	.2794	.2823	.2852
0.8	.2881	.2910	.2939	.2967	.2995	.3023	.3051	.3078	.3106	.3133
0.9	.3159	.3186	.3212	.3238	.3264	.3289	.3315	.3340	.3365	.3389
1.0	.3413	.3438	.3461	.3485	.3508	.3531	.3554	.3577	.3599	.3621
1.1	.3643	.3665	.3686	.3708	.3729	.3749	.3770	.3790	.3810	.3830
1.2	.3849	.3869	.3888	.3907	.3925	.3944	.3962	.3980	.3997	.4015
1.3	.4032	.4049	.4066	.4082	.4099	.4115	.4131	.4147	.4162	.4177
1.4	.4192	.4207	.4222	.4236	.4251	.4265	.4279	.4292	.4306	.4319
1.5	.4332	.4345	.4357	.4370	.4382	.4394	.4406	.4418	.4429	.4441
1.6	.4452	.4463	.4474	.4484	.4495	.4505	.4515	.4525	.4535	.4545
1.7	.4554	.4564	.4573	.4582	.4591	.4599	.4608	.4616	.4625	.4633
1.8	.4641	.4649	.4656	.4664	.4671	.4678	.4686	.4693	.4699	.4706
1.9	.4713	.4719	.4726	.4732	.4738	.4744	.4750	.4756	.4761	.4767
2.0	.4772	.4778	.4783	.4788	.4793	.4798	.4803	.4808	.4812	.4817

Figure 6.15

Finding $z_{.025}$

EXAMPLE 6.3

Find the following.

a $z_{.05}$

b $z_{.01}$

Section 6.3 NORMAL DISTRIBUTION

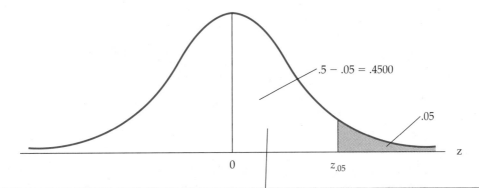

z	.00	.01	.02	.03	.04	.05	.06	.07	.08	.09
0.0	.0000	.0040	.0080	.0120	.0160	.0199	.0239	.0279	.0319	.0359
0.1	.0398	.0438	.0478	.0517	.0557	.0596	.0636	.0675	.0714	.0753
0.2	.0793	.0832	.0871	.0910	.0948	.0987	.1026	.1064	.1103	.1141
0.3	.1179	.1217	.1255	.1293	.1331	.1368	.1406	.1443	.1480	.1517
0.4	.1554	.1591	.1628	.1664	.1700	.1736	.1772	.1808	.1844	.1879
0.5	.1915	.1950	.1985	.2019	.2054	.2088	.2123	.2157	.2190	.2224
0.6	.2257	.2291	.2324	.2357	.2389	.2422	.2454	.2486	.2517	.2549
0.7	.2580	.2611	.2642	.2673	.2704	.2734	.2764	.2794	.2823	.2852
0.8	.2881	.2910	.2939	.2967	.2995	.3023	.3051	.3078	.3106	.3133
0.9	.3159	.3186	.3212	.3238	.3264	.3289	.3315	.3340	.3365	.3389
1.0	.3413	.3438	.3461	.3485	.3508	.3531	.3554	.3577	.3599	.3621
1.1	.3643	.3665	.3686	.3708	.3729	.3749	.3770	.3790	.3810	.3830
1.2	.3849	.3869	.3888	.3907	.3925	.3944	.3962	.3980	.3997	.4015
1.3	.4032	.4049	.4066	.4082	.4099	.4115	.4131	.4147	.4162	.4177
1.4	.4192	.4207	.4222	.4236	.4251	.4265	.4279	.4292	.4306	.4319
1.5	.4332	.4345	.4357	.4370	.4382	.4394	.4406	.4418	.4429	.4441
1.6	.4452	.4463	.4474	.4484	.4495	.4505	.4515	.4525	.4535	.4545
1.7	.4554	.4564	.4573	.4582	.4591	.4599	.4608	.4616	.4625	.4633
1.8	.4641	.4649	.4656	.4664	.4671	.4678	.4686	.4693	.4699	.4706
1.9	.4713	.4719	.4726	.4732	.4738	.4744	.4750	.4756	.4761	.4767
2.0	.4772	.4778	.4783	.4788	.4793	.4798	.4803	.4808	.4812	.4817

Figure 6.16

Finding $z_{.05}$

Solution

a Figure 6.16 depicts the normal curve and $z_{.05}$. If .05 is the area in the tail, then the area between 0 and $z_{.05}$ must be .45. To find $z_{.05}$, we search the table looking for the probability .4500. We don't find this probability, but we find two values that are close: .4495 and .4505. The z-values associated with these probabilities are 1.64 and 1.65, respectively. The average is taken as $z_{.05}$. Thus, $z_{.05} = 1.645$.

164 Chapter 6 CONTINUOUS PROBABILITY DISTRIBUTIONS

z	.00	.01	.02	.03	.04	.05	.06	.07	.08	.09
0.0	.0000	.0040	.0080	.0120	.0160	.0199	.0239	.0279	.0319	.0359
0.1	.0398	.0438	.0478	.0517	.0557	.0596	.0636	.0675	.0714	.0753
0.2	.0793	.0832	.0871	.0910	.0948	.0987	.1026	.1064	.1103	.1141
0.3	.1179	.1217	.1255	.1293	.1331	.1368	.1406	.1443	.1480	.1517
0.4	.1554	.1591	.1628	.1664	.1700	.1736	.1772	.1808	.1844	.1879
0.5	.1915	.1950	.1985	.2019	.2054	.2088	.2123	.2157	.2190	.2224
0.6	.2257	.2291	.2324	.2357	.2389	.2422	.2454	.2486	.2517	.2549
0.7	.2580	.2611	.2642	.2673	.2704	.2734	.2764	.2794	.2823	.2852
0.8	.2881	.2910	.2939	.2967	.2995	.3023	.3051	.3078	.3106	.3133
0.9	.3159	.3186	.3212	.3238	.3264	.3289	.3315	.3340	.3365	.3389
1.0	.3413	.3438	.3461	.3485	.3508	.3531	.3554	.3577	.3599	.3621
1.1	.3643	.3665	.3686	.3708	.3729	.3749	.3770	.3790	.3810	.3830
1.2	.3849	.3869	.3888	.3907	.3925	.3944	.3962	.3980	.3997	.4015
1.3	.4032	.4049	.4066	.4082	.4099	.4115	.4131	.4147	.4162	.4177
1.4	.4192	.4207	.4222	.4236	.4251	.4265	.4279	.4292	.4306	.4319
1.5	.4332	.4345	.4357	.4370	.4382	.4394	.4406	.4418	.4429	.4441
1.6	.4452	.4463	.4474	.4484	.4495	.4505	.4515	.4525	.4535	.4545
1.7	.4554	.4564	.4573	.4582	.4591	.4599	.4608	.4616	.4625	.4633
1.8	.4641	.4649	.4656	.4664	.4671	.4678	.4686	.4693	.4699	.4706
1.9	.4713	.4719	.4726	.4732	.4738	.4744	.4750	.4756	.4761	.4767
2.0	.4772	.4778	.4783	.4788	.4793	.4798	.4803	.4808	.4812	.4817
2.1	.4821	.4826	.4830	.4834	.4838	.4842	.4846	.4850	.4854	.4857
2.2	.4861	.4864	.4868	.4871	.4875	.4878	.4881	.4884	.4887	.4890
2.3	.4893	.4896	.4898	.4901	.4904	.4906	.4909	.4911	.4913	.4916
2.4	.4918	.4920	.4922	.4925	.4927	.4929	.4931	.4932	.4934	.4936
2.5	.4938	.4940	.4941	.4943	.4945	.4946	.4948	.4949	.4951	.4952

Figure 6.17

Finding $z_{.01}$

b To find $z_{.01}$, we seek .4900 in the table. The closest probability is .4901, whose z-value is 2.33. See Figure 6.17.

EXAMPLE 6.4

Mensa is an organization whose members possess IQs that are in the top 2% of the population. It is known that IQs are normally distributed with a mean of 100 and a standard deviation of 16. Find the minimum IQ needed to be a Mensa member.

Solution The minimum IQ to be a Mensa member is the value of X (labeled $x_{.02}$) such that

$$P(X > x_{.02}) = .02$$

Figure 6.18 shows the normal distribution with $x_{.02}$. Below the normal curve, we depict the standard normal curve and $z_{.02}$. We can determine the value of $z_{.02}$ as we did in the example above. In the standard normal table we find $.5 - .02 = .4800$ (its closest value in the table is .4798) and the z-value 2.05. Thus, the standardized value of $x_{.02}$ is $z_{.02} = 2.05$. To find $x_{.02}$ we must *unstandardize* $z_{.02}$. We do so by solving for $x_{.02}$ in the equation

$$z_{.02} = \frac{x_{.02} - \mu}{\sigma}$$

Substituting $z_{.02} = 2.05$, $\mu = 100$, and $\sigma = 16$, we find

$$2.05 = \frac{x_{.02} - 100}{16}$$

Solving, we get

$$x_{.02} = 2.05(16) + 100 = 132.08$$

Rounding up (IQs are integers), we find that the minimum IQ to be a Mensa member is 133.

z_A AND PERCENTILES

In Chapter 3 we introduced percentiles, which are measures of relative standing. The values of z_A are the $100(1 - A)$th percentiles of a standard normal random variable. For example, $z_{.05} = 1.645$, which means that 1.645 is the 95th percentile; 95% of all values of z are below it, and 5% are above it. We interpret other values of z_A similarly.

166 Chapter 6 CONTINUOUS PROBABILITY DISTRIBUTIONS

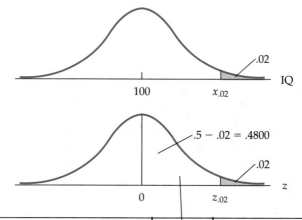

z	.00	.01	.02	.03	.04	.05	.06	.07	.08	.09
0.0	.0000	.0040	.0080	.0120	.0160	.0199	.0239	.0279	.0319	.0359
0.1	.0398	.0438	.0478	.0517	.0557	.0596	.0636	.0675	.0714	.0753
0.2	.0793	.0832	.0871	.0910	.0948	.0987	.1026	.1064	.1103	.1141
0.3	.1179	.1217	.1255	.1293	.1331	.1368	.1406	.1443	.1480	.1517
0.4	.1554	.1591	.1628	.1664	.1700	.1736	.1772	.1808	.1844	.1879
0.5	.1915	.1950	.1985	.2019	.2054	.2088	.2123	.2157	.2190	.2224
0.6	.2257	.2291	.2324	.2357	.2389	.2422	.2454	.2486	.2517	.2549
0.7	.2580	.2611	.2642	.2673	.2704	.2734	.2764	.2794	.2823	.2852
0.8	.2881	.2910	.2939	.2967	.2995	.3023	.3051	.3078	.3106	.3133
0.9	.3159	.3186	.3212	.3238	.3264	.3289	.3315	.3340	.3365	.3389
1.0	.3413	.3438	.3461	.3485	.3508	.3531	.3554	.3577	.3599	.3621
1.1	.3643	.3665	.3686	.3708	.3729	.3749	.3770	.3790	.3810	.3830
1.2	.3849	.3869	.3888	.3907	.3925	.3944	.3962	.3980	.3997	.4015
1.3	.4032	.4049	.4066	.4082	.4099	.4115	.4131	.4147	.4162	.4177
1.4	.4192	.4207	.4222	.4236	.4251	.4265	.4279	.4292	.4306	.4319
1.5	.4332	.4345	.4357	.4370	.4382	.4394	.4406	.4418	.4429	.4441
1.6	.4452	.4463	.4474	.4484	.4495	.4505	.4515	.4525	.4535	.4545
1.7	.4554	.4564	.4573	.4582	.4591	.4599	.4608	.4616	.4625	.4633
1.8	.4641	.4649	.4656	.4664	.4671	.4678	.4686	.4693	.4699	.4706
1.9	.4713	.4719	.4726	.4732	.4738	.4744	.4750	.4756	.4761	.4767
2.0	.4772	.4778	.4783	.4788	.4793	.4798	.4803	.4808	.4812	.4817
2.1	.4821	.4826	.4830	.4834	.4838	.4842	.4846	.4850	.4854	.4857
2.2	.4861	.4864	.4868	.4871	.4875	.4878	.4881	.4884	.4887	.4890

Figure 6.18

Finding $x_{.02}$ in Example 6.4

USING MICROSOFT EXCEL

We can employ Excel to compute probabilities and values of X and Z. To compute cumulative normal probabilities

$P(X < x)$

proceed as follows.

COMMANDS

1 Click f_x, **Function category: Statistical,** and **Function name: NORMDIST.** Click **OK.**
2 Type the value of x **(X),** the mean **(Mean),** the standard deviation **(Standard_dev),** and **true (Cumulative)** to yield a cumulative probability. (Typing **false** will produce the value of the density function, a number with little meaning.) The probability appears on the right side of the dialog box. Clicking **OK** will print the probability in the active cell.
3 Alternatively, type (in any cell)

= NORMDIST([X], [Mean], [Standard_dev], true)

If you type 0 for **Mean** and 1 for **Standard_dev** you will yield standard normal probabilities. Alternatively, click **NORMSDIST** instead of **NORMDIST.** Type the value of z **(Z),** and click **OK.**

For example,

NORMDIST(6000,4500,1000,TRUE) = .9332

and

NORMDIST(1.5,0,1,TRUE) = NORMSDIST(1.5) = .9332

To determine a value of X, given a cumulative probability, follow these instructions:

COMMANDS

1 Click f_x, **Function category: Statistical,** and **Function name: NORMINV.** Click **OK.**
2 Type the cumulative probability **(Probability),** the mean **(Mean),** and the standard deviation **(Standard_dev).**
3 Alternatively, type

=NORMINV([Probability], [Mean], [Standard_dev])

If you type 0 for **Mean** and 1 for **Standard_dev,** you will yield the standard normal random variable. Alternatively, click **NORMSINV** instead of **NORMINV.** Type the cumulative probability **(Probability).**

For example,

NORMINV(.98,100,16) = 132.86

and

NORMSINV(.98) = 2.054

EXERCISES

In Exercises 6.5 to 6.20 find the following probabilities.

6.5 $P(0 < Z < 1.5)$

6.6 $P(0 < Z < 1.51)$

6.7 $P(0 < Z < 1.55)$

6.8 $P(0 < Z < 1.59)$

6.9 $P(0 < Z < 1.6)$

6.10 $P(0 < Z < 2.3)$

6.11 $P(-1.4 < Z < 0.6)$

6.12 $P(Z > -1.44)$

6.13 $P(Z < 2.03)$

6.14 $P(Z > 1.67)$

6.15 $P(Z < 2.84)$

6.16 $P(1.14 < Z < 2.43)$

6.17 $P(-0.91 < Z < -0.33)$

6.18 $P(Z > 3.09)$

6.19 $P(Z > 0)$

6.20 $P(Z > 4.0)$

6.21 Find $z_{.02}$.

6.22 Find $z_{.045}$.

6.23 Find $z_{.20}$.

6.24 X is normally distributed with mean 100 and standard deviation 20. What is the probability that X is greater than 145?

6.25 X is normally distributed with mean 250 and standard deviation 40. What value of X does only the top 15% exceed?

6.26 X is normally distributed with mean 1,000 and standard deviation 250. What is the probability that X lies between 800 and 1,100?

6.27 X is normally distributed with mean 50 and standard deviation 8. What value of X is such that only 8% of values are below it?

6.28 The long distance calls made by the employees of a company are normally distributed with a mean of 6.3 minutes and a standard deviation of 2.2 minutes. Find the probability that a call

  a lasts between 5 and 10 minutes.
  b lasts more than 7 minutes.
  c lasts less than 4 minutes.

6.29 Refer to Exercise 6.28. How long does the longest 10% of calls last?

6.30 The lifetimes of lightbulbs that are advertised to last for 5,000 hours are normally distributed with a mean of 5,100 hours and a standard deviation of 200 hours. What is the probability that a bulb lasts longer than the advertised figure?

6.31 Refer to Exercise 6.30. If we wanted to be sure that 98% of all bulbs last longer than the advertised figure, what figure should be advertised?

6.32 Travelbyus is an Internet-based travel agency wherein customers can see videos of the cities they plan to visit. The number of hits daily is a normally distributed random variable with a mean of 10,000 and a standard deviation of 2,400.

  a What is the probability of getting more than 12,000 hits?
  b What is the probability of getting fewer than 9,000 hits?

6.33 Refer to Exercise 6.32. Some Internet sites have bandwidths that are not sufficient to handle all their traffic, often causing the system to crash. Bandwidth can be measured by the number of hits it can handle. How large a bandwidth should Travelbyus have in order to handle 99.9% of daily traffic?

6.34 The lifetimes of televisions produced by the Hishobi company are normally distributed with a mean of 75 months and a standard deviation of 8 months. If the manufacturer wants to have to replace only 1% of its televisions, what should its warranty be?

6.35 A new car that is a hybrid gas-powered and electric-powered has recently hit the market. The distance traveled on one gallon of fuel is normally distributed with a mean of 65 miles and a standard deviation of 4 miles. Find the probability of the following events.

  a The car travels more than 70 miles per gallon.
  b The car travels less than 60 miles per gallon.
  c The car travels between 55 and 70 miles per gallon.

6.36 The top-selling Red and Voss tire is rated 70,000 miles, which means nothing. In fact, the distance the tires can run until wear-out is a normally distributed random variable with a mean of 82,000 miles with a standard deviation of 6,400 miles.

  a What is the probability that the tire wears out before 70,000 miles?
  b What is the probability that a tire lasts more than 100,000 miles?

6.37 The heights of children 2 years old are normally distributed with a mean of 32 inches and a standard deviation of 1.5 inches. Pediatricians regularly measure the heights of toddlers to determine if there is a problem. There may be a problem when a child is in the top or bottom 5% of heights. Determine the heights of 2-year-old children that could be a problem.

6.38 Refer to Exercise 6.37. Find the probability of the events below.

  a A 2-year-old child is taller than 36 inches.
  b A 2-year-old child is shorter than 34 inches.
  c A 2-year-old child is between 30 and 33 inches tall.

6.39 University and college students average 7.2 hours of sleep per night, with a standard deviation of 40 minutes. If the amount of sleep is normally distributed, what proportion of university and college students sleep for more than 8 hours?

6.40 Refer to Exercise 6.39. Find the amount of sleep that is exceeded by only 25% of students.

6.41 The amount of time devoted to studying statistics each week by students who achieve a grade of A in the course is a normally distributed random variable with a mean of 7.5 hours and a standard deviation of 2.1 hours.

  a What proportion of A students study for more than 10 hours per week?
  b Find the probability that an A student spends between 7 and 9 hours studying.
  c What proportion of A students spend less than 3 hours studying?

6.42 Refer to Exercise 6.41. What is the amount of time below which only 5% of A students spend studying?

6.43 The number of pages printed before replacing the cartridge in a laser printer is normally distributed with a mean of 11,500 pages and a standard deviation of 800 pages. A new cartridge has just been installed.

   a What is the probability that the printer produces more than 12,000 pages before this cartridge needs to be replaced?

   b What is the probability that the printer produces fewer than 10,000 pages?

6.44 Refer to Exercise 6.43. The manufacturer wants to provide guidelines to potential customers advising them of the minimum number of pages they can expect from each cartridge. How many pages should it advertise if the company wants to be correct 99% of the time?

6.45 Battery manufacturers compete on the basis of the amount of time their products last in cameras and toys. A manufacturer of alkaline batteries has observed that its batteries last for an average of 26 hours when used in a toy racing car. The amount of time is normally distributed with a standard deviation of 2.5 hours.

   a What is the probability that the battery lasts between 24 and 28 hours?

   b What is the probability that the battery lasts longer than 28 hours?

   c What is the probability that the battery lasts less than 24 hours?

6.46 Because of the relatively high interest rates, most consumers attempt to pay off their credit card bills promptly. However, this is not always possible. An analysis of the amount of interest paid monthly by a bank's Visa card holders reveals that the amount is normally distributed with a mean of $27 and a standard deviation of $7. What proportion of the bank's Visa card holders

   a pay more than $30 in interest?
   b pay more than $40 in interest?
   c pay less than $15 in interest?

6.47 Refer to Exercise 6.46. What interest payment is exceeded by only 20% of the bank's Visa card holders?

6.48 It is said that sufferers of a cold virus experience symptoms for 7 days. However, the amount of time is actually a normally distributed random variable whose mean is 7.5 days and whose standard deviation is 1.2 days.

   a What proportion of cold sufferers experiences fewer than 4 days of symptoms?

   b What proportion of cold sufferers experiences symptoms for between 7 and 10 days?

6.49 How much money does a typical family of four spend at McDonald's restaurants per visit? The amount is a normally distributed random variable whose mean is $16.40 and whose standard deviation is $2.75. Find the probability that a family of four spends less than $10.

6.50 Refer to Exercise 6.49. What is the amount below which only 10% of families of four spend at McDonald's?

6.51 The final marks in a statistics course are normally distributed with a mean of 70 and a standard deviation of 10. The professor must convert all marks to letter grades. She decides that she wants 10% A's, 30% B's, 40% C's, 15% D's, and 5% F's. Determine the cutoffs for each letter grade.

6.4 OTHER CONTINUOUS DISTRIBUTIONS

In this section we introduce three more continuous distributions. We will employ them extensively in statistical inference.

STUDENT t DISTRIBUTION

The Student t distribution was first derived by William S. Gosset in 1908. (Gosset published his findings under the pseudonym "Student" and used the letter t to represent the random variable, hence the **Student t distribution.**) It is very commonly used in statistical inference, where its applications will be introduced in Chapters 11, 12, 14, 17, and 18.

Student t Density Function

The density function of the Student t distribution is as follows.

$$f(t) = \frac{[(\nu-1)/2]!}{\sqrt{\nu\pi}[(\nu-2)/2]!}\left[1+\frac{t^2}{\nu}\right]^{-(\nu+1)/2} \quad -\infty < t < \infty$$

where $\pi = 3.14159\ldots$ and ν (Greek letter *nu*) is the parameter of the Student t distribution called the **degrees of freedom.**

The mean and variance of a Student random variable are

$$E(t) = 0$$

and

$$V(t) = \frac{\nu}{\nu - 2} \quad \text{for } \nu > 2$$

Figure 6.19 depicts the Student t distribution. As you can see, this distribution is similar to the standard normal distribution. Both are symmetrical about zero. (Both random variables have a mean of zero.) We describe the Student t distribution as mound shaped, whereas the normal distribution is bell shaped.

Figure 6.20 shows both a Student t and the standard normal distributions. The former is more widely spread out than the latter. The variance of a standard normal random variable is 1, while the variance of a Student t random variable is $\nu/(\nu - 2)$, which is greater than 1 for all ν.

Figure 6.21 depicts Student t distributions with several different degrees of freedom. Notice that for larger degrees of freedom, the Student t distribution's dispersion is smaller. For example, when $\nu = 10$, $V(t) = 1.25$; when $\nu = 50$, $V(t) = 1.042$; and when $\nu = 200$, $V(t) = 1.010$. As ν grows larger, the Student t distribution approaches the standard normal distribution.

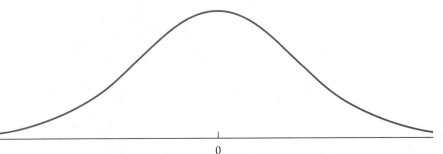

Figure 6.19

Student t distribution

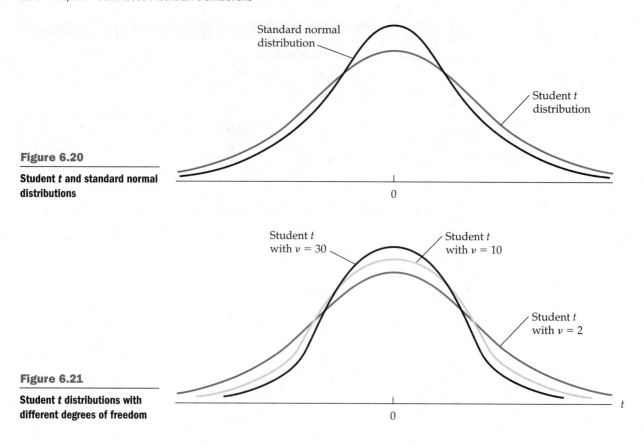

Figure 6.20

Student t and standard normal distributions

Figure 6.21

Student t distributions with different degrees of freedom

DETERMINING STUDENT t VALUES

For each value of ν (the number of degrees of freedom), there is a different Student t distribution. If we wanted to calculate probabilities of the Student t random variable, as we did for the normal random variable, we would need a different table for each ν, which is not feasible. The only way to determine probabilities associated with a Student t random variable is to use Microsoft Excel. The instructions appear below.

As you will discover later in this book, the Student t distribution is employed extensively in statistical inference. And, for inferential methods, we often need to find values of the random variable. To determine values of a normal random variable we used Table 6.1 backwards. Finding values of a Student t random variable is considerably easier. Table 4 in Appendix B (reproduced as Table 6.3) lists values of $t_{A,\nu}$, which are the values of a Student t random variable with ν degrees of freedom such that

$$P(t > t_{A,\nu}) = A$$

Figure 6.22 depicts this notation.

Observe that $t_{A,\nu}$ is provided for degrees of freedom ranging from 1 to 200 and ∞. To read this table, simply identify the degrees of freedom, and find that value or the closest number to it if it is not listed. Then locate the column representing the t_A value you wish. For example, if we want the value of t with 10 degrees of freedom such that the area to its right under the Student t curve is .05, we locate 10 in the first

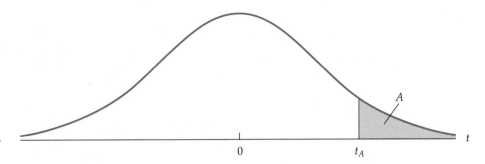

Figure 6.22

Student t distribution with t_A

column and move across this row until we locate the number under the heading $t_{.05}$. We find (see Table 6.4)

$$t_{.05,10} = 1.812$$

Table 6.3 Critical Values of t

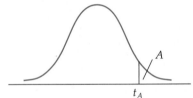

Degrees of Freedom	$t_{.10}$	$t_{.05}$	$t_{.025}$	$t_{.01}$	$t_{.005}$	Degrees of Freedom	$t_{.10}$	$t_{.05}$	$t_{.025}$	$t_{.01}$	$t_{.005}$
1	3.078	6.314	12.706	31.821	63.657	24	1.318	1.711	2.064	2.492	2.797
2	1.886	2.920	4.303	6.965	9.925	25	1.316	1.708	2.060	2.485	2.787
3	1.638	2.353	3.182	4.541	5.841	26	1.315	1.706	2.056	2.479	2.779
4	1.533	2.132	2.776	3.747	4.604	27	1.314	1.703	2.052	2.473	2.771
5	1.476	2.015	2.571	3.365	4.032	28	1.313	1.701	2.048	2.467	2.763
6	1.440	1.943	2.447	3.143	3.707	29	1.311	1.699	2.045	2.462	2.756
7	1.415	1.895	2.365	2.998	3.499	30	1.310	1.697	2.042	2.457	2.750
8	1.397	1.860	2.306	2.896	3.355	35	1.306	1.690	2.030	2.438	2.724
9	1.383	1.833	2.262	2.821	3.250	40	1.303	1.684	2.021	2.423	2.705
10	1.372	1.812	2.228	2.764	3.169	45	1.301	1.679	2.014	2.412	2.690
11	1.363	1.796	2.201	2.718	3.106	50	1.299	1.676	2.009	2.403	2.678
12	1.356	1.782	2.179	2.681	3.055	60	1.296	1.671	2.000	2.390	2.660
13	1.350	1.771	2.160	2.650	3.012	70	1.294	1.667	1.994	2.381	2.648
14	1.345	1.761	2.145	2.624	2.977	80	1.292	1.664	1.990	2.374	2.639
15	1.341	1.753	2.131	2.602	2.947	90	1.291	1.662	1.987	2.369	2.632
16	1.337	1.746	2.120	2.583	2.921	100	1.290	1.660	1.984	2.364	2.626
17	1.333	1.740	2.110	2.567	2.898	120	1.289	1.658	1.980	2.358	2.617
18	1.330	1.734	2.101	2.552	2.878	140	1.288	1.656	1.977	2.353	2.611
19	1.328	1.729	2.093	2.539	2.861	160	1.287	1.654	1.975	2.350	2.607
20	1.325	1.725	2.086	2.528	2.845	180	1.286	1.653	1.973	2.347	2.603
21	1.323	1.721	2.080	2.518	2.831	200	1.286	1.653	1.972	2.345	2.601
22	1.321	1.717	2.074	2.508	2.819	∞	1.282	1.645	1.960	2.326	2.576
23	1.319	1.714	2.069	2.500	2.807						

Source: From M. Merrington, "Table of Percentage Points of the t-Distribution," *Biometrika* 32 (1941): 300. Reproduced by permission of the Biometrika Trustees.

Table 6.4 Finding $t_{.05,10}$

Degrees of Freedom	$t_{.10}$	$t_{.05}$	$t_{.025}$	$t_{.01}$	$t_{.005}$
1	3.078	6.314	12.706	31.821	63.657
2	1.886	2.920	4.303	6.965	9.925
3	1.638	2.353	3.182	4.541	5.841
4	1.533	2.132	2.776	3.747	4.604
5	1.476	2.015	2.571	3.365	4.032
6	1.440	1.943	2.447	3.143	3.707
7	1.415	1.895	2.365	2.998	3.499
8	1.397	1.860	2.306	2.896	3.355
9	1.383	1.833	2.262	2.821	3.250
10	1.372	1.812	2.228	2.764	3.169
11	1.363	1.796	2.201	2.718	3.106
12	1.356	1.782	2.179	2.681	3.055

If the number of degrees of freedom is 25, we determine (see Table 6.5)

$$t_{.05,25} = 1.708$$

If the number of degrees of freedom is 74, we find the number of degrees of freedom closest to 74 listed in the table, which is 70, and determine (see Table 6.6)

$$t_{.05,74} \approx t_{.05,70} = 1.667$$

Because the Student t distribution is symmetric about zero, the value of t such that the area to its *left* is A is $-t_{A,\nu}$. For example, the value of t with 10 degrees of freedom such that the area to its left is .05 is

$$-t_{.05,10} = -1.812$$

Table 6.5 Finding $t_{.05,25}$

Degrees of Freedom	$t_{.10}$	$t_{.05}$	$t_{.025}$	$t_{.01}$	$t_{.005}$
1	3.078	6.314	12.706	31.821	63.657
2	1.886	2.920	4.303	6.965	9.925
3	1.638	2.353	3.182	4.541	5.841
4	1.533	2.132	2.776	3.347	4.604
5	1.476	2.015	2.571	3.365	4.032
.					
.					
.					
21	1.323	1.721	2.080	2.518	2.831
22	1.321	1.717	2.074	2.508	2.819
23	1.319	1.714	2.069	2.500	2.807
24	1.318	1.711	2.064	2.492	2.797
25	1.316	1.708	2.060	2.485	2.787
26	1.315	1.706	2.056	2.479	2.779

Table 6.6 **Finding $t_{.05,70}$**

Degrees of Freedom	$t_{.10}$	$t_{.05}$	$t_{.025}$	$t_{.01}$	$t_{.005}$
1	3.078	6.314	12.706	31.821	63.657
2	1.886	2.920	4.303	6.965	9.925
3	1.638	2.353	3.182	4.541	5.841
4	1.533	2.132	2.776	3.747	4.604
5	1.476	2.015	2.571	3.365	4.032
⋮		⋮			
45	1.301	1.679	2.014	2.412	2.690
50	1.299	1.676	2.009	2.403	2.678
60	1.296	1.671	2.000	2.390	2.660
70	1.294	1.667	1.994	2.381	2.648
80	1.292	1.664	1.990	2.374	2.639
90	1.291	1.662	1.987	2.369	2.632
100	1.290	1.660	1.984	2.364	2.626
120	1.289	1.658	1.980	2.358	2.617
140	1.288	1.656	1.977	2.353	2.611
160	1.287	1.654	1.975	2.350	2.607
180	1.286	1.653	1.973	2.347	2.603
200	1.286	1.653	1.972	2.345	2.601
∞	1.282	1.645	1.960	2.326	2.576

Notice the last row in the Student t table. The number of degrees of freedom is infinite, and the t values are identical (except for the number of decimal places) to the values of z. For example,

$t_{.10,\infty} = 1.282$

$t_{.05,\infty} = 1.645$

$t_{.025,\infty} = 1.960$

$t_{.01,\infty} = 2.326$

$t_{.005,\infty} = 2.576$

In the previous section we showed (or showed how we determine) that

$z_{.10} = 1.28$

$z_{.05} = 1.645$

$z_{.025} = 1.96$

$z_{.01} = 2.33$

$z_{.005} = 2.575$

USING MICROSOFT EXCEL

We can employ Excel to compute probabilities and values of a Student t random variable. To compute Student t probabilities, proceed as follows.

COMMANDS

1 Click f_x, **Function category: Statistical,** and **Function name: TDIST.** Click **OK.**
2 Type the value of x, where x must be positive **(X)**, the degrees of freedom **(Deg_freedom),** and the number of tails **(Tails).** Typing **1** for **Tails** produces the area to the right of x. Typing **2** for **Tails** produces the area to the right of x plus the area to the left of $-x$.
3 Alternatively, type into any cell

   = TDIST([X], [Degrees of freedom], [Tails])

For example,

TDIST(2,50,1) = .025474

and

TDIST(2,50,2) = .050947

To determine a value of a Student t random variable, follow these instructions.

COMMANDS

1 Click f_x, **Function category: Statistical,** and **Function name: TINV.** Click **OK.**
2 Type the probability **(Probability)** and the degrees of freedom **(Deg_freedom).**
3 Alternatively, type into any cell

   = TINV([Probability], [Degrees of freedom])

The result is the value of t such that the area to its right is half the probability. The other half of the probability is located to the left of $-t$. For example,

TINV(.05,200) = 1.972

which means that $P(t > 1.972) + P(t < -1.972) = .025 + .025 = .05$.

CHI-SQUARED DISTRIBUTION

The density function of another very useful random variable is exhibited next.

> **Chi-Squared Density Function**
>
> The chi-squared density function is
>
> $$f(\chi^2) = \frac{1}{[(\nu/2) - 1]!} \frac{1}{2^{\nu/2}} (\chi^2)^{(\nu/2)-1} e^{-\chi^2/2} \quad \chi^2 > 0$$
>
> The parameter ν is the number of degrees of freedom, which, like the degrees of freedom of the Student t distribution, affects the shape.

Figure 6.23 depicts a **chi-squared distribution.** As you can see, it is positively skewed, ranging between 0 and ∞. Like that of the Student t distribution, its shape depends on its number of degrees of freedom. The effect of increasing the degrees of freedom is seen in Figure 6.24.

DETERMINING CHI-SQUARED VALUES

The value of χ^2 with ν degrees of freedom such that the area to its right under the chi-squared curve is equal to A is denoted $\chi^2_{A,\nu}$. We cannot use $-\chi^2_{A,\nu}$ to represent the point such that the area to its *left* is A (as we did with the standard normal and Student t values), because χ^2 is always greater than zero. To represent left-tail critical values, we note that if the area to the left of a point is A, the area to its right must

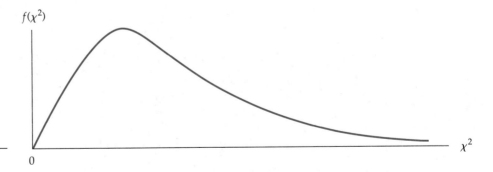

Figure 6.23

Chi-squared distribution

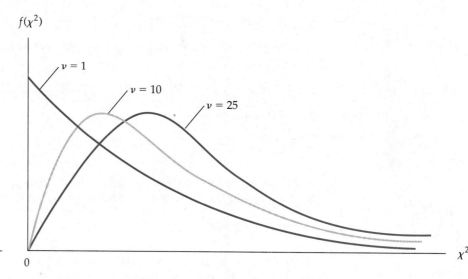

Figure 6.24

Chi-square distributions

be $1 - A$, because the entire area under the chi-squared curve (as well as all continuous distributions) must equal 1. Thus $\chi^2_{1-A,\nu}$ denotes the point such that the area to its left is A. See Figure 6.25.

Table 5 in Appendix B (reproduced as Table 6.7) lists critical values of the chi-squared distribution for degrees of freedom equal to 1 to 30, 40, 50, 60, 70, 80, 90, and 100. For example, to find the point in a chi-squared distribution with 8 degrees of freedom such that the area to its right is .05, locate 8 degrees of freedom in the left column and $\chi^2_{.050}$ across the top. The intersection of the row and column contains the number we seek, as shown in Table 6.8. That is,

$$\chi^2_{.050,8} = 15.5073$$

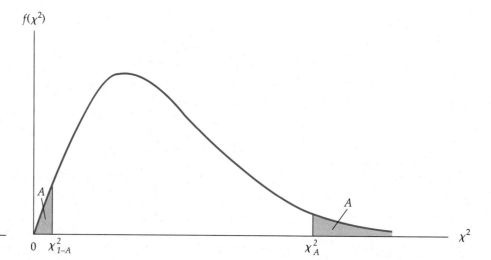

Figure 6.25

χ^2_A and χ^2_{1-A}

Table 6.7 Critical Values of χ^2

Degrees of Freedom	$\chi^2_{.995}$	$\chi^2_{.990}$	$\chi^2_{.975}$	$\chi^2_{.950}$	$\chi^2_{.900}$	$\chi^2_{.100}$	$\chi^2_{.050}$	$\chi^2_{.025}$	$\chi^2_{.010}$	$\chi^2_{.005}$
1	0.0000393	0.0001571	0.0009821	0.0039321	0.0157908	2.70554	3.84146	5.02389	6.63490	7.87944
2	0.0100251	0.0201007	0.0506356	0.102587	0.210720	4.60517	5.99147	7.37776	9.21034	10.5966
3	0.0717212	0.114832	0.215795	0.351846	0.584375	6.25139	7.81473	9.34840	11.3449	12.8381
4	0.206990	0.297110	0.484419	0.710721	1.063623	7.77944	9.48773	11.1433	13.2767	14.8602
5	0.411740	0.55430	0.831211	1.145476	1.61031	9.23635	11.0705	12.8325	15.0863	16.7496
6	0.675727	0.872085	1.237347	1.63539	2.20413	10.6446	12.5916	14.4494	16.8119	18.5476
7	0.989265	1.239043	1.68987	2.16735	2.83311	12.0170	14.0671	16.0128	18.4753	20.2777
8	1.344419	1.646482	2.17973	2.73264	3.48954	13.3616	15.5073	17.5346	20.0902	21.9550
9	1.734926	2.087912	2.70039	3.32511	4.16816	14.6837	16.9190	19.0228	21.6660	23.5893
10	2.15585	2.55821	3.24697	3.94030	4.86518	15.9871	18.3070	20.4831	23.2093	25.1882
11	2.60321	3.05347	3.81575	4.57481	5.57779	17.2750	19.6751	21.9200	24.7250	26.7569
12	3.07382	3.57056	4.40379	5.22603	6.30380	18.5494	21.0261	23.3367	26.2170	28.2995
13	3.56503	4.10691	5.00874	5.89186	7.04150	19.8119	22.3621	24.7356	27.6883	29.8194
14	4.07468	4.66043	5.62872	6.57063	7.78953	21.0642	23.6848	26.1190	29.1413	31.3193
15	4.60094	5.22935	6.26214	7.26094	8.54675	22.3072	24.9958	27.4884	30.5779	32.8013
16	5.14224	5.81221	6.90766	7.96164	9.31223	23.5418	26.2962	28.8454	31.9999	34.2672
17	5.69724	6.40776	7.56418	8.67176	10.0852	24.7690	27.5871	30.1910	33.4087	35.7185
18	6.26481	7.01491	8.23075	9.39046	10.8649	25.9894	28.8693	31.5264	34.8053	37.1564
19	6.84398	7.63273	8.90655	10.1170	11.6509	27.2036	30.1435	32.8523	36.1908	38.5822
20	7.43386	8.26040	9.59083	10.8508	12.4426	28.4120	31.4104	34.1696	37.5662	39.9968
21	8.03366	8.89720	10.28293	11.5913	13.2396	29.6151	32.6705	35.4789	38.9321	41.4010
22	8.64272	9.54249	10.9823	12.3380	14.0415	30.8133	33.9244	36.7807	40.2894	42.7956
23	9.26042	10.19567	11.6885	13.0905	14.8479	32.0069	35.1725	38.0757	41.6384	44.1813
24	9.88623	10.8564	12.4011	13.8484	15.6587	33.1963	36.4151	39.3641	42.9798	45.5585
25	10.5197	11.5240	13.1197	14.6114	16.4734	34.3816	37.6525	40.6465	44.3141	46.9278
26	11.1603	12.1981	13.8439	15.3791	17.2919	35.5631	38.8852	41.9232	45.6417	48.2899
27	11.8076	12.8786	14.5733	16.1513	18.1138	36.7412	40.1133	43.1944	46.9630	49.6449
28	12.4613	13.5648	15.3079	16.9279	18.9392	37.9159	41.3372	44.4607	48.2782	50.9933
29	13.1211	14.2565	16.0471	17.7083	19.7677	39.0875	42.5569	45.7222	49.5879	52.3356
30	13.7867	14.9535	16.7908	18.4926	20.5992	40.2560	43.7729	46.9792	50.8922	53.6720
40	20.7065	22.1643	24.4331	26.5093	29.0505	51.8050	55.7585	59.3417	63.6907	66.7659
50	27.9907	29.7067	32.3574	34.7642	37.6886	63.1671	67.5048	71.4202	76.1539	79.4900
60	35.5346	37.4848	40.4817	43.1879	46.4589	74.3970	79.0819	83.2976	88.3794	91.9517
70	43.2752	45.4418	48.7576	51.7393	55.3290	85.5271	90.5312	95.0231	100.425	104.215
80	51.1720	53.5400	57.1532	60.3915	64.2778	96.5782	101.879	106.629	112.329	116.321
90	59.1963	61.7541	65.6466	69.1260	73.2912	107.565	113.145	118.136	124.116	128.299
100	67.3276	70.0648	74.2219	77.9295	82.3581	118.498	124.342	129.561	135.807	140.169

Source: From C. M. Thompson, "Tables of the Percentage Points of the χ^2-Distribution," *Biometrika* 32 (1941): 188–89. Reproduced by permission of the Biometrika trustees.

To find the point in the same distribution such that the area to its *left* is .05, find the point such that the area to its *right* is .95. Locate $\chi^2_{.950}$ across the top row and 8 degrees of freedom down the left column (also shown in Table 6.8). You should see that

$$\chi^2_{.950,8} = 2.73264$$

Table 6.8 Finding $\chi^2_{.050,8}$ and $\chi^2_{.950,8}$

Degrees of Freedom	$\chi^2_{.995}$	$\chi^2_{.990}$	$\chi^2_{.975}$	$\chi^2_{.950}$	$\chi^2_{.900}$	$\chi^2_{.100}$	$\chi^2_{.050}$	$\chi^2_{.025}$	$\chi^2_{.010}$	$\chi^2_{.005}$
1	0.0000393	0.0001571	0.0009821	0.0039321	0.0157908	2.70554	3.84146	5.02389	6.63490	7.87944
2	0.0100251	0.0201007	0.0506356	0.102587	0.210720	4.60517	5.99147	7.37776	9.21034	10.5966
3	0.0717212	0.114832	0.215795	0.351846	0.584375	6.25139	7.81473	9.34840	11.3449	12.8381
4	0.206990	0.297110	0.484419	0.710721	1.063623	7.77944	9.48773	11.1433	13.2767	14.8602
5	0.411740	0.55430	0.831211	1.145476	1.61031	9.23635	11.0705	12.8325	15.0863	16.7496
6	0.675727	0.872085	1.237347	1.63539	2.20413	10.6446	12.5916	14.4494	16.8119	18.5476
7	0.989265	1.239043	1.68987	2.16735	2.83311	12.0170	14.0671	16.0128	18.4753	20.2777
8	1.344419	1.646482	2.17973	2.73264	3.48954	13.3616	15.5073	17.5346	20.0902	21.9550
9	1.734926	2.087912	2.70039	3.32511	4.16816	14.6837	16.9190	19.0228	21.6660	23.5893
10	2.15585	2.55821	3.24697	3.94030	4.86518	15.9871	18.3070	20.4831	23.2093	25.1882
11	2.60321	3.05347	3.81575	4.57481	5.57779	17.2750	19.6751	21.9200	24.7250	26.7569

MATHEMATICAL DERIVATION (OPTIONAL)

The chi-squared distribution is derived mathematically by squaring the standard normal random variable. That is,

$$\chi^2 = Z^2$$

The mathematical derivation and other topics are stored in a Word file on the CD that accompanies this book. Squaring a standard normal random variable produces a chi-squared distribution with 1 degree of freedom. To demonstrate, we remind you that

$$z_{.025} = 1.96$$

and

$$-z_{.025} = -1.96$$

Figure 6.26 depicts the standard normal curve. If we square every value of z and plot the probability density function, we produce a chi-squared distribution with 1 degree of freedom. The area to the right of 1.96 and the area to the left of -1.96 are combined in the chi-squared distribution. As a result,

$$z^2_{.025} = 1.96^2 = 3.8416$$

From Table 5 we find $\chi^2_{.05,1} = 3.8416$. Similarly

$$z^2_{.05} = 1.645^2 = 2.706025, \quad \chi^2_{.10,1} = 2.70554$$

and

$$z^2_{.005} = 2.575^2 = 6.630625 \quad \chi^2_{.01,1} = 6.63490$$

The imprecise determination of z values (other than 1.96) is the cause of the slight differences between z-squared and chi-squared values.

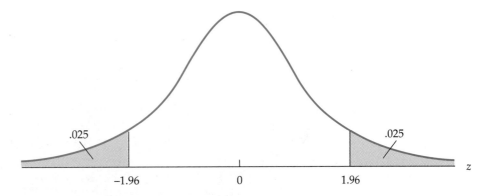
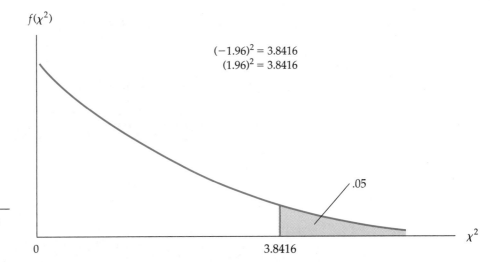

Figure 6.26

Standard normal distribution and chi-squared distribution with $\nu = 1$

USING MICROSOFT EXCEL

We can employ Excel to compute probabilities and values of a chi-squared random variable. To compute the probability to the right of any chi-squared value, proceed as follows.

> COMMANDS
> 1 Click f_x, **Function category: Statistical,** and **Function name: CHIDIST**. Click **OK**.
> 2 Type the value of X **(X)** and the degrees of freedom **(Deg_freedom)**.
> 3 Alternatively, type into any cell
>
>   = CHIDIST([X], [Degrees of freedom])

For example, CHIDIST(6.25139,3) = .1000

To determine a value of a chi-squared random variable, follow these instructions.

> **COMMANDS**
>
> 1 Click f_x, **Function category: Statistical**, and **Function name: CHIINV**. Click **OK**.
> 2 Type the cumulative probability **(Probability)** and the degrees of freedom **(Deg_freedom)**.
> 3 Alternatively, type into any cell
>
> = CHIINV([Probability], [Degrees of freedom])

For example, CHIINV(.10,3) = 6.25139

F DISTRIBUTION

The density function of the **F distribution** is

> **F Density Function**
>
> $$f(F) = \frac{\left(\frac{v_1 + v_2 - 2}{2}\right)!}{\left(\frac{v_1 - 2}{2}\right)!\left(\frac{v_2 - 2}{2}\right)!}\left(\frac{v_1}{v_2}\right)^{\frac{v_1}{2}} \frac{F^{\frac{v_1 - 2}{2}}}{\left(1 + \frac{v_1 F}{v_2}\right)^{\frac{v_1 + v_2}{2}}} \quad F > 0$$
>
> where F ranges from 0 to ∞ and v_1 and v_2 are the parameters of the distribution called degrees of freedom. For reasons that will become clearer in Chapter 12, we call v_1 the *numerator degrees of freedom* and v_2 the *denominator degrees of freedom*.

Figure 6.27 describes the density function when it is graphed. As you can see, the F distribution is positively skewed. Its actual shape depends on the two degrees of freedom.

DETERMINING VALUES OF F

We define F_{A,v_1,v_2} as the value of F with v_1 and v_2 degrees of freedom such that the area to its right under the curve is A. That is,

$P(F > F_{A,v_1,v_2}) = A$

Because the F random variable, like the chi-squared, can equal only positive values, we define F_{1-A,v_1,v_2} as the value such that the area to its left is A. Figure 6.28 depicts

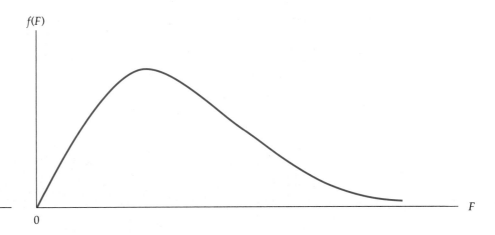

Figure 6.27
F distribution

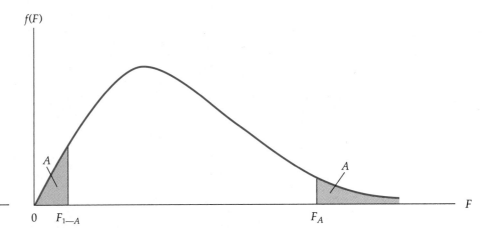

Figure 6.28
F_A and F_{1-A}

this notation. Table 6 in Appendix B provides values of F_{A,v_1,v_2} for A = .05, .025, and .01. Part of Table 6 is reproduced below as Table 6.9.

Values of F_{1-A,v_1,v_2} are unavailable. However, we do not need them, because we can determine F_{1-A,v_1,v_2} from F_{A,v_1,v_2}. That is, statisticians can show that

$$F_{1-A,v_1,v_2} = \frac{1}{F_{A,v_2,v_1}}$$

To determine any critical value, find the numerator degrees of freedom v_1 across the top of Table 6.9 and the denominator degrees of freedom v_2 down the left column. The intersection of the row and column contains the number we seek. To illustrate, suppose that we want to find $F_{.05,5,7}$. Table 6.10 shows how this point is found. Locate the numerator degrees of freedom, 5, across the top and the denominator degrees of freedom, 7, down the left column. The intersection is 3.97. Thus,

$$F_{.05,5,7} = 3.97$$

Note that the order in which the degrees of freedom appear is important. To find $F_{.05,7,5}$ (numerator degrees of freedom = 7 and denominator degrees of freedom = 5),

Table 6.9 Percentile Points of the F Distribution A = .05

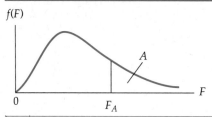

ν_2 \ ν_1	1	2	3	4	5	6	7	8	9
1	161.4	199.5	215.7	224.6	230.2	234.0	236.8	238.9	240.5
2	18.51	19.00	19.16	19.25	19.30	19.33	19.35	19.37	19.38
3	10.13	9.55	9.28	9.12	9.01	8.94	8.89	8.85	8.81
4	7.71	6.94	6.59	6.39	6.26	6.16	6.09	6.04	6.00
5	6.61	5.79	5.41	5.19	5.05	4.95	4.88	4.82	4.77
6	5.99	5.14	4.76	4.53	4.39	4.28	4.21	4.15	4.10
7	5.59	4.74	4.35	4.12	3.97	3.87	3.79	3.73	3.68
8	5.32	4.46	4.07	3.84	3.69	3.58	3.50	3.44	3.39
9	5.12	4.26	3.86	3.63	3.48	3.37	3.29	3.23	3.18
10	4.96	4.10	3.71	3.48	3.33	3.22	3.14	3.07	3.02
11	4.84	3.98	3.59	3.36	3.20	3.09	3.01	2.95	2.90
12	4.75	3.89	3.49	3.26	3.11	3.00	2.91	2.85	2.80
13	4.67	3.81	3.41	3.18	3.03	2.92	2.83	2.77	2.71
14	4.60	3.74	3.34	3.11	2.96	2.85	2.76	2.70	2.65
15	4.54	3.68	3.29	3.06	2.90	2.79	2.71	2.64	2.59
16	4.49	3.63	3.24	3.01	2.85	2.74	2.66	2.59	2.54
17	4.45	3.59	3.20	2.96	2.81	2.70	2.61	2.55	2.49
18	4.41	3.55	3.16	2.93	2.77	2.66	2.58	2.51	2.46
19	4.38	3.52	3.13	2.90	2.74	2.63	2.54	2.48	2.42
20	4.35	3.49	3.10	2.87	2.71	2.60	2.51	2.45	2.39
21	4.32	3.47	3.07	2.84	2.68	2.57	2.49	2.42	2.37
22	4.30	3.44	3.05	2.82	2.66	2.55	2.46	2.40	2.34
23	4.28	3.42	3.03	2.80	2.64	2.53	2.44	2.37	2.32
24	4.26	3.40	3.01	2.78	2.62	2.51	2.42	2.36	2.30
25	4.24	3.39	2.99	2.76	2.60	2.49	2.40	2.34	2.28
26	4.23	3.37	2.98	2.74	2.59	2.47	2.39	2.32	2.27
27	4.21	3.35	2.96	2.73	2.57	2.46	2.37	2.31	2.25
28	4.20	3.34	2.95	2.71	2.56	2.45	2.36	2.29	2.24
29	4.18	3.33	2.93	2.70	2.55	2.43	2.35	2.28	2.22
30	4.17	3.32	2.92	2.69	2.53	2.42	2.33	2.27	2.21
40	4.08	3.23	2.84	2.61	2.45	2.34	2.25	2.18	2.12
60	4.00	3.15	2.76	2.53	2.37	2.25	2.17	2.10	2.04
120	3.92	3.07	2.68	2.45	2.29	2.17	2.09	2.02	1.96
∞	3.84	3.00	2.60	2.37	2.21	2.10	2.01	1.94	1.88

Numerator Degrees of Freedom (columns); Denominator Degrees of Freedom (rows).

Source: From M. Merrington and C. M. Thompson, "Tables of Percentage Points of the Inverted Beta (F)-Distribution," *Biometrika 33* (1943): 73–88. Reproduced by permission of the Biometrika trustees.

Table 6.10 Finding $F_{.05,5,7}$

ν_2 \ ν_1	1	2	3	4	5	6	7	8	9
1	161.4	199.5	215.7	224.6	230.2	234.0	236.8	238.9	240.5
2	18.51	19.00	19.16	19.25	19.30	19.33	19.35	19.37	19.38
3	10.13	9.55	9.28	9.12	9.01	8.94	8.89	8.85	8.81
4	7.71	6.94	6.59	6.39	6.26	6.16	6.09	6.04	6.00
5	6.61	5.79	5.41	5.19	5.05	4.95	4.88	4.82	4.77
6	5.99	5.14	4.76	4.53	4.39	4.28	4.21	4.15	4.10
7	5.59	4.74	4.35	4.12	3.97	3.87	3.79	3.73	3.68
8	5.32	4.46	4.07	3.84	3.69	3.58	3.50	3.44	3.39
9	5.12	4.26	3.86	3.63	3.48	3.37	3.29	3.23	3.18
10	4.96	4.10	3.71	3.48	3.33	3.22	3.14	3.07	3.02
11	4.84	3.98	3.59	3.36	3.20	3.09	3.01	2.95	2.90
12	4.75	3.89	3.49	3.26	3.11	3.00	2.91	2.85	2.80
13	4.67	3.81	3.41	3.18	3.03	2.92	2.83	2.77	2.71
14	4.60	3.74	3.34	3.11	2.96	2.85	2.76	2.70	2.65
15	4.54	3.68	3.29	3.06	2.90	2.79	2.71	2.64	2.59

Numerator Degrees of Freedom (across top); Denominator Degrees of Freedom (down side).

we locate 7 across the top and 5 down the side. The intersection is

$$F_{.05,7,5} = 4.88$$

Suppose that we want to determine the point in an F distribution with $\nu_1 = 4$ and $\nu_2 = 8$ such that the area to its left is .95.

$$F_{.95,4,8} = \frac{1}{F_{.05,8,4}} = \frac{1}{6.04} = .166$$

USING MICROSOFT EXCEL

We can use Excel to compute probabilities and values of an F random variable. To compute the probability to the right of any F value, proceed as follows.

> **COMMANDS**
>
> 1 Click f_x, **Function category: Statistical**, and **Function name: FDIST**. Click **OK**.
> 2 Type the value of x (**X**), numerator degrees of freedom (**Deg_freedom 1**), and denominator degrees of freedom (**Deg_freedom 2**).
> 3 Alternately, type
>
>   = FDIST([X], [Numerator degrees of freedom], [Denominator degrees of freedom])

For example, FDIST(3.97, 5, 7) = .05.

To determine a value of an F random variable, follow these instructions.

> **COMMANDS**
>
> 1 Click f_x, **Function category: Statistical,** and **Function name: FINV.** Click **OK.**
> 2 Type the probability to the right of the value **(Probability),** the numerator degrees of freedom **(Deg_freedom 1),** and the denominator degrees of freedom **(Deg_freedom 2).**
> 3 Alternatively, type
>
> = FINV([Probability], [Numerator degrees of freedom], [Denominator degrees of freedom])

For example, FINV(.05, 5, 7) = 3.97.

EXERCISES

6.52 Use the t table (Table 4) to find the following values of t.
 a $t_{.10,15}$ b $t_{.10,23}$ c $t_{.025,83}$ d $t_{.05,195}$

6.53 Use the t table (Table 4) to find the following values of t.
 a $t_{.005,33}$ b $t_{.10,600}$ c $t_{.05,4}$ d $t_{.01,20}$

6.54 Use Microsoft Excel to find the following values of t.
 a $t_{.10,15}$ b $t_{.10,23}$ c $t_{.025,83}$ d $t_{.05,195}$

6.55 Use Microsoft Excel to find the following values of t.
 a $t_{.05,143}$ b $t_{.01,12}$ c $t_{.025,\infty}$ d $t_{.05,100}$

6.56 Use Microsoft Excel to find the following probabilities.
 a $P(t_{64} > 2.12)$ b $P(t_{27} > 1.90)$
 c $P(t_{159} > 1.33)$ d $P(t_{550} > 1.85)$

6.57 Use Microsoft Excel to find the following probabilities.
 a $P(t_{141} > .94)$ b $P(t_{421} > 2.00)$
 c $P(t_{1000} > 1.96)$ d $P(t_{82} > 1.96)$

6.58 Use the χ^2 table (Table 5) to find the following values of χ^2.
 a $\chi^2_{.10,5}$ b $\chi^2_{.01,100}$ c $\chi^2_{.95,18}$ d $\chi^2_{.99,60}$

6.59 Use the χ^2 table (Table 5) to find the following values of χ^2.
 a $\chi^2_{.90,26}$ b $\chi^2_{.01,30}$ c $\chi^2_{.10,1}$ d $\chi^2_{.80,78}$

6.60 Use Microsoft Excel to find the following values of χ^2.
 a $\chi^2_{.25,66}$ b $\chi^2_{.40,100}$ c $\chi^2_{.50,17}$ d $\chi^2_{.10,17}$

6.61 Use Microsoft Excel to find the following values of χ^2.
 a $\chi^2_{.99,55}$ b $\chi^2_{.05,800}$ c $\chi^2_{.99,43}$ d $\chi^2_{.10,233}$

6.62 Use Microsoft Excel to find the following probabilities.
 a $P(\chi^2_{73} > 80)$ b $P(\chi^2_{200} > 125)$
 c $P(\chi^2_{88} > 60)$ d $P(\chi^2_{1000} > 450)$

6.63 Use Microsoft Excel to find the following probabilities.
 a $P(\chi^2_{250} > 250)$ b $P(\chi^2_{36} > 25)$
 c $P(\chi^2_{600} > 500)$ d $P(\chi^2_{120} > 100)$

6.64 Use the F table (Table 6) to find the following values of F.
 a $F_{.05,3,7}$ b $F_{.05,7,3}$ c $F_{.025,5,20}$ d $F_{.01,12,60}$

6.65 Use the F table (Table 6) to find the following values of F.
 a $F_{.025,8,22}$ b $F_{.05,20,30}$ c $F_{.01,9,18}$ d $F_{.025,24,10}$

6.66 Use Microsoft Excel to find the following values of F.
 a $F_{.05,70,70}$ b $F_{.01,45,100}$
 c $F_{.025,36,50}$ d $F_{.05,500,500}$

6.67 Use Microsoft Excel to find the following values of F.
 a $F_{.01,100,150}$ b $F_{.05,25,125}$
 c $F_{.01,11,33}$ d $F_{.05,300,800}$

6.68 Use Microsoft Excel to find the following probabilities.
 a $P(F_{7,20} > 2.5)$ b $P(F_{18,63} > 1.4)$
 c $P(F_{34,62} > 1.8)$ d $P(F_{200,400} > 1.1)$

6.69 Use Microsoft Excel to find the following probabilities.
 a $P(F_{600,800} > 1.1)$ b $P(F_{35,100} > 1.3)$
 c $P(F_{66,148} > 2.1)$ d $P(F_{17,37} > 2.8)$

6.5 SUMMARY

This chapter dealt with continuous random variables and their distributions. Because a continuous random variable can assume an infinite number of values, the probability that the random variable equals any single value is zero. Consequently, we address the problem of computing the probability of a range of values. We showed that the probability of any interval is the area in the interval under the curve representing the density function.

We introduced the most important distribution in statistics and showed how to compute the probability that a normal random variable falls into any interval. Additionally, we demonstrated how to use the normal table backwards to find values of a normal random variable given a probability. Finally we presented three more continuous random variables and their probability density functions. The Student t, chi-squared, and F distributions will be employed extensively in statistical inference.

IMPORTANT TERMS

Probability density function
Uniform probability distribution
Rectangular probability distribution
Normal distribution
Normal random variable
Standard normal random variable
Student t distribution
Degrees of freedom
Chi-squared distribution
F distribution

SYMBOLS

Symbol	Pronounced	Represents
π	pi	3.14159 ...
z_A	z-sub-A or z-A	Value of Z such that area to its right is A
ν	nu	Degrees of freedom
t_A	t-sub-A or t-A	Value of t such that area to its right is A
χ_A^2	chi-squared-sub-A or chi-squared-A	Value of chi-squared such that area to its right is A
F_A	F-sub-A or F-A	Value of F such that area to its right is A
ν_1	nu-sub-one or nu-one	Numerator degrees of freedom
ν_2	nu-sub-two or nu-two	Denominator degrees of freedom

MICROSOFT EXCEL OUTPUT AND INSTRUCTIONS

Probability/Random Variable	Page
Normal probability	167
Normal random variable	167
Student t probability	176
Student t random variable	176
Chi-squared probability	181
Chi-squared random variable	182
F probability	185
F random variable	186

Chapter 7

Sampling and Sampling Plans

7.1 Introduction

7.2 Sampling

7.3 Sampling Plans

7.4 Errors Involved in Sampling

7.5 Summary

7.1 INTRODUCTION

In Chapter 1, we briefly introduced the concept of statistical inference—the process of inferring information about a population from a sample. Because information about populations can usually be described by parameters, the statistical technique used generally deals with drawing inferences about population parameters from sample statistics. (Recall that a parameter is a measurement about a population, and a statistic is a measurement about a sample.)

In Chapters 5 and 6 we assumed that population parameters are known. In real life, however, calculating parameters is difficult because populations tend to be quite large. As a result, most population parameters are unknown. For example, in order to determine the mean annual income of North American blue-collar workers, we would have to ask each North American blue-collar worker what his or her income is and then calculate the mean of all the responses. Because this population consists of several million people, the task is both expensive and impractical. If we are willing to accept less than 100% accuracy, we can use statistical inference to obtain an estimate. Rather than investigate the entire population, we select a sample of workers, determine the annual income of the workers in this group, and calculate the sample mean. In this chapter, we will discuss the basic concepts and techniques of sampling and several different sampling plans.

7.2 SAMPLING

The chief motive for examining a sample rather than a population is cost. Statistical inference permits us to draw conclusions about a population parameter based on a sample that is quite small in comparison to the size of the population. For example, television executives want to know the proportion of television viewers who watch a network's programs. Because 100 million people may be watching television in the United States on a given evening, determining the actual proportion of the population that is watching certain programs is impractical and prohibitively expensive. The Nielsen ratings provide approximations of the desired information by observing what is watched by a sample of 1,000 television viewers. The proportion of households watching a particular program can be calculated for the households in the Nielsen sample. This sample proportion is then used as an **estimate** of the proportion of all households (the population proportion) that watched the program.

Another illustration of sampling can be taken from the field of quality control. To ensure that a production process is operating properly, the operations manager needs to know what proportion of items being produced is defective. If the quality control technician must destroy the item in order to determine whether it is defective, then there is no alternative to sampling: A complete inspection of the product population would destroy the entire output of the production process.

We know that the sample proportion of television viewers or of defective items is probably not exactly equal to the population proportion we want to estimate. Nonetheless, the sample statistic can come quite close to the parameter it is designed to estimate if the **target population** (the population about which we want to draw inferences) and the **sampled population** (the actual population from which the sample has been taken) are the same. In practice, these may not be the same, as the following example illustrates.

NIELSEN RATINGS

The Nielsen ratings are supposed to provide information about the television shows that all Americans are watching. Hence, the target population is the television viewers of the United States. If the sample of 1,000 viewers was drawn exclusively from the state of New York, however, the sampled population would be the television viewers of New York. In this case, the target population and the sampled population are not the same, and no valid inferences about the target population can be drawn. To allow proper estimation of the proportion of all American television viewers watching a specific program, the sample should contain men and women of varying ages, incomes, occupations, and residences in a pattern similar to that of the target population. The importance of sampling from the target population cannot be overestimated; the consequences of drawing conclusions from improperly selected sample can be costly. One of the most spectacular examples of how not to conduct a survey was the *Literary Digest* poll of 1936.

LITERARY DIGEST POLL

The *Literary Digest* was a popular magazine of the 1920s and 1930s that had correctly predicted the outcomes of several presidential elections. In 1936, the *Digest* predicted that the Republican candidate, Alfred Landon, would defeat the Democratic incumbent, Franklin D. Roosevelt, by a 3 to 2 margin. But in that election, Roosevelt defeated Landon in a landslide victory, garnering the support of 62% of the electorate. The source of this blunder was the sampling procedure, and there were two distinct mistakes. First, the *Digest* sent out 10 million sample ballots to prospective voters. However, most of the names of these people were taken from the *Digest*'s subscription list and from telephone directories. Subscribers to the magazine and people who owned telephones tended to be wealthier than average, and such people then, as today, tended to vote Republican. Additionally, only 2.3 million ballots were returned, resulting in a self-selected sample.

Self-selected samples are almost always biased, because the individuals who participate in them are more keenly interested in the issue than are the other members of the population. You often find similar surveys conducted today when radio and television stations ask people to call and give their opinion on an issue of interest. Again, only listeners who are concerned about the topic and have enough patience to get through to the station will be included in the sample. Hence, the sampled population is comprised entirely of people who are interested in the issue, whereas the target population is made up of all the people within the listening radius of the radio station. As a result, the conclusions drawn from such surveys are frequently wrong.

An excellent example of this phenomenon occurred on ABC's *Nightline* in 1984. Viewers were given a 900 number (cost: 50 cents) and asked to phone in their responses to the question of whether the United Nations should continue to be located in the United States. More than 186,000 people called, with 67% responding "no." At the same time, a (more scientific) market research poll of 500 people revealed that 72% wanted the United Nations to remain in the United States. In general, because the true value of the parameter being estimated is never known, these surveys give the impression of providing useful information. In fact, the results of such surveys

are likely to be no more accurate than the results of the 1936 *Literary Digest* poll*
or *Nightline*'s phone-in show. Statisticians have coined two terms to describe these
polls: SLOP (self-selected opinion poll) and *oy vey* (from the Yiddish lament), both
of which convey the contempt that statisticians have for such data-gathering processes.

EXERCISES

 7.1 For each of the following sampling plans, indicate why the target population and the sampled population are not the same.

a To determine the opinions and attitudes of customers who regularly shop at a particular mall, a surveyor stands outside a large department store in the mall and randomly selects people to participate in the survey.

b A library wants to estimate the proportion of its books that have been damaged. The librarians decide to select one book per shelf as a sample by measuring 12 inches from the left edge of each shelf and selecting the book in that location.

c Political surveyors visit 200 residences during one afternoon to ask eligible voters present in the house at the time whom they intend to vote for.

 7.2 a Describe why the *Literary Digest* poll of 1936 has become infamous.

b What caused this poll to be so wrong?

 7.3 a What is meant by a self-selected sample?

b Give an example of a recent poll that involved a self-selected sample.

c Why are self-selected samples not desirable?

7.3 SAMPLING PLANS

Our objective in this section is to introduce three different sampling plans: simple random sampling, stratified random sampling, and cluster sampling. We begin our presentation with the most basic design.

SIMPLE RANDOM SAMPLING

> **Simple Random Sample**
>
> A **simple random sample** is a sample selected in such a way that every possible sample with the same number of observations is equally likely to be chosen.

One way to conduct a simple random sample is to assign a number to each element in the population, write these numbers on individual slips of paper, toss them into a hat, and draw the required number of slips (the sample size, n) from the hat. This is the kind of procedure that occurs in raffles, when all the ticket stubs go into a large rotating drum from which the winners are selected.

---

*Many statisticians ascribe the *Literary Digest*'s statistical debacle to the wrong causes. For a better understanding of what really happened, read Maurice C. Bryson, "The Literary Digest Poll: Making of a Statistical Myth," *American Statistician* 30(4) (November 1976): 184–85.

Sometimes the elements of the population are already numbered. For example, virtually all adults have Social Security numbers (in the United States) or Social Insurance numbers (in Canada); all employees of large corporations have employee numbers; many people have driver's license numbers, medical plan numbers, student numbers, and so on. In such cases, choosing which sampling procedure to use is simply a matter of deciding how to select from among these numbers.

In other cases, the existing form of numbering has built-in flaws that make it inappropriate as a source of samples. Not everyone has a phone number, for example, so the telephone book does not list all the people in a given area. Many households have two (or more) adults, but only one phone listing. Couples often list the phone number under the man's name, so telephone listings are likely to be disproportionately male. Some people do not have phones, some have unlisted phone numbers, and some have more than one phone; these differences mean that each element of the population does not have an equal probability of being selected.

After each element of the chosen population has been assigned a unique number, sample numbers can be selected at random. A random number table can be used to select these sample numbers. (See, for example, *CRC Standard Management Tables*, W. H. Beyer, Ed., Boca Raton: CRC Press.) Alternatively, we can use Excel to perform this function.

▼ EXAMPLE 7.1

A government income tax auditor has been given responsibility for 1,000 tax returns. A computer is used to check the arithmetic of each return. However, to determine if the returns have been completed honestly, the auditor must check each entry and confirm its veracity. Because it takes, on average, one hour to completely audit a return and she has only one week to complete the task, the auditor has decided to randomly select 40 returns. The returns are numbered from 1 to 1,000. Use a computer random number generator to select the sample for the auditor.

Solution We generated 50 numbers between 1 and 1,000 even though we needed only 40 numbers. We did so because it is likely that there will be some duplicates. We will use the first 40 unique random numbers to select our sample.

MICROSOFT EXCEL RANDOM NUMBERS

383	246	372	952	75
101	46	356	54	199
597	33	911	706	65
900	165	467	817	359
885	220	427	973	488
959	18	304	467	512
15	286	976	301	374
408	344	807	751	986
864	554	992	352	41
139	358	257	776	231

GENERAL COMMANDS	EXAMPLE 7.1
1 Click **Tools, Data Analysis...**, and **Random Number Generation.**	
2 Specify the **Number of Variables**	1
3 Specify the **Number of Random Numbers.**	50
4 Select **Uniform Distribution.**	
5 Specify the range of the uniform distribution (**Parameters**). Click **OK.** Column A will fill with 50 numbers that range between 0 and 1.	0 and 1
6 Multiply Column A by 1,000, and store the products in Column B.	
7 Make Cell C1 active, and click f_x, **Math&Trig, ROUNDUP,** and **Next>.**	
8 Specify the first number to be rounded.	B1
9 Type the **number of digits** (decimal places). Click **Finish.**	0
10 Complete Column C.	

The first five steps command Excel to generate 50 uniformly distributed random numbers between 0 and 1 to be stored in column A. Steps 6 through 10 convert these random numbers to integers between 1 and 1,000. Each number has the same probability (1/1,000 = .001) of being selected. Thus, each member of the population is equally likely to be included in the sample.

INTERPRETING THE RESULTS

The auditor would examine the tax returns selected by the computer. She would pick returns numbered 383, 101, 597, ..., 751, 352, and 776 (the first 40 unique numbers). Each of these would be audited to determine if it was fraudulent. If the objective is to audit these 40 returns, no statistical procedure would be employed. However, if the objective is to estimate the proportion of all 1,000 returns that were dishonest, she would use one of the inferential techniques presented later in this book. ▲

STRATIFIED RANDOM SAMPLING

In making inferences about a population, we attempt to extract as much information as possible from a sample. The basic sampling plan, simple random sampling, often accomplishes this goal at low cost. Other methods, however, can be used to increase the amount of information about the population. One such procedure is stratified random sampling.

> ### Stratified Random Sample
> A **stratified random sample** is obtained by separating the population into mutually exclusive sets, or strata, and then drawing simple random samples from each stratum.

Examples of criteria for separating a population into strata (and of the strata themselves) follow.

1. Sex
   male
   female
2. Age
   under 20
   20–30
   31–40
   41–50
   51–60
   over 60
3. Occupation
   professional
   clerical
   blue-collar
   other
4. Household income
   under $15,000
   $15,000–$29,999
   $30,000–$50,000
   over $50,000

   To illustrate, suppose a public opinion survey is to be conducted in order to determine how many people favor a tax increase. A stratified random sample could be obtained by selecting a random sample of people from each of the four income groups described above. We usually stratify in a way that enables us to obtain particular kinds of information. In this example, we would like to know if people in the different income categories differ in their opinions about the proposed tax increase, because the tax increase will affect the strata differently. We avoid stratifying when there is no connection between the survey and the strata. For example, little purpose is served in trying to determine if people within religious strata have divergent opinions about the tax increase.
   One advantage of stratification is that, besides acquiring information about the entire population, we can also make inferences within each stratum or compare strata.

For instance, we can estimate what proportion of the lowest income group favors the tax increase, or we can compare the highest and lowest income groups to determine if they differ in their support of the tax increase.

Any stratification must be done in such a way that the strata are mutually exclusive: Each member of the population must be assigned to exactly one stratum. After the population has been stratified in this way, we can employ simple random sampling to generate the complete sample. There are several ways to do this. For example, we can draw random samples from each of the four income groups according to their proportions in the population. Thus, if in the population the relative frequencies of the four groups are as listed below, our sample will be stratified in the same proportions. If a total sample of 1,000 is to be drawn, we will randomly select 250 from stratum 1, 400 from stratum 2, 300 from stratum 3, and 50 from stratum 4.

Stratum	Income Categories	Population Proportions
1	under $15,000	25%
2	$15,000–$29,999	40
3	$30,000–$50,000	30
4	over $50,000	5

The problem with this approach, however, is that if we want to make inferences about the last stratum, a sample of 50 may be too small to produce useful information. In such cases, we usually increase the sample size of the smallest stratum to ensure that the sample data provide enough information for our purposes. An adjustment must then be made before we attempt to draw inferences about the entire population. This procedure is beyond the level of this book. We recommend that anyone planning such a survey consult an expert statistician or a reference book on the subject. Better still, become an expert statistician yourself by taking additional statistics courses.

CLUSTER SAMPLING

> **Cluster Sample**
>
> A **cluster sample** is a simple random sample of groups or clusters of elements.

Cluster sampling is particularly useful when it is difficult or costly to develop a complete list of the population members (making it difficult and costly to generate a simple random sample). It is also useful whenever the population elements are widely dispersed geographically. For example, suppose we wanted to estimate the average annual household income in a large city. To use simple random sampling, we would need a complete list of households in the city from which to sample. To use stratified random sampling, we would need the list of households, and we would also need to have each household categorized by some other variable (such as age of household head) in order to develop the strata. A less expensive alternative would be to let each block within the city represent a cluster. A sample of clusters could then be randomly selected, and every household within these clusters could be questioned to determine income. By reducing the distances the surveyor must cover to gather data, cluster sampling reduces the cost.

But cluster sampling also increases sampling error (see Section 7.4), because households belonging to the same cluster are likely to be similar in many respects, including household income. This can be partially offset by using some of the cost savings to choose a larger sample than would be used for a simple random sample.

SAMPLE SIZE

Whichever type of sampling plan you select, you still have to decide what size sample to use. Determining the appropriate sample size will be addressed in detail in Chapters 9 and 11. Until then, we can rely on our intuition, which tells us that the larger the sample size is, the more accurate we can expect the sample estimates to be.

EXERCISES

7.4 A statistics practitioner would like to conduct a survey to ask people their views on a proposed new shopping mall in their community. According to the latest census, there are 500 households in the community. The statistician has numbered each household (from 1 to 500), and she would like to randomly select 25 of these households to participate in the study. Use Excel to generate the sample.

7.5 A safety expert wants to determine the proportion of cars in his state with worn tire treads. The state license plate contains six digits. Use Excel to generate a sample of 20 cars to be examined.

7.6 The operations manager of a large plant with four departments wants to estimate the person-hours lost per month due to accidents. Describe a sampling plan that would be suitable for estimating the plantwide loss and for comparing departments.

7.7 A statistics practitioner wants to estimate the mean age of children in his city. Unfortunately, he does not have a complete list of households. Describe a sampling plan that would be suitable for his purposes.

7.4 ERRORS INVOLVED IN SAMPLING

Two major types of errors can arise when a sample of observations is taken from a population: sampling error and nonsampling error. Anyone reviewing the results of sample surveys and studies, as well as statistics practitioners conducting surveys and applying statistical techniques, should understand the sources of these errors.

SAMPLING ERROR

Sampling error refers to differences between the sample and the population that exist only because of the observations that happened to be selected for the sample. Sampling error is an error that we expect to occur when we make a statement about a population that is based only on the observations contained in a sample taken from the population. To illustrate, consider again the illustration described in Section 7.1 in which we wish to determine the mean annual income of North American blue-collar workers. As we pointed out, we can use statistical inference to estimate the mean income (μ) of the population if we are willing to accept less than 100% accuracy. If we record the incomes of a sample of the workers and find the mean ($\bar{x}$) of this sample of incomes, this sample mean is an estimate of the desired population mean. But the value of $\bar{x}$ will deviate from the population mean (μ) simply by chance, because the value of the sample mean depends on which incomes just happened to be selected for the sample. The difference between the true (unknown) value of the population mean (μ) and its sample estimate $\bar{x}$ is the sampling error. The size of this deviation may be large simply due to bad luck—bad luck that a particularly unrepresentative sample happened to be selected. The only way we can reduce the expected size of this error is to take a larger sample.

Given a fixed sample size, the best we can do is to state the probability that the sampling error is less than a certain amount (as we will discuss in Chapter 9). It is common today for such a statement to accompany the results of an opinion poll. If an opinion poll states that, based on sample results, the incumbent candidate for mayor has the support of 54% of eligible voters in an upcoming election, that statement may be accompanied by the following explanatory note: This percentage is correct to within three percentage points, 19 times out of 20. This statement means that we estimate that the actual level of support for the candidate is between 51% and 57%, and that in the long run this type of procedure is correct 95% of the time.

NONSAMPLING ERROR

Nonsampling error is more serious than sampling error, because taking a larger sample won't diminish the size, or the possibility of occurrence, of this error. Even a census can (and probably will) contain nonsampling errors. **Nonsampling errors** are due to mistakes made in the acquisition of data or due to the sample observations being selected improperly.

1 *Errors in data acquisition.* These types of errors arise from the recording of incorrect responses. This may be the result of incorrect measurements being taken because of faulty equipment, mistakes made during transcription from primary sources, inaccurate recording of data due to misinterpretation of terms, or inaccurate responses to questions concerning sensitive issues such as sexual activity or possible tax evasion.

2 *Nonresponse error.* **Nonresponse error** refers to error (or **bias**) introduced when responses are not obtained from some members of the sample. When this happens, the sample observations that are collected may not be representative of the target population, resulting in biased results (as was discussed in Section 7.2). Nonresponse can occur for a number of reasons. An interviewer may not be able to contact a person listed in the sample, or the sampled person may refuse to respond for some reason. In either case, responses are not obtained from a sampled person, and bias is introduced. The problem of nonresponse is even greater when self-administered questionnaires are used rather than an interviewer, who can attempt to reduce the nonresponse rate by means of callbacks. As noted earlier, the *Literary Digest* fiasco was largely due to a high nonresponse rate, resulting in a biased, self-selected sample.

3 *Selection bias.* **Selection bias** occurs when the sampling plan is such that some members of the target population cannot possibly be selected for inclusion in the sample. Together with nonresponse error, selection bias played a role in the *Literary Digest* poll being so wrong, as voters without telephones or without a subscription to *Literary Digest* were excluded from possible inclusion in the sample taken.

EXERCISES

7.8 a Explain the difference between sampling error and nonsampling error.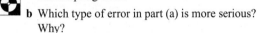
b Which type of error in part (a) is more serious? Why?

7.9 Briefly describe three types of nonsampling errors.

7.10 Is it possible for a sample to yield better results than a census? Explain.

7.5 SUMMARY

Because most populations are very large, it is extremely costly and impractical to investigate each member of the population to determine the values of the parameters. As a practical alternative, we take a sample from the population and use the sample statistics to draw inferences about the parameters. Care must be taken to ensure that the sampled population is the same as the target population.

We can choose from among several different sampling plans, including simple random sampling, stratified random sampling, and cluster sampling. Whatever sampling plan is used, it is important to realize that both sampling error and nonsampling error will occur and to understand what the sources of these errors are.

IMPORTANT TERMS

Estimate
Target population
Sampled population
Self-selected sample
Simple random sample
Stratified random sample

Cluster sample
Sampling error
Nonsampling error
Nonresponse error (bias)
Selection bias

Chapter 8

Sampling Distributions

8.1 Introduction

8.2 Sampling Distribution of the Mean

8.3 Creating the Sampling Distribution by Computer Simulation (Optional)

8.4 Sampling Distribution of a Proportion

8.5 Sampling Distribution of the Difference between Two Means

8.6 From Here to Inference

8.7 Summary

8.1 INTRODUCTION

This chapter introduces the sampling distribution, a fundamental element in statistical inference. We remind you that statistical inference is the process of converting data into information. Here are the parts of the process we have thus far discussed:

1. Parameters describe populations.
2. Parameters are almost always unknown.
3. We take a random sample of a population to obtain the necessary data.
4. We calculate one or more statistics from the data.

For example, to estimate a population mean we compute the sample mean. While there is very little chance that the sample mean and the population mean are identical, we would expect them to be quite close. However, for the purposes of statistical inference, we need to be able to measure how close. The sampling distribution provides this service. It plays a critical role in the process, because the measure of proximity it provides is the key to statistical inference.

8.2 SAMPLING DISTRIBUTION OF THE MEAN

To grasp the idea of a **sampling distribution,** consider the population created by throwing a fair die infinitely many times, with the random variable X indicating the number of spots showing on any one throw. The probability distribution of the random variable X is as follows.

x	1	2	3	4	5	6
$p(x)$	1/6	1/6	1/6	1/6	1/6	1/6

The population is infinitely large, because we can throw the die infinitely many times (or at least imagine doing so). From the definitions of expected value and variance presented in Chapter 5, we calculate the population mean, variance, and standard deviation.

Figure 8.1

Drawing samples of size 2 from a population

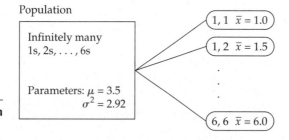

Population mean:
$$\mu = \sum xp(x)$$
$$= 1\left(\frac{1}{6}\right) + 2\left(\frac{1}{6}\right) + 3\left(\frac{1}{6}\right) + 4\left(\frac{1}{6}\right) + 5\left(\frac{1}{6}\right) + 6\left(\frac{1}{6}\right)$$
$$= 3.5$$

Population variance:
$$\sigma^2 = \sum (x - \mu)^2 p(x)$$
$$= (1 - 3.5)^2\left(\frac{1}{6}\right) + (2 - 3.5)^2\left(\frac{1}{6}\right) + (3 - 3.5)^2\left(\frac{1}{6}\right) + (4 - 3.5)^2\left(\frac{1}{6}\right)$$
$$+ (5 - 3.5)^2\left(\frac{1}{6}\right) + (6 - 3.5)^2\left(\frac{1}{6}\right)$$
$$= 2.92$$

Population standard deviation:
$$\sigma = \sqrt{\sigma^2} = \sqrt{2.92} = 1.71$$

Now suppose that μ is unknown and that we want to estimate its value by using the sample mean $\bar{X}$* calculated from a sample of size $n = 2$. In actual practice, only one sample would be drawn, and hence there would be only one value of $\bar{X}$, but, to assess how closely $\bar{X}$ estimates the value of μ, we will develop the sampling distribution of $\bar{X}$ by evaluating every possible sample of size 2.

Consider all the possible different samples of size 2 that could be drawn from the population of die tosses. Figure 8.1 depicts this process. For each sample, we compute the mean. Because the value of the sample mean varies randomly from sample to sample, we can regard $\bar{X}$ as a new random variable created by sampling. Table 8.1 lists all the possible samples and their corresponding values of $\bar{x}$.

Table 8.1 All Samples of Size 2 and Their Means

Sample	$\bar{x}$	Sample	$\bar{x}$	Sample	$\bar{x}$
1, 1	1.0	3, 1	2.0	5, 1	3.0
1, 2	1.5	3, 2	2.5	5, 2	3.5
1, 3	2.0	3, 3	3.0	5, 3	4.0
1, 4	2.5	3, 4	3.5	5, 4	4.5
1, 5	3.0	3, 5	4.0	5, 5	5.0
1, 6	3.5	3, 6	4.5	5, 6	5.5
2, 1	1.5	4, 1	2.5	6, 1	3.5
2, 2	2.0	4, 2	3.0	6, 2	4.0
2, 3	2.5	4, 3	3.5	6, 3	4.5
2, 4	3.0	4, 4	4.0	5, 4	5.0
2, 5	3.5	4, 5	4.5	5, 5	5.5
2, 6	4.0	4, 6	5.0	6, 6	6.0

*Recall our convention introduced in Chapter 5. Uppercase letters represent the random variable, and their lowercase counterparts represent their values. In this chapter, we introduce the sampling distribution of the sample mean. When we refer to the random variable we will use $\bar{X}$; the values of this random variable are represented by $\bar{x}$.

There are 36 different possible samples of size 2; because each sample is equally likely, the probability of any one sample being selected is 1/36. However, $\overline{X}$ can assume only 11 different possible values: 1.0, 1.5, 2.0, ..., 6.0, with certain values of $\overline{X}$ occurring more frequently than others. The value $\overline{x} = 1.0$ occurs only once, so its probability is 1/36. The value $\overline{x} = 1.5$ can occur in two ways; hence, $p(1.5) = 2/36$. The probabilities of the other values of $\overline{x}$ are determined in similar fashion, and the resulting sampling distribution of $\overline{X}$ is shown in Table 8.2.

The most interesting aspect of the sampling distribution of $\overline{X}$ is how different it is from the distribution of X, as can be seen in Figure 8.2.

We can also compute the mean, variance, and standard deviation of the sampling distribution. Once again using the definitions of expected value and variance, we determine the following parameters of the sampling distribution.

Mean of the sampling distribution of $\overline{X}$:

$$\mu_{\overline{x}} = \sum \overline{x} p(\overline{x})$$

$$= 1.0\left(\frac{1}{36}\right) + 1.5\left(\frac{2}{36}\right) + \cdots + 6.0\left(\frac{1}{36}\right)$$

$$= 3.5$$

Variance of the sampling distribution of $\overline{X}$:

$$\sigma_{\overline{x}}^2 = \sum (\overline{x} - \mu_{\overline{x}})^2 p(\overline{x})$$

$$= (1.0 - 3.5)^2 \left(\frac{1}{36}\right) + (1.5 - 3.5)^2 \left(\frac{2}{36}\right) + \cdots + (6.0 - 3.5)^2 \left(\frac{1}{36}\right)$$

$$= 1.46$$

Standard deviation of the sampling distribution of $\overline{X}$:

$$\sigma_{\overline{x}} = \sqrt{\sigma_{\overline{x}}^2} = \sqrt{1.46} = 1.21$$

It is important to recognize that the distribution of $\overline{X}$ is different from the distribution of X. Figure 8.2 shows that the shapes of the two distributions differ. From our previous calculations, we know that the mean of the sampling distribution of $\overline{X}$ is equal to the mean of the distribution of X; that is, $\mu_{\overline{x}} = \mu$. However, the variance

Table 8.2 Sampling Distribution of $\overline{X}$

$\overline{x}$	$p(\overline{x})$
1.0	1/36
1.5	2/36
2.0	3/36
2.5	4/36
3.0	5/36
3.5	6/36
4.0	5/36
4.5	4/36
5.0	3/36
5.5	2/36
6.0	1/36

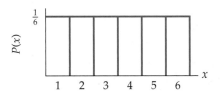
a Distribution of X

b Sampling Distribution of $\overline{X}$

Figure 8.2

Distributions of X and $\overline{X}$

of $\overline{X}$ is not equal to the variance of X; we calculated $\sigma^2 = 2.92$, while $\sigma_{\overline{x}}^2 = 1.46$. It is no coincidence that the variance of $\overline{X}$ is exactly half the variance of X, as we will see shortly.

Don't get lost in the terminology and notation. Remember that μ and σ^2 are the parameters of the population of X. To create the sampling distribution of $\overline{X}$, we repeatedly drew samples of size 2 from the population and calculated $\bar{x}$ for each sample. Thus, we treat $\overline{X}$ as a brand-new random variable, with its own distribution, mean, and variance. The mean is denoted $\mu_{\overline{x}}$, and the variance is denoted $\sigma_{\overline{x}}^2$.

If we now repeat the sampling process with the same population but with other values of n, we produce somewhat different sampling distributions of $\overline{X}$. Figure 8.3 shows the sampling distributions of $\overline{X}$ when $n = 5$, 10, and 25. Observe that in each case $\mu_{\overline{x}} = \mu$, which is equal to 3.5.

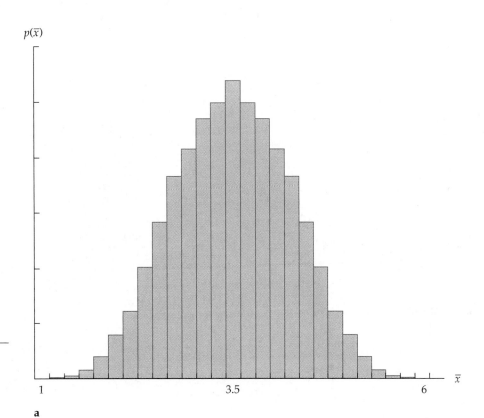

Figure 8.3

(a) Sampling distribution of $\overline{X}$: $n = 5$; (b) sampling distribution of $\overline{X}$: $n = 10$; (c) sampling distribution of $\overline{X}$: $n = 25$

a

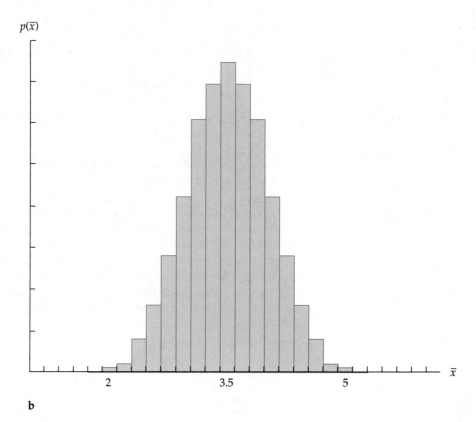

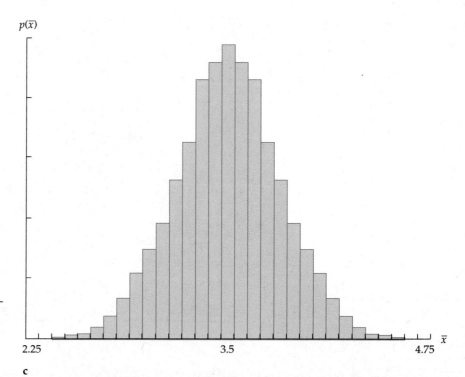

Figure 8.3 (continued)
(a) Sampling distribution of $\bar{X}$: $n = 5$; (b) sampling distribution of $\bar{X}$: $n = 10$; (c) sampling distribution of $\bar{X}$: $n = 25$

Notice that in each case the variance of the sampling distribution is less than that of the parent population; that is, $\sigma_{\bar{x}}^2 < \sigma^2$. This means that a randomly selected value of $\bar{X}$ (the mean of the number of spots observed in, say, five throws of the die) is likely to be closer to the mean value of 3.5 than is a randomly selected value of X (the number of spots observed in one throw). Indeed, this is what you would expect, because in five throws of the die you are likely to get some fives and sixes and some ones and twos, which will tend to offset one another in the averaging process and produce a sample mean reasonably close to 3.5. As the number of throws of the die increases, the probability that the sample mean will be close to 3.5 also increases. Thus, we observe in Figure 8.3 that the sampling distribution of $\bar{X}$ becomes narrower (or more concentrated about the mean) as n increases.

Another thing that happens as n gets larger is that the sampling distribution of $\bar{X}$ becomes increasingly bell shaped. This phenomenon is summarized in the **central limit theorem.**

> **Central Limit Theorem**
>
> The sampling distribution of the mean of a random sample drawn from any population is approximately normal for a sufficiently large sample size. The larger the sample size, the more closely the sampling distribution of $\bar{X}$ will resemble a normal distribution.

The accuracy of the approximation alluded to in the central limit theorem depends on the probability distribution of the parent population and on the sample size. If the population is normal, then $\bar{X}$ is normally distributed for all values of n. If the population is nonnormal, then $\bar{X}$ is approximately normal only for larger values of n. In many practical situations, a sample size of 30 may be sufficiently large to allow us to use the normal distribution as an approximation for the sampling distribution of $\bar{X}$. However, if the population is extremely nonnormal (examples of extremely nonnormal populations include bimodal and highly skewed distributions), the sampling distribution will also be nonnormal even for moderately large values of n.

In our calculation of the mean and variance of X and of $\bar{X}$ when $n = 2$, we found that $\mu_{\bar{x}} = \mu$ and that $\sigma_{\bar{x}}^2 = \sigma^2/2$. Statisticians have shown that, for all sample sizes, the mean of $\bar{X}$ is equal to the mean of X and that the variance of $\bar{X}$ is equal to the variance of X divided by the sample size. That is, $\sigma_{\bar{x}}^2 = \sigma^2/n$.* We can now summarize what we know about the **sampling distribution of the sample mean.**

---

*The variance of $\bar{X}$ is σ^2/n if the population from which we're sampling is infinitely large. If the population is finite, the variance of $\bar{X}$ is

$$\sigma_{\bar{x}}^2 = \left(\frac{\sigma^2}{n}\right)\left(\frac{N-n}{N-1}\right)$$

where N is the population size and $(N - n)/(N - 1)$ is the finite population correction factor. In most practical situations (including all examples and exercises in this book), the target population is finite but very large relative to the sample size (e.g., the population of television viewers in North America). In such cases, the finite population correction factor is so close to 1 that we can ignore it. As a general rule, include the finite population correction factor only if the sample size is greater than 1% of the population size.

> **Sampling Distribution of the Sample Mean**
>
> 1. $\mu_{\bar{x}} = \mu$
> 2. $\sigma_{\bar{x}}^2 = \sigma^2/n$, and $\sigma_{\bar{x}} = \sigma/\sqrt{n}$ (The standard deviation of $\bar{X}$ is called the **standard error of the mean**.)
> 3. If X is normal, $\bar{X}$ is normal. If X is nonnormal, $\bar{X}$ is approximately normal for sufficiently large sample sizes. The definition of "sufficiently large" depends on the extent of nonnormality of X.

CREATING THE SAMPLING DISTRIBUTION EMPIRICALLY

In the analysis above, we created the sampling distribution of the mean theoretically. We did so by listing all of the possible samples of size 2 and their probabilities. (They were all equally likely with probability 1/36.) From this distribution, we produced the sampling distribution. We could also create the distribution empirically by actually tossing two fair dice repeatedly, calculating the sample mean for each sample, counting the number or times each value of $\bar{X}$ occurs, and computing the relative frequencies to estimate the theoretical probabilities. If we toss the two dice a large enough number of times, the relative frequencies and theoretical probabilities (computed above) will be similar. Try it yourself. Toss two dice 500 times, count the number of times each sample mean occurs, and construct the sampling distribution. Obviously this approach is far from ideal because of the excessive amount of time required to toss the dice enough times to make the relative frequencies good approximations for the theoretical probabilities. However, we can use the computer to quickly "simulate" tossing dice many times.

In Section 8.3 we introduce simulation experiments, which will enable you to create sampling distributions. We show how to use Excel to generate a large number of samples to construct sampling distributions empirically. The experiments will deal with the effect of the population distribution and the sample size on the sampling distribution of the mean. Other simulation experiments will let you discover for yourself some of the key statistical concepts that we discuss in this textbook.

▼ EXAMPLE 8.1

The foreman of a bottling plant has observed that the amount of soda pop in each "32-ounce" bottle is actually a normally distributed random variable, with a mean of 32.2 ounces and a standard deviation of 0.3 ounces.

1. If a customer buys one bottle, what is the probability that the bottle will contain more than 32 ounces?

2. If a customer buys a carton of four bottles, what is the probability that the mean of the four will be greater than 32 ounces?

Solution

1 Because the random variable is the amount of soda in one bottle, we want to find $P(X > 32)$, where X is normally distributed, $\mu = 32.2$, and $\sigma = 0.3$. Hence,

$$P(X > 32) = P\left(\frac{X - \mu}{\sigma} > \frac{32 - 32.2}{.3}\right)$$

$$= P(Z > -.67)$$

$$= .5 + .2486 = .7486$$

2 Now we want to find the probability that the mean of four filled bottles exceeds 32 ounces. That is, we want $P(\bar{X} > 32)$. From our previous analysis and from the central limit theorem, we know the following.

1 $\bar{X}$ is normally distributed.
2 $\mu_{\bar{x}} = \mu = 32.2$
3 $\sigma_{\bar{x}} = \sigma/\sqrt{n} = .3/\sqrt{4} = .15$

Hence,

$$P(\bar{X} > 32) = P\left(\frac{\bar{X} - \mu_{\bar{x}}}{\sigma_{\bar{x}}} > \frac{32 - 32.2}{.15}\right)$$

$$= P(Z > -1.33)$$

$$= .5 + .4082 = .9082$$

Figure 8.4 illustrates the distributions used in this example.

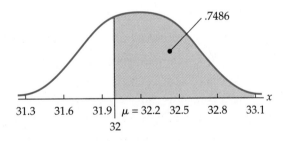

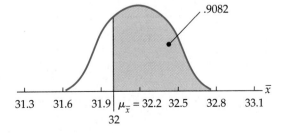

Figure 8.4

Distribution of X and sampling distribution of $\bar{X}$ in Example 8.1

In Example 8.1(2), we began with the assumption that both μ and σ were known. Then, using the sampling distribution, we made a probability statement about $\bar{X}$. Unfortunately, the values of μ and σ are not usually known, so an analysis such as that in Example 8.1 cannot usually be conducted. However, we can use the sampling distribution to infer something about an unknown value of μ on the basis of a sample mean.

▼ EXAMPLE 8.2

The dean of a business school claims that the average weekly income of his school's BBA graduates one year after graduation is $600.

1. If the dean's claim is correct, and if the distribution of weekly incomes has a standard deviation of $100, what is the probability that 25 randomly selected graduates have an average weekly income of less than $550?
2. If a random sample of 25 graduates had an average weekly income of $550, what would you conclude about the validity of the dean's claim?

Solution

1. We want to find $P(\bar{X} < 550)$. The distribution of X, the weekly income, is likely to be positively skewed, but not sufficiently so to make the distribution of $\bar{X}$ nonnormal. As a result, we may assume that $\bar{X}$ is normal with mean $\mu_{\bar{x}} = \mu = 600$ and standard deviation $\sigma_{\bar{x}} = \sigma/\sqrt{n} = 100/\sqrt{25} = 20$. Thus,

$$P(\bar{X} < 550) = P\left(\frac{\bar{X} - \mu_{\bar{x}}}{\sigma_{\bar{x}}} < \frac{550 - 600}{20}\right)$$

$$= P(Z < -2.5)$$

$$= .0062$$

Figure 8.5 describes this calcuation.

2. The probability of observing a sample mean as low as $550 when the population mean is $600 is extremely small. Because this event is quite unlikely, we would have to conclude that the dean's claim is not justified.

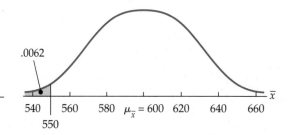

Figure 8.5

Sampling distribution of $\bar{X}$ for Example 8.2

USING THE SAMPLING DISTRIBUTION FOR INFERENCE

Our conclusion in part 2 of Example 8.2 illustrates how the sampling distribution can be used to make inferences about population parameters. The first form of inference is estimation, which we introduce in the next chapter. In preparation for this momentous occasion, we'll present another way of expressing the probability associated with the sampling distribution.

Recall the notation introduced in Chapter 6 (see page 160). We defined z_A to be the value of z such that the area to the right of z_A under the standard normal curve is equal to A. We also showed that $z_{.025} = 1.96$. Because the standard normal distribution is symmetric about 0, the area to the left of -1.96 is also .025. The area between -1.96 and 1.96 is .95. Figure 8.6 depicts this notation. We can express the notation algebraically as

$$P(-1.96 < Z < 1.96) = .95$$

In this section we established that

$$Z = \frac{\bar{X} - \mu}{\sigma/\sqrt{n}}$$

is standard normally distributed. Substituting this form of Z into the probability statement above, we produce

$$P\left(-1.96 < \frac{\bar{X} - \mu}{\sigma/\sqrt{n}} < 1.96\right) = .95$$

With a little algebraic manipulation (multiply all three terms by $\sigma/\sqrt{n}$ and add μ to all three terms), we determine

$$P\left(\mu - 1.96\frac{\sigma}{\sqrt{n}} < \bar{X} < \mu + 1.96\frac{\sigma}{\sqrt{n}}\right) = .95$$

Returning to Example 8.2, where $\mu = 600$, $\sigma = 100$, and $n = 25$, we compute

$$P\left(600 - 1.96\frac{100}{\sqrt{25}} < \bar{X} < 600 + 1.96\frac{100}{\sqrt{25}}\right) = .95$$

Thus, we can say that

$$P(560.8 < \bar{X} < 639.2) = .95$$

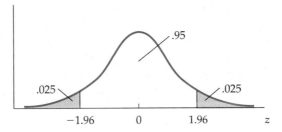

Figure 8.6

$P(-1.96 < Z < 1.96) = .95$

This tells us that there is a 95% probability that a sample mean will fall between 560.8 and 639.2. Because the sample mean was computed to be $550, we would have to conclude that the dean's claim is not supported by the statistic.

Changing the probability from .95 to .90 changes the probability statement to

$$P\left(\mu - 1.645\frac{\sigma}{\sqrt{n}} < \bar{X} < \mu + 1.645\frac{\sigma}{\sqrt{n}}\right) = .90$$

We can also produce a general form of this statement.

$$P\left(\mu - z_{\alpha/2}\frac{\sigma}{\sqrt{n}} < \bar{X} < \mu + z_{\alpha/2}\frac{\sigma}{\sqrt{n}}\right) = 1 - \alpha$$

In this formula α (Greek letter *alpha*) is the probability that $\bar{X}$ does not fall into the interval. To apply this formula, all we need do is substitute the values for μ, σ, n, and α. For example with $\mu = 600$, $\sigma = 100$, $n = 25$, and $\alpha = .01$, we produce

$$P\left(\mu - z_{.005}\frac{\sigma}{\sqrt{n}} < \bar{X} < \mu + z_{.005}\frac{\sigma}{\sqrt{n}}\right) = 1 - .01$$

$$P\left(600 - 2.575\frac{100}{\sqrt{25}} < \bar{X} < 600 + 2.575\frac{100}{\sqrt{25}}\right) = .99$$

$$P(548.5 < \bar{X} < 651.5) = .99$$

which is another probability statement about $\bar{X}$. In Section 9.3 we will use a similar type of probability statement to derive the first statistical inference technique.

EXERCISES

8.1 A normally distributed population has a mean of 40 and a standard deviation of 12. What does the central limit theorem say about the sampling distribution of the mean if samples of size 100 are drawn from this population?

8.2 Refer to Exercise 8.1. Suppose that the population is not normally distributed. Does this change your answer? Explain.

8.3 A sample of $n = 16$ observations is drawn from a normal population with $\mu = 1,000$ and $\sigma = 200$. Find the following.
  a $P(\bar{X} > 1,050)$
  b $P(\bar{X} < 960)$
  c $P(\bar{X} > 1,100)$

8.4 Repeat Exercise 8.3 with $n = 25$.

8.5 Repeat Exercise 8.3 with $n = 100$.

8.6 Given a normal population whose mean is 50 and whose standard deviation is 5:
  a Find the probability that a random sample of 4 has a mean between 49 and 52.
  b Find the probability that a random sample of 16 has a mean between 49 and 52.
  c Find the probability that a random sample of 25 has a mean between 49 and 52.

8.7 Repeat Exercise 8.6 for a standard deviation of 10.

8.8 Repeat Exercise 8.6 for a standard deviation of 20.

8.9 The heights of North American women are normally distributed with a mean of 64 inches and a standard deviation of 2 inches.

  a What is the probability that a randomly selected woman is taller than 66 inches?
  b A random sample of four women is selected. What is the probability that the sample mean is greater than 66 inches?
  c What is the probability that the mean height of a random sample of 100 is greater than 66 inches?

8.10 Refer to Exercise 8.9. If the population of women's heights is not normally distributed, which, if any, of the questions can you answer? Explain.

8.11 An automatic machine in a manufacturing process is operating properly if the lengths of an important subcomponent are normally distributed with mean = 117 cm and standard deviation = 5.2 cm.

  a Find the probability that one selected subcomponent is longer than 120 cm.
  b Find the probability that if four subcomponents are randomly selected, their mean length exceeds 120 cm.
  c Find the probability that if four subcomponents are randomly selected, all four have lengths that exceed 120 cm.

8.12 The amount of time the university professors devote to their jobs per week is normally distributed with a mean of 52 hours and a standard deviation of 6 hours.

  a What is the probability that a professor works for more than 60 hours per week?
  b. Find the probability that the mean amount of work per week for three randomly selected professors is more than 60 hours.
  c Find the probability that, if three professors are randomly selected, all three work for more than 60 hours per week.

8.13 The number of pizzas consumed per month by university students is normally distributed with a mean of 10 and a standard deviation of 3.

  a What proportion of students consume more than 12 pizzas per month?
  b What is the probability that in a random sample of 25 students more than 275 pizzas are consumed? (*Hint:* What is the mean number of pizzas consumed by the sample of 25 students?)

8.14 The marks on a statistics midterm test are normally distributed with a mean of 78 and a standard deviation of 6.

  a What proportion of the class has a midterm mark of less than 75?
  b What is the probability that a class of 50 has an average midterm mark that is less than 75?

8.15 The amount of time spent by North American adults watching television per day is normally distributed with a mean of 6 hours and a standard deviation of 1.5 hours.

  a What is the probability that a randomly selected North American adult watches television for more than 7 hours per day?
  b What is the probability that the average number of hours watching by a random sample of five North American adults is more than 7 hours?
  c What is the probability that in a random sample of five North American adults all watch television for more than 7 hours per day?

8.16 The manufacturer of cans of salmon that are supposed to have a net weight of 6 ounces tells you that the net weight is actually a normal random variable with a mean of 6.05 ounces and a standard deviation of .18 ounces. Suppose that you draw a random sample of 36 cans.

  a Find the probability that the mean weight of the sample is less than 5.97 ounces.
  b Suppose your random sample of 36 cans of salmon produced a mean weight that is less than 5.97 ounces. Comment on the statement made by the manufacturer.

8.17 The number of customers who enter a supermarket each hour is normally distributed with a mean of 600 and a standard deviation of 200. The supermarket is open 16 hours per day. What is the probability that the total number of customers who enter the supermarket in one day is greater than 10,000? (*Hint:* Calculate the average hourly number of customers necessary to exceed 10,000 in one 16-hour day.)

8.18 The sign on the elevator in the Peters Building, which houses the School of Business and Economics at Wilfrid Laurier University, states, "Maximum Capacity 1,140 kilograms (2,500 pounds) or 16 Persons." A professor of statistics wonders what the probability is that 16 persons would weigh more than 1,140 kilograms. Discuss what the professor needs (besides the ability to perform the calculations) in order to satisfy his curiosity.

8.19 Refer to Exercise 8.18. Suppose that the professor discovers that people who use the elevator weigh on average 75 kilograms with a standard deviation of 10 kilograms and that their weights are normally distributed. Calculate the probability that the professor seeks.

8.3 CREATING THE SAMPLING DISTRIBUTION BY COMPUTER SIMULATION (OPTIONAL)

In Section 8.2 we created the sampling distribution of the mean theoretically by listing *all* of the possible samples of size 2 and their probabilities. In each sample, we computed the sample mean, and, collecting like values and their probabilities, we constructed the sampling distribution. Statisticians can create sampling distributions by using calculus, a method we will do you the favor of not presenting. However, there is another way of at least approximating the sampling distribution that may appeal to students of applied statistics.

We can create the sampling distribution of the mean of two dice by tossing two balanced dice repeatedly. For each toss we would record the mean and determine the frequency distribution of these means. You should realize, however, that the frequency distribution so produced would only approximate the theoretical distribution. That's because the theoretical distribution is based on throwing the dice an infinite number of times. (Recall that probability is defined as relative frequency over an infinite number of experiments.) If we toss the dice, say, 100 times, the approximation is likely to be relatively poor. If we toss them 1,000 times, the frequency distribution will be closer to the theoretical, yet not a perfect match. Obviously, we can never throw the dice enough times to achieve perfection. Moreover, even 1,000 tosses would be very time consuming, not to mention mind numbing. Fortunately, there is a way to generate the empirical frequency distribution more quickly, by computer simulation.

Excel possesses a random number generator that can produce data from a variety of distributions. In Chapter 7, we discussed how to generate random numbers from a discrete uniform distribution to help select a random sample. We will use this feature to simulate repeated sampling to illustrate how sampling distributions are created.

SAMPLING FROM A NONNORMAL DISTRIBUTION

We'll begin with the dice-tossing experiment described in Section 8.2. Below we describe the commands that will create two columns of 1,000 numbers where the numbers are drawn from a discrete uniform distribution whose integers fall between 1 and 6 (just like the toss of a balanced die). Excel will be instructed to treat each *row* of the range as a sample of size 2. We then compute the row means and store them in a third column. We can examine the resulting sampling distribution by drawing the histogram and calculating the mean and standard deviation.

▼ **SIMULATION EXPERIMENT 8.1**

Excel Instructions

1 In column A store the numbers 1, 2, 3, 4, 5, 6
2 In cell B1 type

$=1/6$

and drag to fill cells B2 through B6. (Do not type .1667 or any other versions of 1/6, because the sum of the probabilities will not equal 1, causing Excel to issue an error warning.)

3 Click **Tools, Data Analysis...**, and **Random Number Generation.**
4 Type 2 to specify the **Number of Variables** and 1,000 to specify the **Number of Random Variables.** This will create an array of two columns and 1,000 rows.
5 Click **Discrete** distribution. In the parameters box type A1:B6 to specify the **Value and Probability Input Range.**
6 Specify **New Worksheet Ply** (the default). Columns A and B of a new worksheet will fill with the random numbers.
7 Move to Cell C1 and type

=AVERAGE(A1:B1)

8 Drag to fill the rest of column C. Column C will now contain the values of the sample means.
9 Draw the histogram using bins 1.0, 1.5, 2.0, ..., 6.0.
10 Calculate the mean and standard deviation of the sample means in column C. In Cells D1 and D2, respectively, type

=AVERAGE(C1:C1000)

=STDEV(C1:C1000)

We suggest that you write a brief report of the results of the simulation experiment.

Report for Simulation Experiment 8.1

1 Compare the histogram that was drawn to the theoretical sampling distribution. Are they similar?
2 What did you anticipate seeing when the computer printed the mean and standard deviation of the sampling distribution of the mean? What did the computer print?
3 Discuss the significance of your experiment.

▲

INCREASING THE SAMPLE SIZE

To examine the effect of increasing the sample size, repeat the experiment with $n = 5$, 10, and 25.

SIMULATION EXPERIMENT 8.2

Excel Instructions In step 4, type 5 (for $n = 5$), 10, or 25. Change the instructions in steps 7, 8, and 10 to store the sample (row) means in the next available column.

Write a report for each sample size providing the information you gave in Simulation Experiment 8.1.

SAMPLING FROM A NORMAL POPULATION

The central limit theorem states that the sampling distribution of the mean is normal when the population is normal. The following experiment is designed to confirm this part of the theorem.

SIMULATION EXPERIMENT 8.3

In this experiment, we sample from a normal population whose mean is 100 and whose standard deviation is 25. We use a sample size of 4.

Excel Instructions

1. Click **Tools, Data Analysis...**, and **Random Number Generation.**
2. Type 4 to specify the **Number of Variables** and 1000 to specify the **Number of Random Variables.**
3. Click **Normal** distribution. Specify **Mean** 100 and **Standard Deviation** 25.
4. Specify **New Worksheet Ply** (the default).
5. Move to Cell E1 and type

    =AVERAGE(A1:D1)

6. Drag to fill the rest of column E.
7. Draw the histogram, and calculate the mean and standard deviation of the sample means in column E.

Report for Simulation Experiment 8.3

1. Does it appear that the sampling distribution is normal?
2. What did you anticipate seeing when the computer printed the mean and standard deviation of the sampling distribution of the mean? What did the computer print?
3. Discuss the significance of your experiment.

▼ SIMULATION EXPERIMENT 8.4

Repeat Simulation Experiment 8.3, using a sample size of 9. Write a report describing your results.

▼ SIMULATION EXPERIMENT 8.5

Repeat Simulation Experiment 8.3, using a sample size of 25. Write a report describing your results.

8.4 SAMPLING DISTRIBUTION OF A PROPORTION

When the data that we're dealing with in a particular problem are nominal, the parameter of interest is the proportion of times a particular outcome occurs. Using the terminology introduced in Chapter 5, we call these outcomes successes. The estimator of a population proportion of successes is the sample proportion. That is, we count the number of successes in a sample and compute

$$\hat{P} = \frac{X}{n}$$

($\hat{P}$ is read as *p-hat*) where X is the number of successes and n is the sample size. Recall that X is binomially distributed and thus the probability of any value of $\hat{P}$ can be calculated from its value of X. For example, suppose that we have a binomial experiment with $n = 20$ and $p = .5$. To find the probability that the sample proportion $\hat{P}$ is less than or equal to .60 we find the probability that X is less than or equal to 12 (because 12/20 = .60). From Table 1 in Appendix B we find with $n = 20$ and $p = .5$:

$$P(\hat{P} \leq .60) = P(X \leq 12) = .868$$

We calculate the probability of other values of $\hat{P}$ similarly.

NORMAL APPROXIMATION TO THE BINOMIAL

Discrete distributions like the binomial do not lend themselves easily to the kinds of calculation needed for inference. And inference is the reason we need sampling distributions. Fortunately, we can approximate the binomial distribution using the normal distribution. The **normal approximation to the binomial distribution** works best when the number of trials is large and when the binomial distribution is symmetrical (like the normal). The binomial distribution is symmetrical when the probability of a success equals 50% ($p = .5$). The farther p is from .5, the larger n must be in order for a good approximation to result. The value of n must be sufficiently large so that np and $n(1 - p)$ are both greater than 5.

To see how the approximation works, consider a binomial random variable with $n = 20$ and $p = .5$, the graph of which is shown in Figure 8.7. We will approximate the binomial probabilities using a normal distribution whose mean and standard deviation are equal to the mean and standard deviation of the binomial. Recall that the mean and standard deviation of a binomial random variable are np, and $\sqrt{np(1 - p)}$ (see page 134.). For $n = 20$ and $p = .5$, we calculate

$$\mu = np = 20(.5) = 10$$

$$\sigma = \sqrt{np(1 - p)} = \sqrt{20(.5)(1 - .5)} = \sqrt{5} = 2.24$$

Suppose we wish to determine the probability that $X = 10$. We can compute the exact binomial distribution using the binomial formula or Table 1. Both yield

$$P(X = 10) = .176$$

To use the normal approximation we draw (or imagine drawing) the binomial distribution, and we fit a normal curve over it (see Figure 8.8). The area in the rectangle whose base is the interval 9.5 to 10.5 is the exact binomial probability. Notice that to draw a binomial distribution, which is discrete, it was necessary to draw rectangles whose bases were constructed by adding and subtracting 0.5 to the values of X. The 0.5 is called the **continuity correction factor**.

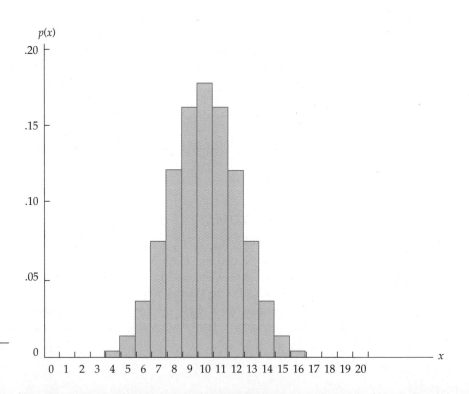

Figure 8.7

Binomial distribution with $n = 20$ and $p = .5$

To calculate the probability that $X = 10$ using the normal distribution requires that we find the area under the normal curve between 9.5 and 10.5. That is,

$$P(X = 10) \approx P(9.5 < Y < 10.5)$$

where Y is a normal random variable approximating the binomial random variable X. We standardize Y (by subtracting its mean and dividing by its standard deviation) and employ Table 3.

$$P(9.5 < Y < 10.5) = P\left(\frac{9.5 - 10}{2.24} < \frac{Y - \mu}{\sigma} < \frac{10.5 - 10}{2.24}\right)$$

$$= P(-.22 < Z < .22) = 2(.0871) = .1742$$

As you can see, the actual value of $P(X = 10)$ and the normal approximation are similar.

The approximation for any other value of X would proceed in the same manner. In general, the binomial probability $P(X = x)$ is approximated by the area under a normal curve between $x - .5$ and $x + .5$. To find the binomial probability $P(X \leq x)$, we calculate the area under the normal curve to the left of $x + .5$. For the same binomial random variable, the probability that its value is less than or equal to 8 is $P(X \leq 8) = .252$. The normal approximation is

$$P(X \leq 8) \approx P(Y < 8.5) = P\left(\frac{Y - \mu}{\sigma} < \frac{8.5 - 10}{2.24}\right) = P(Z < -.67) = .2514$$

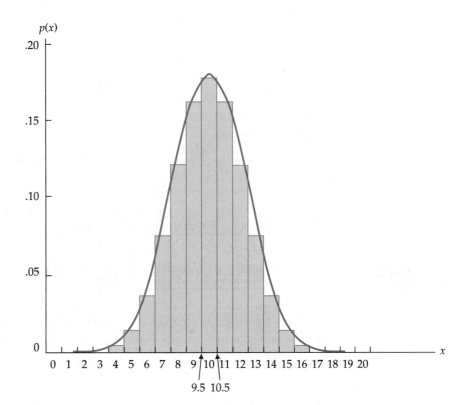

Figure 8.8

Binomial distribution with $n = 20$ and $p = .5$ and normal approximation

We find the area under the normal curve to the right of $x - .5$ to determine the binomial probability $P(X \geq x)$. To illustrate, the probability that the binomial random variable (with $n = 20$ and $p = .5$) is greater than or equal to 14 is $P(X \geq 14) = .058$. The normal approximation is

$$P(X \geq 14) \approx P(Y > 13.5) = P\left(\frac{Y - \mu}{\sigma} > \frac{13.5 - 10}{2.24}\right) = P(Z > 1.56) = .0594$$

For large values of n, the effect of the continuity correction factor is negligible (because each rectangle representing the binomial probability is quite small) and effectively can be omitted. When we use the normal approximation of the binomial in inferential statistics, the values of n will be large enough to ignore the correction factor.

APPROXIMATE SAMPLING DISTRIBUTION OF A SAMPLE PROPORTION

We have established that, for sufficiently large n, a binomial distribution can be approximated by a normal distribution. Thus, the number of successes in n identical independent trials X is approximately normally distributed with mean np and standard deviation $\sqrt{np(1-p)}$. We now turn our attention to the approximate **sampling distribution of a sample proportion** of successes $\hat{P}$.

Using the laws of expected value and variance (for a review, turn to page 126), we can determine the mean, variance, and standard deviation of $\hat{P}$. (The standard deviation of $\hat{P}$ is called the **standard error of the proportion**.) That is,

$$E(\hat{P}) = E\left(\frac{X}{n}\right) = \frac{E(X)}{n} = \frac{np}{n} = p$$

$$V(\hat{P}) = \sigma_{\hat{p}}^2 = V\left(\frac{X}{n}\right) = \frac{V(X)}{n^2} = \frac{np(1-p)}{n^2} = \frac{p(1-p)}{n}$$

$$\sigma_{\hat{p}} = \sqrt{p(1-p)/n}$$

If np and $n(1 - p)$ are both greater than 5, the variable

$$Z = \frac{\hat{P} - p}{\sqrt{p(1-p)/n}}$$

is approximately standard normally distributed.

▼ EXAMPLE 8.3

In the last election, a state representative received 52% of the votes cast. One year after the election, the representative organized a survey that asked a random sample of 300 people whether they would vote for him in the next election. Assuming that his popularity has not changed, what is the probability that more than half of the sample would vote for him?

Solution The number of respondents who would vote for the representative is a binomial random variable with $n = 300$ and $p = .52$. We want to determine the probability that the sample proportion is greater than 50%. That is, we want to find $P(\hat{P} > .50)$.

We now know that the sample proportion $\hat{P}$ is approximately normally distributed with mean $p = .52$ and standard deviation $= \sqrt{p(1-p)/n} = \sqrt{(.52)(.48)/300} = .0288$.

Thus, we calculate

$$P(\hat{P} > .50) = P\left(\frac{\hat{P} - p}{\sqrt{p(1-p)/n}} > \frac{.50 - .52}{.0288}\right) = P(Z > -.69)$$

$$= .5 + .2549 = .7549$$

Assuming that the level of support remains at 52%, the probability that more than half the sample of 300 people would vote for the representative is 75.49%.

EXERCISES

8.20 Given a binomial random variable with $n = 25$ and $p = .3$, find the (exact) probabilities of the following events and their normal approximations.
  a $X = 9$
  b $X \geq 5$
  c $X \leq 11$

8.21 A binomial experiment was performed with 15 identical trials and the probability of success equal to .5. Find the following binomial probabilities and their normal approximations.
  a $P(X = 10)$
  b $P(X \leq 6)$
  c $P(X \geq 11)$

8.22 Use Excel to find the (exact) probabilities and their normal approximations of the following binomial random variables.
  a $X \leq 225, n = 1200, p = .18$
  b $X \leq 130, n = 600, p = .20$
  c $X \geq 150, n = 400, p = .4$
  d $X \geq 220, n = 700, p = .3$
  e $X = 240, n = 500, p = .5$

Use the normal approximation (without the correction factor) to find the probabilities in the following exercises.

8.23 The probability of success on any trial of a binomial experiment is 25%. Find the probability that the proportion of successes in a sample of 500 is less than 22%.

8.24 Repeat Exercise 8.23, given $n = 800$.

8.25 Repeat Exercise 8.23, given $n = 1000$.

8.26 The proportion of eligible voters in the next election who will vote for the incumbent is assumed to be 55%. What is the probability that in a random sample of 500 voters less than 49% say they will vote for the incumbent?

8.27 The assembly line that produces an electronic component of a missile system has historically resulted in a 2% defective rate. A random sample of 800 components is drawn. What is the probability that the defective rate is greater than 4%? Suppose that in the random sample the defective rate is 4%. What does that suggest about the assembly line defective rate?

8.28 The manufacturer of aspirin claims that the proportion of headache sufferers who get relief with just two aspirins is 53%. What is the probability that, in a random sample of 400 headache sufferers, less than 50% obtain relief? If 50% of the sample actually obtained relief, what does this suggest about the manufacturer's claim?

8.29 Repeat Exercise 8.28, using a sample of 1,000.

8.30 A commercial for a household appliances manufacturer claims that less than 5% of all its products require a service call in the first year. A consumer protection association wants to check the claim by surveying 400 households that recently purchased one of the company's appliances. What is the probability that more than 10% require a service call within the first year? What would you say about the commercial's honesty if, in a random sample of 400 households, 10% report at least one service call?

8.31 The Laurier Company's brand has a market share of 30%. Suppose that, in a survey, 1,000 consumers of the product are asked which brand they prefer. What is the probability that more than 32% of the respondents say they prefer Laurier brand?

8.32 A university bookstore claims that 50% of its customers are satisfied with the service and prices.

  a If this true, what is the probability that, in a random sample of 100 customers, less than 45% are satisfied?

  b Suppose that, in a random sample of 100 customers, 45 express satisfaction with the bookstore. What does this tell you about the bookstore's claim?

8.33 A psychologist believes that 80% of male drivers, when lost, continue to drive hoping to find the location they seek rather than ask directions. To examine this belief he took a random sample of male drivers and asked each what he did when lost. If the belief is true, determine the probability that less than 60% said they continue driving.

8.34 The Red Lobster restaurant chain regularly surveys its customers. On the basis of these surveys, the management of the chain claims that 75% of its customers rate the food as excellent. A consumer testing service wants to examine the claim by asking 50 customers to rate the food. What is the probability that less that 60% rate the food as excellent?

8.5 SAMPLING DISTRIBUTION OF THE DIFFERENCE BETWEEN TWO MEANS

Another sampling distribution that you will soon encounter is that of the **difference between two sample means.** The sampling plan calls for independent random samples drawn from each of two normal populations. The samples are said to be independent if the selection of the members of one sample is independent of the selection of the members of the second sample. We will expand upon this discussion in Chapter 12. We are interested in the sampling distribution of the difference between the two sample means.

In Section 8.2, we introduced the central limit theorem, which stated that, in repeated sampling from a normal population whose mean is μ and whose standard deviation is σ, the sampling distribution of the sample mean is normal with mean μ and standard deviation $\sigma/\sqrt{n}$. Statisticians have shown that the difference between two independent normal random variables is also normally distributed. Thus, the difference between two sample means $\overline{X}_1 - \overline{X}_2$ is normally distributed if both populations are normal.

In Chapter 5 we presented the laws of expected value and variance, where we asserted the following.

$$E(X - Y) = E(X) - E(Y)$$

and

$$V(X - Y) = V(X) + V(Y) \quad \text{if } X \text{ and } Y \text{ are independent}$$

Applying these two laws, we determine the mean and variance of $\bar{X}_1 - \bar{X}_2$:

$$\mu_{\bar{x}_1-\bar{x}_2} = E(\bar{X}_1 - \bar{X}_2) = E(\bar{X}_1) - E(\bar{X}_2) = \mu_1 - \mu_2$$

and

$$\sigma^2_{\bar{x}_1-\bar{x}_2} = V(\bar{X}_1 - \bar{X}_2) = V(\bar{X}_1) + V(\bar{X}_2) = \frac{\sigma_1^2}{n_1} + \frac{\sigma_2^2}{n_2}$$

Thus, it follows that in repeated independent sampling from two populations with means μ_1 and μ_2 and standard deviations σ_1 and σ_2, respectively, the sampling distribution of $\bar{X}_1 - \bar{X}_2$ is normal with mean

$$\mu_{\bar{x}_1-\bar{x}_2} = \mu_1 - \mu_2$$

and standard deviation, which is the **standard error of the difference between two means:**

$$\sigma_{\bar{x}_1-\bar{x}_2} = \sqrt{\frac{\sigma_1^2}{n_1} + \frac{\sigma_2^2}{n_2}}$$

If the populations are nonnormal, then the sampling distribution is only approximately normal for large sample sizes. The required sample sizes depend on the extent of nonnormality. However, for most populations, sample sizes of 30 or more are sufficient. Figure 8.9 depicts the sampling distribution of the difference between two means.

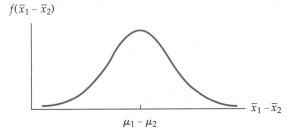

Figure 8.9

Sampling distribution of $\bar{X}_1 - \bar{X}_2$

▼ EXAMPLE 8.4

Suppose that the starting salaries of MBAs at Wilfrid Laurier University (WLU) are normally distributed with a mean of $62,000 and a standard deviation of $14,500. The starting salaries of MBAs at the University of Western Ontario (UWO) are normally distributed with a mean of $60,000 and a standard deviation of $18,300. If a random sample of 50 WLU MBAs and a random sample of 60 UWO MBAs are selected, what is the probability that the sample mean of WLU graduates will exceed that of the UWO graduates?

Solution We want to determine $P(\bar{X}_1 - \bar{X}_2 > 0)$. We know that $\bar{X}_1 - \bar{X}_2$ is normally distributed with mean $\mu_1 - \mu_2 = 62{,}000 - 60{,}000 = 2{,}000$ and standard deviation

$$\sqrt{\frac{\sigma_1^2}{n_1} + \frac{\sigma_2^2}{n_2}} = \sqrt{\frac{14{,}500^2}{50} + \frac{18{,}300^2}{60}} = 3{,}128$$

We can standardize the variable and refer to Table 3:

$$P(\bar{X}_1 - \bar{X}_2 > 0) = P\left(\frac{(\bar{X}_1 - \bar{X}_2) - (\mu_1 - \mu_2)}{\sqrt{\frac{\sigma_1^2}{n_1} + \frac{\sigma_2^2}{n_2}}} > \frac{0 - 2{,}000}{3{,}128}\right)$$

$$= P(Z > -.64) = .5 + .2389 = .7389$$

There is a 73.89% probability that, in drawing a sample of size 50 from the WLU graduates and a sample of size 60 of UWO graduates, the sample mean of WLU graduates will exceed the sample mean of UWO graduates. Note that this means that even though the population mean of WLU graduates is $2,000 more than that of the UWO graduates, there is a 26.11% probability (calculated from $1 - .7389$) that the sample mean of UWO graduates would be greater than the sample mean of WLU graduates.

▲

EXERCISES

8.35 Independent random samples of 10 observations each are drawn from normal populations. The parameters of these populations are

Population 1: $\mu = 280$ $\sigma = 25$
Population 2: $\mu = 270$ $\sigma = 30$

Find the probability that the mean of sample 1 is greater than the mean of sample 2 by more than 25.

8.36 Repeat Exercise 8.35 with samples of size 50.

8.37 Repeat Exercise 8.35 with samples of size 100.

8.38 Suppose that we have two normal populations with means and standard deviations listed below. If random samples of size 25 are drawn from each population, what is the probability that the mean of sample 1 is greater than the mean of sample 2?

Population 1: $\mu = 40$ $\sigma = 6$
Population 2: $\mu = 38$ $\sigma = 8$

8.39 Repeat Exercise 8.38 assuming that the standard deviations are 12 and 16, respectively.

8.40 Repeat Exercise 8.38 assuming that the means are 140 and 138, respectively.

8.41 A widget factory's worker productivity is normally distributed. One worker produces an average of 75 widgets per day with a standard deviation of 20. Another worker produces at an average rate of 65 per day with a standard deviation of 21.

a What is the probability that in any single day worker 1 will outproduce worker 2?

b What is the probability that during one week (5 working days) worker 1 will outproduce worker 2?

8.42 A professor of statistics noticed that the marks in his course are normally distributed. He has also noticed that his morning classes average 73% with a standard deviation of 12% on their final exams. His afternoon classes average 77% with a standard deviation of 10%.

a What is the probability that a randomly selected student in the morning class has a higher final exam mark than a randomly selected student from an afternoon class?

b What is the probability that the mean mark of four randomly selected students from a morning class is greater than the average mark of four randomly selected students from an afternoon class?

8.43 The manager of a restaurant believes that waiters and waitresses who introduce themselves by telling customers their names will get larger tips than those who don't. In fact, she claims that the average tip for the former group is 18% while that of the latter is only 15%. If tips are normally distributed with a standard deviation of 3%, what is the probability that, in a random sample of 10 tips recorded from waiters and waitresses who introduce themselves and 10 tips from waiters and waitresses who don't, the mean of the former will exceed that of the latter?

8.44 The average North American loses an average of 15 days per year due to colds and flu. The natural remedy echinacea is reputed to boost the immune system. One manufacturer of echinacea pills claims that consumers of its product will reduce the number of days lost to colds and flu by one third. To test the claim, a random sample of 50 people was drawn. Half took echinacea, and the other half took placebos. If we assume that the standard deviation of the number of days lost to colds and flu with and without echinacea is 3 days, find the probability that the mean number of days lost for echinacea users is less than that for nonusers.

8.6 FROM HERE TO INFERENCE

The primary function of the sampling distribution is statistical inference. To see how the sampling distribution contributes to the development of inferential methods, we need to briefly review how we got to this point.

In Chapters 5 and 6 we introduced probability distributions, which allowed us to make probability statements about values of the random variable. A prerequisite of this calculation is knowledge of the distribution and the relevant parameters. In Example 5.4, we needed to know that the probability that Pat Statsdud guesses the correct answer is 20% ($p = .2$) and that the number of correct answers (successes) in 10 questions (trials) is a binomial random variable. We then could compute the probability of any number of successes. In Example 6.2, we needed to know that the amount of money students earn in the summer is normally distributed with a mean of $4,500 and a standard deviation of $1,000. These three bits of information allowed us to calculate the probability of various values of the random variable.

Figure 8.10 symbolically represents the use of probability distributions. Simply put, knowledge of the population and its parameter(s) allows us to use the probability distribution to make probability statements about individual members of the population.

In this chapter, we developed the sampling distribution, wherein knowledge of the parameter(s) and some information about the distribution allows us to make probability statements about a sample statistic. In Example 8.2, knowing the population mean and standard deviation and assuming that the population is not extremely nonnormal enabled us to calculate a probability statement about a sample mean. Figure 8.11 describes the application of sampling distributions.

Notice that, in applying both probability distributions and sampling distributions, we need to know the value of the relevant parameters, a highly unlikely circumstance. In the real world, parameters are almost always unknown, because they represent descriptive measurements about extremely large populations. Statistical inference addresses this problem. It does so by reversing the direction of the flow of knowledge in Figure 8.11. In Figure 8.12 we display the character of statistical inference.

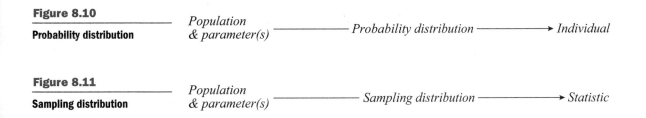

Figure 8.10
Probability distribution

Population & parameter(s) ———— Probability distribution ————▶ Individual

Figure 8.11
Sampling distribution

Population & parameter(s) ———— Sampling distribution ————▶ Statistic

Figure 8.12
Sampling distribution in inference

$$\textit{Statistic}(s) \longleftarrow \textit{Sampling distribution} \longrightarrow \textit{Parameter}(s)$$

Starting in Chapter 9, we will assume that most population parameters are unknown, but that sample statistics are known. The sampling distribution will enable us to draw inferences about the parameter(s) from the statistic(s).

You may be surprised to learn that by and large that is all we do in the remainder of this book. Why then do we need another 11 chapters? Because there are many more parameter and sampling distribution combinations that define the inferential procedures to be presented in an introductory statistics course. However, they all work in the same way. If you understand how one procedure is evolved, you will be likely to understand all of them. Our task in the next two chapters is to ensure that you understand the first inferential method. Your job is identical.

8.7 SUMMARY

The sampling distribution of a statistic is created by repeated sampling from one population. In this chapter, we introduced the sampling distribution of the mean, the proportion, and the difference between two means. We described how these distributions are created theoretically and empirically.

IMPORTANT TERMS

Sampling distribution
Central limit theorem
Sampling distribution of the sample mean
Standard error of the mean
Normal approximation of the binomial distribution
Continuity correction factor
Sampling distribution of a sample proportion
Standard error of the proportion
Sampling distribution of the difference between two sample means
Standard error of the difference between two means

SYMBOLS

Symbol	Pronounced	Represents
$\mu_{\bar{x}}$	mu-x-bar	Mean of the sampling distribution of the sample mean
$\sigma^2_{\bar{x}}$	sigma-squared-x-bar	Variance of the sampling distribution of the sample mean

$\sigma_{\bar{x}}$	sigma-x-bar	Standard deviation of the sampling distribution of the sample mean
α	alpha	Probability
$\hat{P}$	p-hat	Sample proportion
$\sigma_{\hat{p}}^2$	sigma-squared-p-hat	Variance of the sampling distribution of the sample proportion
$\sigma_{\hat{p}}$	sigma-p-hat	Standard deviation of the sampling distribution of the sample proportion
$\mu_{\bar{x}_1 - \bar{x}_2}$	mu-x-bar-1-minus-x-bar-2	Mean of the sampling distribution of the difference between two sample means
$\sigma_{\bar{x}_1 - \bar{x}_2}^2$	sigma-squared-x-bar-1-minus-x-bar-2	Variance of the sampling distribution of the difference between two sample means
$\sigma_{\bar{x}_1 - \bar{x}_2}$	sigma-x-bar-1-minus-x-bar-2	Standard deviation of the sampling distribution of the difference between two sample means

FORMULAS

Expected value of the sample mean

$$E(\overline{X}) = \mu_{\bar{x}} = \mu$$

Variance of the sample mean

$$V(\overline{X}) = \sigma_{\bar{x}}^2 = \frac{\sigma^2}{n}$$

Standard deviation (standard error) of the sample mean

$$\sigma_{\bar{x}} = \frac{\sigma}{\sqrt{n}}$$

Standardizing the sample mean

$$Z = \frac{\overline{X} - \mu}{\sigma/\sqrt{n}}$$

Expected value of the sample proportion

$$E(\hat{P}) = \mu_{\hat{p}} = p$$

Variance of the sample proportion

$$V(\hat{P}) = \sigma_{\hat{p}}^2 = \frac{p(1-p)}{n}$$

Standard deviation (standard error) of the sample proportion

$$\sigma_{\hat{p}} = \sqrt{\frac{p(1-p)}{n}}$$

Standardizing the sample proportion

$$Z = \frac{\hat{P} - p}{\sqrt{p(1-p)/n}}$$

Expected value of the difference between two means

$$E(\overline{X}_1 - \overline{X}_2) = \mu_{\bar{x}_1 - \bar{x}_2} = \mu_1 - \mu_2$$

Variance of the difference between two means

$$V(\overline{X}_1 - \overline{X}_2) = \sigma^2_{\bar{x}_1 - \bar{x}_2} = \frac{\sigma_1^2}{n_1} + \frac{\sigma_2^2}{n_2}$$

Standard deviation (standard error) of the difference between two means

$$\sigma_{\bar{x}_1 - \bar{x}_2} = \sqrt{\frac{\sigma_1^2}{n_1} + \frac{\sigma_2^2}{n_2}}$$

Standardizing the difference between two sample means

$$Z = \frac{(\overline{X}_1 - \overline{X}_2) - (\mu_1 - \mu_2)}{\sqrt{\frac{\sigma_1^2}{n_1} + \frac{\sigma_2^2}{n_2}}}$$

Chapter 9

Introduction to Estimation

9.1 Introduction

9.2 Concepts of Estimation

9.3 Estimating the Population Mean when the Population Standard Deviation Is Known.

9.4 Selecting the Sample Size

9.5 Simulation Experiments (Optional)

9.6 Summary

9.1 INTRODUCTION

Having discussed descriptive statistics (Chapter 3), probability distributions (Chapters 5 and 6), sampling (Chapter 7), and sampling distributions (Chapter 8), we are ready to tackle statistical inference. As we explained in Chapter 1, statistical inference is the process by which we acquire information about populations from samples. There are two general procedures for making inferences about populations: estimation and hypothesis testing. In this chapter, we introduce the concepts and foundations of estimation and demonstrate them with simple examples. In Chapter 10, we will describe the fundamentals of hypothesis testing. Because most of what we do in the remainder of this book applies the concepts of estimation and hypothesis testing, understanding Chapters 9 and 10 is vital to your development as a statistics practitioner.

9.2 CONCEPTS OF ESTIMATION

As its name suggests, the objective of estimation is to determine the approximate value of a population parameter on the basis of a sample statistic. For example, the sample mean is employed to estimate the population mean. We refer to the sample mean as the **estimator** of the population mean. Once the sample mean has been computed, its value is called the estimate.

POINT AND INTERVAL ESTIMATORS

We can use sample data to estimate a population parameter in two ways. First, we can compute the value of the estimator and consider that value as the estimate of the parameter. Such an estimator is called a **point estimator.**

> **Point Estimator**
>
> A point estimator draws inferences about a population by estimating the value of an unknown parameter using a single value or point.

In drawing inferences about a population, it is intuitively reasonable to expect that a large sample will produce more accurate results, because it contains more potential information than a smaller sample does. But point estimators don't have the capacity to reflect the effects of larger sample sizes. The second way of estimating a population parameter is to use an **interval estimator.**

> **Interval Estimator**
>
> An interval estimator draws inferences about a population by estimating the value of an unknown parameter using an interval.

As you will see, the interval estimator is affected by the sample size; because it possesses this feature, we will deal mostly with interval estimators in this text.

To illustrate the difference between point and interval estimators, suppose that a statistics professor wants to estimate the mean summer income of university students. Selecting 25 students at random, he calculates the sample mean weekly income to be $400. The point estimate is the sample mean. That is, he estimates the mean weekly summer income of all university students to be $400. Using the technique described below, he may instead use an interval estimate; he estimates that the average amount of money university students earn per week during the summer lies between $380 and $420.

Numerous applications of estimation occur in the real world. For example, television network executives want to know the proportion of television viewers who are tuned in to their networks; a medical researcher would like to determine the mean number of days to recover after taking a new drug; and an economist wants to know the mean income of university graduates. In each of these cases, in order to accomplish the objective exactly, the statistics practitioner would have to examine each member of the population and then calculate the parameter of interest. For instance, network executives would have to ask each person in the country what he or she is watching to determine the proportion of people who are watching their shows. Because there are millions of television viewers, the task is both impractical and prohibitively expensive. An alternative would be to take a random sample from this population, calculate the sample proportion, and use that as an estimator of the population proportion. The use of the sample proportion to estimate the population proportion seems logical. The selection of the sample statistic to be used as an estimator, however, depends on the characteristics of that statistic. Naturally, we want to use the statistic with the most desirable qualities for our purposes.

One desirable quality of an estimator is unbiasedness.

> **Unbiased Estimator**
>
> An **unbiased estimator** of a population parameter is an estimator whose expected value is equal to that parameter.

This means that, if you were to take an infinite number of samples and calculate the value of the estimator in each sample, the average value of the estimators would equal the parameter. This is equivalent to saying that, on average, the sample statistic is equal to the parameter.

We know that the sample mean $\overline{X}$ is an unbiased estimator of the population mean μ. In presenting the sampling distribution of $\overline{X}$ in Chapter 8 we demonstrated that $E(\overline{X}) = \mu$. We also know that the sample proportion is an unbiased estimator of the population proportion, because $E(\hat{P}) = p$, and that the difference between two sample means is an unbiased estimator of the difference between two population means, because $E(\overline{X}_1 - \overline{X}_2) = \mu_1 - \mu_2$.

Recall that in Chapter 3 we defined the sample variance as

$$s^2 = \frac{\sum(x_i - \overline{x})^2}{n - 1}$$

At the time, it seemed odd that we divided by $n - 1$ rather than by n. The reason for choosing $n - 1$ was to make $E(S^2) = \sigma^2$, so that this definition makes the sample variance an unbiased estimator of the population variance. (The proof of this statement requires about a page of algebraic manipulation, which is more than we would be comfortable in presenting here. Later in this chapter we describe a simulation experiment that demonstrates this point.) Had we defined the sample variance using n in the denominator, the resulting statistic would be a biased estimator of the population variance, one whose expected value is less than the parameter.

Knowing that an estimator is unbiased only assures us that its expected value equals the parameter; it does not tell us how close the estimator is to the parameter. Another desirable quality is that, as the sample size grows larger, the sample statistic should come closer to the population parameter. This quality is called **consistency.**

Consistency

An unbiased estimator is said to be consistent if the difference between the estimator and the parameter grows smaller as the sample size grows larger.

The measure we use to gauge closeness is the variance (or the standard deviation). Thus, $\overline{X}$ is a consistent estimator of μ, because the variance of $\overline{X}$ is σ^2/n. This implies that as n grows larger, the variance of $\overline{X}$ grows smaller. As a consequence, an increasing proportion of sample means falls close to μ.

Figure 9.1 depicts two sampling distributions of $\overline{X}$ when samples are drawn from a normal population whose mean is 0 and whose standard deviation is 10. One sampling distribution is based on samples of size 25, and the other is based on samples of size 100. The former is more spread out than the latter.

Similarly, $\hat{P}$ is a consistent estimator of p because it is unbiased and the variance of $\hat{P}$ is $p(1 - p)/n$, which grows smaller as n grows larger.

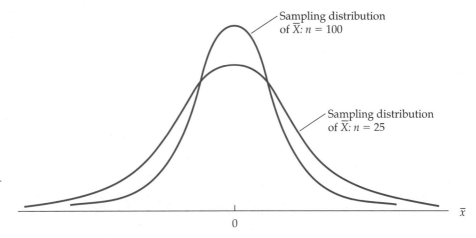

Figure 9.1

Sampling distributions of $\bar{X}$ with $n = 25$ and $n = 100$

A third desirable quality is **relative efficiency,** which compares two unbiased estimators of a parameter.

> **Relative Efficiency**
>
> If there are two unbiased estimators of a parameter, the one whose variance is smaller is said to be relatively efficient.

We have already seen that the sample mean is an unbiased estimator of the population mean and that its variance is σ^2/n. Statisticians have established that (sampling from a normal population) the sample median is also an unbiased estimator of the population mean, but that its variance is $1.57\sigma^2/n$. Consequently, we say that the sample mean is relatively more efficient than the sample median. Not surprisingly, the sample mean will be our first choice in drawing inferences about a population mean.

Over the remaining chapters of this book, we will present the statistical inference of a number of different population parameters. In each case, we will select a sample statistic that is unbiased and consistent, and where there is more than one such statistic we will choose the one that is relatively efficient to serve as the estimator.

DEVELOPING AN UNDERSTANDING OF STATISTICAL CONCEPTS

In this section we described three desirable characteristics of estimators: unbiasedness, consistency, and relative efficiency. An understanding of statistics requires that you know that there are several potential estimators for each parameter but that we chose the estimators used in this book because they possess these characteristics.

EXERCISES

9.1 Define unbiasedness.

9.2 Draw a diagram of the sampling distribution of an unbiased estimator.

9.3 Draw a diagram depicting the sampling distribution of a biased estimator.

9.4 Define consistency.

9.5 Is the sample median a consistent estimator of the population mean? Explain.

9.6 Draw diagrams representing what happens to the sampling distribution of a consistent estimator when the sample size increases.

9.7 Define relative efficiency.

9.8 Draw a diagram representing the sampling distribution representing two unbiased estimators, one of which is relatively efficient.

9.3 ESTIMATING THE POPULATION MEAN WHEN THE POPULATION STANDARD DEVIATION IS KNOWN

We now describe how an interval estimator is produced from a sampling distribution. We choose to demonstrate estimation with an example that is unrealistic. However, this liability is offset by the example's simplicity. When you understand more about estimation, you will be able to apply the technique to more realistic situations.

Suppose we have a population with mean μ and standard deviation σ. The population mean is assumed to be unknown, and our task is to estimate its value. As we just discussed, the estimation procedure requires the statistics practitioner to draw a random sample of size n and calculate the sample mean $\overline{X}$.

In Chapter 8, we established that $\overline{X}$ is normally distributed if X is normally distributed, or approximately normally distributed if X is nonnormal and n is sufficiently large. This means that the variable

$$Z = \frac{\overline{X} - \mu}{\sigma/\sqrt{n}}$$

is standard normally distributed (or approximately so). In Section 8.2 (page 210), we developed the following probability statement associated with the sampling distribution of the sample mean:

$$P\left(\mu - z_{\alpha/2}\frac{\sigma}{\sqrt{n}} < \overline{X} < \mu + z_{\alpha/2}\frac{\sigma}{\sqrt{n}}\right) = 1 - \alpha$$

which was derived from

$$P\left(-z_{\alpha/2} < \frac{\overline{X} - \mu}{\sigma/\sqrt{n}} < z_{\alpha/2}\right) = 1 - \alpha$$

Using a similar algebraic manipulation, we can express the probability in a slightly different form. That is,

$$P\left(\overline{X} - z_{\alpha/2}\frac{\sigma}{\sqrt{n}} < \mu < \overline{X} + z_{\alpha/2}\frac{\sigma}{\sqrt{n}}\right) = 1 - \alpha$$

Notice that in this form the population mean is in the center of the interval created by adding and subtracting $z_{\alpha/2}$ standard errors to the sample mean. It is important for you to understand that this is merely another form of probability statement about the sample mean. This equation says that, with repeated sampling from this population, the proportion of values of $\overline{X}$ for which the interval

$$\overline{X} - z_{\alpha/2}\sigma/\sqrt{n}, \quad \overline{X} + z_{\alpha/2}\sigma/\sqrt{n}$$

includes the population mean μ is equal to $1 - \alpha$. However, this form of probability statement is very useful to us because it is the **interval estimator of μ**.

Interval Estimator of μ*

$$\bar{x} - z_{\alpha/2}\sigma/\sqrt{n}, \quad \bar{x} + z_{\alpha/2}\sigma/\sqrt{n}$$

The probability $1 - \alpha$ is called the **confidence level.**

$\bar{x} - z_{\alpha/2}\sigma/\sqrt{n}$, is called the **lower confidence limit (LCL).**

$\bar{x} + z_{\alpha/2}\sigma/\sqrt{n}$, is called the **upper confidence limit (UCL).**

We often represent the interval estimator as

$$\bar{x} \pm z_{\alpha/2}\sigma/\sqrt{n}$$

where the minus sign defines the lower confidence limit and the plus sign defines the upper confidence limit.

To apply this formula we specify the confidence level $1 - \alpha$, from which we determine α, $\alpha/2$, and $z_{\alpha/2}$ (recall from Chapter 6 that we calculate $z_{\alpha/2}$ by using the normal table—Table 3 in Appendix B—backwards). Because the confidence level is the probability that the interval includes the actual value of μ, we generally set $1 - \alpha$ close to 1 (usually between .90 and .99).

In Table 9.1, we list four commonly used confidence levels and their associated values of $z_{\alpha/2}$. For example, if the confidence level is $1 - \alpha = .95$, $\alpha = .05$,

---

*Since Chapter 5 we've been using the convention whereby an uppercase letter (usually X) represents a random variable and a lowercase letter (usually x) represents one of its values. However, in the formulas used in statistical inference, the distinction between the variable and its value becomes blurred. Accordingly, we will discontinue the notational convention and simply use lowercase letters except when we wish to make a probability statement.

Table 9.1 Four Commonly Used Confidence Levels and $z_{\alpha/2}$

Confidence level: $1 - \alpha$	α	$\alpha/2$	
.90	.10	.05	$z_{.05} = 1.645$
.95	.05	.025	$z_{.025} = 1.96$
.98	.02	.01	$z_{.01} = 2.33$
.99	.01	.005	$z_{.005} = 2.575$

$\alpha/2 = .025$, and $z_{\alpha/2} = z_{.025} = 1.96$. The resulting interval estimator is then called the 95% confidence interval estimator of μ.

As an illustration, suppose we want to estimate the mean value of the distribution resulting from the throw of a fair die. Because we know the distribution, we also know that $\mu = 3.5$ and $\sigma = 1.71$. Pretend now that we know only that $\sigma = 1.71$, that μ is unknown, and that we want to estimate its value. To estimate μ we draw a sample of size $n = 100$ and calculate $\bar{x}$. The interval estimator of μ is

$$\bar{x} \pm z_{\alpha/2} \frac{\sigma}{\sqrt{n}}$$

The 90% confidence interval estimator is

$$\bar{x} \pm z_{\alpha/2} \frac{\sigma}{\sqrt{n}} = \bar{x} \pm 1.645 \frac{1.71}{\sqrt{100}} = \bar{x} \pm .28$$

This notation means that, if we repeatedly draw samples of size 100 from this population, 90% of the values of $\bar{x}$ will be such that μ would lie somewhere between $\bar{x} - .28$ and $\bar{x} + .28$, and 10% of the values of $\bar{x}$ will produce intervals that would not include μ. To illustrate this point, imagine that we draw 40 samples of 100 observations each. The values of $\bar{x}$ and the resulting interval estimates of μ are shown in Table 9.2. Notice that not all the intervals include the true value of the parameter. Samples 5, 16, 22, and 34 produce values of $\bar{x}$ that in turn produce intervals that exclude μ.

Students often react to this situation by asking, "What went wrong with samples 5, 16, 22, and 34?" The answer is nothing. Statistics does not promise 100% certainty. In fact, in this illustration, we expected 90% of the intervals to include μ and 10% to exclude μ. Because we produced 40 intervals, we expected that four (10% of 40) intervals would not contain $\mu = 3.5$.* It is important to understand that, even when the statistics practitioner performs experiments properly, a certain proportion (in this example, 10%) of the experiments will produce incorrect estimates by random chance.

We can improve the confidence associated with the interval estimate. If we let the confidence level $1 - \alpha$ equal .95, the interval estimator is

$$\bar{x} \pm z_{\alpha/2} \frac{\sigma}{\sqrt{n}} = \bar{x} \pm 1.96 \frac{1.71}{\sqrt{100}} = \bar{x} \pm .34$$

---

*In this illustration, exactly 10% of the 40 sample means produced interval estimates that excluded the value of μ, but this will not always be the case. Remember, we expect 10% of the sample means in the long run to result in intervals excluding μ. This group of 40 sample means does not constitute "the long run."

Table 9.2 90% Confidence Interval Estimates of μ

Sample	$\bar{x}$	LCL = $\bar{x}$ − .28	UCL = $\bar{x}$ − .28	Does Interval Include $\mu = 3.5$?
1	3.55	3.27	3.83	Yes
2	3.61	3.33	3.89	Yes
3	3.47	3.19	3.75	Yes
4	3.48	3.20	3.76	Yes
5	3.80	3.52	4.08	No
6	3.37	3.09	3.65	Yes
7	3.48	3.20	3.76	Yes
8	3.52	3.24	3.80	Yes
9	3.74	3.46	4.02	Yes
10	3.51	3.23	3.79	Yes
11	3.23	2.95	3.51	Yes
12	3.45	3.17	3.73	Yes
13	3.57	3.29	3.85	Yes
14	3.77	3.49	4.05	Yes
15	3.31	3.03	3.59	Yes
16	3.10	2.82	3.38	No
17	3.50	3.22	3.78	Yes
18	3.55	3.27	3.83	Yes
19	3.65	3.37	3.93	Yes
20	3.28	3.00	3.56	Yes
21	3.40	3.12	3.68	Yes
22	3.88	3.60	4.16	No
23	3.76	3.48	4.04	Yes
24	3.40	3.12	3.68	Yes
25	3.34	3.06	3.62	Yes
26	3.65	3.37	3.93	Yes
27	3.45	3.17	3.73	Yes
28	3.47	3.19	3.75	Yes
29	3.58	3.30	3.86	Yes
30	3.36	3.08	3.64	Yes
31	3.71	3.43	3.99	Yes
32	3.51	3.23	3.79	Yes
33	3.42	3.14	3.70	Yes
34	3.11	2.83	3.39	No
35	3.29	3.01	3.57	Yes
36	3.64	3.36	3.92	Yes
37	3.39	3.11	3.67	Yes
38	3.75	3.47	4.03	Yes
39	3.26	2.98	3.54	Yes
40	3.54	3.26	3.82	Yes

Because this interval is wider, it is more likely to include the value of μ. If you redo Table 9.2, this time using a 95% confidence interval estimator, only samples 16, 22, and 34 will produce intervals that do not include μ. (Notice that we expected 5% of the intervals to exclude μ and that we actually observed 3/40 = 7.5%.)

The 99% confidence interval estimator is

$$\bar{x} \pm z_{\alpha/2}\frac{\sigma}{\sqrt{n}} = \bar{x} \pm 2.575\frac{1.71}{\sqrt{100}} = \bar{x} \pm .44$$

Applying this interval estimate to the sample means listed in Table 9.2 would result in having all 40 interval estimates include the population mean $\mu = 3.5$. (We expected 1% of the intervals to exclude μ; we observed 0/40 = 0%.)

In Section 9.5 we describe two simulation experiments that generate interval estimators and compute the proportion of intervals that include the mean. You can satisfy yourself that the interval estimators function as we describe here.

ERROR OF ESTIMATION

In Chapter 7 we discussed the sampling error and described it as the difference between the sample and the population caused by chance. We can refine this concept by defining the **error of estimation,** which is defined as the absolute difference between the sample statistic and the population parameter. In this case the error of estimation is

$$|\bar{x} - \mu|$$

We don't know the value of this error (because we don't know the value of μ). However, we can make probability statements about the error of estimation. If we apply a different algebraic manipulation to the following statement

$$P\left(-z_{\alpha/2} < \frac{\bar{X} - \mu}{\sigma/\sqrt{n}} < z_{\alpha/2}\right) = 1 - \alpha$$

we produce

$$P\left(|\bar{X} - \mu| < z_{\alpha/2}\frac{\sigma}{\sqrt{n}}\right) = 1 - \alpha$$

This states that the probability that the error of estimation is less than $z_{\alpha/2}\sigma/\sqrt{n}$ is $(1 - \alpha)$. The quantity $z_{\alpha/2}\sigma/\sqrt{n}$ is called the **bound on the error of estimation** or more simply the **bound,** which we denote as B. It represents how large the error of estimation is likely (where "likely" is defined as probability $1 - \alpha$) to be. Notice that the bound is simply the part of the interval estimator that follows the plus/minus sign.

To illustrate, the bound on the error of estimation of the mean toss of a die when the confidence level is 90% is

$$B = z_{\alpha/2}\frac{\sigma}{\sqrt{n}} = 1.645\frac{1.71}{\sqrt{100}} = .28$$

This figure tells us that 90% of all the sample means will lie within .28 of the actual value of the population mean. Only 10% of the means will be such that the error of estimation is more than .28. Altering the confidence level changes the bound.

In actual practice, only one sample will be drawn, and thus only one value of $\bar{x}$ will be calculated. The resulting interval estimate will either correctly include the parameter or incorrectly exclude it. Unfortunately, statistics practitioners do not know whether in each case they are correct; they know only that, in the long run, they will incorrectly estimate the parameter some of the time. Statistics practitioners accept that as a fact of life.

Section 9.3 ESTIMATING THE POPULATION MEAN WHEN THE POPULATION STANDARD DEVIATION IS KNOWN **237**

STATISTICS PRACTITIONERS AT WORK

The following example illustrates how estimation techniques are applied. It also illustrates how we intend to solve problems in the rest of this book. The solution process that we advocate and use throughout this book is by and large the same one that statistics practitioners use. The process is divided into three stages. The first stage is to identify the correct statistical technique. Of course, for this example you will have no difficulty identifying the technique, because at this point you only know one.

The second stage is to compute the statistics we need. We will employ Microsoft Excel as our primary method. However, to help you understand the concepts we will also calculate the statistics manually in this and the next chapter.

In the third and last stage of the process we interpret the results and deal with the question that began the problem. This may be more difficult than it appears, because to be capable of properly interpreting statistical results one needs to have an understanding of the fundamental principles underlying statistical inference.

▼ EXAMPLE 9.1

Psychologists have been concerned about the amount of time children spend watching television. It is generally believed that children watch far too much violence on television and that they spend little time on more useful activities. A psychologist wanted to know the amount of time children spend watching television. From this information she could determine the amount of violence witnessed. She decided to survey 100 North American children and ask them to keep track of the number of hours of television they watch each week. The data were recorded and appear below. (They are also stored on the data disk in file Xm09-01.) From past experience, it is known that the population standard deviation of the weekly amount of television watched is $\sigma = 8.0$ hours. The psychologist wants an estimate of the mean amount of television watched by all North American children. A confidence level of 95% is judged to be appropriate.

TIME SPENT WATCHING TELEVISION PER WEEK

39.7	23.6	20.6	21.3	33	20.3	35.3	30.8	22.1	24
28.4	20.5	19.5	32.8	27.2	37.1	37.8	21.9	43.8	16.4
15	31.6	20.6	43.9	29.3	22	23.4	23.8	24.5	50.3
26.5	23.4	15.6	35.5	13.9	23.4	26.9	30	34.9	20.6
46.4	9.5	15.5	28.7	23.2	35.7	27.1	15.1	27	34.7
15.1	31.2	36.8	14.7	38.6	26.9	22.8	22.9	25.2	30.2
38.7	40.6	29.1	21.8	23	15.8	21.6	24.4	23.4	29.9
28.9	41.3	29.3	42.4	11.3	19.7	24.4	22.7	30.8	32.9
21.5	38.9	38	33.5	21	36.1	17	27.7	29	39.4
29.8	24.1	26.5	17	21	32.5	26.5	18.3	19.5	31.3

Solution

IDENTIFY

The parameter to be estimated is μ, the mean amount of television watched by all North American children. The estimator is the only one we have presented thus far. It is

$$\bar{x} \pm z_{\alpha/2} \frac{\sigma}{\sqrt{n}}$$

COMPUTE

Manually

We need four figures to manually construct the interval estimate of the mean. They are

$\bar{x}, z_{\alpha/2}, \sigma,$ and n

Using a calculator, we determine the sum of the 100 numbers in the sample:

$$\sum x_i = 2,719.1$$

Dividing the sum by 100, we produce the sample mean:

$$\bar{x} = \frac{\sum x_i}{n} = \frac{2,719.1}{100} = 27.191$$

The confidence level has been specified as .95, which means that $1 - \alpha = .95$. Thus, $\alpha = .05$ and $\alpha/2 = .025$. From Table 3 in Appendix B, or more simply, Table 9.1, we find

$$z_{\alpha/2} = z_{.025} = 1.96$$

We're told that the population standard deviation is $\sigma = 8.0$. The last component is the sample size, which is $n = 100$.

Substituting $\bar{x}, z_{\alpha/2}, \sigma,$ and n into the formula for the interval estimator yields the following.

$$\bar{x} \pm z_{\alpha/2}\frac{\sigma}{\sqrt{n}} = 27.191 \pm 1.96\frac{8.0}{\sqrt{100}} = 27.191 \pm 1.568$$

Thus, the lower and upper confidence limits are

$$\text{LCL} = 27.191 - 1.568 = 25.623$$

and

$$\text{UCL} = 27.191 + 1.568 = 28.759$$

Microsoft Excel Output for Example 9.1

z-Estimate: Mean

	Time
Mean	27.191
Standard Deviation	8.373
Observations	100
SIGMA	8
Standard Error	0.8
Bound	1.568
LCL	25.623
UCL	28.759

The sample mean is $\bar{x} = 27.191$, and the sample standard deviation is $s = 8.373$. (The sample standard deviation is not used in the calculation of the interval estimate; it is provided for information only.) The standard error is

$$\frac{\sigma}{\sqrt{n}} = \frac{8.0}{\sqrt{100}} = 0.8$$

The bound on the error of estimation is

$$B = z_{\alpha/2}\frac{\sigma}{\sqrt{n}} = 1.96\frac{8.0}{\sqrt{100}} = 1.568$$

The lower and upper limits are 25.623 and 28.759, respectively.

GENERAL COMMANDS	EXAMPLE 9.1
1 Type or import the data into one column.	Open file **Xm09-01**.
2 Click **Tools, Data Analysis Plus**, and **z-Estimate: Mean**.	
3 Specify the **Input Range:** (Either highlight the data or type the block coordinates.)	A1:A101
4 Type the value of the population standard deviation **SIGMA**.	8.0
5 Click **Labels** if the first row of the input range contains the name of the variable.	
6 Specify the value of α (**Alpha**) and click **OK**.	.05

There is another way to produce the interval estimate for this problem. If you (or someone else) have calculated the sample mean and know the sample size and population standard deviation, you need not employ the data set and the macro described above. Instead open the file **Stats-Summary.** (It should be in the same directory as the data files.) This workbook contains 18 sheets, each showing the solution to an example in the book. Find the sheet titled **z-Estimate of a Mean.** The worksheet that will be opened represents the solution to Example 9.1. We typed the values of $\bar{x}$ (27.191), σ (8), and n (100) in cells B3, B4, and B5, respectively, and the confidence level in cell B7 (0.95). The bound and the lower and upper confidence limits are calculated in cells B9, B10, and B11, respectively. The complete output is shown below.

Microsoft Excel Output for Example 9.1

z-Estimate of a Mean	
Sample mean	27.191
Population standard deviation	8
Sample size	100
Confidence level	0.95
Bound	1.568
Lower confidence limit	25.623
Upper confidence limit	28.759

There are several ways to use this sheet. First, to solve other problems, simply type in the new values of $\bar{x}$, σ, n, and $1 - \alpha$ in cells B3, B4, B5, and B7, respectively. Do not change any other cells. We recommend you *not* save any of these files in order to avoid inadvertently altering the calculations in cells B9, B10, and B11.

Second, you can perform a "what-if" analysis. That is, this worksheet provides you the opportunity to learn how changing some of the inputs affects the estimate. For example, type 0.99 in cell B7 to see what happens to the size of the interval when you increase the confidence level. Type 1000 in cell B5 to examine the effect of increasing the sample size. Type 1 in cell B4 and see what happens when the population standard deviation is smaller.

In the next three chapters we will describe the other 17 sheets in this workbook. They are all designed in the same way. They can complete the calculation of various techniques from summary statistics and perform what-if analyses.

INTERPRET

The psychologist estimates that the mean amount of time spent watching television by North American children lies somewhere between 25.623 and 28.759 hours per week. From this she may estimate the minimum (as well as maximum) amount of violence. Other pieces of information may be produced from the estimate.

Of course, the point estimate $\bar{x} = 27.191$, alone, would not provide enough information. The psychologist needs to know the lower and upper limits of the estimate to give her enough precision to extract the information she needs.

▲

INTERPRETING THE INTERVAL ESTIMATE

In Example 9.1, we found the 95% confidence interval estimate of the mean number of hours that children watch television per week to be LCL = 25.623 and UCL = 28.759. Some people erroneously interpret this interval to mean that there is a 95% probability that the population mean lies between 25.623 and 28.759. This interpretation is wrong, because it implies that the population mean is a variable about which we can make probability statements. In fact, the population mean is a fixed but unknown quantity. Consequently, we cannot interpret the interval estimate of μ as a probability statement about μ.

To interpret the interval estimate properly, we must remember that it was derived from the sampling distribution of the mean. In Chapter 8, we showed that the sample mean is a random variable with mean μ and standard deviation $\sigma/\sqrt{n}$. It follows that the lower confidence limit and the upper confidence limit are themselves random variables. That is,

$$\text{LCL} = \bar{x} - z_{\alpha/2}\frac{\sigma}{\sqrt{n}}$$

is approximately normally distributed with mean $\mu - z_{\alpha/2}\frac{\sigma}{\sqrt{n}}$ and standard deviation $\sigma/\sqrt{n}$, and

$$\text{UCL} = \bar{x} + z_{\alpha/2}\frac{\sigma}{\sqrt{n}}$$

is approximately normally distributed with mean $\mu + z_{\alpha/2}\frac{\sigma}{\sqrt{n}}$ and standard deviation $\sigma/\sqrt{n}$. Figure 9.2 depicts the sampling distributions of LCL and UCL.

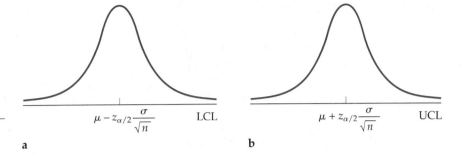

Figure 9.2

Sampling distributions of LCL and UCL

To elaborate further, let's return to the die-tossing illustration we used earlier. The mean and standard deviation of the population of die tosses is $\mu = 3.5$ and $\sigma = 1.71$, respectively. With $n = 100$ and $1 - \alpha = .90$, the following statements are equivalent.

1. $\bar{x}$ is approximately normally distributed with mean $\mu = 3.5$ and standard deviation $\sigma/\sqrt{n} = 1.71/\sqrt{100} = .171$.

2. $\text{CL} = \bar{x} - z_{\alpha/2}\dfrac{\sigma}{\sqrt{n}} = \bar{x} - 1.645\dfrac{1.71}{\sqrt{100}} = \bar{x} - .28$, which is approximately normally distributed with mean $3.5 - .28 = 3.22$ and standard deviation $\sigma/\sqrt{n} = .171$.

3. $\text{CL} = \bar{x} + z_{\alpha/2}\dfrac{\sigma}{\sqrt{n}} = \bar{x} + 1.645\dfrac{1.71}{\sqrt{100}} = \bar{x} + .28$, which is approximately normally distributed with mean $3.5 + .28 = 3.78$ and standard deviation $\sigma/\sqrt{n} = .171$.

Figure 9.3 describes the sampling distributions of the confidence limits.

It must be understood that LCL and UCL are related, because both are based on the value of the sample mean. Thus, if LCL lies between 2.94 (3.22 − .28) and 3.5, UCL will lie between 3.5 and 4.06 (3.78 + .28). In that case, the interval estimate is correct. This will occur for 90% of the sample means. That is, 90% of the sample means will produce an LCL that is less than 3.5 and a UCL that is greater than 3.5.

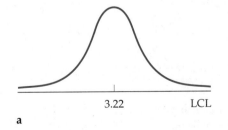 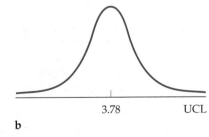

Figure 9.3

Sampling distributions of LCL and UCL (die tossing)

INFORMATION AND THE WIDTH OF THE INTERVAL

Interval estimation, like all other statistical techniques, is designed to convert data into information. However, a wide interval provides little information. For example, suppose that as a result of a statistical study we estimate with 95% confidence that the average starting salary of a newly graduated teacher lies between $10,000 and $60,000. This interval is so wide that very little information was derived from the data. Suppose, however, that the interval estimate was $28,000 to $30,000. This interval is much narrower, providing education students more precise information about the average starting salary.

The width of the interval estimate is a function of the population standard deviation, the confidence level, and the sample size. Consider Example 9.1, where σ was assumed to be 8.0. The interval estimate was 27.191 ± 1.568. Had σ equaled 16.0, the estimate would change to 27.191 ± 3.136. Thus doubling the population standard deviation has the effect of doubling the width of the interval estimate. This result is quite logical. If there is a great deal of variation in the random variable (reflected by a large standard deviation), it is more difficult to accurately estimate the population mean. That difficulty is translated into a wider interval.

While we have no control over the value of σ, we do have the power to select values for the other two elements. In Example 9.1 we chose a 95% confidence level. Had we chosen 99% instead, the interval estimate would be 27.191 ± 2.06. A 90% confidence level results in the interval 27.191 ± 1.316. Decreasing the confidence level will narrow the interval, increasing it widens the interval. But a large confidence level is generally desirable, because that means a larger proportion of interval estimates that will be correct in the long run. There is a direct relationship between the width of the interval and the confidence level. This is because, in order to be more confident in the estimate, we need to widen the interval. (The analogy is that, to be more likely to capture a butterfly, we need a larger butterfly net.) The trade-off between increased confidence and the resulting wider interval estimates must be resolved by the statistics practitioner. As a general rule, however, 95% confidence is considered "standard."

The third element is the sample size. Had the sample size been 400 instead of 100, the interval would have been $27.191 \pm .784$. Increasing the sample size fourfold decreases the width of the interval by half. A larger sample size provides more potential information. The increased amount of information is reflected in a narrower interval. However, there is another trade-off: Increasing the sample size increases the sampling cost. We will discuss these issues when we present sample size selection in Section 9.4.

DEVELOPING AN UNDERSTANDING OF STATISTICAL CONCEPTS

The interval estimator is derived directly from the sampling distribution, an algebraic manipulation that will be repeated throughout this book. In Chapter 8, we used the sampling distribution to make probability statements about the sample mean. Although the form has changed, the interval estimator is also a probability statement about the sample mean. It states that there is $1 - \alpha$ probability that the sample mean will be equal to a value such that the interval $\bar{x} - z_{\alpha/2}\sigma/\sqrt{n}$ to $\bar{x} + z_{\alpha/2}\sigma/\sqrt{n}$ will include the population mean. Once the sample mean is computed, the interval acts as the lower and upper limits of the interval estimate of the population mean.

EXERCISES

Exercises 9.9 to 9.32 are "what-if" analyses designed to determine what happens to the interval estimate when the confidence level, sample size, and standard deviation change. These problems can be solved manually or using Excel's **Stats-Summary** *workbook.*

9.9 Suppose that the amount of time teenagers spend working at part-time jobs is normally distributed with a standard deviation of 20 minutes. A random sample of 100 observations is drawn, and the sample mean computed as 125 minutes. Determine the 90% confidence interval estimate of the population mean.

9.10 Repeat Exercise 9.9 using a 95% confidence level.

9.11 Repeat Exercise 9.9 using a 99% confidence level.

9.12 Repeat Exercise 9.9 changing the population standard deviation to 10.

9.13 Repeat Exercise 9.9 changing the population standard deviation to 40.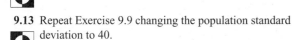

9.14 Repeat Exercise 9.9 with a sample size of 25.

9.15 Repeat Exercise 9.9 with a sample size of 400.

9.16 Summarize Exercises 9.9 to 9.15 by describing what happens to the width of the interval estimate when each of the following happens.
   a The confidence level increases.
   b The standard deviation decreases.
   c The sample size increases.

9.17 One of the few negative side effects of quitting smoking is weight gain. Suppose that the weight gain in the 12 months following a cessation in smoking is normally distributed with a standard deviation of 6 pounds. To estimate the mean weight gain, a random sample of 50 quitters was drawn and the sample mean computed. It was found to be 25 pounds. Determine the 90% confidence interval estimate of the mean 12-month weight gain for all quitters.

9.18 Repeat Exercise 9.17 using a 95% confidence level.

9.19 Repeat Exercise 9.17 using a 99% confidence level.

9.20 Repeat Exercise 9.17 changing the population standard deviation to 10.

9.21 Repeat Exercise 9.17 changing the population standard deviation to 14.

9.22 Repeat Exercise 9.17 with a sample size of 100.

9.23 Repeat Exercise 9.17 with a sample size of 200.

9.24 Summarize Exercises 9.17 to 9.23 by describing what happens to the width of the interval estimate when each of the following happens.
   a The confidence level decreases.
   b The standard deviation increases.
   c The sample size decreases.

9.25 Because of different sales ability, experience, and devotion, the income of real estate agents varies considerably. Suppose that in a large city the annual income is normally distributed with a standard deviation of $7,500. A random sample of 16 real estate agents reveals that the mean annual income is $52,000. Determine the 99% confidence interval estimate of the mean annual income of all real estate agents in the city.

9.26 Repeat Exercise 9.25 using a 95% confidence level.

9.27 Repeat Exercise 9.25 using a 90% confidence level.

9.28 Repeat Exercise 9.25 changing the population standard deviation to $15,000.

9.29 Repeat Exercise 9.25 changing the population standard deviation to $1,000.

9.30 Repeat Exercise 9.25 with a sample size of 4.

9.31 Repeat Exercise 9.25 with a sample size of 1,000.

9.32 Summarize Exercises 9.25 to 9.31 by describing what happens to the width of the interval estimate when each of the following happens.
  a The confidence level decreases.
  b The standard deviation increases.
  c The sample size decreases.

9.33 The following data represent a random sample of nine marks on a statistics quiz. The marks are normally distributed with a standard deviation of 2. Estimate the population mean with 90% confidence.

7, 9, 7, 5, 4, 8, 3, 10, 9

9.34 The following observations are the ages of a random sample of eight men in a bar. It is known that the ages are normally distributed with a standard deviation of 10. Determine the 95% confidence interval estimate of the population mean. Interpret what the interval estimate tells you.

52, 68, 22, 35, 30, 56, 39, 48

9.35 How many rounds of golf do physicians (who play the game) play per year? A survey of 12 physicians revealed the following numbers:

3, 41, 17, 1, 33, 37, 18, 15, 27, 12, 29, 51

Estimate with 95% confidence the mean number of rounds per year played by physicians, assuming the number of rounds is normally distributed with a standard deviation of 7.

The following exercises require the use of a computer and software. The answers may be calculated manually. See Appendix A for the sample statistics.

9.36 Among the most exciting aspects of a university professor's life are departmental meetings where such critical issues as the color the walls will be painted and who gets a new desk are decided. A sample of 20 professors was asked how many hours per year are devoted to these meetings. The responses are listed below (and stored in file Xr09-36). Assuming that this variable is normally distributed with a standard deviation of 8 hours, estimate the mean number of hours spent at departmental meetings by all professors. Use a confidence level of 90%.

14, 17, 3, 6, 17, 3, 8, 4, 20, 15,
7, 9, 0, 5, 11, 15, 18, 13, 8, 4

9.37 The number of cars sold annually by used-car salespeople is normally distributed with a standard deviation of 15. A random sample of 400 salespeople was taken, and the number of cars sold annually for each was recorded and stored in file Xr09-37. Find the 95% confidence interval estimate of the population mean. Interpret what the interval estimate tells you.

9.38 It is known that the amount of time needed to change the oil on a car is normally distributed with a standard deviation of 5 minutes. A random sample of the time needed for 100 oil changes was stored in file Xr09-38. Determine the 99% confidence interval estimate of the mean of the population.

9.39 A survey of 400 statistics professors was undertaken. Each was asked how much time was devoted to teaching graphical techniques. We believe that the times are normally distributed with a standard deviation of 30 minutes. The data are stored in file Xr09-39. Estimate the population mean with 95% confidence.

9.40 In a survey conducted to determine, among other things, the cost of vacations, 64 individuals were randomly sampled. Each person was asked to compute the cost of her or his most recent vacation. The data are stored in file Xr09-40. Assuming that the standard deviation is $400, estimate with 95% confidence the average cost of all vacations.

9.41 In an article about disinflation, various investments were examined. The investments included stocks, bonds, and real estate. Suppose that a random sample of 200 rates of return on real estate investments were computed and stored in file Xr09-41. Assuming that the standard deviation of all rates of return on real estate investments is 2.1%, estimate the mean rate of return on all real estate investments with 90% confidence. Interpret the estimate.

9.42 The dean of students is in the process of investigating how many classes university students miss each semester. To help answer this question, she took a random sample of 100 university students and asked each to report how many classes he or she had missed in the previous semester. These data are stored in file Xr09-42. Estimate the mean number of classes missed by all students at the university. Use a 99% confidence level, and assume that the population standard deviation is known to be 2.2 classes.

9.43 As part of a project to develop better lawn fertilizers, a research chemist wanted to determine the mean weekly growth rate of Kentucky bluegrass, a common type of grass. A sample of 250 blades of grass was measured, and the amount of growth in one week was recorded. These data are stored in file Xr09-43. Assuming that weekly growth is normally distributed with a standard deviation of .10 inches, estimate with 99% confidence the mean weekly growth of Kentucky bluegrass. Briefly interpret what the interval estimate tells you about the growth of Kentucky bluegrass.

9.44 A time study of a large production facility was undertaken to determine the mean time required assembling a widget. A random sample of the times to assemble 50 widgets was recorded and stored in file Xr09-44. An analysis of the assembly times reveals that they are normally distributed with a standard deviation of 1.3 minutes. Estimate with 95% confidence the mean assembly time for all widgets. What do your results tell you about the assembly times?

9.4 SELECTING THE SAMPLE SIZE

As we discussed in the previous section, if the interval estimate is very wide it provides little information. Fortunately, statistics practitioners can control the width of the interval by determining the sample size necessary to produce narrow intervals. They do so by specifying the bound on the error of estimation. Recall that we define the bound as

$$B = z_{\alpha/2} \frac{\sigma}{\sqrt{n}}$$

With a little algebra we solve for n and produce the **sample size to estimate μ**.

Sample Size to Estimate a Mean

$$n = \left(\frac{z_{\alpha/2}\sigma}{B}\right)^2$$

In this chapter we have assumed that we know the value of the population standard deviation. In practice this is seldom the case. (In Chapter 11 we introduce a more realistic interval estimator of the population mean.) To use the formula above, it is frequently necessary to "guesstimate" the value of σ. That is, we must use our knowledge of the variable with which we're dealing to assign some value to σ. Unfortunately, we cannot be very precise in this guess. However, in guesstimating the value of σ, we prefer to err on the high side. To understand why, consider the following example.

▼ EXAMPLE 9.2

Lumber companies need to be able to estimate the amount of lumber that they can harvest in a tract of land to determine whether the effort will be profitable. To do so, they must estimate the mean diameter of the trees. It has been decided to estimate that parameter to within one inch with 99% confidence. A forester familiar with the territory guesses that the diameters of the trees are normally distributed with a standard deviation of 6 inches. How large a sample should be taken?

Solution The phrase "to within one inch" is interpreted as specifying the bound to be 1 inch. The confidence level is 99% ($1 - \alpha = .99$). Thus $\alpha = .01$ and $\alpha/2 = .005$. It follows that $z_{\alpha/2} = 2.575$. Substituting this quantity, $B = 1$, and $\sigma = 6$, we compute

$$n = \left(\frac{z_{\alpha/2}\sigma}{B}\right)^2 = \left(\frac{2.575 \times 6}{1}\right)^2 = 239$$

To estimate with 99% confidence the mean of a normal population whose standard deviation is assumed to be 6 requires a random sample of 239 trees. From the data, the sample mean will be computed, and ultimately the interval estimator will be produced. If the standard deviation is actually 6 inches, the interval estimate will be $\bar{x} \pm 1$.

However, if the standard deviation is actually a number larger than 6, the interval estimator will be wider than planned and thus less precise and useful. To illustrate, suppose that after sampling the trees we discover that σ is actually 12 inches. The interval estimator becomes

$$\bar{x} \pm z_{\alpha/2}\frac{\sigma}{\sqrt{n}} = \bar{x} \pm 2.575\frac{12}{\sqrt{239}} = \bar{x} \pm 2$$

which is twice the width we had planned.

If we discover that the standard deviation is less than we assumed when we determined the sample size, the interval estimator will be narrower and therefore more precise. In the example above, if σ is actually 3 inches, the interval estimator becomes

$$\bar{x} \pm z_{\alpha/2}\frac{\sigma}{\sqrt{n}} = \bar{x} \pm 2.575\frac{3}{\sqrt{239}} = \bar{x} \pm 0.5$$

Although this means that we have sampled more trees than needed, the additional cost is relatively low when compared to the value of the information derived.

▲

EXERCISES

9.45 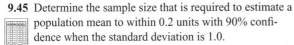 Determine the sample size that is required to estimate a population mean to within 0.2 units with 90% confidence when the standard deviation is 1.0.

9.46 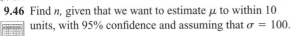 Find n, given that we want to estimate μ to within 10 units, with 95% confidence and assuming that $\sigma = 100$.

9.47 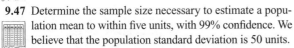 Determine the sample size necessary to estimate a population mean to within five units, with 99% confidence. We believe that the population standard deviation is 50 units.

9.48 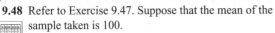 Refer to Exercise 9.47. Suppose that the mean of the sample taken is 100.
  a Determine the 99% confidence interval estimate of the mean.
  b You should have answered part (a) in less than five seconds. Why?

9.49 Refer to Exercise 9.47. Suppose that after taking the sample you find that σ is 70 and that $\bar{x} = 100$. Find the 99% confidence interval estimate.

9.50 Refer to Exercise 9.47. Suppose that, after taking the sample, you find that σ is 20 and that $\bar{x} = 100$. Find the 99% confidence interval estimate.

9.51 A medical statistician wants to estimate the average weight loss of people who are on a new diet plan. In a preliminary study, he guesses that the standard deviation of the population of weight losses is about 10 pounds. How large a sample should he take to estimate the mean weight loss to within two pounds, with 90% confidence?

9.52 The operations manager of a large production plant would like to estimate the average amount of time a worker takes to assemble a new electronic component. After observing a number of workers assembling similar devices, she guesses that the standard deviation is 6 minutes. How large a sample of workers should she take if she wishes to estimate the mean assembly time to within 20 seconds? Assume that the confidence level is to be 99%.

9.53 A statistics professor wants to compare today's students with those from 25 years ago. All of his current students' marks are stored on a computer so that he can easily determine the population mean. However, the marks from 25 years ago reside only in his musty files. He does not want to retrieve all the marks and will be satisfied with a 95% confidence interval estimate of the mean mark 25 years ago. If he assumes that the population standard deviation is 12, how large a sample should he take to estimate the mean to within 2 marks?

9.54 A medical researcher wants to investigate the amount of time it takes for patients' headache pain to be relieved after taking a new prescription painkiller. She plans to use statistical methods to estimate the mean of the population of relief times. She believes that the population is normally distributed with a standard deviation of 20 minutes. How large a sample should she take to estimate the mean time to within 1 minute with 90% confidence?

9.5 SIMULATION EXPERIMENTS (OPTIONAL)

The simulation experiments in this section are designed to reinforce some of the concepts introduced earlier in this chapter. We begin by recycling some of the experiments described in Chapter 8.

UNBIASEDNESS

▼ **SIMULATION EXPERIMENT 9.1**

You can demonstrate that the sample mean is an unbiased estimator of the population mean by repeating Experiment 8.4 and noting the average value of the sample means.

Report for Experiment 9.1

1. What is the mean of the population you have sampled?
2. If the sample mean is an unbiased estimator of the population mean, what value did you expect to be output when you computed the mean of the sample means?
3. What conclusion can you draw from this experiment?

▼ **SIMULATION EXPERIMENT 9.2**

To demonstrate that the sample variance is an unbiased estimator of the population variance, repeat Experiment 8.4, but calculate the row variances instead of the row means. At step 5 in cell J1 type

= VAR(A1:I1)

to calculate the variance of samples of size 9. Don't draw the histogram—simply output the mean of the sample variances.

Report for Experiment 9.2

1. What is the variance of the population you have sampled?
2. If the sample variance is an unbiased estimator of the population variance, what value did you expect to be output when you computed the mean of the sample variances?
3. What conclusion can you draw from this experiment?

RELATIVE EFFICIENCY

▼ **SIMULATION EXPERIMENT 9.3**

In discussing relative efficiency, we pointed out that the median is an unbiased estimator of the population mean. Repeat Experiment 8.4 again, but compute the median instead of the mean or variance. Type the following in cell J1, and calculate the median for each row:

= MEDIAN(A1:I1)

Report for Experiment 9.3

1. Does it appear that the sampling distribution of the median is normal?
2. What are the mean and standard deviation of the sampling distribution of the median?
3. Compare the mean and standard deviation of the sampling distribution of the median with the mean and standard deviation of the sampling distribution of the mean (Experiment 8.4).

4 What does this experiment tell you about the relationship between the sampling distribution of the mean and the sampling distribution of the median when sampling from the same normal population and identical sample sizes?

▲

INTERVAL ESTIMATES

One of the fundamental concepts of statistical inference is that the interval estimate does not always include the value of the parameter we're trying to estimate. That means, for example, that the process that produces 95% confidence intervals will produce intervals that contain the true value of the parameter 95% of the time. The remaining 5% of the time, the interval estimate will be incorrect. The objective of the next experiment is to show that interval estimates are correct most, but not all, of the time.

We will generate 1,000 samples of 9 observations from a normal population with mean 100 and standard deviation 25. For each sample we will compute the 90% confidence interval estimator, which is

$$\bar{x} \pm 1.645 \frac{\sigma}{\sqrt{n}} = \bar{x} \pm 1.645 \frac{25}{\sqrt{9}} = \bar{x} \pm 13.71$$

The computer will then be instructed to count the number of intervals that include the true value of the population mean, which is $\mu = 100$.

▼ SIMULATION EXPERIMENT 9.4

Repeat the first six steps of Experiment 8.4. At this point you should have the row means computed in column J.

7 In column K, calculate the lower confidence limit LCL = $\bar{x}$ − 13.71. In cell K1 type

= J1−13.71

and drag to fill the column.

8 In column L, calculate the upper confidence limit UCL = $\bar{x}$ + 13.71. In cell L1 type

= J1+13.71

and drag to fill the column.

9 In cell M1 type

= AND(K1 < 100, L1 > 100)

and drag to fill the column. If K1 (lower confidence limit) is less than 100 and L1 (upper confidence limit) is greater than 100, the word TRUE will be recorded in cell M1. If not, the word FALSE will be recorded.

10 Click f_x, **Statistical**, **COUNTIF**, and **Next>**.

11 Specify the range of the column where the words TRUE and FALSE are stored: M1:M1000

12 Specify the criteria: **TRUE**

13 Click **Finish**.

Steps 9 through 13 count the number of times the word TRUE appears in the last column. Thus, it counts the number of intervals that include the true value of μ.

Report for Experiment 9.4

1 What is the number of intervals containing the true population mean you anticipated seeing?

2 What is the number of intervals containing the true population mean observed in this experiment?

3 What conclusions can you draw from this experiment?

▼ SIMULATION EXPERIMENT 9.5

Repeat Experiment 9.4 using a confidence level of 95%. The 95% confidence interval estimator is

$$\bar{x} \pm 1.96 \frac{\sigma}{\sqrt{n}} = \bar{x} \pm 1.96 \frac{25}{\sqrt{9}} = \bar{x} \pm 16.33$$

Write a brief report describing your findings.

9.6 SUMMARY

This chapter introduced the concepts of estimation and the estimator of a population mean when the population standard deviation is known. It also presented a formula to calculate the sample size necessary to estimate a population mean.

IMPORTANT TERMS

Estimator
Point estimator
Interval estimator
Unbiased estimator
Consistency
Relative efficiency
Interval estimator of μ

Confidence level
Lower confidence limit (LCL)
Upper confidence limit (UCL)
Error of estimation
Bound on the error of estimation
Sample size to estimate μ

SYMBOLS

Symbol	Pronounced	Represents
α	alpha	Probability that interval estimator excludes the parameter
$1 - \alpha$	one-minus-alpha	Probability that interval estimator includes the parameter
B		Bound on the error of estimation

FORMULAS

Interval estimator of μ

$$\bar{x} \pm z_{\alpha/2} \frac{\sigma}{\sqrt{n}}$$

Sample size

$$n = \left(\frac{z_{\alpha/2}\sigma}{B}\right)^2$$

MICROSOFT EXCEL OUTPUT AND INSTRUCTIONS

Technique	Page
Interval estimator of μ (raw data)	238–239
Interval estimator of μ (Stats-Summary)	239

SUPPLEMENTARY EXERCISES

9.55 The stereotypical image of the Japanese manager is that of a workaholic with little or no leisure time. In a survey, a random sample of 250 Japanese middle managers was asked how many hours per week they spent in leisure activities (e.g., sports, movies, television). The results of the survey are stored in file Xr09-55. Assuming that the population standard deviation is 6 hours, estimate with 90% confidence the mean leisure time per week for all Japanese middle managers. What do these results tell you?

9.56 One measure of physical fitness is the amount of time it takes for the pulse rate to return to normal after exercise. A random sample of 100 women aged 40–50 exercised on stationary bicycles for 30 minutes. The amount of time it took for their pulse rates to return to pre-exercise levels was measured and recorded. The data are stored in file Xr09-56. If the times are normally distributed with a standard deviation of 2.3 minutes, estimate with 99% confidence the true mean pulse-recovery time for all 40- to 50-year-old women. Interpret the results.

9.57 A survey of 20 randomly selected companies asked them to report the annual income of their presidents. These data are stored in file Xr09-57. Assuming that incomes are normally distributed with a standard deviation of $30,000, determine the 90% confidence interval estimate of the mean annual income of all company presidents. Interpret the statistical results.

9.58 The operations manager of a plant making cellular telephones has proposed rearranging the production process to be more efficient. She wants to estimate the time to assemble the telephone using the new arrangement. She believes that the population standard deviation is 15 seconds. How large a sample of workers should she take to estimate the mean assembly time to within 2 seconds with 95% confidence?

9.59 A random sample of 300 university professors was asked how many vacation days they take annually. The results are stored in file Xr09-59. Estimate with 95% confidence the mean number of vacation days all university professors take annually if the population standard deviation is known to be five days. What do the statistics tell you about the population in question?

9.60 To help make a decision about expansion plans, the president of a music company needs to know how many compact discs teenagers buy annually. Accordingly, he commissions a survey of 250 teenagers. Each is asked to report how many CDs he or she purchased in the previous 12 months. The responses are stored in file Xr09-60. Estimate with 90% confidence the mean annual number of CDs purchased by all teenagers. Assume that the population standard deviation is 3 CDs.

9.61 The label on one-gallon cans of paint states that the amount of paint in the can is sufficient to paint 400 square feet. However, this number is quite variable. In fact the amount of coverage is known to be approximately normally distributed with standard deviation of 25 square feet. How large a sample should be taken to estimate the true mean coverage of all one-gallon cans to within 5 square feet with 95% confidence?

9.62 Refer to Exercise 9.61. Suppose that a random sample of 100 one-gallon cans of paint was drawn and used to paint walls. The amount of coverage for each can was recorded and stored in file Xr09-62. Estimate the mean coverage with 95% confidence.

9.63 The Doll Computer Company makes its own computers and delivers them directly to customers who order them via the Internet. Doll competes primarily on price and speed of delivery. To achieve its objective of speed, Doll makes each of its five most popular computers and transports them to warehouses across the country. The computers are stored in the warehouses, from which it generally takes one day to deliver a computer to the customer. This strategy requires high levels of inventory that add considerably to the cost. To lower these costs, the operations manager wants to employ an inventory model. He notes that both daily demand and lead time are random variables. He concludes that demand during lead time is normally distributed, and he needs to know the mean in order to compute the optimum inventory level. He observes 60 lead-time periods and records the demand during each period. These data are stored in file Xr09-63. The manager would like a 99% confidence interval estimate of the mean demand during lead time. Assume that the manager knows that the standard deviation is 50 computers.

Chapter 10

Introduction to Hypothesis Testing

10.1 Introduction

10.2 Concepts of Hypothesis Testing

10.3 Testing the Population Mean when the Population Standard Deviation is Known

10.4 Calculating the Probability of a Type II Error

10.5 The Road Ahead

10.6 Summary

10.1 INTRODUCTION

In Chapter 9, we introduced estimation and showed how it is used. Now we're going to present the second general procedure of making inferences about a population—hypothesis testing. The purpose of this type of inference is to determine whether enough statistical evidence exists to enable us to conclude that a belief or hypothesis about a parameter is reasonable. Illustrations of hypothesis testing include the following.

Illustration 1

Thousands of new products are developed every year. For a variety of reasons, most never reach the market. Products that reach the final stages are evaluated by marketing managers, who attempt to predict how well a product will sell. Statistics in general, and hypothesis testing specifically, often help in this assessment. Suppose that a company has developed a new product that it hopes will be very successful. After a complete financial analysis, the company's directors have determined that if more than 12% of potential customers buy the product, the company will make a profit. A random sample of potential customers is asked whether they would buy the product. The sampling procedure and data collection would be as described in Chapter 7. Statistical techniques would convert the raw data into information that would permit the marketing manager to decide whether to proceed. The parameter is the proportion of customers who would buy the product. The hypothesis to test is that the proportion is greater than 12%. In Chapter 11, we present the technique that would be used in this illustration.

Illustration 2

Scientific and medical discoveries are tested to determine whether they are improvements over current technology. When a new drug is developed, it is usually tested to determine whether it is effective. The test is conducted by selecting a random sample of patients suffering from the disease that the drug is intended to cure. Half the sample is given the new drug and the other half is given a placebo, which looks like a drug but contains no medication. The experiment is called a "double-blind" experiment, because neither the patients nor the doctors know which patients are taking the drug and which the placebo. The effectiveness of each is then measured. We can measure the degree of improvement and compute means, or we can count the number of patients cured and determine proportions. In either case, we compare the two samples, and, using methods presented in Chapter 12, we can infer whether the new drug works.

Illustration 3

Forecasting is one of the most important topics in statistics. Managers often need forecasts of product demand, commodity prices, and general economic activity to help make decisions. Economists and statisticians create equations called mathematical models that are designed to forecast. Hypothesis testing is used extensively in evaluating how well the model is likely to perform. Suppose that a stock analyst wants to predict the return on a particular investment. The analyst would collect data for a variety of variables that she thinks are related to return on investment. Using techniques introduced in Chapters 17 and 18, she would develop the model. She would conduct various tests of hypotheses to determine whether the model is likely to produce accurate forecasts.

In the next section, we will introduce the concepts of hypothesis testing, and in Section 10.3 we will develop the method employed to test a hypothesis about a population mean when the population standard deviation is known. The rest of the chapter deals with related topics.

10.2 CONCEPTS OF HYPOTHESIS TESTING

In Chapter 9 we introduced estimation, the first form of statistical inference. In this chapter we present the second form, **hypothesis testing.** The term is likely to be new to most readers, but the concepts underlying hypothesis testing are quite familiar. There are a variety of nonstatistical applications of hypothesis testing, the best known of which is a criminal trial.

When a person is accused of a crime, he or she faces a trial. The prosecution presents its case, and a jury must make a decision on the basis of the evidence presented. In fact, the jury conducts a test of hypothesis. There are actually two hypotheses that are tested. The first is called the **null hypothesis,** represented by H_0 (pronounced *H-nought*: *nought* is a British term for zero). It is

H_0: The defendant is innocent.

The second is called the **alternative** or **research hypothesis,** denoted H_1. In a criminal trial it is

H_1: The defendant is guilty.

Of course, the jury does not know which hypothesis is correct. The members of the jury must make a decision on the basis of evidence presented by both the prosecution and defense. There are only two possible decisions: Convict or acquit the defendant. In statistical parlance, convicting the defendant is equivalent to *rejecting the null hypothesis in favor of the alternative hypothesis.* When a jury acquits a defendant, a statistician says that we *don't reject the null hypothesis.* Notice that we do not say that we accept the null hypothesis. In a criminal trial that would be interpreted as finding the defendant innocent. Our justice system does not allow this decision.

When testing hypotheses there are two possible errors. A **Type I error** occurs when we reject a true null hypothesis. A **Type II error** is defined as not rejecting a false null hypothesis. In the criminal trial, a Type I error is made when an innocent person is wrongly convicted. A Type II error occurs when a guilty defendant is acquitted. The probability of a Type I error is denoted by α, which is called the **significance level.** The probability of a Type II error is denoted β (Greek letter *beta*). α and β are inversely related, which means that reducing one will increase the other.

In our justice system, Type I errors are regarded as more serious. As a consequence, the system is set up so that the probability of a Type I error is small. This is arranged by placing the burden of proof on the prosecution (the prosecution must prove guilt—the defense need not prove anything) and by having judges instruct the jury to find the defendant guilty only if there is "evidence beyond a reasonable doubt."

In the absence of enough evidence, the jury must acquit, even though there may be some evidence of guilt. The consequence of this arrangement is that the probability of acquitting guilty people is relatively large. Oliver Wendall Holmes, a United States Supreme Court Justice, once phrased the relationship between the probabilities of Type I and Type II errors in the following way: "Better to acquit 100 guilty men than convict one innocent one." In Justice Holmes's opinion the probability of a Type I error should be 1/100 of the probability of a Type II error.

The critical concepts are these.

1 There are two hypotheses. One is called the null hypotheses and the other the alternative or research hypothesis.

2 The testing procedure assumes that the null hypothesis is true.

3 The goal of the process is to determine whether there is enough evidence to infer that the alternative hypothesis is true.

4 There are two possible decisions:

Reject the null hypothesis in favor of the alternative hypothesis.

Do not reject the null hypothesis in favor of the alternative hypothesis.

5 Two possible errors can be made in any test. Type I errors occur when we reject a true null hypothesis, and Type II errors occur when we don't reject a false null hypothesis.

Let's extend these concepts to statistical hypothesis testing.

In statistics, we frequently test hypotheses about parameters. The hypotheses we test are generated by questions that statistics practitioners need to answer. To illustrate, suppose that in Example 9.1 the psychologist did not want to estimate the mean amount of time spent watching television but instead wanted to know whether the mean exceeds 25 hours, which may be the point at which the child sees some critical amount of violence. That is, the psychologist wants to determine whether she can infer that $\mu > 25$. As was the case with the criminal trial, whatever we're investigating is specified as the alternative (research) hypothesis. Thus,

$H_1 : \mu > 25$

Had the psychologist wanted to determine whether the mean is *less than* 25, she would have set up the alternative hypothesis as

$H_1 : \mu < 25$

If she wanted to know whether the mean *differs from* 25, the alternative hypothesis would be

$H_1 : \mu \neq 25$

To test the alternative hypothesis, we employ the sampling distribution of the mean. We do so by assuming that the mean television time is equal to 25. (In the criminal trial, we assume that the defendant is innocent.) This is represented by the null hypothesis. That is,

$H_0 : \mu = 25$

Why do we need the null hypothesis if we want to know whether the alternative hypothesis is true? Because the sampling distribution requires us to assume that μ is a specific value. In this case we will assume that its value is 25.

The next element in the procedure is to randomly sample the population and calculate the sample mean. This is called the **test statistic.** The test statistic is the criterion upon which we base our decision about the hypotheses. (In the criminal trial analogy, this is equivalent to the evidence presented in the case.) The test statistic is based on the best estimator of the parameter. In Chapter 9, we stated that the best estimator of a population mean is the sample mean.

If the test statistic's value is inconsistent with the null hypothesis, we reject it and infer that the alternative hypothesis is true. For example, if we're trying to decide whether the mean is greater than 25, a large value of $\bar{x}$ (say, 35) would provide evidence that the null hypothesis is false and that the alternative is true. If the test statistic value is consistent with the null hypothesis, we do not reject the null. For instance, if $\bar{x}$ is close to 25 (say, 25.4), we could not say that this provides a great deal of evidence to infer that the population mean is greater than 25. In the absence of sufficient evidence we do not reject the null hypothesis in favor of the alternative. (In the absence of sufficient evidence of guilt, a jury finds the defendant not guilty.)

In a criminal trial "sufficient evidence" is defined as "evidence beyond a reasonable doubt." In statistics we need to use the test statistic's sampling distribution to define "sufficient evidence." We will do so in the next section.

10.3 TESTING THE POPULATION MEAN WHEN THE POPULATION STANDARD DEVIATION IS KNOWN

To illustrate the process, consider the following example.

▼ EXAMPLE 10.1

The manager of a department store is thinking about establishing a new billing system for the store's credit customers. After a thorough financial analysis, he determines that the new system will be cost effective only if the mean monthly account is more than $170. A random sample of 400 monthly accounts is drawn for which the sample mean is $178. (The data are stored in file Xm10-01.) The manager knows that the accounts are approximately normally distributed with a standard deviation of $65. Can he conclude from this that the new system will be cost effective?

Solution This example deals with the population of the credit accounts at the store. To conclude that the system will be cost effective requires the manager to show that the mean account for all customers is greater than $170. Consequently, we set up the alternative hypothesis to express this circumstance:

$$H_1: \mu > 170$$

The null hypothesis must state that the parameter is equal to a specific value. As a result, we express the null hypothesis as

$H_0: \mu = 170$

As we pointed out above, the test statistic is the best estimator of the parameter. In Chapter 9, we used the sample mean to estimate the population mean. To conduct this test we ask and answer the following question. "Is a sample mean of 178 sufficiently greater than 170 to allow us to confidently infer that the population mean is greater than 170?"

There are two approaches to answering this question. The first is called the rejection region method. It can be used in conjunction with the computer, but it is mostly used by those computing statistics manually. The second is the *p*-value approach, which in most applications can be employed only in conjunction with a computer and statistical software. We recommend, however, that users of statistical software be familiar with both approaches.

REJECTION REGION METHOD

It seems reasonable to reject the null hypothesis if the value of the sample mean is large relative to 170. If we had calculated the sample mean to be, say, 500, it would be quite apparent that the null hypothesis is false, and we would reject it. On the other hand, values of $\bar{x}$ close to 170, such as 171, do not allow us to reject the null hypothesis, because it is entirely possible to observe a sample mean of 171 from a population whose mean is 170. Unfortunately, the decision is not always so obvious. In this example, the sample mean was calculated to be 178, a value apparently neither very far away from nor very close to 170. In order to make a decision about this sample mean we set up the **rejection region.**

> **Rejection Region**
>
> The rejection region is a range of values such that, if the test statistic falls into that range, we decide to reject the null hypothesis in favor of the alternative hypothesis.

Suppose we define the value of the sample mean that is just large enough to reject the null hypothesis as $\bar{x}_L$. The rejection region is

$\bar{x} > \bar{x}_L$

Because a Type I error is defined as rejecting a true null hypothesis, and the probability of committing a Type I error is α, it follows that

$\alpha = P(\text{rejecting } H_0, \text{given that } H_0 \text{ is true})$

$= P(\bar{x} > \bar{x}_L, \text{given that } H_0 \text{ is true})$

Figure 10.1 depicts the sampling distribution and the rejection region.

Figure 10.1
Sampling distribution for Example 10.1

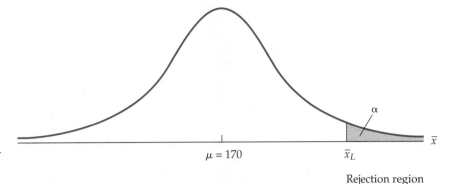

From Section 8.2, we know that the sampling distribution of $\bar{x}$ is normal or approximately normal, with mean μ and standard deviation $\sigma/\sqrt{n}$. As a result, we can standardize $\bar{x}$ and obtain the following probability:

$$P\left(\frac{\bar{x} - \mu}{\sigma/\sqrt{n}} > \frac{\bar{x}_L - \mu}{\sigma/\sqrt{n}}\right) = P\left(Z > \frac{\bar{x}_L - \mu}{\sigma/\sqrt{n}}\right) = \alpha$$

In Section 6.3, we defined the notation z_α, which is the value of Z such that

$$P(Z > z_\alpha) = \alpha$$

Because both probability statements involve the same distribution (standard normal) and the same probability (α), it follows that the limits are identical. Thus,

$$\frac{\bar{x}_L - \mu}{\sigma/\sqrt{n}} = z_\alpha$$

We know that $\sigma = 65$ and $n = 400$. Because the probabilities defined above are conditional upon the null hypothesis being true, we have $\mu = 170$. (This is the reason why the null hypothesis must state that the parameter is equal to a specific value.) To calculate the rejection region, we need a value of α. Suppose that the supervisor chose α to be 5%. It follows that $z_\alpha = z_{.05} = 1.645$. We can now find the value $\bar{x}_L$.

$$\frac{\bar{x}_L - \mu}{\sigma/\sqrt{n}} = z_\alpha$$

$$\frac{\bar{x}_L - 170}{65/\sqrt{400}} = 1.645$$

$$\bar{x}_L = 175.34$$

Therefore the rejection region is

$$\bar{x} > 175.34$$

The sample mean was computed to be 178. Because the test statistic (sample mean) is in the rejection region (it is greater than 175.34), we reject the null hypothesis. Thus, there is sufficient evidence to infer that the mean monthly account is greater than 170.

Our calculations determined that any value of $\bar{x}$ above 175.34 represents an event that is quite unlikely to occur when sampling (with $n = 400$) from a population whose mean is 170 (and whose standard deviation is 65). This suggests that the assumption that the null hypothesis is true is incorrect, and consequently we reject the null hypothesis in favor of the alternative hypothesis.

▲

THE STANDARDIZED TEST STATISTIC

The preceding test used the test statistic $\bar{x}$. As a result, the rejection region had to be set up in terms of $\bar{x}$. To do so, we solved an equation to determine the value of $\bar{x}_L$. An easier method to use specifies that the test statistic be the standardized value of $\bar{x}$. That is, we use the **standardized test statistic**

$$z = \frac{\bar{x} - \mu}{\sigma/\sqrt{n}}$$

and the rejection region consists of all values of z that are greater than z_α. Algebraically, the rejection region is

$$z > z_\alpha$$

We can redo Example 10.1 using the standardized test statistic.

$H_0 : \mu = 170$

$H_1 : \mu > 170$

Test statistic: $z = \dfrac{\bar{x} - \mu}{\sigma/\sqrt{n}}$

Rejection region: $z > z_\alpha = z_{.05} = 1.645$

Value of the test statistic: $z = \dfrac{\bar{x} - \mu}{\sigma/\sqrt{n}} = \dfrac{178 - 170}{65/\sqrt{400}} = 2.46$

Conclusion: Because 2.46 is greater than 1.645, reject the null hypothesis, and conclude that there is enough evidence to infer that the mean monthly account is greater than $170.

As you can see, the conclusions we draw from using the test statistic $\bar{x}$ and the standardized test statistic z are identical. Figures 10.2 and 10.3 depict the two sampling distributions, highlighting the equivalence of the two tests.

Because of the convenience and because statistical software packages employ it, the standardized test statistic will be used throughout this book. For simplicity we will refer to the standardized test statistic simply as the "test statistic."

Incidentally, when a null hypothesis is rejected the test is said to be *statistically significant* at whatever significance level the test was conducted at. Summarizing Example 10.1, we would say that the test was significant at the 5% significance level.

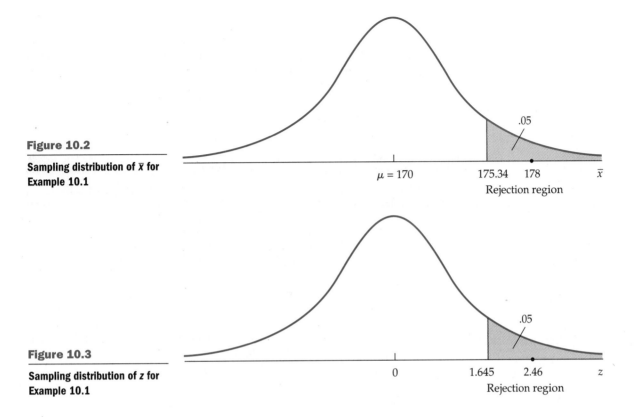

Figure 10.2

Sampling distribution of $\bar{x}$ for Example 10.1

Figure 10.3

Sampling distribution of z for Example 10.1

INTERPRETING THE RESULTS OF A TEST

In Example 10.1, we rejected the null hypothesis. Does this prove that the alternative hypothesis is true? The answer is no: Because our conclusion is based on sample data (and not on the entire population), we can never prove anything by using statistical inference. Consequently, we summarize the test by stating that there is enough statistical evidence to infer that the null hypothesis is false and that the alternative hypothesis is true.

Now suppose that $\bar{x}$ had equaled 174 instead of 178. We would then have calculated $z = 1.23$, which is not in the rejection region. Could we conclude on this basis that there is enough statistical evidence to infer that the null hypothesis is true and hence that $\mu = 170$? Again, the answer is no, because it is absurd to suggest that a sample mean of 174 provides enough evidence to infer that the population mean is 170. Because we're testing a single value of the parameter under the null hypothesis, we can never have enough statistical evidence to establish that the null hypothesis is true (unless we sample the entire population).

Consequently, if the value of the test statistic does not fall into the rejection region, rather than say we accept the null hypothesis (which implies that we're stating that the null hypothesis is true), we state that we do not reject the null hypothesis, and we conclude that not enough evidence exists to show that the alternative hypothesis is true. While it may appear that we're being overly technical, such is not the case. Your ability to set up tests of hypotheses properly and to interpret their results correctly very much depends on your understanding of this point. The point is that the conclusion is based on the alternative hypothesis. In the final analysis, there are only two possible conclusions of a test of hypothesis.

> **Conclusions of a Test of Hypothesis**
>
> If we reject the null hypothesis, we conclude that there is enough statistical evidence to infer that the alternative hypothesis is true.
>
> If we do not reject the null hypothesis, we conclude that there is not enough statistical evidence to infer that the alternative hypothesis is true.

Observe that the conclusion is based on the alternative hypothesis, because it represents what we are investigating. That is why it is also called the research hypothesis. Whatever we're trying to show statistically must be represented by the alternative hypothesis (bearing in mind that we have only three choices for the alternative hypothesis—the parameter is greater than, less than, and not equal to some specified value).

When we introduced statistical inference in Chapter 9, we pointed out that the first step in the solution is to identify the technique. Part of this process when the problem involves hypothesis testing is the specification of the hypotheses. Because the alternative hypothesis represents the circumstance we're researching, we will identify it first. The null hypothesis automatically follows because the null hypothesis must specify equality. However, by tradition, when we list the two hypotheses, the null hypothesis comes first, followed by the alternative hypothesis. All examples in this book will follow that format.

p-VALUE METHOD

There are several drawbacks to the rejection region method. Foremost among them is the type of information provided by the result of the test. The rejection region method produces a yes or no response to the question "Is there sufficient statistical evidence to infer that the alternative hypothesis is true?" The implication is that the result of the test of hypothesis will be converted automatically into one of two possible courses of action: one action as a result of rejecting the null hypothesis in favor of the alternative and another as a result of not rejecting the null hypothesis in favor of the alternative. In Example 10.1, the rejection of the null hypothesis seems to imply that the new billing system will be installed.

In fact, this is not the way in which the result of a statistical analysis is utilized. The statistical procedure is only one of several factors considered by a statistics practitioner when making a decision. In Example 10.1, the manager discovered that there was enough statistical evidence to conclude that the mean monthly account is greater than 170. However, before taking any action, the manager would like to consider a number of factors, including the cost and feasibility of restructuring the billing system and the possibility of making an error, in this case a Type I error.

What is needed to take full advantage of the information available from the test result and make a better decision is a measure of the amount of statistical evidence supporting the alternative hypothesis so that it can be weighed in relation to the other factors, especially the economic ones. The *p*-value of a test provides this measure.

p-Value

The *p*-value of a test is the probability of observing a test statistic at least as extreme as the one computed given that the null hypothesis is true.

In Example 10.1, the *p*-value is the probability of observing a sample mean at least as large as 178 when the population mean is 170. Thus,

$$p\text{-value} = P(\bar{x} > 178) = P\left(\frac{\bar{x} - \mu}{\sigma/\sqrt{n}} > \frac{178 - 170}{65/\sqrt{400}}\right) = P(Z > 2.46) = .0069$$

Figure 10.4 describes this calculation.

INTERPRETING THE *p*-VALUE

To properly interpret the results of an inferential procedure, you must remember that the technique is based on the sampling distribution. And the sampling distribution allows us to make probability statements about a sample statistic, assuming knowledge of the population parameter. Thus, the probability of observing a sample mean at least as large as 178 from a population whose mean is 170 is .0069, which is very small. In other words, we have just observed an unlikely event, an event so unlikely

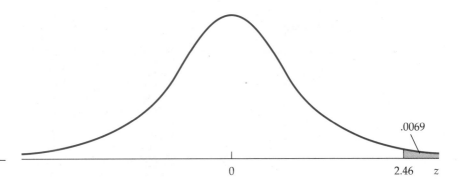

Figure 10.4

p-value for Example 10.1

that we must doubt the assumption that began the process. Recall that we assume that the null hypothesis is true in order to calculate the value of the test statistic. Consequently, we have reason to reject the null hypothesis and support the alternative hypothesis.

You may be tempted to simplify the interpretation by stating that the *p*-value is the probability that the null hypothesis is true. Don't! As was the case with interpreting the interval estimator, you cannot make a probability statement about a parameter. It is not a random variable.

The *p*-value of a test provides valuable information because it measures the amount of statistical evidence that supports the alternative hypothesis. To understand this interpretation fully, refer to Table 10.1, where we list several values of $\bar{x}$, their *z*-statistics, and *p*-values for Example 10.1. Notice that the closer $\bar{x}$ is to the hypothesized mean, 170, the larger the *p*-value is. The farther $\bar{x}$ is above 170, the smaller the *p*-value is. Values of $\bar{x}$ far above 170 tend to indicate that the alternative hypothesis is true. Thus, the smaller the *p*-value, the more statistical evidence exists to support the alternative hypothesis. Figure 10.5 graphically depicts the information in Table 10.1.

This raises the question: How small does the *p*-value have to be to infer that the alternative hypothesis is true? In general, the answer depends on a number of factors, including the costs of making Type I and Type II errors. In Example 10.1, a Type I error would occur if the manager adopts the new billing system when it is not cost effective. If the cost of this action is high, we attempt to minimize its probability. In the rejection region method we do so by setting the significance level quite low, say 1%. Using the *p*-value method we would insist that the *p*-value be quite small, providing sufficient evidence to infer that the mean monthly account is greater than 170 before proceeding with the new billing system.

DESCRIBING THE *p*-VALUE

It is often useful to describe a *p*-value in nonquantitative language. This may be necessary when attempting to express the results of a test to another person not fortunate enough to have taken a statistics course.

Table 10.1 **Test Statistics and *p*-Values for Example 10.1**

Sample Mean $\bar{x}$	Test Statistic $z = \dfrac{\bar{x} - \mu}{\sigma/\sqrt{n}} = \dfrac{\bar{x} - 170}{65/\sqrt{400}}$	*p*-value
170	0	.5000
171	0.3077	.3792
172	0.6154	.2692
173	0.9231	.1780
174	1.2308	.1092
175	1.5385	.0620
176	1.8462	.0324
177	2.1538	.0156
178	2.4615	.0069
179	2.7692	.0028
180	3.0769	.0010

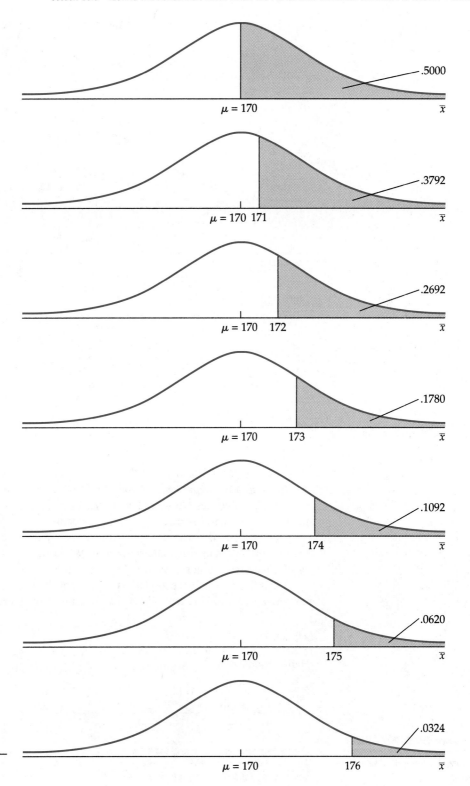

Figure 10.5

p-values for Example 10.1

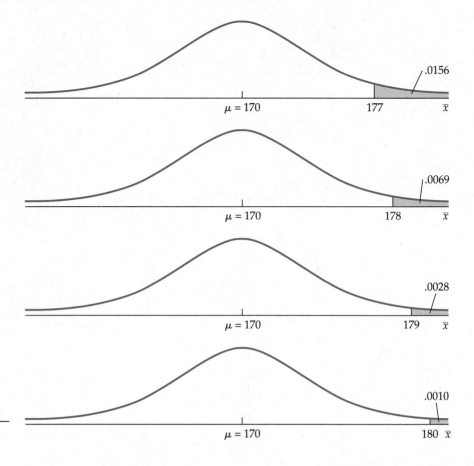

Figure 10.5 (continued)

p-values for Example 10.1

If the *p*-value is less than 1%, we say that there is overwhelming evidence to infer that the alternative hypothesis is true. We also say that the test is **highly significant.**

If the *p*-value lies between 1 and 5%, there is strong evidence to infer that the alternative hypothesis is true. The result is deemed to be **significant.**

If the *p*-value is between 5 and 10%, we say that there is weak evidence to indicate that the alternative hypothesis is true. When the *p*-value is greater than 5%, we say that the result is not statistically significant.

When the *p*-value exceeds 10%, we say that there is no evidence to infer that the alternative hypothesis is true. Figure 10.6 summarizes these terms.

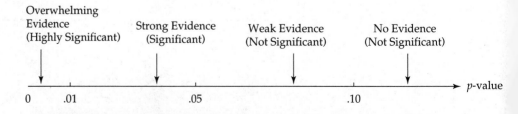

Figure 10.6

Describing the *p*-value of a test

GUIDELINES FOR CONDUCTING TESTS OF HYPOTHESES IN THIS BOOK AND BEYOND

In virtually all applications of statistical hypothesis testing, the statistics practitioner reads the *p*-value from the computer printout and interprets it. If it is clear that the cost of making a Type I error greatly exceeds the cost of a Type II error, the practitioner would not reject the null hypothesis in favor of the alternative unless the *p*-value is small, generally below 1%. That is, we would require overwhelming evidence to support the alternative hypothesis (and its resulting course of action).

If the cost of a Type II error is relatively large, *p*-values as large as 10% may provide sufficient evidence to reject the null hypothesis.

When the costs of the two types of error are similar or when it is difficult to evaluate the consequences of both types of error, we reject the null hypothesis when the *p*-value is less than 5%, which we use as a "standard."

In the exercises and cases in this book where we do not specify a significance level, you are expected to judge the size of the *p*-value using the guidelines above. It will be necessary for you to consider the consequences of each type of error. In some cases, the relative costs will be obvious. In others, a degree of subjective assessment is necessary.

In Example 10.1, the cost of making a Type I error (concluding that the system will be cost effective when in fact it is not) is likely much higher than the cost of a Type II error (not concluding the system will be cost effective when it is). The consequence of a Type I error is that the company will incur the cost of installing the system (perhaps hundreds of thousands of dollars) but will not gain much benefit from it. The cost of a Type II error is much lower, because we're likely to continue searching or developing another, perhaps better, system. See Section 10.4, which discusses how to calculate the probability of a Type II error.

We'll complete Example 10.1 by producing the Excel printout from the data in file Xm10-01.

Microsoft Excel Output for Example 10.1

Z-Test: Mean

	Accounts
Mean	178.00
Standard Deviation	68.367
Observations	400
Hypothesized Mean	170
SIGMA	65
z Stat	2.46
P(Z<=z) one-tail	0.0069
z Critical one-tail	1.6449
P(Z<=z) two-tail	0.0138
z Critical two-tail	1.96

The value of the test statistic is $z = 2.46$. The quantity **P(Z<=z) one-tail** is the probability in the tail region of the sampling distribution. It is the probability of observing as extreme a value as the test statistic. In this case, it is the *p*-value of the test. The next figure, **z Critical one-tail,** is the critical value defining the rejection region. The value 1.6449 is the result of specifying $\alpha = .05$ in the dialogue box. (See the instructions below.) We'll explain **P(Z<=z) two-tail** and **z Critical two-tail** later in this section, when we discuss two-tail tests.

COMMANDS	COMMANDS FOR EXAMPLE 10.1
1 Type or import the data.	Open file **Xm10-01.**
2 Click **Tools, Data Analysis Plus,** and z-Test: Mean.	
3 Specify the **Input Range:**.	A1:A401
4 Type the value of the **Hypothesized Mean:**.	170
5 Specify the population **Standard Deviation (Sigma).**	65
6 Click **Labels** if the first row of the input range contains the name of the variable.	
7 Type the value of α **(Alpha),** and click **OK.**	.05

The printout above was produced from the raw data. That is, we input the 400 observations in the data set, and the computer calculated the value of the test statistic and the *p*-value. As was the case with estimation, we can use Excel in another way. Open the file **Stats-Summary.xls.** Find the sheet **z-Test of a Mean.** You can input the components of the test $\bar{x}, \sigma, n$, and the hypothesized value of μ. The value of z and the one- and two-tail *p*-values (discussed below) will be output. We can also conduct a what-if analysis. Try changing any of the inputs to discover their effects on the test statistic and *p*-value.

▼ **EXAMPLE 10.2**

There are a variety of government agencies devoted to ensuring that food producers package their products in such a way that the weight or volume of the contents listed on the label is correct. For example, bottles of catsup whose labels state that the contents have a net weight of 16 ounces must have a net weight of at least 16 ounces. However, it is impossible to check all packages sold in the country. As a result, statistical techniques are used. A random sample of the product is selected and its contents measured. If the mean of the sample provides sufficient evidence to infer that the mean weight of all bottles is less than 16 ounces, the product label is deemed to be unacceptable. Suppose that a government inspector weighs the contents of a random sample of 25 bottles of catsup labeled "Net weight: 16 ounces" and records the measurements below. (The data are also stored on the data disk in file Xm10-02.) The inspector knows from previous experiments that the weights of all catsup bottles are normally distributed with a standard deviation of 0.4 ounces. Can the inspector conclude that the product label is unacceptable?

NET WEIGHT OF "16-OUNCE" CATSUP BOTTLES

15.8	16.1	15.6	16	15.7
16	16.2	15.9	16.8	15.8
16.2	17.3	16	15.7	15.6
15.7	15	16.2	15.6	15.5
15.4	16.8	15.6	15.3	15.7

Solution

IDENTIFY

The objective of the study is to draw a conclusion about the mean weight of all catsup bottles. Thus, the parameter to be tested is the population mean μ. We want to know if there is enough statistical evidence to show that the population mean is less than 16 ounces. Thus, the alternative hypothesis is

$$H_1 : \mu < 16$$

The null hypothesis automatically follows:

$$H_0 : \mu = 16$$

The test statistic is the only one we've presented thus far. It is

$$z = \frac{\bar{x} - \mu}{\sigma/\sqrt{n}}$$

COMPUTE

Manually

To solve this problem employing the rejection region method we require a significance level. A Type I error occurs when we reject a true null hypothesis. In this case, this means that the labels are valid, but the inspector concludes that they are not acceptable. The cost to the company would be considerable. As a consequence, a small significance level is recommended. Thus we set $\alpha = .01$.

We wish to reject the null hypothesis in favor of the alternative only if the sample mean and hence the value of the test statistic are small enough. As a result we locate the rejection region in the left tail of the sampling distribution. To understand why, remember that we're trying to decide if there is enough statistical evidence to infer that the mean is less than 16 (which is the alternative hypothesis). If we observe a large sample mean (and hence a large value of z), do we want to reject the null hypothesis in favor of the alternative? The answer is an emphatic no. It is absurd to conclude that if the sample mean is say, 20, there is enough evidence to conclude that the mean of all bottles is less than 16. Consequently, we want to reject the null hypothesis only if the sample mean (and hence the value of z) is small. How small is small enough? The answer is determined by the significance level and the rejection region. Thus, we set up the rejection region as

$$z < -z_\alpha = -z_{.01} = -2.33$$

Note that the direction of the inequality in the rejection region ($z < -z_\alpha$) matches the direction of the inequality in the alternative hypothesis ($\mu < 16$). Also note the negative sign, because the rejection region is in the left tail (containing values of z less than zero) of the sampling distribution.

From the data, we compute the sample mean. It is

$$\bar{x} = 15.90$$

Because the population standard deviation is known to be $\sigma = 0.4$, the sample size is $n = 25$, and the value of μ is hypothesized to be 16, we compute the value of the test statistic as

$$z = \frac{\bar{x} - \mu}{\sigma/\sqrt{n}} = \frac{15.90 - 16}{0.4/\sqrt{25}} = -1.25$$

Because the value of the test statistic, $z = -1.25$, is not less than -2.33, we do not reject the null hypothesis in favor of the alternative hypothesis. There is insufficient evidence to infer that the mean is less than 16 ounces.

We can determine the *p*-value of the test. It is

$$p\text{-value} = P(Z < -1.25) = .1056$$

In this type of one-tail (left-tail) test of hypothesis we calculate the *p*-value as $P(Z < z)$ where z is the actual value of the test statistic. Figure 10.7 depicts the sampling distribution, test statistic, and *p*-value.

Microsoft Excel Output for Example 10.2

Z-Test: Mean	
	Catsup
Mean	15.9
Standard Deviation	0.5017
Observations	25
Hypothesized Mean	16
SIGMA	0.4
z Stat	−1.25
P(Z<=z) one-tail	0.1056
z Critical one-tail	2.3263
P(Z<=z) two-tail	0.2112
z Critical two-tail	2.5758

The value of the test statistic is $z = -1.25$ and the *p*-value of the test is .1056.

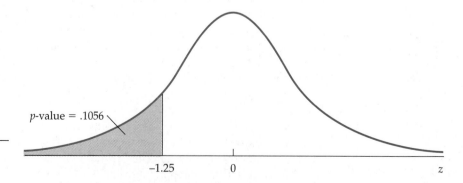

Figure 10.7

Sampling distribution for Example 10.2

INTERPRET

Because we were not able to reject the null hypothesis, we say that there is not enough evidence to infer that the mean weight of all catsup bottles is less than 16 ounces. Note that there was some evidence to indicate that the mean of the entire population of weights of catsup bottles is less than 16 ounces. We did calculate the sample mean to be 15.9. However, to reject the null hypothesis we need enough statistical evidence, and in this case we simply did not have enough reason to reject the null hypothesis in favor of the alternative. In the absence of evidence to show that the mean weight of all catsup bottles is less than 16 ounces, the inspector would not find the labels to be unacceptable.

▲

The statistical tests conducted in Examples 10.1 and 10.2 are called **one-tail tests** because the rejection region is located in only one tail of the sampling distribution. The right tail in Example 10.1 is the important one, because the alternative hypothesis specifies that the mean is *greater than* 170. In Example 10.2, the left tail is critical because the alternative hypothesis specifies that the mean is *less than* 16.

We'll now present an example that requires a **two-tail test.**

▼ EXAMPLE 10.3

The supervisor of a production line that assembles computer keyboards has been experiencing problems since a new process was instituted. He notes that there has been an increase in the number of defective units and occasional backlogs when one station's productivity does not match that of the other stations. Upon reviewing the process, the supervisor discovered that the management scientists who developed the production process assumed that the amount of time to complete a critical part of the process is normally distributed with a mean of 130 seconds and a standard deviation of 15 seconds. He is satisfied that the process times are normally distributed with a standard deviation of 15 seconds, but he is unsure about the mean assembly time. In order to determine what if any adjustments should be made to the line, he measures the times for 100 randomly selected assemblies of the critical part. (These data are listed below and stored in file Xm10-03.) Can the supervisor conclude that the management scientists' belief that the mean assembly time is 130 seconds is incorrect?

TIMES TO COMPLETE CRITICAL PART

145	126	142	147	96	130	135	152	130	107
128	129	129	132	103	138	140	120	117	131
153	106	115	112	119	113	134	144	99	130
141	118	147	133	141	104	141	117	106	140
119	122	148	126	130	116	131	142	121	121
101	133	139	114	127	132	133	134	114	140
136	124	108	136	130	136	126	125	121	134
104	146	147	112	148	149	133	130	120	157
154	136	141	127	125	105	126	120	107	103
106	113	132	116	138	104	111	138	117	106

Solution

IDENTIFY

In this problem we want to know whether the mean assembly time μ is different from 130 seconds. Consequently, we set up the alternative hypothesis to express this condition:

$$H_1 : \mu \neq 130$$

The null hypothesis specifies that the mean is equal to the value specified under the alternative hypothesis. Hence,

$$H_0 : \mu = 130$$

COMPUTE

Manually

To set up the rejection region we need to realize that we can reject the null hypothesis when the test statistic is large or when it is small. That is, we must set up a two-tail rejection region. Because the total area in the rejection region must be α, we divide this probability by 2. Thus, the rejection region* is

$$z < -z_{\alpha/2} \quad \text{or} \quad z > z_{\alpha/2}$$

Because we're uncertain whether a Type I or Type II error is more costly, we will set the significance level at 5%. Hence, $\alpha = .05$, $\alpha/2 = .025$, and $z_{\alpha/2} = z_{.025} = 1.96$. It follows that the rejection region is

$$z < -1.96 \quad \text{or} \quad z > 1.96$$

From the data we compute

$$\bar{x} = 126.8$$

The value of the test statistic is

$$z = \frac{\bar{x} - \mu}{\sigma/\sqrt{n}} = \frac{126.8 - 130}{15/\sqrt{100}} = -2.13$$

Because -2.13 is less than -1.96, we reject the null hypothesis.

We can also calculate the p-value of the test. Because it is a two-tail test, we determine the p-value by finding the area in both tails. That is,

$$p\text{-value} = P(Z < -2.13) + P(Z > 2.13) = .0166 + .0166 = .0332$$

Or, more simply, multiply the probability in one tail by 2.

In general, the p-value in a two-tail test is determined by

$$p\text{-value} = 2P(Z > |z|)$$

where z is the actual value of the test statistic and $|z|$ is its absolute value. Figure 10.8 exhibits the test statistic's sampling distribution and p-value.

---

* Statisticians often represent this rejection region as $|z| > z_{\alpha/2}$, which reads, "the *absolute* value of z is greater than $z_{\alpha/2}$." We prefer our method because it is clear that we are performing a two-tail test.

Figure 10.8

Sampling distribution for Example 10.3

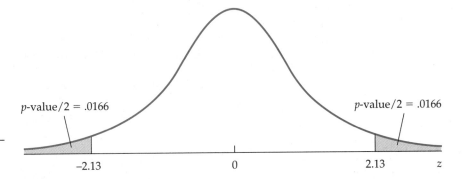

Microsoft Excel Output for Example 10.3

```
Z-Test: Mean
                        Time
Mean                    126.8
Standard Deviation      14.482
Observations            100
Hypothesized Mean       130
SIGMA                   15
z Stat                  −2.133
P(Z<=z) one-tail        0.0164
z Critical one-tail     1.6449
P(Z<=z) two-tail        0.0328
z Critical two-tail     1.96
```

The value of the test statistic is $z = -2.133$. The rejection region is defined by **z Critical two-tail**, which is 1.96. We read the *p*-value of the test from **P(Z<=z)two-tail**, which is .0328. (This differs slightly from the *p*-value calculated manually, because of rounding.)

INTERPRET

There is evidence to infer that the mean time to complete this stage of the production process is not equal to 130 seconds. The operations manager will have to juggle several issues in deciding how to proceed. First, how likely is it that the mean time is actually some other value than 130 seconds? The *p*-value indicates that there is statistical evidence, but it is hardly overwhelming. Second, the problem of bottlenecks and increased defects may be attributable to some other cause or causes. Third, even if the problems are isolated to this stage of the production process, it may not be possible or financially feasible to correct the process. It is clear that more investigating is necessary. However, the statistical analysis allows the manager to start the search for a solution.

TESTING HYPOTHESES AND INTERVAL ESTIMATORS

In this chapter and the previous one we showed that the test statistic and the interval estimator are both derived from the sampling distribution. It shouldn't be a surprise, then, that we can use the interval estimator to test hypotheses. To illustrate, consider Example 10.3. The 95% confidence interval estimate of the population mean is

$$\bar{x} \pm z_{\alpha/2} \frac{\sigma}{\sqrt{n}} = 126.8 \pm 1.96 \frac{15}{\sqrt{100}} = 126.8 \pm 2.94$$

LCL = 123.86   and   UCL = 129.74

We estimate that μ lies between 123.86 and 129.74. Because the interval does not include 130, we can conclude that there is sufficient evidence to infer that the population mean differs from 130.

The 95% confidence interval estimator for Example 10.2 is LCL = 15.74 and UCL = 16.06. Because this interval includes 16, we cannot infer that the labels are unacceptable.

In Example 10.1, the 95% confidence interval estimate is LCL = 171.63 and UCL = 184.37. The interval estimate excludes 170, allowing us to conclude that the population mean account is not equal to $170.

As you can see, the interval estimator can be employed to conduct tests of hypotheses. This process is equivalent to the rejection region approach. But, instead of finding the critical values of the rejection region and determining whether the test statistic falls into the rejection region, we compute the interval estimate and determine whether the hypothesized value of the mean falls into the interval.

Using the interval estimator to test hypotheses has the advantage of simplicity. Apparently, we don't need the formula for the test statistic; we only need the interval estimator. However, there are two serious drawbacks.

First, when conducting a one-tail test, our conclusion may not answer the question. In Example 10.1 we wanted to know whether there was enough evidence to infer that the mean is *greater than* 170. The interval estimate concludes that the mean *differs from* 170. You may be tempted to say that because the entire interval is greater than 170 there is enough statistical evidence to infer that the population mean is greater than 170. However, in attempting to draw this conclusion, we run into the problem of determining the procedure's significance level. Is it 5% or is it 2.5%? We may be able to overcome this problem through the use of **one-sided interval estimators**. However, if the purpose of using interval estimators instead of test statistics is simplicity, one-sided estimators are a contradiction.

Second, the interval estimator does not yield a *p*-value, which we have argued is the better way to draw inferences about a parameter. Using the interval estimator to test hypotheses forces the statistics practitioner into making a reject/don't reject decision rather than providing information about how much statistical evidence exists to be judged with other factors in the decision process. Furthermore, we only postpone the point in time when a test of hypothesis must be used. In later chapters we will present problems where only a test produces the information we need to make decisions.

DEVELOPING AN UNDERSTANDING OF STATISTICAL CONCEPTS 1

As is the case with the interval estimator, the test of hypothesis is based on the sampling distribution of the sample statistic. The result of a test of hypothesis is a probability statement about the sample statistic. We assume that the population mean is specified by the null hypothesis. We then compute the test statistic and determine how likely it is to observe this large (or small) a value when the null hypothesis is true. If the probability is small, we conclude that the assumption that the null hypothesis is true is unfounded, and we reject it.

DEVELOPING AN UNDERSTANDING OF STATISTICAL CONCEPTS 2

When we (or the computer) calculate the value of the test statistic

$$z = \frac{\bar{x} - \mu}{\sigma/\sqrt{n}}$$

we're also measuring the difference between the sample statistic $\bar{x}$ and the hypothesized value of the parameter μ. The unit of measurement of the difference is the standard error $\sigma/\sqrt{n}$. In Example 10.3, we found that the difference between the sample mean and the hypothesized mean was -2.133. That is, the sample mean was 2.133 standard errors below the hypothesized value of 130. The standard normal probability table told us that this value is considered unlikely. As a result we rejected the null hypothesis.

The concept of measuring the difference between the sample statistic and the hypothesized value of the parameter is one that will be used frequently throughout this book.

EXERCISES

The next five exercises feature nonstatistical applications of hypothesis testing. For each, identify the hypotheses, define Type I and Type II errors, and discuss the consequences of each error. In setting up the hypotheses you will have to consider where to place the "burden of proof."

10.1 It is the responsibility of the federal government to judge the safety and effectiveness of new drugs. There are two possible decisions: Approve the drug or disapprove the drug.

10.2 You are contemplating earning a Ph.D. in your favorite subject and then going on to a faculty position. If you succeed, a life of fame, fortune, and happiness awaits you. If you fail, you've wasted five years of your life. Should you go for it?

10.3 You are the center fielder of the New York Yankees. It is the bottom of the ninth inning of the seventh game of the World Series. The Yanks lead by 2 with 2 out and men on second and third. The batter is known to hit for high average and runs very well but has mediocre power. A single will tie the game, and a hit over your head will likely result in the Yanks losing. Do you play shallow?

10.4 You are faced with two investments. One is very risky, but the potential returns are high. The other is safe, but the potential is quite limited. Pick one.

10.5 (At the time this edition was being prepared Swissair Flight 111 had just crashed off the coast of Nova Scotia after the plane caught on fire.) You are the pilot of a jumbo jet. You smell smoke in the cockpit. The nearest airport is less than 5 minutes away. Should you land the plane immediately?

10.6 Several years ago in a high-profile case, a defendant was acquitted in a double-murder trial but was subsequently found responsible for the deaths in a civil trial. (Guess the name of the defendant—the answer is in Appendix C.) In a civil trial the plaintiffs (the victims' relatives) are required only to show that the preponderance of evidence points to the guilt of the defendant. Aside from the other issues in the cases, discuss why these results are logical.

For each of the next six exercises, calculate the value of the test statistic, set up the rejection region, determine the p-value and interpret the result.

10.7

$H_0: \mu = 1000$
$H_1: \mu \neq 1000$
$\sigma = 200, n = 100, \bar{x} = 980, \alpha = .01$

10.8

$H_0: \mu = 50$
$H_1: \mu > 50$
$\sigma = 5, n = 9, \bar{x} = 51, \alpha = .03$

10.9

$H_0: \mu = 15$
$H_1: \mu < 15$
$\sigma = 2, n = 25, \bar{x} = 14.3, \alpha = .10$

10.10

$H_0: \mu = 100$
$H_1: \mu \neq 100$
$\sigma = 10, n = 100, \bar{x} = 100, \alpha = .05$

10.11

$H_0: \mu = 70$
$H_1: \mu > 70$
$\sigma = 20, n = 100, \bar{x} = 80, \alpha = .01$

10.12

$H_0: \mu = 50$
$H_1: \mu < 50$
$\sigma = 15, n = 100, \bar{x} = 48, \alpha = .05$

*Exercises 10.13 to 10.30 are "what-if" analyses, designed to determine what happens to the test statistic and p-value when the sample size, standard deviation, and sample mean change. These problems can be solved manually or using the z-test of a Mean worksheet in **Stats-Summary**.*

10.13 A random sample of 50 young adult men (20 to 30 years old) was interviewed. Each person was asked how many minutes of sports he watched on television daily. The sample mean was found to be $\bar{x} = 64$. Suppose that $\sigma = 20$. Test to determine at the 5% significance level whether there is enough statistical evidence to infer that the mean amount of television watched by all young adult men is greater that 60 minutes.

10.14 Repeat Exercise 10.13 with $n = 25$.

10.15 Repeat Exercise 10.13 with $n = 100$.

10.16 Repeat Exercise 10.13 with $\sigma = 10$.

10.17 Repeat Exercise 10.13 with $\sigma = 40$.

10.18 Repeat Exercise 10.13 with $\bar{x} = 62$.

10.19 Repeat Exercise 10.13 with $\bar{x} = 68$.

10.20 Summarize Exercises 10.13 to 10.19 by describing what happens to the value of the test statistic when each of the following happens

a The sample size increases.
b The standard deviation decreases.
c The value of $\bar{x}$ increases.

10.21 A random sample of 50 second-year university students enrolled in a statistics course was drawn. At the course's completion, each student was asked how many hours he or she spent doing homework in statistics. The sample mean was computed to be $\bar{x} = 33$. It is known that the population standard deviation is $\sigma = 8.0$. The instructor has recommended that students devote 3 hours per week for the duration of the 12-week semester, which totals to 36 hours. Test to determine whether there is evidence that the average student spent less than the recommended amount of time. Compute the *p*-value of the test.

10.22 Repeat Exercise 10.21 with $n = 25$.

10.23 Repeat Exercise 10.21 with $n = 100$.

10.24 Repeat Exercise 10.21 with $\sigma = 5$.

10.25 Repeat Exercise 10.21 with $\sigma = 12$.

10.26 Repeat Exercise 10.21 with $\bar{x} = 30$.

10.27 Repeat Exercise 10.21 with $\bar{x} = 35$.

10.28 Summarize Exercises 10.21 to 10.27 by describing what happens to the *p*-value of the test when each of the following happens

    a The sample size increases.
    b The standard deviation decreases.
    c The value of $\bar{x}$ decreases.

10.29 Perform a what-if analysis on Example 10.1. Change each of the following, and determine the effect.

    a Sample size becomes 100.
    b The standard deviation is 100.
    c The value of the sample mean is 176.

10.30 Perform a what-if analysis on Example 10.2. Change each of the following, and determine the effect.

    a Sample size becomes 100.
    b The standard deviation is 1.
    c The value of the sample mean is 15.8.

10.31 Determine if there is enough statistical evidence at the 1% significance level to infer from the following information that the population mean is less than 250.

$$\bar{x} = 247, \ \sigma = 40, \ n = 400$$

10.32 A random sample of 200 observations from a normal population whose standard deviation is 100 produced a mean of 150. Does this statistic provide sufficient evidence at the 5% significance level to infer that the population mean is less than 160?

10.33 Determine if there is enough statistical evidence at the 10% significance level to infer that the population mean is not equal to 50, given that $\bar{x} = 56$, $\sigma = 10$, and $n = 25$.

10.34 Suppose that the following observations were drawn from a normal population whose standard deviation is 10. Test with $\alpha = .10$ to determine whether there is enough evidence to conclude that the population mean differs from 25.

21   37   33   47   28   16   29   37   41   20

10.35 Given the following data drawn from a population whose standard deviation is known to be 1, test to determine if there is enough evidence at the 5% significance level to infer that the population mean is greater than 5.

9   8   4   8   7   8   5   9   7   4   8
5   6   3   9   4   7   4   6   3   9

10.36 A machine that produces ball bearings is set so that the average diameter is .50 inch. In a sample of 100 ball bearings, it was found that $\bar{x} = .51$ inch. Assuming that the standard deviation is .05 inch, can we conclude at the 5% significance level that the mean diameter is not .50 inch?

The following exercises require the use of a computer and software. The answers may be calculated manually. See Appendix A for the sample statistics.

10.37 A manufacturer of lightbulbs advertises that, on average, its long-life bulb will last more than 5,000 hours. To test the claim, a statistician took a random sample of 100 bulbs and measured the amount of time until each bulb burned out. The data are stored in file Xr10-37.

    a If we assume that the lifetime of this type of bulb has a standard deviation of 400 hours, can we conclude at the 5% significance level that the claim is true?
    b What is the *p*-value of the test?

10.38 In the midst of labor–management negotiations, the president of a company argues that the company's blue-collar workers, who are paid an average of $30,000 per year, are well paid because the mean annual income of all blue-collar workers in the country is less than $30,000. That figure is disputed by the union, which does not believe that the mean blue-collar income is less than $30,000. To test the company president's belief, an arbitrator draws a random sample of 350 blue-collar workers from across the country and asks each to report his or her annual income. The results are stored in file Xr10-38. If the arbitrator assumes that the blue-collar incomes are distributed with a standard deviation of $8,000, can it be inferred at the 10% significance level that the company president is correct?

10.39 The academic vice-president of a large university claims that the SAT scores of applicants to the university have increased during the past five years. Five years ago, the mean and standard deviation of SAT scores of high school applicants were 560 and 50, respectively. Twenty applications for this year's program were randomly selected and the SAT scores recorded. These are stored in file Xr10-39. If we assume that the distribution of SAT scores of this year's applicants is the same as that of five years ago, with the possible exception of the mean, can we conclude at the 1% significance level that the dean's claim is true?

10.40 A study in the *Academy of Management Journal* (D. R. Woods and R. L. LaForge, "The Impact of Comprehensive Planning on Financial Performance," *Academy of Management Journal* 22 [3] [1979]: 516–526) reported that the average annual return on investment for American banks was 10.2% with a standard deviation of 0.8%. The article hypothesized that banks that exercised comprehensive planning would outperform the average bank. A random sample of 26 banks that exercised comprehensive planning was drawn, and the return on investment for each was calculated and stored in file Xr10-40. Assuming that the return on investment is normally distributed with a standard deviation of 0.8%, can we conclude that the article's hypothesis is correct? Report the p-value of the text and judge its size.

10.41 Past experience indicates that the monthly long-distance telephone bill is normally distributed with a mean of $17.85 and a standard deviation of $3.87. After an advertising campaign aimed at increasing long-distance telephone usage, a random sample of 25 household bills was taken. The results are stored in file Xr10-41.

a Calculate the p-value of a test to determine whether we can infer that the campaign was successful. What does the p-value tell you? Assume that the standard deviation of the bills after the campaign is the same as before.

b What are the consequences of Type I and Type II errors? Which is more expensive?

10.42 Many Alpine ski centers base their projections of revenues and profits on the assumption that the average Alpine skier skis four times per year. To investigate the validity of this assumption, a random sample of skiers is drawn, and each is asked to report the number of times he or she skied the previous year. The responses are stored in file Xr10-42. If we assume that the standard deviation is 2, can we infer at the 10% significance level that the assumption is wrong?

10.43 Companies that send out bills to their customers would like them to be paid more quickly. An accountant for an electricity company suggested that changing the color of the paper that the invoice is printed on might improve speed of payment. The current mean time to pay bills is 22 days with a standard deviation of 6 days. An experiment was organized wherein a random sample of 500 customers was selected and sent an invoice colored blue. The number of days until payment was received was recorded in file Xr10-43.

a Find the p-value of the test to determine whether there is enough evidence to infer that the blue invoice improves speed of payment. What does the p-value tell you?

b What assumption was made to answer part (a)?

c Discuss the consequences of Type I and Type II errors. Which is more expensive?

10.44 The golf professional at a private course claims that members who have taken lessons from him lowered their handicap by more than 5 strokes. The club manager decides to test the claim by randomly sampling 25 members who have had lessons and asking each to report the reduction in handicap. These data are stored in file Xr10-44, where a negative number indicates an increase in the handicap. Assuming that the reduction in handicap is approximately normally distributed with a standard deviation of 2 strokes, test the golf professional's claim using a 10% significance level.

10.45 In an attempt to reduce the number of person-hours lost as a result of industrial accidents, a large production plant installed new safety equipment. In a test of the effectiveness of the equipment, a random sample of 50 departments was chosen. The number of person-hours lost in the month prior to and the month after the installation of the safety equipment was recorded. The percentage change was calculated, and the data stored in file Xr10-45. Assume that the population standard deviation is 5. What conclusion can you draw? (Interpret the p-value, or set up the rejection region using what you judge to be an appropriate significance level.)

10.46 A highway patrol officer believes that the average speed of cars traveling over a certain stretch of highway exceeds the posted limit of 55 mph. The speeds of a random sample of 200 cars were recorded and stored in file Xr10-46. Do these data provide sufficient evidence at the 1% significance level to support the officer's belief? What is the p-value of the test? (Assume that the standard deviation is known to be 5.)

10.47 An automotive expert claims that the large number of self-serve gasoline stations has resulted in poor automobile maintenance, and that the average tire pressure is more than 4 psi (pounds per square inch) below its manufacturer's specification. As a quick test, 50 tires are examined, and the number of psi each tire is below specification is recorded and stored in file Xr10-47. If we assume that tire pressure is normally distributed with $\sigma = 1.5$ psi, can we infer at the 10% significance level that the expert is correct? What is the *p*-value?

10.48 For the past few years, the number of customers of a drive-up bank in New York has averaged 20 per hour, with a standard deviation of 3 per hour. This year, another bank one mile away opened a drive-up window. The manager of the first bank believes that this will result in a decrease in the number of customers. The number of customers who arrived during 36 randomly selected hours was recorded and stored in file Xr10-48. Can we conclude that the manager is correct? What is the *p*-value? (Judge the *p*-value by considering the costs of Type I and Type II errors.)

10.49 A fast-food franchiser is considering building a restaurant at a certain location. Based on financial analyses, a site is acceptable only if the number of pedestrians passing the location averages more than 100 per hour. The number of pedestrians observed for each of 40 hours was recorded and stored in file Xr10-49.

  a Assuming that the population standard deviation is known to be 12, can we conclude that the site is acceptable?

  b Determine the *p*-value of the test. Interpret its value relative to the costs of Type I and Type II errors.

10.50 In recent years, a number of companies have been formed that offer competition to AT&T in long-distance calls. All advertise that their rates are lower than AT&T's, and as a result their bills will be lower. AT&T has responded by arguing that for the average consumer there will be no difference in billing. Suppose that a statistician working for AT&T determines that the mean and standard deviation of monthly long-distance bills for all its residential customers are $17.85 and $3.87, respectively. He then takes a random sample of 100 customers and recalculates their last month's bill using the rates quoted by a leading competitor. These data are stored in file Xr10-50. Assuming that the standard deviation of this population is the same as for AT&T, can we conclude at the 5% significance level that there is a difference between AT&T's bills and those of the leading competitor? What is the *p*-value of the test?

10.51 The current no-smoking regulations in office buildings require workers who smoke to take breaks and leave the building in order to satisfy their habits. A study indicates that such workers average 32 minutes per day taking smoking breaks. The standard deviation is 6 minutes. To help reduce the average, break rooms with powerful exhausts were installed in the buildings. To see if these rooms serve their designed purpose, a random sample of smokers was taken. The total amount of time away from their desks was measured for one day. These data are stored in file Xr10-51. Test to determine whether there has been a decrease in the mean time away from their desks. Compute the *p*-value, and judge it relative to the costs of Type I and Type II errors.

10.52 A low-handicap golfer who uses Titleist brand golf balls observed that his average drive is 230 yards with a standard deviation of 10 yards. Nike has just introduced a new ball, which has been endorsed by Tiger Woods. Nike claims that the ball will travel further than Titleist. To test the claim the golfer hits 100 drives with a Nike ball and measures the distance. These data are stored in file Xr10-52. Conduct a test to determine whether Nike is correct. Compute the *p*-value, and interpret its value relative to the costs of Type I and Type II errors.

10.4 CALCULATING THE PROBABILITY OF A TYPE II ERROR

To properly interpret the results of a test of hypothesis requires that you be able to specify an appropriate significance level or to judge the *p*-value of a test. However, it also requires that you have an understanding of the relationship between Type I and Type II errors. In this section, we describe how the probability of a Type II error is computed and interpreted.

Recall Example 10.1, where we conducted the test using the sample mean as the test statistic and we computed the rejection region (with $\alpha = .05$) as

$$\bar{x} > 175.34$$

A Type II error occurs when a false null hypothesis is not rejected. Thus, in Example 10.1, if $\bar{x}$ is less than 175.34, we will not reject the null hypothesis. If we do not reject the null hypothesis, we will not install the new billing system. Thus, the consequence of a Type II error in this example is that we will not install the new system when it would be cost effective. The probability of this occurring is the probability of a Type II error. It is defined as

$$\beta = P(\bar{x} < 175.34, \text{ given that the null hypothesis is false})$$

The condition that the null hypothesis is false only tells us that the mean is not equal to 170. If we want to compute β, we need to specify a value for μ. Suppose that, when the mean account is at least \$180, the new billing system's savings become so attractive that the manager would hate to make the mistake of not installing it. As a result he would like to determine the probability of not installing the new system when it would produce large cost savings. Because calculating sampling distribution (approximately normal) probabilities requires us to substitute one value of μ (as well as σ and n), we will calculate the probability of not installing the new system when μ is *equal* to 180.

$$\beta = P(\bar{x} < 175.34, \text{ given that } \mu = 180)$$

We know that $\bar{x}$ is approximately normally distributed with mean μ and standard deviation $\sigma/\sqrt{n}$. To proceed, we standardize $\bar{x}$ and use the standard normal table (Table 3 in Appendix B).

$$\beta = P\left(\frac{\bar{x} - \mu}{\sigma/\sqrt{n}} < \frac{175.34 - 180}{65/\sqrt{400}}\right) = P(Z < -1.43) = .0764$$

This tells us that when the mean account is actually \$180, the probability of incorrectly not rejecting the null hypothesis is .0764. Figure 10.9 graphically depicts how the calculation was performed. Notice that in order to calculate the probability of a Type II error, we had to express the rejection region in terms of the unstandardized

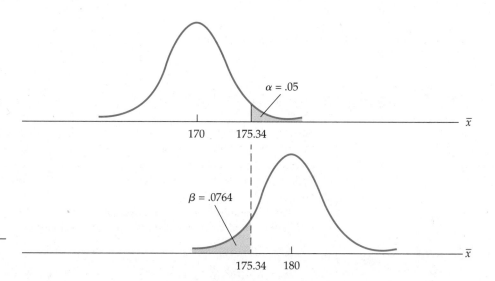

Figure 10.9

Calculating β for $\mu = 180$, $\alpha = .05$, and $n = 400$

test statistic $\bar{x}$, and we had to specify a value for μ other than the one shown in the null hypothesis. In this illustration the value of μ used was based on a financial analysis indicating that when μ is at least $180, the cost savings would be very attractive.

EFFECT ON β OF CHANGING μ

We can see the effect of using different values of μ in the calculation of the probability of a Type II error. For example, suppose that instead of using $180 we use $185. That is, we wish to compute

$$\beta = P(\bar{x} < 175.34, \text{ given that } \mu = 185)$$

Algebraically we get

$$\beta = P\left(\frac{\bar{x} - \mu}{\sigma/\sqrt{n}} < \frac{175.34 - 185}{65/\sqrt{400}}\right) = P(Z < -2.97) = .0015$$

As you can see, by increasing the value of μ the probability of a Type II error decreases. This is logical, because when the actual value of μ is further away from the value of μ specified under the null hypothesis, it is easier to make the correct decision. Making it easier to choose correctly is reflected in a smaller probability of a Type II error.

Figure 10.10 depicts this calculation. Notice that, by moving μ to the right, we shift the entire distribution to the right, which results in a smaller value of β. If we use a smaller value of μ the distribution shifts left. For example, if $\mu = 172$, the probability of of a Type II error is computed as

$$\beta = P(\bar{x} < 175.34, \text{ given that } \mu = 172)$$

$$\beta = P\left(\frac{\bar{x} - \mu}{\sigma/\sqrt{n}} < \frac{175.34 - 172}{65/\sqrt{400}}\right) = P(Z < 1.03) = .5 + 3485 = .8485$$

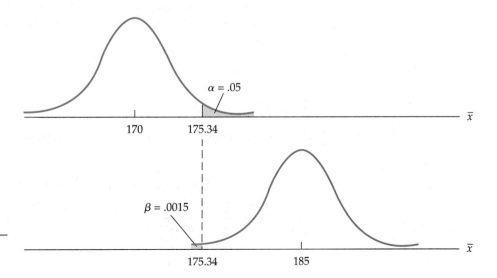

Figure 10.10

Calculating β for $\mu = 185$, $\alpha = .05$, and $n = 400$

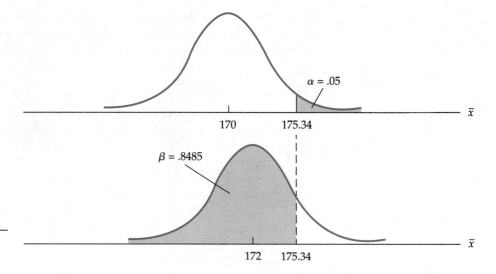

Figure 10.11

Calculating β for $\mu = 172$, $\alpha = .05$, and $n = 400$

See Figure 10.11. Now it is more difficult to make the correct decision, which is reflected by a larger value of β.

It should be noted that, because the alternative hypothesis specifies $\mu > 170$, we have no interest in using a value of μ that is less than 170. Although such a calculation is possible, it has little meaning.

EFFECT ON β OF CHANGING α

Suppose that, in the illustration above (where we assumed that the actual value of μ is 180), we had used a significance level of 1% instead of 5%. The rejection region expressed in terms of the standardized test statistic would be

$$z > z_{.01} = 2.33$$

or

$$\frac{\bar{x} - 170}{65/\sqrt{400}} > 2.33$$

Solving for $\bar{x}$, we find the rejection region in terms of the unstandardized test statistic:

$$\bar{x} > 177.57$$

The probability of a Type II error when $\mu = 180$ is

$$\beta = P\left(\frac{\bar{x} - \mu}{\sigma/\sqrt{n}} < \frac{177.57 - 180}{65/\sqrt{400}}\right) = P(Z < -.74) = .2296$$

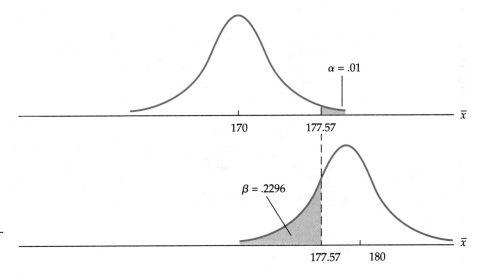

Figure 10.12

Calculating β for $\mu = 180$, $\alpha = .01$, and $n = 400$

Figure 10.12 depicts this calculation. Compare this figure with Figure 10.9. As you can see, by decreasing the significance level from 5% to 1% we have shifted the critical value of the rejection region to the right and thus enlarged the area where the null hypothesis is not rejected. The probability of a Type II error increases from .0764 to .2296.

This calculation illustrates the inverse relationship between the probabilities of Type I and Type II errors alluded to in Section 10.2. It is important to understand this relationship. From a practical point of view, it tells us that if you want to decrease the probability of a Type I error (by specifying a small value of α), you increase the probability of a Type II error.

In applications where the cost of a Type I error is considerably larger than the cost of a Type II error, this is appropriate. In fact, a significance level of 1% or less is probably justified. However, when the cost of a Type II error is relatively large, a significance level of 5% or more may be appropriate.

Unfortunately, there is no simple formula to determine what the significance level should be. It is necessary for the statistics practitioner to consider the costs of both mistakes in deciding what to do. Judgment and knowledge of the factors in the decision are critical.

JUDGING THE TEST

There is another important concept to be derived from this section. A statistical test of hypothesis is effectively defined by the significance level and the sample size, both of which are selected by the statistics practitioner. We can judge how well the test functions by calculating the probability of a Type II error at some value of the parameter. To illustrate, in Example 10.1 the manager chose a sample size of 400 and a 5% significance level on which to base his decision. With those selections we found β to be .0764 when the actual mean is 180. If we believe that the cost of a Type II error is high, and thus that the probability is too large, we have two ways to reduce the probability. We can increase the value of α. However, this would result in an increase in the chances of making a Type I error, which is very costly. Or, we can increase the sample size.

Suppose that the manager chose a sample size of 1,000. We'll now recalculate β with $n = 1,000$ (and $\alpha = .05$ and the actual value of $\mu = 180$). The rejection region is

$$z > z_{.05} = 1.645$$

or

$$\frac{\bar{x} - 170}{65/\sqrt{1000}} > 1.645$$

which yields

$$\bar{x} > 173.38$$

The probability of a Type II error is

$$\beta = P\left(\frac{\bar{x} - \mu}{\sigma/\sqrt{n}} < \frac{173.38 - 180}{65/\sqrt{1000}}\right) = P(Z < -3.22) = 0 \text{ (approximately)}$$

In this case, we left $\alpha = .05$, but we reduced the probability of not installing the system when the actual mean account is $180 to virtually zero.

DEVELOPING AN UNDERSTANDING OF STATISTICAL CONCEPTS: LARGER SAMPLE SIZE EQUALS MORE INFORMATION EQUALS BETTER DECISIONS

Figure 10.13 displays the calculation of β when $\mu = 180$, $\alpha = .05$, and $n = 1,000$. As compared to Figure 10.9 (calculation of β when $\mu = 180$, $\alpha = .05$, and $n = 400$), we can see that the sampling distribution of the mean is narrower because the standard error of the mean $\sigma/\sqrt{n}$ becomes smaller as n increases. Narrower distributions represent more information. The increased information is reflected in a smaller probability of a Type II error.

The calculation of the probability of a Type II error for $n = 400$ and for $n = 1,000$ illustrates a concept whose importance cannot be overstated. By increasing the sample size, we reduce the probability of a Type II error. By reducing the probability of a Type II error, we make this type of error less frequently, and hence we make better decisions in the long run. This finding lies at the heart of applied statistical analysis and reinforces the book's first sentence: "Statistics is a way to get information from data."

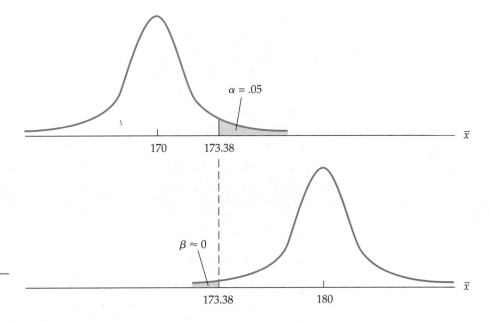

Figure 10.13

Calculating β for $\mu = 180$, $\alpha = .05$, and $n = 1,000$

POWER OF A TEST

Another way of expressing how well a test performs is to report its power—the probability of its leading us to rejecting the null hypothesis when it is false. Thus, the power of a test is $1 - \beta$.

When more than one test can be performed in a given situation, we would naturally prefer to use the test that is correct more frequently. If (given the same alternative hypothesis, sample size, and significance level) one test has a higher power than a second test, the first test is said to be more powerful.

Using Microsoft Excel to Determine β

We have made it possible to utilize Excel to calculate β for any test of hypothesis. To do so, follow these instructions.

Open **Stats-Summary.xls.** Find the worksheet titled **Beta.** In this worksheet, we have calculated the probability of a Type II error for three different tests. Column B performs a one-tail (right-tail) test; Column C, a one-tail (left-tail) test; and Column D, a two-tail test. The worksheet shows the value of β for Example 10.1 (with $\mu = 180$), Example 10.2 (with $\mu = 15.8$), and Example 10.3 (with $\mu = 135$). To use the worksheet, go to the column representing the test you are conducting, and substitute new values for μ under the null hypothesis, σ, n, α, and the value of μ under the alternative hypothesis in rows 5 through 9, respectively. In row 11, Excel computes the critical value of the rejection region for the two one-tail tests and in rows 11 and 12 for the two-tail test. Rows 14 and 16 list the value of β and the power of the test.

Microsoft Excel Output of the Probability of a Type II Error and the Power of the Test

We show the current form of the worksheet below.

Probability of a Type II error

	Right-tail Test	Left-tail Test	Two-tail Test
H0: MU	170	16	130
SIGMA	65	0.4	15
Sample size	400	25	100
ALPHA	0.05	0.05	0.05
H1: MU	180	15.8	135
Critical value(s)	175.35	15.87	127.06
			132.94
Prob(Type II error)	0.0761	0.1962	0.0848
Power of the test	0.9239	0.8038	0.9152

In Column C, Excel computes the probability of a Type II error for Example 10.2 when the mean weight of all catsup bottles is actually 15.8 ounces. The probability of not finding the labels unacceptable when the mean weight is 15.8 is .1962. If regulators decide that this probability is too large, they would have to change the test. If we increase the significance level, we would increase the percentage of tests where we erroneously conclude that the labels are unacceptable when in fact they are acceptable. We can also increase the sample size. (See Exercise 10.53.)

Column D shows the calculation of β for Example 10.3 when $\mu = 135$. Note that in a two-tail test, there are two critical values that define the rejection region. The probability of a Type II error is the area between these two values, assuming that population mean is 135 (in this illustration).

SETTING UP THE ALTERNATIVE HYPOTHESIS TO DEFINE TYPE I AND TYPE II ERRORS

We've already discussed how the alternative hypothesis is set up. It represents the condition we're investigating. In Example 10.2, we wanted to know whether there was sufficient statistical evidence to infer that the labels are unacceptable. That is, that the mean of all catsup bottles is less than 16 ounces. In this textbook, you will encounter many problems using similar phraseology. Your job will be to conduct the test that answers the question.

In real life, however, the statistics practitioner (that's you, five years from now) will be asking and answering the question. In general, you will find that the question can be posed in two ways. In Example 10.2, we asked whether there was evidence to conclude that the labels are unacceptable. Another way of investigating the issue is to determine whether there is sufficient evidence to infer that the labels are acceptable. We remind you of the criminal trial analogy. In a criminal trial, the burden of proof falls on the prosecution to prove that the defendant is guilty. In other countries with less emphasis on individual rights, the defendant is required to prove his or her innocence. In the United States and Canada (and in other countries), we chose the former because we consider the conviction of an innocent defendant to be the greater error. Thus the test is set up with the null and alternative hypotheses as described in Section 10.2.

In a statistical test where we are responsible for asking the question, as well as answering it, we must ask the question so that we directly control the error that is more costly. As you have already seen, we control the probability of a Type I error by specifying its value (the significance level). Consider Example 10.2 once again. There are two possible errors: conclude that the labels are unacceptable when they are acceptable, and conclude that the labels are acceptable when they are not. Which error is more costly? If the government regulators conclude that the labels are unacceptable, the catsup manufacturer will have to take some remedial action. If, in reality, the labels are accurate, the company will incur a high cost for no valid reason. On the other hand, if the regulators conclude that the labels are acceptable (by not taking any action) but in reality they are not, consumers will receive less catsup than they are entitled to. The cost to individual consumers is small. Consequently, the error we wish to avoid is the erroneous conclusion that the labels are unacceptable. We define this as a Type I error. As a result the "burden of proof" is placed on government regulators to deliver sufficient statistical evidence that the mean net weight is less than 16 ounces. The alternative hypothesis is formulated accordingly.

EXERCISES

10.53 Calculate the probability of a Type II error in Example 10.2 when $\mu = 15.80$ and the sample size is 100.

10.54 Calculate the probability of a Type II error in Example 10.2 if the true mean weight of all catsup bottles is 15.7 ounces.

10.55 Find the probability of a Type II error in Example 10.3 if $\mu = 132$.

10.56 Draw a figure similar to Figure 10.9 depicting the calculation in Exercise 10.54.

10.57 Draw a figure similar to Figure 10.9 depicting the calculation in Exercise 10.55.

10.58 Calculate the probability of a Type II error for the following test of hypothesis, given that $\mu = 203$:

$H_0: \mu = 200$

$H_1: \mu \neq 200$

$\alpha = .05, \sigma = 10, n = 100$

10.59 Find the probability of a Type II error for the following test of hypothesis, given that $\mu = 1050$:

$H_0: \mu = 1{,}000$

$H_1: \mu > 1{,}000$

$\alpha = .01, \sigma = 50, n = 25$

10.60 Determine β for the following test of hypothesis, given that $\mu = 48$:

$H_0: \mu = 50$

$H_1: \mu < 50$

$\alpha = .05, \sigma = 10, n = 40$

10.61 For the test of hypothesis

$H_0: \mu = 1{,}000$

$H_1: \mu \neq 1{,}000$

$\alpha = .05, \sigma = 200, n = 100$

find β when $\mu = 900, 940, 980, 1{,}020, 1{,}060, 1{,}100$.

10.62 For Exercise 10.61, graph μ (on the horizontal axis) versus β (on the vertical axis). If necessary, calculate β for additional values of μ. The resulting graph is called the *operating characteristic (OC) curve*.

10.63 For Exercise 10.61, graph μ versus $1 - \beta$. Recall that $1 - \beta$ is called the power of the test; as a consequence, the graph is called the *power curve*.

10.64 Repeat Exercises 10.61 through 10.63 with $n = 25$.

10.65 What do you notice about the graphs in Exercises 10.63 and 10.64? What are the implications of your observations?

10.66 Refer to Exercise 10.39. Find the probability of erroneously concluding that there is not enough evidence to support the claim when, in fact, the true mean GMAT score is 600.

10.67 Refer to Exercise 10.41. Find the probability of a Type II error when the true mean is $18.00. (Use $\alpha = .01$.)

10.68 Refer to Exercise 10.43. Find the probability of concluding that the blue invoices do not improve the speed of payment when the mean is really 21 days. (Use $\alpha = .05$.)

10.69 Refer to Exercise 10.49. What is the probability of committing a Type II error when the mean number of pedestrians is 105? (Use $\alpha = .05$.)

10.70 Suppose that in Example 10.1 we wanted to determine whether there was sufficient evidence to conclude that the new system would *not* be cost effective. Set up the null and alternative hypotheses, and discuss the consequences of Type I and Type II errors. Conduct the test. Is your conclusion the same as the one reached in Example 10.1? Explain.

10.71 A school-board administrator believes that the average number of days absent per year among students is fewer than 10 days. From past experience, he knows that the population standard deviation is 3 days. In testing to determine whether his belief is true, he could use any of the following plans:

    i $n = 100$, $\alpha = .01$
    ii $n = 75$, $\alpha = .05$
    iii $n = 50$, $\alpha = .10$

Which plan has the lower probability of a Type II error, given that the true population average is 9 days?

10.5 THE ROAD AHEAD

We had two principal goals to accomplish in Chapters 9 and 10. First, we wanted to present the concepts of estimation and hypothesis testing. Second, we wanted to show how to produce interval estimates and conduct tests of hypotheses. The importance of both of these goals should not be underestimated. Almost everything that follows this chapter will involve either estimating a parameter or testing a set of hypotheses. Consequently, Sections 9.3 and 10.3 set the pattern for the way in which statistical techniques are applied. It is no exaggeration to state that if you understand how to produce and use interval estimates and how to conduct and interpret hypothesis tests, then you are well on your way to the ultimate goal of being a competent statistics practitioner. It is fair for you to ask what more you must accomplish to achieve this goal. The answer, simply put, is much more of the same.

In the chapters that follow, we plan to present about three dozen different statistical techniques that can be (and frequently are) employed by statistics practitioners. To calculate manually the value of test statistics and interval estimates requires nothing more than the ability to add subtract, multiply, divide, and compute square roots. If you intend to use the computer, all you need to know are the commands. The key, then, to applying statistics is knowing which formula to calculate or which set of commands to issue. Thus, the real challenge of the subject lies in being able to define the problem and identify which statistical method is the most appropriate one to use.

Most students have some difficulty recognizing the particular kind of statistical problem they are addressing unless, of course, the problem appears among the exercises at the end of a section that just introduced the technique needed. Unfortunately, in practice, statistical problems do not appear already so identified. Consequently, we have adopted an approach to teaching statistics that is designed to help identify the statistical technique.

A number of factors determine which statistical method should be used, but two are especially important: the type of data and the purpose of the statistical inference.

In Chapter 2, we pointed out that there are effectively three types of data—nominal, ordinal, and interval. Recall that nominal data represent categories such as marital status, occupation, and gender. Statistics practitioners often record nominal data by assigning numbers to the responses (e.g., 1 = single; 2 = married; 3 = divorced; 4 = widowed). Because these numbers are assigned completely arbitrarily, any calculations performed on them are meaningless. All that we can do with nominal data is count the number of times each category is observed. Ordinal data are obtained from questions whose answers represent a rating or ranking system. For example, if students are asked to rate a university professor, the responses may be excellent, good, fair, or poor. To draw inferences about such data, we convert the responses to numbers. Any numbering system is valid as long as the order of the responses is preserved. Thus "4 = excellent; 3 = good; 2 = fair; 1 = poor" is just as valid as "10 = excellent; 5 = good; 2 = fair; 1 = poor." Because of this feature, the most appropriate statistical procedures for ordinal data are ones based on a ranking process.

Interval data are real numbers such as those representing income, age, height, weight, and volume. Computation of means and variances is permissible.

The second key factor in determining the statistical technique is the purpose of doing the work. Every statistical method has some specific objective. There are five such objectives addressed in this book.

PROBLEM OBJECTIVES

1 **Describe a single population.** Our objective here is to describe some property of a population of interest. The decision about which property to describe is generally dictated by the type of data. For example, suppose the population of interest consists of all purchasers of home computers. If we are interested in the purchasers' incomes (for which the data are interval), we may calculate the mean or the variance to describe that aspect of the population. But if we are interested in the brand of computer that has been bought (for which the data are nominal), all we can do is compute the proportion of the population that purchases each brand.

2 **Compare two populations.** In this case, our goal is to compare a property of one population with a corresponding property of a second population. For example, suppose the populations of interest are male and female purchasers of computers. We could compare the means of their incomes, or compare the proportion of each population that purchases a given brand. Once again, the data type generally determines what kinds of properties we compare.

3 **Compare two or more populations.** We might want to compare the average income in each of several locations in order (for example) to decide where to build a new shopping center. Or we might want to compare the proportions of voters who support the incumbent candidate in different constituencies in order to determine where the campaign should focus its attention. In each case, the problem objective involves comparing two or more populations.

4 **Analyze the relationship between two variables.** There are numerous situations in which we want to know how one variable is related to another. Governments need to know what effect rising interest rates have on the unemployment rate. Companies want to investigate how the sizes of their advertising

budgets influence sales volume. In most of the problems in this introductory text, the two variables to be analyzed will be of the same type; we will not attempt to cover the fairly large body of statistical techniques that has been developed to deal with two variables of different types.

5 **Analyze the relationship among two or more variables.** Our objective here is usually to forecast one variable (called the dependent variable) on the basis of several other variables (called independent variables). We will deal with this problem only in situations in which all variables are interval.

Table 10.2 lists the types of data and the five **problem objectives.** For each combination, the table specifies the chapter and/or section where the appropriate statistical technique is presented. For your convenience, a more detailed version of this table is reproduced inside the front cover of this book.

A BRIEF COMMENT ABOUT DERIVATIONS

Because this book is about statistical applications, we assume that our readers have little interest in the mathematical derivations of the techniques described. However, it might be helpful for you to have some understanding about the process that produces the formulas.

As described above, factors such as the problem objective and the type of data determine the parameter to be estimated and tested. For each parameter, statisticians have determined which statistic to use. That statistic has a sampling distribution that can usually be expressed as a formula. For example, in this chapter, the parameter of interest was the population mean μ, whose best estimator is the sample mean $\bar{x}$. Assuming that the population standard deviation σ is known, the sampling distribution of $\bar{x}$ is normal (or approximately so) with mean μ and standard deviation $\sigma/\sqrt{n}$. The sampling distribution can be described by the formula

$$z = \frac{\bar{x} - \mu}{\sigma/\sqrt{n}}$$

Table 10.2 Guide to Statistical Inference Showing Where Each Technique Is Introduced

	Data Type		
Problem Objective	Nominal	Ordinal	Interval
Describe a single population	Sec 11.4, 15.2	Not covered	Sec 11.2, 11.3
Compare two populations	Sec 12.6, 15.3	Sec 16.2, 16.3	Sec 12.2, 12.4, 12.5, 16.2, 16.3
Compare two or more populations	Sec 15.3	Sec 16.4, 16.5	Chapter 14 Sec 16.4, 16.5
Analyze the relationship between two variables	Sec 15.3	Sec 17.7	Chapter 17
Analyze the relationship among two or more variables	Not covered	Not covered	Chapter 18

This formula also describes the test statistic for μ with σ known. With a little algebra, we were able to derive (in Section 9.3) the interval estimator of μ.

In future chapters, we will repeat this process, which in several cases involves the introduction of another sampling distribution. The sampling distributions we will use are the Student t, chi-squared, and F distributions, which were presented in Section 6.4. While its shape and formula will differ from the sampling distribution used in this chapter, the pattern will be the same. In general, the formula that expresses the sampling distribution will describe the test statistic. Then some algebraic manipulation (which we will not show) produces the interval estimator. Consequently, we will reverse the order of presentation of the two techniques. That is, we will present the test of hypothesis first, followed by the interval estimator.

10.6 SUMMARY

In this chapter, we introduced the concepts of hypothesis testing and applied them to testing hypotheses about a population mean. We showed how to specify the null and alternative hypotheses, set up the rejection region, compute the value of the test statistic, and, finally, to make a decision. Equally as important, we discussed how to interpret the test results. This chapter also demonstrated another way to make decisions—by calculating and using the p-value of the test. To help interpret test results, we showed how to calculate the probability of a Type II error. Finally, we provided a road map of how we plan to present statistical techniques. We complete this summary by listing the formulas described in this chapter.

IMPORTANT TERMS

Hypothesis testing
Null hypothesis
Alternative or research hypothesis
Type I error
Type II error
Significance level
Test statistic
Rejection region

Standardized test statistic
p-value of the test
Highly significant
Significant
One-tail test
Two-tail test
One-sided interval estimator
Problem objective

SYMBOLS

Symbol	Pronounced	Represents
H_0	H-nought	Null hypothesis
H_1	H-one	Alternative hypothesis
α	alpha	Probability of a Type I error
β	beta	Probability of a Type II error
$\bar{x}_L$	X-bar-sub L or X-bar-L	Value of $\bar{x}$ large enough to reject H_0

FORMULAS

Test statistic

$$z = \frac{\bar{x} - \mu}{\sigma/\sqrt{n}}$$

MICROSOFT EXCEL OUTPUT AND INSTRUCTIONS

Technique	Page
Test of μ (Raw data)	267–268
β	286

Chapter 11

Inference about a Single Population

11.1 Introduction

11.2 Inference about a Population Mean when the Standard Deviation Is Unknown

11.3 Inference about a Population Variance

11.4 Inference about a Population Proportion

11.5 Summary

11.1 INTRODUCTION

In the previous two chapters, we introduced the concepts of statistical inference and showed how to estimate and test a population mean. However, the illustration we chose is unrealistic because the techniques require us to use the population standard deviation σ, which, in general, is unknown. The purpose, then, of Chapters 9 and 10 was to create the model for the way in which we plan to present other statistical techniques. That is, we will begin by identifying the parameter to be estimated or tested. We will then specify the parameter's estimator (each parameter has an estimator chosen because of the characteristics we discussed at the beginning of Chapter 9) and its sampling distribution. Using simple mathematics, statisticians have derived the interval estimator and the test statistic. This pattern will be used repeatedly as we introduce new techniques.

In Section 10.5, we described the five problem objectives addressed in this book, and we laid out the order of presentation of the statistical methods. In this chapter, we will present techniques employed when the problem objective is to describe a single population. When the data are interval, the parameters of interest are the population mean μ and the population variance σ^2. In Section 11.2, we describe how to make inferences about the population mean under the more realistic assumption that the population standard deviation is unknown. In Section 11.3, we continue to deal with interval data, but our parameter of interest becomes the population variance.

In Chapter 2 and in Section 10.5, we pointed out that when the data are nominal, the only computation that makes sense is determining the proportion of times each value occurs. Section 11.4 discusses inference about the proportion p.

Here are three illustrations of inference about a single population.

Illustration 1

When an election for political office takes place, the television networks cancel regular programming and instead provide election coverage. When the ballots are counted, the results are reported. However, for important offices such as president or senator in large states, the networks actively compete to see which will be the first to predict a winner. The way in which this is done is through **exit polls,** where a random sample of voters who exit the polling booth are asked whom they voted for. From the data, the sample proportion of voters supporting the candidates is computed. A statistical technique is applied to determine whether there is enough evidence to infer that the leading candidate will garner enough votes to win.

Illustration 2

Bottlenecks in a production line can occur for a variety of reasons. For example, if the amount of time taken to complete a certain task is always greater than some others on an assembly line, several workers and/or machines will be idle while others will be overworked. In designing the way in which products will be produced, managers must consider a number of factors in order to avoid bottlenecks. Suppose that in an experiment to measure the amount of time required to complete a task on an assembly line, a random sample of workers' times is measured. The problem objective is to describe the population of workers' times—data that are interval. The managers would like to draw inferences about the central location of the times, in which case the parameter to be estimated or tested is the population mean. Equally important in this scenario is the variability of the times, because a great deal of variation can cause bottleneck problems even when the mean assembly times of different tasks are identical. Consequently, management will also draw inferences about the population variance. Several different production designs will be examined before a final decision is made.

Illustration 3

The profits of television networks depend greatly on the number of viewers who are tuned into each network. In most North American cities, viewers can choose from among the major networks and an assortment of independent stations, cable, and pay TV. Because the population of television viewers is so large (over 100 million in the United States and over 10 million in Canada), statistical techniques are used to draw inferences about it. In North America, this service is provided by several firms, including A. C. Nielsen, and the results are known as the Nielsen ratings. The problem objective is to describe the population of television viewers. Each respondent would be asked (among other things) which programs he or she watches at particular times. The data are nominal. Consequently, the parameter of interest is the proportion of viewers who watch each program.

MANUAL CALCULATIONS AND MICROSOFT EXCEL OUTPUT

When we've introduced statistical techniques, we demonstrated their use by performing the calculations manually and by showing Excel printouts and instructions. The purpose of showing the manual calculations was to assist you in understanding both the mathematics and the statistical concepts. Calculating means, variances, standard deviations, z-statistics, and z-estimators makes the statistics more easily understood.

However, we now cease showing how calculations are performed manually. There are several reasons for this decision. First, although the calculations require only basic arithmetic, they will become more complicated and time-consuming. In many cases we need shortcut methods, whose application does not enhance anyone's understanding of the subject. Second, to appreciate the power of statistics, students need to be exposed to realistic applications, which require large sample sizes. And large sample sizes and manual calculations are not compatible. Third, we emphasize the importance of identifying the correct technique. If we supply part of the summary statistics such as means and variances, we will make the job of determining the appropriate method to use unrealistically simple.

At the "compute" stage of all examples we will employ only Microsoft Excel. In all exercises and cases we supply only the raw data.

11.2 INFERENCE ABOUT A POPULATION MEAN WHEN THE STANDARD DEVIATION IS UNKNOWN

In Sections 9.3 and 10.3, we demonstrated how to estimate and test the population mean when the population standard deviation is known. The interval estimator and the test statistic were derived from the sampling distribution of the sample mean with σ known, expressed as

$$z = \frac{\bar{x} - \mu}{\sigma/\sqrt{n}}$$

In this section, we assume that σ is unknown. Consequently, the sampling distribution above cannot be used. Instead, we substitute the sample standard deviation s in place of the unknown population standard deviation σ. The result is called a ***t*-statistic** because that is what mathematician William S. Gosset called it. In 1908, Gosset showed that the *t*-statistic defined as

$$t = \frac{\bar{x} - \mu}{s/\sqrt{n}}$$

is Student *t* distributed when the sampled population is normally distributed. The number of degrees of freedom is $v = n - 1$. Recall that we introduced the **Student *t* distribution** in Section 6.4.

With exactly the same logic used to develop the test statistic in Section 10.3 and the interval estimator in Section 9.3, we derive the following inferential methods.

Test Statistic for μ when σ Is Unknown

When the population standard deviation is unknown and the population is normal, the test statistic for testing hypotheses about μ is

$$t = \frac{\bar{x} - \mu}{s/\sqrt{n}}$$

which is Student *t* distributed with $v = n - 1$ degrees of freedom.

Interval Estimator of μ when σ Is Unknown

The interval estimator of a mean of a normal population when the population standard deviation is unknown is

$$\bar{x} \pm t_{\alpha/2} \frac{s}{\sqrt{n}} \quad v = n - 1$$

The bound on the error of estimation is

$$B = t_{\alpha/2} \frac{s}{\sqrt{n}}$$

Note that the use of the *t*-statistic is valid provided that the population is normal. We will discuss how to check the required condition later.

We have now presented two different test statistics and two different interval estimators for a population mean. In Chapters 9 and 10 we drew inference about μ under the assumption that the population standard deviation σ was known. In that case we employed the test statistic

Section 11.2 INFERENCE ABOUT A POPULATION MEAN WHEN THE STANDARD DEVIATION IS UNKNOWN

$$z = \frac{\bar{x} - \mu}{\sigma/\sqrt{n}}$$

and interval estimator

$$\bar{x} \pm z_{\alpha/2} \frac{\sigma}{\sqrt{n}}$$

However, as we've already discussed, the application of these methods is quite rare in practice. Consequently, henceforth, when we test hypotheses about a population mean, we will use the Student t test statistic and interval estimator summarized in the boxes above.

▼ EXAMPLE 11.1

Ecologists have long advocated recycling newspapers as a way of saving trees and reducing landfills. In recent years, a number of companies have gone into the business of collecting used newspapers from households and recycling them. An analysis of the costs of picking up the newspapers and recycling them and the prices recycled paper sells for indicates that recycling is viable only if the mean weekly newspaper collection from each household exceeds two pounds. In a study to determine the feasibility of a recycling plant, a random sample of 240 households was drawn, and the weekly weight of newspapers discarded for recycling for each household was recorded and listed below as well as stored in file Xm11-01. Do these data provide sufficient evidence to conclude that recycling newspapers is a viable project?

AMOUNT OF DISCARDED NEWSPAPER FOR RECYCLING

1.84	2.19	2.54	2.05	1.89	2.13	2.59	2.12	2.59	1.62
2.24	2.24	1.86	1.51	2.2	1.51	1.73	2.44	1.32	2
2.51	1.96	2.5	2.81	1.93	2.36	1.86	2.32	2.62	2.53
1.55	1.79	2.26	2.37	2.22	2.79	2.14	2.19	2.23	2.13
2.97	3.27	3.13	2.21	1.68	2.47	1.81	1.95	1.61	1.95
2.48	2.03	1.13	1.91	2.35	2.38	2.1	2.42	2.42	1.8
2.22	1.05	1.71	2.45	2.01	1.24	1.28	2.2	2.09	2.15
2.6	2.61	1.69	2.38	2.1	1.75	2.22	2.53	2.73	2.52
1.93	1.66	2.31	2.8	2.6	1.91	1.89	1.68	2.3	2.28
1.83	1.71	2.01	2.31	2.51	1.87	2.2	2.36	1.81	1.94
2.41	1.58	2.34	1.3	2.8	1.8	1.99	2.01	2.07	1.68
1.71	1.61	1.78	2.45	2.62	1.89	2.25	1.97	2.35	2.11
1.55	2	2.71	2.76	2.11	2.66	2.28	1.81	2.48	2.43
1.99	1.73	2.23	2.2	2.13	1.92	2.65	1.31	1.47	2.28
2.27	2.29	2.04	1.84	2.24	2.21	1.64	1.57	2.3	1.86
2.33	2.07	1.89	2.3	1.04	2.58	1.59	1.89	2.34	2.18
1.6	2.12	1.98	1.52	2.68	2.19	1.43	2.21	2.38	2.57
2.06	2	1.68	1.27	1.89	1.79	1.51	1.14	2.36	2.41
1.82	2.57	2.08	2.55	1.56	1.56	2.34	1.91	1.96	1.71
1.92	1.85	2.48	2.65	1.1	2.84	2.45	2.5	1.76	1.75
1.8	1.89	2.38	2.83	1.9	2.34	2.31	2.04	1.82	2.35
2.09	2.5	2.37	1.95	2.24	2.75	2.35	2.93	1.6	2.24
1.61	2.92	1.63	2.32	2.46	2.35	2.31	2.19	2.06	1.46
2.16	1.37	1.6	2.39	2.18	2.25	1.93	1.28	2.12	1.2

Solution

IDENTIFY

The problem objective is to describe the population of the amount of newspaper discarded for recycling by the households in this city. The data are interval, indicating that the parameter to be tested is the population mean. Because the statistics practitioner wants to know whether recycling is viable, and recycling is viable if the mean amount of discarded newspaper per household exceeds 2 pounds, the alternative hypothesis is

$$H_1 : \mu > 2$$

The null hypothesis automatically follows

$$H_0 : \mu = 2$$

The test statistic is

$$t = \frac{\bar{x} - \mu}{s/\sqrt{n}} \quad v = n - 1$$

COMPUTE

Microsoft Excel Output for Example 11.1

t-Test and Estimate: Mean

Newspaper

Mean	2.090
Standard Deviation	0.415
Hypothesized Mean	2
df	239
t Stat	3.377
P(T<=t) one-tail	0.0004
t Critical one-tail	2.342
P(T<=t) two-tail	0.0008
t Critical two-tail	2.5966
Standard Error	0.0268
Bound	0.0696
LCL	2.021
UCL	2.160

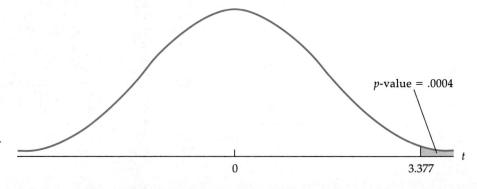

Figure 11.1

Sampling distribution of the test statistic for Example 11.1

The value of the test statistic is 3.377 The *p*-value is the value of **P(T<=t) one-tail,** which is .0004. The critical value of the rejection region **(t Critical one-tail)** is 2.342. Notice that the output includes the interval estimate, including the standard error, bound, and lower and upper limits. The confidence level is 99%, because we specified α = .01. We estimate that the mean amount of discarded newspaper per household lies between 2.021 and 2.160. We can use this estimate to predict how successful recycling will be.

GENERAL COMMANDS	EXAMPLE 11.1
1 Type or import the data into one column.	Open file **Xm11-01.**
2 Click **Tools, Data Analysis Plus, and t-Test and Estimate: Mean.**	
3 Specify the **Input Range.**	A1:A241
4 Type the value of the **Hypothesized Mean:.** (You must type some value here even if you only wish to estimate a mean.)	2
5 Click **Labels** if applicable.	
6 Specify the value of α **(Alpha),** and click **OK.**	.01

To conduct this test from statistics or perform a what-if analysis, open file **Stats-Summary.xls.** Activate the worksheet **t-Test of a Mean.**

INTERPRET

A Type I error is made when we conclude that recycling is viable when in fact the mean household amount of discarded newspaper is not greater than 2 pounds. Because this cost is high, we demand a small significance level, 1%.

Because the *p*-value = .0004, we have overwhelming statistical evidence to infer that the mean amount of newspaper discarded for recycling is greater than 2 pounds per household. If the analysis of costs is correct, recycling is viable. The mean amount of newspaper discarded for recycling is estimated to lie between 2.021 and 2.160 pounds per week.

Figure 11.1 describes the sampling distribution and *p*-value for this example.

▲

CHECKING THE REQUIRED CONDITIONS

The use of the *t*-statistic and *t*-estimator require that the population is normal. However, the techniques introduced in this section and throughout the book are **robust,** which means that if the population is nonnormal, the techniques are still valid provided that the population is not *extremely* nonnormal. The easiest way to check to determine whether the variable is extremely nonnormal is to draw a histogram. Figure 11.2 depicts the Excel histogram for Example 11.1, which suggests that amounts of discarded newspaper may be normal or at least not extremely nonnormal.

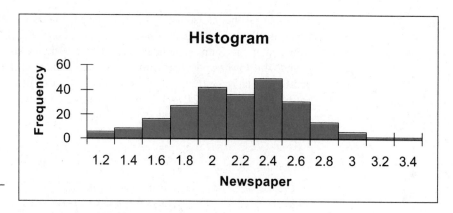

Figure 11.2

Histogram for Example 11.1

DEVELOPING AN UNDERSTANDING OF STATISTICAL CONCEPTS 1

Notice that the *t*-statistic, like the *z*-statistic, measures the difference between the sample mean $\bar{x}$ and the hypothesized value of μ in terms of the number of standard errors. However, when the population standard deviation σ is unknown, we estimate the standard error as $s/\sqrt{n}$.

DEVELOPING AN UNDERSTANDING OF STATISTICAL CONCEPTS 2

When we introduced the Student *t* distribution in Section 6.4, we pointed out that it is more widely spread out than the standard normal. This circumstance is logical. The only variable in the *z*-statistic is the sample mean $\bar{x}$, which will vary from sample to sample. The *t*-statistic has two variables, the sample mean $\bar{x}$ and the sample standard deviation *s*, both of which will vary from sample to sample. Because of this feature, the *t*-statistic will display greater variability. To demonstrate these points, suppose that a sample of size $n = 8$ produced a mean of $\bar{x} = 10$ and $s = 2$. Using the **Stats-Summary** workbook and the worksheet **t-Estimate of a Mean,** we determined the 95% confidence interval estimate of μ.

Microsoft Excel t-Interval Estimator

t-Estimate of a Mean	
Sample mean	10
Sample standard deviation	2
Sample size	8
Confidence level	0.95
Bound	1.6720
Lower confidence limit	8.3280
Upper confidence limit	11.6720

That is, the interval estimate is

$$10 \pm 1.6720$$

Now assume that we know the value of the population standard deviation, and it is $\sigma = 2$. Turning to the worksheet **z-Estimate of a Mean,** we determined the 95% confidence interval estimate of μ.

Microsoft Excel z-Interval Estimator

z-Estimate of a Mean	
Sample mean	10
Population standard deviation	2
Sample size	8
Confidence level	0.95
Bound	1.3859
Lower confidence limit	8.6141
Upper confidence limit	11.3859

This interval estimate is

10 ± 1.3859

Notice that the interval estimate assuming σ is known is narrower than the interval estimate with σ unknown. This point reinforces a principle we discussed in Chapter 9. That is, with more information we produce narrower (which we remind you means better) interval estimates. In this illustration knowledge of the value of the population standard deviation σ allows us to be more precise.

Factors That Identify the t-Test and Estimator of μ

1 Problem objective: describe a single population.

2 Data type: interval.

3 Descriptive measurement: central location.

EXERCISES

In exercises requiring a test of hypothesis where we do not specify a significance level, we expect you to use the guidelines provided in Chapter 10 to judge the size of the p-value.

*Exercises 11.1 to 11.52 are "what-if" analyses designed to determine what happens to the test statistics and interval estimates when elements of the statistical inference change. These problems can be solved manually or using one or more worksheets in **Stats-Summary.xls.***

11.1 A random sample of 50 was drawn from a population. The sample mean and standard deviation are $\bar{x} = 500$ and $s = 125$. Estimate μ with 95% confidence.

11.2 Repeat Exercise 11.1 with $n = 100$.

11.3 Repeat Exercise 11.1 with $n = 25$.

11.4 Repeat Exercise 11.1 with $s = 200$.

11.5 Repeat Exercise 11.1 with $s = 75$.

11.6 Repeat Exercise 11.1 with a 90% confidence level.

11.7 Repeat Exercise 11.1 with a 99% confidence level.

11.8 Repeat Exercise 11.1 with $\bar{x} = 200$.

11.9 Repeat Exercise 11.1 with $\bar{x} = 1{,}000$.

11.10 Review the results of Exercises 11.1 to 11.9.
    a What is the effect of decreasing the sample size?
    b What is the effect of increasing s?
    c What is the effect of increasing the confidence level?
    d What is the effect of increasing $\bar{x}$?

11.11 A random sample of 100 was drawn from a population. The sample mean and standard deviation are $\bar{x} = 33$ and $s = 5$. Estimate μ with 90% confidence.

11.12 Repeat Exercise 11.11 with $n = 25$.

11.13 Repeat Exercise 11.11 with $n = 400$.

11.14 Repeat Exercise 11.11 with $s = 2$.

11.15 Repeat Exercise 11.11 with $s = 15$.

11.16 Repeat Exercise 11.11 with a 95% confidence level.

11.17 Repeat Exercise 11.11 with a 99% confidence level.

11.18 Repeat Exercise 11.11 with $\bar{x} = 133$.

11.19 Repeat Exercise 11.11 with $\bar{x} = 233$.

11.20 Review the results of Exercises 11.11 to 11.19.
    a What is the effect of decreasing the sample size?
    b What is the effect of increasing s?
    c What is the effect of increasing the confidence level?
    d What is the effect of increasing $\bar{x}$?

11.21 The sample mean and standard deviation from a random sample of 20 observations from a normal population were computed as $\bar{x} = 23$ and $s = 9$. Can we infer that the population mean is greater than 20?

11.22 Repeat Exercise 11.21 with $n = 10$.

11.23 Repeat Exercise 11.21 with $n = 50$

11.24 Repeat Exercise 11.21 with $s = 5$.

11.25 Repeat Exercise 11.21 with $s = 20$.

11.26 Repeat Exercise 11.21 with $\bar{x} = 21$.

11.27 Repeat Exercise 11.21 with $\bar{x} = 26$.

11.28 Review the results of Exercises 11.21 to 11.27.
    a What is the effect of decreasing the sample size?
    b What is the effect of increasing s?
    c What is the effect of increasing $\bar{x}$?

11.29 A random sample of 10 observations was drawn from a normal population. The sample mean and sample standard deviation are $\bar{x} = 50$ and $s = 15$. Estimate the population mean with 95% confidence.

11.30 Repeat Exercise 11.29 assuming that you know that the population standard deviation is $\sigma = 15$.

11.31 Review Exercises 11.29 and 11.30. Explain why the interval estimate produced in Exercise 11.30 is narrower than that in Exercise 11.29.

11.32 A random sample of six observations was drawn from a normal population. The sample mean and sample standard deviation are $\bar{x} = 1{,}000$ and $s = 100$. Estimate the population mean with 95% confidence.

11.33 Repeat Exercise 11.32, assuming that you know that the population standard deviation is $\sigma = 100$.

11.34 Review Exercises 11.32 and 11.33. Explain why the interval estimate produced in Exercise 11.33 is narrower than that in Exercise 11.32.

11.35 After sampling 1,000 members of a normal population, you find $\bar{x} = 15,500$ and $s = 9,950$. Estimate the population mean with 90% confidence.

11.36 Repeat Exercise 11.35, assuming that you know that the population standard deviation is $\sigma = 9,950$.

11.37 Review Exercises 11.35 and 11.36. Explain why the interval estimates were virtually identical.

11.38 From a random sample of $n = 200$ observations you calculate $\bar{x} = 150$ and $s = 25$. Estimate the population mean with 90% confidence.

11.39 Repeat Exercise 11.38, assuming that you know that the population standard deviation is $\sigma = 25$.

11.40 Review Exercises 11.38 and 11.39. Explain why the interval estimates were virtually identical.

11.41 A random sample of eight observations was taken from a normal population. The sample mean and standard deviation are $\bar{x} = 75$ and $s = 50$. Use Excel to determine the p-value of a test to determine whether there is enough evidence to infer that the population mean is less than 100.

11.42 Repeat Exercise 11.41, assuming that you know that the population standard deviation is $\sigma = 50$.

11.43 Review Exercises 11.41 and 11.42. Explain why the p-value in Exercise 11.42 is smaller than that of Exercise 11.41.

11.44 A random sample of five observations was taken from a normal population. The sample mean and standard deviation are $\bar{x} = 12$ and $s = 4$. Use Excel to determine the p-value of a test to determine whether there is enough evidence to infer that the population mean is greater than 10.

11.45 Repeat Exercise 11.44, assuming that you know that the population standard deviation is $\sigma = 4$.

11.46 Review Exercises 11.43 and 11.44. Explain why the p-value in Exercise 11.44 is smaller than that of Exercise 11.43.

11.47 A random sample of 400 observations was taken from a normal population. The sample mean and standard deviation are $\bar{x} = 5,100$ and $s = 750$. Use Excel to determine the p-value of a test to determine whether there is enough evidence to infer that the population mean is greater than 5,000.

11.48 Repeat Exercise 11.47, assuming that you know that the population standard deviation is $\sigma = 750$.

11.49 Review Exercises 11.47 and 11.48. Explain why the p-value in Exercise 11.48 is only slightly smaller than that of Exercise 11.47.

11.50 A random sample of 250 observations was taken from a normal population. The sample mean and standard deviation are $\bar{x} = 295$ and $s = 50$. Use Excel to determine the p-value of a test to determine whether there is enough evidence to infer that the population mean is less than 300.

11.51 Repeat Exercise 11.50, assuming that you know that the population standard deviation is $\sigma = 50$.

11.52 Review Exercises 11.50 and 11.51. Explain why the p-value in Exercise 11.51 is only slightly smaller than that of Exercise 11.50.

11.53 A random sample of 11 observations was drawn from a normal population. These are

7 1 2 8 4 9 3 4 9 5 2

Estimate the population mean with 90% confidence.

11.54 The following sample was randomly drawn from a normal population. Estimate the population mean with 95% confidence.

21 45 33 27 36 19
23 31 37 42 20 18

11.55 Given the following observations, test to determine whether the mean differs from 15.

12 23 11 25 28 7 16 27 11 8 4 14

11.56 The following observations were drawn from a large population (the data are also stored in file Xr11-56).

22 18 25 28 19 20 24 26 19
26 27 22 23 25 25 18 20
26 18 26 27 24 20 19 18

a Estimate the population mean with 95% confidence.
b Test to determine if we can infer that the population mean is greater than 20.
c What is the required condition of the techniques used in parts (a) and (b)? Use a graphical technique to check to see if that required condition is satisfied.

11.57 A random sample of 75 observations from a normal population is stored in file Xr11-57. Test to determine whether we can conclude that the population mean is not equal to 103.

11.58 A growing concern for educators in the United States is the number of teenagers who have part-time jobs while they attend high school. It is generally believed that the amount of time teenagers spend working is deducted from the amount of time devoted to schoolwork. To investigate this problem, a school guidance counselor took a random sample of 200 15-year-old high school students and asked how many hours per week each worked at a part-time job. The results were recorded and stored in file Xr11-58. Estimate with 95% confidence the mean amount of time all 15-year-old high school students devote per week to part-time jobs.

11.59 A federal agency responsible for enforcing laws governing weights and measures routinely inspects packages to determine if the weight of the contents is at least as great as that advertised on the package. A random sample of 50 containers whose packaging states that the contents weigh 8 ounces was drawn. The contents were weighed, and the results (to the nearest tenth) are stored in file Xr11-59. Estimate the mean weight of all the containers with 99% confidence.

11.60 A diet doctor claims that the average North American is more than 20 pounds overweight. To test his claim, a random sample of 100 North Americans was weighed, and the difference between their actual weight and their ideal weight was calculated. The data are stored in file Xr11-60. Do these data allow us to infer that the doctor's claim is true?

11.61 A courier service advertises that its average delivery time is less than six hours for local deliveries. A random sample of times for 50 deliveries to an address across town was stored in file Xr11-61.
   a Is this sufficient evidence to support the courier's advertisement?
   b What assumption must be made in order to answer part (a)? Use whatever graphical technique you deem appropriate to confirm that the required condition is satisfied.

11.62 A manufacturer of a brand of designer jeans has pitched her advertising to develop an expensive and classy image. The suggested retail price is $75. However, she is concerned that retailers are undermining her image by offering the jeans at discount prices. To better understand what is happening, she randomly samples 30 retailers who sell her product and determines the price. The results are stored in file Xr11-62. She would like an estimate of the mean selling price of the jeans at all retail stores.
   a Determine the 95% confidence interval estimate.
   b What assumption must be made to be sure that the estimate produced in part (a) is valid? Use a graphical technique to check the required condition.

11.63 Couriers like UPS and FedEx compete on service and price. One way to reduce costs is to keep labor costs low by hiring and laying off workers to meet demand. This strategy requires managers to hire and train new workers. But newly hired and trained workers are not as productive as more experienced ones. Thus, determining the number of workers required and the work schedule is difficult. The current work schedule is based on the belief that trainees will achieve more than 90% of the level of experienced workers within one week of hiring. To determine the accuracy of this number, an operations manager conducted an experiment. Fifty trainees were observed for one hour, and the number of packages processed and routed was recorded. These data appear below and are stored in file Xr11-63. It is known that experienced workers process an average of 500 packages per hour. The manager is concerned that if he concludes that the mean is greater than 450 when it isn't, the result will be many late deliveries, a disaster for a courier. Can the manager conclude from the data that the belief is correct?

11.64 During the last decade a number of institutions dedicated to improving the quality of products and services in the United States have been formed. Many of these groups annually give awards to companies that produce high-quality goods and services. An investor believes that publicly traded companies that win awards are likely to outperform companies that do not win such awards. To help determine his return on investment in such companies he took a random sample of 50 firms that won quality awards the previous year and computed the annual return had he invested. These data are stored in file Xr11-64. The investor would like an estimate of the returns he can expect. A 95% confidence level is deemed appropriate.

11.65 The natural remedy echinacea is reputed to boost the immune system, which will reduce flu and colds. A six-month study was undertaken to determine whether the remedy works. A random sample of 200 Seattle-area residents who frequently experienced cold and flu symptoms was taken. All were given daily doses of the herbal remedy. The number of days that each person was ill with a respiratory infection for the next 12 months was recorded and stored in file Xr11-65. Estimate with 95% confidence the mean number of days of illness for all Seattle residents who take echinacea.

11.66 Refer to Exercise 11.65. If the average number of days of illness with respiratory infection per year in Seattle is 15.5 days, can we infer that echinacea is effective?

11.3 INFERENCE ABOUT A POPULATION VARIANCE

In Section 11.2, where we presented the inferential methods about a population mean, we were interested in acquiring information about the central location of the population. As a result, we tested and estimated the population mean. If we are interested instead in making an inference about the variability, the parameter we need to investigate is the population variance σ^2. Inference about the variance can be used in a variety of applications. For example, quality-control engineers must ensure that their company's products meet specifications. One way of judging the consistency of a production process is to compute the variance of the size, weight, or volume of the product. That is, if the variation in product size, weight, or volume is large, it is likely that an unsatisfactorily large number of products will lie outside the specifications for that product. Another example comes from the subject area of finance. Investors use the variance of the returns on a portfolio of stocks, bonds, or other investments as a measure of the uncertainty and risk inherent in that portfolio. Investors often go to great lengths to avoid risky investments.

The task of deriving the test statistic and the interval estimator provides us with another opportunity to show how statistical techniques in general are developed. We begin by identifying the best estimator. That estimator has a sampling distribution, from which we produce the test statistic and the interval estimator.

POINT ESTIMATOR OF σ^2

The point estimator for σ^2 is the sample variance introduced in Section 3.3 and used repeatedly, most recently in Section 11.2, to test and estimate a population mean. The statistic s^2 has the desirable characteristics presented in Section 9.2; that is, s^2 is an unbiased consistent estimator of σ^2.

SAMPLING DISTRIBUTION OF s^2

Statisticians have shown that the sum of squared differences $\sum(x_i - \bar{x})^2$ [which is equal to $(n-1)s^2$] divided by the population variance is chi-squared distributed with $n-1$ degrees of freedom provided that the population is normal. The statistic

$$\chi^2 = \frac{(n-1)s^2}{\sigma^2}$$

is called the **chi-squared statistic (χ^2-statistic).**

It is for this reason that the chi-squared distribution was introduced in Section 6.4.

TESTING AND ESTIMATING THE POPULATION VARIANCE

As we discussed in Section 10.5, the formula that describes the sampling distribution is the formula of the test statistic.

> **Test Statistic for σ^2**
>
> The test statistic used to test hypotheses about σ^2, the variance of a normal population, is
>
> $$\chi^2 = \frac{(n-1)s^2}{\sigma^2}$$
>
> which is chi-squared distributed with $v = n - 1$ degrees of freedom.

Using the notation developed in Section 6.4, we can make the following probability statement:

$$P(\chi^2_{1-\alpha/2} < \chi^2 < \chi^2_{\alpha/2}) = 1 - \alpha$$

Substituting

$$\chi^2 = \frac{(n-1)s^2}{\sigma^2}$$

and with some algebraic manipulation we have

> **Interval Estimator of σ^2**
>
> Lower confidence limit (LCL) = $\dfrac{(n-1)s^2}{\chi^2_{\alpha/2}}$
>
> Upper confidence limit (UCL) = $\dfrac{(n-1)s^2}{\chi^2_{1-\alpha/2}}$

EXAMPLE 11.2

Container-filling machines are used to package a variety of liquids, including milk, soft drinks, and paint. Ideally, the amount of liquid should vary only slightly, because large variations will cause some containers to be underfilled (cheating the customer) and some to be overfilled (resulting in costly waste). The president of a company that developed a new type of machine boasts that this machine can fill 1-liter (1,000 cubic centimeters) containers so consistently that the variance of the fills will be less than 1 cc². To examine the veracity of the claim, a random sample of 115 l-liter fills was taken and the results recorded (in cc). These data are listed below and also stored in file Xm11-02. Do these data support the president's claim?

ONE-LITER FILLS

999.21	999.92	1000.88	1000.89	1000.47	1000.75	1000.39	999.78
999.31	1001.57	999.04	999.63	1000.34	999.38	999.08	998.36
999.57	999.91	1000.68	1001	998.33	1001.25	999.7	1001.34
1001.46	1000.35	1000.14	1000.53	999.59	1001.06	1000.19	1000.34
1000.59	1001.68	1000.32	1000.02	1000.08	999.6	999.66	1000.05
998.8	998.52	1000.69	999.94	999.88	999.57	1001.11	999.56
1000.32	999.43	1000.55	1001.29	1000.48	999.61	999.5	999.84
1001.49	1000.51	999.66	1001.06	998.35	999.62	1001.03	1000.16
999.92	1001.18	1001.63	1001.45	999.37	1000.05	999.81	1000.11
999.61	999.62	1000.25	1000.34	999.52	998.35	1000.73	999.05
1000.45	1000.02	998.66	999.47	999.5	997.93	1000.48	
1001.4	1001.49	1000.38	998.9	999.76	999.28	998.83	
1000.46	998.49	1000.88	1000.41	998.97	997.88	1002.51	
1000.32	1000.56	1001.78	998.13	1002.05	1000.4	1000.18	
999.97	1000.61	997.87	1000.6	1000.45	999.13	1000.39	

Solution

IDENTIFY

The problem objective is to describe the population of l-liter fills from this machine. The data are interval, and we're interested in the variability of the fills. It follows that the parameter of interest is the population variance. Because we want to determine whether there is enough evidence to support the claim that the variance is less than 1, the alternative hypothesis is

$$H_1 : \sigma^2 < 1$$

The null hypothesis automatically follows

$$H_0 : \sigma^2 = 1$$

The test statistic is

$$\chi^2 = \frac{(n-1)s^2}{\sigma^2}$$

COMPUTE

Microsoft Excel Output for Example 11.2

Chi-Squared Test and Estimate: Variance		
Fills		
Sample Variance		0.8933
Hypothesized Variance		1
df		114
chi-squared Stat		101.8396
P(CHI<=chi) one-tail		0.2144
chi-squared Critical one tail	Left-tail	90.3511
	Right-tail	139.9207
P(CHI<=chi) two-tail		0.4288
chi-squared Critical two tail	Left-tail	86.3425
	Right-tail	145.4413
LCL		0.7002
UCL		1.1795

The value of the test statistic is 101.8396, and its *p*-value is .2144. The output also includes the interval estimate of the variance.

GENERAL COMMANDS	EXAMPLE 11.2
1 Type or import the data into one column.	Open file **Xm11-02.**
2 Click **Tools, Data Analysis Plus,** and **Chi-squared Test and Estimate: Variance.**	
3 Specify the **Input Range:**.	A1:A116
4 Type the value of the **Hypothesized Variance:**. (You must type some value here even if you only wish to estimate a variance.)	1
5 Click **Labels,** if applicable.	
6 Specify the value of α **(Alpha),** and click **OK.**	.05

To conduct this test from statistics or perform a what-if analysis, open file **Stats-Summary.xls** and activate the worksheet **Chi-Sq Test of a Variance.** You may also activate the worksheet **Chi-Sq Estimate of a Variance.**

INTERPRET

The two types of error are: concluding that the claim is true when it isn't (Type I) and not concluding that the claim is true when it is (Type II). The first error is likely to be more costly, although it depends on the current machine's variance. It seems reasonable to set the significance level at a point below 5%.

The *p*-value is .2144 [**P(CHI<=chi) one-tail**]. There is not enough evidence to infer that the claim is true. There is little support for the president's claim and likely little reason to switch to his product.

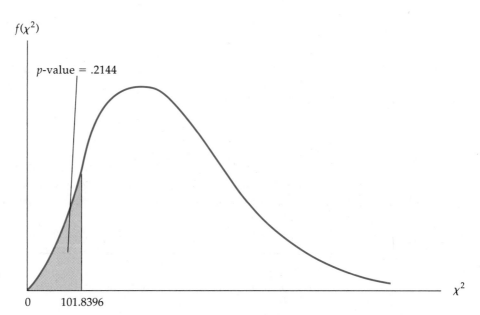

Figure 11.3

Sampling distribution of the test statistic for Example 11.2

Figure 11.3 depicts the sampling distribution and *p*-value.

CHECKING THE REQUIRED CONDITIONS

The chi-squared test of σ^2, like most of the other inferential methods, is robust. However, it is somewhat more sensitive to violations of the normality requirement than the *t*-test of μ. Consequently, we must be more careful to ensure that the fills are not extremely nonnormally distributed.

Figure 11.4 exhibits the histogram of the fills, indicating an obvious bell shape.

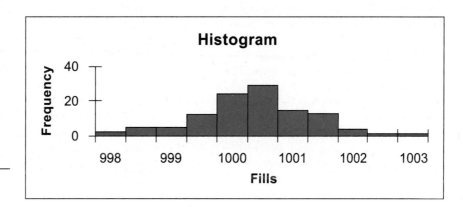

Figure 11.4

Histogram of fills in Example 11.2

Factors That Identify the Chi-Squared Test and Estimator of σ^2

1 Problem objective: describe a single population.
2 Data type: interval.
3 Descriptive measurement: variability.

EXERCISES

In exercises requiring a test of hypothesis where we do not specify a significance level, we expect you to use the guidelines provided in Chapter 10 to judge the size of the p-value.

Exercises 11.67 to 11.74 are "what-if" analyses designed to determine what happens to the test statistics and interval estimates when elements of the statistical inference change. These problems can be solved manually or using a worksheet in **Stats-Summary.xls.**

11.67 A random sample of 100 observations was drawn from a normal population. The sample variance was calculated to be $s^2 = 220$. Test to determine whether we can infer that the population variance differs from 300.

11.68 Repeat Exercise 11.67, changing the sample size to 50.

11.69 Repeat Exercise 11.67, changing the sample size to 400.

11.70 Review the interval estimates created in Exercises 11.67 to 11.69. What is the effect of increasing the sample size?

11.71 The sample variance of a random sample of ten observations from a normal population was found to be $s^2 = 80$. Use Excel to compute the p-value of a test to determine whether we can infer that σ^2 is less than 100.

11.72 Repeat Exercise 11.71 with $n = 100$.

11.73 Repeat Exercise 11.71 with $n = 500$.

11.74 What is the effect on the p-value of increasing the sample size in Exercises 11.71 to 11.73?

11.75 Test the hypotheses below
$H_0: \sigma^2 = 100$
$H_1: \sigma^2 > 100$
given the data:
85 59 66 81 35 57 55 63 66.

11.76 The following data were drawn from a normal population:
92 93 54 58 74 53 63 83 64 51 103
Test to determine if there is enough evidence to conclude that the population variance is less than 500.

11.77 A random sample of 12 observations was taken from a normal population. These data are listed below. Estimate the population variance with 90% confidence.
45 66 38 49 21 64
51 58 47 46 29 35

11.78 Given the following sample, estimate the population variance with 95% confidence:
4 5 6 5 5 6 5 6

11.79 The following observations were drawn from a normal population:
497 511 498 494 479 526
510 515 489 488 491
Estimate the population variance with 90% confidence.

11.80 One important factor in inventory control is the variance of the daily demand for the product. A management scientist has developed the optimal order quantity and reorder point, assuming that the variance is equal to 250. Recently, the company has experienced some inventory problems, which induced the operations manager to doubt the assumption. To examine the problem, the manager took a sample of 25 daily demands and stored them in file Xr11-80. Do these data provide sufficient evidence to infer that the management scientist's assumption about the variance is wrong?

11.81 Refer to Example 11.80. What are the smallest and largest values that σ^2 is likely to assume? (Define "likely" as 95% confidence.)

11.82 Some traffic experts believe that the major cause of highway collisions is the differing speeds of cars. That is, when some cars are driven slowly while others are driven at speeds well in excess of the speed limit, cars tend to congregate in bunches, increasing the probability of accidents. Thus, the greater the variation in speeds, the greater is the number of collisions that occur. Suppose that one expert believes that when the variance exceeds 18 $(mph)^2$, the number of accidents will be unacceptably high. A random sample of the speeds of 245 cars on a highway with one of the highest accident rates in the country is taken. These data are stored in file Xr11-82. Can we conclude that the variance in speeds exceeds 18 $(mph)^2$?

11.83 One problem facing the managers of maintenance departments is when to change the bulbs in streetlamps. If bulbs are changed only when they burn out, it is quite costly to send crews out to change only one bulb at a time. This method also requires someone to report the problem, and in the meantime, the light is off. If each bulb lasts approximately the same amount of time, they can all be replaced periodically, producing significant cost savings in maintenance. Suppose that a financial analysis of the lights at Yankee Stadium has concluded that it will pay to replace all of the lightbulbs at the same time if the variance of the lives of the bulbs is less than 200 $hours^2$. The length of life of the last 100 bulbs was recorded and stored in file Xr11-83. What conclusion can be drawn from these data?

11.4 INFERENCE ABOUT A POPULATION PROPORTION

In this section, we continue to address the problem of describing a single population. However, we shift our attention to populations of nominal data, which means that the population consists of categorical values. For example, in a political survey, the statistics practitioner asks potential voters for which (of five) candidates for mayor they intend to vote. The responses are the names of the five candidates, which could be represented by their names, letters (A, B, C, D, and E), or by numbers (1, 2, 3, 4, and 5). When numbers are used, it should be understood that the numbers only represent the name of the candidate, are completely arbitrarily assigned, and cannot be treated as real numbers. That is, we cannot calculate means and variances.

PARAMETER

Recall our earlier discussion in Chapter 2. When the data are nominal, all that we are permitted to do to describe the population or sample is to count the number of occurrences of each value. From the counts, we calculate proportions. Thus, the parameter of interest in describing a single population of nominal data is the population proportion p. In Section 5.4, this parameter was used to calculate probabilities based on the binomial experiment. One of the characteristics of the binomial experiment is that there are only two possible outcomes per trial. Most practical applications of inference about p involve more than two outcomes. However, in most cases we're interested in only one outcome, which we label a success. All other outcomes are labeled as failures. For example, in brand-preference surveys we are interested in our company's brand. In political surveys we wish to estimate or test the proportion of voters who will vote for one particular candidate—likely the one who has paid for the survey.

STATISTIC AND SAMPLING DISTRIBUTION

The logical statistic employed to estimate and test the population proportion is the sample proportion defined as

$$\hat{p} = \frac{x}{n}$$

where x is the number of successes in the sample and n is the sample size. In Section 8.4, we presented the approximate sampling distribution of $\hat{p}$. (The actual distribution is based on the binomial distribution, which does not lend itself to statistical inference.) The sampling distribution of $\hat{p}$ is approximately normal with mean p and standard deviation $\sqrt{p(1-p)/n}$ (provided that np and $n(1-p)$ are greater than 5). We express this sampling distribution as

$$z = \frac{\hat{p} - p}{\sqrt{p(1-p)/n}}$$

As you have already seen, the formula that summarizes the sampling distribution also represents the test statistic.

Test Statistic for p

$$z = \frac{\hat{p} - p}{\sqrt{p(1-p)/n}}$$

which is approximately normal for np and $n(1-p)$ greater than 5.

As was the case with the z-statistic (Chapter 10) and the t-statistic (Chapter 11), this test statistic measures the number of standard errors between the sample statistic and the value of the parameter specified by the null hypothesis. The statistic is $\hat{p}$, the parameter is p, and the standard error is

$$\sqrt{p(1-p)/n}$$

INTERVAL ESTIMATOR OF p

Using the same algebra employed in Sections 9.3 and 11.2, we attempt to derive the interval estimator of p from the sampling distribution. The result is

$$\hat{p} \pm z_{\alpha/2}\sqrt{p(1-p)/n}$$

This formula, although technically correct, is useless. To understand why, examine the standard error of the sampling distribution $\sqrt{p(1-p)/n}$. To produce the interval estimate, we must compute the standard error, which requires us to know the value of p, which is the parameter we wish to estimate. This is the first of several statistical techniques where we face the same problem, how to determine the value of the standard error. In this application, the problem is easily and logically solved; simply estimate the value of p with $\hat{p}$. Thus, we estimate the standard error with $\sqrt{\hat{p}(1-\hat{p})/n}$. The interval estimator follows.

> **Interval Estimator of p**
>
> $$\hat{p} \pm z_{\alpha/2}\sqrt{\frac{\hat{p}(1-\hat{p})}{n}}$$
>
> which is valid provided that $n\hat{p}$ and $n(1-\hat{p})$ are greater than 5.

The bound on the error of estimation is

$$B = z_{\alpha/2}\sqrt{\frac{\hat{p}(1-\hat{p})}{n}}.$$

▼ EXAMPLE 11.3

As we described in the introduction to this chapter, television networks employ exit polls to allow them to predict an election winner shortly after the polls close. Suppose that during an election for a Senate seat, NBC conducted an exit poll where voters leaving the polling booth were asked for whom they voted. The responses were

1 = Democrat

2 = Republican

The responses of a sample of 926 voters are stored in file Xm11-03. (We will not list the values; they are just a bunch of 1s and 2s.) Do these data allow NBC to confidently predict the election winner?

Solution

IDENTIFY

The objective is to describe the population of voters in this state. The values of the variable are "Democrat" and "Republican," stored in the file as "1" and "2." The data are nominal. The parameter employed to describe a single population of nominal data is the population proportion. We will arbitrarily define "1" as a success. NBC can only predict a winner if there is enough evidence to infer that the proportion of successes is either greater than .5 (the Democrat wins) or less than .5 (the Republican wins). Thus the alternative hypothesis is

$H_1: p \neq .5$

The null hypothesis is

$H_0: p = .5$

The test statistic is

$$z = \frac{\hat{p} - p}{\sqrt{p(1-p)/n}}$$

314 Chapter 11 INFERENCE ABOUT A SINGLE POPULATION

COMPUTE

Microsoft Excel Output for Example 11.3

z-Test and Estimate: Proportion

	Votes
Sample Proportion	0.5346
Observations	926
Hypothesized Proportion	0.5
z Stat	2.1032
P(Z<=z) one-tail	0.0177
z Critical one-tail	2.3263
P(Z<=z) two-tail	0.0354
z Critical two-tail	2.5758
Standard Error	0.0164
Bound	0.0422
LCL	0.4923
UCL	0.5768

GENERAL COMMANDS	EXAMPLE 11.3
1 Type or import the data into one column.	Open file **Xm11-03**.
2 Click **Tools, Data Analysis Plus**, and **z-Test and Estimate: Proportion**.	
3 Specify the **Input Range:**.	**A1:A927**
4 Specify the **Code for Success:**.	**1**
5 Type the value of the **Hypothesized Proportion:**. (You must type some value even if you wish only to estimate the proportion.)	**.5**
6 Click **Labels**, if applicable.	
7 Type a value for α (**Alpha**), and click **OK**.	**.01**

To complete the technique from the sample proportion or to conduct a what-if analysis, use the **z-Test of a Proportion** worksheet in the **Stats-Summary.xls** file.

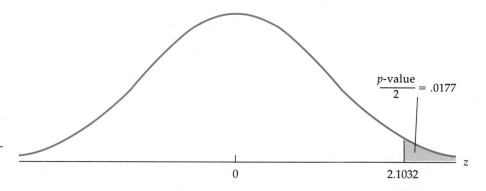

Figure 11.5

Sampling distribution for the test statistic for Example 11.3

INTERPRET

If NBC makes a prediction that later turns out to be incorrect (Type I error), the network news will be embarrassed. To minimize the chances of a Type I error, we will set the significance level at 1%. The value of the test statistic is $z = 2.1032$, which has a *p*-value of .0354. Although there is some evidence that the proportion of votes will be different from 50%, the evidence is not strong enough to make a prediction. NBC should simply wait for the actual votes to be counted. Figure 11.5 depicts the sampling distribution.

The next example features nominal data with more than two categories and whose objective is to estimate the total number of successes in the population.

EXAMPLE 11.4

Statistical techniques play a vital role in helping advertisers determine how many viewers watch the shows that they sponsor. There are several companies that sample television viewers to determine what shows they watch, the best known of which is the A. C. Nielsen firm. The Nielsen Ratings are based on a sample of randomly selected families. A device attached to the family television keeps track of the channels the television receives. The ratings then produce the proportions of each show from which sponsors can determine the number of viewers and the potential value of any commercials.

Suppose that the results of a survey of 2,000 television viewers at 11:40 P.M. on Monday, September 28, 1998, were stored in file Xm11-04 using the following codes:

1 *The Tonight Show with Jay Leno* (NBC)
2 *Late Show with David Letterman* (CBS)
3 *Nightline* (ABC)
4 Other
5 Television turned off

If there are 100 million potential television sets in the population, estimate with 95% confidence the number of televisions tuned to *The Tonight Show*.

Solution

IDENTIFY

The problem objective is to describe the population of television viewers, and the data are nominal. The parameter to be estimated is the proportion of televisions tuned to *The Tonight Show*. The interval estimator is

$$\hat{p} \pm z_{\alpha/2}\sqrt{\frac{\hat{p}(1-\hat{p})}{n}}$$

COMPUTE

Microsoft Excel Output for Example 11.4

z-Test and Estimate: Proportion	
Viewers	
Sample Proportion	0.113
Observations	2000
Hypothesized Proportion	0.5
z Stat	−34.61
P(Z<=z) one-tail	0
z Critical one-tail	1.6449
P(Z<=z) two-tail	0
z Critical two-tail	1.96
Standard Error	0.0071
Bound	0.0139
LCL	0.0991
UCL	0.1269

We typed **1** as the **Code for Success** and **.05** for the value of α, and we arbitrarily specified **.5** for the value of the **Hypothesized Proportion**. (Note that if you leave this box empty, an error message will appear.) We will ignore all the statistics pertaining to a test of hypothesis. We're only interested in the estimate of the proportion of *Tonight Show* viewers. The lower and upper limits of the estimate of p is LCL = .0991 and UCL = .1269.

To complete the technique from the sample proportion or to conduct a what-if analysis, activate the **z-Estimate of a Proportion** worksheet in **Stats-Summary.xls**.

INTERPRET

We estimate that between 9.91% and 12.69% of all television sets had received *The Tonight Show*. If we multiply these figures by the total number of televisions, 100 million, we produce an interval estimate of the number of televisions tuned to *The Tonight Show*. That is, we estimate that figure to lie between 9.91 million and 12.69 million. Sponsoring companies can then determine the value of any commercials that appeared on the show.

DEVELOPING AN UNDERSTANDING OF STATISTICAL CONCEPTS 1

The test statistic introduced in this section follows the pattern of the z-test and the t-test of μ. The value of the test statistic is the difference between the sample proportion and the hypothesized proportion measured in terms of the standard error.

DEVELOPING AN UNDERSTANDING OF STATISTICAL CONCEPTS 2

The interval estimator of a proportion requires us to estimate the standard error by using the sample data. We will encounter the same problem again and resolve it in a similar manner.

MISSING DATA

When statistics practitioners conduct experiments to collect data, they often encounter nonresponses. This is quite common in political surveys where we ask voters for whom they intend to vote in the next election. Some people will answer that they haven't decided or that they will refuse to answer. This is a troublesome issue for statistics practitioners. We can't force people to answer our questions. However, if the number of nonresponses is high, the results of our analysis may be invalid because the sample is no longer truly random. To understand why, suppose that people who are in the top quarter of household incomes regularly refuse to answer questions about their incomes. The estimate of the population household income mean will be lower than the actual value. The issue can be complicated. There are several ways to compensate for nonresponses. The simplest method is simply to eliminate the nonresponses. To illustrate, suppose that in a political survey respondents are asked for whom they intend to vote in a two-candidate race. Surveyors record the results as 1 = Candidate A, 2 = Candidate B, 3 = "Don't know," and 4 = "Refuse to say." If we wish to infer something about the proportion of decided voters who will vote for Candidate A, we can simply omit codes 3 and 4. If we're doing the work manually, we will count the number of voters who prefer Candidate A and the number who prefer Candidate B. The sum of these two numbers is the total sample size.

In the language of statistical software, nonresponses that we wish to eliminate are collectively called "missing data." Excel recognizes blank cells as missing data. However, the macros in Data Analysis Plus do not. For these techniques, the easiest way to omit missing data is to simply delete the cells containing the missing data. Sort the data (Click **Data** and **Sort...**) in order, highlight the cells you wish to delete, and with one keystroke delete the missing data.

RECODING DATA

To recode data we employ a logical function. Click f_x, **Logical (Function category:)**, and **IF (Function name:)**. To illustrate, suppose that in column A you have stored data consisting of codes 1 through 6, and you wish to convert all 4s, 5s, and 6s to 9s. Activate cell B1, and type

=IF(A1>=4,9,A1)

This logical function determines whether the value in cell A1 is greater than or equal to 4. If so, Excel places a 9 in cell B1. If A1 is less than 4, B1 = A1. Dragging to fill in column B coverts all 4s, 5s, and 6s to 9s and stores the results in column B.

If 4s, 5s, and 6s represent nonresponses, you can replace these codes with a blank. Type in cell B1

=IF(A1>=4," ",A1)

SELECTING THE SAMPLE SIZE TO ESTIMATE THE PROPORTION

When we introduced the sample size selection method to estimate a mean in Section 9.4, we pointed out that the sample size depends on the confidence level and the bound on the error of estimation. When the parameter to be estimated is a proportion the bound is

$$B = z_{\alpha/2}\sqrt{\frac{\hat{p}(1-\hat{p})}{n}}$$

Solving for n we produce:

Sample Size to Estimate a Proportion

$$n = \left(\frac{z_{\alpha/2}\sqrt{\hat{p}(1-\hat{p})}}{B}\right)^2$$

To illustrate the use of this formula, suppose that in a brand-preference survey we want to estimate the proportion of consumers who prefer our company's brand to within .03 with 95% confidence. This means that when the sample is taken and the calculations completed, the interval estimate is to be $\hat{p} \pm .03$. Thus, $B = .03$. Because $1 - \alpha = .95$, $\alpha = .05$, $\alpha/2 = .025$, and $z_{\alpha/2} = z_{.025} = 1.96$. Therefore,

$$n = \left(\frac{1.96\sqrt{\hat{p}(1-\hat{p})}}{.03}\right)^2$$

To solve for n, we need to know $\hat{p}$. Unfortunately, this value is unknown, because the sample has not yet been taken. At this point, we can use either of two methods to solve for n.

Method 1

If we have no knowledge of even the approximate values of $\hat{p}$, we let $\hat{p} = .5$. We choose $\hat{p} = .5$ because the product $\hat{p}(1 - \hat{p})$ equals its maximum value at $\hat{p} = .5$. (Figure 11.6 illustrates this point.) This, in turn, results in a conservative value of n, and as a result, the confidence interval will be no wider than the interval $\hat{p} \pm .03$. If, when the sample is drawn, $\hat{p}$ does not equal .5, the interval estimate will be better (that is, narrower) than planned. Thus,

$$n = \left(\frac{1.96\sqrt{(.5)(.5)}}{.03}\right)^2 = (32.67)^2 = 1{,}068$$

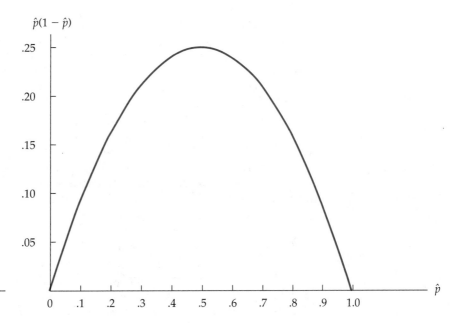

Figure 11.6

Graph of $\hat{p}$ versus $\hat{p}(1 - \hat{p})$

If it turns out that $\hat{p} = .5$, the interval estimate is $.5 \pm .03$. If not, the interval estimate will be narrower. For instance, if it turns out that $\hat{p} = .2$, the estimate is $.2 \pm .024$, which is better than we had planned.

Method 2

If we have some idea about the value of $\hat{p}$, we can use that quantity to determine n. For example, if we believe that $\hat{p}$ will turn out to be approximately .2, we can solve for n as follows:

$$n = \left(\frac{1.96\sqrt{(.2)(.8)}}{.03}\right)^2 = (26.13)^2 = 683$$

Notice that this produces a smaller value of n (thus reducing sampling costs) than does method 1. If $\hat{p}$ actually lies between .2 and .8, however, the estimate will not be as good as we wanted, because the interval will be wider than desired.

Method 1 is often used to determine the sample size used in public opinion surveys reported by newspapers, magazines, television, and radio. These polls usually estimate proportions to within 3%, with 95% confidence. (The media often state the confidence level as "19 times out of 20.") If you've ever wondered why opinion polls almost always estimate proportions to within 3%, consider the sample size required to estimate a proportion to within 1%:

$$n = \left(\frac{1.96\sqrt{(.5)(.5)}}{.01}\right)^2 = (98)^2 = 9,605$$

The sample size 9,605 is 9 times the sample size needed to estimate a proportion to within 3%. Thus, to divide the width of the interval by 3 requires multiplying the sample size by 9. The cost would also increase considerably. For most applications, the increase in accuracy (created by decreasing the width of the interval estimate) does not overcome the increased cost. Interval estimates with 5% or 10% bounds (sample sizes 385 and 97, respectively) are generally considered too wide to be useful. Thus, the 3% bound is the happy compromise between cost and accuracy.

TESTING HYPOTHESES ABOUT PROPORTIONS USING THE BINOMIAL DISTRIBUTION (OPTIONAL)

As we've noted several times, the actual distribution of $\hat{p}$ is related to the binomial distribution. But we use the normal approximation of $\hat{p}$ because the binomial is a discrete random variable, and discrete distributions make it difficult for us to determine the critical values of an interval estimator and the rejection region and p-value of a test of hypothesis. However, when we use the computer to perform the calculations, we can use the exact binomial distribution. Excel users can produce the p-value of the test, but they cannot find the interval estimate using the binomial distribution.

To illustrate how Excel can produce the p-value refer to Example 11.3, where we wanted to know whether there was sufficient evidence to infer that the proportion of voters was different from 50%. In a sample of 926, we found $x = 495$ voters who supported candidate A. The exact (binomial) p-value of the test is

p-value $= 2P(X \geq 495$, given that $n = 926$ and $p = .5)$

Note that the p-value is twice (because we conducted a two-tail test) the probability that X is greater than or equal to 495 (the number of successes in the sample) when the sample size is 926, and the probability of success is .5 (the value specified by the null hypothesis). For smaller values of n, the binomial formula or table (Table 1 in Appendix B) may be used. For larger values of n, we must use the computer.

To determine the number of successes, in the Xm11-03 worksheet, type into any empty cell

=COUNTIF(A2:A927,1)

which counts the number of 1s and reports that $x = 495$.

To find the $P(X \geq 495)$, we need to calculate the cumulative probability

$P(X \leq 494$, given that $n = 926$ and $p = .5)$

Using the instructions provided on page 133, we typed the following into an empty cell in any Excel worksheet.

=BINOMDIST(494, 926, .5, TRUE)

which yielded .9808. Thus, the exact p-value is

p-value $= 2P(X \geq 495) = 2[1 - P(X \leq 494)] = 2[1 - .9808] = .0384$

Note that the approximate p-value (based on the normal distribution) was found to be .0354.

We complete this section by reviewing the factors that tell us when to test and estimate a population proportion.

Factors That Identify the z-Test and Interval Estimator of p

1 Problem objective: describe a single population
2 Data type: nominal.

EXERCISES

*Exercises 11.84 to 11.107 are "what-if" analyses designed to determine what happens to the test statistics and interval estimates when elements of the statistical inference change. These problems can be solved manually or using **Stats-Summary.xls.***

11.84 In a random sample of 500 observations, we found the proportion of successes to be 48%. Estimate with 95% confidence the population proportion of successes.

11.85 Repeat Exercise 11.84 with $n = 200$.

11.86 Repeat Exercise 11.84 with $n = 1,000$.

11.87 Repeat Exercise 11.84 with $\hat{p} = 33\%$.

11.88 Repeat Exercise 11.84 with $\hat{p} = 10\%$.

11.89 Review Exercises 11.84 to 11.88.
  a Discuss the effect on the width of the interval estimate of reducing the sample size.
  b Discuss the effect on the width of the interval estimate of reducing the sample proportion.

11.90 In a random sample of 50 observations, we found the proportion of successes to be 10%. Estimate with 90% confidence the population proportion of successes.

11.91 Repeat Exercise 11.90 with $n = 100$.

11.92 Repeat Exercise 11.90 with $n = 400$.

11.93 Repeat Exercise 11.90 with $\hat{p} = 50\%$.

11.94 Repeat Exercise 11.90 with $\hat{p} = 90\%$.

11.95 Review Exercises 11.90 to 11.94.
  a Discuss the effect on the width of the interval estimate of increasing the sample size.
  b Discuss the effect on the width of the interval estimate of changing the sample proportion.

11.96 Given that $\hat{p} = .84$ and $n = 600$, estimate p with 90% confidence.

11.97 In a random sample of 250, we found 75 successes. Estimate the population proportion of success, with 99% confidence.

11.98 If $\hat{p} = .59$ and $n = 100$, can we conclude that the population proportion p is greater than .50?

11.99 Suppose that, in a sample of 200, we observe 140 successes. Is this sufficient evidence at the 1% significance level to indicate that the population proportion of successes is greater than 65%?

11.100 Manually calculate the *p*-value of the test in Exercise 11.99.

11.101 Determine the sample size necessary to estimate a population proportion to within .03 with 90% confidence, assuming you have no knowledge of the approximate value of the sample proportion.

11.102 Suppose that you used the sample size calculated in Exercise 11.101 and found $\hat{p} = .5$.
  a Estimate the population proportion with 90% confidence.
  b Is this the result you expected? Explain.

11.103 Suppose that you used the sample size calculated in Exercise 11.101 and found $\hat{p} = .75$.

  a Estimate the population proportion with 90% confidence.

  b Is this the result you expected? Explain.

  c If you were hired to conduct this analysis would the person who hired you be satisfied with the interval estimate you produced? Explain.

11.104 Redo Exercise 11.101, assuming that you know that the sample proportion will be no less than .75.

11.105 Suppose that you used the sample size calculated in Exercise 11.104 and found $\hat{p} = .75$.

  a Estimate the population proportion with 90% confidence.

  b Is this the result you expected? Explain.

11.106 Suppose that you used the sample size calculated in Exercise 11.104 and found $\hat{p} = .92$.

  a Estimate the population proportion with 90% confidence.

  b Is this the result you expected? Explain.

  c If you were hired to conduct this analysis would the person who hired you be satisfied with the interval estimate you produced? Explain.

11.107 Suppose that you used the sample size calculated in Exercise 11.104 and found $\hat{p} = .5$.

  a Estimate the population proportion with 90% confidence.

  b Is this the result you expected? Explain.

  c If you were hired to conduct this analysis, would the person who hired you be satisfied with the interval estimate you produced? Explain.

11.108 In a television commercial, the manufacturer of a toothpaste claims that more than four out of five dentists recommend the ingredients in his product. To test that claim, a consumer-protection group randomly samples 400 dentists and asks each one whether he or she would recommend toothpaste that contained the ingredients. The responses are 1 = No and 2 = Yes. The responses are stored in file Xr11-108. Can the consumer group infer that the claim is true?

11.109 A professor of statistics recently adopted a new textbook. At the completion of the course, 100 randomly selected students were asked to assess the book. The responses are as follows:

  Excellent (1), Good (2), Adequate (3), Poor (4)

The results are stored in file Xr11-109 using the codes in parentheses.

  a Do these results allow us to conclude that more than 50% of all business students would rate it as excellent?

  b Do these results allow us to conclude that more than 90% of all business students would rate it as at least adequate?

11.110 Refer to Example 11.4. Estimate with 95% confidence the number of televisions that had received *Late Show with David Letterman.*

11.111 The wine industry is an important part of the economy of western New York State and of southern Ontario. The wine is made from grapes grown on vines. During the winter, some of the grape wines die from the extreme cold that is common in this part of the continent. In the spring, the vines are pruned. If the vine is brown, it means that the plant is dead; green indicates a healthy vine. To test how well a vineyard has survived the winter, a random sample of vines is selected. The results of the latest pruning are stored in file Xr11-111 (2 = dead, and 1 = alive). Estimate with 90% confidence the degree of winter kill for this vine.

11.112 The Wilfrid Laurier University bookstore conducts annual surveys of its customers. One question asks respondents to rate the prices of textbooks. The wording is

  "The bookstore's prices of textbooks are reasonable."

The manager of the bookstore claims that more than half of her customers agree with the statement. A statistics professor doubts the claim and decides to apply his skill to determine whether the claim is true. A random sample of 100 students is drawn and each is asked whether he or she agrees with the statement. The responses (1 = yes, and 2 = no) are stored in file Xr11-112. Is there sufficient evidence to infer that the claim is true?

11.113 In 1991 the Canadian government passed legislation that allowed a man to take a paid leave (paid from employment insurance) from work following the birth of his child. Women have had that right for several decades. A government economist wanted to know how many men are taking advantage of the law and staying home to care for the baby (instead of the mother). A random sample of 300 men whose wives gave birth the previous week was asked whether they took leave to care for the baby. The responses (1 = no, and 2 = yes) are stored in file Xr11-113. Estimate with 95% confidence the proportion of all men who take advantage of the new law.

11.114 When homeowners hire a real estate agent to sell their homes, the sales agreement specifies a commission (usually 5%). However, in the negotiation of the sales price it is common for the agent to reduce the commission in order to complete the sales of the house. A homeowner who is about to list his house is told by an agent that a reduction in the commission is quite rare, occurring less than 25% of the time. The homeowner is skeptical and decides to test the claim. A random sample of 200 recently sold houses was drawn, and whether the commission was reduced was recorded and stored in file Xr11-114 (1 = reduction, and 2 = no reduction). Can we infer from these data that the agent is not telling the truth?

11.5 SUMMARY

The inferential methods presented in this chapter address the problem of describing a single population. When the data are interval, the parameters of interest are the population mean μ and the population variance σ^2. The Student t distribution is used to test and estimate the mean when the population standard deviation is unknown. The chi-squared distribution is used to make inferences about a population variance. When the data are nominal, the parameter to be tested and estimated is the population proportion p. The sample proportion follows an approximate normal distribution, which produces the test statistic and the interval estimator. We also discussed how to determine the sample size required to estimate a population proportion.

IMPORTANT TERMS

Exit polls
t-statistic
Student t distribution
Robust
Degrees of freedom
Chi-squared statistic

SYMBOLS

Symbol	Pronounced	Represents
v	nu	Degrees of freedom
χ^2	chi-squared	Chi-squared statistic
$\hat{p}$	p-hat	Sample proportion

FORMULAS

Test statistic for μ

$$t = \frac{\bar{x} - \mu}{s/\sqrt{n}}$$

Interval estimator of μ

$$\bar{x} \pm t_{\alpha/2} \frac{s}{\sqrt{n}}$$

Test statistic for σ^2

$$\chi^2 = \frac{(n-1)s^2}{\sigma^2}$$

Interval Estimator of σ^2

$$\text{LCL} = \frac{(n-1)s^2}{\chi^2_{\alpha/2}}$$

$$\text{UCL} = \frac{(n-1)s^2}{\chi^2_{1-\alpha/2}}$$

Test statistic for p

$$z = \frac{\hat{p} - p}{\sqrt{p(1-p)/n}}$$

Interval estimator of p

$$\hat{p} \pm z_{\alpha/2}\sqrt{\frac{\hat{p}(1-\hat{p})}{n}}$$

MICROSOFT EXCEL OUTPUT AND INSTRUCTIONS

Technique	Page
t-test and estimator of μ	298–299
Chi-squared test and estimator of σ^2	308
z-test and estimator of p	314

SUPPLEMENTARY EXERCISES

11.115 One of the issues that came up in a recent municipal election was the high cost of housing. A candidate seeking to unseat an incumbent claimed that the average family spends more than 30% of its annual income on housing. A housing expert was asked to investigate the claim. A random sample of 125 households was drawn, and each household was asked to report the percentage of household income spent on housing costs. The data are stored in file Xr11-115.

  a Is there enough evidence to infer that the candidate is correct?

  b Using a confidence level of 95%, estimate the mean percentage spent on housing by all households.

  c What is the required condition for the techniques used in parts (a) and (b)? Use a graphical technique to check whether it is satisfied.

11.116 To help forecast the winner in a Democratic senate primary, a survey was conducted. A random sample of 681 registered Democrats was asked for whom they intended to vote in the primary to be held the following day. The results are stored in file Xr11-116 using the following codes.

  Barbara Jones (1), Bill Smith (2), Pat Jackson (3)

  a Estimate with 90% confidence the proportion of all voters who will vote for Barbara Jones.

  b Can we conclude that Bill Smith will receive more than 20% of the vote?

11.117 The "just-in-time" policy of inventory control (developed by the Japanese) is growing in popularity. For example, General Motors recently spent $2 billion on its Oshawa, Ontario, plant so that it will be less than one hour from most suppliers. Suppose that an automobile parts supplier claims to deliver parts to any manufacturer in an average time of less than one hour. In an effort to test the claim, a manufacturer recorded the times (in minutes) of 24 deliveries from this supplier. These data are stored in file Xr11-117. Can we conclude that the supplier's assertion is correct?

11.118 Robots are being used with increasing frequency on production lines to perform monotonous tasks. To determine whether a robot welder should replace human welders in producing automobiles, an experiment was performed. The time for the robot to complete a series of welds was found to be 38 seconds. A random sample of 20 workers was taken, and the time for each worker to complete the welds was measured and stored in file Xr11-118. The mean was calculated to be 38 seconds, the same as the robot's time. However, the robot's time did not vary, whereas there was variation among the workers' times. An analysis of the production line revealed that if the variance exceeds 17 seconds^2, there will be problems. Perform an analysis of the data, and determine whether problems using human welders are likely.

11.119 The television networks often compete on the evening of an election day to be the first to identify the winner of the election correctly. One commonly used technique is the random sampling of voters as they exit the polling booths. Suppose that, in a two-candidate race, 500 voters were asked for whom they voted. The results were stored in file Xr11-119 using the code 1 = Democrat and 2 = Republican. Can we conclude that the Republican candidate will win?

11.120 Suppose that, in a large state university (with numerous campuses), the marks in an introductory statistics course are normally distributed with a mean of 68%. To determine the effect of requiring students to pass a calculus test (which at present is not a prerequisite), a random sample of 50 students who have taken calculus is given a statistics course. The marks out of 100 are stored in file Xr11-120.
   a Estimate with 95% confidence the mean statistics mark for all students who have taken calculus.
   b Do these data provide evidence to infer that students with a calculus background would perform better in statistics than students with no calculus?

11.121 Duplicate bridge is a game in which players compete for master points. When a player receives 300 master points, he or she becomes a life master. Because that title comes with a year's free subscription to the American Contract Bridge League (ACBL) monthly bulletin, the ACBL is interested in knowing the status of non–life masters. Suppose that a random sample of 80 non–life masters was asked how many master points they have. The results are stored in file Xr11-121. The ACBL would like an estimate of the mean number of master points held by all non–life masters. A confidence level of 90% is considered adequate in this case.

11.122 A national health care system was an issue in the recent presidential election campaign and is likely to be a subject of debate for many years. The issue arose because of the large number of Americans who have no health insurance. Under the present system, free health care is available to poor people, while relatively well-off Americans buy their own health insurance. Those who are considered working poor and who are in the lower-middle-class economic stratum appear to be most unlikely to have adequate medical insurance. To investigate this problem, a statistician surveyed 250 families whose gross income last year was between $15,000 and $25,000. Family heads were asked whether they have medical insurance coverage. The answers were stored in file Xr11-122 (2 = Has medical insurance, and 1 = Doesn't have medical insurance). The statistician wanted an estimate of the fraction of all families whose incomes are in the range of $15,000 to $25,000 who have medical insurance. Perform the necessary calculations to produce an interval estimate with 90% confidence.

11.123 The routes of postal deliverers are carefully planned so that each deliverer works between 7 and 7.5 hours per shift. The planned routes assume an average walking speed of 2 miles per hour and no shortcuts across lawns. In an experiment to examine the amount of time deliverers actually spend completing their shifts, a random sample of 75 postal deliverers was secretly timed. The data from the survey are stored in file Xr11-123.
   a Estimate with 99% confidence the mean shift time for all postal deliverers.
   b Check to determine if the required condition for this statistical inference is satisfied.
   c Is there enough evidence to conclude that postal workers are on average spending less than seven hours per day doing their jobs?

11.124 As you can easily appreciate, the number of Internet users is rapidly increasingly. A recent survey reveals that there are about 30 million Internet users in North America. Suppose that a survey of 200 of these people asked them to report the number of hours they spent on the Internet last week. The results are stored in file Xr11-124. Estimate with 95% confidence the mean weekly amount of time spent by all North Americans on the Internet.

11.125 The manager of a branch of a major bank wants to improve service. She is thinking about giving one dollar to any customer who waits in line for a period of time that is considered excessive. (The bank ultimately decided that more than eight minutes is excessive.) However, to get a better idea about the level of current service, she undertakes a survey of customers. A student is hired to measure the time spent waiting in line by a random sample of 50 customers. Using a stopwatch, the student determined the amount of time between the time the customer joined the line and the time he or she reached the teller. The times were recorded and are stored in file Xr11-125. Construct a 90% confidence interval estimate of the mean waiting time for the bank's customers.

11.126 In an examination of consumer loyalty in the travel business, 72 first-time visitors to a tourist attraction were asked whether they planned to return. The responses were stored in file Xr11-126 where 2 = Yes and 1 = No. Estimate with 95% confidence the proportion of all first-time visitors who planned to return to the same destination.

11.127 Engineers who are in charge of the production of springs used to make car seats are concerned about the variability of the springs. The springs are designed to be 500 mm long. When the springs are too long, they will loosen and fall out. When they are too short, they will not fit into the frames. The springs that are too long and too short must be reworked at considerable additional cost. The engineers have calculated that a standard deviation of 2 mm will result in an acceptable number of springs that must be reworked. A random sample of 100 springs was measured. The data are stored in file Xr11-127. Can we infer that the number of springs requiring reworking is unacceptably large?

11.128 Refer to Exercise 11.127. Suppose the engineers recoded the data so that springs that were the correct length were recorded as 1, springs that were too long were recorded as 2, and springs that were too short were recorded as 3. These data are stored in file Xr11-128. Can we infer that less than 90% of the springs are the correct length?

11.129 An advertisement for a major home appliance manufacturer claims that its repair personnel are the loneliest in the world because its appliances require the smallest number of service calls. To examine this claim, a researcher drew a random sample of 100 owners of five-year-old washing machines. The number of service calls made in the five-year period was recorded and stored in file Xr11-129. Find the 90% confidence interval estimate of the mean number of service calls for all five-year-old washing machines.

11.130 An oil company sends out monthly statements to its customers who purchased gasoline and other items using the company's credit card. Until now, the company has not included a preaddressed envelope for returning payments. The average and the standard deviation of the number of days before payment is received are 9.8 and 3.2, respectively. As an experiment to determine whether enclosing preaddressed envelopes speeds up payment, 150 customers selected at random were sent preaddressed envelopes with their bills. The number of days to payment was recorded and stored in file Xr11-130.

a Do the data provide sufficient evidence to establish that enclosure of preaddressed envelopes improves the average speed of payments?

b Can we conclude that the variability in payment speeds decreases when a preaddressed envelope is sent?

11.131 A rock promoter is in the process of deciding whether to book a new band for a rock concert. He knows that this band appeals almost exclusively to teenagers. According to the latest census, there are 400,000 teenagers in the area. The promoter decides to do a survey to try to estimate the proportion of teenagers who will attend the concert. How large a sample should be taken in order to estimate the proportion to within .02 with 99% confidence?

11.132 In Exercise 11.131, suppose that the promoter decided to draw a sample of size 600 (because of financial considerations). Each teenager was asked whether he or she would attend the concert. The answers were stored in file Xr11-132 using the following codes: 2 = Yes, I will attend; 1 = No, I will not attend. Estimate with 95% confidence the number of teenagers who will attend the concert.

11.133 The number of automatic teller machines (ATMs) has grown dramatically in the last several years. It has been estimated that for each transaction performed by an ATM instead of a teller, a bank saves more than one dollar. However, because of the cost of installation and maintenance, the number and placement of ATMs must be carefully planned. To help determine potential savings and thus whether a particular site should have an ATM, a bank conducted a survey to estimate the frequency with which potential ATM users will actually use the ATM annually. The responses are stored in file Xr11-133. Estimate with 95% confidence the mean frequency of use annually per person.

11.134 The owner of a downtown parking lot suspects that the person she hired to run the lot is stealing some money. The receipts as provided by the employee indicate that the average number of cars parked in the lot is 125 per day and that, on average, each car is parked for three and a half hours. In order to determine whether the employee is stealing, the owner watches the lot for five days. On those days, the number of cars parked is as follows:

120 130 124 127 128

The time spent on the lot for the 629 cars that the owner observed during the five days was stored in the file Xr11-134. Can the owner conclude that the employee is stealing? (Hint: Because there are two ways to steal, two tests should be performed.)

Case 11.1 Pepsi's Exclusivity Agreement with a University

In the last few years, colleges and universities have signed exclusivity agreements with a variety of private companies. These agreements bind the university to sell that company's products exclusively on the campus. Many of the agreements involved food and beverage firms.

A large university with a total enrolment of about 50,000 students has offered Pepsi-Cola an exclusivity agreement, which would give Pepsi exclusive rights to sell their products at all university facilities for the next year and an option for future years. In return, the university would receive 35% of the on-campus revenues and an additional lump sum of $200,000 per year. Pepsi has been given two weeks to respond.

The management at Pepsi quickly reviews what they know. The market for soft drinks is measured in terms of the equivalent of 10-ounce cans. Pepsi currently sells an average of 22,000 cans or their equivalents per week (over the 40 weeks of the year that the university operates). The cans sell for an average of 75 cents each. The costs, including labor, amount to 20 cents per can. Pepsi is unsure of its market share but suspects it is considerably less than 50%. A quick analysis reveals that if its current market share were 25%, then, with an exclusivity agreement, Pepsi would sell 88,000 cans per week or 3,520,000 cans per year (calculated as 88,000 cans per week × 40 weeks). The gross revenue would be computed as follows:

Gross revenue = 3,520,000 cans × $.75 revenue/can = $2,640,000

This figure must be multiplied by 65%, because the university would rake in 35% of the gross. Thus,

65% × $2,640,000 = $1,716,000

The total cost of 20 cents per can (or $704,000) and the annual payment to the university of $200,000 is subtracted to obtain the net profit:

Net profit = $1,716,000 − $704,000 − $200,000 = $812,000

Pepsi's current annual profit is

Current profit = 40 weeks × 22,000 cans/week × $.55/can = $484,000

If the current market share is 25%, the potential gain from the agreement is

$812,000 − $484,000 = $328,000

The only problem with this analysis is that Pepsi does not know how many soft drinks are sold weekly at the university. Coke is not likely to supply Pepsi with information about the sales of its brands, which together with Pepsi's line of products constitutes virtually the entire market.

A recent graduate of a business program volunteers that a survey of the university's students can supply the missing information. Accordingly, she organizes a survey that asks 500 students to keep track of the number of soft drinks they purchase on campus over the next seven days. The responses are stored in file C11-01.

Perform a statistical analysis to extract the needed information from the data. Estimate with 95% confidence the parameter that is at the core of the decision problem. Use the estimate to compute estimates of the annual profit. Assume that Coke and Pepsi drinkers would be willing to buy either product in the absence of their first choice.

On the basis of maximizing profits from sales of soft drinks at the university, should Pepsi agree to the exclusivity agreement?

Case 11.2 Pepsi's Exclusivity Agreement with a University: The Coke Side of the Equation

While the executives of Pepsi-Cola are trying to decide what to do, the university informs them that a similar offer has gone out to the Coca-Cola Company. Furthermore, if both companies want exclusive rights, then a bidding war will take place. The executives at Pepsi would like to know how likely is it that Coke will want exclusive rights under the conditions outlined by the university.

Perform a similar analysis to the one you did in Case 11.1, but this time from Coke's point of view.

Is it likely that Coke will want to conclude an exclusivity agreement with the university? Discuss the reasons for your conclusions.

Case 11.3 Number of Uninsured Motorists*

A number of years ago the Michigan legislature passed a law requiring insurance for all drivers. Prior to this event drivers did not have to be covered by insurance. The law was challenged on the grounds that it discriminated against poor people who would not be able legally to drive. At issue at the trial was the number of Michigan motorists who would be coerced by the law into buying insurance. To determine this, it was nec-

---

*Adapted from L. Katz, "Presentation of a Confidence Interval Estimate as Evidence in a Legal Proceeding," Department of Statistics, Michigan State University (1974).

essary to count the number of uninsured motorists. (These would be the people who would be forced by law to buy insurance.) There were a total of 4,505,665 license plates for passenger vehicles registered in Michigan at the time. An investigation of each one of these to determine whether the drivers had insurance coverage would be prohibitively expensive and time-consuming. It was decided that the state would draw a random sample of motorists and estimate the number of Michigan's driving population who were uninsured from the sample data. A random sample of 249 license plates was drawn using statistically sound sampling methods. Each was investigated to determine its insurance status. The license plates sampled were placed in one of three categories. The categories and the code on the disk are as follows.

1 Insured

2 Not insured

3 Missing

(License plates that were drawn for the sample but for which investigators were unable to find the car or its owner were classified as missing.) The data are stored in file C11–03.

Your job is to estimate the proportion of all Michigan passenger vehicles that are not insured. Provide two methods for dealing with the missing data. From each method, determine the upper and lower limits for the estimated number or motorists who would have been forced by law to buy insurance. Discuss which method is more reasonable.

Chapter 12

Inference about Two Populations

12.1 Introduction

12.2 Inference about the Difference between Two Means: Independent Samples

12.3 Observational and Experimental Data

12.4 Inference about the Difference between Two Means: Matched Pairs Experiment

12.5 Inference about the Ratio of Two Variances

12.6 Inference about the Difference between Two Population Proportions

12.7 Summary

12.1 INTRODUCTION

We can compare learning how to apply statistical techniques to learning how to drive a car. We began by describing what you are going to do in this course (Chapter 1), followed by a presentation of the essential background material (Chapters 2 through 8). Learning the concepts of statistical inference and applying them the way we did in Chapters 9 and 10 is akin to driving a car in an empty parking lot. You're driving, but it's not a realistic experience. Learning Chapter 11 is like driving on a quiet side street with little traffic. The experience represents real driving, but much of the difficulty has been postponed. In this chapter, you begin to drive for real, with many of the actual problems faced by licensed drivers, and the experience prepares you to tackle the next difficulty.

In this chapter, we present a variety of techniques whose objective is to compare two populations. In Sections 12.2 and 12.4, we deal with interval data; the parameter of interest is the difference between two means. The difference between these two sections introduces yet another factor that determines the correct statistical method—the design of the experiment used to gather the data. In Section 12.2, the samples are independently drawn, whereas in Section 12.4, the samples are taken in a matched pairs experiment. In Section 12.3, we discuss the difference between observational and experimental data, a distinction that is critical to the way in which we interpret statistical results.

Section 12.5 presents the procedures employed to infer whether two population variances differ. The parameter is the ratio σ_1^2/σ_2^2. (When comparing two variances, we use the ratio rather than the difference because of the sampling distribution.)

Section 12.6 addresses the problem of comparing two populations of nominal data. The parameter to be tested and estimated is the difference between two proportions.

Following are illlustrations of situations in which the problem objective is to compare two populations.

Illustration 1

Medical scientists are involved in various research projects. A number of projects are examining ways to reduce cholesterol levels, because high levels of cholesterol are linked to heart attacks and strokes. One method of testing the effectiveness of a new drug is to give the drug to one group of people and a placebo (a pill with no medicine) to another group of people. To judge how well the drug works, the reduction in cholesterol would be measured for each person. The mean reduction for those taking the drug could be compared with the mean reduction for those taking the placebo. The objective is to test to see if the former is greater than the latter. The parameter is the difference between two means $\mu_1 - \mu_2$.

Illustration 2

Politicians are constantly concerned about how the voting public perceives their actions and behaviors. Politicians are particularly concerned with the extent to which constituents approve of their behavior and the ways in which that approval changes over time. As a result, they frequently poll the public to determine the proportion of voters who support them and whether that support has changed since the previous survey. The parameter of interest to them is the difference between two proportions $p_1 - p_2$, where p_1 is the proportion of support at present and p_2 is the proportion of support at the time of the previous survey.

Illustration 3

Operations managers of production facilities are always looking for ways to improve productivity in their plants. This can be accomplished by rearranging sequences of operations, acquiring new technology, or improving the training of workers. When one or more such changes are made, their effect on the operation of the entire plant is of interest. The manager can measure the effect by comparing productivity after the innovation with productivity before the innovation. Because productivity is often measured by the mean number of units produced per hour, the parameter of interest is the difference between two means $\mu_1 - \mu_2$. We may also be interested in comparing the consistency before and after the innovation; the parameter to be tested or estimated is σ_1^2/σ_2^2.

Illustration 4

Market managers and advertisers are eager to know which segments of the population are buying their products. If they can determine these groups, they can target their advertising messages and tailor their products to these customers. For example, if advertisers determine that the decision to purchase a particular household product is made more frequently by men than by women, the interests and concerns of men will be the focus of most commercial messages. The advertising media also depend on whether the product is of greater interest to men or to women. The most common way of measuring this circumstance is to find the difference in the proportions of men and women buying the product. In these situations, the parameter to be tested or estimated is the difference between two proportions $p_1 - p_2$.

12.2 INFERENCE ABOUT THE DIFFERENCE BETWEEN TWO MEANS: INDEPENDENT SAMPLES

In order to test and estimate the difference between two population means, the statistics practitioner draws random samples from each of two populations. In this section, we discuss independent samples. In Section 12.4, where we present the matched pairs experiment, the distinction between independent samples and matched pairs will be made clear. For now, we define independent samples as samples completely unrelated to one another.

Figure 12.1 depicts the sampling process. Observe that we draw a sample of size n_1 from population 1 and a sample of size n_2 from population 2. For each sample, we compute the sample means and sample variances.

The best estimator of the difference between two population means $\mu_1 - \mu_2$ is the difference between two sample means $\bar{x}_1 - \bar{x}_2$. In Chapter 8 we presented the sampling distribution of $\bar{x}_1 - \bar{x}_2$.

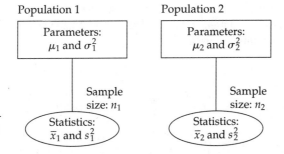

Figure 12.1

Independent samples from two populations

Thus,

$$z = \frac{(\bar{x}_1 - \bar{x}_2) - (\mu_1 - \mu_2)}{\sqrt{\frac{\sigma_1^2}{n_1} + \frac{\sigma_2^2}{n_2}}}$$

is a standard normal (or approximately standard normal) random variable. It follows that this will be the test statistic.

The interval estimator is

$$(\bar{x}_1 - \bar{x}_2) \pm z_{\alpha/2}\sqrt{\frac{\sigma_1^2}{n_1} + \frac{\sigma_2^2}{n_2}}$$

However, these formulas are rarely used because the population variances σ_1^2 and σ_2^2 are almost always unknown. Consequently, it is necessary to estimate the standard

---

Sampling Distribution of $\bar{x}_1 - \bar{x}_2$

1. $\bar{x}_1 - \bar{x}_2$ is normally distributed if the populations are normal and approximately normal if the populations are nonnormal and the sample sizes are large.

2. The expected value of is $\bar{x}_1 - \bar{x}_2$ is

$$E(\bar{x}_1 - \bar{x}_2) = \mu_1 - \mu_2$$

3. The variance of $\bar{x}_1 - \bar{x}_2$ is

$$V(\bar{x}_1 - \bar{x}_2) = \frac{\sigma_1^2}{n_1} + \frac{\sigma_2^2}{n_2}$$

The standard error of $\bar{x}_1 - \bar{x}_2$ is

$$\sqrt{\frac{\sigma_1^2}{n_1} + \frac{\sigma_2^2}{n_2}}$$

error of the sampling distribution. The way to do this depends on whether the two unknown population variances are equal. When they are equal, the test statistic and interval estimator are as follows.

Test Statistic for $\mu_1 - \mu_2$ when $\sigma_1^2 = \sigma_2^2$

$$t = \frac{(\bar{x}_1 - \bar{x}_2) - (\mu_1 - \mu_2)}{\sqrt{s_p^2 \left(\frac{1}{n_1} + \frac{1}{n_2}\right)}} \qquad \nu = n_1 + n_2 - 2$$

where

$$s_p^2 = \frac{(n_1 - 1)s_1^2 + (n_2 - 1)s_2^2}{n_1 + n_2 - 2}$$

The quantity s_p^2 is called the **pooled variance estimator.** It is the weighted average of the two sample variances. The requirement that the population variances be equal makes this calculation feasible, because we need only one estimate of the common value of σ_1^2 and σ_2^2. It makes sense for us to use the pooled variance estimator because, in combining both samples, we produce a better estimate.

The test statistic is Student t distributed with $n_1 + n_2 - 2$ degrees of freedom, provided that the two populations are normal. The interval estimator is derived by mathematics that by now has become routine.

Interval Estimator of $\mu_1 - \mu_2$ when $\sigma_1^2 = \sigma_2^2$

$$(\bar{x}_1 - \bar{x}_2) \pm t_{\alpha/2} \sqrt{s_p^2 \left(\frac{1}{n_1} + \frac{1}{n_2}\right)} \qquad \nu = n_1 + n_2 - 2$$

We will refer to the above formulas as the **equal-variances test statistic** and **equal-variances interval estimator,** respectively.

The question naturally arises, "How do we know when the population variances are equal?" The answer is that because σ_1^2 and σ_2^2 are unknown, we can't know for certain whether they're equal. However, we can use the sample variances s_1^2 and s_2^2 to make inferences about the population variances. In Section 12.5, we will present a statistical technique that will allow us to test for equality. However, for now we will simply examine the sample variances and informally judge their relative values to determine whether we can assume that the population variances are equal.

When the population variances are unequal, we cannot use the pooled variance estimate. Instead, we estimate each population variance with its sample variance. Unfortunately, the sampling distribution of the resulting statistic

$$\frac{(\bar{x}_1 - \bar{x}_2) - (\mu_1 - \mu_2)}{\sqrt{\dfrac{s_1^2}{n_1} + \dfrac{s_2^2}{n_2}}}$$

is neither normal nor Student t. However, it can be approximated by a Student t distribution with degrees of freedom equal to

$$\nu = \frac{(s_1^2/n_1 + s_2^2/n_2)^2}{\dfrac{(s_1^2/n_1)^2}{n_1 - 1} + \dfrac{(s_2^2/n_2)^2}{n_2 - 1}}$$

The test statistic and interval estimator are easily derived from the sampling distribution.

Test Statistic for $\mu_1 - \mu_2$ when $\sigma_1^2 \neq \sigma_2^2$

$$t = \frac{(\bar{x}_1 - \bar{x}_2) - (\mu_1 - \mu_2)}{\sqrt{\left(\dfrac{s_1^2}{n_1} + \dfrac{s_2^2}{n_2}\right)}} \qquad \nu = \frac{(s_1^2/n_1 + s_2^2/n_2)^2}{\dfrac{(s_1^2/n_1)^2}{n_1 - 1} + \dfrac{(s_2^2/n_2)^2}{n_2 - 1}}$$

Interval Estimator of $\mu_1 - \mu_2$ when $\sigma_1^2 \neq \sigma_2^2$

$$(\bar{x}_1 - \bar{x}_2) \pm t_{\alpha/2}\sqrt{\left(\dfrac{s_1^2}{n_1} + \dfrac{s_2^2}{n_2}\right)} \qquad \nu = \frac{(s_1^2/n_1 + s_2^2/n_2)^2}{\dfrac{(s_1^2/n_1)^2}{n_1 - 1} + \dfrac{(s_2^2/n_2)^2}{n_2 - 1}}$$

We will refer to the above formulas as the **unequal-variances test statistic** and **unequal-variances interval estimator,** respectively.

DECISION RULE: EQUAL-VARIANCES OR UNEQUAL-VARIANCES t-TESTS AND ESTIMATORS

Statisticians have shown that the number of degrees of freedom associated with the equal-variances test statistic and interval estimator is always greater than or equal to the number of degrees of freedom for the unequal-variances test statistic and interval estimator. Larger degrees of freedom have the same effect as having larger sample sizes. And we have seen that larger sample sizes yield more information, which is reflected in more powerful tests (lower probabilities of Type II errors) and narrower interval estimates. As a result, we would prefer that the population variances be equal. Accordingly, we adopt the following rule. We will use the equal-variances test statistic and interval estimator unless there is evidence (based on the sample variances) to indicate that the population variances are unequal, in which case we will use the unequal-variances *t*-test and estimator.

▼ EXAMPLE 12.1

Despite some controversy, scientists generally agree that high-fiber cereals reduce the likelihood of various forms of cancer. However, one scientist claims that people who eat high-fiber cereal for breakfast will consume, on average, fewer calories for lunch than people who don't eat high-fiber cereal for breakfast (*Toronto Star,* 2 July 1991). If this is true, we can add weight-reduction to the other reasons to eat a high-fiber cereal for breakfast. As a preliminary test of the claim, 150 people were randomly selected and asked what they regularly eat for breakfast and lunch. Each person was identified as either a consumer or a nonconsumer of high-fiber cereal, and the number of calories consumed at lunch was measured and recorded. These data are listed below and stored in columns A and B of file Xm12-01. Can the scientist conclude that his belief is correct?

CALORIES CONSUMED AT LUNCH

Consumers of High-Fiber Cereal

568	646	607	555	530	714	593	647	650
498	636	529	565	566	639	551	580	629
589	739	637	568	687	693	683	532	651
681	539	617	584	694	556	667	467	
540	596	633	607	566	473	649	622	

Nonconsumers of High-Fiber Cereal

705	754	740	569	593	637	563	421	514	536
819	741	688	547	723	553	733	812	580	833
706	628	539	710	730	620	664	547	624	644
509	537	725	679	701	679	625	643	566	594
613	748	711	674	672	599	655	693	709	596
582	663	607	505	685	566	466	624	518	750
601	526	816	527	800	484	462	549	554	582
608	541	426	679	663	739	603	726	623	788
787	462	773	830	369	717	646	645	747	
573	719	480	602	596	642	588	794	583	
428	754	632	765	758	663	476	490	573	

Section 12.2 INFERENCE ABOUT THE DIFFERENCE BETWEEN TWO MEANS: INDEPENDENT SAMPLES

Solution

IDENTIFY

To assess the claim, the scientist needs to compare the population of consumers of high-fiber cereal to the population of nonconsumers. The data are interval (obviously, we've recorded real numbers). This problem-objective/data-type combination tells us that the parameter to be tested is the difference between two means $\mu_1 - \mu_2$. The claim to be tested is that the mean caloric intake of consumers (μ_1) is less than that of nonconsumers (μ_2). Hence the alternative hypothesis is

$$H_1 : (\mu_1 - \mu_2) < 0$$

To identify the test statistic, the scientist instructs the computer to output the sample variances. They are

$$s_1^2 = 4{,}103 \text{ and } s_2^2 = 10{,}670$$

There is reason to believe that the population variances are unequal. Thus, we use the unequal-variances test statistic.

The complete test follows.

$$H_0 : (\mu_1 - \mu_2) = 0$$

$$H_1 : (\mu_1 - \mu_2) < 0$$

Test statistic: $t = \dfrac{(\bar{x}_1 - \bar{x}_2) - (\mu_1 - \mu_2)}{\sqrt{\dfrac{s_1^2}{n_1} + \dfrac{s_2^2}{n_2}}}$ $\quad \nu = \dfrac{(s_1^2/n_1 + s_2^2/n_2)^2}{\dfrac{(s_1^2/n_1)^2}{n_1 - 1} + \dfrac{(s_2^2/n_2)^2}{n_2 - 1}}$

COMPUTE

Microsoft Excel Output for Example 12.1

t-Test: Two-Sample Assuming Unequal Variances

	Consumers	Nonconsumers
Mean	604.02	633.23
Variance	4102.98	10669.77
Observations	43	107
Hypothesized Mean Difference	0	
df	123	
t Stat	−2.0911	
P(T<=t) one-tail	0.0193	
t Critical one-tail	1.6573	
P(T<=t) two-tail	0.0386	
t Critical two-tail	1.9794	

The value of the test statistic (**t Stat**) is −2.0911. The one-tail *p*-value (**P(T<=t) one-tail**) is .0193.

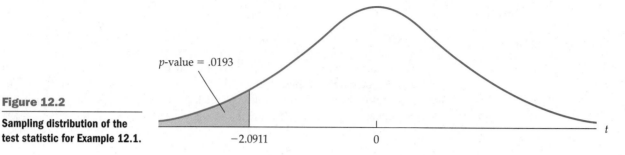

Figure 12.2

Sampling distribution of the test statistic for Example 12.1.

GENERAL COMMANDS	EXAMPLE 12.1
1 Type or import the data into two columns.	Open file **Xm12-01**.
2 Click **Tools, Data Analysis...**, and **t-test: Two-Sample Assuming Unequal Variances**.	
3 Specify the **Variable 1 Range**.	A1:A44
4 Specify the **Variable 2 Range**.	B1:B108
5 Type the value of the **Hypothesized Mean Difference**.	0
6 Click **Labels** if applicable.	
7 Specify the value of α **(Alpha:)**, and click **OK**.	.05

Figure 12.2 depicts the Student t sampling distribution and the p-value of the test.

To conduct this test from means and standard deviations or to perform a what-if analysis, activate the **t-test of 2 Means (Uneq-Var)** worksheet in **Stats-Summary.xls**.

INTERPRET

A Type I error in this example occurs when we conclude that consumers of high-fiber cereal consume fewer calories at lunch than do nonconsumers when in fact there is no difference between the two groups. A Type II error occurs when we erroneously fail to conclude that consumers of high-fiber cereal eat less at lunch than nonconsumers. It is difficult to judge which error is more costly. As a result we take a neutral position and suggest that any p-value less than 5% should be interpreted as providing enough evidence to reject the null hypothesis.

The p-value of the test is .0193. As a result, we conclude that there is sufficient evidence to infer that consumers of high-fiber cereal do eat fewer calories at lunch than do nonconsumers. However, there are two reasons to be cautious about concluding that high-fiber cereals constitute an effective contribution to weight loss. First, the data were likely self-reported, which means that each person determined the number of calories recorded that he or she consumed. Such data are often unreliable. Ideally, a less subjective method of counting calories should be used. Second, the way in which the experiment was performed may lead to several contradictory interpretations of the data. We will discuss this important issue in the next section.

In addition to testing to determine whether a difference exists, we can estimate the difference in mean caloric intake.

The interval estimator of the difference between two means with unequal population variances is

$$(\bar{x}_1 - \bar{x}_2) \pm t_{\alpha/2} \sqrt{\frac{s_1^2}{n_1} + \frac{s_2^2}{n_2}}$$

COMPUTE

The Excel output does not include the interval estimate. However, you can use the **t-Estimate of 2 Means (Uneq-Var)** worksheet of **Stats-Summary.xls.** Simply plug in the values of the sample means, sample standard deviations, and sample sizes, as well as the confidence level. The result of this action appears below.

Microsoft Excel Output of the 95% confidence Interval Estimate: Example 12.1

t-Estimate of the Difference between Two Means (Unequal-Variances)

Sample 1
Sample mean	604.02
Sample standard deviation	64.05
Sample size	43

Sample 2
Sample mean	633.2
Sample standard deviation	103.3
Sample size	107

Confidence level	0.95
Degrees of freedom	122.60
Difference between means	−29.2
Bound	27.65
Lower confidence limit	−56.86
Upper confidence limit	−1.56

INTERPRET

We estimate that the average consumer of high-fiber cereal eats between 1.56 and 56.86 fewer calories at lunch than does the average nonconsumer of high-fiber cereal.

▼ EXAMPLE 12.2

The plant manager of a company that manufactures office equipment is attempting to determine the process that will be used to assemble a new ergonomic chair. The material, machines, and workforce have already been decided. However there are two methods under consideration. The methods differ by the order in which the separate operations are performed. To help decide which should be used, an experiment was performed. Twenty-five randomly selected workers each assembled the chair using method A, and another 25 workers each assembled the chair using method B. The assembly times in minutes were recorded and are exhibited below and stored in file Xm12-02. The plant manager would like to know whether the assembly times of the two methods differ.

ASSEMBLY TIMES: METHOD A					ASSEMBLY TIMES: METHOD B				
6.8	5	6.1	7.1	6.4	5.2	6.5	4.2	5.9	5.9
5	5.9	6.2	6.1	6.1	6.7	5.9	4.5	7.1	4.9
7.9	5.2	7.1	5	6.6	5.7	6.7	5.3	5.8	5.3
5.2	6.5	4.6	6.3	7.7	6.6	6.6	7.9	7	4.2
7.6	7.4	6	7	6.4	8.5	4.2	7	5.7	7.1

Solution

IDENTIFY

The data are interval, and the objective of the experiment is to compare the two populations of assembly times. The parameter of interest is the difference between two population means $\mu_1 - \mu_2$. The plant manager wants to determine whether a difference between the two methods exists. As a result the alternative hypothesis is

$$H_1 : (\mu_1 - \mu_2) \neq 0$$

To identify the correct test statistic, we need to calculate the sample variances. They are

$$s_1^2 = .8478 \text{ and } s_2^2 = 1.3031$$

Because the sample variances are similar, we will assume that the populations variances are equal and use the equal-variances test statistic.

$$H_0 : (\mu_1 - \mu_2) = 0$$
$$H_1 : (\mu_1 - \mu_2) \neq 0$$

Test statistic: $t = \dfrac{(\bar{x}_1 - \bar{x}_2) - (\mu_1 - \mu_2)}{\sqrt{s_p^2 \left(\dfrac{1}{n_1} + \dfrac{1}{n_2}\right)}} \quad v = n_1 + n_2 - 2$

COMPUTE

Microsoft Excel Output for Example 12.2

t-test: Two-Sample Assuming Equal Variances

	Method-A	Method-B
Mean	6.288	6.016
Variance	0.8478	1.3031
Observations	25	25
Pooled Variance	1.0754	
Hypothesized Mean Difference	0	
df	48	
t Stat	0.9273	
P(T<=t) one-tail	0.1792	
t Critical one-tail	1.6772	
P(T<=t) two-tail	0.3584	
t Critical two-tail	2.0106	

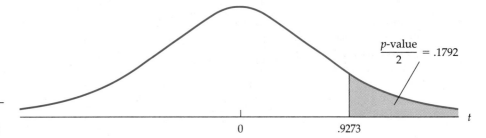

Figure 12.3

Sampling distribution of the test statistic for Example 12.2

GENERAL COMMANDS	EXAMPLE 12.2
1 Type or import the data into two columns.	Open file **Xm12-02**.
2 Click **Tools, Data Analysis...**, and **t-Test: Two-Sample Assuming Equal Variances.**	
3 Specify the **Variable 1 Range.**	A1:A26
4 Specify the **Variable 2 Range.**	B1:B26
5 Type the value of the **Hypothesized Mean Difference.**	0
6 Click **Labels** if applicable.	
7 Specify the value of α **(Alpha:)**, and click **OK**.	.05

Use the **t-Test of 2 Means (Eq-Var)** worksheet to complete this test from the sample statistics and to perform a what-if analysis.

INTERPRET

The *p*-value is .3584. (Note that this is a two-tail test; see Figure 12.3.) We conclude that there is little evidence to infer that the mean times differ. Once again, the manager should determine that all of the required conditions are satisfied (see the following subsection) and that there are no other factors that need to be considered. For example, is the quality of the finished product identical using the two designs? Is it possible that one method is better than another, but this experiment failed to demonstrate it because it takes longer to adapt to the new production design than this experiment allowed? If the conclusion stands, however, the manager should choose the method using some other criterion, such as worker preference.

Figure 12.3 depicts the sampling distribution.

▲

To estimate the difference between two means with equal population variances, activate the **t-Estimate of 2 Means (Eq-Var)** worksheet in **Stats-Summary.xls** and substitute the sample statistics and confidence level. We produced the following output.

Microsoft Excel Output of the 95% Confidence Interval Estimate: Example 12.2

t-Estimate of the Difference Between Two Means (Equal-Variances)	
Sample 1	
Sample mean	6.2880
Sample standard deviation	0.921
Sample size	25
Sample 2	
Sample mean	6.0160
Sample standard deviation	1.1420
Sample size	25
Confidence level	0.95
Pooled Variance estimate	1.0762
Difference between means	0.2720
Bound	0.5900
Lower confidence limit	−0.3180
Upper confidence limit	0.8620

CHECKING THE REQUIRED CONDITION

Both the equal-variances and unequal-variances techniques require that the populations be normally distributed. As before, we can check to see if the requirement is satisfied by drawing the histograms of the data. To illustrate, we used Excel to create the histograms for Examples 12.1 (Figures 12.4 and 12.5) and 12.2 (Figures 12.6 and 12.7). Although the histograms are not bell shaped, it appears that in both examples the data are at least approximately normal. Because this technique is robust, we can be confident in the validity of the results.

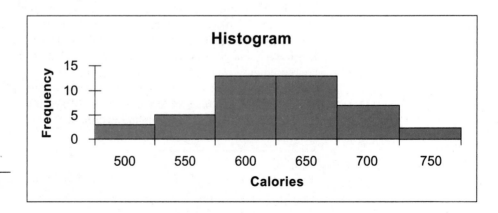

Figure 12.4

Histogram for Example 12.1: Consumers

Figure 12.5

Histogram for Example 12.1: Nonconsumers

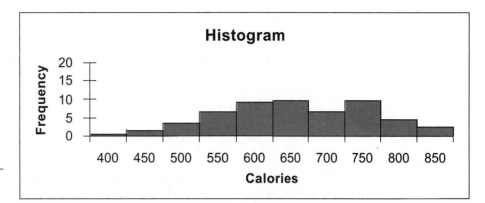

Figure 12.6

Histogram for Example 12.2: Method A

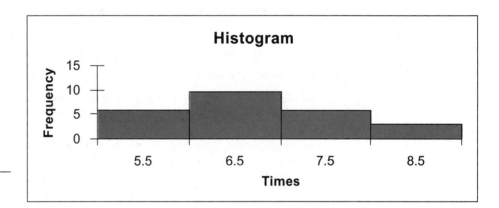

Figure 12.7

Histogram for Example 12.2: Method B

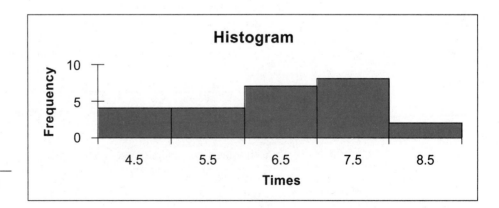

VIOLATION OF THE REQUIRED CONDITION

When the populations are extremely nonnormal, we will use a nonparametric technique—the Wilcoxon rank sum test (Chapter 16)—to replace the equal-variances test of $\mu_1 - \mu_2$. We have no alternative to the unequal-variances test of $\mu_1 - \mu_2$ when the populations are very nonnormal.

DEVELOPING AN UNDERSTANDING OF STATISTICAL CONCEPTS 1

The formulas above are relatively complicated. However, conceptually both test statistics are based on the concepts we introduced in Chapter 10 and repeated in Chapter 11. That is, the value of the test statistic is the difference between the statistic $\bar{x}_1 - \bar{x}_2$ and the hypothesized value of the parameter $\mu_1 - \mu_2$ measured in terms of the standard error. As was the case with the interval estimator of p, the standard error must be estimated from the data for all inferential procedures introduced here.

DEVELOPING AN UNDERSTANDING OF STATISTICAL CONCEPTS 2

The method we use to compute the standard error of the sampling distribution of $\bar{x}_1 - \bar{x}_2$ depends on whether the population variances are equal. When they are equal we calculate and use the pooled variance estimator s_p^2. An important principle is being applied here and will be again in this chapter (Section 12.6) and in later chapters. The principle can be loosely stated as follows: Where possible, it is advantageous to pool sample data to estimate the standard error. In the application of the equal-variances t-test and estimator, we are able to pool because we assume that the two samples were drawn from populations with a common variance. Combining both samples increases the accuracy of the estimate. Thus, s_p^2 is a better estimator of the common variance than either s_1^2 or s_2^2 separately.

When the two population variances are unequal, we cannot pool the data and produce a common estimator. We must compute s_1^2 and s_2^2 and use them to estimate σ_1^2 and σ_2^2, respectively.

Here is a summary of how we recognize the techniques presented in this section.

Factors That Identify the Equal-Variances t-Test and Estimator of $\mu_1 - \mu_2$

1 Problem objective: compare two populations.
2 Data type: interval.
3 Descriptive measurement: central location.
4 Experimental design: independent samples.
5 Population variances: equal.

Factors That Identify the Unequal-Variances t-Test and Estimator of $\mu_1 - \mu_2$

1. Problem objective: compare two populations.
2. Data type: interval.
3. Descriptive measurement: central location.
4. Experimental design: independent samples.
5. Population variances: unequal.

EXERCISES

*Exercises 12.1 to 12.18 are "what-if" analyses designed to determine what happens to the test statistics and interval estimates when elements of the statistical inference change. These problems can be solved manually or using **Stats-Summary.xls**.*

12.1 In random samples of 25 from two normal populations we found the following statistics:

$\bar{x}_1 = 524 \quad s_1 = 129$
$\bar{x}_2 = 469 \quad s_2 = 150$

  a Test to determine whether we can infer that the population means differ.
  b Estimate the difference between the two population means with 95% confidence.

12.2 Repeat Exercise 12.1, increasing the standard deviations to $s_1 = 255$ and $s_2 = 307$.

12.3 Review Exercises 12.1 and 12.2. Describe what happens when the sample standard deviations get larger.

12.4 Repeat Exercise 12.1 with samples of size 100.

12.5 Review Exercises 12.1 and 12.4. Discuss the effects of increasing the sample size.

12.6 Random sampling from two normal populations produced the following results:

$\bar{x}_1 = 63 \quad s_1 = 8 \quad n_1 = 50$
$\bar{x}_2 = 60 \quad s_2 = 7 \quad n_2 = 45$

  a Can we infer that μ_1 is greater than μ_2?
  b Estimate with 90% confidence the difference between the two population means.

12.7 Repeat Exercise 12.6, changing the sample standard deviations to 25 and 22, respectively.

12.8 Review Exercises 12.6 and 12.7. What happens when the sample standard deviations increase?

12.9 Repeat Exercise 12.6, doubling the sample sizes.

12.10 Review Exercises 12.6 and 12.9. Describe the effects of increasing the sample sizes.

12.11 After drawing random samples from two normal populations a statistics practitioner produced the following statistics:

$\bar{x}_1 = 884 \quad s_1 = 302 \quad n_1 = 38$
$\bar{x}_2 = 963 \quad s_2 = 49 \quad n_2 = 24$

  a Can we infer that μ_1 is less than μ_2?
  b Estimate with 95% confidence the difference between population means.

12.12 Repeat Exercise 12.11, increasing the sample standard deviations to 588 and 103, respectively.

12.13 Review Exercises 12.11 and 12.12. Discuss the effect of increasing the sample standard deviations.

12.14 Repeat Exercise 12.11, increasing the sample sizes to 80 and 75, respectively.

12.15 Review Exercises 12.11 and 12.14. Describe what happens when the sample sizes increase.

12.16 Repeat Exercise 12.11, changing $\bar{x}_1$ to 850.

12.17 Repeat Exercise 12.11, changing $\bar{x}_1$ to 900.

12.18 Review Exercises 12.11, 12.16, and 12.17. Discuss the effect of increasing and decreasing $\bar{x}_1$.

In exercises requiring a test of hypothesis where we do not specify a significance level we expect you to use the guidelines provided in Chapter 10 to judge the size of the p-value.

12.19 The data obtained from sampling from two populations are stored in columns A and B, respectively, in file Xr12-19.
  a Conduct a test at the 5% significance level to determine whether the population means differ.
  b Estimate the difference between population means with 95% confidence.
  c What is the required condition(s) of the techniques employed in parts (a) and (b)?
  d Check to ensure that the required condition(s) is satisfied.

12.20 Random samples were drawn from each of two populations. The data are stored in columns A and B, respectively, in file Xr12-20.
  a Is there sufficient evidence at the 1% significance level to infer that the mean of population 1 is greater than the mean of population 2?
  b Estimate with 90% confidence the difference between the two population means.
  c What is the required condition(s) of the techniques employed in parts (a) and (b)?
  d Check to ensure that the required condition(s) is satisfied.

12.21 Two samples of 40 observations from each of two populations were taken. The data are stored in columns A and B in file Xr12-21. Do these data provide sufficient evidence to infer that the mean of population 2 is greater than the mean of population 1?

12.22 A statistics practitioner gathered data from two populations and stored them in columns A (sample 1) and B (sample 2) in file Xr12-22. Can the statistics practitioner infer at the 10% significance level that the mean of population 1 is less than the mean of population 2?

12.23 The president of Tastee Inc., a baby-food producer, claims that her company's product is superior to that of her leading competitor, because babies gain weight faster with her product. (This is a good thing for babies.) To test the claim, a survey was undertaken. Mothers of babies were asked which baby food they intended to feed their babies. Those who responded Tastee or the leading competitor were asked to keep track of their babies' weight gains over the next two months. There were 15 mothers who indicated that they would feed their babies Tastee and 25 who responded that they would feed their babies the product of the leading competitor. Each baby's weight gain (in ounces) was recorded and stored in column A (Tastee) and column B (leading competitor) in file Xr12-23.
  a Can we conclude that, using weight gain as our criterion, Tastee baby food is indeed superior?
  b Estimate with 95% confidence the difference between the mean weight gains of the two products.
  c Check to ensure that the required condition(s) is satisfied.

12.24 Medical experts advocate the use of vitamin and mineral supplements to help fight infections. A study, undertaken by Dr. Ranjit Schneider of Memorial University (reported in the British journal *Lancet*, November 1992), recruited 96 men and women age 65 and older. Half of them received daily supplements of vitamins and minerals, while the other half received placebos. The supplements contained the daily recommended amounts of 18 vitamins and minerals. These included vitamins B-6, B-12, C, D, and E, thiamine, riboflavin, niacin, calcium, copper, iodine, iron, selenium, magnesium, and zinc. The doses of vitamins A and E were slightly less than the daily requirements. The supplements included four times the amount of beta-carotene that the average person ingests daily. The number of days of illness from infections (ranging from colds to pneumonia) was recorded for each person. The data are stored in columns A (supplements) and B (placebo) in file Xr12-24. Can we infer that taking vitamin and mineral supplements daily increases the body's immune system?

12.25 A sociologist is attempting to determine whether differences exist between 17- to 21-year-old female and male drivers. To help shed light on this issue 100 male and 100 female drivers were surveyed. Each was asked how many miles he or she drove in the past year. The distances (in thousands of miles) are stored in column A (males) and column B (females) in file Xr12-25.
  a Can we conclude that male and female drivers differ in the numbers of miles driven per year?
  b Estimate with 95% confidence the difference in mean distance driven by male and female drivers.
  c Check to ensure that the required condition(s) of the techniques used in parts (a) and (b) is satisfied.

12.26 The president of a company that manufactures automobile air conditioners is considering switching his supplier of condensers. Supplier A, the current producer of condensers for the manufacturer, prices its product 5% higher than supplier B does. Because the president wants to maintain his company's reputation for quality he wants to be sure that supplier B's condensers last at least as long as supplier A's. After a careful analysis, the president decided to retain supplier A if there is sufficient statistical evidence that supplier A's condensers last longer on the average than supplier B's condensers. In an experiment, 30 midsize cars were equipped with air conditioners using type A condensers while another 30 midsize cars were equipped with type B condensers. The number of miles (in thousands) driven by each car before the condenser broke down was recorded and the data stored in column A (supplier A) and column B (supplier B) in file Xr12-26. Should the president retain supplier A?

12.27 High blood pressure is a leading cause of strokes. Medical researchers are constantly seeking ways to treat patients suffering from this condition. A specialist in hypertension claims that regular aerobic exercise can reduce high blood pressure just as successfully as drugs, with none of the adverse side effects. To test the claim, 50 patients who suffer from high blood pressure were chosen to participate in an experiment. For 60 days, half the sample exercised three times per week for one hour; the other half took the standard medication. The percentage reduction in blood pressure was recorded for each individual, and the resulting data are stored in file Xr12-27 (column A = exercise and column B = drug).

   a Can we conclude that exercise is more effective than medication in reducing hypertension?
   b Estimate with 95% confidence the difference in mean percentage reduction in blood pressure between drugs and exercise programs.
   c Check to ensure that the required condition(s) of the techniques used in parts (a) and (b) is satisfied.

12.28 The program director of an FM radio station is developing this year's program. One of the factors that determine the schedule is the age of the station's listeners. To examine the issue, the director commissioned a survey. One objective was to measure the difference in listening habits between the 18 to 34 and the 35 to 50 age groups. The survey asked 250 people in each age category how much time they spent listening to FM radio per day. The results (in minutes) were recorded and stored in file Xr12-28 (column A = listening times for 18 to 34 age group and column B = listening times for 35 to 50 age group).

   a Can we conclude at the 5% significance level that a difference exists between the two groups?
   b Estimate with 95% confidence the difference in mean time listening to FM radio between the two age groups.
   c Are the required conditions satisfied for the techniques you used in parts (a) and (b)?

12.29 A statistics professor is about to select a statistical software package for her course. One of the most important features, according to the professor, is the ease with which students learn to use the software. She has narrowed the selection to two possibilities: software A, a menu-driven statistical package with some high-powered techniques; and software B, a spreadsheet that has the capability of performing most techniques. To help make her decision, she asks 40 statistics students selected at random to choose one of the two packages. She gives each student a statistics problem to solve by computer and the appropriate manual. The amount of time (in minutes) each student needs to complete the assignment was recorded and stored in file Xr12-29 (column A = package A and column B = package B).

   a Can the professor conclude from these data that the two software packages differ in the amount of time needed to learn how to use them?
   b Estimate with 95% confidence the difference in the mean amount of time needed to learn to use the two packages.
   c What are the required conditions for the techniques used in parts (a) and (b)?
   d Check to see if the required conditions are satisfied.

12.30 One factor in low productivity is the amount of time wasted by workers. Wasted time includes time spent cleaning up mistakes, waiting for more material and equipment, and performing any other activity not related to production. In a project designed to examine the problem, an operations management consultant took a survey of 200 workers in companies that were classified as successful (on the basis of their latest annual profits) and another 200 workers from unsuccessful companies. The amount of time (in hours) wasted during a standard 40-hour workweek was recorded for each worker. These data are stored in columns A (successful companies) and B (unsuccessful companies) in file Xr12-30.

   a Do these data provide enough evidence to infer that the amount of time wasted in unsuccessful firms exceeds that of successful ones?
   b Estimate with 95% confidence how much more time is wasted in unsuccessful firms than in successful ones.

12.3 OBSERVATIONAL AND EXPERIMENTAL DATA

As we've pointed out several times, the ability to properly interpret the results of a statistical technique is a critical skill for you to develop. This ability is dependent on your understanding of Type I and Type II errors and the fundamental concepts that are part of statistical inference. However, there is another component that needs to be understood: the difference between **observational data** and **experimental data.** To explain this difference, we will reexamine Examples 12.1 and 12.2 and analyze the way the data were obtained in each example.

In Example 12.1, we randomly selected 150 people and, on the basis of *their* responses, assigned them to one of two groups: high-fiber consumers and nonconsumers. We then recorded the number of calories consumed at lunch for the members of each group. Such data are called observational. Now examine Example 12.2, where the data were gathered by randomly assigning 25 workers to assemble chairs using method A and another 25 workers to assemble chairs using method B. Data produced in this manner are said to be experimental or controlled. The statistical technique to be applied is not affected by whether the data are observational or experimental. However, the interpretation of the results may be affected.

In Example 12.1, we found that there was evidence to infer that people who eat high-fiber cereal for breakfast consume fewer calories at lunch than do nonconsumers of high-fiber cereal. From this result, we're inclined to believe that eating a high-fiber cereal at breakfast may be a way to reduce weight. However, other interpretations are possible. For example, people who eat fewer calories are probably more health conscious, and such people are more likely to eat high-fiber cereal as part of a healthy breakfast. In this interpretation, high-fiber cereals do not necessarily lead to fewer calories at lunch. Instead another factor, general health consciousness, leads to both fewer calories at lunch and high-fiber cereal for breakfast. Notice that the conclusion of the statistical procedure is unchanged. On average, people who eat high-fiber cereal consume fewer calories at lunch. However, because of the way the data were gathered, we have more difficulty interpreting this result.

Suppose that we redo Example 12.1 using the experimental approach. We randomly select 150 people to participate in the experiment. We randomly assign 75 (or 43, as we had in the original experiment) to eat high-fiber cereal for breakfast and the other 75 to eat something else. We then record the number of calories each person consumes at lunch. Ideally, in this experiment both groups will be similar in all other dimensions, including health consciousness. (Larger sample sizes increase the likelihood that the two groups will be similar.) If the statistical result is about the same as in Example 12.1, we may have some valid reason to believe that high-fiber cereal leads to a decrease in caloric intake.

Experimental data are usually more expensive to obtain because of the planning required to set up the experiment; observational data usually require less work to gather. Furthermore, in many situations it is impossible to conduct a controlled experiment. For example, suppose that we want to determine if engineering students outperform arts students in MBA programs. In a controlled experiment we would randomly assign some students to achieve a degree in engineering and other students to obtain an arts degree. We would then make them sign up for an MBA program where we would record their grades. Unfortunately for statistical despots (and fortunately for the rest of us), we live in a democratic society, which makes the coercion necessary to perform this controlled experiment impossible.

To answer our question about the relative performance of engineering and arts students, we have no choice but to obtain our data by observational methods. We would take a random sample of engineering students and arts students who have already entered MBA programs and record their grades. If we find that engineering students do better, we may tend to conclude that an engineering background better prepares students for an MBA program. However, it may be true that better students tend to choose engineering as their undergraduate major, and that better students achieve higher grades in all programs, including the MBA program.

Although we've discussed observational and experimental data in the context of the test of the difference between two means, you should be aware that the issue of how the data are obtained is relevant to the interpretation of all the techniques that follow

EXERCISES

12.31 Examine Exercises 12.23 to 12.30. Which of the data sets were obtained by observational methods, and which were obtained through controlled experiments? Explain the reasons for your choices.

12.32 Provide two interpretations of the results you produced in Exercise 12.23.

12.33 Discuss how the data in Exercise 12.23 could have been obtained through a controlled experiment.

12.34 Suppose that you are analyzing one of the hundreds of statistical studies linking smoking with lung cancer. The study analyzed thousands of randomly selected people, some of whom had lung cancer. The statistics indicate that those who have lung cancer smoked on average significantly more than those who did not have lung cancer.

a Explain how you know that the data are observational.

b Is there another interpretation of the statistics besides the obvious one that smoking causes lung cancer? If so, what is it? (Students who produce the best answers will be eligible for a job in the public relations department of a tobacco company.)

c Is it possible to conduct a controlled experiment to produce data that address the question of the relationship between smoking and lung cancer? If so, describe the experiment.

12.4 INFERENCE ABOUT THE DIFFERENCE BETWEEN TWO MEANS: MATCHED PAIRS EXPERIMENT

We continue our presentation of statistical techniques that address the problem of comparing two populations of interval data. In Section 12.2, the parameter of interest was the difference between two population means, where the data were generated from independent samples. In this section, the data are gathered from a matched pairs experiment. To illustrate why matched pairs experiments are needed and how we deal with data produced in this way, consider the following example.

▼ EXAMPLE 12.3

Tire manufacturers are constantly researching ways to produce tires that last longer. New innovations are tested by professional drivers on racetracks. However, any promising inventions are also test-driven by ordinary drivers. The latter tests are closer to what the tire company's customers will actually experience. Suppose that to determine whether a new steel-belted radial tire lasts longer than the company's current model, two new-design tires were installed on the rear wheels of 20 randomly selected cars and two current-design tires were installed on the rear wheels of another 20 cars. All drivers were told to drive in their usual way until the tires wore out. The number of miles driven by each driver was recorded and is shown below, as well its being stored in file Xm12-03. Can the company infer that the new tire will last on average longer than the current tire?

Distance (in thousands of miles) until Wear-Out

New-Design Tire	Current-Design Tire
70 83 78 46 74 56 74 52 99 57	46 64 58 60 74 64 72 84 96 83
77 84 72 98 81 63 88 69 54 97	71 38 71 90 63 62 78 73 75 42

Solution

IDENTIFY

The objective is to compare two populations of interval data. The parameter is the difference between two means $\mu_1 - \mu_2$ (μ_1 = mean distance to wear-out for the new-design tire, and μ_2 = mean distance to wear-out for the current-design tire). Because we want to determine whether the new tire lasts longer, the alternative hypothesis will specify that μ_1 is greater than μ_2. Calculation of the sample variances allows us to use the equal-variances test statistic.

$$H_0 : (\mu_1 - \mu_2) = 0$$

$$H_1 : (\mu_1 - \mu_2) > 0$$

Test statistic: $t = \dfrac{(\bar{x}_1 - \bar{x}_2) - (\mu_1 - \mu_2)}{\sqrt{s_p^2\left(\dfrac{1}{n_1} + \dfrac{1}{n_2}\right)}}$

COMPUTE

Microsoft Excel Output for Example 12.3

t-Test: Two-Sample Assuming Equal Variances

	New Design	Current Design
Mean	73.6	68.2
Variance	243.4	226.8
Observations	20	20
Pooled Variance	235.1	
Hypothesized Mean Difference	0	
df	38	
t Stat	1.114	
(T<=t) one-tail	0.1362	
t Critical one-tail	1.6860	
(T<=t) two-tail	0.2724	
t Critical two-tail	2.0244	

INTERPRET

The value of the test statistic ($t = 1.114$) and its p-value (.1362) indicate that there is very little evidence to support the hypothesis that the new-design tire lasts longer on average than the current-design tire.

▲

As was the case with some earlier examples, we have some evidence to support the alternative hypothesis, but not enough. Note that the difference in sample means is $(\bar{x}_1 - \bar{x}_2) = (73.6 - 68.2) = 5.4$. However, we judge the difference in sample means in relation to the standard error. From the Excel printout, we see that $s_p^2 = 235.1$. We can manually calculate the standard error of this test. It is

$$\sqrt{s_p^2 \left(\frac{1}{n_1} + \frac{1}{n_2}\right)} = \sqrt{235.1 \left(\frac{1}{20} + \frac{1}{20}\right)} = 4.85$$

Consequently, the value of the test statistic is $t = 5.4/4.85 = 1.114$, a value that does not allow us to reject the null hypothesis. We can see that although the difference between the sample means was quite large, the variability of the data, as measured by s_p^2, was also large, resulting in a small test statistic value.

▼ EXAMPLE 12.4

Suppose now we redo the experiment in the following way. On 20 randomly selected cars, one of each type of tire is installed on the rear wheels and, as above, the cars are driven until the tires wear out. The number of miles until wear-out occurred is shown below and stored in file Xm12-04. Can we conclude from these data that the new tire is superior?

Distance (in thousands of miles) until Wear-Out

Car	New Design	Current Design
1	57	47
2	66	52
3	102	85
4	62	56
5	81	78
6	87	75
7	61	50
8	65	49
9	74	70
10	62	64
11	100	98
12	90	86
13	83	78
14	84	90
15	86	94
16	62	58
17	67	59
18	40	41
19	71	61
20	77	81

Solution

The experiment described in Example 12.3 is one where the samples are independent. That is, there was no relationship between the observations in one sample and the observations in the second sample. However, in this example, the experiment was designed in such a way that each observation in one sample is matched with an observation in the other sample. The matching is conducted by using the same set of cars for each sample. Thus, it is logical to compare the distance until wear-out for both types of tires for each car. This type of experiment is called **matched pairs.** Here is how we conduct the test.

For each car, we calculate the matched pairs difference between the distances obtained with each type of tire.

Car	New Design	Current Design	Difference
1	57	47	10
2	66	52	14
3	102	85	17
4	62	56	6
5	81	78	3
6	87	75	12
7	61	50	11
8	65	49	16
9	74	70	4
10	62	64	−2
11	100	98	2
12	90	86	4
13	83	78	5
14	84	90	−6
15	86	94	−8
16	62	58	4
17	67	59	8
18	40	41	−1
19	71	61	10
20	77	81	−4

The experimental design tells us that the parameter of interest is the mean of the population of differences, which we label μ_D. Note that $\mu_1 - \mu_2 = \mu_D$, but that we test μ_D because of the way the experiment was performed. The hypotheses to be tested are

$H_0: \mu_D = 0$

$H_1: \mu_D > 0$

We have already presented inferential techniques about a population mean. Recall that in Chapter 11 we introduced the *t*-test of μ. Thus, to test hypotheses about μ_D, we use the following test statistic.

Test Statistic for μ_D

$$t = \frac{\bar{x}_D - \mu_D}{s_D/\sqrt{n_D}}$$

which is Student t distributed with $v = n_D - 1$ degrees of freedom, provided that the differences are normally distributed.

Aside from the subscript D, this test statistic is identical to the one presented in Chapter 11.

COMPUTE

Microsoft Excel Output for Example 12.4

t-Test: Paired Two Sample for Means

	New Design	Current Design
Mean	73.85	68.60
Variance	237.7	294.1
Observations	20	20
Pearson Correlation	0.9100	
Hypothesized Mean Difference	0	
df	19	
t Stat	3.300	
P(T<=t) one-tail	0.0019	
t Critical one-tail	1.7291	
P(T<=t) two-tail	0.0038	
t Critical two-tail	2.0930	

GENERAL COMMANDS	EXAMPLE 12.4
1 Type or import the data into two columns.	Open file **Xm12-04.**
2 Click **Tools, Data Analysis...**, and **t-Test: PairedTwo-Sample for Means.**	
3 Specify the **Variable 1 Range.**	A1:A21
4 Specify the **Variable 2 Range.**	B1:B21
5 Type the value of the **Hypothesized Mean Difference.**	0
6 Click **Labels** if applicable.	
7 Specify the value of α (**Alpha:**) and click **OK.**	.05

Figure 12.8
Sampling distribution of test statistic for Example 12.4

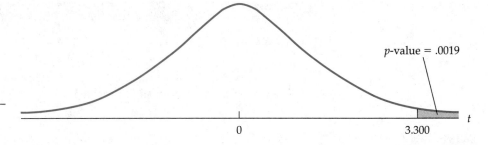

| INTERPRET | The value of the test statistic is $t = 3.300$ with a p-value of .0019. Figure 12.8 exhibits the sampling distribution and p-value. There is now overwhelming evidence to infer that the new-design tire lasts longer on average than the current design. By redoing the experiment as matched pairs, we were able to extract this information from the data. |

ESTIMATING THE MEAN DIFFERENCE

Applying the usual algebra we derive the interval estimator of μ_D.

Interval Estimator of μ_D

$$\bar{x}_D \pm t_{\alpha/2}\frac{s_D}{\sqrt{n_D}} \quad v = n_D - 1$$

| COMPUTE |

Microsoft Excel Output of the Interval Estimator for Example 12.4

t-Test and Estimate: Mean

	Difference
Mean	5.25
Standard Deviation	7.115
Hypothesized Mean	0
df	19
t Stat	3.300
P(T<=t) one-tail	0.0019
t Critical one-tail	1.7291
P(T<=t) two-tail	0.0038
t Critical two-tail	2.093
Standard Error	1.591
Bound	3.330
LCL	1.920
UCL	8.580

GENERAL COMMANDS	EXAMPLE 12.4
1 Type or import the data into two columns.	Open file **Xm12-04**.
2 In another column (we chose column C), calculate the matched pairs differences. If you wish, type the name of the variable in row 1.	**Difference**
3 Click **Tools, Data Analysis Plus**, and **t-Test and Estimate: Mean**.	
4 Specify the **Input Range:**.	C1:C21
5 Type the value of the **Hypothesized Mean**.	0
6 Click **Labels**, if applicable.	
7 Specify the value of α (**Alpha:**), and click **OK**.	.05

Notice that the test results in this printout are the same as the printout above. That is, $t = 3.300$ and p-value $= .0019$. The last part of the output tells us that the estimated mean difference lies between 1.920 and 8.580 thousand miles. That is, the new-design tires last on average between 1,920 and 8,580 miles more than the current-design tires.

INDEPENDENT SAMPLES OR MATCHED PAIRS: WHICH EXPERIMENTAL DESIGN IS BETTER?

Examples 12.3 and 12.4 demonstrated that the experimental design is an important factor in statistical inference. However, these two examples raise several questions about experimental designs.

1 Why does the matched pairs experiment result in rejecting the null hypothesis, whereas the independent samples experiment could not?

2 Should we always use the matched pairs experiment? In particular, are there disadvantages to its use?

3 How do we recognize when a matched pairs experiment has been performed?

Here are our answers.

1 The matched pairs experiment worked in Example 12.4 by reducing the variation in the data. To understand this point, examine the statistics from both examples. In Example 12.3, we found $\bar{x}_1 - \bar{x}_2 = 5.4$. In Example 12.4, we computed $\bar{x}_D = 5.25$. Thus, the numerators of the two test statistics were almost identical. However, the reason that the test statistic in Example 12.3 was so much smaller than in Example 12.4 was because of the standard errors. In Example 12.3, we calculated

$$s_p^2 = 235.1 \quad \text{and} \quad \sqrt{s_p^2\left(\frac{1}{n_1} + \frac{1}{n_2}\right)} = 4.85$$

Example 12.4 produced

$$s_D = 7.115 \quad \text{and} \quad \frac{s_D}{\sqrt{n_D}} = 1.591$$

As you see, the difference in the test statistics was caused not by the numerator but by the denominator. This raises another question: Why was the variation in the data of Example 12.3 so much greater than the variation in the data of Example 12.4? If you examine the data and statistics from Example 12.3, you will find that there was a great deal of variation between the cars. That is, some drivers drove in a way that extended the life of the tires, while others drove faster and braked harder, resulting in shorter tire lives. This high level of variation made the difference between the sample means appear to be small. As a result, we could not reject the null hypothesis.

Looking at the data from Example 12.4, we see that there is very little variation among the paired differences. Now the variation caused by different driving habits has been markedly decreased. The smaller variation causes the value of the test statistic to be larger. Consequently, we reject the null hypothesis.

2 Will the matched pairs experiment always produce a larger test statistic than the independent samples experiment? The answer is not necessarily. Suppose that in our example we found that most drivers drove in about the same way and that there was very little difference among drivers in the distances until tire wear-out. In such circumstances, the matched pairs experiment would result in no significant decrease in variation when compared to independent samples. It is possible that the matched pairs experiment may be less likely to reject the null hypothesis than the independent samples experiment. Calculating the degrees of freedom provides the reason. In Example 12.3, the number of degrees of freedom was 38, whereas in Example 12.4, it was 19. Even though we had the same number of observations (20 in each sample), the matched pairs experiment had half the number of degrees of freedom as the equivalent independent samples experiment. For exactly the same value of the test statistic, a smaller number of degrees of freedom in a Student t distributed test statistic yields a larger p-value. What this means is that if there is little reduction in variation to be achieved by the matched pairs experiment, the statistics practitioner should choose instead to conduct the experiment with independent samples.

3 As you've seen, in this book we deal with questions arising from experiments that have already been conducted. Thus, one of your tasks is to determine the appropriate test statistic. In the case of comparing two populations of interval data, you must decide whether the samples are independent (in which case the parameter is $\mu_1 - \mu_2$) or matched pairs (in which case the parameter is μ_D) in order to select the correct test statistic. To help you do so, we suggest you ask and answer the following question: "Does some natural relationship exist between each pair of observations that provides a logical reason to compare the first observation of sample 1 with the first observation of sample 2, the second observation of sample 1 with the second observation of sample 2, and so on?" If so, the experiment was conducted by matched pairs. If not, it was conducted using independent samples.

OBSERVATIONAL AND EXPERIMENTAL DATA

The points we made in Section 12.3 are also valid in this section. That is, we can design a matched pairs experiment where the data are gathered using a controlled experiment or by observation. The data in Examples 12.3 and 12.4 are experimental, which means the statistics practitioner randomly assigned the tires to the cars. As a consequence, when we established that the new-design tire lasted longer, we were able to conclude that the new tire is indeed superior. Because very few cars would be equipped with different brands of tires on their rear wheels, in this type of problem only experimental data are available. In most applications of the matched pairs experiment, the data are experimental because of the control required to conduct such experimental designs.

CHECKING THE REQUIRED CONDITION

The required condition is that the differences are normally distributed. However, the technique is robust, so that only extreme departures from normality invalidate the results. As you can see from the histogram (Figure 12.9) the differences appear to be normally distributed, removing any doubts about the test's validity.

VIOLATION OF THE REQUIRED CONDITION

If the differences are very nonnormal, we cannot use the t-test of μ_D. We can, however, employ a nonparametric technique—the Wilcoxon signed rank sum test, which we present in Chapter 16.

DEVELOPING AN UNDERSTANDING OF STATISTICAL CONCEPTS 1

Two of the most important principles in statistics were applied in this section. The first is the concept of analyzing sources of variation. In Examples 12.3 and 12.4, we showed that, by reducing the variation among drivers, we were able to detect a real

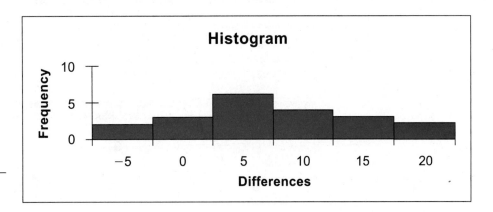

Figure 12.9

Histogram of Differences in Example 12.4

difference between tire designs. This was an application of the more general procedure of analyzing data and attributing some fraction of the variation to several sources. The two sources of variation were the car drivers and the tire designs. However, we were not interested in the variation among drivers because we weren't interested in determining whether drivers actually differ. Instead we merely wanted to eliminate that source of variation, making it easier to determine if tire designs represented a real source of variation and thus that one tire design is superior to another tire design.

In Chapter 14, we will introduce a technique called the analysis of variance, which does what its name suggests; it analyzes sources of variation in an attempt to detect real differences. In most applications of this procedure, we will be interested in each source of variation and not simply in reducing one source. We refer to the process as *explaining the variation.* The concept of explaining variation also will be applied in Chapters 17 and 18.

DEVELOPING AN UNDERSTANDING OF STATISTICAL CONCEPTS 2

The second principle demonstrated in this section is that statistics practitioners can design data-gathering procedures in such a way that they can analyze sources of variation. Before conducting the experiment in Example 12.4, the statistics practitioner suspected that there were large differences among drivers in the way they wear tires out. Consequently, the experiment was organized so that the effects of those differences were mostly eliminated. It is also possible to design experiments that allow for easy detection of real differences and minimize the costs of data gathering. Unfortunately, we will not present this topic. However, you should understand that the entire subject of the design of experiments is an important one, because statistics practitioners often need to be able to analyze data to detect differences, and the cost is almost always a factor.

Here is a summary of the test statistic, interval estimator, and how we determine when to use these techniques.

Test Statistic for μ_D

$$t = \frac{\bar{x}_D - \mu_D}{s_D/\sqrt{n_D}} \quad \nu = n_D - 1$$

Interval Estimator for μ_D

$$\bar{x}_D \pm t_{\alpha/2} \frac{s_D}{\sqrt{n_D}} \quad \nu = n_D - 1$$

Factors That Identify the t-Test and Estimator of μ_D

1. Problem objective: compare two populations.
2. Data type: interval.
3. Descriptive measurement: central location.
4. Experimental design: matched pairs.

EXERCISES

12.35 Given the following data generated from a matched pairs experiment, test to determine if we can infer that the mean of population 1 exceeds the mean of population 2.

Pair:	1	2	3	4	5
Sample 1:	20	23	15	18	19
Sample 2:	17	16	9	19	15

12.36 The data below and stored in file Xr12-36 were produced from a matched pairs experiment. Determine whether these data are sufficient to infer that the two population means differ.

Pair:	1	2	3	4	5	6	7	8	9	10
Sample 1:	7	12	19	17	22	18	30	33	40	48
Sample 2:	10	13	18	21	25	19	31	31	44	47

12.37 The following data were generated from a matched pairs experiment. (The data are stored in columns A and B of file Xr12-37.)

a Estimate with 90% confidence the mean difference.
b Briefly describe what the interval estimate in part (a) tells you.

Pair:	1	2	3	4	5	6	7	8	9	10
Sample 1:	3	12	15	10	17	14	10	9	16	8
Sample 2:	7	13	14	14	23	13	12	12	18	9

12.38 Samples of size 12 were drawn independently from two normal populations. These data are listed below and stored in columns A and B of file Xr12-38. A matched pairs experiment was then conducted; 12 pairs of observations were drawn from the same populations. These data are also shown below and stored in columns C and D of file Xr12-38.

a Using the data taken from independent samples, test to determine whether the means of the two populations differ.
b Using the data taken from independent samples, estimate with 95% confidence the difference between the two population means.
c Repeat part (a) using the matched pairs data.
d Repeat part (b) using the matched pairs data.
e Describe the differences between parts (a) and (c) and between (b) and (d). Discuss why these differences occurred.

Independent Samples

| Sample 1: | 66 | 19 | 88 | 72 | 61 | 32 | 75 | 61 | 71 | 54 | 79 | 40 |
| Sample 2: | 69 | 37 | 66 | 59 | 27 | 18 | 47 | 67 | 83 | 61 | 32 | 37 |

Matched Pairs

Pair:	1	2	3	4	5	6	7	8	9	10	11	12
Sample 1:	55	45	52	87	78	42	62	90	23	60	67	53
Sample 2:	48	37	43	75	78	35	45	79	12	53	59	37

12.39 Repeat Exercise 12.38 using the data below, which are also stored in columns A through D of file Xr12-39.

Independent Samples

| Sample 1: | 199 | 261 | 295 | 183 | 161 | 104 | 199 | 248 | 105 | 197 | 249 | 218 |
| Sample 2: | 286 | 211 | 121 | 134 | 210 | 68 | 166 | 157 | 258 | 184 | 116 | 203 |

Matched Pairs

Pair:	1	2	3	4	5	6	7	8	9	10	11	12
Sample 1:	218	144	286	208	234	256	133	87	224	212	256	133
Sample 2:	154	160	239	198	211	241	136	39	192	183	215	117

12.40 Repeat Exercise 12.38 using the data below, which are also stored in columns A through D of file Xr12-40.

Independent Samples

| Sample 1: | 103 | 78 | 101 | 112 | 111 | 100 | 97 | 105 | 119 | 89 | 104 | 99 |
| Sample 2: | 71 | 86 | 100 | 89 | 92 | 105 | 85 | 85 | 97 | 98 | 107 | 96 |

Matched Pairs

Pair:	1	2	3	4	5	6	7	8	9	10	11	12
Sample 1:	91	120	97	94	107	107	91	118	94	101	87	102
Sample 2:	88	75	108	84	97	92	92	76	86	97	107	98

12.41 Discuss what you have discovered from Exercises 12.38 to 12.40.

12.42 In an effort to determine whether or not a new type of fertilizer is more effective than the type currently in use, researchers took 12 two-acre plots of land scattered throughout the county. Each plot was divided into two equal-sized subplots, one of which was treated with the current fertilizer and the other of which was treated with the new fertilizer. Wheat was planted, and the crop yields were measured. These data are stored in file Xr12-42 and listed below.

  a Can we conclude that the new fertilizer is more effective than the current one?
  b Estimate with 95% confidence the difference in mean crop yields between the two fertilizers.
  c What is the required condition(s) for the validity of the results obtained in parts (a) and (b)?
  d Is the required condition(s) satisfied?
  e Are these data experimental or observational? Explain.
  f How should the experiment be conducted if the researchers believed that the land throughout the county was pretty much the same?

Plot:	1	2	3	4	5	6	7	8	9	10	11	12
Current fertilizer:	56	45	68	72	61	69	57	55	60	72	75	66
New fertilizer:	60	49	66	73	59	77	61	60	58	75	72	71

12.43 The president of a large company is in the process of deciding whether to adopt a lunchtime exercise program. The purpose of such programs is to improve the health of workers and, in so doing, reduce medical expenses. To get more information, she instituted an exercise program for the employees in one office. The president knows that during the winter months medical expenses are relatively high because of the incidence of colds and flu. Consequently, she decides to use a matched pairs design by recording medical expenses for the 12 months before the program and for 12 months after the program. The "before" and "after" expenses (in thousands of dollars) are compared on a month-to-month basis and shown below. (These data are stored in columns A and B of file Xr12-43.)

  a Do the data indicate that exercise programs reduce medical expenses?
  b Estimate with 95% confidence the mean savings produced by exercise programs.
  c Was it appropriate to conduct a matched pairs experiment? Explain.

Month:	Jan	Feb	Mar	Apr	May	Jun	Jul	Aug	Sep	Oct	Nov	Dec
Before program:	68	44	30	58	35	33	52	69	23	69	48	30
After program:	59	42	20	62	25	30	56	62	25	75	40	26

12.44 Do waiters or waitresses earn larger tips? To answer this question, a restaurant consultant undertook a preliminary study. The study involved measuring the percentage of the total bill left as a tip for one randomly selected waiter and one randomly selected waitress in each of 50 restaurants during a one-week period. The data are stored in columns A and B of file Xr12-44. What conclusions can be drawn from these data?

12.45 In order to determine the effect of advertising in the Yellow Pages, Bell Telephone took a sample of 40 retail stores that did not advertise in the Yellow Pages last year but did so this year. The annual sales (in thousands of dollars) for each store in both years were recorded and stored in file Xr12-45.

  a Estimate with 90% confidence the improvement in sales between the two years.
  b Can we infer that advertising in the Yellow Pages improves sales?
  c Check to ensure that the required condition(s) of the techniques above is satisfied.
  d Would it be advantageous to perform this experiment with independent samples? Explain why or why not.

12.46 Research scientists at a pharmaceutical company have recently developed a new nonprescription sleeping pill. They decide to test its effectiveness by measuring the time it takes for people to fall asleep after taking the pill. Preliminary analysis indicates that the time to fall asleep varies considerably from one person to another. Consequently, they organize the experiment in the following way. A random sample of 100 volunteers who regularly suffer from insomnia is chosen. Each person is given one pill containing the newly developed drug and one placebo. (A placebo is a pill that contains absolutely no medication.) Participants are told to take one pill one night and the second pill one night a week later. (They do not know whether the pill they are taking is the placebo or the real thing, and the order of use is random.) Each participant is fitted with a device that measures the time until sleep occurs. The data are stored in columns A and B of file Xr12-46. Can we conclude that the new drug is effective?

12.5 INFERENCE ABOUT THE RATIO OF TWO VARIANCES

In Sections 12.2 and 12.4, we dealt with statistical inference concerning the difference between two population means. The problem objective in each case was to compare two populations of interval data, and our interest was in comparing measures of central location. This section discusses the statistical techniques to use when the problem objective and the data type are the same as in Sections 12.2 and 12.4, but our interest is in comparing variability. Here we will study the ratio of two population variances σ_1^2/σ_2^2. We make inferences about the ratio because the sampling distribution concerns ratios rather than differences.

In the previous chapter, we presented the procedures used to draw inferences about a single population variance. We pointed out that variance can be used to address problems where we need to know the variance in order to judge the consistency of a production process. We also use variance to measure the risk associated with a portfolio of investments. In this section we compare two variances, enabling us to compare the consistency of two production processes. We can also compare the relative risks of two sets of investments.

There is one other important use of the statistical methods to be presented in this section. One of the factors that determine the correct technique when testing or estimating the difference between two means from independent samples is whether the two unknown population variances are equal. Statistics practitioners often test for the equality of σ_1^2/σ_2^2 before deciding which of the two procedures introduced in Section 12.2 is to be used. We will proceed in a manner that is probably becoming quite familiar.

PARAMETER

As you will see shortly, we compare two population variances by determining the ratio. Consequently, the parameter is σ_1^2/σ_2^2.

POINT ESTIMATOR OF σ_1^2/σ_2^2

We have previously noted that the sample variance (defined in Chapter 3) is an unbiased and consistent estimator of the population variance. Not surprisingly, the estimator of the parameter σ_1^2/σ_2^2 is the ratio of the two sample variances drawn from their respective populations. The point estimator is s_1^2/s_2^2.

SAMPLING DISTRIBUTION OF s_1^2/s_2^2

Statisticians have shown that the ratio of two independent chi-squared variables divided by their degrees of freedom is F distributed. (The **F distribution** was presented in Section 6.4.) The degrees of freedom of the F distribution are identical to the degrees of freedom for the two chi-squared distributions. In Section 11.3, we pointed out that $(n-1)s^2/\sigma^2$ is chi-squared distributed provided that the sampled population is normal. If we have independent samples drawn from two normal populations, then both $(n_1 - 1)s_1^2/\sigma_1^2$ and $(n_2 - 1)s_2^2/\sigma_2^2$ are chi-squared distributed. If we divide them by their respective numbers of degrees of freedom and take the ratio, we produce

$$\frac{\frac{(n_1 - 1)s_1^2/\sigma_1^2}{(n_1 - 1)}}{\frac{(n_2 - 1)s_2^2/\sigma_2^2}{(n_2 - 1)}}$$

which simplifies to

$$\frac{s_1^2/\sigma_1^2}{s_2^2/\sigma_2^2}$$

This statistic is F distributed with $v_1 = n_1 - 1$ and $v_2 = n_2 - 1$ degrees of freedom. Recall that v_1 is called the **numerator degrees of freedom** and v_2 is called the **denominator degrees of freedom**.

TESTING σ_1^2/σ_2^2

In this book, our null hypothesis will always specify that the two variances are equal. As a result, the ratio will equal 1. Thus, the null hypothesis will always be expressed as

$$H_0 : \sigma_1^2/\sigma_2^2 = 1$$

The alternative hypothesis can state that the ratio σ_1^2/σ_2^2 is either not equal to 1, greater than 1, or less than 1. Technically, the test statistic is

$$F = \frac{s_1^2/\sigma_1^2}{s_2^2/\sigma_2^2}$$

However, under the null hypothesis, which states that $\sigma_1^2/\sigma_2^2 = 1$, the test statistic becomes as follows.

Test Statistic for σ_1^2/σ_2^2

The test statistic employed to test that σ_1^2/σ_2^2 is equal to one is

$$F = \frac{s_1^2}{s_2^2}$$

which is F distributed with $v_1 = n_1 - 1$ and $v_2 = n_2 - 1$ degrees of freedom, provided that the populations are normal.

EXAMPLE 12.5

In Example 12.1, we applied the unequal-variances t-test of $\mu_1 - \mu_2$. We chose that test statistic after computing the variance of the sample of consumers of high-fiber cereal to be 4,103.0 and the variance of the sample of nonconsumers of high-fiber cereal to be 10,669.8. The difference between the two sample variances appears to indicate that the population variances differ. Test to determine whether that decision was correct.

Solution

IDENTIFY

We need to conduct the F-test of σ_1^2/σ_2^2 to determine whether the two population variances differ. The test proceeds as follows.

$$H_0: \sigma_1^2/\sigma_2^2 = 1$$
$$H_1: \sigma_1^2/\sigma_2^2 \neq 1$$

Test statistic: $F = \dfrac{s_1^2}{s_2^2}$

COMPUTE

Microsoft Excel Output for Example 12.5

F-Test Two-Sample for Variances

	Consumers	Nonconsumers
Mean	604.0	633.2
Variance	4103.0	10669.8
Observations	43	107
df	42	106
F	.3845	
P(F<=f) one-tail	0.0004	
F Critical one-tail	0.6371	

The value of the test statistic is $F = .3845$. Excel outputs the one-tail p-value .0004. Because we're conducting a two-tail test, we double that value. Thus, the p-value of the test we're conducting is $2 \times .0004 = .0008$.

GENERAL COMMANDS	EXAMPLE 12.5
1 Type or import the data into two columns.	Open file **Xm12-01**.
2 Click **Tools, Data Analysis...**, and **F-Test Two-Sample for Variances.**	
3 Specify the **Variable 1 Range:**.	A1:A44
4 Specify the **Variable 2 Range:**.	B1:B108
5 Click **Labels** if applicable.	
6 Specify the value of α (**Alpha:**), and click **OK**.	.05

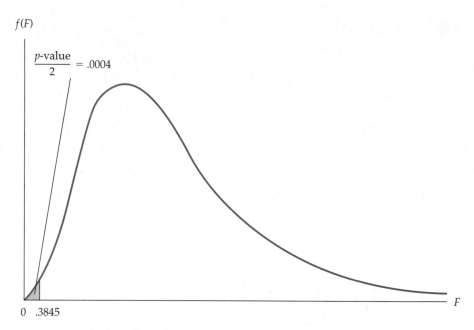

Figure 12.10

Sampling distribution of the test statistic for Example 12.5

Use the **F-Test of 2 Variances** worksheet in **Stats-Summary.xls** to complete this test from the sample variances and to perform a what-if analysis.

INTERPRET

There is strong evidence to infer that the population variances differ. It follows that we were justified in using the unequal-variances t-test in Example 12.1. We're confident that the normality requirement for this test is satisfied. It is the same requirement for the t-test, which we checked when we drew the histograms. (See Figures 12.4 and 12.5.)

Figure 12.10 depicts the F sampling distribution in this example.

DECISION RULE: EQUAL-VARIANCES OR UNEQUAL-VARIANCES TEST STATISTIC AND ESTIMATOR OF $\mu_1 - \mu_2$

Following up on our discussion in Section 12.2, we now have a simple decision rule. If there is sufficient evidence from the F-test to infer that the population variances are unequal, use the unequal-variances test statistic and interval estimator. In the absence of such evidence, use the equal-variances formulas.

ESTIMATING THE RATIO OF TWO POPULATION VARIANCES

Another way of expressing the notation associated with the F distribution is

$$P(F_{1-\alpha/2} < F < F_{\alpha/2}) = 1 - \alpha$$

This states that the probability that an F distributed random variable falls between $F_{1-\alpha/2}$ and $F_{\alpha/2}$ is $1 - \alpha$. If we substitute the F statistic

$$F = \frac{s_1^2/\sigma_1^2}{s_2^2/\sigma_2^2}$$

into the equation, we produce

$$P\left(F_{1-\alpha/2} < \frac{s_1^2/\sigma_1^2}{s_2^2/\sigma_2^2} < F_{\alpha/2}\right) = 1 - \alpha$$

Applying algebra, we isolate σ_1^2/σ_2^2 in the center of the probability statement and produce the interval estimator.

Interval Estimator of σ_1^2/σ_2^2

$$\text{LCL} = \left(\frac{s_1^2}{s_2^2}\right)\frac{1}{F_{\alpha/2,v_1,v_2}}$$

$$\text{UCL} = \left(\frac{s_1^2}{s_2^2}\right)\frac{1}{F_{1-\alpha/2,v_1,v_2}}$$

where $v_1 = n_1 - 1$ and $v_2 = n_2 - 1$.

▼ EXAMPLE 12.6

Determine the 95% confidence interval estimate of the ratio of the two population variances in Example 12.1.

Solution

IDENTIFY

The interval estimator is

$$\text{LCL} = \left(\frac{s_1^2}{s_2^2}\right)\frac{1}{F_{\alpha/2,v_1,v_2}}$$

$$\text{UCL} = \left(\frac{s_1^2}{s_2^2}\right)\frac{1}{F_{1-\alpha/2,v_1,v_2}}$$

We substituted the sample variances, sample sizes, and the confidence level into the **F-Estimate of 2 Variances** worksheet in **Stats-Summary.xls** to produce the printout below.

COMPUTE

Microsoft Excel Output for Example 12.6

F-Estimate of the Ratio of Two Variances	
Sample 1	
Sample variance	4103.0
Sample size	43
Sample 2	
Sample variance	10669.8
Sample size	107
Confidence level	0.95
Lower confidence limit	0.2374
Upper confidence limit	0.6594

INTERPRET

As we pointed out in Chapter 10, we can often use an interval estimator to test hypotheses. In this example the interval estimate excludes the value of 1. Consequently, we can draw the same conclusion as we did in Example 12.5; the appropriate technique to test the data in Example 12.1 was the unequal-variances t-test of $\mu_1 - \mu_2$.

▲

Factors That Identify the F-Test and Estimator of σ_1^2 / σ_2^2

1. Problem objective: compare two populations.
2. Data type: interval.
3. Descriptive measurement: variability.

EXERCISES

12.47 Random samples from two normal populations produced the following statistics.

$$s_1^2 = 350 \quad n_1 = 30 \quad s_2^2 = 700 \quad n_2 = 30$$

Can we infer that the two population variances differ?

12.48 Refer to Exercise 12.47. Estimate with 95% confidence the ratio of the two population variances.

12.49 Given the following statistics, test to determine whether the variance of population 1 is larger than the variance of population 2.

$$s_1^2 = 60 \quad n_1 = 20 \quad s_2^2 = 25 \quad n_2 = 20$$

12.50 Given the data below, test the following hypotheses.

$$H_0: \sigma_1^2/\sigma_2^2 = 1$$
$$H_1: \sigma_1^2/\sigma_2^2 \neq 1$$

Sample 1:	7	4	9	12	8	6	9	14		
Sample 2:	10	7	13	18	4	8	21	20	5	8

12.51 Random samples from two normal populations produced the following results. Is there enough evidence to infer that the population variances differ?

Sample 1:	27	52	41	20	33	59	41	28	29	51
Sample 2:	18	15	19	31	49	12	48	29	45	50

12.52 Can we conclude from the data stored in columns A and B of file Xr12-52 that the variance of population A is less than that of population B?

12.53 Refer to Exercise 12.52. Estimate the ratio of population variances with 95% confidence.

12.54 Test to determine whether the t-test of $\mu_1 - \mu_2$ you applied in Exercise 12.23 was justified.

12.55 Determine whether the test you used to answer Exercise 12.24 was appropriate.

12.56 Did you use the correct technique when you answered the question posed in Exercise 12.30?

12.57 The weekly returns of two portfolios were recorded for one year with the results stored in columns A and B of file Xr12-57. Can we conclude that portfolio B is riskier than portfolio A?

12.58 An important statistical measurement in service facilities (such as restaurants and banks) is the variability in service times. As an experiment, two bank tellers were observed, and the service times for each of 100 customers were recorded and stored in columns A and B of file Xr12-58. Do these data allow us to infer that the variance in service times differs between the two tellers?

12.6 INFERENCE ABOUT THE DIFFERENCE BETWEEN TWO POPULATION PROPORTIONS

In this section, we present the procedures for drawing inferences about the difference between populations whose data are nominal. When data are nominal, the only meaningful computation is to count the number of occurrences of each type of outcome and calculate proportions. Consequently, the parameter to be tested and estimated in this section is the difference between two population proportions, $p_1 - p_2$.

To draw inferences about $p_1 - p_2$, we take a sample of size n_1 from population 1 and a sample of size n_2 from population 2 (Figure 12.11 depicts the sampling process). For each sample, we count the number of successes (recall that we call

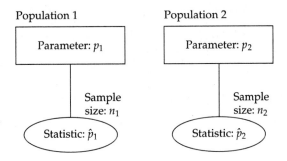

Figure 12.11

Sampling from two populations of nominal data

anything we're looking for a success) in each sample, which we label x_1 and x_2, respectively. The sample proportions are then computed.

$$\hat{p}_1 = \frac{x_1}{n_1} \quad \text{and} \quad \hat{p}_2 = \frac{x_2}{n_2}$$

Statisticians have proved that the statistic $\hat{p}_1 - \hat{p}_2$ is an unbiased consistent estimator of the parameter $p_1 - p_2$.

SAMPLING DISTRIBUTION OF $\hat{p}_1 - \hat{p}_2$

Using the same mathematics as in Chapter 8 to derive the sampling distribution of the sample proportion $\hat{p}$, we determine the sampling distribution of the difference between two sample proportions.

Sampling Distribution of $\hat{p}_1 - \hat{p}_2$

1 The statistic $\hat{p}_1 - \hat{p}_2$ is approximately normally distributed, provided that the sample sizes are large enough so that $n_1 p_1$, $n_1(1 - p_1)$, $n_2 p_2$, and $n_2(1 - p_2)$ are all greater than or equal to 5. (Because p_1 and p_2 are unknown, we express the sample size requirement as $n_1 \hat{p}_1$, $n_1(1 - \hat{p}_1)$, $n_2 \hat{p}_2$, and $n_2(1 - \hat{p}_2)$, are greater than or equal to 5.

2 The mean of $\hat{p}_1 - \hat{p}_2$ is

$$E(\hat{p}_1 - \hat{p}_2) = p_1 - p_2$$

3 The variance of $\hat{p}_1 - \hat{p}_2$ is

$$V(\hat{p}_1 - \hat{p}_2) = \frac{p_1(1 - p_1)}{n_1} + \frac{p_2(1 - p_2)}{n_2}$$

The standard error is

$$\sigma_{\hat{p}_1 - \hat{p}_2} = \sqrt{\frac{p_1(1 - p_1)}{n_1} + \frac{p_2(1 - p_2)}{n_2}}$$

Thus, the variable

$$z = \frac{(\hat{p}_1 - \hat{p}_2) - (p_1 - p_2)}{\sqrt{\frac{p_1(1 - p_1)}{n_1} + \frac{p_2(1 - p_2)}{n_2}}}$$

is approximately standard normally distributed.

TESTING THE DIFFERENCE BETWEEN TWO PROPORTIONS

We would like to use the z-statistic just described as our test statistic; however, the standard error of $\hat{p}_1 - \hat{p}_2$, which is

$$\sigma_{\hat{p}_1 - \hat{p}_2} = \sqrt{\frac{p_1(1 - p_1)}{n_1} + \frac{p_2(1 - p_2)}{n_2}}$$

is unknown, because both p_1 and p_2 are unknown. As a result, the standard error must be estimated from the sample data. There are two different estimators of the standard error, and the determination of which one to use depends on the null hypothesis. If the null hypothesis states that $p_1 - p_2 = 0$, the hypothesized equality of the two population proportions allows us to pool the data from the two samples. Thus, the estimated standard error of $\hat{p}_1 - \hat{p}_2$ is

$$\sqrt{\frac{\hat{p}(1 - \hat{p})}{n_1} + \frac{\hat{p}(1 - \hat{p})}{n_2}} = \sqrt{\hat{p}(1 - \hat{p})\left(\frac{1}{n_1} + \frac{1}{n_2}\right)}$$

where $\hat{p}$ (no subscript) is the **pooled proportion estimator,** defined as

$$\hat{p} = \frac{x_1 + x_2}{n_1 + n_2}$$

The principle used in estimating the standard error of $\hat{p}_1 - \hat{p}_2$ is analogous to that applied in Section 12.2 to produce the pooled variance estimator s_p^2, which is used to test $\mu_1 - \mu_2$ with σ_1^2 and σ_2^2 unknown but equal. That principle roughly states that, where possible, pooling data from two samples produces a better estimator of the standard error. Here, pooling is made possible by hypothesizing (under the null hypothesis) that $p_1 = p_2$. (In Section 12.2, we used the pooled variance estimator because we assumed that $\sigma_1^2 = \sigma_2^2$.) We will call this application Case 1.

Test Statistic for $p_1 - p_2$: Case 1

If the null hypothesis specifies

$$H_0: (p_1 - p_2) = 0$$

the test statistic is

$$z = \frac{(\hat{p}_1 - \hat{p}_2) - (p_1 - p_2)}{\sqrt{\hat{p}(1 - \hat{p})\left(\frac{1}{n_1} + \frac{1}{n_2}\right)}}$$

Because we hypothesize that $p_1 - p_2 = 0$, we simplify the test statistic to

$$z = \frac{(\hat{p}_1 - \hat{p}_2)}{\sqrt{\hat{p}(1 - \hat{p})\left(\frac{1}{n_1} + \frac{1}{n_2}\right)}}$$

The second case applies when, under the null hypothesis, we state that $p_1 - p_2 = D$, where D is some value other than zero. Under such circumstances, we cannot pool the sample data to estimate the standard error of $\hat{p}_1 - \hat{p}_2$. The appropriate test statistic is described next as Case 2.

> **Test Statistic for $p_1 - p_2$: Case 2**
>
> If the null hypothesis specifies
>
> $H_0 : (p_1 - p_2) = D$
>
> where $(D \neq 0)$, the test statistic is
>
> $$z = \frac{(\hat{p}_1 - \hat{p}_2) - (p_1 - p_2)}{\sqrt{\dfrac{\hat{p}_1(1 - \hat{p}_1)}{n_1} + \dfrac{\hat{p}_2(1 - \hat{p}_2)}{n_2}}}$$

Notice that in Case 2 the standard error is calculated by simply substituting the sample statistics $\hat{p}_1$ and $\hat{p}_2$ in place of the population parameters p_1 and p_2.

You will find that, in most practical applications (including the exercises and cases in this book), Case 1 applies. In most problems, we want to know if the two population proportions differ; that is,

$H_1 : (p_1 - p_2) \neq 0$

or if one proportion exceeds the other; that is,

$H_1 : (p_1 - p_2) > 0 \quad \text{or} \quad H_1 : (p_1 - p_2) < 0$

In some other problems, however, the objective is to determine if one proportion exceeds the other by a specific nonzero quantity. In such situations, Case 2 applies.

ESTIMATING THE DIFFERENCE BETWEEN TWO POPULATION PROPORTIONS

We derive the interval estimator of $p_1 - p_2$ in the same manner we have been using since Chapter 9.

> **Interval Estimator of $p_1 - p_2$**
>
> $$(\hat{p}_1 - \hat{p}_2) \pm z_{\alpha/2} \sqrt{\dfrac{\hat{p}_1(1 - \hat{p}_1)}{n_1} + \dfrac{\hat{p}_2(1 - \hat{p}_2)}{n_2}}$$
>
> This technique is valid when $n_1\hat{p}_1$, $n_1(1 - \hat{p}_1)$, $n_2\hat{p}_2$, and $n_2(1 - \hat{p}_2)$ are greater than or equal to 5.

Notice that that the standard error is estimated using the individual sample proportions rather than the pooled proportion. In this procedure we cannot assume that the population proportions are equal as we did in the Case 1 test statistic.

The bound on the error of estimation is

$$B = z_{\alpha/2}\sqrt{\frac{\hat{p}_1(1-\hat{p}_1)}{n_1} + \frac{\hat{p}_2(1-\hat{p}_2)}{n_2}}$$

▼ EXAMPLE 12.7

In a study that was highly publicized, doctors discovered that aspirin seems to help prevent heart attacks. The research project, which was scheduled to last for five years, employed 22,000 American physicians (all male). Half took an aspirin tablet three times per week, while the other half took a placebo on the same schedule. The researchers determined whether the physician suffered a heart attack. If so, a "2" was recorded. If not, a "1" was recorded. These data are stored in columns A (aspirin takers) and B (placebo takers) in file Xm12-07. Determine whether these results indicate that aspirin is effective in reducing the incidence of heart attacks.

Solution

IDENTIFY

The problem objective is to compare two populations. The first is the population of men who take aspirin regularly, and the second is the population of men who do not regularly take aspirin. The data are nominal because the values are "the man suffered a heart attack" and "the man did not suffer a heart attack." These two factors tell us that the parameter to be tested is the difference between two population proportions $p_1 - p_2$ (where p_1 = proportion of all men who regularly take aspirin who suffer a heart attack, and p_2 = proportion of all men who do not take aspirin who suffer a heart attack). Because we want to know if aspirin is effective in reducing heart attacks, the alternative hypothesis is

$$H_1 : (p_1 - p_2) < 0$$

The null hypothesis must be

$$H_0 : (p_1 - p_2) = 0$$

which tells us that this is an application of Case 1. Thus, the test statistic is

$$z = \frac{(\hat{p}_1 - \hat{p}_2)}{\sqrt{\hat{p}(1-\hat{p})\left(\frac{1}{n_1} + \frac{1}{n_2}\right)}}$$

Microsoft Excel Output for Example 12.7

> COMPUTE

z-Test and Estimate: Two Proportions		
	Aspirin	Placebo
Sample Proportions	0.0095	0.0172
Observations	11000	11000
Hypothesized Difference	0	
z Stat	−4.9992	
P(Z<=z) one-tail	0	
z Critical one-tail	1.6449	
P(Z<=z) two-tail	0	
z Critical two-tail	1.96	
Standard Error	0.0015	
Bound	0.0029	
LCL	−0.0107	
UCL	−0.0048	

GENERAL COMMANDS	EXAMPLE 12.7
1 Type or import the data into two adjacent columns.	Open file **Xm12-07**.
2 Click **Tools, Data Analysis Plus,** and **z-Test and Estimate: Two Proportions.**	
3 Specify the **Variable 1 Range:**.	A1:A11001
4 Specify the **Variable 2 Range:**.	B1:B11001
5 Type the value of **Code for Success:**.	2
6 Type the value of **Hypothesized Difference:**. (You must type some value even if you only wish an estimate.)	0
7 Click **Labels**, if applicable.	
8 Type the value of α (**Alpha**), and click **OK**.	.05

To conduct this procedure from the sample proportions or to perform a what-if analysis, activate the **z-Test of 2 Proportions (Case 1)** worksheet in the **Stats-Summary.xls** file.

> INTERPRET

The value of the test statistic is $z = -4.9992$ with a p-value of 0. There is overwhelming evidence to infer that aspirin reduces the incidence of heart attacks among men. In fact, the evidence was so strong that the experiment, which was originally scheduled to run for five years, was cut short after only three years so that the results could be made public.

The interval estimate tells us that the proportion of heart attacks among men who take aspirin is estimated to be between 0.48% to 1.07% lower than those who do not. If 100 million men take aspirin regularly, we can avoid between 480,000 and 1,070,000 heart attacks.

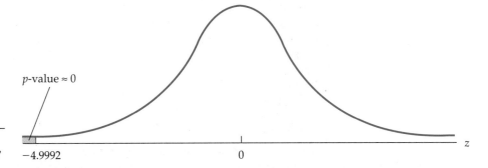

Figure 12.12

Sampling distribution of the test statistic for Example 12.7

One of the flaws in the aspirin study is that all of the subjects were male physicians. Consequently, the effect of taking aspirin regularly on women is unknown. Moreover, to claim that aspirin reduces the frequency of heart attacks among all men is based on the assumption that male physicians are similar to all men, a contention that can be disputed. Nevertheless, medical researchers appear to be satisfied in advising middle-aged people who are not adversely affected by aspirin to take aspirin regularly.

Figure 12.12 describes the sampling distribution.

▼ **EXAMPLE 12.8**

A soap manufacturer is hoping to improve sales with the introduction of more attractive packaging but can't decide which of two new designs to adopt. Having decided to conduct an experiment to collect data to help make his decision, the marketing manager selects two communities known to be similar in terms of sales and preferences. Soap packaged with the new design A is distributed to one community, and soap packaged with the new design B is distributed to the other community. In each community, soap in the old packaging will continue to be available and will be placed next to the soap packaged with one of the new designs. The purchases in each community were stored in file Xm12-08. Column A stores the results (2 = purchased soap packaged with new design and 1 = purchased soap packaged in old design) for community 1, and column B contains the codes for community 2. Because design A is considerably more costly than design B, management has decided that design A will have to outsell design B by more than 3% to be financially viable. Conduct a test to help management decide which type of packaging to use.

Solution

INTERPRET

The problem objective is to compare two populations, each population consisting of the values "consumer purchased product in the new packaging" and "consumer purchased product in the old packaging." The data are obviously nominal, making the parameter to be tested the difference between two population proportions $p_1 - p_2$, where p_1 is the proportion of consumers in community 1 who purchased the product with new packaging A and p_2 is the proportion of consumers in community 2 who purchased the product with new packaging B.

We want to determine whether we can infer that p_1 is more than 3% larger than p_2. Hence we set up the alternative hypothesis accordingly:

$$H_1 : (p_1 - p_2) > .03$$

The null hypothesis meekly follows.

$$H_0 : (p_1 - p_2) = .03$$

Because the null hypothesis specifies a value for the parameter that is not zero, we identify this procedure as Case 2. The test statistic is

$$z = \frac{(\hat{p}_1 - \hat{p}_2) - (p_1 - p_2)}{\sqrt{\frac{\hat{p}_1(1 - \hat{p}_1)}{n_1} + \frac{\hat{p}_2(1 - \hat{p}_2)}{n_2}}}$$

COMPUTE

Microsoft Excel Output for Example 12.8

z-Test and Estimate: Two Proportions

	Design A	Design B
Sample Proportions	0.6416	0.5774
Observations	904	1046
Hypothesized Difference	0.03	
z Stat	1.5467	
P(Z<=z) one-tail	0.061	
z Critical one-tail	1.6449	
P(Z<=z) two-tail	0.122	
z Critical two-tail	1.96	
Standard Error	0.0221	
Bound	0.0433	
LCL	0.0208	
UCL	0.1075	

GENERAL COMMANDS   **EXAMPLE 12.8**

Follow the instructions provided for Example 12.7.
At step 6 type the value of **Hypothesized Difference:.**   .03

As was the case with previous Excel computations, starting in Chapter 9, we can complete the test using sample proportions as well as what-if analyses. Activate the **z-Test of 2 Proportions (Case 2)** worksheet in the **Stats-Summary.xls** file.

Figure 12.13

Sampling distribution of the test statistic for Example 12.8

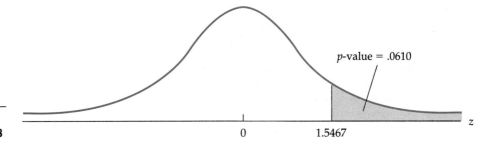

INTERPRET

Because the costs of Type I and Type II errors are similar, a 5% significance level is appropriate. As a result, there is not enough evidence to infer that the proportion of customers who buy the product with the design A packaging is more than 3% higher than the proportion of customers who buy the product with the design B packaging. In the absence of sufficient evidence, the analysis suggests that the product should be packaged using design B.

Figure 12.13 exhibits the sampling distribution.

The test statistics, interval estimator, and the critical factors that identify their use are listed below.

Test Statistics for $p_1 - p_2$

Case 1: $z = \dfrac{(\hat{p}_1 - \hat{p}_2)}{\sqrt{\hat{p}(1 - \hat{p})\left(\dfrac{1}{n_1} + \dfrac{1}{n_2}\right)}}$

Case 2: $z = \dfrac{(\hat{p}_1 - \hat{p}_2) - (p_1 - p_2)}{\sqrt{\dfrac{\hat{p}_1(1 - \hat{p}_1)}{n_1} + \dfrac{\hat{p}_2(1 - \hat{p}_2)}{n_2}}}$

Interval Estimator of $p_1 - p_2$

$(\hat{p}_1 - \hat{p}_2) \pm z_{\alpha/2} \sqrt{\dfrac{\hat{p}_1(1 - \hat{p}_1)}{n_1} + \dfrac{\hat{p}_2(1 - \hat{p}_2)}{n_2}}$

Factors That Identify the z-Test and Estimator of $p_1 - p_2$

1 Problem objective: compare two populations.

2 Data type: nominal.

EXERCISES

*Exercises 12.59 to 12.74 are "what-if" analyses designed to determine what happens to the test statistics and interval estimates when elements of the statistical inference change. These problems can be solved manually or using a worksheet in the **Stats-Summary** file.*

12.59 Random samples from two binomial populations yielded the following statistics:

$\hat{p}_1 = .44 \quad n_1 = 50$
$\hat{p}_2 = .38 \quad n_2 = 50$

Estimate the difference between the two population proportions with 95% confidence.

12.60 Repeat Exercise 12.59, increasing the sample sizes to 100.

12.61 Repeat Exercise 12.59, increasing the sample sizes to 400.

12.62 Review Exercises 12.59 to 12.61. Describe what happens when the sample sizes increase.

12.63 Random samples from two binomial populations yielded the following statistics:

$\hat{p}_1 = .16 \quad n_1 = 25$
$\hat{p}_2 = .12 \quad n_2 = 25$

Estimate the difference between the two population proportions with 90% confidence.

12.64 Repeat Exercise 12.63, with $\hat{p}_1 = .36$ and $\hat{p}_2 = .32$.

12.65 Repeat Exercise 12.63, with $\hat{p}_1 = .56$ and $\hat{p}_2 = .52$.

12.66 Review Exercises 12.63 to 12.65. Describe what happens when the sample proportions increase.

12.67 After sampling from two binomial populations we found the following:

$\hat{p}_1 = .32 \quad n_1 = 100$
$\hat{p}_2 = .28 \quad n_2 = 100$

Calculate the *p*-value of the test to determine whether we can infer that p_1 is greater than p_2.

12.68 Repeat Exercise 12.67, increasing the sample sizes to 400.

12.69 Repeat Exercise 12.67, increasing the sample sizes to 1,000.

12.70 Review Exercises 12.67 to 12.69. Describe what happens when the sample sizes increase.

12.71 The following statistics were calculated.

$\hat{p}_1 = .90 \quad n_1 = 100$
$\hat{p}_2 = .98 \quad n_2 = 100$

Calculate the *p*-value of the test to determine whether we can infer that p_1 is less than p_2.

12.72 Repeat Exercise 12.71, with $\hat{p}_1 = .70$ and $\hat{p}_2 = .78$.

12.73 Repeat Exercise 12.71, with $\hat{p}_1 = .50$ and $\hat{p}_2 = .58$.

12.74 Review Exercises 12.71 to 12.73. Describe what happens when the sample proportions decrease.

12.75 Cold and allergy medicines have been available for a number of years. One serious side effect of these medications is that they cause drowsiness, which makes them dangerous for industrial workers. In recent years, a nondrowsy cold and allergy medicine has been developed. One such product, Hismanal, is claimed by its manufacturer to be the first once-a-day nondrowsy allergy medicine. The nondrowsy part of the claim is based on a clinical experiment in which 1,604 patients were given Hismanal and 1,109 patients were given a placebo. Each person was asked whether he or she felt drowsy. The responses (2 = yes and 1 = no) are stored in columns A and B, respectively, in file Xr12-75. Do these data allow us to infer that Hismanal's claim is false?

12.76 Surveys have been widely used by politicians around the world as a way of monitoring the opinions of the electorate. Six months ago, a survey was undertaken to determine the degree of support for a national party leader. This month, another survey was taken. The responses (2 = support the leader and 1 = do not support the leader) are stored in columns A and B in file Xr12-76.

a Can we infer that the national leader's popularity has decreased?

b Can we infer that the national leader's popularity has decreased by more than 5%?

c Estimate with 95% confidence the decrease in percentage support between now and six months ago.

12.77 The process that is used to produce a complex component used in medical instruments typically results in defective rates in the 40% range. Recently, two innovative processes have been developed to replace the existing process. Process 1 appears to be more promising, but it is considerably more expensive to purchase and operate than process 2. After a thorough analysis of the costs, management decides that it will adopt process 1 only if the proportion of defective components it produces is more than 8% smaller than that produced by process 2. In a test to guide the decision, both processes were used to produce 300 components. A defective component was recorded as 1 and a nondefective as 2. The data are stored in columns A and B in file Xr12-77. Conduct a test to help management make a decision.

12.78 An insurance company is thinking about offering discounts on its life insurance policies to nonsmokers. As part of its analysis, it randomly selects 200 men who are 60 years old and asks them if they smoke at least one pack of cigarettes per day and if they have ever suffered from heart disease. The results are stored in file Xr12-78 using the following format.

Column A: sample of smokers: 2 = suffer from heart disease; 1 = do not suffer from heart disease

Column B: sample of nonsmokers: 2 = suffer from heart disease; 1 = do not suffer from heart disease

a Can the company conclude that smokers have a higher incidence of heart disease than nonsmokers?

b Estimate with 90% confidence the difference in the fraction of men suffering from heart disease between smokers and nonsmokers.

12.79 The impact of the accumulation of carbon dioxide in the atmosphere caused by burning fossil fuels such as oil, coal, and natural gas has been hotly debated for more than a decade. Some environmentalists and scientists have predicted that the excess carbon dioxide will increase the earth's temperature over the next 50 to 100 years with disastrous consequences. This belief is often called the "greenhouse effect." Other scientists claim that we don't know what the effect will be, and yet others believe that the earth's temperature is likely to decrease. Given the debate among scientists, it is not surprising that the general population is confused. To gauge the public's opinion on the subject, two years ago a random sample of 400 people was asked if they believed in the greenhouse effect. This year, 500 people were asked the same question. The results are stored in file Xr12-79 using the following format.

Column A: results two years ago: 2 = believe greenhouse effect; 1 = do not believe greenhouse effect

Column B: results this year: 2 = believe greenhouse effect; 1 = do not believe greenhouse effect

a Can we infer that there has been a decrease in belief in the greenhouse effect?

b Estimate the real change in the public's opinion about the subject. Use a 90% confidence level.

12.80 One of the questions that appears in the professor/course ratings that students fill out at the end of each course at a university is whether the student would recommend the course to a friend. A psychology professor believes that the answer to this question depends on, among other things, how well the student is doing. To test his belief, the professor asked his students to indicate on the questionnaire whether their mark at that point in the course was B or better, or C or worse. The data have been stored in columns A and B in file Xr12-80, where 1 = would not recommend course to a friend and 2 = would recommend course to a friend. Column A contains the responses for those whose grades are B or better. Column B contains the responses for students performing at C or worse. Does it appear that students performing at a grade of B or better are more likely to respond affirmatively to the question about recommending the course to a friend?

12.81 Angina is a common problem for many middle-aged and senior North Americans. Angina is caused by a buildup of fatty tissue in the arteries causing the flow of blood to the heart muscle to diminish. The disease is treated in a variety of ways, including a procedure called angioplasty. In the procedure a wire is inserted into the artery and a balloon is inflated, which squashes the blockage. The problem with this procedure is that the blockage tends to reappear. Surgeons attempt to overcome the problem by inserting a stent at the location of the blockage. To determine whether this innovation is effective, a sample of 40- to 50-year-old men and women who required an angioplasty was recruited. Half the sample had the procedure without the stent; the remainder had the stent inserted. Statistics practitioners then observed whether the procedure needs to be redone within five years. The results are stored in columns A (no stent) and B (stent, where 1 = does not require procedure within five years and 2 = does require procedure) in file Xr12-81.

a Can we infer that inserting a stent reduces the proportion of people who need an angioplasty within five years?

b Suppose that a risk analysis indicates that the stent is worth the added risk if the proportion of people who need a second angioplasty is more than 5% greater for those without the stent. Can we infer from the data that the stent is worthwhile?

12.82 Numerous critics of television, including child psychologists, believe that television induces violence among its viewers. In particular, preteens are affected by the violence they watch. An experiment was conducted to produce information on the subject. A random sample of children between the ages of 10 and 13 was drawn. Half the sample watched a one-hour television show devoid of violence. The other half watched a one-hour show that contained several violent scenes. At the end of the show, the children were asked several questions whose answers identified each child as either aggressive or not aggressive. The data are stored in columns A (nonviolent show) and B (violent show), where 1 = not aggressive and 2 = aggressive in file Xr12-82. Can we conclude from these data that children who watch violence on television are more likely to be aggressive?

12.83 Statisticians can and have shown that flying large commercial airlines is the safest form of transportation. Yet, many people have a fear of flying; some refuse to fly at all, and others fly but suffer doing so. Efforts to "cure" these people of their phobia have taken many different forms. However, some psychologists suggest that for many fear of flying is "caused" by the lack of control one endures by flying. To investigate, a sample of people was asked to complete a questionnaire that identified them as individuals who are either comfortable or uncomfortable not being in control. All were asked whether they fear flying. The responses are stored in file Xr12-83 using the following format.

Column A: Comfortable not being in control;
2 = fear flying and 1 = no fear
Column B: Uncomfortable not being in control;
2 = fear flying and 1 = no fear

Do the data allow us to infer that those who are uncomfortable not being in control are more likely to fear flying?

12.7 SUMMARY

In this chapter, we presented a variety of techniques that allow statistics practitioners to compare two populations. When the data are interval and we are interested in measures of central location, we encountered two more factors that must be considered when choosing the appropriate technique. When the samples are independent, we can use either the equal-variances or unequal-variances formulas. When the samples are matched pairs, we have only one set of formulas. We introduced the F distribution, which is used to make inferences about two population variances. When the data are nominal, the parameter of interest is the difference between two proportions. For this parameter we had two test statistics and one interval estimator. Finally, we discussed observational and experimental data, important concepts in attempting to interpret statistical findings.

IMPORTANT TERMS

Pooled variance estimator
Equal-variances test statistic and interval estimator
Unequal-variances test statistic and interval estimator
Observational data
Experimental data
Independent samples
Matched pairs experiment
F distribution
Numerator degrees of freedom
Denominator degrees of freedom
Pooled proportion estimator

FORMULAS

Equal-variances t-test of $\mu_1 - \mu_2$

$$t = \frac{(\bar{x}_1 - \bar{x}_2) - (\mu_1 - \mu_2)}{\sqrt{s_p^2\left(\frac{1}{n_1} + \frac{1}{n_2}\right)}} \quad \nu = n_1 + n_2 - 2$$

Equal-variances interval estimator of $\mu_1 - \mu_2$

$$(\bar{x}_1 - \bar{x}_2) \pm t_{\alpha/2}\sqrt{s_p^2\left(\frac{1}{n_1} + \frac{1}{n_2}\right)}$$

Unequal-variances t-test of $\mu_1 - \mu_2$

$$t = \frac{(\bar{x}_1 - \bar{x}_2) - (\mu_1 - \mu_2)}{\sqrt{\left(\frac{s_1^2}{n_1} + \frac{s_2^2}{n_2}\right)}} \quad \nu = \frac{(s_1^2/n_1 + s_2^2/n_2)^2}{\left(\frac{(s_1^2/n_1)^2}{n_1 - 1} + \frac{(s_2^2/n_2)^2}{n_2 - 1}\right)}$$

Unequal-variances interval estimator of $\mu_1 - \mu_2$

$$(\bar{x}_1 - \bar{x}_2) \pm t_{\alpha/2}\sqrt{\frac{s_1^2}{n_1} + \frac{s_2^2}{n_2}}$$

t-Test of μ_D

$$t = \frac{\bar{x}_D - \mu_D}{s_D/\sqrt{n_D}} \quad \nu = n_D - 1$$

t-Estimator of μ_D

$$\bar{x}_D \pm t_{\alpha/2}\frac{s_D}{\sqrt{n_D}}$$

F-test of σ_1^2/σ_2^2

$$F = \frac{s_1^2}{s_2^2} \quad \nu_1 = n_1 - 1 \quad \text{and} \quad \nu_2 = n_2 - 1$$

F-estimator of σ_1^2/σ_2^2

$$\text{LCL} = \left(\frac{s_1^2}{s_2^2}\right)\frac{1}{F_{\alpha/2, v_1, v_2}}$$

$$\text{UCL} = \left(\frac{s_1^2}{s_2^2}\right)\frac{1}{F_{1-\alpha/2, v_1, v_2}}$$

z-Test and estimator of $p_1 - p_2$

Case 1: $z = \dfrac{(\hat{p}_1 - \hat{p}_2)}{\sqrt{\hat{p}(1 - \hat{p})\left(\dfrac{1}{n_1} + \dfrac{1}{n_2}\right)}}$

Case 2: $z = \dfrac{(\hat{p}_1 - \hat{p}_2) - (p_1 - p_2)}{\sqrt{\dfrac{\hat{p}_1(1 - \hat{p}_1)}{n_1} + \dfrac{\hat{p}_2(1 - \hat{p}_2)}{n_2}}}$

z-Interval estimator of $p_1 - p_2$

$$(\hat{p}_1 - \hat{p}_2) \pm z_{\alpha/2}\sqrt{\frac{\hat{p}_1(1 - \hat{p}_1)}{n_1} + \frac{\hat{p}_2(1 - \hat{p}_2)}{n_2}}$$

MICROSOFT EXCEL OUTPUT AND INSTRUCTION

Technique	Page
Equal-variances t-test of $\mu_1 - \mu_2$	340–341
Equal-variances estimator of $\mu_1 - \mu_2$	342
Unequal-variances t-test of $\mu_1 - \mu_2$	337–338
Unequal-variances estimator of $\mu_1 - \mu_2$	339
t-Test of μ_D	353
t-Estimator of μ_D	354–355
F-test of σ_1^2/σ_2^2	363
F-estimator of σ_1^2/σ_2^2	366
z-Test and estimator of $p_1 - p_2$ (Case 1)	372
z-Test and estimator of $p_1 - p_2$ (Case 2)	374

SUPPLEMENTARY EXERCISES

12.84 Is eating oat bran an effective way to reduce cholesterol? Early studies indicated that eating oat bran daily reduces cholesterol levels by 5 to 10%. Reports of this study resulted in the introduction of many new breakfast cereals with various percentages of oat bran as an ingredient. However, a January 1990 experiment performed by medical researchers in Boston, Massachusetts, cast doubt on the effectiveness of oat bran. In that study, 120 volunteers ate oat bran for breakfast, and another 120 volunteers ate another grain cereal for breakfast. At the end of six weeks, the percentage of cholesterol reduction was computed for both groups. These data are stored in columns A (% cholesterol reduction with oat bran) and B (% cholesterol reduction with other cereal) in file Xr12-84. Can we infer that oat bran is different from other cereals in terms of cholesterol reduction?

12.85 An inspector for the Atlantic City Gaming Commission suspects that a particular blackjack dealer may be cheating when he deals at expensive tables. To test her belief, she observed 500 hands each at the $100-limit table and the $3,000-limit table. For each hand, she recorded whether the dealer won (code = 2) or lost (code = 1). When a tie occurs, there is no winner or loser. These data are stored in file Xr12-85. (Column A stores the outcomes for the $100-limit table, and column B stores the outcomes for the $3,000-limit table.) Can the inspector conclude that the dealer is cheating in favor of the casino at the more expensive table?

12.86 A restaurant located in an office building decides to adopt a new strategy for attracting customers. Every week it advertises in the city newspaper. To measure how well the advertising is working, the restaurant owner recorded the weekly gross sales for the 15 weeks after the campaign began and the weekly gross sales for the 24 weeks immediately prior to the campaign. These data are stored in columns A (during campaign) and B (before campaign) in file Xr12-86.

  a Can the restaurateur conclude that the advertising campaign is successful?

  b Assume that the profit is 20% of the gross. If the ads cost $50 per week, can the restaurateur conclude that the ads are profitable?

  c What are the required conditions for the test results to be valid in parts (a) and (b)?

12.87 Because of the high cost of energy, homeowners in northern climates need to find ways to cut their heating costs. A building contractor wanted to investigate the effect on heating costs of increasing the insulation. As an experiment, he located a large subdevelopment built around 1970 with minimal insulation. His plan was to insulate some of the houses and compare the heating costs in the insulated homes with those that remained uninsulated. However, it was clear to him that the size of the house was a critical factor in determining heating costs. Consequently, he found 15 pairs of identical-sized houses ranging from about 1,200 to 2,800 square feet. He insulated one house in each pair (levels of R20 in the walls and R32 in the attic) and left the other house unchanged. The heating cost for the following winter season was recorded for each house. The data are stored in file Xr12-87. (Column A = size of the house, column B = heating cost of uninsulated house, column C = heating cost of insulated house.)

  a Do these data allow the contractor to infer that the heating cost for the insulated house is less than that for the uninsulated one?

  b Estimate with 95% confidence the mean savings due to insulating the house.

  c What is the required condition for the use of the techniques in parts (a) and (b)?

12.88 In recent years, a number of state governments have passed mandatory seat-belt laws. Although the use of seat belts is known to save lives and reduce serious injuries, compliance with seat-belt laws is not universal. In an effort to increase the use of seat belts, a government agency sponsored a two-year study. Among its objectives was to determine if there was enough evidence to infer that seat-belt usage increased between last year and this year. To test this belief, random samples of drivers last year and this year were asked whether they always used their seat belts. The responses were stored in file Xr12-88 in the following way:

Column A: last year's survey responses: 2 = wear seat belt; 1 = do not wear seat belt

Column B: this year's survey responses: 2 = wear seat belt; 1 = do not wear seat belt

What conclusions can be drawn from the results?

12.89 An important component of the cost of living is the amount of money spent on housing. Housing costs include rent (for tenants), mortgage payments and property tax (for home owners), heating, electricity, and water. An economist undertook a five-year study to determine how housing costs have changed. Five years ago, he took a random sample of 200 households and recorded the percentage of total income spent on housing. This year, he took another sample of 200 households. The data are stored in columns A (five years ago) and B (this year) in file Xr12-89.

  a Conduct a test to determine whether the economist can infer that housing cost as a percentage of total income has increased over the last five years.

  b Use whatever statistical method you deem appropriate to check the required condition(s) of the test used in part (a).

12.90 In designing advertising campaigns to sell magazines, it is important to know how much time each of a number of demographic groups spends reading magazines. In a preliminary study, 40 people were randomly selected. Each was asked how much time per week he or she spent reading magazines; additionally, each was categorized by gender. The data are stored in file Xr12-90 in the following way: column A = time spent reading magazines per week in minutes for men; column B = time spent reading magazines per week in minutes for women. Is there sufficient evidence to conclude that men and women differ in the amount of time spent reading magazines?

12.91 In a study to determine whether gender affects salary offers for graduating MBA students, 25 pairs of students were selected. Each pair consisted of a female and a male student who were matched according to their grade-point averages, courses taken, ages, and previous work experience. The highest salary offered (in thousands of dollars) to each graduate was recorded and stored in file Xr12-91. (Column A = salary offer for females and column B = salary offer for males.)

  a Is there enough evidence to infer that gender is a factor in salary offers?
  b Discuss why the experiment was organized in the way it was.
  c Is the required condition for the test in part (a) satisfied?

12.92 Have North Americans grown to distrust television and newspaper journalists? A study was conducted this year to compare what Americans currently thought of the press versus what they said three years ago. The survey asked respondents whether they agreed that press tends to favor one side when reporting on political and social issues. A random sample of people were asked to participate in this year's survey. The results of a survey of another random sample taken three years ago are also available. The responses are stored in file Xr12-92 using the following format:

Column A: this year's survey response: 2 = agree; 1 = disagree

Column B: three-years-ago survey response: 2 = agree; 1 = disagree

Can we conclude that Americans have become more distrustful of television and newspaper reporting this year than they were three years ago?

12.93 Before deciding which of two types of stamping machines should be purchased, the plant manager of an automotive parts manufacturer wants to determine the number of units that each produces. The two machines differ in cost, reliability, and productivity. The firm's accountant has calculated that machine A must produce 25 more nondefective units per hour than machine B to warrant buying machine A. To help decide, both machines were operated for 24 hours. The total number of units and the number of defective units produced by each machine per hour were recorded. These data are stored in file Xr12-93 (column A = total number of units produced by machine A; column B = number of defectives produced by machine A; column C = total number of units produced by machine B; column D = number of defectives produced by machine B). Determine which machine should be purchased.

12.94 Refer to Exercise 12.93. Can we conclude that the defective rate differs between the two machines?

12.95 The growing use of bicycles to commute to work has caused many cities to create exclusive bicycle lanes. These lanes are usually created by disallowing parking on streets that formerly allowed curbside parking. Merchants on such streets complain that the removal of parking will cause their businesses to suffer. To examine this problem, the mayor of a large city decided to launch an experiment on one busy street that had one-hour parking meters. The meters were removed, and a bicycle lane was created. The mayor asked the three businesses (a dry cleaner, a doughnut shop, and a convenience store) in one block to record daily sales for two complete weeks (Sunday to Saturday) prior to the change and two complete weeks after the change. The data are stored in file Xr12-95 (column A = day of the week, column B = sales before change for dry cleaner, column C = sales after change for dry cleaner, column D = sales before change for doughnut shop, column E = sales after change for doughnut shop, column F = sales before change for convenience store, and column G = sales after change for convenience store). What conclusions can you draw from these data?

12.96 There may be a new health concern—too much iron in our bodies. An article in the *Wall Street Journal* (17 January 1992) reported that some scientists have implicated iron as a factor in various diseases, including cancer. Part of the problem, it is believed, is that iron builds up in the body over many years. To examine the issue, a random sample of 20-year-old men and women and 40-year-old men and women was drawn. The amount of iron in their bodies was measured and recorded. The results are stored in file Xr12-96 in the following way: column A shows the amount of stored iron in men (in milligrams); column B indicates the men's ages; column C shows the amount of stored iron in women; column D lists the women's ages. (Note: You will have to sort the data before applying statistical techniques.)

  a Can we infer that 40-year-old men have more iron in their bodies than do 20-year-old men?
  b Repeat part (a) for women.

12.97 It is known that clinical depression is linked to several other diseases. Scientists at Johns Hopkins University undertook a study to determine whether heart disease is one of these. A group of 1,190 male medical students were tracked over a 40-year period. Of these, 132 had suffered clinically diagnosed depression. For each student the scientists recorded whether the student died of a heart attack (code = 2) or did not (code = 1). These data are stored in columns A (clinically depressed) and B (not clinically depressed) in file Xr12-97.

a Can we infer that men who are clinically depressed are more likely to die from heart diseases?
b If the answer above is yes, can you interpret this to mean that depression causes heart disease? Explain.

12.98 Most English professors complain that students don't write very well. In particular, they point out that students often confuse quality and quantity. A study at the University of Texas examined this claim. In the study undergraduate students were asked to compare the cost benefits of Japanese and American cars. All wrote their analyses on computers. Unbeknownst to the students the computers were rigged so that some students would have to type twice as many words to fill a single page. The number of words used by each student was recorded and stored in file Xr12-98. (Column A contains the number of words written by students who were allotted a small space and column B contains the number of words written by students who were allotted a large space.) Can we conclude that students write in such a way as to fill the allotted space?

12.99 Approximately 20 million Americans work for themselves. Most run single-person businesses out of their homes. One-quarter of these individuals use personal computers in their businesses. A market research firm, Computer Intelligence InfoCorp, wanted to know whether single-person businesses that use personal computers are more successful than those with no computer. They surveyed 150 single-person firms and recorded their annual incomes. These data are stored in file Xr12-99. (Column A stores the incomes of businesses that use a computer, and column B stores the incomes of businesses that do not). Can we infer that single-person businesses that use a personal computer earn more than those that do not?

12.100 Many small retailers advertise in their neighborhoods by sending out flyers. People who are paid according to the number of flyers delivered deliver these to homes. Each deliverer is given several streets whose homes become their responsibility. One of the ways retailers use to check the performance of deliverers is to randomly sample some of the homes and ask the homeowner whether he or she received the flyer. Recently university students started a new delivery service. They have promised better service at a competitive price. A retailer wanted to know whether the new company's delivery rate is better than that of the existing firm. She had both companies deliver her flyers. Random samples of homes were drawn, and each was asked whether he or she received the flyer (2 = yes and 1 = no). These data are stored in columns A (new company) and B (older company) in file Xr12-100. Can the retailer conclude that the new company is better?

12.101 A study in a journal published by the *American Academy of Pediatrics* (as reported in the *Miami Herald*, January 14, 1997) seems to suggest that preschoolers who drink more than a cup and half of fruit juice tend to be heavier and shorter than children who do not. The study involved 168 healthy youngsters of whom 59 drank on average more than 12 ounces of fruit juice per day. Each child was weighed and measured. These measurements were converted to percentage of normal height and weight for a child of that age and gender. For example, a weight percentage of 1.1 indicates that that child is 10% heavier than the average. The data are stored in columns A, B, C, and D in file Xr12-101. Column A stores the weight percentages for children who drank more than 12 ounces of fruit juice per day; column B contains the weight percentages for children who drank less than 12 ounces of fruit juice per day. Column C contains the height percentages for children who drank more than 12 ounces of fruit juice per day, and column D stores the height percentages for children who drank less than 12 ounces of fruit juice per day.

a Can we conclude that children who drink more than 12 ounces of fruit juice per day are heavier than those who do not?
b Can we conclude that children who drink more than 12 ounces of fruit juice per day are shorter than those who do not?
c Are these data experimental or observational?
d If the answers to questions (a) and (b) are affirmative, does this indicate that drinking too much fruit juice causes the preschoolers to be heavier and shorter? Explain. Is there another explanation for your results?

12.102 Toyota manufactures a variety of cars over a wide range of prices. At the low end, the company makes Tercels and Corollas. At the high end, they make Avalons and Solaras. The prevailing view of the marketing managers is that, as a general rule, buyers of Tercels are younger than buyers of Avalons. To determine whether the belief is correct, the company randomly sampled 100 owners of Tercels and 100 owners of Avalons bought within the past month. The age of each buyer was recorded and stored in columns A and B, respectively, in file Xr12-102. Is there sufficient evidence to conclude that buyers of Avalons are older than buyers of Tercels?

12.103 Refer to Exercise 12.102. As part of the survey, respondents were also asked to report their annual household incomes. These data (in $1,000s) are stored in columns A (incomes of Tercel buyers) and B (incomes of Avalon buyers) in file Xr12-103. Can we conclude that Avalon owners have higher household incomes than do Tercel buyers?

12.104 We know that physical exercise improves the body, but can it also improve the mind? Scientists at The Salk Institute at La Jolla, California, took a random sample of adult mice and divided them into two groups. Both groups were trained to find a platform in a maze filled with cloudy water. (Mice hate swimming, so they seek the platform as a refuge.) The first group exercised on a running wheel, about 5 kilometers per day. The second group remained inactive. After about a month the mice were placed in the maze and timed until each reached the platform. The times (in seconds) are stored in columns A (exercisers) and B (inactive) in file Xr12-104. Can we infer that exercise improves memory in adult mice?

12.105 In 1991 the Canadian government passed legislation that allowed men to take a paid leave (paid from employment insurance) from work following the birth of a child. Women have had that right for several decades. An economist working for a feminist organization wanted to know whether there has been an increase in the number of men taking advantage of the law and staying home (instead of the mother) to care for the baby. In 1993, a random sample of 300 men whose wives gave birth the previous week was asked whether they took leave to care for the baby. In 1999, the survey was repeated with a sample of 350 men. The results of both surveys are stored in columns A (1993) and B (1999) in file Xr12-105, where 1 = took the leave and 2 = did not take the leave. Can we infer that there has been an increase in the incidence of men taking the leave rather than their wives?

12.106 Is it a good strategy to "churn" stocks? Churning occurs when investors buy a stock, sell it shortly thereafter, and buy another stock. In a recent study, a random sample of investors was drawn. When the investor sold a stock and bought another, both stocks were tracked. The one-month return on the sold and bought stocks was recorded and stored in columns A and B, respectively in file Xr12-106. Do these data allow us to infer that the stocks the average investor sold performed better than the one he or she bought?

12.107 Benjamin Franklin wrote ". . . early to rise, makes a man healthy, wealthy, and wise." Recent research may cast doubt on the healthy part. Researchers at the University of Westminster in London, England, studied the saliva of 142 volunteers who woke up between 5:30 and 10:30 A.M. The amount of cortisol was measured. (Cortisol is a hormone that affects temperament and concentration.) The amount of cortisol of those who awoke between 5:30 and 7:00 is stored in column A; the amount for the others is stored in column B. The data are stored in file Xr12-107. Can we infer that early risers have less cortisol than late risers?

12.108 Police and politicians are constantly seeking ways to slow drivers. The city of Vaughan experimented with a new idea. At a cost of $2,500 a solid line was painted three feet from the curb the entire length of a road, and the space was filled with diagonal lines. The lines made the road look narrower. A random sample of car speeds was taken before and after the lines were drawn. The speeds are stored in columns A (before) and B (after) in file Xr12-108. Can we infer that the painted lines reduce speeds?

12.109 Some psychologists believe that people have a built-in bias for selecting the first thing on a list of alternatives. Two psychologists at Ohio State University recorded the vote from the 1992 general election. In Ohio the order of names on the ballot is rotated from one precinct to another. The votes from two precincts are stored in columns A and B in file Xr12-109. Column A contains the votes for five candidates from a precinct where candidate 1 was listed first. Column B contains the votes from a second precinct where candidate 1 was listed last. Can we conclude that candidates receive more votes when their names are listed first than when they are listed last?

12.110 The Department of Business at Wilfrid Laurier University (WLU) offers a BBA degree. Students who desire may take a concentration in accounting and in so doing prepare themselves for the Uniform Final Examination (UFE), which is a stepping stone to the CA (Chartered Accountant) designation. In addition to the degree, WLU offers a diploma in accounting, which requires fewer courses than the degree. To determine the quality of the diploma students, a random sample of students who graduated with a BBA (with a concentration in accounting) and those who received the diploma was drawn. The UFE marks were recorded and stored in columns A and B, respectively, in file Xr12-110. Can we infer that BBA graduates (with accounting concentration) outperform graduates of the accounting diploma program?

12.111 As part of the Physicians Health Study reported in this chapter, physicians also reported whether they were losing their hair at the crown. For each physician who suffered a heart attack a "2" was recorded, and a "1" was recorded if the physician did not suffer a heart attack. The responses are stored in columns A (balding at the crown) and B (not balding at the crown) in file Xr12-111. Can we infer that men who are losing their hair at the crown are more likely to suffer a heart attack?

12.112 Several years ago the National Hockey League was concerned about the large number of tie games. Consequently, they instituted a five-minute, sudden-death overtime period in games that were tied after regulation 60 minutes. However, by early 1999 league governors were still unhappy. It appears that most teams are satisfied with the one point garnered from the tie and thus play a very conservative (and uninteresting) brand of hockey in the overtime period. Up to that point, only 26% of overtime games produced a winner (74% remain tied) and the number of shots per minute was 0.93. As a result, another change was instituted prior to the 1999–2000 season. The rule changed the number of players from six (five skaters and a goalie) to five. It was called "four-on-four" hockey. To determine whether the new rule improved offensive play, the results of a number of games were recorded. In column A in file Xr12-112, a statistics practitioner recorded the number of shots per minute in overtime games prior to the 1999–2000 season. Column B stores the number of shots per minute in overtime games in the 1999–2000 season. Can we conclude from these data that "four-on-four" hockey increases action?

12.113 An increasing number of public elementary, middle, and high schools are requiring their students to wear uniforms. Proponents argue that clothing rules eliminate the baggy gang-inspired look and make it easier to spot intruders. It is also hoped that uniforms would improve morale, behavior, and performance. To examine the issue, a random sample of students from a school requiring uniforms (instituted this year) and another school with no clothing regulations was drawn. The marks in reading and arithmetic were recorded last year (before the uniform rule) and this year. These data are stored in the following format in file Xr12-113.

Column A: School requiring uniforms last year's marks in reading

Column B: School requiring uniforms this year's marks in reading

Column C: School requiring uniforms last year's marks in arithmetic

Column D: School requiring uniforms this year's marks in arithmetic

Column E: School with no clothing requirement last year's marks in reading

Column F: School with no clothing requirement this year's marks in reading

Column G: School with no clothing requirement last year's marks in arithmetic

Column H: School with no clothing requirement this year's marks in arithmetic

a Can we infer that the requirement of uniforms improves reading?
b Can we infer that the requirement of uniforms improves arithmetic?
c How can the data from schools with no clothing requirement be used to draw a conclusion?

12.114 To help determine the value of early childhood education programs, a sample of 363 children was drawn. Some of the children attended early childhood programs when they were two or three years old, and the rest did not. Vocabulary tests were conducted when each child was six years old. These data are stored in columns A (attended early childhood education program) and B (did not attend early childhood education program) in file Xr12-114. Do these data allow us to infer that children who attend early childhood education programs have larger vocabularies than those who do not?

12.115 Refer to Exercise 12.114. The children were also classified according to whether their parents had postsecondary education. The results of the vocabulary tests are stored in columns A and B in file Xr12-115. Can we conclude that children whose parents have postsecondary education have larger vocabularies than children of parents who do not have a postsecondary education?

12.116 Refer to Exercise 12.114. In another related study, children were sampled and divided into two groups. Group 1 children were defined as those whose parents read to them regularly, and group 2 children had parents who did not. Their vocabulary scores are stored in columns A and B, respectively, in file Xr12-116. What can we infer from these data?

Case 12.1 Bonanza International*

Bonanza International is one of the top 15 fast-food franchisers in the United States. Like McDonald's, Burger King, and most others, Bonanza uses a menu board to inform customers about its products. One of Bonanza's bright young executives believes that not all positions on the board are equal; specifically, that the position of the menu item on the board influences sales. If this hypothesis is true, Bonanza would be well advised to place its high-profit items in the positions that produce the highest sales.

After watching the eye movements of several people, the executive determined that customers first look at the upper right-hand corner, then cross the top row toward the left-hand side, then move down to the lower left-hand corner, and finally scan across the bottom toward the right.

This analysis suggests that items listed in the upper right-hand corner may achieve higher sales than items listed in the lower left-hand corner. In order to test this hypothesis, 10 stores with similar characteristics were selected as test restaurants, and two moderately popular items were selected as test menu items. During weeks one and three (of a four-week study), item A was placed in the upper right-hand corner, and item B was placed in the lower left-hand corner. During weeks two and four, the positions were reversed. The number of sales of each item was recorded, and the results are summarized in Tables A and B. These data are also stored in columns A through D of file C12-01.

On the basis of the data, what can you conclude regarding the executive's belief?

Table A Sales of Item A

Store	Sales with Item A in Upper Right-Hand Corner	Sales with Item A in Lower Left-Hand Corner
1	642	485
2	912	681
3	221	138
4	312	237
5	295	258
6	775	725
7	511	553
8	726	524
9	476	384
10	570	529

*Adapted from M. G. Sobol and T. E. Barry, "Item Positioning for Profits: Menu Boards at Bonanza International," *Interfaces* (February 1980): 55–60.

Table B Sales of Item B

Store	Sales with Item B in Upper Right-Hand Corner	Sales with Item B in Lower Left-Hand Corner
1	372	351
2	334	312
3	160	136
4	285	305
5	271	189
6	464	430
7	327	310
8	642	557
9	213	215
10	493	446

Case 12.2 Accounting Course Exemptions*

One of the problems encountered in teaching accounting in a business program is the issue of what to do with students who have taken one or more accounting courses in high school. Should these students be exempted from the introductory accounting course usually offered in the first or second year of the business program? Some professors have argued that high school courses do not have the breadth or depth of university courses and that, as a consequence, high school accounting students should not be exempted. Others think that the high school accounting course coverage is sufficiently close to that of the university course and that forcing students with high school accounting to "retake" the course is a waste of time and resources.

In order to examine the problem, students who were enrolled in the third year of the Bachelor of Commerce program at St. Mary's University were sampled. In the second year of this program, two introductory accounting half-credits are required: ACT 241 and ACT 242. Of the 638 students enrolled in ACT 241 in the fall semester, 374 were selected because of the similarities in their educational backgrounds (excluding high school accounting). Student files were examined for all 374 students, of whom 275 continued on to ACT 242 in the winter semester. For each student, researchers recorded the mark (out of 100) in ACT 241 and the mark in ACT 242 (if it was taken), as well as the number of high school accounting courses (either 0, 1, or 2). The results are stored in the file C12–02. Columns A, B, and C contain the marks in ACT 241 for students who have taken 0, 1, and 2 high school accounting courses, respectively. Columns D, E, and F contain the marks in ACT 242 for students who have taken 0, 1, and 2 high school accounting courses, respectively.

The researchers would like to know whether students with one high school accounting course outperform those with no high school accounting and whether students with two high school accounting courses outperform those with no high school accounting. What exemption policy should be adopted?

*Adapted from E. Morass, G. Walsh, and N. M. Young, "Accounting for Performance: An Analysis of the Relationship Between Success in Introductory Accounting in University and Prior Study of Accounting in High School," *Proceedings of the 14th Annual Atlantic Schools of Business Conference* (1984): 13–44.

Chapter 13

Statistical Inference: Review of Chapters 11 and 12

13.1 Introduction

13.2 Guide to Identifying the Correct Technique: Chapters 11 and 12

13.1 INTRODUCTION

This chapter is more than just a review of the previous two chapters. It is a critical part of your development as a statistics practitioner. When you solved problems at the end of each section in the preceding chapters (you *have* been solving problems at the end of each section covered, haven't you?), you probably had no great difficulty identifying the correct technique to use. You used the statistical technique introduced in that section. While those exercises provided practice in setting up hypotheses, producing computer output of tests of hypothesis and interval estimators, and interpreting the results, you did not address a fundamental question faced by statistics practitioners: which technique to use. If you still do not appreciate the dimension of this problem, consider the following, which lists all the inferential methods covered thus far.

t-test and estimator of μ

χ^2-test and estimator of σ^2

z-test and estimator of p

t-test and estimator of $\mu_1 - \mu_2$ (equal variances formulas)

t-test and estimator of $\mu_1 - \mu_2$ (unequal variances formulas)

t-test and estimator of μ_D

F-test and estimator of σ_1^2/σ_2^2

z-test (cases 1 and 2) and estimator of $p_1 - p_2$

Counting tests and interval estimators of a parameter as two different techniques, a total of 17 statistical procedures have been presented thus far, and there is much left to be done. Faced with statistical problems that require the use of some of these techniques (such as in real-world applications or on a midterm test), most students need some assistance in identifying the appropriate method. In the next section, we discuss in greater detail how to make this decision. At the end of this chapter you will have the opportunity to practice your decision skills; we've provided exercises and cases that cumulatively require all of the inferential techniques introduced in Chapters 11 and 12. Solving these problems will require you to do what statistics practitioners must do. You must analyze the problem, identify the technique or techniques, employ statistical software and a computer to yield the required statistics, and interpret the results.

13.2 GUIDE TO IDENTIFYING THE CORRECT TECHNIQUE: CHAPTERS 11 AND 12

As you've probably already discovered, the two most important factors in determining the correct statistical technique are the problem objective and the data type. In some situations, once these have been recognized, the technique automatically follows. In other cases, however, several additional factors must be identified before you can proceed. For example, when the problem objective is to compare two populations and the data are interval, three other significant issues must be addressed: the descriptive measurement (central location or variability), whether the samples are independently drawn, and, if so, whether the unknown population variances are equal.

The flowchart in Figure 13.1 represents the logical process that leads to the identification of the appropriate method. We've also included a more detailed guide (Table 13.1) to the statistical techniques that lists the formulas of the test statistics, the interval estimators, and the required conditions.

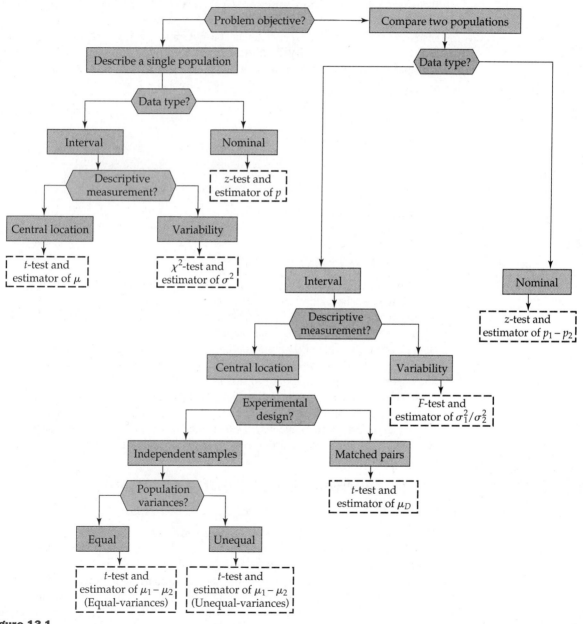

Figure 13.1

Flowchart of techniques: Chapters 11 and 12

Table 13.1 Summary of Statistical Inference: Chapters 11 and 12

Problem objective: Describe a single population.
  Data type: Interval
    Descriptive measurement: Central location
      Parameter: μ
      Test statistic: $t = \dfrac{\bar{x} - \mu}{s/\sqrt{n}}$
      Interval estimator: $\bar{x} \pm t_{\alpha/2} \dfrac{s}{\sqrt{n}}$
      Required condition: Population is normal.
    Descriptive measurement: Variability
      Parameter: σ^2
      Test statistic: $\chi^2 = \dfrac{(n-1)s^2}{\sigma^2}$
      Interval estimator: $\text{LCL} = \dfrac{(n-1)s^2}{\chi^2_{\alpha/2}}$ $\text{UCL} = \dfrac{(n-1)s^2}{\chi^2_{1-\alpha/2}}$
      Required condition: Population is normal.
  Data type: Nominal
    Parameter: p
    Test statistic: $z = \dfrac{\hat{p} - p}{\sqrt{p(1-p)/n}}$
    Interval estimator: $\hat{p} \pm z_{\alpha/2} \sqrt{\dfrac{\hat{p}(1-\hat{p})}{n}}$
    Required condition: $np \geq 5$ and $n(1-p) \geq 5$ (for test)
                        $n\hat{p} \geq 5$ and $n(1-\hat{p}) \geq 5$ (for estimate)

Problem objective: Compare two populations.
  Data type: Interval
    Descriptive measurement: Central location
      Experimental design: Independent samples
        Population variances: $\sigma_1^2 = \sigma_2^2$
        Parameter: $\mu_1 - \mu_2$
        Test statistic: $t = \dfrac{(\bar{x}_1 - \bar{x}_2) - (\mu_1 - \mu_2)}{\sqrt{s_p^2 \left(\dfrac{1}{n_1} + \dfrac{1}{n_2}\right)}}$
        Interval estimator: $(\bar{x}_1 - \bar{x}_2) \pm t_{\alpha/2} \sqrt{s_p^2 \left(\dfrac{1}{n_1} + \dfrac{1}{n_2}\right)}$
        Required condition: Populations are normal.
        Population variances: $\sigma_1^2 \neq \sigma_2^2$
        Parameter: $\mu_1 - \mu_2$
        Test statistic: $t = \dfrac{(\bar{x}_1 - \bar{x}_2) - (\mu_1 - \mu_2)}{\sqrt{\left(\dfrac{s_1^2}{n_1} + \dfrac{s_2^2}{n_2}\right)}}$
        Interval estimator: $(\bar{x}_1 - \bar{x}_2) \pm t_{\alpha/2} \sqrt{\left(\dfrac{s_1^2}{n_1} + \dfrac{s_2^2}{n_2}\right)}$
        Required condition: Populations are normal.
      Experimental design: Matched pairs
        Parameter: μ_D
        Test statistic: $t = \dfrac{\bar{x}_D - \mu_D}{s_D/\sqrt{n_D}}$
        Interval estimator: $\bar{x}_D \pm t_{\alpha/2} \dfrac{s_D}{\sqrt{n_D}}$
        Required condition: Differences are normal.

Descriptive measurement: Variability

Parameter: σ_1^2/σ_2^2

Test statistic: $F = s_1^2/s_2^2$

Interval estimator: $\text{LCL} = \left(\dfrac{s_1^2}{s_2^2}\right)\dfrac{1}{F_{\alpha/2,\nu_1,\nu_2}}$

$\text{UCL} = \left(\dfrac{s_1^2}{s_2^2}\right) F_{\alpha/2,\nu_2,\nu_1}$

Required condition: Populations are normal.

Data type: Nominal

Parameter: $p_1 - p_2$

Test statistic:

Case 1: $H_0: (p_1 - p_2) = 0$

$$z = \dfrac{(\hat{p}_1 - \hat{p}_2)}{\sqrt{\hat{p}(1-\hat{p})\left(\dfrac{1}{n_1} + \dfrac{1}{n_2}\right)}}$$

Case 2: $H_0: (p_1 - p_2) = D \quad (D \ne 0)$

$$z = \dfrac{(\hat{p}_1 - \hat{p}_2) - (p_1 - p_2)}{\sqrt{\dfrac{\hat{p}_1(1-\hat{p}_1)}{n_1} + \dfrac{\hat{p}_2(1-\hat{p}_2)}{n_2}}}$$

Interval estimator: $(\hat{p}_1 - \hat{p}_1) \pm z_{\alpha/2}\sqrt{\dfrac{\hat{p}_1(1-\hat{p}_1)}{n_1} + \dfrac{\hat{p}_2(1-\hat{p}_2)}{n_2}}$

Required conditions: $n_1\hat{p}_1$, $n_1(1-\hat{p}_1)$, $n_2\hat{p}_2$, and $n_2(1-\hat{p}_2) \ge 5$

▼ EXAMPLE 13.1

Is the antilock braking system (ABS), now available as a standard feature on many cars, really effective? The ABS works by automatically pumping brakes extremely quickly on slippery surfaces so the brakes do not lock, avoiding an uncontrollable skid. If ABS is effective, we would expect that cars equipped with ABS would have fewer accidents, and the costs of repairs for the accidents that do occur would be smaller. To investigate the effectiveness of ABS, the Highway Loss Data Institute gathered data on a random sample of 500 1991 General Motors cars that did not have ABS and 500 1992 GM cars that were equipped with ABS. For each year, the institute recorded whether the car was involved in an accident and, if so, the cost of making repairs. Forty-two 1991 cars and 38 1992 cars were involved in accidents. The costs of repairs were stored in columns A (1991 cars) and B (1992 cars) in file Xm13-01. Using frequency of accidents and cost of repairs as measures of effectiveness, can we conclude that ABS is effective? If so, estimate how much better are cars equipped with ABS compared to cars without ABS.

Solution

This is a typical illustration of the work that statistics practitioners perform and the way they do it. The Highway Loss Data Institute wants to determine whether ABS is effective. Even before the data are gathered, the statistics practitioner must decide which techniques to apply. To do so requires the statistics practitioner to frame the questions so that tests of hypotheses or interval estimators can he specified. Simply asking whether ABS works is not sufficiently well defined. Because there are several ways to measure the effectiveness of ABS, the following questions were posed.

a Is there sufficient evidence to infer that the accident rate is lower in ABS-equipped cars than in cars without ABS? (If ABS is effective, we would expect a lower accident rate in ABS-equipped cars.)

b Is there sufficient evidence to infer that the cost of repairing accident damage in ABS-equipped cars is less than that of cars without ABS? (When accidents do occur we expect the severity of accidents to be lower in ABS-equipped cars, assuming that ABS is effective.)

c Assuming that we discover that ABS-equipped cars suffer less damage in accidents, estimate how much cheaper they are to repair on average than cars without ABS.

These questions allow the statistics practitioner to select the appropriate techniques. We will proceed through the flowchart to illustrate how this is done. When the data are gathered and stored in the computer, the statistics practitioner executes the commands to output the results. The results are interpreted to answer the central question: Is ABS effective?

Question (a)

IDENTIFY

The first factor to identify in the flowchart is the problem objective. In question (a), the problem objective is to compare two populations: 1991 model results and 1992 model results. Next, we're asked to determine the data type. The data are nominal. (The values of the random variable are "accident occurred" and "no accident occurred.") The flowchart (Figure 13.1) identifies the technique as the z-test and estimator of $p_1 - p_2$. We define

p_1 = proportion of 1991 model cars without ABS involved in an accident

p_2 = proportion of 1992 model cars with ABS involved in an accident

Because we want to know whether ABS brakes are effective in reducing accidents, we specify the alternative hypothesis as

$$H_1 : (p_1 - p_2) > 0$$

The null hypothesis automatically becomes

$$H_0 : (p_1 - p_2) = 0$$

which indicates the use of the case 1 test statistic.

$$\text{Test statistic: } z = \frac{(\hat{p}_1 - \hat{p}_2)}{\sqrt{\hat{p}(1 - \hat{p})\left(\frac{1}{n_1} + \frac{1}{n_2}\right)}}$$

COMPUTE

The sample proportion of accidents for 1991 cars is

$$\hat{p}_1 = \frac{42}{500} = .084$$

The accident rate for 1992 cars is

$$\hat{p}_2 = \frac{38}{500} = .076$$

The sample proportions and sample sizes were input into the **z-Test of 2 Proportions (Case 1)** worksheet of **Stats-Summary.xls** with the resulting printout shown below.

Microsoft Excel Output for Example 13.1a

z-Test of the Difference between Two Proportions (Case 1)	
Sample 1	
Sample proportion	0.084
Sample size	500
Sample 2	
Sample proportion	0.076
Sample size	500
z Stat	0.4663
P(Z<=z) one-tail	0.3205
P(Z<=z) two-tail	0.6410

The value of the test statistic and its *p*-value are .4663 and .3205, respectively. There is not enough evidence to infer that ABS-equipped cars have fewer accidents than cars without ABS.

Question (b)

IDENTIFY

The problem objective is to compare two populations. The data are interval, because we measure the cost of repairs, which is a real number. The flowchart now asks about the descriptive measurement, which we identify as central location. (We want to know whether the repair costs in one population are in general larger than the repair costs in the second population.) The next question asks us to identify the experimental design. Because there is no relationship between the two samples, we know that the samples are independent. The next factor we need to specify is whether the population variances are equal. To make this decision we apply the *F*-test of σ_1^2/σ_2^2. The results (see the printout below), $F = 1.1535$ and *p*-value = $2(.3313) = .6626$, indicate that there is not enough evidence to infer that the variances differ. Putting all the factors together we identify the equal-variances *t*-test of $\mu_1 - \mu_2$.

μ_1 = mean cost of repairing 1991 model cars damaged in accidents

μ_2 = mean cost of repairing 1992 model cars damaged in accidents

Because we want to know whether μ_1 is greater than μ_2, we specify the alternative hypothesis as

$$H_1 : (\mu_1 - \mu_2) > 0$$

and the null hypothesis is

$$H_0 : (\mu_1 - \mu_2) = 0$$

The test statistic is

$$t = \frac{(\bar{x}_1 - \bar{x}_2) - (\mu_1 - \mu_2)}{\sqrt{s_p^2\left(\dfrac{1}{n_1} + \dfrac{1}{n_2}\right)}}$$

COMPUTE

Microsoft Excel Output of the F-Test in Example 13.1b

F-Test Two-Sample for Variances

	Cost 1991	Cost 1992
Mean	2075.0	1714.5
Variance	450343	390409
Observations	42	38
df	41	37
F	1.1535	
P(F<=f) one-tail	0.3313	
F Critical one-tail	1.7129	

Microsoft Excel Output for the t-Test in Example 13.1b

t-Test: Two-Sample Assuming Equal Variances

	Cost 1991	Cost 1992
Mean	2075.0	1714.5
Variance	450343	390409
Observations	42	38
Pooled Variance	421913	
Hypothesized Mean Difference	0	
df	78	
t Stat	2.479	
P(T<=t) one-tail	0.0077	
t Critical one-tail	1.6646	
P(T<=t) two-tail	0.0153	
t Critical two-tail	1.9908	

The value of the test statistic is $z = 2.479$. The p-value is .0077. The t-test of $\mu_1 - \mu_2$ indicates that the cost of repairs is less for ABS-equipped cars than cars without ABS.

Question (c)

IDENTIFY

To measure how much better off a car owner is with ABS, we determine the 95% confidence interval estimator of the difference between the two mean costs. The interval estimator is

$$(\bar{x}_1 - \bar{x}_2) \pm t_{\alpha/2}\sqrt{s_p^2\left(\dfrac{1}{n_1} + \dfrac{1}{n_2}\right)}$$

COMPUTE

We used the **t-Estimate of 2 Means (Eq-Var)** worksheet in **Stats-Summary.xls** with the result shown next.

Microsoft Excel Output for Example 13.1 c

t-Estimate of the Difference between Two Means (Equal-Variances)	
Sample 1	
Sample mean	2075.0
Sample standard deviation	671.1
Sample size	42
Sample 2	
Sample mean	1714.5
Sample standard deviation	624.8
Sample size	38
Confidence level	0.95
Pooled Variance estimate	421914
Difference between means	360.5
Bound	289.52
Lower confidence limit	70.98
Upper confidence limit	650.02

The 95% confidence interval estimate of $\mu_1 - \mu_2$ is

$$\text{LCL} = 70.98 \quad \text{and} \quad \text{UCL} = 650.02$$

CHECKING THE REQUIRED CONDITIONS

Figures 13.2 and 13.3 depict the histograms of the costs of repairs of the 1991 and 1992 cars, respectively. While the histograms are not bell shaped, it appears that costs of repairs in both years are not extremely nonnormal. The conclusions we reached in the statistical procedures above are valid.

INTERPRETING ALL OF THE STATISTICAL RESULTS

The data indicate that the accident rate in ABS-equipped cars may be no lower than that of cars without ABS. However, the cost of repairing the accident damage is less for the former group. We estimate that the average repair bill for an ABS-equipped car is between $70.98 and $650.02 less than a car not equipped with ABS. Can we now say conclusively that ABS is effective? Unfortunately, this is only one interpretation of the results.

Because the experiment uses observational data, we must be careful about the meaning of the tests. It is possible that poor drivers will buy ABS-equipped cars and better drivers will not. If so, the results would tend to indicate that ABS is either ineffective, as in part (a), or not as effective as it really is, as in parts (b) and (c). Experimental data may have been able to overcome this problem. Experimental data could be gathered by randomly selecting people to drive either ABS-equipped cars or cars without ABS for one year and recording the data. In this way the drivers in both groups should be quite similar, making the comparison more definitive.

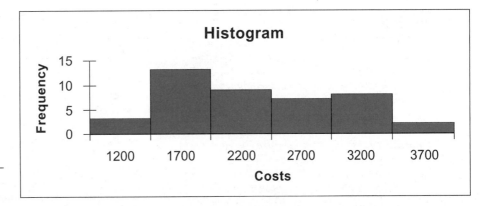

Figure 13.2

Histogram of repair costs of 1991 cars

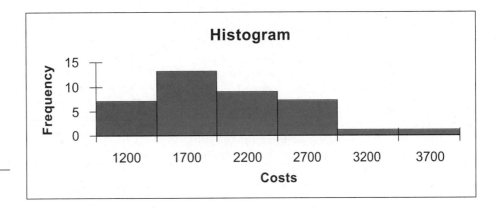

Figure 13.3

Histogram of repair costs of 1992 cars

Another problem in interpreting the results to proclaim that ABS is effective is that it is possible that driving ABS-equipped cars changes the behavior of the drivers. They may drive more dangerously in the mistaken belief that ABS will save them. It may be possible to remedy this problem by not telling the drivers which type they have been assigned to drive. However, most drivers will likely know from the feel and performance of the brakes. Another experiment can be undertaken to determine whether driving behavior is indeed altered by ABS. (See Exercise 13.23.)

Yet another difficulty arose because the experiment was performed using different model years. The ABS-equipped cars were all 1992 models, and the cars without ABS were all 1991 models. The results we observed may be due to differences either in the repair costs or the performance of the models between 1991 and 1992 cars. Undoubtedly, it would have been better to compare 1992 cars with and without ABS. (Note that we are merely reporting the way the Highway Loss Data Institute actually conducted the study; we are not endorsing their methods. We may have gathered the data in a different way.)

Besides teaching you how to identify the appropriate statistical technique, this example also highlights the issues that must be considered when interpreting the results.

EXERCISES

The purpose of the exercises that follow is twofold. First, the exercises provide you with practice in the critical skill of identifying the correct technique. Second, they allow you to improve your ability to determine the statistics needed to answer the question and interpret the results. We believe that the first skill is underdeveloped, because up to now you have had little practice. The exercises you've worked on have appeared at the end of sections and chapters where the correct techniques have just been presented. Determining the correct technique should not have been difficult. Because the exercises that follow were selected from the types that you have already encountered at the ends of Chapters 11 and 12, they will help you develop your technique-identification skills.

You will note that in the exercises that require a test of hypothesis we do not specify a significance level. We have left this decision to you. After analyzing the issues raised in the exercise, use your own judgment to determine whether the p-value is small enough to reject the null hypothesis.

13.1 Shopping malls are more than places where we buy things. We go to malls to watch movies; buy breakfast, lunch, and dinner; exercise; meet friends; and, in general, to socialize. To study the trends a sociologist took a random sample of 100 mall shoppers and asked a variety of questions. This survey was first conducted three years ago with another sample of 100 shoppers. In both surveys respondents were asked to report the number of hours they spend in malls during an average week The results are stored in columns A (this year's survey) and B (survey three years ago) in file Xr13-01. Can we conclude that the amount of time spent at malls has decreased over the past three years?

13.2 It is often useful for retailers to determine why their potential customers chose to visit their store. Possible reasons include advertising, advice from a friend, or previous experience. To determine the effect of full-page advertisements in the local newspaper, the owner of an electronic-equipment store asked 200 randomly selected people who visited the store whether they had seen the ad. He also determined whether the customers had bought anything, and, if so, how much they spent. There were 113 respondents who saw the ad. Of these, 49 made a purchase. Of the 87 respondents who did not see the ad, 21 made a purchase. The amounts spent were stored in file Xr13-02 in the following way.

Column A = amount spent at store among purchasers who saw the advertisement

Column B = amount spent at store among purchasers who did not see the advertisement

a Can the owner conclude that customers who see the ad are more likely to make a purchase than those who do not see the ad?

b Can the owner conclude that customers who see the ad spend more than those who do not see the ad (among those who make a purchase)?

c Estimate with 95% confidence the proportion of all customers who see the ad and who then make a purchase.

d Estimate with 95% confidence the mean amount spent by customers who see the ad and make a purchase.

13.3 In an attempt to reduce the number of person-hours lost as a result of industrial accidents, a large multi-plant corporation installed new safety equipment in all departments and all plants. To test the effectiveness of the equipment, a random sample of 25 plants was drawn. The number of person-hours lost in the month prior to installation of the safety equipment and in the month after installation were recorded. The results are stored in columns A (number of person-hours lost before installation), and B (number of person-hours lost after installation) of file Xr13-03. Can we conclude that the equipment is effective?

13.4 The United States Postal Service (USPS) offers a service called Priority Mail that originally promised two-day delivery. At that time, it cost about $3.00 to send a letter by Priority Mail within the United States. A spokesperson for the USPS claimed that it had a success rate of more than 95% in delivering letters within the two-day deadline. Station WARY in Miami (as reported in their newscast of December 24, 1992) decided to conduct an experiment to determine whether the $3.00 cost is worthwhile. Letters were sent by Priority Mail and by ordinary mail (at that time, a 29-cent stamp) from New York City to Cleveland, Ohio. Letters that arrived within the deadline were recorded with a 2; letters that were late were recorded with a 1. The data are stored in file Xr13-04. (Column A stores the results of Priority Mail, and column B stores the data for ordinary mail.)

a Do these data provide sufficient evidence to support the spokesperson's claim?

b Do these data provide sufficient evidence to infer that Priority Mail delivered letters within two days more frequently than did ordinary mail?

13.5 The electric company is considering an incentive plan to encourage its customers to pay their bills promptly. The plan is to discount the bills 1% if the customer pays within five days, as opposed to the usual 25 days. As an experiment, 50 customers are offered the discount on their September bill. The amount of time each takes to pay his or her bill is recorded. The amount of time a random sample of 50 customers not offered the discount take to pay their bills is also recorded. Both sets of data are stored in file Xr13-05. (Column A represents the first set of customers, and column B represents the second set.) Do these data allow us to infer that the discount plan works?

13.6 Traffic experts are always looking for ways to control automobile speeds. Some communities have experimented with "traffic-calming" techniques. These include speed bumps and various obstructions that force cars to slow to drive around them. Critics point out that the techniques are counterproductive because they cause drivers to speed on other parts of these roads. In an analysis of the effectiveness of speed bumps, a statistics practitioner organized a study over a one-mile stretch of city road that had ten stop signs. He then took a random sample of 100 cars and recorded their average speed (the speed limit was 30 mph) and the number of proper stops at the stop signs. He repeated the observations for another sample of 100 cars after speed bumps were placed on the road. These data were stored in columns A and B (average speeds before and after the speed bumps) and columns C and D (number of proper stops before and after the speed bumps) in file Xr13-06. Do these data allow the statistics practitioner to conclude that the speed bumps are effective?

13.7 The proliferation of self-serve pumps at gas stations has generally resulted in poorer automobile maintenance. One feature of poor maintenance is low tire pressure, which results in shorter tire life and higher gasoline consumption. To examine this problem, an automotive expert took a random sample of cars across the country and measured the tire pressure. The difference between the recommended tire pressure and the observed tire pressure was recorded and stored in file Xr13-07. (A recording of 8 means that that tire is 8 pounds per square inch [PSI] less than the amount recommended by the tire manufacturer.) Suppose that for each PSI below recommendation, tire life decreases by 100 miles and gasoline consumption increases by 0.1 gallons per mile. Estimate with 95% confidence the effect on tire life and gasoline consumption.

13.8 Many North American cities encourage the use of bicycles as a way to reduce pollution and traffic congestion. So many people now regularly use the bicycle to get to work and for exercise that some jurisdictions have enacted bicycle helmet laws, which specify that all bicycle riders must wear helmets to protect against head injuries. Critics of these laws complain that it is a violation of individual freedom and that helmet laws tend to discourage bicycle usage. To examine this issue, a researcher randomly sampled 50 bicycle users and asked each to record the number of miles he or she rode weekly. Several weeks later the helmet law was enacted. The number of miles each of the 50 bicycle riders rode weekly was recorded for the week after the law was passed. These data are stored in columns A (miles ridden before law) and B (miles ridden after law) in file Xr13-08. Can we infer from these data that the law discourages bicycle usage?

13.9 Cardizem CD is a prescription drug that is used to treat high blood pressure and angina. One common side effect of such drugs is the occurrence of headaches and dizziness. To determine whether their drug has the same side effects the drug's manufacturer, Marion Merrell Dow, Inc., undertook a study. A random sample of 908 high blood pressure sufferers was recruited; 607 took Cardizem CD and 301 took a placebo. Each reported whether she or he suffered from headaches and/or dizziness (2 = yes and 1 = no). The responses were recorded in columns A (Cardizem users) and B (placebo) in file Xr13-09. Can the pharmaceutical company scientist infer that Cardizem CD users are more likely to suffer headache and dizziness side effects than nonusers?

13.10 A fast-food franchiser is considering building a restaurant at a downtown location. Based on a financial analysis, a site is acceptable only if the number of pedestrians passing the location during the work day averages more than 200 per hour. To help decide whether to build on the site, a statistics practitioner observes the number of pedestrians who pass the site each hour over a 40-hour work week. These data are stored in file Xr13-10. Should the franchiser build on this site?

13.11 There has been much debate about the effects of secondhand smoke. A U.S. government study (*Globe and Mail*, 20 June 1991) observed samples of households with children living with at least one smoker and households with children living with no smokers. Each child's health was measured. The data from this study are stored in file Xr13-11 in the following way:

Column A: children living with at least one smoker; 2 = child is in fair to poor health; 1 = child is healthy

Column B: children living with no smokers; 2 = child is in fair to poor health; 1 = child is healthy

a Can we infer that children in smoke-free households are less likely to be in fair to poor health than children in households with at least one smoker?

b Assuming that there are 10 million children living in homes with at least one smoker, estimate with 95% confidence the number of children who are in fair to poor health living in a home with at least one smoker.

13.12 An actual U.S. government-funded study surveyed people to determine how they eat spaghetti. The study recorded whether respondents consume spaghetti by winding it on a fork or cutting the noodles into small pieces. Not included in the study, evidently, are those who slurp the noodles directly from their plates without using dining implements at all. The responses are stored in file Xr13-12 (2 = wind the strands and 1 = cut the strands). Can we conclude that more Americans eat their spaghetti by winding on a fork than by cutting the strands?

13.13 Most automobile repair shops now charge according to a schedule that is claimed to be based on average times. This means that instead of determining the actual time to make a repair and multiplying this value by their hourly rate, repair shops determine the cost from a schedule that is calculated from average times. A critic of this policy is examining how closely this schedule adheres to the actual time to complete a job. He randomly selects five jobs. According to the schedule these jobs should take 45 minutes, 60 minutes, 80 minutes, 100 minutes, and 125 minutes, respectively. The critic then takes a random sample of repair shops and records the actual times for each of 20 cars for each job. The times are stored in columns A through E, respectively, in file Xr13-13. For each job, can we infer that the time specified by the schedule is greater than the actual time?

13.14 Most people who quit smoking cigarettes do so for health reasons. However, some quitters find that they gain weight after quitting, and scientists estimate that the health risks of smoking two packs of cigarettes per day and of carrying 65 extra pounds of weight are about equivalent. In an attempt to learn more about the effects of quitting smoking, the U.S. Centers for Disease Control conducted a study (reported in *Time,* 25 March 1991). A sample of 1,885 smokers was taken. During the course of the experiment, some of the smokers quit their habit. The amount of weight gained by all of the subjects was recorded and stored in file Xr13-14. The file is organized in the following way:

Column A: weight gain of smokers who quit in the study

Column B: weight gain of smokers who continued smoking in the study

Do these data allow us to conclude that quitting smoking results in weight gains?

13.15 Golf-equipment manufacturers compete against one another by offering a bewildering array of new products and innovations. Oversized clubs, square grooves, and graphite shafts are examples of such innovations. The effect of these new products on the average golfer is, however, much in doubt. One product, a perimeter weighted iron, was designed to increase the consistency of distance and accuracy. The most important aspect of irons is consistency, which means that ideally there should be no variation in distance from shot to shot. To examine the relative merits of two brands of perimeter-weighted irons, an average golfer used the 7-iron, hitting 100 shots using each of two brands. The distance in yards was recorded and stored in columns A (Brand A) and B (Brand B) in file Xr13-15. Can the golfer conclude that Brand B is superior to Brand A?

13.16 No one disputes the value of physical exercise. Regular exercise has been proven to prolong life and decrease the incidence of certain diseases. But what about exercise of the mind? Are there ways in which one can exercise one's intellect without resorting to the mental equivalent of boring calisthenics? The answer may lie in the game of bridge. In a study undertaken at Scripps College in California, researchers tested 50 bridge players and 50 nonplayers aged between 55 and 91 (as reported in the *ACBL Bulletin,* July 1992). The test measured working memory, reasoning, reaction time, and vocabulary. The results of the tests are stored in file Xr13-16 (columns A and B = working memory for players and nonplayers; columns C and D = reasoning for players and nonplayers; columns E and F = reaction time for players and nonplayers; columns G and H = vocabulary for players and nonplayers). Bearing in mind that the game of bridge places demands on memory and reasoning but requires only 15 words and can be played quite slowly, can we infer that playing bridge improves the memory and reasoning of seniors but does not affect their reaction time and vocabulary?

13.17 Advertising is critical in the residential real estate industry. Agents are always seeking ways to increase sales through improved advertising methods. A particular agent believes that he can increase the number of inquiries (and thus the probability of making a sale) by describing the house for sale without indicating its asking price. To support his belief, he conducted an experiment in which 100 houses for sale were advertised in two ways—with and without the asking price. The number of inquiries for each house was recorded as well as whether the customer saw the ad with or without the asking price shown. The number of inquiries for each house is stored in file Xr13-17 in the following way:

Column A: number of inquiries from customers who saw ad with the asking price shown

Column B: number of inquiries from customers who saw ad without the asking price shown

Do these data allow the real estate agent to infer that ads with no price shown are more effective in generating interest in a house?

13.18 In most offices, the copier is the most frequently used and abused machine. Consequently, buyers of copiers need to know how frequently service will be required before a decision to buy is made. Prior to making a major purchase, the general manager of a large company asks 150 recent buyers of this copier if they required maintenance in the first year and, if so, how frequently. The number of service calls is stored in file Xr13-18. If the president plans to buy 1,000 copiers, estimate with 95% confidence the number of service calls he expects in the first year.

13.19 Throughout the day there are a number of exercise shows appearing on television. These usually feature attractive and fit men and women performing various exercises who urge viewers to duplicate the activity at home. Some viewers are exercisers. However, some people like to watch the shows without exercising (which explains why they use attractive people as demonstrators). Various companies sponsor the shows, and there are commercial breaks. One sponsor wanted to determine whether there are differences between exercisers and nonexercisers in terms of how well they remember the sponsor's name. A random sample of viewers was selected and called after the exercise show was over. Each was asked to report whether he or she exercised or only watched. They were also asked to name the sponsor's brand name (2 = yes, they could, and 1 = no, they couldn't). These results are stored in columns A (exercisers) and B (watchers) in file Xr13-19. Can the sponsor conclude that exercisers are more likely to remember the sponsor's brand name than those who do not exercise?

13.20 A professor of statistics hands back his graded midterms in class by calling out the name of each student and personally handing the exam over to its owner. At the end of the process he notes that there are several exams left over, the result of students missing that class. He forms the theory that the absence is caused by a poor performance by those students on the test. If the theory is correct, the leftover papers will have lower marks than those papers handed back. He records the marks (out of 100) for the leftover papers in column A and the marks of the returned papers in column B. The data are stored in file Xr13-20. Do the data support the professor's theory?

13.21 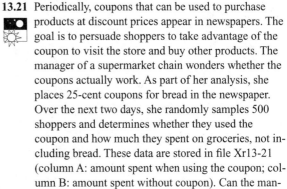 Periodically, coupons that can be used to purchase products at discount prices appear in newspapers. The goal is to persuade shoppers to take advantage of the coupon to visit the store and buy other products. The manager of a supermarket chain wonders whether the coupons actually work. As part of her analysis, she places 25-cent coupons for bread in the newspaper. Over the next two days, she randomly samples 500 shoppers and determines whether they used the coupon and how much they spent on groceries, not including bread. These data are stored in file Xr13-21 (column A: amount spent when using the coupon; column B: amount spent without coupon). Can the manager conclude that coupon users spend more money on groceries than do non–coupon-users?

13.22 According to the latest census, the number of households in a large metropolitan area is 425,000. The home-delivery department of the local newspaper reports that there are 104,320 households that receive daily home delivery. To increase home delivery sales, the marketing department launches an expensive advertising campaign. A financial analyst tells the publisher that for the campaign to be successful, home delivery sales must increase to more than 110,000 households. Anxious to see if the campaign is working, the publisher authorizes a telephone survey of 400 households within one week of the beginning of the campaign and asks each household head whether or not he or she has the newspaper delivered. The responses are stored in file Xr13-22 (2 = yes; 1 = no).

a Do these data indicate that the campaign will increase home delivery sales?
b Do these data allow the publisher to conclude that the campaign will be successful?

13.23 Does driving an ABS-equipped car change the behavior of drivers? To help answer this question, the following experiment was undertaken. A random sample of 200 drivers who currently operate cars without ABS was selected. Each person was given an identical car to drive for one year. Half the sample were given cars that had ABS, and the other half were given cars with standard-equipment brakes. Computers on the cars recorded the average speed (in miles per hour) during the year. These data are stored in file Xr13-23. Column A contains the average speeds of the drivers who were given ABS-equipped cars, and column B stores the speeds of the drivers who were given cars with standard brakes. Can we infer that operating an ABS-equipped car changes the behavior of the driver?

13.24 Health care costs in the United States and Canada are concerns for citizens and politicians. The question is how can we devise a system wherein people's medical bills are covered but yet individuals attempt to reduce costs. An American company has come up with a possible solution. Golden Rule is an insurance company in Indiana with 1,300 employees. The company offered its employees a choice of programs. One choice was a medical savings account (MSA) plan. Here's how it works. To ensure that a major illness or accident does not financially destroy an employee, Golden Rule offers catastrophic insurance—a policy that covers all expenses above $2,000 per year. At the beginning of the year the company deposits $1,000 (for a single employee) and $2,000 (for an employee with a family) into the MSA. For minor expenses, the employee pays from his or her MSA. As an incentive for the employee to spend wisely, any money left in the MSA at the end of the year can be withdrawn by the employee. To determine how well it works a random sample of employees who opted for the medical savings account plan was compared to employees who chose the regular plan. At the end of the year the medical expenses for each employee were recorded and stored in columns A (MSA plan) and B (regular plan) in file Xr13-24. Critics of MSA say that the plan leads to poorer health care, and as a result employees are less likely to be in excellent health. To address this issue, each employee was examined. The results of the examination are stored in columns C (MSA plan) and D (regular plan) where 1 = excellent health and 2 = not in excellent health.

a Can we infer from these data that MSA is effective in reducing costs?

b Can we infer that the critics of MSA are correct?

13.25 Is there a connection between your initials and your lifespan? Researchers at the University of California at San Diego looked at the approximately 5 million people who died in California between 1969 and 1997 (results reported in the *Miami Herald,* March 28, 1998). The researchers found 2,287 men whose initials were judged to be negative. Examples include ILL, DED, and SIK. They also found 1,186 men whose initials were deemed to be good. Included in this list are WIN, JOY, and LOV. The lifespan of each man was stored in columns A and B in file Xr13-25. Can we infer that those with good initials live longer than those with negative initials?

13.26 Refer to Exercise 13.25. Six suicides were reported among those with good initials while those with negative initials had 79 suicides. Can we conclude that a man with negative initials is more likely to commit suicide?

13.27 Does prayer help sick people recover? This question was addressed by researchers who published their findings in the *Archives of Internal Medicine*. The researchers randomly sampled 990 patients at the Mid-America Heart Institute in Kansas City, Mo. All had serious life-threatening cardiac conditions. The sample was divided into two equal groups (based on the last digit of their medical records). For 28 days volunteers outside the hospital prayed for a speedy recovery for a specific patient in group 1. Group 2 patients were not prayed for. Neither the patients nor their doctors were told which group each patient was assigned to. At the end of the experiment the researchers observed that 43 group 1 patients and 54 group 2 patients died. Can we infer from these results that praying helps people survive?

13.28 Refer to Exercise 13.27. For those who recovered, the number of days to recover was recorded and stored in columns A (group 1) and B (group 2) in file Xr13-28. Do these data allow us to conclude that prayer speeds recovery?

13.29 Is there a relationship between testosterone levels and behavior? To answer this question medical researchers measured the testosterone level of 250 penitentiary inmates. They also recorded whether they were convicted of murder. The testosterone level of murderers was stored in column A in file Xr13-29. That of other felons was stored in column B. Can we infer that murderers have higher levels of testosterone than do other felons?

13.30 Refer to Exercise 13.29. The levels of testosterone of trial lawyers and other lawyers (such as corporate and tax attorneys) was recorded and stored in columns A (trial lawyers) and B (other lawyers) in file Xr13-30. Do these data allow us to conclude that trial lawyers have higher levels of testosterone than do other lawyers?

Case 13.1   Quebec Separation: *Oui ou non?*

Since the 1960s, there has been an ongoing campaign among Quebecers to separate from Canada and form an independent nation. Should Quebec separate, the ramifications for the rest of Canada, American states that border Quebec, the North American Free Trade Agreement, and numerous multinational corporations would be enormous. In the 1993 federal election, the prosovereigntist *Bloc Quebecois* won 54 of Quebec's 75 seats in the House of Commons. In 1994, the separatist *Parti Quebecois* formed the provincial government in Quebec and promised to hold a referendum on separation. Like most political issues, polling plays an important role in trying to influence voters and to predict the outcome of the referendum vote. Shortly after the 1993 federal election, *The Financial Post Magazine,* in cooperation with several polling companies, conducted a survey of Quebecers.

A total of 641 adult Quebecers were interviewed. They were asked the following question. (Francophones were asked the questions in French.) The pollsters also recorded the language (English or French) the respondent answered in.

> If a referendum were held today on Quebec's sovereignty with the following question, "Do you want Quebec to separate from Canada and become an independent country?" would you vote yes or no?
>
> 2 Yes
>
> 1 No

The responses are stored in columns A (planned referendum vote for Francophones) and B (planned referendum vote for Anglophones) in file C13-01.

Infer from the data:

a If the referendum was held on the day of the survey, would Quebec vote to remain in Canada?

b Estimate with 95% confidence the difference between French- and English-speaking Quebecers in their support for separation.

Case 13.2   Host Selling and Announcer Commercials*

A study was undertaken to compare the effects of host selling commercials and announcer commercials on children. Announcer commercials are straightforward commercials in which the announcer describes to viewers why they should buy a particular product. Host selling commercials feature a children's show personality or television character who extols the virtues of the product. In 1975, the National Association of Broadcasters prohibited using show characters to advertise products during the

---

*Adapted from J. H. Miller, "An Empirical Evaluation of the Host Selling Commercial and the Announcer Commercial When Used on Children," *Developments in Marketing Science* 8 (1985): 276–78.

same program in which the characters appear. However, this prohibition was overturned in 1982 by a judge's decree.

The objective of the study was to determine whether the two types of advertisements have different effects on children watching them. Specifically, the researchers wanted to know whether children watching host selling commercials would remember more details about the commercial and would be more likely to buy the advertised product than children watching announcer commercials. The experiment consisted of two groups of children ranging in age from 6 to 10. One group of 121 children watched a program in which two host selling commercials appeared. The commercials tried to sell Canary Crunch, a breakfast cereal. A second group of 121 children watched the same program but was exposed to two announcer commercials for the same product. Immediately after the show, the children were given a questionnaire that tested their memory concerning the commercials they had watched.

Each child was marked (out of 10) on his or her ability to remember details of the commercial. In addition, each child was offered a free box of cereal. The children were shown four different brands of cereal—Froot Loops (FL), Boo Berries (BE), Kangaroo Hops (KH), and Canary Crunch (CC; the advertised cereal)—and asked to pick the one they wanted. The results are stored in file C13-02 in the following way:

Column A: recall test mark for the children who watched the host commercial

Column B: recall test mark for the children who watched the announcer commercial

Column C: children's choice of cereal (where 1 = Froot Loops, 2 = Boo Berries, 3 = Kangaroo Hops, and 4 = Canary Crunch) for children who watched the host commercial

Column D: children's choice of cereal for children who watched the announcer commercial

Are there differences in memory test marks and children's choice of advertised cereal between the two groups of children?

Chapter 14

Analysis of Variance

14.1 Introduction

14.2 Single-Factor (One-Way) Analysis of Variance: Independent Samples

14.3 Analysis of Variance Experimental Designs

14.4 Single-Factor Analysis of Variance: Randomized Blocks

14.5 Two-Factor Analysis of Variance: Independent Samples

14.6 Multiple Comparisons

14.7 Bartlett's Test

14.8 Summary

14.1 INTRODUCTION

The technique presented in this chapter allows statistics practitioners to compare two or more populations of interval data. The technique is called the **analysis of variance** and it is an extremely powerful and commonly used procedure. The analysis of variance determines whether differences exist between population means. Ironically, the procedure works by analyzing the sample variance, hence the name. We will examine several different forms of the technique. Illustrations of problems where the statistical methods introduced in this chapter would be applied follow.

Illustration 1 A farm products manufacturer wants to determine if the yields of a crop differ when the soil is treated with various fertilizers. Similar plots of land are planted with the same type of seed but are fertilized differently. At the end of the growing season, the crop yields from the different plots are recorded. The analysis of variance procedure is applied, and the mean yields per acre for each fertilizer are computed. Historically, this type of experiment was one of the first to employ the analysis of variance, and the terminology of the original experiment is still used. No matter what the experiment, the test is designed to determine whether there are significant differences between the treatment means.

Illustration 2 Medical researchers have developed several promising drugs to treat influenza. It is decided to test four of these drugs on human patients. A group of volunteers who have the flu are each given one of the drugs, and the number of days the flu persists is measured. The researchers would conduct an analysis of variance to determine whether differences exist between the days until recovery. The parameters are μ_1, μ_2, μ_3, and μ_4, the mean number of days the flu lasts for each drug. If differences exist, the researchers can use additional techniques to determine which are best.

Illustration 3 Golf equipment manufacturers are constantly researching new designs and materials with the goal of producing golf clubs that are capable of hitting longer distances. When new products are produced they are tested in several ways. In one experiment, average golfers hit golf balls with the new clubs and with their older clubs. Suppose that 100 golfers are asked to use their own drivers and two newly designed ones. The distances that the balls travel are recorded. The analysis of variance technique is applied to determine whether there are differences between the mean distances for each of the three drivers. If differences exist, further research is conducted to determine which designs and/or materials are best.

In each of the illustrations above we are able to classify the populations using only one criterion or **factor.** Each population is called a factor **level.** In Illustration 1, the factor that defines the populations is the type of fertilizer, and the levels represent the different fertilizers being tested. The drug is the factor in Illustration 2, and there are four levels. In Illustration 3, the factor is the design of the golf club, and there are three levels of this factor.

To illustrate a problem where there are two factors that describe the populations, suppose that in Illustration 2 we vary the way the drug is delivered, by injection, orally, or through the skin. In this case, there are two factors. Factor 1 is the drug, which has 4 levels, and factor 2 is the delivery method, which has 3 levels. In all, there are 12 populations, or treatments.

14.2 SINGLE-FACTOR (ONE-WAY) ANALYSIS OF VARIANCE: INDEPENDENT SAMPLES

The analysis of variance is a procedure that tests to determine whether differences exist between two or more population means. The name of the technique derives from the way in which the calculations are performed. That is, the technique analyzes the variance of the data to determine whether we can infer that the population means differ. As in Chapter 12, the experimental design is a determinant in identifying the proper method to use. In this section, we describe the procedure to apply when the samples are independently drawn.

Figure 14.1 depicts the sampling process for drawing independent samples. The mean and variance of population $j(j = 1, 2, \ldots, k)$ are labeled μ_j and σ_j^2, respectively. Both parameters are unknown. For each population, we draw independent random samples. For each sample, we can compute the mean $\bar{x}_j$ and the variance s_j^2.

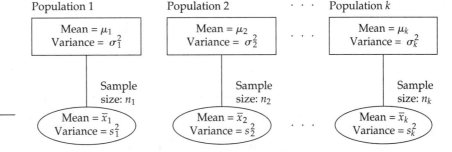

Figure 14.1

Sampling scheme for independent samples

▼ EXAMPLE 14.1

An apple juice manufacturer has developed a new product—a liquid concentrate that, when mixed with water, produces 1 liter of apple juice. The product has several attractive features. First, it is more convenient than canned apple juice, which is the way apple juice is currently sold. Second, because the apple juice that is sold in cans is actually made from concentrate, the quality of the new product is at least as high as canned apple juice. Third, the cost of the new product is slightly lower than canned apple juice. The marketing manager has to decide how to market the new product. She can create advertising that emphasizes convenience, quality, or price. To facilitate a decision, she conducts an experiment. In three different cities that are similar in size and demographic makeup, she launches the product with advertising stressing the convenience of the liquid concentrate (e.g., easy to carry from store to home and takes up less room in the freezer) in one city. In the second city, the advertisements emphasize the quality of the product ("average" shoppers are depicted discussing how good the apple juice tastes). Advertising that highlights the relatively low cost of the liquid concentrate is used in the third city. The number of packages sold weekly is recorded for the 20 weeks following the beginning of the campaign. These data are stored in file Xm14-01 and are listed in the accompanying table. The marketing manager wants to know if differences in sales exist between the three advertising strategies.

Weekly Sales in the Three Cities

City 1 (Convenience)	City 2 (Quality)	City 3 (Price)
529	804	672
658	630	531
793	774	443
514	717	596
663	679	602
719	604	502
711	620	659
606	697	689
461	706	675
529	615	512
498	492	691
663	719	733
604	787	698
495	699	776
485	572	561
557	523	572
353	584	469
557	634	581
542	580	679
614	624	532

Solution You should confirm that the data are interval and that the problem objective is to compare three populations (sales of the liquid concentrate in the three cities). Following the pattern that we have used repeatedly in this book, we introduce the statistical technique by specifying the null and alternative hypotheses. The null hypothesis will state that there are no differences between the population means. Hence,

$$H_0: \mu_1 = \mu_2 = \mu_3$$

The analysis of variance determines whether there is enough statistical evidence to show that the null hypothesis is false. Consequently, the alternative hypothesis will always specify the following.

H_1 : At least two means differ

The next step is to determine the test statistic, which is somewhat more involved than the test statistics we have introduced thus far. The process of performing the analysis of variance is facilitated by the notation in Table 14.1.

The variable X is called the **response variable,** and its values are called **responses.** The unit that we measure is called an **experimental unit.** In this example,

Table 14.1 Notation for the Single-Factor Analysis of Variance: Independent Samples

Independent Samples From k Populations (Treatments)

	Treatment			
	1	**2**	**j**	**k**
	x_{11}	x_{12}	x_{1j}	x_{1k}
	x_{21}	x_{22}	x_{2j}	x_{2k}
	.	.	. . .	.
	.	.	. . .	.
	.	.	. . .	.
	$x_{n_1 1}$	$x_{n_2 2}$	$x_{n_j j}$	$x_{n_k k}$
For Each Treatment (Column)				
Sample Size	n_1	n_2	n_j	n_k
Sample Mean	$\bar{x}_1$	$\bar{x}_2$	$\bar{x}_j$	$\bar{x}_k$

x_{ij} = ith observation of the jth sample

n_j = number of observations in the sample taken from the jth population

$\bar{x}_j$ = mean of the jth sample = $\dfrac{\sum_{i=1}^{n_j} x_{ij}}{n_j}$

$\bar{\bar{x}}$ = grand mean of all the observations = $\dfrac{\sum_{j=1}^{k}\sum_{i=1}^{n_j} x_{ij}}{n}$

where $n = n_1 + n_2 + \cdots + n_k$ and k is the number of populations. Notice that we allow the sample sizes to be different.

the response variable is weekly sales, and the experimental units are the weeks in the three cities when we record sales figures. The sales figures are the responses. As you can see, there is only one factor, advertising approach, that defines the populations, and there are three levels of this factor. They are advertising that emphasizes convenience, advertising that emphasizes quality, and advertising that emphasizes price.

TEST STATISTIC

The test statistic is computed in accordance with the following rationale. If the null hypothesis is true, the population means would all be equal. We would then expect that the sample means would be close to one another. If the alternative hypothesis is true, however, there would be large differences between some of the sample means. The statistic that measures the proximity of the sample means to each other is called the **between-treatments variation,** denoted **SST,** which stands for **sum of squares for treatments.**

Sum of Squares for Treatments

$$\text{SST} = \sum_{j=1}^{k} n_j (\bar{x}_j - \bar{\bar{x}})^2$$

As you can deduce from this formula, if the sample means are close to each other, all of the sample means would be close to the grand mean, and as a result, SST would be small. In fact, SST achieves its smallest value (zero) when all the sample means are equal. That is, if

$$\bar{x}_1 = \bar{x}_2 = \cdots = \bar{x}_k$$

then

$$SST = 0$$

It follows that a small value of SST supports the null hypothesis. In this example, we compute the sample means and the grand mean as

$$\bar{x}_1 = 577.55$$
$$\bar{x}_2 = 653.00$$
$$\bar{x}_3 = 608.65$$
$$\bar{\bar{x}} = 613.07$$

Then

$$SST = \sum_{i=1}^{n} n_j(\bar{x}_j - \bar{\bar{x}})^2$$
$$= 20(577.55 - 613.07)^2 + 20(653.00 - 613.07)^2 + 20(608.65 - 613.07)^2$$
$$= 57,512.23$$

If large differences exist between the sample means, at least some sample means differ considerably from the grand mean, producing a large value of SST. It is then reasonable to reject the null hypothesis in favor of the alternative hypothesis. The key question to be answered in this test (as in all other statistical tests) is, "How large does the statistic have to be for us to justify rejecting the null hypothesis?" In our example, SST = 57,512.23. Is this value large enough to indicate that the population means differ? To answer this question, we need to know how much variation exists in the weekly sales, which is measured by the **within-treatments variation,** which is denoted by **SSE (sum of squares for error).** The within-treatments variation provides a measure of the amount of variation we can expect from the random variable we've observed.

Sum of Squares for Error

$$SSE = \sum_{j=1}^{k} \sum_{i=1}^{n_j} (x_{ij} - \bar{x}_j)^2$$

To understand this concept, examine Tables 14.2 and 14.3 and Figures 14.2 and 14.3. Table 14.2 and Figure 14.2 describe an example in which, because the variation within each sample is quite small, SST is judged to be a large number. That is, this random variable displays very little variation. Consequently, the differences between

Table 14.2 Relatively Large Variation between Samples

	Treatment	
1	**2**	**3**
10	15	20
10	16	20
11	14	20
10	16	20
9	14	20
$\bar{x}_1 = 10$	$\bar{x}_2 = 15$	$\bar{x}_3 = 20$

Table 14.3 Relatively Small Variation between Samples

	Treatment	
1	**2**	**3**
1	19	5
12	31	33
20	4	20
10	9	12
7	12	30
$\bar{x}_1 = 10$	$\bar{x}_2 = 15$	$\bar{x}_3 = 20$

the sample means appear to be caused by real differences between the population means. Contrast this example with the one depicted in Table 14.3. The value of SST in Table 14.3 is equal to that in Table 14.2. However, the variation within the samples is large, which tells us that this random variable features a great deal of variation. By comparison, SST is small, and we would conclude that the differences between the sample means do not allow us to infer that the population means differ.

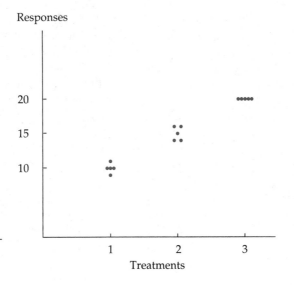

Figure 14.2

Relatively large variation between samples

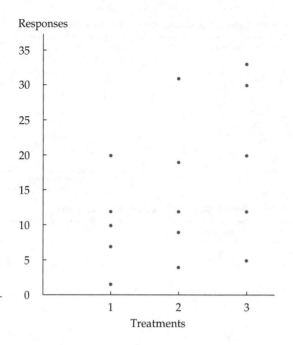

Figure 14.3

Relatively small variation between samples

When SSE is partially expanded, we get

$$\text{SSE} = \sum_{i=1}^{n_1}(x_{i1} - \bar{x}_1)^2 + \sum_{i=1}^{n_2}(x_{i2} - \bar{x}_2)^2 + \cdots + \sum_{i=1}^{n_k}(x_{ik} - \bar{x}_k)^2$$

If you examine each of the k components of SSE, you'll see that each is a measure of the variability of that sample. If we divide each component by $n_j - 1$, we compute the sample variances. We can express this by rewriting SSE as

$$\text{SSE} = (n_1 - 1)s_1^2 + (n_2 - 1)s_2^2 + \cdots + (n_k - 1)s_k^2$$

where s_j^2 is the sample variance of sample j. SSE is thus the combined or pooled variation of the k samples. This is an extension of a calculation we made in Section 12.2, where we tested and estimated the difference between two means using the pooled estimate of the common population variance (denoted s_p^2). One of the required conditions for that statistical technique is that the population variances are equal. That same condition is now necessary for us to use SSE. That is, we require that

$$\sigma_1^2 = \sigma_2^2 = \cdots = \sigma_k^2$$

Returning to our example, we calculate the sample variances as follows.

$s_1^2 = 10{,}775.00$

$s_2^2 = 7{,}238.11$

$s_3^2 = 8{,}670.24$

Thus,

$$\text{SSE} = (n_1 - 1)s_1^2 + (n_2 - 1)s_2^2 + (n_3 - 1)s_3^2$$
$$= 19(10{,}775.00) + 19(7{,}238.61) + 19(8{,}670.24)$$
$$= 506{,}983.50$$

The next step is to compute quantities called the **mean squares.** The **mean square for treatments** is computed by dividing SST by the number of treatments minus 1.

Mean Square for Treatments

$$\text{MSE} = \frac{\text{SST}}{k - 1}$$

The **mean square for error** is determined by dividing SSE by the total sample size (labeled n) minus the number of treatments.

Mean Square for Error

$$\text{MSE} = \frac{\text{SSE}}{n - k}$$

Finally, the test statistic is defined as the ratio of the two mean squares.

Test Statistic

$$F = \frac{\text{MST}}{\text{MSE}}$$

SAMPLING DISTRIBUTION OF THE TEST STATISTIC

The test statistic is F-distributed with $k - 1$ and $n - k$ degrees of freedom, provided that the response variable is normally distributed. In Section 6.4, we introduced the F distribution, and we used it to test and estimate the ratio of two population variances in Chapter 12. The test statistic in that application was the ratio of two sample variances s_1^2 and s_2^2. If you examine the definitions of SST and SSE, you will see that both measure variation similar to the numerator in the formula used to calculate the sample variance s^2 used throughout this book. When we divide SST by $k - 1$ and SSE by $n - k$ to calculate MST and MSE, respectively, we're actually computing variance estimators. Thus, the ratio $F = \text{MST}/\text{MSE}$ is the ratio of two sample variances. The degrees of freedom for this application are the denominators in the mean squares. That is, $v_1 = k - 1$ and $v_2 = n - k$. For Example 14.1, the degrees of freedom are

$$v_1 = k - 1 = 3 - 1 = 2$$

$$v_2 = n - k = 60 - 3 = 57$$

In our example, we found

$$\text{MST} = \frac{\text{SST}}{k - 1} = \frac{57{,}512.23}{2} = 28{,}756.12$$

$$\text{MSE} = \frac{\text{SSE}}{n - 1} = \frac{506{,}983.50}{57} = 8{,}894.45$$

$$F = \frac{\text{MST}}{\text{MSE}} = \frac{28{,}756.12}{8{,}894.45} = 3.23$$

The purpose of calculating the F-statistic is to determine whether the value of SST is large enough to reject the null hypothesis. As you can see, if SST is large, F will be large. We judge the magnitude of the test statistic by computing the p-value of the test, which is

$$P(F > 3.23) = .0468$$

Alternatively, we can set up the rejection region, which is

$$F > F_{\alpha, k-1, n-k}$$

Figure 14.4 depicts the sampling distribution for Example 14.1.

The results of the analysis of variance are usually reported in an **analysis of variance (ANOVA) table.** Table 14.4 shows the general organization of the ANOVA table, while Table 14.5 shows the ANOVA table for Example 14.1.

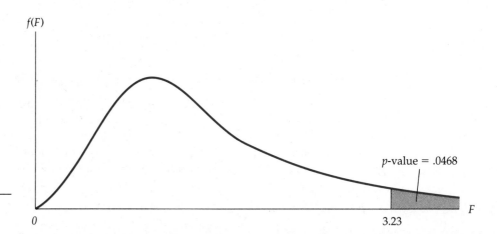

Figure 14.4

Sampling distribution for Example 14.1

Table 14.4 **ANOVA Table for the Single-Factor Analysis of Variance: Independent Samples**

Source of Variation	Degrees of Freedom	Sums of Squares	Mean Squares	F-Statistic
Treatments	$k - 1$	SST	$MST = \dfrac{SST}{(k-1)}$	$F = \dfrac{MST}{MSE}$
Error	$n - k$	SSE	$MSE = \dfrac{SSE}{(n-k)}$	
Total	$n - 1$	SS(Total)		

Table 14.5 **ANOVA Table for Example 14.1**

Source of Variation	Degrees of Freedom	Sums of Squares	Mean Squares	F-Statistic
Treatments	2	57,512.23	28,756.12	F = 3.23
Error	57	506,983.50	8,894.45	
Total	59	564,495.73		

The terminology used in the ANOVA table (and for that matter, in the test itself) is based on the partitioning of the sum of squares. Such partitioning is derived from the following equation (whose validity can be demonstrated by using the rules of summation).

$$\sum_{j=1}^{k}\sum_{i=1}^{n_j}(x_{ij} - \bar{\bar{x}}_j)^2 = \sum_{j=1}^{k} n_j(\bar{x}_j - \bar{\bar{x}})^2 + \sum_{j=1}^{k}\sum_{i=1}^{n_j}(x_{ij} - \bar{x}_j)^2$$

The term on the left represents the total variation of all the data. This expression is denoted **SS(Total).** If we divide SS(Total) by the total sample size minus 1 (that is, by $n - 1$), we would compute the sample variance (assuming that the null hypothesis is true). The first term on the right of the equal sign is SST, and the second term is SSE. As you can see, the total variation SS(Total) is partitioned into two sources of variation. The sum of squares for treatments (SST) is the variation attributed to the differences between the treatment means, while the sum of squares for error (SSE) measures the variation within the samples. The preceding equation can be restated as

SS(TOTAL) = SST + SSE

The test is then based on the comparison of SST and SSE.

Recall that in discussing the advantages and disadvantages of the matched pairs experiment in Section 12.4, we pointed out that statistics practitioners frequently seek ways to reduce or explain the variation in a random variable. In the analysis of variance introduced in this section, the sum of squares for treatments explains some of the variation. The sum of squares for error measures the amount of variation that is unexplained. If SST explains a significant portion of the total variation, we conclude that the population means differ. In Sections 14.4 and 14.5, we will introduce other experimental designs of the analysis of variance, ones that attempt to reduce or explain even more of the variation.

COMPUTE

Microsoft Excel Output for Example 14.1

Anova: Single Factor

SUMMARY

Groups	Count	Sum	Average	Variance
Convenience	20	11551	577.55	10775
Quality	20	13060	653	7238.11
Price	20	12173	608.65	8670.24

ANOVA

Source of Variation	SS	df	MS	F	P-value	F crit
Between Groups	57512.23	2	28756.12	3.23	0.0468	3.16
Within Groups	506983.5	57	8894.45			
Total	564495.7	59				

The value of the test statistic is $F = 3.23$. The p-value is .0468. The critical value of the rejection region is **(F Crit)** 3.16.

GENERAL COMMANDS	EXAMPLE 14.1
1 Type or import the data into adjacent columns.	Open file **Xm14-01**.
2 Click **Tools, Data Analysis...**, and **Anova: Single Factor**.	
3 Specify the **Input Range:**.	A1:C21
4 Click **Labels in First Row**, if applicable.	
5 Specify the value of α (**Alpha**), and click **OK**.	.05

INTERPRET

The p-value is .0468, which means there is evidence to infer that mean weekly sales of the apple juice concentrate are different in at least two of the cities. Can we conclude that the effects of the advertising approaches differ? Recall that it is easier to answer this type of question when the data are obtained through a controlled experiment. In this example, the marketing manager randomly assigned an advertising approach to each city. Thus, the data are experimental. As a result, we are quite confident that the approach used to advertise the product will produce different sales figures.

Incidentally, when the data are obtained through a controlled experiment in the single-factor analysis of variance, we call the experimental design the **completely randomized design** of the analysis of variance.

CHECKING THE REQUIRED CONDITIONS

The *F*-test of the analysis of variance requires that the random variable be normally distributed with equal variances. The technique, though, is robust, so that only extreme nonnormality will invalidate the test results. This is easily checked graphically by producing the histograms for each sample. From the histograms below (Figures 14.5, 14.6, and 14.7), we can see that there is no reason to believe that the weekly sales are extremely nonnormally distributed.

The equality of variances is examined by printing the sample variances, which are output by Excel. The similarity of sample variances allows us to assume that the population variances are equal. In Section 14.7 we present Bartlett's test, a statistical procedure designed to test for the equality of variances.

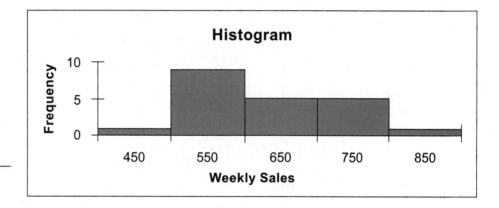

Figure 14.5

Histogram of weekly sales in City 1

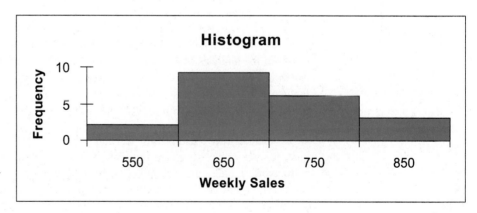

Figure 14.6

Histogram of weekly sales in City 2

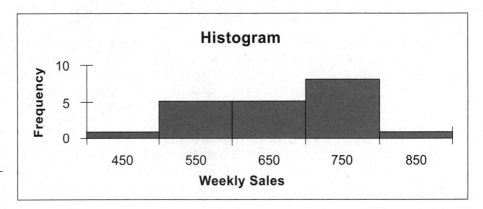

Figure 14.7

Histogram of weekly sales in City 3

VIOLATION OF THE REQUIRED CONDITIONS

If the data are very nonnormal, we can replace the independent samples single-factor model of the analysis of variance with its nonparametric counterpart, which is the Kruskal-Wallis test. (See Section 16.4.) If the population variances are unequal, we can use several methods to correct the problem. However, these corrective measures are beyond the level of this book.

CAN WE USE THE *t*-TESTS OF THE DIFFERENCE BETWEEN TWO MEANS INSTEAD OF THE ANALYSIS OF VARIANCE?

The analysis of variance tests to determine whether there are differences between two or more population means. The *t*-test of $\mu_1 - \mu_2$ determines whether there is a difference between two population means. The question arises: Can we use *t*-tests instead of the analysis of variance? That is, instead of testing all the means in one test as in the analysis of variance, why not test each pair of means? In Example 14.1, we would test $\mu_1 - \mu_2$, $\mu_1 - \mu_3$, and $\mu_2 - \mu_3$. If we found no evidence of a difference in each test, we would conclude that none of the means differ. If there was evidence of a difference in at least one test, we would conclude that some of the means differ.

There are two reasons why we don't use multiple *t*-tests instead of one *F*-test. First, we would have to perform many more calculations. Even with a computer, this extra work is tedious. Second, and more importantly, conducting multiple tests increases the probability of making Type I errors. To understand why, consider a problem where we want to compare six populations, all of which are identical. If we conduct an analysis of variance where we set the significance level at 5%, there is a 5% chance that we would reject the true null hypothesis. That is, there is a 5% chance that we would conclude that differences exist when in fact they don't.

To replace the F-test, we would perform 15 t-tests. (This number is derived from the number of combinations of pairs of means to test, which is $C_2^6 = 6 \times 5/2 = 15$.) Each test would have a 5% probability of erroneously rejecting the null hypothesis. The probability of committing one or more Type I errors is 53.7%.*

One remedy for this problem is to decrease the significance level. In this illustration, we would perform the t-tests with $\alpha = .05/15$, which is equal to .0033. (We will use this procedure in Section 14.6 when we discuss multiple comparisons.) Unfortunately, this would increase the probability of a Type II error. Regardless of the significance level, performing multiple t-tests increases the likelihood of making mistakes. Consequently, when we want to compare more than two populations of interval data, we use the analysis of variance.

Now that we've argued that the t-tests cannot replace the analysis of variance, we need to argue that the analysis of variance cannot replace the t-test.

CAN WE USE THE ANALYSIS OF VARIANCE INSTEAD OF THE t-TEST OF $\mu_1 - \mu_2$?

The analysis of variance is the first of several techniques that allow us to compare two or more populations. Most of the examples and exercises deal with more than two populations. However, it should be noted that like all other techniques whose objective is to compare two or more populations, we can use the analysis of variance to compare only two populations. If that's the case, why do we need techniques to compare exactly two populations? Specifically, why do we need the t-test of $\mu_1 - \mu_2$ when the analysis of variance can be used to test two population means?

To understand why we still need the t-test to make inferences about $\mu_1 - \mu_2$, suppose that we plan to use the analysis of variance to test two population means. The null and alternative hypotheses are

$H_0: \mu_1 = \mu_2$

$H_1:$ At least two means differ

Of course, the alternative hypothesis specifies that $\mu_1 \neq \mu_2$. However, if we want to determine whether μ_1 is greater than μ_2 (or vice versa), we cannot use the analysis of variance because this technique only allows us to test for a difference. Thus, if we want to test to determine if one population mean exceeds the other, we must use the t-test of $\mu_1 - \mu_2$ (with $\sigma_1^2 = \sigma_2^2$). Moreover, the analysis of variance requires that the population variances are equal. If they are not, we must use the unequal variances test statistic.

---

*The probability of committing at least one Type I error is computed from a binomial distribution with $n = 15$ and $p = .05$. Thus,

$P(X \geq 1) = 1 - P(X = 0) = 1 - .463 = .537$

RELATIONSHIP BETWEEN THE F-STATISTIC AND THE t-STATISTIC

It is probably useful for you to understand the relationship between the t-statistic and the F-statistic. The test statistic for testing hypotheses about $\mu_1 - \mu_2$ with equal variances is

$$t = \frac{(\bar{x}_1 - \bar{x}_2) - (\mu_1 - \mu_2)}{\sqrt{s_p^2\left(\frac{1}{n_1} + \frac{1}{n_2}\right)}}$$

If we square this quantity, the result is the F-statistic. That is, $F = t^2$. To illustrate this point, we'll redo Example 12.2 using the analysis of variance. If you reexamine Example 12.2, you'll see that the null and alternative hypotheses were

$H_0: \mu_1 = \mu_2$

$H_1: \mu_1 \neq \mu_2$

The value of the test statistic was $t = .9273$ with a p-value of .3584. Using the analysis of variance (the Excel output is shown below), we find that the value of the test statistic is $F = .860$, which is $(.9273)^2$, and that the p-value is .3584. Thus, we draw exactly the same conclusion using the analysis of variance as we did when we applied the t-test of $\mu_1 - \mu_2$.

Microsoft Excel Analysis of Variance Output for Example 12.2

Anova: Single Factor

SUMMARY

Groups	Count	Sum	Average	Variance
Design-A	25	157.2	6.288	0.8478
Design-B	25	150.4	6.016	1.3031

ANOVA

Source of Variation	SS	df	MS	F	P-value	F crit
Between Groups	0.92	1	0.925	0.860	0.3584	4.043
Within Groups	51.62	48	1.075			
Total	52.54	49				

DEVELOPING AN UNDERSTANDING OF STATISTICAL CONCEPTS 1

Although the formula for the F-statistic is more complex than any we have seen thus far, it is conceptually and mathematically similar to many of the test statistics we have used before. Like the z-test and t-test of μ (Chapters 10 and 11) and the two t-tests of $\mu_1 - \mu_2$ (Chapter 12), the numerator measures the differences between the sample means and the denominator measures the expected variability of the data.

DEVELOPING AN UNDERSTANDING OF STATISTICAL CONCEPTS 2

The analysis of variance partitions the total sum of squares, which enables us to measure how much variation is attributable to differences between populations and how much variation is attributable to differences within populations. As we pointed out in Section 12.4, explaining the variation is an extremely important topic, one that will be seen again in other models of the analysis of variance and in regression analysis (Chapters 17 and 18).

Let's review how we recognize the need to use this model of the analysis of variance.

Factors That Identify the Independent Samples Single-Factor Analysis of Variance

1 Problem objective: compare two or more populations.

2 Data type: interval.

3 Experimental design: independent samples.

EXERCISES

14.1 Provide an example with $k = 4$ where SST = 0.

14.2 Provide an example with $k = 4$ where SSE = 0.

14.3 Random samples of 25 were taken from each of three populations. The data are stored in columns A, B, and C in file Xr14-03.

    a Can we infer that the three population means differ?
    b What are the required conditions for the test in part (a)?
    c Use whatever techniques you deem necessary to check the required conditions.

14.4 The data in file Xr14-04 were generated by drawing random samples from five populations. (Columns A through E are used.)

    a Is there sufficient evidence to infer that differences exist between the population means?
    b What are the required conditions for the technique used in part (a)?
    c Are the conditions satisfied?

14.5 Because there are no national or regional standards, it is difficult for university admission committees to compare graduates of different high schools. University administrators have noted that an 80% average at a high school with low standards may be equivalent to a 70% average at another school with higher standards of grading. In an effort to more equitably compare applications, a pilot study was initiated. Random samples of students who were admitted the previous year were drawn. All of the students entered the business program with averages between 74% and 76% from a random sample of four local high schools. Their average grades in the first year at the university were computed and stored in columns A through D of file Xr14-05.

    a Can the university admissions officer conclude that there are differences in grading standards between the four high schools?
    b What are the required conditions for the test conducted in part (a)?
    c Does it appear that the required conditions of the test in part (a) are satisfied?

14.6 The friendly folks at the Internal Revenue Service (IRS) are always looking for ways to improve the wording and format of its tax return forms. Three new forms have been developed recently. To determine which, if any, are superior to the current form, 120 individuals were asked to participate in an experiment. Each of the three new forms and the currently used form were filled out by 30 different people. The amount of time (in minutes) taken by each person to complete the task was recorded and stored in columns A through D (forms A through D, respectively) in file Xr14-06.

  a What conclusions can be drawn from these data?
  b What are the required conditions for the test conducted in part (a)?
  c Does it appear that the required conditions of the test in part (a) are satisfied?

14.7 A manufacturer of outdoor brass lamps and mailboxes has received numerous complaints about premature corrosion. The manufacturer has identified the cause of the problem as being the low-quality lacquer used to coat the brass. He decides to replace his current lacquer supplier with one of five possible alternatives. In order to judge which is best, he uses each of the five lacquers to coat 25 brass mailboxes and puts all 125 mailboxes outside. He records, for each, the number of days until the first sign of corrosion is observed. The results are stored in columns A through E of file Xr14-07.

  a Is there sufficient evidence to allow the manufacturer to conclude that differences exist between the five lacquers?
  b What are the required conditions for the test conducted in part (a)?
  c Does it appear that the required conditions of the test in part (a) are satisfied?

14.8 A study performed by a Columbia University professor (described in *Report on Business,* August 1991) counted the number of times per minute professors from three different departments said "uh" or "ah" during lectures to fill gaps between words. The data derived from observing 100 minutes from each of the three departments are stored in file Xr14-08. (Column A contains the data for the English department, column B stores the data for the mathematics department, and column C stores the data for the political science department.) If we assume that the more frequent use of "uh" and "ah" results in more boring lectures, can we conclude that some departments' professors are more boring than others?

14.9 In the introduction to this chapter we mentioned that the first use of the analysis of variance was in the 1920s. It was employed to determine whether different amounts of fertilizer yielded different amounts of crop. Suppose that a scientist at an agricultural college wanted to redo the original experiment using three different types of fertilizer. Accordingly, he applied fertilizer A to 20 one-acre plots of land, fertilizer B to another 20 plots, and fertilizer C to yet another 20 plots of land. At the end of the growing season, the crop yields were recorded and stored in columns A through C, respectively, in file Xr14-09. Can the scientist infer that differences exist between the crop yields?

14.10 In 1994 the chief executive officers of the major tobacco companies testified before a Senate subcommittee. One of the accusations made was that tobacco firms added nicotine to their cigarettes, which made them even more addictive to smokers. Company scientists argued that the amount of nicotine in cigarettes depended completely on the size of the tobacco leaf. That is, during poor growing seasons the tobacco leaves would be smaller than in normal or good growing seasons. However, because the amount of nicotine in a leaf is a fixed quantity, smaller leaves would result in cigarettes having more nicotine (because a greater fraction of the leaf would be used to make a cigarette). To examine the issue, a university chemist took random samples of tobacco leaves that were grown in greenhouses where the amount of water was allowed to vary. Three different groups of tobacco leaves were grown. Group 1 leaves were grown with about an average season's rainfall. Group 2 leaves were given about 67% of group 1's water, and group 3 leaves were given 33% of group 1's water. The size of the leaf (in grams) and the amount of nicotine in each leaf were measured and stored in file Xr14-l0. Columns A, B, and C contain the leaf sizes for groups 1, 2, and 3, respectively. Columns D, E, and F contain the amounts of nicotine (in milligrams) for groups 1, 2, and 3, respectively.

  a Test to determine whether the leaf sizes differ between the three groups.
  b Test to determine whether the amounts of nicotine differ in the three groups.
  c What conclusions about the scientists' claim can you draw from these data?

14.11 There is a bewildering number of breakfast cereals on the market. Each company produces several different products in the belief that there are distinct markets. For example, there are the market composed primarily of children and the markets for diet-conscious adults and for health-conscious adults. Each cereal the companies produce has at least one market as its target. However, consumers make their own decisions, which may or may not match the target predicted by the cereal maker. In an attempt to distinguish between consumers, a survey of adults between the ages of 25 and 65 was undertaken. All were asked several questions, including age and which brand of cereal they consumed most frequently. The cereal choices are

1 Sugar Smacks, a children's cereal
2 Special K, a cereal aimed at dieters
3 Fiber One, a cereal that is designed and advertised as healthy
4 Cheerios, a combination of healthy and tasty

The results of the survey are stored in file Xr14-11, where columns A through D contain the ages of the adults who selected Sugar Smacks, Special K, Fiber One, and Cheerios, respectively. Determine whether there are differences between the ages of the consumers of the four breakfast cereals.

14.12 Applicants to law schools must take the Law School Admission Test (LSAT). There are several companies that offer assistance in preparing for the test. To determine whether they work, and if so, which one is best, an experiment was conducted. Several hundred law school applicants were surveyed and asked to report their LSAT score and which if any LSAT preparation course they took. The scores are stored in file Xr14-12 using the following format:

Column A: LSAT score for applicants with preparatory course A

Column B: LSAT score for applicants with preparatory course B

Column C: LSAT score for applicants with preparatory course C

Column D: LSAT score for applicants with no preparatory course

Do these data allow us to infer that there are differences between the four groups of LSAT scores?

14.3 ANALYSIS OF VARIANCE EXPERIMENTAL DESIGNS

Since we introduced the matched pairs experiment in Section 12.4, the experimental design has been one of the factors that determine which technique we use. As we pointed out in that section, statistics practitioners often design experiments to help extract the information they need to assist them in making decisions. The independent samples single-factor experiment is only one of many different experimental designs of the analysis of variance. For each experimental design, we can describe the behavior of the response variable using a mathematical expression or model. Although we will not exhibit the mathematical expressions (we introduce models in Chapter 17) in this chapter, we think it is useful for you to be aware of the elements that distinguish one model or experimental design from another. In this section, we present some of these elements, and in so doing, we introduce two of the experimental designs that will be presented later in this chapter.

SINGLE-FACTOR AND MULTIFACTOR EXPERIMENTS

Recall that the group of treatments or populations is called a factor. The experiment described in Section 14.2 is a **single-factor experiment,** because it addresses the problem of comparing two or more populations defined on the basis of only one factor. A **multifactor experiment** is one where there are two or more factors that define the treatments. The experiment described in Example 14.1 is a single-factor experiment because the treatments were the three advertising approaches. That is, the factor is the advertising approach, and the three levels are advertising that emphasizes convenience, advertising that emphasizes quality, and advertising that emphasizes price.

Suppose that in another study, the medium used to advertise also varied: We can advertise on television or in newspapers. We would then develop a **two-factor** analysis of variance experiment where the first factor, advertising approach, has three levels, and the second factor, advertising medium, has two levels. We will discuss two-factor experiments in Section 14.5.

INDEPENDENT SAMPLES AND BLOCKS

In Section 12.4, we introduced statistical techniques where the data were gathered from a matched pairs experiment. As we pointed out in Section 12.4, this type of experimental design reduces the variation within the samples, making it easier to detect differences between the two populations. When the problem objective is to compare more than two populations, the experimental design that is the counterpart of the matched pairs experiment is called the randomized block experiment. The term *block* refers to a matched group of observations from each population. Here is an example:

> To determine whether incentive pay plans are effective, a statistics practitioner selected three groups of five workers who assemble electronic equipment. Each group will be offered a different incentive plan. The treatments are the incentive plans, the response variable is the number of units produced in one day, and the experimental units are the workers. If we obtain data from independent samples, we may not be able to detect differences between the pay plans because of variation between workers. If there are differences between workers, we need to identify the source of the differences. Suppose, for example, that we know that more experienced workers produce more units no matter what the pay plan. We could improve the experiment if we were to block the workers into five groups of three according to their experience. The three workers with the most experience will represent block 1, the next three will constitute block 2, and so on. As a result, the workers in each block will have approximately the same amount of experience. By designing the experiment in this way, the statistics practitioner removes the effect of different amounts of experience on the response variable. By doing so, we improve the chances of detecting real differences between pay incentives.

We can also perform a blocked experiment by using the same subject (person, plant, or store) for each treatment. For example, we can determine whether sleeping pills are effective by giving three brands of pills to the same group of people to measure the effects. Such applications are called **repeated measures** designs. Technically, this is a different design than the randomized block. However, the single-factor model is analyzed in the same way for both experimental designs. Hence, we will treat repeated measures experiments as randomized block experiments.

In Section 14.4, we introduce the technique used to calculate the test statistic for this type of experiment.

FIXED- AND RANDOM-EFFECTS EXPERIMENTAL DESIGNS

If our analysis includes all possible levels of a factor, the technique is called a **fixed-effects** experiment of the analysis of variance. If the levels included in the study represent a random sample of all the levels that exist, the technique is called a **random-effects** experiment. In Example 14.1, there were only three possible advertising approaches. Consequently, the study is a fixed-effects experiment. However, if there were other advertising approaches besides the three described in the example, and we wanted to know whether there were differences in sales between all the advertising approaches, the application would be a random-effects experiment. Here's another example.

> To determine if there is a difference in the number of units produced by the machines in a large factory, 4 machines out of 50 in the plant are randomly selected for study. The number of units each produces per day for 10 days will be recorded. This experiment is a random-effects experiment because the statistical results will allow us to determine whether there are differences between the 50 machines.

In some experiments, there are no differences in calculations of the test statistic between fixed and random effects. However, in others, including the two-factor experiment presented in Section 14.5, the calculations are different.

14.4 SINGLE-FACTOR ANALYSIS OF VARIANCE: RANDOMIZED BLOCKS

The purpose of designing a randomized block experiment is to reduce the within-treatments variation to more easily detect differences between the treatment means. In the independent samples single-factor analysis of variance, we partitioned the total variation into the between-treatments and the within-treatments variation. That is,

SS(Total) = SST + SSE

In the randomized block experiment, we partition the total variation into three sources of variation:

SS(Total) = SST + SSB + SSE

where SSB, the **sum of squares for blocks,** measures the variation between the blocks. When the variation associated with the blocks is removed SSE is reduced, making it easier to determine if differences exist between the treatment means.

To help you understand the formulas, we will use the following notation.

$\bar{x}[T]_j$ = mean of the observations in the jth treatment

$\bar{x}[B]_i$ = mean of the observations in the ith block

b = number of blocks

Table 14.6 summarizes the notation we use in this model.

The definitions of SS(Total) and SST in the randomized block experiment are identical to those in the independent samples design. SSE in the independent samples design is equal to the sum of SSB and SSE in the randomized block design.

Table 14.6 Notation for the Randomized Block Design of the Analysis of Variance

Blocked Samples from k Populations (Treatments)

Block	Treatment 1	2	...	k	Block Mean
1	x_{11}	x_{12}	...	x_{1k}	$\bar{x}[B]_1$
2	x_{21}	x_{22}	...	x_{2k}	$\bar{x}[B]_2$
.	.	.	...	.	.
.	.	.	...	.	.
.	.	.	...	.	.
b	x_{b1}	x_{b2}	...	x_{bk}	$\bar{x}[B]_b$
Treatment mean	$\bar{x}[T]_1$	$\bar{x}[T]_2$	...	$\bar{x}[T]_k$	

Sums of Squares in the Randomized Block Experiment

$$\text{SS(Total)} = \sum_{j=1}^{k} \sum_{i=1}^{b} (x_{ij} - \bar{\bar{x}})^2$$

$$\text{SST} = \sum_{j=1}^{k} b(\bar{x}[T]_j - \bar{\bar{x}})^2$$

$$\text{SSB} = \sum_{i=1}^{b} k(\bar{x}[B]_i - \bar{\bar{x}})^2$$

$$\text{SSE} = \sum_{j=1}^{k} \sum_{i=1}^{b} (x_{ij} - \bar{x}[T]_j - \bar{x}[B]_i + \bar{\bar{x}})^2$$

The test is conducted by determining the mean squares, which are computed by dividing the sums of squares by their respective degrees of freedom.

Mean Squares for the Randomized Block Experiment

$$\text{MST} = \frac{\text{SST}}{k-1}$$

$$\text{MSB} = \frac{\text{SSB}}{b-1}$$

$$\text{MSE} = \frac{\text{SSE}}{n-k-b+1}$$

Finally, the test statistic is

Test Statistic for Treatments in the Randomized Block Experiment

$$F = \frac{\text{MST}}{\text{MSE}}$$

which is F-distributed with degrees of freedom $v_1 = k - 1$ and $v_2 = n - k - b + 1$.

An interesting, and sometimes useful, by-product of the test of the treatment means is that we can also test to determine if the block means differ. This will allow us to determine whether the experiment should have been conducted as a randomized block experiment. (If there are no differences between the blocks, the randomized block experiment is less likely to detect real differences between the treatment means.) Such a discovery could be useful in future similar experiments. The test of the block means is similar to that of the treatment means.

Test Statistic for Blocks in the Randomized Block Experiment

$$F = \frac{\text{MSB}}{\text{MSE}}$$

which is F-distributed with degrees of freedom $v_1 = b - 1$ and $v_2 = n - k - b + 1$.

Like the independent samples experiment, the statistics generated in the randomized block experiment are summarized in an ANOVA table, whose general form is exhibited in Table 14.7.

Table 14.7 ANOVA Table for the Randomized Block Design

Source of Variation	Degrees of Freedom	Sums of Squares	Mean Squares	F-Statistics
Treatments	$k - 1$	SST	$\text{MST} = \dfrac{\text{SST}}{(k - 1)}$	$F = \dfrac{\text{MST}}{\text{MSE}}$
Blocks	$b - 1$	SSB	$\text{MSB} = \dfrac{\text{SSB}}{(b - 1)}$	$F = \dfrac{\text{MSB}}{\text{MSE}}$
Error	$n - k - b + 1$	SSE	$\text{MSE} = \dfrac{\text{SSE}}{(n - k - b + 1)}$	
Total	$n - 1$	SS (Total)		

EXAMPLE 14.2

Many North Americans suffer from high levels of cholesterol, which can lead to heart attacks. For those with very high levels (over 280), doctors prescribe drugs to reduce cholesterol levels. A pharmaceutical company has recently developed four such drugs. To determine if any differences exist in their benefits, an experiment was organized. The company selected 25 groups of four men, each of whom had levels in excess of 280. In each group, the men were matched according to age and weight. The drugs were administered over a two-month period, and the reduction in cholesterol was recorded. The data are listed below and stored in columns B, C, D, and E in file Xm14-02 (Column A stores the code representing the group number). Do these results allow the company to conclude that differences exist between the four new drugs?

GROUP	DRUG 1	DRUG 2	DRUG 3	DRUG 4
1	6.6	12.6	2.7	8.7
2	7.1	3.5	2.4	9.3
3	7.5	4.4	6.5	10
4	9.9	7.5	16.2	12.6
5	13.8	6.4	8.3	10.6
6	13.9	13.5	5.4	15.4
7	15.9	16.9	15.4	16.3
8	14.3	11.4	17.1	18.9
9	16	16.9	7.7	13.7
10	16.3	14.8	16.1	19.4
11	14.6	18.6	9	18.5
12	18.7	21.2	24.3	21.1
13	17.3	10	9.3	19.3
14	19.6	17	19.2	21.9
15	20.7	21	18.7	22.1
16	18.4	27.2	18.9	19.4
17	21.5	26.8	7.9	25.4
18	20.4	28	23.8	26.5
19	21.9	31.7	8.8	22.2
20	22.5	11.9	26.7	23.5
21	21.5	28.7	25.2	19.6
22	25.2	29.5	27.3	30.1
23	23	22.2	17.6	26.6
24	23.7	19.5	25.6	24.5
25	28.4	31.2	26.1	27.4

Solution

IDENTIFY

The problem objective is to compare four populations, and the data are interval. Because the researchers recorded the cholesterol reduction for each drug for each member of the similar groups of men, we identify the experimental design as randomized block. The response variable is the cholesterol reduction, the treatments are the drugs, and the blocks are the 25 similar groups of men. The complete test is as follows:

$H_0: \mu_1 = \mu_2 = \mu_3 = \mu_4$

$H_1:$ At least two means differ

Test statistic: $F = \dfrac{\text{MST}}{\text{MSE}}$

COMPUTE

Microsoft Excel Output for Example 14.2

ANOVA						
Source of Variation	SS	df	MS	F	P-value	F crit
Rows	3848.7	24	160.361	10.11	0.0000	1.67
Columns	196.0	3	65.318	4.12	0.0094	2.73
Error	1142.6	72	15.869			
Total	5187.2	99				

The output includes block and treatment statistics (sums, averages, and variances, which are not shown here), and the ANOVA table. The F-statistic to determine if differences exist between the four drugs **(Columns)** is 4.12. Its p-value is .0094. The other F-statistic, which is 10.11 (p-value = 0) indicates that there are differences between the groups of men **(Rows)**. Notice that the printout includes the critical values of both rejection regions **(F crit)**.

GENERAL COMMANDS	EXAMPLE 14.2
1 Type or import the data into adjacent columns. 2 Click **Tools, Data Analysis...**, and **Anova: Two-Factor Without Replication.** 3 Specify the **Input Range:**. Click **Labels**, if applicable. If you do, both the treatments and blocks must be labeled (as in Xm14-02). 4 Specify the value of α **(Alpha)**, and click **OK**.	Open file **Xm14-02**. A1:E26 .05

INTERPRET

A Type I error occurs when you conclude that differences exist when, in fact, they do not. A Type II error is committed when the test reveals no difference when at least two means differ. It would appear that both errors are equally costly. Accordingly, we judge the p-value against a standard of 5%. Because the p-value = .0094, we conclude that there is sufficient evidence to infer that at least two of the drugs differ. An examination reveals that cholesterol reduction is greatest using drugs 2 and 4. Further testing is recommended to determine which is best.

Figure 14.8 depicts the sampling distribution of the F-statistic.

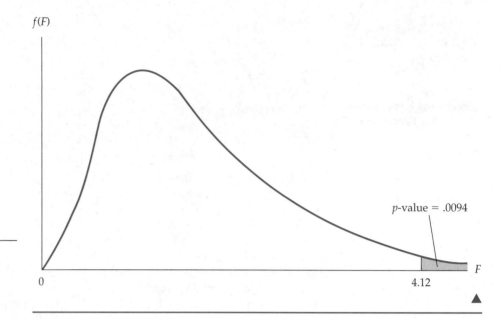

Figure 14.8

Sampling distribution of F-statistic (treatments) for Example 14.2

CHECKING THE REQUIRED CONDITIONS

The F-test of the randomized block experiment has the same requirements as the independent samples experiment. That is, the response variable is normally distributed, and the population variances are equal. The histograms (not shown) appear to be approximately bell shaped. The equality of variances requirement also appears to be met.

VIOLATION OF REQUIRED CONDITIONS

When the response variable is extremely nonnormal, we can replace the randomized block design of the analysis of variance with the Friedman test, which is introduced in Section 16.5.

CRITERIA FOR BLOCKING

In Section 12.4, we listed the advantages and disadvantages of performing a matched pairs experiment. The same comments are valid when we discuss performing a blocked experiment. The purpose of blocking is to reduce the variation caused by differences between the experimental units. By grouping the experimental units into homogeneous blocks with respect to the response variable, the statistics practitioner increases the chances of detecting actual differences between the treatment means. Hence, we need to find criteria for blocking that significantly affect the response variable. For example, suppose that a statistics practitioner wants to determine which of four methods of teaching statistics is best. In an independent samples design, he might take four samples of 10 students, teach each sample by a different method, grade the students at the end of the course, and perform an F-test to determine if differences exist. However, it is likely that there are very large differences between students within each class that may hide differences between classes. To reduce this variation, the statistics practitioner needs to identify variables that are linked to a student's grade in statistics. For example, overall ability of the student, completion of mathematics courses, and exposure to other statistics courses are all related to performance in a statistics course.

The experiment could be performed in the following way. The statistics practitioner selects four students at random whose average grade before statistics is 95 to 100. He then randomly assigns the students to one of the four classes. He repeats the process with students whose average is 90 to 95, 85 to 90, ..., and 50 to 55. The final grades would be used to test for differences between the classes.

Any characteristics that are related to the experimental units are potential blocking criteria. For example, if the experimental units are people, we may block according to age, gender, income, work experience, intelligence, residence (country, county, or city), weight, or height. If the experimental unit is a factory and we're measuring number of units produced hourly, blocking criteria include workforce experience, age of the plant, and quality of suppliers.

DEVELOPING AN UNDERSTANDING OF STATISTICAL CONCEPTS

As we explained above, the randomized block experiment is an extension of the matched pairs experiment discussed in Section 12.4. In the matched pairs experiment, we simply remove the effect of the variation caused by differences between the experimental units. The effect of this removal is seen in the decrease in the value of the standard error (compared to the standard error in the test statistic produced from independent samples) and the increase in the value of the t-statistic. In the randomized block experiment of the analysis of variance, we actually measure the variation between the blocks by computing SSB. The sum of squares for error is reduced by SSB, making it easier to detect differences between the treatments. Additionally, we can test to determine whether the blocks differ—a procedure we were unable to perform in the matched pairs experiment.

To illustrate, let's return to Examples 12.3 and 12.4, which were experiments to determine whether there was a difference between two tire designs. (In fact, we tested to determine whether the new-design tires outlast the current-design tires. However, the analysis of variance can only test for differences.) In Example 12.3 (independent samples), there was insufficient evidence to infer a difference between the two types of tires. In Example 12.4 (matched pairs experiment), there was enough evidence to infer a difference. As we pointed out in Section 12.4, matching cars allowed us to more easily discern a difference between the two types of tires. If we repeat Examples 12.3 and 12.4 using the analysis of variance, we come to the same conclusion. The Excel outputs are shown below.

Microsoft Excel Analysis of Variance Output for Example 12.3

ANOVA

Source of Variation	SS	df	MS	F	P-value	F crit
Between Groups	291.6	1	291.6	1.240	0.2724	4.0982
Within Groups	8934.0	38	235.1			
Total	9225.6	39				

Microsoft Excel Analysis of Variance Output for Example 12.4

ANOVA

Source of Variation	SS	df	MS	F	P-value	F crit
Rows	9624.5	19	506.6	20.01	0.0000	2.1682
Columns	275.6	1	275.6	10.89	0.0038	4.3808
Error	480.9	19	25.3			
Total	10381.0	39				

In Example 12.3, we partition the total sum of squares [SS(Total) = 9,225.6] into two sources of variation: SST = 291.6 and SSE = 8,934. In Example 12.4, the total sum of squares is SS(Total) = 10,381, SST (sum of squares for tires) = 275.6, SSB (sum of squares for cars) = 9,624.5, and SSE = 480.9. As you can see, the total sums of squares for both examples are about the same (9,225.6 and 10,381), and the sums of squares for treatments are also approximately equal (291.6 and 275.6). However, where the two calculations differ is in the sums of squares for error. SSE in Example 12.4 is much smaller than SSE in Example 12.3 because the randomized block experiment allows us to measure and remove the effect of the variation between cars. The sum of squares for blocks is 9,624.5, a statistic that measures how much variation exists between the cars. As a result of removing this variation, SSE is small. Thus, we conclude in Example 12.4 that the tires differ, whereas there was not enough evidence in Example 12.3 to draw the same conclusion.

We'll complete this section by listing the factors that we need to recognize to use this model of the analysis of variance.

> **Factors That Identify the Randomized Block Analysis of Variance**
>
> 1 Problem objective: compare two or more populations.
> 2 Data type: interval.
> 3 Experimental design: blocked samples.

EXERCISES

14.13 The following data were generated from a randomized block experiment.

  a Test to determine whether the treatment means differ.
  b Test to determine whether the block means differ.

		Treatment	
Block	1	2	3
1	7	12	8
2	10	8	9
3	12	16	13
4	9	13	6
5	12	10	11

14.14 A randomized block experiment produced the data below.

  a Can we infer that the treatment means differ?
  b Can we infer that the block means differ?

	Treatment			
Block	1	2	3	4
1	6	5	4	4
2	8	5	5	6
3	7	6	5	6

14.15 Data from a randomized block experiment (three treatments and 10 blocks) are stored in file Xr14-15 in columns B to D. Column A stores the block numbers. Can we conclude that the treatment means differ?

14.16 The data from a randomized block experiment with $k = 4$ and $b = 25$ are stored in file Xr14-16 in columns B to E. Column A contains the block codes.

  a Can we conclude that the treatment means differ?
  b Is there enough evidence to infer that the block means differ?
  c What are the required conditions for the procedure used in part (a)?
  d Are the conditions listed in part (c) satisfied?

14.17 Repeat Exercise 12.44. Do you draw the same conclusion?

14.18 Refer to Exercise 14.9. Despite failing to show that differences in the three types of fertilizer exist, the scientist continued to believe that there were differences and that the differences were masked by the variation between the plots of land. Accordingly, he conducted another experiment. In the second experiment he found 20 three-acre plots of land scattered across the county. He divided each into three plots and applied the three types of fertilizer on each of the one-acre plots. The crop yields were recorded and stored in columns B through D of file Xr14-18. Column A contains the plot numbers.

  a Can the scientist infer that there are differences between the three types of fertilizer?
  b What do these test results reveal about the variation between the plots?

14.19 In recent years, lack of confidence in the Postal Service has led many companies to send all of their correspondence by private courier. A large company is in the process of selecting one of three possible couriers to act as its sole delivery method. To help in making the decision, an experiment was performed whereby letters were sent using each of the three couriers at 12 different times of the day to a delivery point across town. The number of minutes required for delivery was recorded and stored in file Xr14-19 (columns B through D list the delivery times of couriers A, B, and C, and column A contains codes representing the time of day).

  a Can we conclude that there are differences in delivery times between the three couriers?
  b Did the statistics practitioner choose the correct design? Explain.

14.20 Exercise 14.6 described an experiment that involved comparing the completion times associated with four different income tax forms. Suppose the experiment is redone in the following way. Thirty people are asked to fill out all four forms. The completion times (in minutes) are recorded and stored in columns B through E of file Xr14-20. Column A stores the taxpayer number.

  a Is there sufficient evidence to infer that differences in the completion times exist between the four forms?
  b Comment on the suitability of this experimental design in this problem.

14.21 A recruiter for a computer company would like to determine whether there are differences in sales ability between business, arts, and science graduates. She takes a random sample of 20 business graduates who have been working for the company for the past two years. Each is then matched with an arts graduate and a science graduate with similar educational and working experience. The commission earned by each (in thousands of dollars) in the last year was recorded and stored in columns B, C, and D, respectively, in file Xr14-21 with column A containing the block number.

  a Is there sufficient evidence to allow the recruiter to conclude that there are differences in sales ability between the holders of the three types of degrees?
  b Conduct a test to determine whether an independent samples design would have been a better choice.
  c What are the required conditions for the test in part (a)?
  d Are the required conditions satisfied?

14.22 The advertising revenues commanded by a radio station depend on the number of listeners it has. The manager of a station that plays mostly hard rock music wants to learn more about its listeners—mostly teenagers and young adults. In particular, he wants to know if the amount of time they spend listening to radio music varies by the day of the week. If the manager discovers that the mean time per day is about the same, he will schedule the most popular music evenly throughout the week. Otherwise, the top hits will be played mostly on the days that attract the greatest audience. An opinion survey company is hired, and it randomly selects 200 teenagers and asks them to record the amount of time spent listening to music on the radio for each day of the previous week. The data are stored in file Xr14-22 (column A contains the teenagers' identification codes, and columns B through H store the listening times for Sunday through Saturday). What can the manager conclude from these data?

14.5 TWO-FACTOR ANALYSIS OF VARIANCE: INDEPENDENT SAMPLES

In Section 14.2, we addressed problems where the data were generated from single-factor experiments. In Example 14.1, the treatments were the three different advertising approaches. Thus, there were three levels of this factor. In this section, we address the problem where the experiment features two factors. The general term for such data-gathering procedures is **factorial experiments.** In factorial experiments, we can examine the effect on the response variable of two or more factors, although we address the problem of only two factors in this book. We can use the analysis of variance to determine whether the levels of each factor are different from one another.

We will present the technique for the fixed-effects experiments. That means we will address problems where all the levels of the factors are included in the experiment.

▼ EXAMPLE 14.3

Suppose that, in Example 14.1, in addition to varying the advertising approach, the manufacturer also decided to advertise in one of the two media that are available: television and newspapers. As a consequence, the experiment was repeated in the following way. Six similar cities were selected. In City 1, the advertising emphasized convenience, and all the advertising was conducted on television. In City 2, advertising also emphasized convenience, but all the advertising was conducted in the daily newspaper. Quality was emphasized in Cities 3 and 4. City 3 learned about the product from television commercials, and City 4 saw newspaper advertising. Price was the advertising emphasis in Cities 5 and 6. City 5 saw television commercials, and City 6 saw newspaper advertisements. In each city, the weekly sales for each of 10 weeks were recorded. These data are listed in the accompanying table and in file Xm14-03 (columns A to F store the 10 observations for each of the cities). What conclusions can be drawn from these results?

CITY-1	CITY-2	CITY-3	CITY-4	CITY-5	CITY-6
491	464	677	689	575	803
712	559	627	650	614	584
558	759	590	704	706	525
447	557	632	652	484	498
479	528	683	576	478	812
624	670	760	836	650	565
546	534	690	628	583	708
444	657	548	798	536	546
582	557	579	497	579	616
672	474	644	841	795	587

Solution

IDENTIFY

Notice that there are six treatments. However, the treatments are defined by two different factors. One factor is the advertising approach, which has three levels (convenience, quality, and price). The second factor is the advertising medium, which has two levels (television and newspaper). If we assume that there are only three advertising approaches and only two advertising media, we identify this experiment as a fixed-effects design. We can proceed to solve this problem in the same way we did in Section 14.2. That is, we test the following hypotheses.

$H_0 : \mu_1 = \mu_2 = \mu_3 = \mu_4 = \mu_5 = \mu_6$

H_1 : At least two means differ

COMPUTE

Microsoft Excel Output for Example 14.3

Anova: Single Factor

SUMMARY

Groups	Count	Sum	Average	Variance
City-1	10	5555	555.5	8641.4
City-2	10	5759	575.9	8545.9
City-3	10	6430	643.0	3884.7
City-4	10	6871	687.1	12558.5
City-5	10	6000	600.0	9527.6
City-6	10	6244	624.4	12523.8

ANOVA

Source of Variation	SS	df	MS	F	P-value	F crit
Between Groups	113620.3	5	22724.1	2.45	0.0452	2.39
Within Groups	501136.7	54	9280.3			
Total	614757.0	59				

436 Chapter 14 ANALYSIS OF VARIANCE

INTERPRET

The value of the test statistic is $F = 2.45$ with a p-value of .0452. We conclude that there is enough evidence to infer that at least two means differ. This statistical result raises more questions. Namely, can we conclude that the differences in weekly sales between the cities are caused by differences between the advertising approaches? Or are they caused by differences between television and newspaper advertising? Or, perhaps, are there combinations of advertising approach and advertising medium that result in especially high or low sales? To show how we test for each type of difference, we need to develop some terminology.

A **complete factorial experiment** is an experiment in which the data for all possible combinations of the levels of the factors are gathered. That means that in Example 14.3 we measured the sales for all six combinations. This experiment is called a complete 3 × 2 factorial experiment. Had we omitted gathering sales figures for (say) emphasizing price and advertising on television, we would not have a complete factorial experiment.

In general, we will refer to one of the factors as factor A (arbitrarily chosen). The number of levels of this factor will be denoted by a. The other factor is called factor B, and its number of levels is denoted by b. This terminology becomes clearer when we present the data from Example 14.3 in another format. Table 14.8 depicts the layout for a **two-way classification,** which is another name for the complete factorial experiment.

Table 14.8 **Two-Way Classification for Example 14.3**

Weekly Sales of Apple Juice Concentrate

Factor B:	Factor A: Advertising Approach		
Advertising Medium	Convenience	Quality	Price
Television	491	677	575
	712	627	614
	558	590	706
	447	632	484
	479	683	478
	624	760	650
	546	690	583
	444	548	536
	582	579	579
	672	644	795
Newspaper	464	689	803
	559	650	584
	759	704	525
	557	652	498
	528	576	812
	670	836	565
	534	628	708
	657	798	546
	557	497	616
	474	841	587

The number of observations for each combination is called a **replicate.** The number of replicates is denoted by r. In this book, we only address problems in which the number of replicates is the same for each treatment. Such a design is called **balanced.**

Thus, we use a complete factorial experiment where the number of treatments is ab with r replicates per treatment. In Example 14.3, $a = 3$, $b = 2$, and $r = 10$. As a result, we have 10 observations for each of the six treatments.

If there are differences between the treatment means, we would like to know if both factors affect the response. That is, are there differences between the levels of A and differences between the levels of B? If only one factor affects the response, is it A or is it B? If both A and B affect the response, do they do so independently, or do they interact, which means that some combinations of levels of factors A and B result in higher responses and some result in lower responses?

Figures 14.9 to 14.12 graphically depict the possible differences.

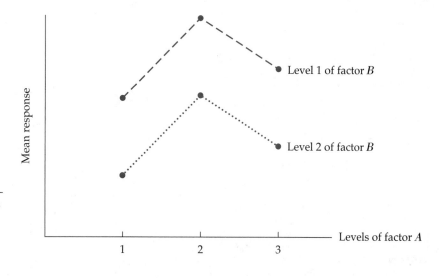

Figure 14.9

Differences between the levels of factor *A* and differences between the levels of factor *B*; no interaction

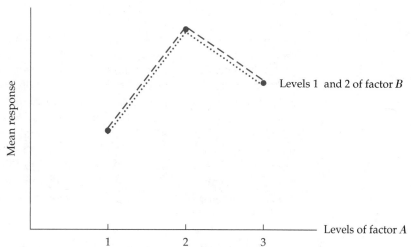

Figure 14.10

Differences between the levels of factor *A* and no difference between the levels of factor *B*; no interaction

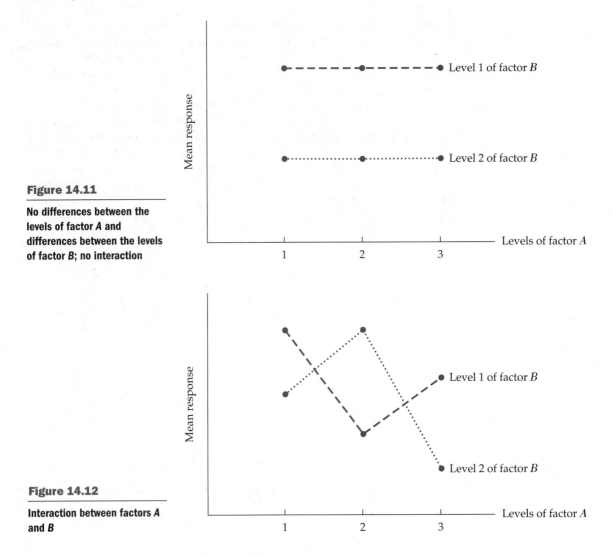

Figure 14.11

No differences between the levels of factor *A* and differences between the levels of factor *B*; no interaction

Figure 14.12

Interaction between factors *A* and *B*

Figure 14.9 graphs the mean weekly sales when there are differences between the levels of *A* as well as differences between the levels of *B*. However, the factors affect sales independently, which means there is no interaction. Figure 14.10 describes the case where there are differences between the levels of *A*, but no difference between the levels of *B*. Figure 14.11 depicts differences between the levels of *B* but no differences between the levels of *A*. Figure 14.12 shows the levels of *A* and *B* interacting.

To test for each possibility, we conduct several *F*-tests similar to the one performed in Section 14.2. Figure 14.13 illustrates the partitioning of the total sum of squares that leads to the *F*-tests. We've included in this figure the partitioning used in the single-factor study. This part of the analysis of variance has already been performed in Example 14.3. When the single-factor analysis of variance allows us to infer that differences between the treatment means exist, we continue our analysis by partitioning the treatment sum of squares into three sources of variation. The first is sum of squares for factor *A*, which we label SS(*A*), which measures the variation between the levels of factor A. Its degrees of freedom are $a - 1$. The second is the sum of squares for factor B, whose degrees of freedom are $b - 1$. SS(B) is the variation between the levels of factor B. The interaction sum of squares is labeled SS(AB), which is a measure of the amount of variation between the combinations of factors A and B. Its degrees of freedom are $(a - 1) \times (b - 1)$. The sum of squares for error is SSE, and its degrees of freedom are $n - ab$. (Recall that n is the total sample size, which in this experiment is $n = abr$.) Notice that SSE and its number of degrees of freedom are identical in both partitions. As in the previous model, SSE is the variation within the treatments.

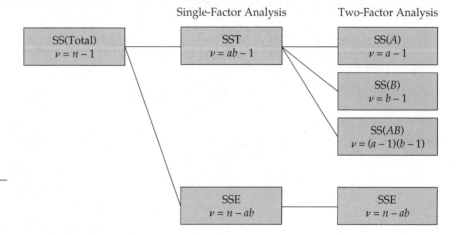

Figure 14.13

Partitioning SS(Total) in single-factor and two-factor analyses of variance

For those whose mathematical confidence is high, we have provided a listing of the notation and the definitions of the sums of squares below. (See Table 14.9 on page 441.) Learning how the sums of squares are calculated is useful but hardly essential to your ability to conduct the tests.

We then perform three F-tests to determine whether the differences between the treatment means are caused by differences between levels of A, levels of B, or interaction. We call this analysis the two-factor or two-way analysis of variance. These tests are summarized below.

To help you understand the formulas, we will use the following notation.

$\bar{x}[AB]_{ij}$ = mean of the responses in the ijth treatment (mean of the treatment when the factor A level is i and the factor B level is j)

$\bar{x}[A]_i$ = mean of the responses when the factor A level is i

$\bar{x}[B]_j$ = mean of the responses when the factor B level is j

$\bar{\bar{x}}$ = mean of all the responses

a = number of factor A levels

b = number of factor B levels

r = number of replicates

In this notation, $\bar{x}[AB]_{11}$ is the mean of the responses for factor A level 1 and factor B level 1. The mean of the responses for factor A level 1 is $\bar{x}[A]_1$. The mean of the responses for factor B level 1 is $\bar{x}[B]_1$. The sums of squares are defined as follows.

Sums of Squares in the Two-Factor Experiment

$$\text{SS(Total)} = \sum_{i=1}^{a} \sum_{j=1}^{b} \sum_{k=1}^{r} (x_{ijk} - \bar{\bar{x}})^2$$

$$\text{SS}(A) = rb \sum_{i=1}^{a} (\bar{x}[A]_i - \bar{\bar{x}})^2$$

$$\text{SS}(B) = ra \sum_{j=1}^{b} (\bar{x}[B]_j - \bar{\bar{x}})^2$$

$$\text{SS}(AB) = r \sum_{i=1}^{a} \sum_{j=1}^{b} (\bar{x}[AB]_{ij} - \bar{x}[A]_i - \bar{x}[B]_j + \bar{\bar{x}})^2$$

$$\text{SSE} = \sum_{i=1}^{a} \sum_{j=1}^{b} \sum_{k=1}^{r} (x_{ijk} - \bar{x}[AB]_{ij})^2$$

To compute SS(A), we calculate the sum of the squared differences between the factor A level means, which are denoted $\bar{x}[A]_i$, and the grand mean $\bar{\bar{x}}$. The sum of squares for factor B, SS(B), is defined similarly. The interaction sum of squares SS(AB) is calculated by taking each treatment mean (a treatment consists of a combination of a level of factor A and a level of factor B), subtracting the factor A level mean, subtracting the factor B level mean, adding the grand mean, squaring this quantity, and adding. The sum of squares for error SSE is calculated by subtracting the treatment means from the observations, squaring, and adding.

F-TESTS CONDUCTED IN TWO-FACTOR ANALYSIS OF VARIANCE

Test for Differences between the Levels of Factor A

H_0 : No difference between the means of the a levels of factor A

H_1 : At least two means differ

Test statistic: $F = \dfrac{\text{MS}(A)}{\text{MSE}}$

Table 14.9 Notation For Two-Factor Model

Factor B	Factor A				
	1	2	...	a	
1	$\begin{matrix} x_{111} \\ x_{112} \\ \cdot \\ \cdot \\ \cdot \\ x_{11r} \end{matrix}$ $\bar{x}[AB]_{11}$	$\begin{matrix} x_{211} \\ x_{212} \\ \cdot \\ \cdot \\ \cdot \\ x_{21r} \end{matrix}$ $\bar{x}[AB]_{21}$		$\begin{matrix} x_{a11} \\ x_{a12} \\ \cdot \\ \cdot \\ \cdot \\ x_{a1r} \end{matrix}$ $\bar{x}[AB]_{a1}$	$\bar{x}[B]_1$
2	$\begin{matrix} x_{121} \\ x_{122} \\ \cdot \\ \cdot \\ \cdot \\ x_{12r} \end{matrix}$ $\bar{x}[AB]_{12}$	$\begin{matrix} x_{221} \\ x_{222} \\ \cdot \\ \cdot \\ \cdot \\ x_{22r} \end{matrix}$ $\bar{x}[AB]_{22}$		$\begin{matrix} x_{a21} \\ x_{a22} \\ \cdot \\ \cdot \\ \cdot \\ x_{a2r} \end{matrix}$ $\bar{x}[AB]_{a2}$	$\bar{x}[B]_2$
$\vdots$					
b	$\begin{matrix} x_{1b1} \\ x_{1b2} \\ \cdot \\ \cdot \\ \cdot \\ x_{1br} \end{matrix}$ $\bar{x}[AB]_{1b}$	$\begin{matrix} x_{2b1} \\ x_{2b2} \\ \cdot \\ \cdot \\ \cdot \\ x_{2br} \end{matrix}$ $\bar{x}[AB]_{2b}$		$\begin{matrix} x_{ab1} \\ x_{ab2} \\ \cdot \\ \cdot \\ \cdot \\ x_{abr} \end{matrix}$ $\bar{x}[AB]_{ab}$	$\bar{x}[B]_b$
	$\bar{x}[A]_1$	$\bar{x}[A]_2$		$\bar{x}[A]_a$	$\bar{\bar{x}}$

Test for Differences between the Levels of Factor B

H_0 : No difference between the means of the b levels of factor B

H_1 : At least two means differ

Test statistic: $F = \dfrac{\text{MS}(B)}{\text{MSE}}$

Test for Interaction between Factors A and B

H_0 : Factors A and B do not interact to affect the mean responses

H_1 : Factors A and B do interact to affect the mean responses

Test statistic: $F = \dfrac{\text{MS}(AB)}{\text{MSE}}$

> **Required Conditions**
>
> 1 The distribution of the response is normally distributed.
> 2 The variance for each treatment is identical.
> 3 The samples are independent.

As in the two previous analysis of variance experimental designs, we summarize the results in an ANOVA table. Table 14.10 depicts the general form of the table for the complete factorial experiment.

Table 14.10 ANOVA Table for the Two-Factor Factorial Experiment with Fixed Effects and Independent Samples

Source of Variability	Degrees of Freedom	Sums of Squares	Mean Squares	F-Ratios
Factor A	$a - 1$	SS(A)	$MS(A) = \dfrac{SS(A)}{(a-1)}$	$F = \dfrac{MS(A)}{MSE}$
Factor B	$b - 1$	SS(B)	$MS(B) = \dfrac{SS(B)}{(b-1)}$	$F = \dfrac{MS(B)}{MSE}$
Interaction	$(a-1)(b-1)$	SS(AB)	$MS(AB) = \dfrac{SS(AB)}{(a-1)(b-1)}$	$F = \dfrac{MS(AB)}{MSE}$
Error	$n - ab$	SSE	$MSE = \dfrac{SSE}{(n-ab)}$	
Total	$n - 1$	SS(Total)		

We'll illustrate the techniques using the data in Example 14.3.

COMPUTE

Microsoft Excel Output for Example 14.3

Anova: Two-Factor with Replication

ANOVA

Source of Variation	SS	df	MS	F	P-value	F crit
Sample	13172.0	1	13172.0	1.419	0.2387	4.020
Columns	98838.6	2	49419.3	5.325	0.0077	3.168
Interaction	1609.6	2	804.8	0.087	0.9171	3.168
Within	501136.7	54	9280.3			
Total	614757.0	59				

The actual output includes a variety of statistics, which we have omitted. In the ANOVA table, **Sample** refers to factor B (medium), and **Columns** refers to factor A (advertising approach). Thus, MS(B) = 13,172.0, MS(A) = 49,419.3, MS(AB) = 804.8, and MSE = 9280.3. The F-statistics are 1.419 (medium), 5.325 (advertising approach), and .087 (interaction).

GENERAL COMMANDS	EXAMPLE 14.3
1 Type or import the data. (See file Xm14–03a for the correct format.	Open file **Xm14-03a**.
2 Click **Tools, Data Analysis...**, and **Anova: Two-Factor with Replication**.	
3 Specify the **Input Range:**.	A1:D21
4 Type the number of replications r (**Rows per sample:**).	10
5 Specify the value of α (**Alpha**), and click **OK**.	.05

INTERPRET

Test for Differences in Mean Weekly Sales between the Three Advertising Approaches

H_0: No difference between the means of the three levels of factor A

H_1: At least two means differ

Test statistic: $F = \dfrac{\text{MS}(A)}{\text{MSE}}$

Value of the test statistic: From the computer output above, we have
MS(A) = 49,419.3, MSE = 9,280.3, and F = 5.3
(p-value = .0077)

There is sufficient evidence to infer that differences in mean weekly sales exist between the different advertising approaches.

Test for Differences in Mean Weekly Sales between the Advertising Media

H_0: No difference between the means of the two levels of factor B

H_1: At least two means differ

Test statistic: $F = \dfrac{\text{MS}(B)}{\text{MSE}}$

Value of the test statistic: From the computer output above, we find
MS(B) = 13,172 and MSE = 9,280.3. Thus, F = 1.419
(p-value = .2387)

There is insufficient evidence to infer that differences in mean weekly sales exist between television and newspaper advertising.

Test for Interaction between Factors A and B

H_0: Factors A and B do not interact to affect mean weekly sales

H_1: Factors A and B do interact to affect mean weekly sales

Test statistic: $F = \dfrac{\text{MS}(AB)}{\text{MSE}}$

Value of the test statistic: From the printout above, MS(AB) = 804.8, MSE = 9,280.3, and $F = .087$ (p-value = .9171)

There is not enough evidence to conclude that there is an interaction between advertising approach and advertising medium that affects mean weekly sales.

Figure 14.14 graphs the mean sales for each factor. As you can see, there are differences between the levels of factor A, no difference between the levels of factor B, and no interaction is apparent. These results indicate that emphasizing quality produces the highest sales and that television and newspapers are equally effective.

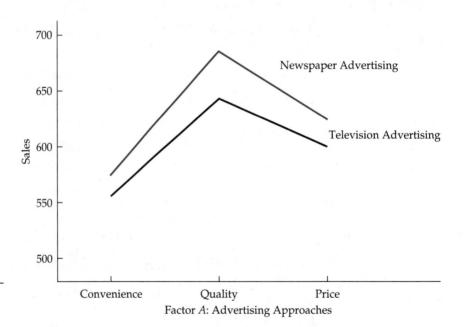

Figure 14.14

Mean responses for factors A and B in Example 14.3

CONDUCTING THE ANALYSIS OF VARIANCE FOR THE COMPLETE FACTORIAL EXPERIMENT

In addressing the problem outlined in Example 14.3, we began by conducting a single-factor analysis of variance to determine if differences existed between the six treatment means. This was done primarily for pedagogical reasons to enable you to see that when the treatment means differ we need to analyze the reasons for the differences. However, in practice, we generally do not conduct this test in the complete factorial experiment (although it should be noted that some statistics practitioners prefer this "two-stage" approach). We recommend that you proceed directly to the two-factor analysis of variance.

You should also note the order in which we conducted the three F-tests in the two-factor analysis of variance. The order of the tests that we performed above in this part of the analysis was based on the way that most software (including Excel) outputs the results of the tests. That is, the computer lists factors A and B as the first two sources of variation. Interaction is shown as the third source of variation. However, the test for interaction should be conducted first. To understand why, examine Figure 14.12, which depicts A and B interacting. When factors A and B interact, certain combinations of the levels of A and levels of B result in different mean responses. In most of these cases, the F-test of the differences between the levels of A and for the levels of B will yield statistically significant results. However, this inference may be wrong because the only differences between the levels of A and/or B may be due to the interaction. Thus, we first test to determine if there is interaction between factors A and B. If evidence to that effect exists, we do not perform the other two tests. If no interaction is evident, we test to determine if there are differences between the levels of A and if there are differences between the levels of B.

DEVELOPING AN UNDERSTANDING OF STATISTICAL CONCEPTS

You may have noticed that there are similarities between the independent samples two-factor experiment and the randomized block experiment (Section 14.4). In fact, when the number of replicates is one, the calculations are identical. This raises the question: What is the difference between a factor in a multifactor experiment and a block in a randomized block experiment? In general, the difference between the two experimental designs is that in the randomized block experiment, blocking is performed specifically to reduce variation, whereas in the two-factor experiment the effect of the factors on the response variable is of interest to the statistics practitioner. The criteria that define the blocks are always characteristics of the experimental units. Consequently, factors that are characteristics of the experimental units will be treated not as factors in a multifactor experiment but as blocks in a randomized block experiment.

Let's review how we recognize the need to use this experiment of the analysis of variance.

Factors That Identify the Independent Samples Two-Factor Analysis of Variance

1 Problem objective: compare two or more populations (populations are defined as combinations of levels of two factors)

2 Data type: interval

3 Experimental design: independent samples

EXERCISES

14.23 The following data were generated from a 2 × 2 factorial experiment with three replicates.

  a Test to determine if factors *A* and *B* interact.
  b Test to determine if differences exist between the levels of factor *A*.
  c Test to determine if differences exist between the levels of factor *B*.

	Factor B	
Factor A	1	2
1	6	12
	9	10
	7	11
2	9	15
	10	14
	5	10

14.24 The data shown below were taken from a 2 × 3 factorial experiment with four replicates.

  a Test to determine if factors *A* and *B* interact.
  b Test to determine if differences exist between the levels of factor *A*.
  c Test to determine if differences exist between the levels of factor *B*.

	Factor B	
Factor A	1	2
1	23	20
	18	17
	17	16
	20	19
2	27	29
	23	23
	21	27
	28	25
3	23	27
	21	19
	24	20
	16	22

14.25 Headaches are one of the most common, but least understood, ailments. Most people get headaches several times per month; over-the-counter medication is usually sufficient to eliminate their pain. However, for a significant proportion of people, headaches are debilitating and make their lives almost unbearable. Many such people have investigated a wide spectrum of possible treatments, including narcotic drugs, hypnosis, biofeedback, and acupuncture, with little or no success. In the last few years, a promising new treatment has been developed. Simply described, the treatment involves a series of injections of a local anesthetic to the occipital nerve (located in the back of the neck). The current treatment procedure is to schedule the injections once a week for four weeks. However, it has been suggested that another procedure may be better, one that features one injection every other day for a total of four injections. Additionally, some physicians recommend other combinations of drugs that may increase the effectiveness of the injections. To analyze the problem, an experiment was organized. It was decided to test for a difference between the two schedules of injection and to determine whether there are differences between four drug mixtures. Because of the possibility of an interaction between the schedule and the drug, a complete factorial experiment was chosen. Five headache patients were randomly selected for each combination of schedule and drug. Forty patients were treated, and each was asked to report the frequency, duration, and severity of his or her headache prior to treatment and for the 30 days following the last injection. An index ranging from 0 to 100 was constructed for each patient, where 0 indicates no headache pain and 100 specifies the worst headache pain. The improvement in the headache index for each patient was recorded and reproduced in the accompanying table. (A negative value indicates a worsening condition.) (The author is grateful to Dr. Lorne Greenspan for his help in writing this example.)

  a What are the factors in this experiment?
  b What is the response variable?
  c Identify the levels of each factor.
  d Can we conclude that differences exist between the two schedules?
  e Can we conclude that differences exist between the four drug mixtures?

f Before answering the questions in parts (d) and (e), what test must you conduct?
g What is the result of the test referred to in part (f)? What does this test tell you?

Improvement in Headache Index

Schedule	Drug Mixture			
	1	2	3	4
One Injection	17	24	14	10
Every Week	6	15	9	−1
(Four Weeks)	10	10	12	0
	12	16	0	3
	14	14	6	−1
One Injection	18	−2	20	−2
Every Two Days	9	0	16	7
(Four Days)	17	17	12	10
	21	2	17	6
	15	6	18	7

14.26 Most college instructors prefer to have their students participate actively in class. Ideally, students will ask their professor questions and answer their professor's questions, making the classroom experience more interesting and useful. Many professors seek ways to encourage their students to participate in class. A statistics professor at a community college in upper New York State believes that there are a number of external factors that affect student participation. He believes that the time of day and the configuration of seats are two such factors. Consequently, he organized the following experiment. Six classes of about 60 students each were scheduled for one semester. Two classes were scheduled at 9:00 AM, two at 1:00 PM, and two at 4:00 PM. At each of the three times, one of the classes was assigned to a room where the seats were arranged in rows of 10 seats. The other class was a U-shaped, tiered room, where students not only face the instructor but face their fellow students as well. In each of the six classrooms, over five days, student participation was measured by counting the number of times students asked and answered questions. These data are displayed in the accompanying table and stored in file Xr14-26 in exactly the same format as the table.

a How many factors are there in this experiment? What are they?
b What is the response variable?
c Identify the levels of each factor.
d What conclusions can the professor draw from these data?

Class Configuration	Time		
	9:00 AM	1:00 PM	4:00 PM
Rows	10	9	7
	7	12	12
	9	12	9
	6	14	20
	8	8	7
U-Shape	15	4	7
	18	4	4
	11	7	9
	13	4	8
	13	6	7

14.27 Detergent manufacturers frequently make claims about the effectiveness of their products. A consumer-protection service decided to test the five best-selling brands of detergent, each of whose manufacturers claims that its product produces the "whitest whites" in all water temperatures. The experiment was conducted in the following way. One hundred fifty white sheets were equally soiled. Thirty sheets were washed in each brand—10 with cold water, 10 with lukewarm water, and 10 with hot water. After washing, the "whiteness" scores for each sheet were measured with laser equipment. The results are stored in file Xr14-27 using the following format.

Column A: water temperature category

Column B: scores for detergent 1 (first 10 rows = cold water, middle 10 rows = lukewarm, and last 10 rows = hot)

Column C: scores for detergent 2 (same format as column B)

Column D: scores for detergent 3 (same format as column B)

Column E: scores for detergent 4 (same format as column B)

Column F: scores for detergent 5 (same format as column B)

a What are the factors in this experiment?
b What is the response variable?
c Identify the levels of each factor.

d Can we conclude that differences exist between the five detergents?

e Can we conclude that differences exist between the three water temperatures?

f Before answering the questions in parts (d) and (e), what test must you conduct?

g What is the result of the test referred to in part (f)? What does this test tell you?

14.28 Refer to Exercise 14.6. Suppose that the experiment is redone in the following way. Thirty taxpayers fill out each of the four forms. However, 10 taxpayers in each group are in the lowest income bracket, 10 are in the next income bracket, and the remaining 10 are in the highest bracket. The amount of time needed to complete the returns is recorded and stored in file Xr14-28 using the following format.

Column A: group number

Column B: times to complete form 1 (first 10 rows = low income, next 10 rows = next income bracket, and last 10 rows = highest bracket)

Column C: times to complete form 2 (same format as column B)

Column D: times to complete form 3 (same format as column B)

Column E: times to complete form 4 (same format as column B)

a How many treatments are there in this experiment?

b How many factors are there? What are they?

c What are the levels of each factor?

d Can we conclude that differences exist between the four forms?

e Can we conclude that taxpayers in different brackets require different amounts of time to complete their tax forms?

f What test must be conducted before answering the questions in parts (d) and (e)?

g What is the result of the test referred to in part (f)? What does this test tell you?

14.29 The headrests on a car's front seats are designed to protect the driver and front-seat passenger from whiplash when the car is hit from behind. The frame of the headrest is made from metal rods. A machine is used to bend the rod into a U-shape exactly 440 millimeters wide. The width is critical; too wide or too narrow, and it won't fit into the holes drilled into the car seat frame. The company has experimented with several different metal alloys in the hope of finding a material that will result in more headrest frames that fit. Another possible source of variation is the machines used. To learn more about the process the operations manager conducts an experiment. Both of the machines are used to produce 10 headrests from each of the five metal alloys now being used. Each frame is measured and the data (in millimeters) stored in file Xr14-29 using the following format.

Column B: Machine 1, rows 1 to 10 alloy A, rows 11 to 20, alloy B, etc.

Column C: Machine 2, rows 1 to 10 alloy A, rows 11 to 20, alloy B, etc.

Analyze the data to determine whether the alloys, machines, or both are sources of variation.

14.30 A paint manufacturer is attempting to improve the process that fills the one-gallon containers. The foreperson has suggested that the nozzle can be made from several different alloys. Furthermore, the way that the process "knows" when to stop the flow of paint can be accomplished in two ways—by setting a predetermined amount or by measuring the amount of paint already in the can. To determine what factors lead to variation, an experiment is conducted. For each of the four alloys that could be used to make the nozzles and the two measuring devices, five cans are filled. The amount of paint in each container is precisely measured. The data in liters are stored in file Xr14-30 in the following way.

Column B: Device 1, rows 1 to 5 alloy A, rows 6 to 10 alloy B, etc.

Column C: Device 2, rows 1 to 5 alloy A, rows 6 to 10 alloy B, etc.

Can we infer that the alloys, the measuring devices, or both are sources of variation?

14.31 In Example 12.2, the operations manager wanted to know whether the two methods used to assemble a new ergonomic chair were different with respect to the amount of time taken to assemble the chair. The manager was now faced with another problem. The marketing department of the firm has ascertained that there is a growing market for a specialized desk that houses the various parts of a computer system. The operations manager is summoned to put together a plan that will produce high-quality desks at low cost. The characteristics of the desk have been dictated by the marketing department, which in turn has specified the material that the desk will be made from and the machines used to produce the parts. However, there are three methods that can be utilized. Moreover, because of the complexity of the operation, the manager realizes that it is possible that different skill levels of the workers can yield different results. Accordingly, he organized an experiment. Workers from each of three skill levels were chosen. These groups were further divided into two subgroups. Each subgroup assembled the desks using methods A and B. The amount of time taken to assemble each of eight desks was recorded and stored in file Xr14-31. (Columns B and C contain the times for methods A and B; rows 1 to 8, 9 to16, and 17 to 24 store the times for the three skill levels.) What can we infer from these data?

14.6 MULTIPLE COMPARISONS

When the null hypothesis in the analysis of variance is rejected, it is often desirable to know which treatment means are responsible for the difference between population means. For example, if an experiment is undertaken to determine whether four local high schools prepare students differently for university, the university's admission officer would like to know which high schools produce better students and which produce poorer ones.

While it may appear that all we need to do is examine the sample means and identify the largest or the smallest to determine which population means are largest or smallest, this is not the case. To illustrate, suppose that in a five-treatment analysis of variance, we discover that differences exist. The sample means are calculated as follows.

$$\bar{x}_1 = 20 \quad \bar{x}_2 = 19 \quad \bar{x}_3 = 25 \quad \bar{x}_4 = 22 \quad \bar{x}_5 = 17$$

The statistics practitioner wants to know which of the following conclusions are valid.

1 μ_3 is larger than the other means.
2 μ_3 and μ_4 are larger than the other means.
3 μ_5 is smaller than the other means.
4 μ_5 and μ_2 are smaller than the other means.
5 μ_3 is larger than the other means, and μ_5 is smaller than the other means.

From the information we have, it is impossible to determine which, if any, of the statements are true. We need a statistical method to make this determination.

There are several statistical inference procedures that deal with this problem. We will present three methods that allow us to determine which population means differ. All three methods apply to the independent samples single-factor analysis of variance.

FISHER'S LEAST SIGNIFICANT DIFFERENCE (LSD) METHOD

Fisher's Least Significant Difference method was briefly introduced in Section 14.2 (page 418). To determine which population means differ, we could perform a series of t-tests of the difference between two means on all of pairs of population means to determine which are significantly different. In Chapter 12 we introduced the equal-variances *t*-test of the difference between two means. The test statistic and interval estimator are, respectively,

$$t = \frac{(\bar{x}_1 - \bar{x}_2) - (\mu_1 - \mu_2)}{\sqrt{s_p^2\left(\dfrac{1}{n_1} + \dfrac{1}{n_2}\right)}}$$

$$(\bar{x}_1 - \bar{x}_2) \pm t_{\alpha/2}\sqrt{s_p^2\left(\dfrac{1}{n_1} + \dfrac{1}{n_2}\right)}$$

with degrees of freedom $v = n_1 + n_2 - 2$.

Recall that s_p^2 is the pooled variance estimator, which is an unbiased estimator of the variance of the two populations. (Recall that in using these formulas we assume that the population variances are equal.) In this section we modify the test statistic and interval estimator.

Statisticians have proven that MSE is an unbiased estimator of the common variance of the populations we're testing. Because MSE is based on all the observations in the k samples, it will be a better estimator than s_p^2 (which is based on only two samples). Thus, we could draw inferences about every pair of means by substituting MSE in place of s_p^2 in the test statistic and interval estimator above. The number of degrees of freedom would also change to $n - k$ (where n is the total sample size). The test statistic to determine whether μ_i and μ_j differ is

$$t = \frac{(\bar{x}_i - \bar{x}_j) - (\mu_i - \mu_j)}{\sqrt{\text{MSE}\left(\dfrac{1}{n_i} + \dfrac{1}{n_j}\right)}}$$

The interval estimator is

$$(\bar{x}_i - \bar{x}_j) \pm t_{\alpha/2}\sqrt{\text{MSE}\left(\dfrac{1}{n_i} + \dfrac{1}{n_j}\right)}$$

and degrees of freedom are $v = n - k$.

A simple way of determining whether differences exist between each pair of population means is to compare the absolute value of the difference between two sample means and

$$t_{\alpha/2}\sqrt{\text{MSE}\left(\dfrac{1}{n_i} + \dfrac{1}{n_j}\right)}$$

If we define the least significant difference (LSD) as

$$\text{LSD} = t_{\alpha/2}\sqrt{\text{MSE}\left(\dfrac{1}{n_i} + \dfrac{1}{n_j}\right)}$$

we will conclude that μ_i and μ_j differ if

$$|\bar{x}_i - \bar{x}_j| > \text{LSD}$$

LSD will be the same for all pairs of means if all k sample sizes are equal. If some sample sizes differ, LSD must be calculated for each combination.

In Section 14.2 we argued that this method is flawed because it will increase the probability of committing a Type I error. That is, it is more likely to conclude that a difference exists in some of the population means when in fact none differ. On page 419 we calculated that if $k = 6$ and all population means are equal, the probability of erroneously inferring at the 5% significance level that at least two means differ is about 53.7%. The 5% figure is now referred to as the *comparisonwise Type I error rate*. The true probability of making at least one Type I error is called the *experimentwise Type I error rate*, denoted α_E. The experimentwise Type I error rate can be calculated as

$$\alpha_E = 1 - (1 - \alpha)^C$$

where C is the number of pairwise comparisons. That is, $C = k(k - 1)/2$. Statisticians can show that

$$\alpha_E \leq C\alpha$$

which means that if we want the probability of making at least one Type I error to be no more than α_E, we simply specify $\alpha = \alpha_E/C$. The resulting procedure is called the Bonferroni adjustment.

BONFERRONI ADJUSTMENT

The **Bonferroni adjustment** is made by dividing the specified experimentwise Type I error rate by the number of combinations of pairs of population means. For example, if $k = 6$, then

$$C = \frac{k(k - 1)}{2} = \frac{6(5)}{2} = 15$$

If we want the true probability of a Type I error to be no more than 5%, we divide this probability by C. Thus,

$$\alpha = \alpha_E/C = .05/15 = .0033$$

To illustrate Fisher's LSD method and the Bonferroni adjustment, consider Example 14.1, where we tested to determine whether three population means differ using a 5% significance level. The three sample means are 577.55, 653.0, and 608.65. The pairwise absolute differences are

$$|\bar{x}_1 - \bar{x}_2| = |577.55 - 653.0| = |-75.45| = 75.45$$
$$|\bar{x}_1 - \bar{x}_3| = |577.55 - 608.65| = |-31.10| = 31.10$$
$$|\bar{x}_2 - \bar{x}_3| = |653.0 - 608.65| = |44.35| = 44.35$$

If we conduct the LSD procedure with $\alpha = .05$ we find $t_{\alpha/2, n-k} = t_{.025, 57} = 2.002$. (This figure was determined from Excel.) Thus,

$$t_{\alpha/2}\sqrt{\text{MSE}\left(\frac{1}{n_i} + \frac{1}{n_j}\right)} = 2.002\sqrt{8894\left(\frac{1}{20} + \frac{1}{20}\right)} = 59.71$$

We can see that only one pair of sample means differ by more than 59.71. That is, $|\bar{x}_1 - \bar{x}_2| = 75.45$, and the other two differences are less than LSD. Consequently, we conclude that only μ_1 and μ_2 differ.

If we perform the LSD procedure with the Bonferroni adjustment, the number of pairwise comparisons is 3 [calculated as $C = k(k-1)/2 = 3(2)/2$]. We set $\alpha = .05/3 = .0167$. Thus, $t_{\alpha/2, n-k} = t_{.0083, 57} = 2.467$ and

$$t_{\alpha/2} \sqrt{\text{MSE}\left(\frac{1}{n_i} + \frac{1}{n_j}\right)} = 2.467 \sqrt{8894\left(\frac{1}{20} + \frac{1}{20}\right)} = 73.54$$

Again we conclude that only μ_1 and μ_2 differ. Notice, however, in the second calculation LSD is larger, reflecting the higher hurdle dictated by the smaller probability.

The drawback to LSD is that we increase the probability of at least one Type I error. The Bonferroni adjustment corrects this problem. However, recall that the probabilities of Type I and Type II errors are inversely related. The Bonferroni adjustment uses a smaller value of α, which results in an increased probability of a Type II error. A Type II error occurs when a difference between population means exists yet we cannot detect it. The next multiple comparison method addresses this problem.

TUKEY'S MULTIPLE COMPARISON METHOD

A more powerful test is **Tukey's multiple comparison method.** This technique determines a critical number such that, if any pair of sample means has a difference greater than this critical number, we conclude that the pair's two corresponding population means are different.

The test is based on the Studentized range, which is defined as the variable

$$q = \frac{\bar{x}_{max} - \bar{x}_{min}}{s/\sqrt{n}}$$

where $\bar{x}_{max}$ and $\bar{x}_{min}$ are the largest and smallest sample means, respectively, assuming that there are no differences between the population means. We can find a critical number ω (Greek letter *omega*) such that if the difference between any pair of sample means exceeds ω, we can take this as sufficient evidence that the pair's corresponding population means differ. We define the number ω as follows.

> **Critical Number ω**
>
> $$\omega = q_\alpha(k,v)\sqrt{\frac{MSE}{n_g}}$$
>
> where
>
> n = number of observations ($n = n_1 + n_2 + \cdots + n_k$)
>
> v = number of degrees of freedom associated with MSE = $n - k$
>
> n_g = number of observations in each of k samples
>
> α = significance level
>
> $q_\alpha(k,v)$ = critical value of the Studentized range

Tables that provide values of $q_\alpha(k,v)$ are available for those who need them. We don't, because we'll let Excel do the calculations.

Theoretically, this procedure requires that all sample sizes be equal. However, if the sample sizes are different, we can use still use this technique provided that the sample sizes are at least similar. The value of n_g used above is the *harmonic mean* of the sample sizes. That is,

$$n_g = \frac{k}{\frac{1}{n_1} + \frac{1}{n_2} + \cdots + \frac{1}{n_k}}$$

Applying Tukey's method to Example 14.1, we compute the following.

COMPUTE

Microsoft Excel Output for Tukey's and Fisher's LSD Method: Example 14.1 ($\alpha = .05$)

Multiple Comparisons

Treatment	Treatment	Difference	LSD Alpha = 0.05	Omega Alpha = 0.05
Convenience	Quality	−75.45	59.721	71.701
	Price	−31.1	59.721	71.701
Quality	Price	44.35	59.721	71.701

Microsoft Excel Output for Tukey and Fisher's LSD with the Bonferroni Adjustment: Example 14.1 ($\alpha = .05/3 = .0167$)

Multiple Comparisons

Treatment	Treatment	Difference	LSD Alpha = 0.0167	Omega Alpha = 0.05
Convenience	Quality	−75.45	73.542	71.701
	Price	−31.1	73.542	71.701
Quality	Price	44.35	73.542	71.701

The printout includes ω (Tukey's method), the differences between sample means for each combination of populations, and Fisher's LSD. (The Bonferroni adjustment is made by specifying another value for α.)

GENERAL COMMANDS	EXAMPLE 14.1
1 Type or import the data into adjacent columns.	Open file **Xm14-01**.
2 Click **Tools, Data Analysis Plus,** and **Multiple Comparisons.**	
3 Specify the **Input Range**.	A1:C21
4 Click **Labels,** if applicable.	
5 Type the value of α. To use the Bonferroni adjustment and divide α by $C = k(k - 1)/2$. For Tukey, Excel computes ω only for $\alpha = .05$. Click **OK**.	.05 (for Fisher's LSD) .0167 for Bonferroni

INTERPRET

Using any of the three multiple comparison methods we discover that only μ_1 and μ_2 differ, and no other pairs differ. This tells the marketing manager that advertising emphasizing quality (City 2) outsells advertising stressing convenience (City 1), and that there is no evidence to infer that advertising emphasizing price (City 3) is any different from the other two. It would appear that the company should launch an advertising campaign that features various ways of describing the high quality of the product. We may also advertise stressing the price of the product.

WHICH MULTIPLE COMPARISON METHOD TO USE

In Example 14.1, all three multiple comparison methods yielded the same results. This will not always be the case. When the results differ, the statistics practitioner must choose which one to use. Bear in mind as well that there are other techniques besides the ones described here. Unfortunately, no one procedure works best in all types of problems. Most statisticians agree with the following guideline.

If you have identified two or three pairwise comparisons that you wish to make before conducting the analysis of variance, use the Bonferroni method. This means that if in a problem there are 10 populations but you're particularly interested in comparing population (say) 3 with 7 and population 5 with 9, use Bonferroni with $C = 2$.

If you plan on comparing all possible combinations, use Tukey.

When do we use Fisher's LSD? If the purpose of the analysis is to point to areas that should be investigated further, Fisher's LSD method is indicated.

Incidentally, to employ Fisher's LSD or the Bonferroni adjustment, you must perform the analysis of variance first. Tukey's method can be employed *instead* of the analysis of variance.

EXERCISES

14.32 Apply Fisher's LSD method with the Bonferroni adjustment to determine which schools differ in Exercise 14.5. Use $\alpha = .05$.

14.33 Repeat Exercise 14.32, applying Tukey's method instead.

14.34 Apply Tukey's multiple comparison method to determine which forms differ in Exercise 14.6. (Use $\alpha = .05$.)

14.35 Repeat Exercise 14.34 applying the Bonferroni adjustment.

14.36 Use Tukey's multiple comparison method with $\alpha = .05$ to determine which lacquers differ in Exercise 14.7.

14.37 Repeat Exercise 14.36 using the Bonferroni adjustment with $\alpha = .10$.

14.38 Police cars, ambulances, and other emergency vehicles are required to carry road flares. One of the most important features of flares is their burning times. To help decide which of four brands on the market to use, a police laboratory technician measured the burning time for a random sample of 10 flares of each brand. The results, recorded to the nearest minute, are stored in columns A through D of file Xr14-38.

  a Can we conclude at the 5% significance level that differences exist between the burning times of the four brands of flares?
  b Apply Fisher's LSD method with the Bonferroni adjustment to determine which flares are better.
  c Repeat part (b) using Tukey's method.

14.39 An engineering student who is about to graduate decided to survey various firms in the Silicon Valley to see which offered the best chance for early promotion and career advancement. He surveyed 30 small firms (size level is based on gross revenues), 30 medium-sized firms, and 30 large firms and determined how much time must elapse before an average engineer can receive a promotion. These data are stored in columns A through C of file Xr14-39.

  a Can the engineering student conclude at the 5% significance level that speed of promotion varies between the three sizes of engineering firms?
  b If differences exist, which of the following is true? Use Tukey's method.
  i Small firms differ from the other two.
  ii Midsized firms differ from the other two.
  iii Large firms differ from the other two.
  iv All three firms differ from one another.
  v Small firms differ from large firms.

14.7 BARTLETT'S TEST

One of the required conditions for the F-test of the analysis of variance is that the population variances are equal. Informally, we can examine the sample variances and judge whether they are close enough to allow us to assume that the population variances are equal. Alternatively, we can conduct a test similar to the F-test of σ_1^2/σ_2^2 we introduced in Chapter 12 to determine which t-test of $\mu_1 - \mu_2$ to use.

There are several statistical tests available for multiple variances. We will introduce only one, **Bartlett's test,** which is applied to the independent samples single-factor experiment.

The hypotheses we test are

$$H_0: \sigma_1^2 = \sigma_2^2 = \cdots = \sigma_k^2$$

$H_1:$ At least one variance differs

The test statistic is

$$B = \frac{1}{C}\left[(n-k)\ln(\text{MSE}) - \sum_{i=1}^{k}(n_i - 1)\ln(s_i^2)\right]$$

where

$$C = 1 + \frac{1}{3(k-1)}\left[\left(\sum_{i=1}^{k}\frac{1}{n_i - 1}\right) - \frac{1}{n - k}\right]$$

(Note: ln represents the natural logarithm.)

When the null hypothesis is true, B will be small. We reject the null hypothesis only when B is large. The p-value of the test is determined from the sampling distribution, which is chi-squared with $k - 1$ degrees of freedom. The required condition is the same as the condition for the analysis of variance; the populations must be normal. We illustrate Bartlett's test using Example 14.1.

COMPUTE

Microsoft Excel Output for Bartlett's Test: Example 14.1

Bartlett's Test

SUMMARY

Groups	Variances
Convenience	10775
Quality	7238.1
Price	8670.2
B Stat	0.7389
df	2
p-value	0.6911
chi-squared Critical	5.9915

Excel prints the sample variances for each group as well as the test statistic and its p-value.

GENERAL COMMANDS	EXAMPLE 14.1
1 Type or import the data into adjacent columns.	Open file **Xm14-01**.
2 Click **Tools**, **Data Analysis Plus**, and **Bartlett's Test**.	
3 Specify the **Input Range**.	A1:C21
4 Click **Labels**, if applicable.	
5 Type the value of α (**Alpha**), and click **OK**.	.05

INTERPRET

There is not enough evidence to infer that the population variances differ. We've already shown that the histograms are approximately bell shaped. We're now confident that the required conditions of the F-test of the analysis of variance are satisfied.

EXERCISES

14.40 Refer to Exercise 14.3. Test to determine whether the equality of variance requirement is satisfied.

14.41 Can we infer that variances are unequal in Exercise 14.4?

14.42 Is there sufficient evidence in Exercise 14.5 to infer that the population variances are unequal?

14.43 Test to determine whether the equality of variance requirement is satisfied in Exercise 14.6.

14.44 Refer to Exercise 14.7. Test to determine whether the equality of variance requirement is satisfied.

14.45 Conduct a test to determine whether the equality of variance requirement is satisfied in Exercise 14.8.

14.46 Can we infer that the equality of variances requirement is violated in Exercise 14.9?

14.8 SUMMARY

The analysis of variance allows us to test for differences between the populations when the data are interval. Three different analysis of variance experimental designs were introduced in this chapter. The first design is the independent samples single-factor model wherein the treatments are defined as the levels of one factor and the samples are independent. The second design also defines the treatments on the basis of one factor. However, the randomized block design uses data gathered by observing the results of a matched or blocked experiment. The third design is the independent samples two-factor model wherein the treatments are defined as the combination of the levels of two factors. The analysis of variance of all experimental designs are based on partitioning the total sum of squares into sources of variation from which the mean squares and F-statistics are computed. We introduced the least significant difference method, the Bonferroni adjustment, and Tukey's multiple comparison method to determine which means differ in the independent samples single-factor model. Finally we presented Bartlett's test, which tests to determine whether the equal-variances requirement is satisfied.

IMPORTANT TERMS

Analysis of variance
Factor
Level
Response variable
Response
Experimental unit
Between-treatments variation
Sum of squares for treatments (SST)
Within-treatments variation
Sum of squares for error (SSE)
Mean squares
Mean square for treatments
Mean square for error
ANOVA table
Total sum of squares SS(Total)
Completely randomized design

Single-factor experiment
Multifactor experiment
Two-factor
Repeated measures
Fixed effects
Random effects
Factorial experiment
Complete factorial experiment
Two-way classification
Replicate
Balanced
Sum of squares for blocks (SSB)
Fisher's Least Significant Difference
Bonferroni adjustment
Tukey's multiple comparison method
Bartlett's test

FORMULAS

Independent Samples Single-Factor Experiment

$$\text{SST} = \sum_{j=1}^{k} n_j (\bar{x}_j - \bar{\bar{x}})^2$$

$$\text{SSE} = \sum_{j=1}^{k} \sum_{i=1}^{n_j} (x_{ij} - \bar{x}_j)^2$$

$$\text{MST} = \frac{\text{SST}}{k-1}$$

$$\text{MSE} = \frac{\text{SSE}}{n-k}$$

$$F = \frac{\text{MST}}{\text{MSE}}$$

Randomized Block Experiment

$$\text{SS(Total)} = \sum_{j=1}^{k} \sum_{i=1}^{b} (x_{ij} - \bar{\bar{x}})^2$$

$$\text{SST} = \sum_{j=1}^{k} b(\bar{x}[T]_j - \bar{\bar{x}})^2$$

$$\text{SSB} = \sum_{i=1}^{b} k(\bar{x}[B]_i - \bar{\bar{x}})^2$$

$$\text{SSE} = \sum_{j=1}^{k} \sum_{i=1}^{b} (x_{ij} - \bar{x}[T]_j - \bar{x}[B]_i + \bar{\bar{x}})^2$$

$$\text{MST} = \frac{\text{SST}}{k-1}$$

$$\text{MSB} = \frac{\text{SSB}}{b-1}$$

$$\text{MSE} = \frac{\text{SSE}}{n - k - b + 1}$$

$$F = \frac{\text{MST}}{\text{MSE}}$$

$$F = \frac{\text{MSB}}{\text{MSE}}$$

Two-Factor Experiment

$$\text{SS(Total)} = \sum_{i=1}^{a} \sum_{j=1}^{b} \sum_{k=1}^{r} (x_{ijk} - \bar{\bar{x}})^2$$

$$\text{SS(A)} = rb \sum_{i=1}^{a} (\bar{x}[A]_i - \bar{\bar{x}})^2$$

$$\text{SS(B)} = ra \sum_{j=1}^{b} (\bar{x}[B]_j - \bar{\bar{x}})^2$$

$$\text{SS(AB)} = r \sum_{i=1}^{a} \sum_{j=1}^{b} (\bar{x}[AB]_{ij} - \bar{x}[A]_i - \bar{x}[B]_j + \bar{\bar{x}})^2$$

$$\text{SSE} = \sum_{i=1}^{a} \sum_{j=1}^{b} \sum_{k=1}^{r} (x_{ijk} - \bar{x}[AB]_{ij})^2$$

$$F = \frac{\text{MS}(A)}{\text{MSE}}$$

$$F = \frac{\text{MS}(B)}{\text{MSE}}$$

$$F = \frac{\text{MS}(AB)}{\text{MSE}}$$

Least Significant Difference Comparison Method

$$\text{LSD} = t_{\alpha/2} \sqrt{\text{MSE}\left(\frac{1}{n_i} + \frac{1}{n_j}\right)}$$

Tukey's Multiple Comparison Method

$$\omega = q_\alpha(k,v) \sqrt{\frac{\text{MSE}}{n_g}}$$

Bartlett's Test

$$B = \frac{1}{C}\left[(n-k)\ln(\text{MSE}) - \sum_{i=1}^{k}(n_i - 1)\ln(s_i^2)\right]$$

$$C = 1 + \frac{1}{3(k-1)}\left[\left(\sum_{i=1}^{k}\frac{1}{n_i - 1}\right) - \frac{1}{n-k}\right]$$

MICROSOFT EXCEL OUTPUT AND INSTRUCTIONS

Technique	Page
Independent samples single-factor ANOVA	416
Randomized block ANOVA	429
Two-factor ANOVA	442–443
Multiple comparisons (LSD, Bonferroni adjustment, and Tukey)	451–452
Bartlett's test	456

SUPPLEMENTARY EXERCISES

14.47 The possible revision of the residential property tax has been a sensitive political issue in a large city that consists of five boroughs. Currently, property tax is based on an assessment system that dates back to 1950. This system has produced numerous inequities whereby newer homes tend to be assessed at higher values than older homes. A new system based on the market value of the house has been proposed. Opponents of the plan argue that residents of some boroughs would have to pay considerably more on the average, while residents of other boroughs would pay less. As part of a study examining this issue, several homes in each borough were assessed under both plans. The percentage increase (a decrease is represented by a negative increase) in each case was recorded and stored in columns A through E of file Xr14-47.

    a Can we conclude that there are differences in the effect the new assessment system would have on the five boroughs?

    b If differences exist, which boroughs differ? Use Tukey's multiple comparison method.

    c What are the required conditions for your conclusions to be valid?

    d Are the required conditions satisfied?

14.48 The editor of the student newspaper was in the process of making some major changes in the newspaper's layout. He was also contemplating changing the typeface of the print used. To help himself make a decision, he set up an experiment in which 20 individuals were asked to read four newspaper pages, with each page printed in a different typeface. If the reading speed differs, the typeface that is read fastest will be used. However, if there is not enough evidence to allow the editor to conclude that such differences exist, the current typeface will be continued. The times (in seconds) to completely read one page were stored in columns A to D of file Xr14-48. What should the editor do?

14.49 Each year thousands of workers are injured while performing their jobs. It is in the interests of both the worker and his or her employer that the worker return to work as quickly as possible. As part of an analysis of the amount of time taken for workers to return to work, a sample of male blue-collar workers aged 35 to 45 who suffered a common wrist fracture was taken. The researchers believed that the mental and physical condition of the individual affects recovery time. Each man was given a questionnaire to complete, which measured whether he tended to be optimistic or pessimistic. His physical condition was also evaluated and categorized as very physically fit, average, or in poor condition. The number of days until the wrist returned to full function was measured for each individual. These data are stored in file Xr14-49 in the following way:

    Column B: time to recover for optimists (rows 1 to 10 = very fit, rows 11 to 20 = in average condition, rows 21 to 30 = poor condition)

    Column C: time to recover for pessimists (same format as column B)

a What are the factors in this experiment? What are the levels of each factor?
b Can we conclude that pessimists and optimists differ in their recovery times?
c Can we conclude that physical condition affects recovery times?

14.50 In the past decade, American companies have spent nearly $1 trillion on computer systems. However, productivity gains have been quite small. During the 1980s, productivity in U.S. service industries (where most computers are used) grew by only 0.7% annually. In the 1990s, this figure rose to 1.5%. (Source: *New York Times Service,* 22 February 1995). The problem of small productivity increases may be caused by employee difficulty in learning how to use the computers. Suppose that in an experiment to examine the problem, 100 firms were studied. Each company had bought a new computer system five years ago. The companies reported their increase in productivity over the five-year period and were also classified as offering extensive employee training, some employee training, little employee training, or no formal employee training in the use of computers. (There were 25 firms in each group.) The results are stored in columns A through D of file Xr14-50.

a Can we conclude that differences in productivity gain exist between the four groups of companies?
b If there are differences, what are they?

14.51 An avid skier wants to find the resort with the shortest lift lines. Because she is a whiz at statistics, she decides to conduct an experiment. She collects data on the amount of time she spends waiting in line at the lift for each run she takes at three different resorts. These results are stored in columns A through C of file Xr14-51.

a Can she conclude that there are differences in waiting times between the three resorts?
b What are the required conditions for the techniques above?
c How would you check to determine that the required conditions are satisfied?

14.52 A popularly held belief about university professors is that they don't work very hard and that the higher their rank, the less work they do. A statistics student decided to determine whether the belief is true. She took a random sample of 20 university instructors in each of the faculties of business, engineering, arts, and sciences. In each sample of 20, five were instructors, five were assistant professors, five were associate professors, and five were full professors. Each professor was surveyed and asked to report confidentially the number of weekly hours of work. These data are stored in file Xr14-52 in the following way:

Column B: hours of work for business professors (first five rows = instructors, next five rows = assistant professors, next five rows = associate professors, and last five rows = full professors)

Column C: hours of work for engineering professors (same format as column B)

Column D: hours of work for arts professors (same format as column B)

Column E: hours of work for science professors (same format as column B)

a If we conduct the test under the single-factor analysis of variance, how many levels are there? What are they?
b Test to determine if differences exist using a single-factor analysis of variance.
c If we conduct tests using the two-factor analysis of variance, what are the factors? What are their levels?
d Are there differences between the four ranks of instructor?
e Are there differences between the four faculties?
f What test must be conducted before attempting to answer parts (d) and (e)?
g What is the result of the test described in part (f)? What does the test result tell you?

14.53 In marketing children's products, it's extremely important to produce television commercials that hold the attention of the children who view them. A psychologist hired by a marketing research firm wants to determine whether differences in attention span exist between children watching advertisements for different types of products. One hundred fifty children under 10 years of age were recruited for an experiment. One-third watched a 60-second commercial for a new computer game, one-third watched a commercial for a breakfast cereal, and another third watched a commercial for children's clothes. Their attention spans were measured. The results (in seconds) were stored in the first three columns of file Xr14-53. Do these data provide enough evidence to conclude that there are differences in attention span between the three products advertised?

14.54 Upon reconsidering the experiment in Exercise 14.53, the psychologist decides that the age of the child may influence the attention span. Consequently, the experiment is redone in the following way. Three 10-year-olds, three 9-year-olds, three 8-year-olds, three 7-year-olds, three 6-year-olds, three 5-year-olds, and three 4-year-olds are randomly assigned to watch one of the commercials, and their attention spans are measured. The data are stored in file Xr14-54. Columns B, C, and D store the times for the three advertisements. Do the results indicate that there are differences in the abilities of the products advertised to hold children's attention?

14.55 North American automobile manufacturers have become more concerned with quality because of foreign competition. One aspect of quality is the cost of repairing damage caused by accidents. A manufacturer is considering several new types of bumpers. In order to test how well they react to low-speed collisions, 40 bumpers of each of five different types were installed on midsize cars, which were then driven into a wall at 5 miles per hour. The cost of repairing the damage in each case was assessed, and the relevant data stored in columns A to E of file Xr14-55.

   a Is there sufficient evidence to infer that the bumpers differ in their reactions to low-speed collisions?
   b If differences exist, which bumpers differ?

14.56 It is important for salespeople to be knowledgeable about how people shop for certain products. Suppose that a new car salesman believes that the age and sex of a car shopper affects the way he or she makes an offer on a car. He records the initial offers made by a group of men and women shoppers on a $20,000 Mercury Sable. Besides the sex of the shopper, the salesman also notes the age category. The amount of money below the asking price that each person offered initially for the car was recorded and stored in file Xr14-56 using the following format. Column B contains the data for the under 30 group; the first 25 rows store the results for female shoppers and the last 25 rows are the male shoppers. Columns C and D store the data for the 30 to 45 age category and over 45 category, respectively. What can we conclude from these data?

14.57 A study was undertaken to investigate whether different training programs and software packages offered by a business college were more effective than others. The study recorded the number of words per minute typed by six groups of 40 students who completed the training programs. The training program and software packages assigned to each group are as described below.

   Group 1: Hands-on training/MS Word software
   Group 2: Computer tutorial/MS Word software
   Group 3: Hands-on training/WordPerfect software
   Group 4: Computer tutorial/WordPerfect software
   Group 5: Hands-on training/AmiPro software
   Group 6: Computer tutorial/AmiPro software

The typing speeds for each student were recorded in columns A through F for groups 1 to 6, respectively, in file Xr14-57. Can we conclude that the typing speeds differ between the six groups of students?

14.58 Refer to Exercise 14.57. Do the data allow us to infer that there are combinations of software packages and training programs that produce faster speeds than other combinations?

14.59 Many of you reading this page probably learned how to read using the whole-language method. This approach maintains that the natural and effective way is to be exposed to whole words in context. Students learn how to read by recognizing words they have seen before. In the past generation, this has been the dominant teaching approach throughout North America. It replaced phonics, wherein children are taught to sound out the letters to form words. The whole-language method was instituted with little or no research and has been severely criticized in the past. A recent study may have resolved the question of which method should be employed. Barbara Foorman, an educational psychologist at the University of Houston, described the experiment at the annual meeting of the American Association for the Advancement of Science. The subjects were 375 low-achieving, poor, first-grade students in Houston schools. The students were divided into three groups. One was educated according to the whole-language philosophy, a second group was taught using a pure phonics approach, and the third was taught employing a mixed or embedded phonics technique. At the end of the term students were asked to read words on a list of 50 words. The number of words each child could read are stored in columns A, B, and C (whole-language, embedded phonics, and pure phonics, respectively) in file Xr14-59.

   a Can we infer that differences exist between the effects of the three teaching approaches?
   b If differences exist, identify which method appears to be best.

14.60 Are babies exposed to music before their birth smarter than those who are not? And, if so, what kind of music is best? Researchers at the University of Wisconsin conducted an experiment with rats. The researchers selected a random sample of pregnant rats and divided the sample into three groups. Mozart's works were played to one group, a second group was exposed to white noise (a steady hum with no musical elements); and the third group listened to Philip Glass's music (very simple compositions). The researchers then trained the young rats to run a maze in search of food. The amount of time for the rats to complete the maze was measured for all three groups. These data are stored in file Xr14-60 (column A = time for "Mozart" rats, column B = "white noise" rats, and column C = time for "Glass" rats).

   a Can we infer from these data that there are differences between the three groups?

   b If there are differences, determine which group is best.

14.61 When the stock market has a large one-day decline, does it bounce back the next day or does the bad news endure? To answer this question, an economist examined a random sample of daily changes to the Toronto Stock Index (TSE). He recorded the percent change. He classified declines as

   Down by less than 0.5%

   Down by 0.5% to 1.5%

   Down by 1.5% to 2.5%

   Down by more than 2.5%

   For each of these days he recorded the percent loss the following day. These data are stored in columns A through D of file Xr14-61. Do these data allow us to infer that there are differences in changes to the TSE depending on the loss the previous day? (This exercise is based on a study undertaken by Tim Whitehead, an economist for Left Bank Economics, a consulting firm near Paris, Ontario.)

14.62 Increasing tuition has resulted in some students being saddled with large debts upon graduation. To examine this issue, a random sample of recent graduates was asked to report whether they had student loans, and if so, how much was the debt at graduation. Each person who reported that he or she owed money was also asked to whether his or her degree was a BA, BSc, BBA, or other. The amounts of debt were stored in columns A through D, respectively, in file Xr14-62. Can we conclude that debt levels differ among the four types of degree?

14.63 Studies indicate that single male investors tend to take the most risk, while married female investors tend to be conservative. This raises the question: Who does best? The risk-adjusted returns for single and married men and for single and married women were recorded and stored in columns A through D, respectively, in file Xr14-63. Can we infer that differences exist among the four groups of investors?

14.64 Like all fine restaurants Ye Olde Steak House in Windsor, Ontario, attempts to have three "seatings" on weekend nights. Three seatings means that each table gets three different customers. Obviously any group that lingers over dessert and coffee may result in the loss of one seating and profit for the restaurant. In an effort to determine which types of groups tend to linger, a random sample of 150 groups was drawn. For each group, the number of members and the length of time that the group stayed were recorded. These data are stored in file Xr14-64 in the following way:

   Columns A: length of time for two people

   Columns B: length of time for three people

   Columns C: length of time for four people

   Columns D: length of time for more than four people

   Do these data allow us to infer that the length of time in the restaurant depends on the size of the party?

Chapter 15

Chi-Squared Tests

15.1 Introduction

15.2 Chi-Squared Goodness-of-Fit Test

15.3 Chi-Squared Test of a Contingency Table

15.4 Summary of Tests on Nominal Data

15.5 Chi-Squared Test for Normality

15.6 Summary

15.1 INTRODUCTION

This chapter introduces two more statistical techniques that involve nominal data. The first is a goodness-of-fit test applied to data produced by a multinomial experiment, a generalization of a binomial experiment. The second uses data arranged in a table (called a contingency table) to determine whether or not two classifications of a population of nominal data are statistically independent; this test can also be interpreted as a comparison of two or more populations. The sampling distribution of the test statistics in both tests is the chi-squared distribution introduced in Chapter 6.

Following are two illustrations of situations in which chi-squared tests could be applied.

Illustration 1 Firms periodically estimate the proportion (or market share) of consumers who prefer their products, as well as the market shares of competitors. These market shares may change over time as a result of advertising campaigns or the introduction of new improved products. To determine whether the actual current market shares are in accord with its beliefs, a firm might sample several consumers and compute, for each of k competing companies, the proportion of consumers sampled who prefer that company's product. Such an experiment, in which each consumer is classified as preferring one of the k companies, is called a multinomial experiment. If only two companies were considered ($k = 2$), we would be dealing with the familiar binomial experiment. After computing the proportion of consumers preferring each of the k companies, a goodness-of-fit test could be conducted to determine whether the sample proportions (or market shares) differ significantly from those hypothesized by the firm. The problem objective is to describe the population of consumers, and the data are nominal.

Illustration 2 During the presidential primaries, candidates in each party strive to win delegates for the nominating convention. Although the rules vary from state to state, it is possible for registered members of one party to "cross over" and vote in another party's primary. It is important for candidates to determine where their support is coming from. Suppose that in one state the people voting in the primary are polled and asked to report for whom they voted and whether they are registered members of the Democratic or Republican parties or are independent. A test could then be conducted to determine whether the two variables are related. Rather than interpreting this test as a test of the relationship between two nominal variables defined over a single population, we could view registered Democrats, Republicans, and independents as representing three different populations. Then we could interpret the test as testing for differences in preferences among the three populations.

15.2 CHI-SQUARED GOODNESS-OF-FIT TEST

This section presents another test designed to describe a single population of nominal data. The first such test was introduced in Section 11.4, where we discussed the statistical procedure employed to test hypotheses about a population proportion. In that case, the nominal variable could assume one of only two possible values, success or failure. Our tests dealt with hypotheses about the proportion of successes in the entire population. Recall that the experiment that produces the data is called a binomial experiment. We now introduce the **multinomial experiment,** which is an extension of the binomial experiment, wherein there are two or more possible outcomes per trial.

> **Multinomial Experiment**
>
> A multinomial experiment is one possessing the following properties:
>
> 1. The experiment consists of a fixed number n of trials.
> 2. The outcome of each trial can be classified into one of k categories, called cells.
> 3. The probability p_i that the outcome will fall into cell i remains constant for each trial. Moreover, $p_1 + p_2 + \cdots + p_k = 1$.
> 4. Each trial of the experiment is independent of the other trials.

As you can see, when $k = 2$, the multinomial experiment is identical to the binomial experiment. Just as we count the number of successes (recall that we label the number of successes x) and failures in a binomial experiment, we count the number of outcomes falling into each of the k cells in a multinomial experiment. In this way, we obtain a set of observed frequencies $f_1, f_2, \ldots, f_k$ where f_i is the observed frequency of outcomes falling into cell i, for $i = 1, 2, \ldots, k$. Because the experiment consists of n trials and an outcome must fall into some cell,

$$f_1 + f_2 + \cdots + f_k = n$$

Just as we used the number of successes x (by calculating the sample proportion $\hat{p}$, which is equal to x/n) to draw inferences about p, so do we use the observed frequencies to draw inferences about the cell probabilities. We'll proceed in what by now has become a standard procedure. We will set up the hypotheses and develop the test statistic and its sampling distribution. We'll demonstrate the process with the following example.

▼ EXAMPLE 15.1

Two companies, A and B, have recently conducted aggressive advertising campaigns in order to maintain and possibly increase their respective shares of the market for fabric softener. These two companies enjoy a dominant position in the market. Before the advertising campaigns began, the market share of company A was 45%, while company B had 40% of the market. Other competitors accounted for the remaining 15%. To determine whether these market shares changed after the advertising campaigns, a marketing analyst solicited the preferences of a random sample of 200 customers of fabric softener. Of the 200 customers, 102 indicated a preference for company A's product, 82 preferred company B's fabric softener, and the remaining 16 preferred the products of one of the competitors. Can the analyst infer that customer preferences have changed from the levels they were at before the advertising campaigns were launched?

Solution

IDENTIFY

The population in question is composed of the brand preferences of the fabric softener customers. The data are nominal because each respondent will choose one of three possible answers—product A, product B, or other. If there were only two categories, or if we were only interested in the proportion of one company's customers (which we would label as successes and label the others as failures), we would identify the technique as the z-test of p. However, in this problem we're interested in the proportions of all three categories. We recognize this experiment as a multinomial experiment, and we identify the technique as the chi-squared goodness-of-fit test.

Because we want to know if the market shares have changed, we specify those precampaign market shares in the null hypothesis.

$$H_0: p_1 = .45, \ p_2 = .40, \ p_3 = .15$$

The alternative hypothesis attempts to answer our question, "Have the proportions changed?" Thus,

H_1 : At least one p_i is not equal to its specified value

TEST STATISTIC

If the null hypothesis is true, we would expect the number of customers selecting Brand A, Brand B, and other to be 200 times the proportions specified under the null hypothesis. That is,

$$e_1 = 200(.45) = 90$$
$$e_2 = 200(.40) = 80$$
$$e_3 = 200(.15) = 30$$

In general, the **expected frequency** for each cell is given by

$$e_i = np_i$$

This expression is derived from the formula for the expected value of a binomial random variable, first seen in Section 5.4.

If the expected frequencies e_i and the observed frequencies f_i are quite different, we would conclude that the null hypothesis is false, and we would reject it. However, if the expected and observed frequencies are similar, we would not reject the null hypothesis. The test statistic we employ to measure the similarity of the expected and observed frequencies is described below.

Test Statistic for the Goodness-of-Fit Test

$$\chi^2 = \sum_{i=1}^{k} \frac{(f_i - e_i)^2}{e_i}$$

The sampling distribution of the test statistic is approximately chi-squared with degrees of freedom $v = k - 1$, provided that the sample size is large.

We will discuss the sample size requirement later in this section. Recall that the chi-squared distribution was introduced in Section 6.4

The table below demonstrates the calculation of the test statistic. Thus, the value of the test statistic is $\chi^2 = 8.18$. As usual we judge the size of this test statistic by determining the *p*-value (or by specifying the rejection region).

Company	Observed Frequency f_i	Expected Frequency e_i	$(f_i - e_i)$	$\dfrac{(f_i - e_i)^2}{e_i}$
A	102	90	12	1.60
B	82	80	2	0.05
Other	16	30	−14	6.53
Total	200	200		$\chi^2 = 8.18$

When the null hypothesis is true, the observed and expected frequencies should be similar, in which case the test statistic will be small. Thus, a small test statistic supports the null hypothesis. If the null hypothesis is untrue, some of the observed and expected frequencies will differ and the test statistic will be large. Consequently, we want to reject the null hypothesis when χ^2 is large. Throughout this book we have judged the size of the test statistic by determining the *p*-value of the test, which is

$$p\text{-value} = P(\chi^2 > 8.18) = .0167$$

(See the Excel printout below.)

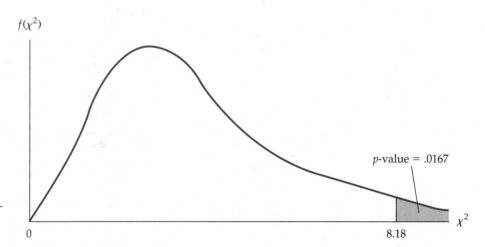

Figure 15.1

Sampling distribution of the test statistic for Example 15.1

COMPUTE

If we have the raw data representing the nominal responses we must first determine the frequency of each category (the observed values). The printouts below were generated from the observed and expected values. You can also perform what-if analyses to determine for yourself the effect of changing some of the observed values and the sample size.

Microsoft Excel Output for Example 15.1

Observed	Expected
102	90
82	80
16	30

0.0167

The output from the commands listed below is the *p*-value of the test. It is .0167.

GENERAL COMMANDS		EXAMPLE 15.1	
1 Type the observed values into one column and the expected values into another column. (If you wish, you can type the cell probabilities specified in the null hypothesis and let Excel convert these into expected values by multiplying by the sample size.)		**Observed** 102 82 16	**Expected** 90 80 30
2 Activate some empty cell, and click f_x, **Statistical,** and **CHITEST.**		Cell A6	
3 Specify the range of the observed values **(Actual_range).** Do not include the cell containing the name of the variable.		A2:A4	
4 Specify the range of the expected values **(Expected_range).** Do not include the cell containing the name of the variable.		B2:B4	
5 Click **FINISH.**			

INTERPRET

There is evidence to infer that the proportions have changed since the advertising campaigns. If the sampling was conducted properly, we can be quite confident in our conclusion. This technique has only one required condition, which is satisfied. (See the rule of five, described below.) It is probably a worthwhile exercise to determine the nature and causes of the changes. The results of this analysis will determine the design and timing of other advertising campaigns.

Figure 15.1 describes the sampling distribution of the test statistic.

RULE OF FIVE

The test statistic used to compare the relative sizes of observed and expected frequencies is

$$\chi^2 = \sum_{i=1}^{k} \frac{(f_i - e_i)^2}{e_i}$$

We previously stated that this test statistic has an approximate chi-squared distribution. In fact, the actual distribution of this test statistic is discrete, but it can be approximated conveniently by using a chi-squared distribution, which is continuous when the sample size n is large, just as we approximated the discrete binomial distribution by using the normal distribution. This approximation may be poor, however, if the expected cell frequencies are small. For the (discrete) distribution of the test statistic to be adequately approximated by the (continuous) chi-squared distribution, the conventional (and conservative) rule—known as the rule of five—is to require that the expected frequency for each cell be at least 5. Where necessary, cells should be combined in order to satisfy this condition. The choice of cells to be combined should be made in such a way that meaningful categories result from the combination.

Consider the following modification of Example 15.1. Suppose that three companies (A, B, and C) have recently conducted aggressive advertising campaigns; the market shares prior to the campaigns were $p_1 = .45$ for company A, $p_2 = .40$ for company B, $p_3 = .13$ for company C, and $p_4 = .02$ for other competitors. In a test to see if market shares changed after the advertising campaigns, the null hypothesis would now be

$$H_0: p_1 = .45, \ p_2 = .40, \ p_3 = .13, \ p_4 = .02$$

Hence, if the preferences of a sample of 200 customers were solicited, the expected frequencies would be

$e_1 = 200(.45) = 90$

$e_2 = 200(.40) = 80$

$e_3 = 200(.13) = 26$

$e_4 = 200(.02) = 4$

Because the expected cell frequency e_4 is less than 5, the rule of five requires that it be combined with one of the other expected frequencies (say, e_3) to obtain a combined cell frequency of (in this case) 30. Although e_4 could have been combined with e_1 or e_2, we have chosen to combine it with e_3, so that we still have a separate category representing each of the two dominant companies (A and B). After this combination is made, the null hypothesis reads

$$H_0: p_1 = .45, \ p_2 = .40, \ p_3 = .15$$

where p_3 now represents the market share of all competitors of companies A and B. Therefore, the appropriate number of degrees of freedom for the chi-squared test statistic would be $k - 1 = 3 - 1 = 2$, where k is the number of cells after some have been combined to satisfy the rule of five.

Let's summarize the factors that allow us to recognize when to use the **chi-squared goodness-of-fit test** for a multinomial experiment.

Factors That Identify the Chi-Squared Goodness-of-Fit Test

1. Problem objective: describe a single population.
2. Data type: nominal.
3. Number of categories: 2 or more.

EXERCISES

15.1 Consider a multinomial experiment involving $n = 300$ trials and $k = 5$ cells. The observed frequencies resulting from the experiment are shown below, and the hypotheses to be tested are as follows.

$H_0: p_1 = .1,\ p_2 = .2,\ p_3 = .3,\ p_4 = .2,\ p_5 = .2$

H_1: At least one p_i is not equal to its specified value.

Test the hypotheses.

Cell	1	2	3	4	5
Frequency	24	64	84	72	56

15.2 Repeat Exercise 15.1 with the following frequencies.

Cell	1	2	3	4	5
Frequency	12	32	42	36	28

15.3 Repeat Exercise 15.1 with the following frequencies.

Cell	1	2	3	4	5
Frequency	6	16	21	18	14

15.4 Review the results of Exercises 15.1 to 15.3. What is the effect of decreasing the sample size?

15.5 Consider a multinomial experiment involving $n = 150$ trials and $k = 4$ cells. The observed frequencies resulting from the experiment are shown below, and the hypotheses to be tested are as follows.

$H_0: p_1 = .3,\ p_2 = .3,\ p_3 = .2,\ p_4 = .2$

H_1: At least one p_i is not equal to its specified value.

Test the hypotheses.

Cell	1	2	3	4
Frequency	38	50	38	24

15.6 For Exercise 15.5, retest the hypotheses, assuming that the experiment involved twice as many trials ($n = 300$) and that the observed frequencies were twice as high as before, as shown below.

Cell	1	2	3	4
Frequency	76	100	76	48

15.7 The results of a multinomial experiment with $k = 5$ are stored in file Xr15-07. Each outcome is identified by the numbers 1 through 5. Test to determine if there is enough evidence to infer that the proportion of each outcome is the same.

15.8 A multinomial experiment was conducted with $k = 4$. Each outcome is stored as an integer from 1 to 4 and the results of a survey are stored in file Xr15-08. Test the following hypotheses.

$H_0: p_1 = .15,\ p_2 = .40,\ p_3 = .35,\ p_4 = .10$

H_1: At least one p_i is not equal to its specified value

15.9 To determine whether a single die is balanced, or fair, the die was rolled 600 times. The outcomes are stored in file Xr15-09. Is there sufficient evidence to allow you to conclude that the die is not fair?

15.10 Grades assigned by an economics instructor have historically followed a symmetrical distribution: 5% A's, 25% B's, 40% C's, 25% D's, and 5% F's. This year, a sample of 150 grades was drawn. The grades (1 = A, 2 = B, 3 = C, 4 = D, and 5 = F) are stored in file Xr15-10. Can you conclude that this year's grades are distributed differently from grades in the past?

15.11 Pat Statsdud is about to write a multiple-choice exam but, as usual, knows absolutely nothing. Pat plans to guess one of the five choices. Pat has been given one of the professor's previous exams with the correct answers marked. The correct choices are stored in file Xr15-11 where 1 = (a), 2 = (b), 3 = (c), 4 = (d), and 5 = (e). Help Pat determine whether this professor does not randomly distribute the correct answer over the five choices. If this is true, how does it affect Pat's strategy?

15.12 Financial managers are interested in the speed with which customers who make purchases on credit pay their bills. In addition to calculating the average number of days that unpaid bills (called accounts receivable) remain outstanding, they often prepare an aging schedule. An aging schedule classifies outstanding accounts receivable according to the time that has elapsed since billing and records the proportion of accounts receivable belonging to each classification. A large firm has determined its aging schedule for the past five years. These results are shown in the accompanying table. During the past few months, however, the economy has taken a downturn. The company would like to know if the recession has affected the aging schedule. A random sample of 250 accounts receivable was drawn and each account was classified (and stored in file Xr15-12) as follows.

1 = 0 to 14 days outstanding

2 = 15 to 29 days outstanding

3 = 30 to 59 days outstanding

4 = 60 or more days outstanding

Number of Days Outstanding	Proportion of Accounts Receivable Past Five Years
0 to 14	.72
15 to 29	.15
30 to 59	.10
60 and over	.03

Determine whether the aging schedule has changed.

15.13 License records in a county reveal that 15% of cars are subcompacts (1), 25% are compacts (2), 40% are midsize (3), and the rest are an assortment of other styles and models (4). A random sample of accidents involving cars licensed in the county was drawn. The type of car was stored in file Xr15-13 using the codes in parentheses. Can we infer that certain sizes of cars are involved in a higher than expected percentage of accidents?

15.14 In an election held last year that was contested by three parties, party A captured 27% of the vote, party B garnered 51%, and party C received the remaining votes. A survey of 1,200 voters asked each to identify the party that he or she would vote for in the next election. These results are stored in file Xr15-14 where 1 = party A, 2 = party B, and 3 = party C. Can we infer that voter support has changed since the election?

15.15 In a number of pharmaceutical studies volunteers who take placebos (but are told they have taken a cold remedy) report the following side effects.

Headache (1)	5%
Drowsiness (2)	7%
Stomach upset (3)	4%
No side effect (4)	84%

A random sample of 250 people who were given a placebo (but who thought they had taken an anti-inflammatory) reported whether they had experienced each of the side effects. These data are stored in file Xr15-15 using the codes in parentheses. Can we infer that the reported side effects of the placebo for an anti-inflammatory differ from those of a cold remedy?

15.3 CHI-SQUARED TEST OF A CONTINGENCY TABLE

We introduce another chi-squared test, this one designed to satisfy two different problem objectives. The **chi-squared test of a contingency table** is used to determine if there is enough evidence to infer that two nominal variables are related and to infer that differences exist among two or more populations of a nominal variable. Completing both objectives entails classifying items according to two different criteria. To see how this is done, consider the following example.

▼ **EXAMPLE 15.2**

Current projections indicate that, with the current strong economy, it is likely that the federal government will be generating budget surpluses for the next 10 years or more. How to spend this money is an issue that will be debated for many years. The major options appear to be

1 Reduce taxes.

2 Pay down the debt.

3 Improve Social Security benefits.

4 Institute a universal Canadian-style health care plan.

Like most issues, politicians need to know which parts of the electorate support these options. Suppose that a random sample of 1,000 people was asked which option they support and their political affiliations. The possible responses to the question about political affiliation were Democrat, Republican, and independent (which included a variety of political persuasions). The responses were summarized in a table called a **contingency table** or **cross-classification table,** shown below. Do these results allow us to conclude that political affiliation and support for the budget surplus options are related?

	Political Affiliation		
Budget Surplus Options	Democrat	Republican	Independent
Reduce taxes	118	193	45
Pay down the debt	83	132	41
Improve Social Security benefits	109	88	31
Institute universal health care	101	31	28

Solution One way to solve the problem is to consider that there are two variables represented by the contingency table. The variables are the political affiliation and the budget surplus option. Both are nominal. The values of political affiliation are Democrat, Republican, and independent. The values of budget surplus option are reduce taxes, pay down the debt, improve Social Security, and institute universal health care. The problem objective is to analyze the relationship between the two variables.

Another way of addressing the problem is to determine whether differences exist among Democrats, Republicans, and independents. In other words, we treat the members of each political group as a separate population. Each population has four possible values, namely, the budget surplus options. (We can also answer the question by treating the budget surplus options as populations and the political affiliations as the values of the random variable.) Here the problem objective is to compare three populations.

As you will shortly discover, both objectives lead to the same test. Consequently, we address both objectives at the same time.

The null hypothesis will specify that there is no relationship between the two variables. We state this in the following way:

H_0 : The two variables are independent

The alternative hypothesis specifies one variable affects the other, expressed as

H_1 : The two variables are dependent

If the null hypothesis is true, political affiliation and budget surplus option are independent of one another. This means that whether a voter is a Democrat, Republican, or an independent does not affect his or her choice of what to do with the surplus. Consequently, there is no difference in budget surplus options among Democrats, Republicans, and independents. If the alternative hypothesis is true, political affiliation is related to the budget surplus option. Thus, there are differences among the three political groups.

TEST STATISTIC

The test statistic is the same as the one employed to test proportions in the goodness-of-fit test. That is, the test statistic is

$$\chi^2 = \sum_{i=1}^{k} \frac{(f_i - e_i)^2}{e_i}$$

where k is the number of cells in the contingency table. If you examine the null hypothesis described in the goodness-of-fit test and the one described above, you will discover a major difference. In the goodness-of-fit test, the null hypothesis lists values for the probabilities p_i. The null hypothesis for the chi-squared test of a contingency table states only that the two variables are independent. However, we need the probabilities to compute the expected values e_i, which in turn are needed to calculate the value of the test statistic. (The entries in the table are the observed values f_i.) The question immediately arises: From where do we get the probabilities? The answer is that they must come from the data after we assume that the null hypothesis is true.

If we consider each political affiliation as a separate population, each column represents a multinomial experiment with four cells. If the null hypothesis is true, the three multinomial populations should have similar proportions in each cell. We can estimate the cell probabilities by calculating the total in each row and dividing by the sample size. Thus,

$$P(\text{Reduce taxes}) = \frac{(118 + 193 + 45)}{1000} = \frac{356}{1000}$$

$$P(\text{Pay down the debt}) = \frac{(83 + 132 + 41)}{1000} = \frac{256}{1000}$$

$$P(\text{Improve Social Security benefits}) = \frac{(109 + 88 + 31)}{1000} = \frac{228}{1000}$$

$$P(\text{Institute universal health care}) = \frac{(101 + 31 + 28)}{1000} = \frac{160}{1000}$$

We can calculate the expected values for each cell in the three multinomial experiments by multiplying these probabilities by the total number of members of each political group. By adding down each column, we find there were 411 Democrats, 444 Republicans, and 145 independents.

Expected Values of the Democrats

Budget Surplus Option	Expected Value
Reduce taxes	$411 \times \frac{356}{1000} = 146.32$
Pay down the debt	$411 \times \frac{256}{1000} = 105.22$
Improve Social Security benefits	$411 \times \frac{228}{1000} = 93.71$
Institute universal health care	$411 \times \frac{160}{1000} = 65.76$

Expected Values of the Republicans

Budget Surplus Option	Expected Value
Reduce taxes	$444 \times \dfrac{356}{1000} = 158.06$
Pay down the debt	$444 \times \dfrac{256}{1000} = 113.66$
Improve Social Security benefits	$444 \times \dfrac{228}{1000} = 101.23$
Institute universal health care	$444 \times \dfrac{160}{1000} = 71.04$

Expected Values of the Independents

Budget Surplus Option	Expected Value
Reduce taxes	$145 \times \dfrac{356}{1000} = 51.62$
Pay down the debt	$145 \times \dfrac{256}{1000} = 37.12$
Improve Social Security benefits	$145 \times \dfrac{228}{1000} = 33.06$
Institute universal health care	$145 \times \dfrac{160}{1000} = 23.20$

Notice that the expected values are computed by multiplying the column total by the row total and dividing by the sample size.

Expected Frequencies for a Contingency Table

The expected frequency of the cell in column j and row i is

$$e_{ij} = \frac{\text{Column } j \text{ total} \times \text{Row } i \text{ total}}{\text{Sample size}}$$

The expected cell frequencies are shown in parentheses in the table below. As in the case of the goodness-of-fit test, the expected cell frequencies should satisfy the rule of five.

Budget Surplus Option	Democrat	Political Affiliation Republican	Independent
Reduce taxes	118 (146.32)	193 (158.06)	45 (51.62)
Pay down debt	83 (105.22)	132 (113.66)	41 (37.12)
Improve Social Security benefits	109 (93.71)	88 (101.23)	31 (33.06)
Institute universal health care	101 (65.76)	31 (71.04)	28 (23.20)

We can now calculate the value of the test statistic.

$$\chi^2 = \sum_{i=1}^{k} \frac{(f_i - e_i)^2}{e_i}$$

$$= \frac{(118 - 146.32)^2}{146.32} + \frac{(83 - 105.22)^2}{105.22} + \frac{(109 - 93.71)^2}{93.71} + \frac{(101 - 65.76)^2}{65.76}$$

$$+ \frac{(193 - 158.06)^2}{158.06} + \frac{(132 - 113.66)^2}{113.66} + \frac{(88 - 101.23)^2}{101.23} + \frac{(31 - 71.04)^2}{71.04}$$

$$+ \frac{(45 - 51.62)^2}{51.62} + \frac{(41 - 37.12)^2}{37.12} + \frac{(31 - 33.06)^2}{33.06} + \frac{(28 - 23.20)^2}{23.20}$$

$$= 68.91$$

The number of degrees of freedom for a contingency table with r rows and c columns is

$$\nu = (r - 1)(c - 1)$$

For this example the number of degrees of freedom is

$$\nu = (r - 1)(c - 1) = (4 - 1)(3 - 1) = 6$$

Because we will reject the null hypothesis if the test statistic is large, the p-value of the test is

$$P(\chi^2 > 68.91)$$

which will be calculated by Excel.

COMPUTE

Excel can produce the chi-squared statistic and its p-value from either a contingency table whose frequencies have already been tallied or from raw data. If you or someone else created the contingency table, the output and instructions are as follows.

Microsoft Excel Output for Example 15.2

Contingency Table

	Dem	Rep	Ind	TOTAL
Reduce	118	193	45	356
Pay	83	132	41	256
Social	109	88	31	228
Health	101	31	28	160
TOTAL	411	444	145	1000

chi-squared Stat   68.9037
df   6
p-value   0
chi-squared Critical 12.5916

| GENERAL COMMANDS (COMPLETED TABLE) | EXAMPLE 15.2 |

1. Type the frequencies into adjacent columns.
2. Click **Tools, Data Analysis Plus,** and **Contingency Table.**

	Dem	Rep	Ind
Reduce	118	193	45
Pay	83	132	41
Social	109	88	31
Health	101	31	28

3. Specify the **Input Range:**. A1:D5
4. Click **Labels,** if the first row and first column of the input range contain the names of the categories.
5. Specify the value of α (**Alpha**), and click **OK**. .05

We also created file Xm15-02, which contains the raw data using the following format.

COLUMN A

Budget Surplus Options

1 = Reduce taxes
2 = Pay down debt
3 = Improve Social Security benefits
4 = Institute universal health care plan

COLUMN B

Political Affiliation

1 = Democrat
2 = Republican
3 = Independent

Microsoft Excel Output for Example 15.2 (Raw Data)

Contingency Table

Affiliation	Option 1	2	3	TOTAL
1	118	193	45	356
2	83	132	41	256
3	109	88	31	228
4	101	31	28	160
TOTAL	411	444	145	1000

chi-squared Stat	68.9037
df	6
p-value	0
chi-squared Critical	12.5916

GENERAL COMMANDS (RAW DATA)	EXAMPLE 15.2
1 Type or import the data into adjacent columns where one column represents the codes for one variable and the second column stores the codes for the second variable. The codes must be positive integers.	Open file **Xm15-02**.
2 Click **Tools, Data Analysis Plus,** and **Contingency Table (Raw Data)**.	
3 Specify the **Input Range:**.	A1:B1001
4 Click **Labels**, if applicable.	
5 Specify α **(Alpha)**, and click **OK**.	.05

INTERPRET

Both outputs are virtually identical. The value of the test statistic is $\chi^2 = 68.9037$. The *p*-value is 0. There is strong evidence to infer that the political affiliation and budget surplus option are related. This information can be used in a variety of ways. Candidates may tailor their policies to keep their supporters happy, or advertising campaigns can be launched stressing the candidate's views on the options.

Figure 15.2 describes the sampling distribution of the test statistic.

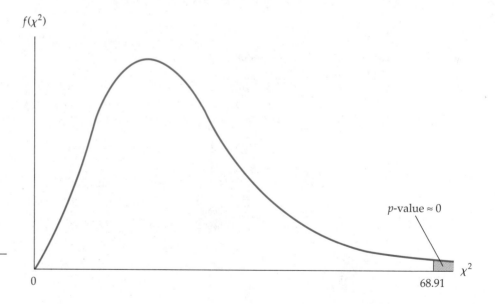

Figure 15.2

Sampling distribution for Example 15.2

RULE OF FIVE

In the previous section, we pointed out that the expected values should be at least five to ensure that the chi-squared distribution provides an adequate approximation of the sampling distribution. In a contingency table where one or more cells have expected values of less than 5, we need to combine rows or columns to satisfy the rule of five. To illustrate, suppose that we want to test for dependence in the following contingency table.

					TOTAL
	10	14	4	|	28
	12	16	7	|	35
	8	8	4	|	20
TOTAL	30	38	15	|	83

The expected values are as follows.

10.1	12.8	5.1
12.7	16.0	6.3
7.2	9.2	3.6

The expected value of the cell in row 3 and column 3 is less than 5. To eliminate the problem, we can add column 3 to one of columns 1 and 2 or add row 3 to either row 1 or row 2. The combining of rows or columns should be done so that the combination forms a logical unit, if possible. For example, if the columns represent the age groups, young (under 40), middle-aged (40 to 65), and senior (over 65), it is logical to combine columns 2 and 3. The observed and expected values are combined to produce the following table (expected values in parentheses).

				TOTAL
	10 (10.1)	18 (17.9)	|	28
	12 (12.7)	23 (22.3)	|	35
	8 (7.2)	12 (12.8)	|	20
TOTAL	30	53	|	83

The degrees of freedom must be changed as well. The number of degrees of freedom of the original contingency table is $(3 - 1) \times (3 - 1) = 4$. The number of degrees of freedom of the combined table is $(3 - 1) \times (2 - 1) = 2$. The rest of the procedure is unchanged.

Here is a summary of the factors that tell us when to apply the chi-squared test of a contingency table. Note that there are two problem objectives satisfied by this statistical procedure.

Factors That Identify the Chi-Squared Test of a Contingency Table

1. Problem objectives: analyze the relationship between two variables; compare two or more populations.
2. Data type: nominal.

EXERCISES

15.16 Conduct a test to determine whether the two classifications L and M are independent, using the data in the accompanying contingency table.

	M_1	M_2
L_1	28	68
L_2	56	36

15.17 Repeat Exercise 15.16 using the following table.

	M_1	M_2
L_1	14	34
L_2	28	18

15.18 Repeat Exercise 15.16 using the following table.

	M_1	M_2
L_1	7	17
L_2	14	9

15.19 Review the results of Exercises 15.16 to 15.18. What is the effect of decreasing the sample size?

15.20 Conduct a test to determine whether the two classifications R and C are independent, using the data in the accompanying contingency table.

	C_1	C_2	C_3
R_1	40	32	48
R_2	30	48	52

15.21 The trustee of a company's pension plan has solicited the opinions of a sample of the company's employees about a proposed revision of the plan. A breakdown of the responses is shown in the accompanying table. Is there evidence to infer that the responses differ among the three groups of employees?

Responses	Blue-Collar Workers	White-Collar Workers	Managers
For	67	32	11
Against	63	18	9

15.22 The operations manager of a company that manufactures shirts wants to determine whether there are differences in the quality of workmanship among the three daily shifts. She randomly selects 600 recently made shirts and carefully inspects them. Each shirt is classified as either perfect or flawed, and the shift that produced it is also recorded. The accompanying table summarizes the number of shirts that fell into each cell. Do these data provide sufficient evidence to infer that there are differences in quality among the three shifts?

	Shift		
Shirt Condition	1	2	3
Perfect	240	191	139
Flawed	10	9	11

15.23 The MBA program was experiencing problems scheduling their courses. The demand for the program's optional courses and majors was quite variable from one year to the next. In one year students seemed to want marketing courses, and in other years accounting or finance were the rage. In desperation, the dean of the business school turned to a statistics professor for assistance. The statistics professor believed that the problem may be the variability in the academic background of the students and that the undergraduate degree affects the choice of major. As a start he took a random sample of last year's MBA students and recorded the undergraduate degree and the major selected in the graduate program. The undergraduate degrees included BA (1), BEng (2), BBA (3), as well as several others (4). There are three possible majors for the MBA students: accounting (1), finance (2), and marketing (3). The results were stored in columns A and B file Xr15-23 using the codes in parentheses. Can the statistics professor conclude that the undergraduate degree affects the choice of major?

15.24 To determine if commercials viewed during happy television programs are more effective than those viewed during sad television programs, a study was conducted in which a random sample of students viewed an upbeat segment from "Real People" with commercials, while another random sample of students viewed a very sad segment from "Sixty Minutes" with commercials. The students were then asked what they were thinking during the final commercial. From their responses, they were categorized as thinking primarily about the commercial (1), thinking primarily about the program (2), or thinking about both (3). The results were stored in file Xr15-24. (Column A lists the program, 1 = "Real People" and 2 = "Sixty Minutes," and column B lists the responses.) Do commercials viewed during happy television programs appear to have a different effect than those viewed during sad television programs? (Source: Marvin E. Goldberg and Gerald J. Corn, "Happy and Sad TV Programs: How They Affect Reactions to Commercials," *Journal of Consumer Research* 14 (1987): 387–403.)

15.25 Acute otitis media, an infection of the middle ear, is a very common childhood illness. Although it is normally treated with amoxicillin, emerging resistance to the antibiotic has promoted the search for an alternative. A recent article discussed the efficacy of one such alternative: trimethoprim-sulfamethoxazole. In this study, 203 "patients were randomly assigned to receive either amoxicillin (1) or trimethoprim-sulfamethoxazole (2) by means of a computer-generated table of random numbers." Each patient was judged to be cured (1), improved (2), or to have no improvement (3). The data are stored in columns A (drug) and B (outcome) of file Xr15-25. Can we conclude from these data that there are differences in outcomes for children treated with amoxicillin and for children treated with trimethoprim-sulfamethoxazole? (Source: William Feldman, Joanne Momy, and Corinne Dulberg, "Trimethoprim-Sulfamethoxazole v. Amoxicillin in the Treatment of Acute Otitis Media," *Canadian Medical Association Journal* 139 (1988): 961–64.)

15.26 An antismoking group recently had a large advertisement published in local newspapers throughout Florida. Several statistical facts and medical details were included, in the hope that the ad would have meaningful impact on smokers. The antismoking group is concerned, however, that smokers might have read less of the advertisement than did nonsmokers. This concern is based on the belief that a reader tends to spend more time reading articles that agree with his or her predisposition. The antismoking group has conducted a survey asking those who saw the advertisement if they read the headline only (1), some detail (2), or most of the advertisement (3). The questionnaire also asks respondents to identify themselves as either a heavy smoker—more than two packs per day (1); a moderate smoker—between one and two packs per day (2); a light smoker—less than one pack per day (3); or a nonsmoker (4). The results are stored in file Xr15-26 (column A: type of smoker; column B: survey responses). Do the data indicate that level of smoking affects how much one reads of an antismoking advertisement?

15.27 An investor who can correctly forecast the direction and size of changes in foreign currency exchange rates is able to reap huge profits in the international currency markets. A knowledgeable reader of the *Wall Street Journal* (in particular, of the currency futures market quotations) can determine the direction of change in various exchange rates that is predicted by all investors, viewed collectively. Predictions from 216 investors, together with the subsequent actual directions of change, are stored in file Xr15-27 (column A: predicted change where 1 = positive and 2 = negative; column B: actual change where 1 = positive and 2 = negative).

a Test the hypothesis that a relationship exists between the predicted and actual directions of change.

b To what extent would you make use of these predictions in formulating your forecasts of future exchange rate changes?

15.28 During the past decade, many cigarette smokers have attempted to quit. Unfortunately, nicotine is highly addictive. There is a large number of different methods that smokers employ to help them quit. These include nicotine patches, hypnosis, and various forms of therapy. A researcher for the Addiction Research Council wanted to determine why some people quit while others attempted to quit but failed. He surveyed 1,000 people who planned to quit smoking. He determined their educational level and whether one year later they continued to smoke. Educational level was recorded in the following way.

1 = did not finish high school
2 = high school graduate
3 = university or college graduate
4 = completed a postgraduate degree

A continuing smoker was recorded as 1; a quitter was recorded as 2. These data are stored in columns A (education level) and B (continuing smoker?) in file Xr15-28. Can we infer that the amount of education is a factor in determining whether a smoker will quit?

15.29 In Exercises 12.102 and 12.103 we applied statistical analysis to help segment the market for Toyota cars on the basis of ages and incomes of customers, respectively. Suppose that, as part of the survey in those exercises, the questionnaire also asked about marital status. Recall that the study involved 100 customers of Tercels and 100 customers of Avalons. The responses to the question about marital status were recorded as

1 Single
2 Married
3 Divorced
4 Widowed

The results are stored in file Xr15-29, where column A contains the responses of all buyers and column B identifies the type of car (1 = Tercel and 2 = Avalon). Can we infer that differences in marital status exist between the buyers of the two Toyotas?

15.30 After a thorough analysis of the market a publisher of business and economics statistics books has divided the market into three general approaches to the teaching of applied statistics. These are (1) use of a computer and statistical software with no manual calculations; (2) traditional teaching of concepts and solution of problems by hand; (3) mathematical approach with emphasis on derivations and proofs. The publisher wanted to know if this market could be segmented on the basis of the educational background of the instructor. As a result the statistics editor organized a survey that asked 195 professors of business and economics statistics to report their approach to teaching and which one of the following categories represents their highest degree.

1 Business (MBA or PhD in business)
2 Economics
3 Mathematics or engineering
4 Other

The responses are stored in columns A (teaching approach) and B (degree) in file Xr15-30. Can the editor infer that there are differences in type of degree among the three teaching approaches? If so, how can the editor use this information?

15.4 SUMMARY OF TESTS ON NOMINAL DATA

At this point in the textbook, we've described four tests that are used when the data are nominal. These are as follows.

1 z-test of p (Section 11.4)
2 z-test of $p_1 - p_2$ (Section 12.6)
3 Chi-squared goodness-of-fit test (Section 15.2)
4 Chi-squared test of a contingency table (Section 15.3)

In the process of presenting these techniques, it was necessary to concentrate on one technique at a time and focus on the kinds of problems each addresses. However, this approach tends to conflict somewhat with our promised goal of emphasizing the "when" of statistical inference. In this section, we summarize the statistical tests on nominal data to help clarify how to select the correct method.

There are two critical factors in identifying the technique used when the data are nominal. The first, of course, is the problem objective. The second is the number of categories that the nominal variable can assume. Table 15.1 provides a guide to help select the correct technique.

Table 15.1 Statistical Techniques for Nominal Data

Problem Objective	Number of Categories	Statistical Technique
Describe a single population	2	z-test of p or the chi-squared goodness-of-fit test
Describe a single population	more than 2	Chi-squared goodness-of-fit test
Compare two populations	2	z-test of $p_1 - p_2$ or the chi-squared test of a contingency table
Compare two populations	more than 2	Chi-squared test of a contingency table
Compare two or more populations	2 or more	Chi-squared test of a contingency table
Analyze the relationship between two variables	2 or more	Chi-squared test of a contingency table

Notice that when we describe a single population of nominal data with exactly two categories, we can use either of two techniques. We can employ the z-test of p or the chi-squared goodness-of-fit test. These two tests are equivalent because, if there are only two categories, the multinomial experiment is actually a binomial experiment (one of the categorical outcomes is labeled success and the other is labeled failure). Statisticians have established that if we square the value of z, the test statistic for the test of p, we produce the χ^2-statistic. That is, $z^2 = \chi^2$. Thus, if we want to conduct a two-tail test of a population proportion, we can employ either technique. However, the chi-squared test of a binomial experiment can only test to determine if the hypothesized values of p_1 (which we can label p) and p_2 (which we call $1 - p$) are not equal to their specified values. Consequently, to perform a one-tail test of a population proportion, we must use the z-test of p. (This issue was discussed in Chapter 14 when we pointed out that we can use either the t-test of $\mu_1 - \mu_2$ or the analysis of variance to conduct a test to determine if two population means differ.)

When we test for differences between two populations of nominal data with two categories, we can also use either of two techniques: the z-test of $p_1 - p_2$ (Case 1) or the chi-squared test of a contingency table. Once again, we can use either technique to perform a two-tail test about $p_1 - p_2$. (Squaring the value of the z-statistic yields the value of χ^2-statistic.) However, one-tail tests must be conducted by the z-test of $p_1 - p_2$. The rest of the table is quite straightforward. Notice that when we want to compare two populations when there are more than two categories, we use the chi-squared test of a contingency table.

Figure 15.3 offers another summary of the tests that deal with nominal data introduced in this book. There are two groups of tests: ones that test hypotheses about single populations and ones that test either for differences or for independence. In the first set, we have the z-test of p, which can be replaced by the chi-squared goodness-of-fit test. The latter test is employed when there are more than two categories.

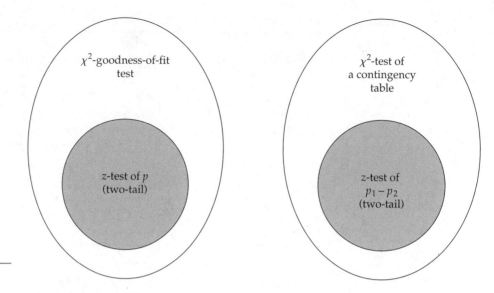

Figure 15.3

Tests on qualitative data

To test for differences between two proportions, we apply the z-test of $p_1 - p_2$. We can use instead the chi-squared test of a contingency table, which can be applied to a variety of other problems.

DEVELOPING AN UNDERSTANDING OF STATISTICAL CONCEPTS

Table 15.1 and Figure 15.3 summarize how we deal with nominal data. We determine the frequency of each category and use these frequencies to compute test statistics. We can then compute proportions to calculate z-statistics or use the frequencies to calculate χ^2-statistics. Because squaring a standard normal random variable produces a chi-squared variable, we can employ either statistic to test for differences. As a consequence, when you encounter nominal data in the problems described in this book (and other introductory applied statistics books), the most logical starting point in selecting the appropriate technique will be either a z-statistic or a χ^2-statistic. However, you should know that there are other statistical procedures that can be applied to nominal data, techniques that are not included in this book.

15.5 CHI-SQUARED TEST FOR NORMALITY

We can use the goodness-of-fit test presented in Section 15.2 in another way. We can test to determine whether data were drawn from any distribution. The most common application of this procedure is a test of normality.

In the examples and exercises shown in Section 15.2, the probabilities specified in the null hypothesis were derived from the question. In Example 15.1, the probabilities p_1, p_2, and p_3 were the market shares before the advertising campaign. To test for normality (or any other distribution), the probabilities must first be calculated using the hypothesized distribution. To illustrate, consider Example 11.1, where we tested the mean amount of discarded newspaper using the Student t distribution. The required condition for this procedure is that the data must be normally distributed. To determine whether the 240 observations in our sample were indeed taken from a normal distribution, we must calculate the theoretical probabilities assuming a normal distribution. To do so, we must first calculate the sample mean and standard deviation. They are $\bar{x} = 2.0904$ and $s = .4148$. Next we find the probabilities of an arbitrary number of intervals. For example, we can find the probabilities of the following intervals.

Interval 1: $X \leq 1.2608$

Interval 2: $1.2608 < X \leq 1.6756$

Interval 3: $1.6756 < X \leq 2.0904$

Interval 4: $2.0904 < X \leq 2.5052$

Interval 5: $2.5052 < X \leq 2.9200$

Interval 6: $X > 2.9200$

(The intervals were chosen to simplify the calculations.) The probabilities are computed using the normal distribution and the values of $\bar{x}$ and s as estimators of μ and σ. Thus,

$$P(X \leq 1.2608) = P\left(\frac{X - \mu}{\sigma} \leq \frac{1.2608 - 2.0904}{.4148}\right) = P(Z \leq -2) = .0228$$

$$P(1.2608 < X \leq 1.6756) = P\left(\frac{1.2608 - 2.0904}{.4148} < \frac{X - \mu}{\sigma} \leq \frac{1.6756 - 2.0904}{.4148}\right)$$
$$= P(-2 < Z \leq -1) = .1359$$

$$P(1.6756 < X \leq 2.0904) = P\left(\frac{1.6756 - 2.0904}{.4148} < \frac{X - \mu}{\sigma} \leq \frac{2.0904 - 2.0904}{.4148}\right)$$
$$= P(-1 < Z \leq 0) = .3413$$

$$P(2.0904 < X \leq 2.5052) = P\left(\frac{2.0904 - 2.0904}{.4148} < \frac{X - \mu}{\sigma} \leq \frac{2.5052 - 2.0904}{.4148}\right)$$
$$= P(0 < Z \leq 1) = .3413$$

$$P(2.5052 < X \leq 2.9200) = P\left(\frac{2.5052 - 2.0904}{.4148} < \frac{X - \mu}{\sigma} \leq \frac{2.9200 - 2.0904}{.4148}\right)$$
$$= P(1 < Z \leq 2) = .1359$$

$$P(X > 2.9200) = P\left(\frac{X - \mu}{\sigma} > \frac{2.9200 - 2.0904}{.4148}\right) = P(Z > 2) = .0228$$

To test for normality is to test the following hypotheses.

$H_0 : p_1 = .0228, \ p_2 = .1359, \ p_3 = .3413, \ p_4 = .3413, \ p_5 = .1359, \ p_6 = .0228$

H_1 : At least one proportion differs from its specified value

We complete the test as we did in Section 15.2, except that the number of degrees of freedom associated with the chi-squared statistic is the number of intervals minus 1 minus the number of parameters estimated, which in this illustration is two. (We estimated the population mean μ and the population standard deviation σ.) Thus, in this case the number of degrees of freedom is $v = 6 - 2 - 1 = 3$. The expected values are

$$e_1 = np_1 = 240(.0228) = 5.47$$
$$e_2 = np_2 = 240(.1359) = 32.62$$
$$e_3 = np_3 = 240(.3413) = 81.91$$
$$e_4 = np_4 = 240(.3413) = 81.91$$
$$e_5 = np_5 = 240(.1359) = 32.62$$
$$e_6 = np_6 = 240(.0228) = 5.47$$

The observed values can be determined manually by counting the number of values in each interval. Thus,

$$f_1 = 7$$
$$f_2 = 31$$
$$f_3 = 76$$
$$f_4 = 90$$
$$f_5 = 32$$
$$f_6 = 4$$

The chi-squared statistic is

$$\chi^2 = \sum_{i=1}^{k} \frac{(f_i - e_i)^2}{e_i} = \frac{(7 - 5.47)^2}{5.47} + \frac{(31 - 32.62)^2}{32.62} + \frac{(76 - 81.91)^2}{81.91}$$
$$+ \frac{(90 - 81.91)^2}{81.91} + \frac{(32 - 32.62)^2}{32.62} + \frac{(4 - 5.47)^2}{5.47}$$
$$= 2.1412$$

CLASS INTERVALS

In practice you can use any intervals you like. We chose the intervals we did to facilitate the calculation of the normal probabilities. The number of intervals was chosen to comply with the rule of five, which requires that all expected values be at least equal to 5. Because the number of degrees of freedom is $k - 3$, the minimum number of intervals is $k = 4$.

COMPUTE

We programmed Excel to calculate the value of the test statistic so that the expected values are at least five (where possible) and the minimum number of intervals is 4. Hence, if the number of observations is more than 220, the intervals and probabilities are

Interval	Probability
$Z \leq -2$	.0228
$-2 < Z \leq -1$	.1359
$-1 < Z \leq 0$	.3413
$0 < Z \leq 1$	.3413
$1 < Z \leq 2$	.1359
$Z > 2$	.0228

If the sample size is less than or equal to 220 and greater than 80, the intervals are

Interval	Probability
$Z \leq -1.5$	.0668
$-1.5 < Z \leq -0.5$	.2417
$-0.5 < Z \leq 0.5$	.3829
$0.5 < Z \leq 1.5$	.2417
$Z > 1.5$	.0668

If the sample size is less than or equal to 80, we employ the minimum number of intervals 4. When the sample size is less than 32, at least one expected value will be less than 5. The intervals are

Interval	Probability
$Z \leq -1$	.1587
$-1 < Z \leq 0$	.3413
$0 < Z \leq 1$	.3413
$Z > 1$	.1587

Microsoft Excel Output of the Chi-Squared Test of Normality of the Data in Example 11.1

Chi-Squared Test of Normality

Newspaper

Mean	2.0904		
Standard deviation	0.4148		
Observations	240		
Intervals	Probability	Expected	Observed
(z <= −2)	0.02275	5.46	7
(−2 < z <= −1)	0.135905	32.6172	31
(−1 < z <= 0)	0.341345	81.9228	76
(0 < z <= 1)	0.341345	81.9228	90
(1 < z <= 2)	0.135905	32.6172	32
(z > 2)	0.02275	5.46	4
chi-squared Stat	2.1412		
df	3		
p-value	0.5436		
chi-squared Critical	7.8147		

GENERAL COMMANDS	EXAMPLE 10.1
1 Type or import the data into one column.	Open file **Xm11–01**.
2 Click **Tools, Data Analysis Plus,** and **Chi-Squared Test of Normality**.	
3 Specify the **Input Range:**.	A1:A241
4 Click **Labels**, if applicable.	
5 Specify the value of α **(Alpha)**, and click **OK**.	.05

INTERPRET

The *p*-value of the test is .5436. We conclude that there is not enough evidence to infer that the variable in Example 11.1 is not normally distributed. The technique we used and the conclusion we drew in Example 11.1 were valid.

Figure 15.4 describes the sampling distribution and *p*-value.

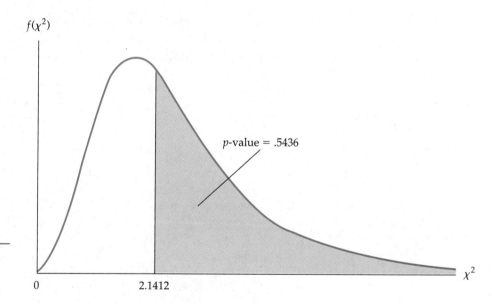

Figure 15.4

Sampling distribution of the chi-squared test statistic for normality for Example 11.1

The goodness-of-fit test can be used to test for any distribution. All we need do is calculate estimates of the distribution's parameters and use them to calculate probabilities of ranges for continuous random variables and individual values for discrete random variables. We then determine the frequencies and compute the chi-squared statistic.

EXERCISES

15.31 One hundred observations were drawn from a population whose distribution is unknown. The data are stored in file Xr15-31. Can we infer that the population is not normal?

15.32 A random sample of 250 observations is stored in file Xr15-32. Test to determine whether the population is nonnormal.

15.33 Test to determine whether the data stored in file Xr15-33 were drawn from a nonnormal population.

The following exercises require you to test for normality in earlier exercises. It should be noted that the variables are required to be normally distributed, but that departures from normality do not necessarily invalidate the test's conclusions.

15.34 Refer to Exercise 11.58. Test to determine whether the amount of time spent working at part-time jobs is nonnormally distributed.

15.35 The test in Exercise 11.59 requires that the weights are normally distributed. Conduct a test to determine whether the required condition is unsatisfied.

15.36 In Exercise 11.64 you were required to compute the t-estimate of the mean return on investment. Test the required condition of this procedure.

15.37 Refer to Exercise 11.83, where the chi-squared test of a variance was applied. Test to determine whether the required condition was unsatisfied.

15.38 In Exercise 12.24, you applied the t-test of the difference between two means. One of the required conditions is that the number of days of illness in both populations is normally distributed. Was this required condition unsatisfied?

15.39 The required condition of the procedure applied in Exercise 12.27 is that the percentage reduction in blood pressure is normally distributed. Test to determine whether this condition is not met.

15.40 Exercise 12.30 required you to conduct a t-test of the difference between two means. Each sample's productivity data are required to be normally distributed. Is that required condition violated?

15.41 In Exercise 12.44, you performed a test of the mean matched pairs difference. The test result depends on the requirement that the differences are normally distributed. Test to determine whether the requirement is violated.

15.42 Can we conclude that the requirement that the differences be normally distributed in Exercise 12.46 was not satisfied?

15.43 The analysis of variance procedure you conducted in Exercise 14.5 requires that the grades be normally distributed. Test this requirement.

15.44 Refer to Exercise 14.12. Were the LSAT scores in each group nonnormally distributed?

15.6 SUMMARY

This chapter introduced three statistical techniques. The first is the chi-squared goodness-of-fit test, which is applied when the problem objective is to describe a single population of nominal data with two or more categories. The second is the chi-squared test of a contingency table. There are two objectives of this test, to analyze the relationship between two nominal variables and to compare two or more populations of nominal data. The last procedure is designed to test for normality.

IMPORTANT TERMS

Multinomial experiment
Expected frequency
Chi-squared goodness-of-fit test

Chi-squared test of a contingency table
Contingency table
Cross-classification table

SYMBOLS

Symbol	Represents
f_i	Frequency of the ith category
e_i	Expected value of the ith category

FORMULAS

Test statistic for all procedures

$$\chi^2 = \sum_{i=1}^{k} \frac{(f_i - e_i)^2}{e_i}$$

MICROSOFT EXCEL OUTPUT AND INSTRUCTIONS

Technique	Page
Chi-squared goodness-of-fit test	469
Chi-squared test of a contingency table	476–477
Chi-squared test of a contingency table (raw data)	477–478
Chi-squared test of normality	487–488

SUPPLEMENTARY EXERCISES

15.45 An organization dedicated to ensuring fairness in television game shows is investigating "Wheel of Fortune." In this show, three contestants are required to solve puzzles by selecting letters. Each contestant gets to select the first letter and continues selecting until he or she chooses a letter that is not in the hidden word, phrase, or name. The order of contestants is random. However, contestant 1 gets to start game 1, contestant 2 starts game 2, and so on. The contestant who wins the most money is declared the winner, and he or she is given an opportunity to win a grand prize. Usually, more than three games are played per show, and as a result it appears that contestant 1 has an advantage: contestant 1 will start two games, whereas contestant 3 will usually start only one game. To see if this is the case, a random sample of 30 shows was taken, and the starting position of the winning contestant for each show was recorded. These are shown in the following table.

Starting Position	Number of Winners
1	14
2	10
3	6

Do the tabulated results allow us to conclude that the game is unfair?

15.46 Econetics Research Corporation, a well-known Montreal-based consulting firm, wants to test how it can influence the proportion of questionnaires returned from surveys. Believing that the inclusion of an inducement to respond may be important, it sends out 1,000 questionnaires: 200 promise to send respondents a summary of the survey results, 300 indicate that 20 respondents (selected by lottery) will be awarded gifts, and 500 are accompanied by no inducements. Of these, 80 questionnaires promising a summary, 100 questionnaires offering gifts, and 120 questionnaires offering no inducements are returned. What can you conclude from these results? (*Hint:* The sample size is 1,000, and your analysis must include the complete sample.)

15.47 It has been estimated that employee absenteeism costs North American companies more than $100 billion per year. As a first step in addressing the rising cost of absenteeism, the personnel department of a large corporation recorded the weekdays during which individuals in a sample of 362 absentees were away over the past several months. Do these data suggest that absenteeism is higher on some days of the week than on others?

Day of the Week	Monday	Tuesday	Wednesday	Thursday	Friday
Number Absent	87	62	71	68	74

15.48 Suppose that the personnel department in Exercise 15.47 continued its investigation by categorizing absentees according to the shift on which they worked, as shown in the accompanying table. Is there sufficient evidence of a relationship between the days on which employees are absent and the shift on which the employees work?

Shift	Monday	Tuesday	Wednesday	Thursday	Friday
Day	52	28	37	31	33
Evening	35	34	34	37	41

15.49 A management behavior analyst has been studying the relationship between male/female supervisory structures in the workplace and the level of employees' job satisfaction. The results of a recent survey are shown in the table below. Conduct a test to determine whether the level of job satisfaction depends on the boss/employee gender relationship.

Level of Satisfaction	Boss/Employee			
	Female/Male	Female/Female	Male/Male	Male/Female
Satisfied	21	25	54	71
Neutral	39	49	50	38
Dissatisfied	31	48	10	11

15.50 During the decade of the 1980s, professional baseball thrived in North America. Attendance rose continuously from 45 million in 1984 to 58 million in 1991. However, in 1992, attendance dropped about 2 million. In addition, the number of television viewers also decreased. In order to examine the popularity of baseball relative to other sports, surveys were performed. In 1985 and again in 1993, a Harris Poll asked a random sample of 500 people to name their favorite sport. The results, which were published in the *Wall Street Journal* (6 July 1993), are stored in file Xr15-50 in the following way. Column A: results from 1985—1 = professional football, 2 = baseball, 3 = professional basketball, 4 = college basketball, 5 = college football, and 6 = other; column B: results from 1993, using the same codes.

  a Do these results indicate that North Americans changed their favorite sport between 1985 and 1993? (*Hint:* You will have to reorganize the data to answer this question.)

  b Do these results indicate that the popularity of baseball has changed between 1985 and 1993?

15.51 According to *NBC News* (March 11, 1994) more than 3,000 Americans quit smoking each day. (Unfortunately, more than 3,000 Americans start smoking each day.) Because nicotine is one of the most addictive drugs, quitting smoking is a difficult and frustrating task. It usually takes several tries before success is achieved. There are various methods, including cold turkey, nicotine patch, hypnosis, and group therapy sessions. In an experiment to determine how these methods differ, a random sample of smokers who have decided to quit is selected. Each smoker has chosen one of the methods listed above. After one year the respondents report whether they have quit (1 = yes and 2 = no) and which method they used (1 = cold turkey; 2 = nicotine patch; 3 = hypnosis; 4 = group therapy sessions). These data are stored in columns A and B, respectively, in file Xr15-51. Is there sufficient evidence to conclude that the four methods differ in their success?

15.52 A newspaper publisher, trying to pinpoint his market's characteristics, wondered whether the way people read a newspaper is related to the reader's educational level. A survey asked adult readers which section of the paper they read first and asked them to report their highest educational level. These data were recorded (column A = first section read, where 1 = front page, 2 = sports, 3 = editorial, and 4 = other; and column B = educational level where 1 = did not complete high school, 2 = high school graduate, 3 = university or college graduate, and 4 = postgraduate degree) and stored in file Xr15-52. What do these data tell the publisher about how educational level affects the way adults read the newspaper?

15.53 Every week the Florida Lottery draws 6 numbers between 1 and 49. Lottery ticket buyers are naturally interested in whether certain numbers are drawn more frequently than others. To assist players, the *Sun-Sentinel* publishes the number of times each of the 49 numbers has been drawn in the past 52 weeks. The numbers and the frequency with which each occurred are stored in columns A and B, respectively, in file Xr15-53. These data are from the Sunday, January 5, 1997, edition.

  a If the numbers are drawn from a uniform distribution, what is the expected frequency for each number?

  b Can we infer that the data were not generated from a uniform distribution?

15.54 Canadians have the option of investing income in registered retirement savings plans (RRSPs). Subject to limits calculated on the basis of income, employer retirement plans, and previous RRSPs, money invested in RRSPs is not taxable. (Money withdrawn from retirement plans is taxable.) Critics argue that RRSPs are a tax loophole for the rich, because only wealthier people are in a position to take advantage of the tax provisions. In a study to determine who uses RRSPs, the Caledon Institute of Social Policy randomly sampled a variety of Canadians in different tax brackets. (Survey results were published in the *Globe and Mail*, February 5, 1994.) For each respondent the researchers recorded the tax bracket (1 = less than $20,000; 2 = $20 to 40,000; 3 = $40 to $60,000; 4 = $60 to 100,000; 5 = over $100,000), and whether they invested in an RRSP this year (2 = yes; 1 = no). These data are stored in columns A and B, respectively in file Xr15-54. Can we infer from these data that there are differences in RRSP positions among the income groups?

15.55 The relationship between drug companies and medical researchers is under scrutiny because of possible conflict of interest. The issue that started the controversy was a 1995 case control study that suggested that the use of calcium-channel blockers to treat hypertension led to an increased risk of heart disease. This led to an intense debate both in technical journals and in the press. Researchers writing in the *New England Journal of Medicine* ("Conflict of Interest in the Debate over Calcium Channel Antagonists," 8 January 1998, p. 101) looked at the 70 reports that appeared during 1996–1997, classifying them as favorable, neutral, or critical toward the drugs. The researchers then contacted the authors of the reports and questioned them about financial ties to drug companies. The results are stored in file Xr15-55 in the following way:

Column A: Results of the scientific study; 1 = favorable, 2 = neutral, 3 = critical

Column B: 1 = financial ties to drug companies, 2 = no ties to drug companies

Do these data allow us to infer that the research findings for calcium-channel blockers are affected by whether the research is funded by drug companies?

15.56 The statistical analysis in Exercise 15.55 led to a further analysis. In addition to recording whether the study was favorable, neutral, or critical, researchers also recorded whether competing companies funded the research. The data for competing companies were stored in column B of file Xr15-56. (Column A is identical to column A of file Xr15-55.) Do these data allow us to infer that the research findings for calcium-channel blockers are affected by whether the research is funded by competing companies?

15.57 Toronto has four daily newspapers. They are the *National Post,* the *Globe and Mail,* the *Sun,* and the *Toronto Star.* A marketing consultant wanted to determine the demographic characteristics of the readers of each newspaper. Accordingly, he organized a survey that asked newspaper readers in Toronto which paper they regularly read and to indicate their occupation. The data were stored in file Xr15-57 in the following way:

Column A: Newspaper where 1 = *National Post*, 2 = *Globe and Mail*, 3 = *Sun*, 4 = *Toronto Star*

Column B: Occupation where 1 = managerial, 2 = blue-collar, 3 = professional, 4 = other

Can we infer that the readerships of the four daily newspapers differ in terms of the occupations of their readers?

15.58 How does level of affluence affect health care? To address one dimension of the problem, a random sample of heart attack victims was drawn. Each was categorized as a low (1), medium (2), or high (3) income earner. Each was also categorized as having survived (1) or died (2). These data are stored in columns A and B, respectively, in file Xr15-58. Can we infer that income level and surviving a heart attack are related?

15.59 Who is more likely to ask for directions when lost, men or women? The conventional wisdom overwhelmingly favors women. However, conventional wisdom may be wrong. Market Facts Inc. conducted a U.S.-wide survey that asked 503 men and 502 women what they do when they are lost while driving a car. The responses are

1. Consult a map
2. Ask someone for directions
3. Continue driving until location or direction determined
4. Other

The responses are stored in column A in file Xr15-59. Column B contains a code for the gender of the respondent (1 = male and 2 = female). Can we infer that men and women differ in their action when lost?

15.60 Researchers at the University of Pennsylvania School of Medicine asked parents of 437 children who are 16 years old whether their children slept with lights on before the age of 2. For each child the researchers recorded a 2 if the child became myopic (near-sighted) or a 1 if the child did not. These results are stored in column A in file Xr15-60. Column B stores the codes for the responses to the question about lights where 1 = night-light, 2 = regular light, and 3 = no light.

Do the data allow us to infer that night lights and myopia are related?

15.61 Ulcers are caused by the *Helicobacter pylori* bacteria. To determine whether drinking alcoholic beverages has an effect on this bacteria, researchers in Germany took a random sample of 1,800 people. Each was classified as a teetotaler (1), drink one drink per day (2), drink two drinks per day (3), or drink more than two drinks per day (4). The incidence of ulcers was recorded as a 2; 1 represents no ulcers. The data are stored in columns A and B, respectively, in file Xr15-61. Can we infer that drinking affects the incidence of ulcers?

Case 15.1 Predicting the Outcomes of Basketball, Baseball, Football, and Hockey Games from Intermediate Results*

Some basketball fans generally believe that it doesn't pay to watch an entire game because the outcome is determined in the last few minutes (some say the last two minutes) of the game. Is this really true, and, if so, is basketball different in this respect from other professional sports played in North America? For example, is it true that the team that leads a baseball game after seven innings almost always wins the game? To address these questions, three researchers tracked basketball, baseball, football, and hockey games. The results of games during the 1990 season (for baseball and football) and during the 1990–1991 season (for basketball and hockey)—whether the early-game leader and whether the late-game leader won—were recorded. Early-game leaders are defined as the teams that are ahead after one quarter of basketball and football, one period of hockey, or three innings of baseball. Late-game leaders are defined as the teams that are ahead after three quarters of basketball and football, two periods of hockey, or seven innings of baseball.

The data are stored in file C15–01 in the following way:

Column A: Results of games where 2 = early-game leader wins and 1 = early-game leader loses

Column B: Early-game-leader game where 1 = basketball, 2 = baseball, 3 = football, and 4 = hockey

Column C: Results of games where 2 = late-game leader wins and 1 = late-game leader loses

Column D: Late-game-leader game where 1 = basketball, 2 = baseball 3 = football, and 4 = hockey

Can we infer from these data that all four professional sports experience the same proportion of early-game leaders winning the game?

Can we infer from these data that all four professional sports experience the same proportion of late-game leaders winning the game?

*Adapted from H. Cooper, K. M. DeNeve, and E Mosteller, "Predicting Professional Sports Game Outcomes from Intermediate Game Scores," *Chance* 5, Nos. 3–4 (1992): 18–22

Case 15.2 Can Exposure to a Code of Professional Ethics Help Make Managers More Ethical?*

In many North American business schools, the issue of whether a course on ethics should be compulsory has been hotly debated. The empirical evidence appears to be far from consistent on the effects of such courses. To help shed more light on the issue, two researchers organized a study in which they took a random sample of 68 accounting students and 132 nonaccounting students. As part of their curriculum, the accounting students were exposed to the American Institute of Certified Public Accountants' code of professional ethics. The nonaccounting business students did not take any course that dealt with issues of ethical behavior.

All 200 students in the study were taking a required senior-level policy course. As part of the course, they were assigned to read the article "Crisis in Conscience at Quasar" by A. Fendrock (*Harvard Business Review,* March–April 1968, 112–20). In the case, Universal, the parent company, learned that the senior managers of one of its subsidiaries, Quasar, deliberately lied about financial conditions in their monthly report to corporate headquarters. Quasar's president, John Kane, and its controller, Hugh Kay, were forced to resign. Universal wanted to know why no one at Quasar provided any information about the true financial conditions, whether any other executives were accomplices to the phony reports, and what could be done to avert such occurrences in the future. Universal sent a fact finder to interview other executives at Quasar—George Kessler, vice president, manufacturing; William Heller, vice president, engineering; Peter Loomis, vice president, marketing; Donald Morgan, chief accountant; and Paul Brown, vice president, industrial relations.

After studying the case, students completed the questionnaire shown below. The results are stored in file C15–02. (The responses to questions 1 to 6 for all students are stored in columns A to F; column G indicates whether the student was an accounting student [1] or a nonaccounting business student [2].)

Does it appear that accounting students exposed to a code of ethics answer the questionnaire differently from nonaccounting business students not exposed to the same code?

1 If you had been John Kane, president of Quasar, do you think you would have been tempted to withhold the bad news from corporate management at the parent company?
2 Yes
1 No

2 Do you think that under the circumstances you would have withheld the bad news?
2 Yes
1 No

*Adapted from W. E. Fulmer and B. R. Cargile, "Ethical Perceptions of Accounting Students: Does Exposure to a Code of Professional Ethics Help?" *Issues in Accounting Education* (Fall 1987): 207–19.

3 Do you think you would have gone around the president and reported the bad news to corporate headquarters at Universal?
2 Yes
1 No

4 Do you think Kane's withholding the bad news was (check one) . . .
_____ 1 practical?
_____ 2 unethical?
_____ 3 poor judgment?

5 Do you think the blame lies with (check just one) . . .
_____ 1 Universal's corporate management?
_____ 2 Quasar's president?
_____ 3 Quasar's controller?
_____ 4 other?

6 Is the problem one of (check just one) . . .
_____ 1 poor organization?
_____ 2 lack of communication?
_____ 3 excessive personal loyalty?
_____ 4 inadequate supervision?
_____ 5 other?

Chapter 16

Nonparametric Statistical Techniques

16.1 Introduction

16.2 Wilcoxon Rank Sum Test

16.3 Sign Test and Wilcoxon Signed Rank Sum Test

16.4 Kruskal-Wallis Test

16.5 Friedman Test

16.6 Summary

16.1 INTRODUCTION

Throughout this book we have presented statistical techniques that are used when the data are either interval or nominal. In this chapter, we introduce statistical techniques that deal with ordinal data. We will introduce three methods that compare two populations and two procedures used to compare two or more populations. As you've seen, when we compare two or more populations of interval data we measure the difference among means. However, as we discussed in Chapter 2, when the data are ordinal, the mean is not the most appropriate measure of location. As a result, the methods in this chapter do not test the difference in population means; instead, they test characteristics of populations without referring to specific parameters. For this reason, these techniques are called **nonparametric statistical techniques.** Rather than testing to determine whether the population means differ, we will test to determine whether the **population locations** differ.

Although nonparametric methods are designed to test ordinal data, they have another area of application. The statistical tests described in Sections 12.2 and 12.4 and Chapter 14 require that the populations be normally distributed. If the data are extremely nonnormal, the t-tests of the difference between two means, the t-test of the mean difference, and the F-tests of the analysis of variance are invalid. Fortunately, nonparametric techniques can be used instead. For this reason, nonparametric procedures are often (perhaps more accurately) called **distribution-free statistics.** The techniques presented here can be used when the data are interval and the populations are extremely nonnormal. In such circumstances, we will treat the interval data as if they were ordinal. Consequently, even when the data are interval and the mean is the appropriate measure of location, we will choose instead to test population locations.

Figure 16.1 depicts the distributions of two populations when their locations are the same. Notice that, because we don't know (or care) anything about the shape of the distributions, we represent them as nonnormal. Figure 16.2 describes a circumstance when the location of population 1 is to the right of the location of population 2. The location of population 1 is to the left of the location of population 2 in Figure 16.3.

498 Chapter 16 NONPARAMETRIC STATISTICAL TECHNIQUES

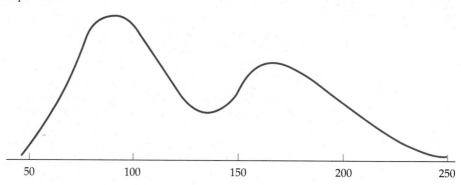

Figure 16.1

Population locations are the same

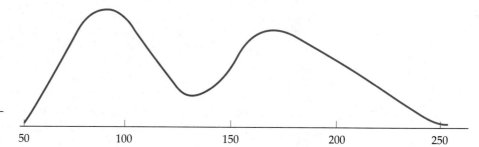

Figure 16.2

Location of population 1 is to the right of the location of population 2

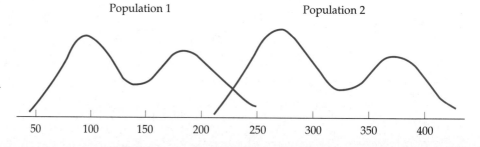

Figure 16.3

Location of population 1 is to the left of the location of population 2

When the problem objective is to compare two populations the null hypothesis will state

H_0 : The two population locations are the same.

The alternative hypothesis can take on any one of the following three forms:

1 If we want to know whether there is sufficient evidence to infer that there is a difference between the two populations, the alternative hypothesis is

H_1 : The location of population 1 is different from the location of population 2.

2 If we want to know whether we can conclude that the random variable in population 1 is larger in general than the random variable in population 2 (see Figure 16.2), the alternative hypothesis is

H_1 : The location of population 1 is to the right of the location of population 2.

3 If we want to know whether we can conclude that the random variable in population 1 is smaller in general than the random variable in population 2 (see Figure 16.3), the alternative hypothesis is

H_1 : The location of population 1 is to the left of the location of population 2.

As usual, the alternative hypothesis specifies whatever we're investigating.

As you will see, nonparametric tests utilize a ranking procedure as an integral part of the calculations. You've actually dealt with such a process already in this book. In Chapter 3, we introduced the median as a measure of central location. The median is computed by placing the observations in order and selecting the observation that falls in the middle. Thus, the appropriate measure of central location of ordinal data is the median, a statistic that is the product of a ranking process.

In the next section, we present the Wilcoxon rank sum test. Section 16.3 introduces the sign test, and the Wilcoxon signed rank sum test. All three of these methods are employed when the objective is to compare two populations. Sections 16.4 and 16.5 introduce the Kruskal-Wallis test and the Friedman test, respectively, procedures that are utilized when the objective is to compare two or more populations.

16.2 WILCOXON RANK SUM TEST

The test we introduce in this section deals with problems with the following characteristics:

1 The problem objective is to compare two populations.
2 The data are either ordinal or interval where the populations are extremely nonnormal.
3 The samples are independent.

To illustrate how to compute the test statistic for the **Wilcoxon rank sum test,** we offer the following simplified example.

EXAMPLE 16.1

Suppose that we want to determine whether the following observations drawn from two populations allow us to conclude at the 5% significance level that the location of population 1 is to the left of the location of population 2.

Sample 1: 22 23 20

Sample 2: 18 27 26

We want to test the following hypotheses.

H_0 : The two population locations are the same.

H_1 : The location of population 1 is to the left of the location of population 2.

TEST STATISTIC

The first step is to rank all six observations with rank 1 assigned to the smallest observation and rank 6 to the largest.

Sample 1	Rank	Sample 2	Rank
22	3	18	1
23	4	27	6
20	2	26	5
	$T_1 = 9$		$T_2 = 12$

Observe that 18 is the smallest number, so it receives a rank of 1; 20 is the second-smallest number, and it receives a rank of 2. We continue until rank 6 is assigned to 27, which is the largest of the observations. In case of ties, we average the ranks of the tied observations. The second step is to calculate the sum of the ranks of each sample. The rank sum of sample 1, denoted T_1, is 9. The rank sum of sample 2, denoted T_2, is 12. (Note that T_1 plus T_2 must equal the sum of the integers from 1 to 6, which is 21.) We can use either rank sum as the test statistic. We arbitrarily select T_1 as the test statistic and label it T. The value of the test statistic in this example is $T = T_1 = 9$.

SAMPLING DISTRIBUTION OF THE TEST STATISTIC

A small value of T indicates that most of the smaller observations are in sample 1 and that most of the larger observations are in sample 2. This would imply that the location of population 1 is to the left of the location of population 2. Therefore, in order for us to conclude statistically that this is the case, we need to show that T is small. The definition of "small" comes from the sampling distribution of T. As we did in Chapter 8 when we derived the sampling distribution of the sample mean, we can derive the sampling distribution of T by listing all possible values of T. In Table 16.1 we show all possible rankings of two samples of size 3 and the rank sums.

If the null hypothesis is true and the two population locations are identical, then it follows that each possible ranking is equally likely. Because there are 20 different possibilities, each value of T has the same probability, namely, 1/20. Notice that there is one value of 6, one value of 7, two values of 8, and so on. Table 16.2 summarizes the values of T and their probabilities.

Table 16.1 All Possible Ranks and Rank Sums of Two Samples of Size 3

Ranks of Sample 1	Rank Sum	Ranks of Sample 2	Rank Sum
1, 2, 3	6	4, 5, 6	15
1, 2, 4	7	3, 5, 6	14
1, 2, 5	8	3, 4, 6	13
1, 2, 6	9	3, 4, 5	12
1, 3, 4	8	2, 5, 6	13
1, 3, 5	9	2, 4, 6	12
1, 3, 6	10	2, 4, 5	11
1, 4, 5	10	2, 3, 6	11
1, 4, 6	11	2, 3, 5	10
1, 5, 6	12	2, 3, 4	9
2, 3, 4	9	1, 5, 6	12
2, 3, 5	10	1, 4, 6	11
2, 3, 6	11	1, 4, 5	10
2, 4, 5	11	1, 3, 6	10
2, 4, 6	12	1, 3, 5	9
2, 5, 6	13	1, 3, 4	8
3, 4, 5	12	1, 2, 6	9
3, 4, 6	13	1, 2, 5	8
3, 5, 6	14	1, 2, 4	7
4, 5, 6	15	1, 2, 3	6

Table 16.2 Sampling Distribution of T with Two Samples of Size 3

T	$P(T)$
6	1/20
7	1/20
8	2/20
9	3/20
10	3/20
11	3/20
12	3/20
13	2/20
14	1/20
15	1/20
Total	1

From this sampling distribution, we can see that $P(T \leq 6) = P(T = 6) = 1/20 = 0.05$. Because we're trying to determine whether the value of the test statistic is small enough for us to reject the null hypothesis at the 5% significance level, we specify the rejection region as $T \leq 6$. Because $T = 9$, we cannot reject the null hypothesis.

We can also compute the *p*-value of the test. It is

$$p\text{-value} = P(T \leq 9) = p(6) + p(7) + p(8) + p(9)$$
$$= .05 + .05 + .10 + .15 = .35$$

Figure 16.4 displays the sampling distribution and the *p*-value of the test.

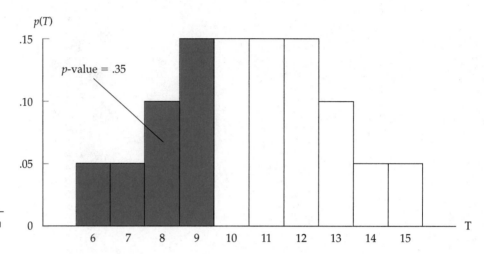

Figure 16.4

Sampling distribution of *T* with $n_1 = 3$ and $n_2 = 3$

Statisticians have determined that the sampling distribution of *T* is approximately normal for sample sizes in excess of 10. (We will not address problems with smaller sample sizes, because sample sizes that small are virtually useless.) The mean $E(T)$ and standard deviation σ_T are

$$E(T) = \frac{n_1(n_1 + n_2 + 1)}{2}$$

and

$$\sigma_T = \sqrt{\frac{n_1 n_2(n_1 + n_2 + 1)}{12}}$$

Thus, the standardized test statistic is

$$z = \frac{T - E(T)}{\sigma_T}$$

Here is a more realistic example, with a considerably larger sample size.

EXAMPLE 16.2

A research scientist has created a new painkiller. In a preliminary experiment to determine its effectiveness, 300 people were randomly selected, of whom 150 were given the new painkiller and 150 were given aspirin. All 300 were told to use the drug when headaches or other minor pains occurred and to indicate which of the following statements most accurately represented the effectiveness of the drug they took.

5 = The drug was extremely effective.

4 = The drug was quite effective.

3 = The drug was somewhat effective.

2 = The drug was slightly effective.

1 = The drug was not at all effective.

The responses are stored in columns A (new painkiller) and B (aspirin) in file Xm16-02 using the codes. A partial listing is provided below. Can we conclude that the new painkiller is perceived to be more effective?

Ratings of New Painkiller	Ratings of Aspirin
3	2
1	3
2	3
.	
.	
.	
4	2
2	1
5	5

Solution

IDENTIFY

The objective is to compare two populations: the perceived effectiveness of the new painkiller and of aspirin. We recognize that the data are ordinal, because except for the order of the codes, the numbers used to record the results are arbitrary. Finally, the samples are independent. These factors tell us that the appropriate technique is the Wilcoxon rank sum test. We denote the effectiveness scores of the new painkiller as population 1, and population 2 represents the effectiveness scores of aspirin. Because we want to know whether the new painkiller is better than aspirin, the alternative hypothesis is

H_1 : The location of population 1 is to the right of the location of population 2.

The complete test follows.

H_0 : The two population locations are the same.

H_1 : The location of population 1 is to the right of the location of population 2.

Test statistic: $z = \dfrac{T - E(T)}{\sigma_T}$

COMPUTE

Microsoft Excel Output for Example 16.2

Wilcoxon Rank Sum Test

	Rank Sum	Observations
New Painkiller	24817.5	150
Aspirin	20332.5	150
z Stat	2.985	
P(Z<=z) one-tail	0.0014	
z Critical one-tail	1.6449	
P(Z<=z) two-tail	0.0028	
z Critical two-tail	1.96	

The rank sums are $T_1 = 24{,}817.5$ and $T_2 = 20{,}332.5$. The value of the test statistic is $z = 2.985$, and, because we're conducting a one-tail test, the p-value is .0014.

GENERAL COMMANDS	EXAMPLE 16.2
1 Type or import the data into two adjacent columns.	Open file **Xm16-02**.
2 Click **Tools, Data Analysis Plus,** and **Wilcoxon Rank Sum Test**.	
3 Specify the **Input Range:**.	A1:B151
4 Click **Labels**, if applicable.	
5 Specify the value of α **(Alpha),** and click **OK**.	.05

INTERPRET

There is overwhelming evidence to infer that the new painkiller is perceived to be more effective than aspirin. We note that the data were generated from a controlled experiment. That is, the subjects were assigned to take either the new painkiller or aspirin. (Recall that when subjects decide for themselves which medication to take, the data are observational.) The controlled experiment helps support the claim that the new painkiller is indeed more effective than aspirin.

Figure 16.5 depicts the approximate sampling distribution.

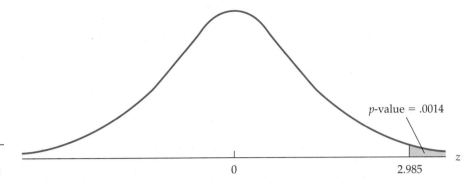

Figure 16.5

Sampling distribution of the test statistic for Example 16.2

APPLYING NONPARAMETRIC TECHNIQUES TO INTERVAL DATA

As we have noted several times in this book, the use of the Student t and F sampling distributions to test for differences between means requires that the populations be normal. However, the techniques are robust, which means that the techniques are valid even when the populations are *somewhat* nonnormal. Only when they are extremely nonnormal must we use the nonparametric methods introduced in this chapter. The question naturally arises: What constitutes extreme nonnormality? Unfortunately, we do not have a simple answer.

The chi-squared test of normality presented in Chapter 15 (and other similar tests not presented in this book) tests to determine whether there is enough statistical evidence to infer that a variable is not normally distributed. However, arriving at this conclusion does not and cannot tell us whether a variable is extremely nonnormal. Consequently, we will revert to a method we used in Chapters 12 and 14 to check the required condition of normality; we will draw histograms.

A bell-shaped histogram is an indication that the variable is normal. Because the bell shape is unimodal and symmetric, we will identify a variable as extremely nonnormal when its histogram exhibits more than one mode and/or when it is highly skewed. The following example is an illustration of this approach. Other examples in this chapter will reinforce this point.

▼ EXAMPLE 16.3

Because of the high cost of hiring and training new employees, employers would like to ensure that they retain highly qualified workers. To help develop a hiring program, the personnel manager of a large company wanted to compare how long business and nonbusiness university graduates worked for the company before quitting to accept a position elsewhere. The manager selected a random sample of 25 business and 20 nonbusiness graduates who had been hired five years ago. The number of months each had worked for the company was recorded. (Those who had not quit were recorded as having worked for 60 months.) The data are stored in column A (business graduates) and column B (nonbusiness graduates) in file Xm16-03 and are listed below. Can the personnel manager conclude that a difference in duration of employment exists between business and nonbusiness graduates?

DURATION OF EMPLOYMENT OF BUSINESS GRADUATES (IN MONTHS)	DURATION OF EMPLOYMENT OF NONBUSINESS GRADUATES (IN MONTHS)
60, 11, 18, 19, 5, 25, 60, 7,	25, 60, 22, 18, 23, 36, 39, 15,
8, 17, 37, 4, 8, 28, 27, 11,	35, 16, 28, 19, 60, 29, 16, 18
60, 25, 5, 13, 22, 11, 17, 9, 4	60, 17, 60, 32

Solution The problem objective is to compare two populations whose data are interval. The samples are independent. Thus, the appropriate parametric technique is the *t*-test of $\mu_1 - \mu_2$, which requires that the populations be normally distributed. However, when the histograms are drawn (see Figures 16.6 and 16.7), it becomes clear that this requirement is unsatisfied. And, in fact, the distributions of duration of employment appear to be extremely nonnormal. It follows that the correct statistical procedure is the Wilcoxon rank sum test.

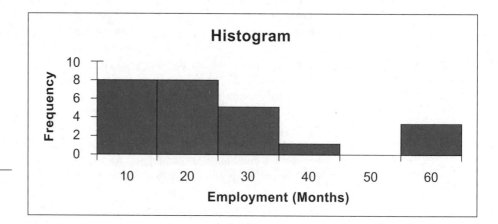

Figure 16.6

Histogram of length of employment of business graduates in Example 16.3

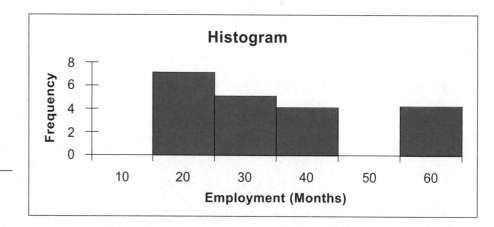

Figure 16.7

Histogram of length of employment of nonbusiness graduates in Example 16.3

H_0: The two population locations are the same.

H_1: The location of population 1 (business graduates) is different from the location of population 2 (nonbusiness graduates).

Test statistic: $z = \dfrac{T - E(T)}{\sigma_T}$

COMPUTE

Microsoft Excel Output for Example 16.3

Wilcoxon Rank Sum Test

	Rank Sum	Observations
Business	459.5	25
Nonbusiness	575.5	20
z Stat	−2.638	
P(Z<=z) one-tail	0.0042	
z Critical one-tail	1.6449	
P(Z<=z) two-tail	0.0084	
z Critical two-tail	1.96	

INTERPRET

The value of the test statistic is $z = -2.638$ and the p-value is .0084. (See Figure 16.8.) There is strong evidence to infer that the duration of employment is different for business and nonbusiness graduates. The data cannot tell us the cause of this conclusion. Moreover, we don't know what the results would have been had we surveyed employees 10 years after they were employed.

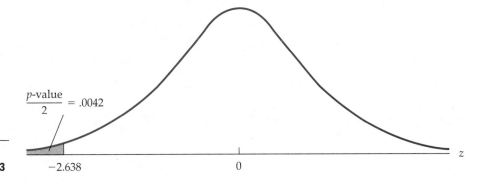

Figure 16.8

Sampling distribution of the test statistic for Example 16.3

REQUIRED CONDITIONS

The Wilcoxon rank sum test (like most of the nonparametric tests presented in this book) actually tests to determine whether the population distributions are identical. This means that it tests not only for identical locations, but for identical spreads (variances) and shapes (distributions) as well. Unfortunately, this means that the rejection of the null hypothesis may not necessarily signify a difference in population locations. The rejection of the null hypothesis may be due instead to a difference in distribution shapes and/or spreads. To avoid this problem, we will require that the two probability distributions be identical except with respect to location, which then becomes the sole focus of the test. This requirement is made for all the nonparametric tests in this chapter.

The histograms in Figures 16.6 and 16.7 appear to confirm that the two distributions are identical except for location and that they are far from bell shaped. The use of the Wilcoxon rank sum test is justified for this example.

DEVELOPING AN UNDERSTANDING OF STATISTICAL CONCEPTS

When applying nonparametric techniques, we do not perform any calculations using the original data. Instead, we perform computations only on the ranks. (We determine the rank sums and use them to make our decision.) As a result, we do not care about the actual distribution of the data (hence the name *distribution-free techniques*), and we do not specify parameters in the hypotheses (hence the name *nonparametric techniques*). Although there are other techniques that do not specify parameters in the hypotheses, we use the term *nonparametric* for procedures that feature these concepts.

Here is a summary of how to identify the Wilcoxon rank sum test.

Factors That Identify the Wilcoxon Rank Sum Test

1 Problem objective: compare two populations.

2 Data type: ordinal or interval and extremely nonnormal.

3 Experimental design: independent samples.

EXERCISES

16.1 In a taste test of a new beer, 25 people rated the new beer and another 25 rated the leading brand on the market. The possible ratings were Poor, Fair, Good, Very good, and Excellent. The responses for the new beer and the leading beer were stored in columns A and B in file Xr16-01 using a 1-2-3-4-5 coding system. Can we infer that the new beer is less highly rated than the leading brand?

16.2 Refer to Exercise 16.1. The responses were recoded so that 3 = Poor, 8 = Fair, 22 = Good, 37 = Very good, and 55 = Excellent. These data are stored in columns A and B in file Xr16-02.

a Can we infer that the new beer is less highly rated than the leading brand?

b Compare your answer in part (a) with that of Exercise 16.1.

16.3 To determine whether the satisfaction rating of an airline differs between business class and economy class, a survey was performed. Random samples of both groups were asked to rate their satisfaction with the quality of service using the following responses.

Very satisfied
Quite satisfied
Somewhat satisfied
Neither satisfied nor dissatisfied
Somewhat dissatisfied
Quite dissatisfied
Very dissatisfied

Using a 7-6-5-4-3-2-1 coding system, the results were stored in columns A (business class) and B (economy class) in file Xr16-03. Can we infer that business and economy class differ in their degree of satisfaction with the service?

16.4 Refer to Exercise 16.3. The responses were recoded using the following values: 88-67-39-36-25-21-18. These data are stored in columns A and B in file Xr16-04.

a Can we infer that business and economy class differ in their degree of satisfaction with the airline?
b Compare your answer in part (a) with that of Exercise 16.3.

16.5 Refer to Example 16.2. Suppose that the responses were coded as follows.

100 = The drug was extremely effective.
60 = The drug was quite effective.
40 = The drug was somewhat effective.
35 = The drug was slightly effective.
10 = The drug was not at all effective.

These data are stored in file Xr16-05.

a Determine whether we can infer that the new painkiller is more effective.
b Compare your answer in part (a) to Example 16.2.

16.6 A survey of statistics professors asked them to rate the importance of teaching nonparametric techniques. The possible responses are

Very important
Quite important
Somewhat important
Not too important
Not important at all

The professors were classified as either a member of the Mathematics Department or a member of some other department. The responses (the codes are 5, 4, 3, 2, and 1, respectively) were stored in columns A (Mathematics Department) and B (other department) in file Xr16-06. Can we infer that members of the Mathematics Department rate nonparametric techniques as more important than do members of other departments?

16.7 In recent years, insurance companies offering medical coverage have given discounts to companies that are committed to improving the health of their employees. To help determine whether this policy is reasonable, the general manager of one large insurance company organized a study of a random sample of 30 workers who regularly participate in their company's lunchtime exercise program and 30 workers who do not. Over a two-year period, he observed the total dollar amount of medical expenses for each individual. These data are stored in columns A (exercisers) and B (non-exercisers) of file Xr16-07. Can the manager conclude that companies that provide exercise programs should be given discounts?

16.8 Feminist organizations often use the issue of who does the housework in two-career families as a gauge of equality. Suppose that a study was undertaken and a random sample of 125 two-career families was taken. The wives were asked to report the number of hours of housework they performed the previous week. The results together with the responses from a survey performed last year (with a different sample of two-career families) are stored in columns A (hours of housework this year) and B (hours of housework last year) in file Xr16-08. Can we conclude that women are doing less housework today than last year?

16.9 The American public's support for the space program is important for the program's continuation and for the financial health of the aerospace industry. In a poll conducted by the Gallup organization last year, a random sample of 100 Americans were asked, "Should the amount of money being spent on the space program be increased or kept at current levels (3), decreased (2), or ended altogether (1)?" The survey was conducted again this year. The results are stored in file Xr16-09 using the codes in parentheses. (Column A = opinions last year, and column B = opinions this year.) Can we conclude that public support decreased between this year and last year?

16.10 Certain drugs differ in their side effects depending on the gender of the patient. In a study to determine whether men or women suffer more serious side effects when taking a powerful penicillin substitute, 50 men and 50 women were given the drug. Each was asked to evaluate the level of stomach upset on a

4-point scale, where 4 = extremely upset, 3 = somewhat upset, 2 = not too upset, and 1 = not upset at all. The results are stored in file Xr16-10 with column A = females' evaluations and column B = males' evaluations. Can we conclude that men and women experience different levels of stomach upset from the drug?

16.11 The president of Tastee Inc., a baby-food producer, claims that her company's product is superior to that of her leading competitor because babies gain weight faster with her product. As an experiment, 40 healthy infants are randomly selected. For two months, 15 of the babies are fed Tastee baby food and the other 25 are fed the competitor's product. Each baby's weight gain (in ounces) is stored in columns A (Tastee) and B (leading competitor) in file Xr16-11. Can we conclude that if we use weight gain as our criterion, Tastee baby food is indeed superior? This exercise is identical to Exercise 12.23 (except for the data).

16.12 Does a woman's size influence the ways that others judge her? This question was addressed by a researcher at Ohio State University (*Working Mother*, April 1992). The experiment consisted of asking women to rate how professional two women looked. One woman wore a size 6 dress and the other wore a size 14. Suppose that the researcher asked 20 women to rate the woman wearing the size 6 dress and another 20 to rate the woman wearing the size 14 dress. The ratings were as follows.

Highly professional
Somewhat professional
Not very professional
Not at all professional

The responses were coded 4, 3, 2, and 1, respectively and stored in columns A (size 6) and B (size 14) in file Xr16-12. Do these data provide sufficient evidence to infer that women perceive another woman wearing a size 6 dress as more professional than one wearing a size 14 dress?

16.13 Medical experts advocate the use of vitamin and mineral supplements to help fight infections. A study, undertaken by Dr. Ranjit Schneider of Memorial University (reported in the British journal *Lancet*, November 1992), recruited 96 men and women age 65 and older. One-half of them received daily supplements of vitamins and minerals, while the other half received placebos. The supplements contained the daily recommended amounts of 18 vitamins and minerals. These included vitamins B-6, B-12, C, D, and E, thiamine, riboflavin, niacin, calcium, copper, iodine, iron, selenium, magnesium, and zinc. The doses of vitamins A and E were slightly less than the daily requirements. The supplements included four times the amount of beta-carotene that the average person ingests daily. The number of days of illness from infections (ranging from colds to pneumonia) was recorded for each person. The data are stored in columns A (supplement) and B (placebo) in file Xr16-13. Can we infer that taking vitamin and mineral supplements daily increases the body's immune system? This exercise is identical to Exercise 12.24 (except for the data).

16.14 When new textbooks are written, they are usually reviewed by instructors who teach the course rather than by students who learn from the book. Suppose that a publisher of a new high school calculus book asked a random sample of 50 teachers and 100 high school students to rate their degree of satisfaction with the book using the following responses.

Very satisfied
Somewhat satisfied
Neither satisfied nor dissatisfied
Somewhat dissatisfied
Very unsatisfied

The responses were coded using a 5-4-3-2-1 system. The codes for the teachers and students are stored in columns A and B, respectively, in file Xr16-14. Can we infer that teachers and students differ in their assessment of the book?

16.15 A statistics professor is about to select a statistical software package for her course. One of the most important features, according to the professor, is the ease with which students learn to use the software. She has narrowed the selection to two possibilities: software A, a menu-driven statistical package with some high-powered techniques; and software B, a spreadsheet that has the capability of performing most techniques. To help make her decision, she asks 40 statistics students selected at random to choose one of the two packages. She gives each student a statistics problem to solve by computer and the appropriate manual. The amount of time (in minutes) each student needs to complete the assignment was recorded and stored in file Xr16-15 (column A = package A and column B = package B). (This exercise is identical to 12.29 a except for the data.) Can the professor conclude from these data that the two software packages differ in the amount of time needed to learn how to use them?

16.3 SIGN TEST AND WILCOXON SIGNED RANK SUM TEST

In the preceding section, we discussed the nonparametric technique for comparing two populations of data that are either ordinal or interval and where the data are independently drawn. In this section, the problem objective and data type remain as they were in Section 16.2, but we will be working with data generated from a matched pairs experiment. We have dealt with this type of experiment before. In Section 12.4, we dealt with the mean of the paired differences represented by the parameter μ_D. In this section, we introduce two nonparametric techniques that test hypotheses in problems with the following characteristics.

1 The problem objective is to compare two populations.
2 The data are either ordinal or interval and extremely nonnormal.
3 The samples are matched pairs.

To extract all the potential information from a matched pairs experiment, we must create the matched pairs differences. Recall that we did so when conducting the t-test and estimate of μ_D. We then calculated the mean and standard deviation of these differences and completed the test statistic and interval estimator. The first step in both nonparametric methods presented here is the same: compute the differences for each pair of observations. However, if the data are ordinal, we cannot perform any calculations on those differences because differences have no meaning.

To understand this point, consider comparing two populations of responses of people rating a product or service. The responses are "excellent," "good," "fair," and "poor." Recall that we can assign any numbering system so long as the order is maintained. The simplest system is 4-3-2-1. However, any other system such as 66-38-25-11 (a set of numbers randomly selected except for maintaining the order) is equally valid. Now suppose that in one matched pair the sample 1 response was "excellent" and the sample 2 response was "good." Calculating the matched pairs difference under the 4-3-2-1 system the difference is $4 - 3 = 1$. Using the 66-38-25-11 system, the difference is $66 - 38 = 28$. If we treat this and other differences as real numbers, we are likely to produce different results depending on which numbering system we used. Thus, we cannot use any method that uses the actual differences. However, we can use the sign of the differences. In fact, when the data are ordinal, that is the only method that is valid. That is, no matter what numbering system is employed, we know that "excellent" is better than "good." In the 4-3-2-1 system the difference between "excellent" and "good" is $+1$. In the 66-38-25-11 system the difference is $+28$. If we ignore the magnitude of the number and only record the sign, the two numbering systems (and all other systems where the rank order is maintained) will produce exactly the same result.

As you will shortly discover, the sign test only uses the sign of the differences. That's why it's called the **sign test.**

When the data are interval, however, differences have real meaning. Although we can use the sign test when the data are interval, doing so results in a loss of potentially useful information. For example, knowing that the difference in sales between two matched used car salespeople is 25 cars is much more informative than simply knowing that the first salesperson sold more cars than a second salesperson. As a result, when the data are interval but extremely nonnormal, we will use the **Wilcoxon signed rank sum test,** which incorporates not only the sign of the difference (hence the name), but the magnitude as well.

SIGN TEST

The sign test is employed in the following situations.

1 The problem objective is to compare two populations.
2 The data are ordinal.
3 The experimental design is matched pairs.

TEST STATISTIC

The sign test is quite simple. For each matched pair, we calculate the difference between the observation in sample 1 and the related observation in sample 2. We then count the number of positive differences and the number of negative differences. If the null hypothesis is true, we expect the number of positive differences to be approximately equal to the number of negative differences. Expressed another way, we expect the number of positive differences and the number of negative differences each to be approximately equal to half the total sample size. If either number is too large or too small, we reject the null hypothesis. By now you know that the determination of what is too large or too small comes from the sampling distribution of the test statistic. We will arbitrarily choose the test statistic to be the number of positive differences, which we denote x. The test statistic x is a binomial random variable, and, under the null hypothesis, the binomial proportion is $p = .5$. Thus, the sign test is none other than the z-test of p first developed in Section 11.4.

Recall from Chapters 5 and 8 that x is binomially distributed and that, for sufficiently large n ($np \geq 5$ and $n(1-p) \geq 5$), x is approximately normally distributed with mean np and standard deviation $\sqrt{np(1-p)}$. Thus, the standardized test statistic is

$$z = \frac{x - np}{\sqrt{np(1-p)}}$$

The null hypothesis

H_0 : The two population locations are the same.

is equivalent to testing

$H_0 : p = .5$

Therefore, the test statistic becomes

$$z = \frac{x - np}{\sqrt{np(1-p)}} = \frac{x - .5n}{\sqrt{n(.5)(.5)}} = \frac{x - .5n}{.5\sqrt{n}}$$

The normal approximation of the binomial distribution is valid when $np \geq 5$ and $n(1 - p) \geq 5$. When $p = .5$,

$$np = n(.5) \geq 5 \quad \text{and} \quad n(1 - p) = n(.5) \geq 5$$

implies that n must be greater than or equal to 10. Thus, this is one of the required conditions of the sign test. However, the quality of the inference with so small a sample size is poor. Larger sample sizes are strongly recommended.

It is common practice in this type of test to eliminate the matched pairs of observations when the differences equal zero. Consequently, n equals the number of nonzero differences in the sample.

▼ EXAMPLE 16.4

In an experiment to determine which of two cars is perceived to have the more comfortable ride, 100 people rode (separately) in the back seat of an expensive European model and also in the back seat of a North American midsized car. Each of the 100 people was asked to rate the ride on the following 5-point scale.

1 = Ride is very uncomfortable.

2 = Ride is quite uncomfortable.

3 = Ride is neither uncomfortable nor comfortable.

4 = Ride is quite comfortable.

5 = Ride is very comfortable.

The results are stored in columns A (European car ratings) and B (North American car ratings) in file Xm16-04. Some of these data are shown below. Do these data allow us to conclude that the European car is perceived to be more comfortable than the North American car?

Rating of European Car	Rating of North American Car
3	4
5	3
4	4
.	
.	
.	
5	5
4	5
3	4

Solution

IDENTIFY

The problem objective is to compare two populations of ordinal data. Because the same 100 people rated both cars, we recognize the experimental design as matched pairs. The sign test is applied, with the following results.

H_0: The two population locations are the same.

H_1: The location of population 1 (European car rating) is to the right of the location of population 2 (North American car rating)

Test statistic: $z = \dfrac{x - .5n}{.5\sqrt{n}}$

COMPUTE

Microsoft Excel Output for Example 16.4

Sign Test	
Difference	European - North American
Positive Differences	42
Negative Differences	27
Zero Differences	31
z Stat	1.806
P(Z<=z) one-tail	0.0355
z Critical one-tail	1.6449
P(Z<=z) two-tail	0.071
z Critical two-tail	1.96

The value of the test statistic is $Z = 1.806$ and the *p*-value of the test is .0355.

GENERAL COMMANDS	EXAMPLE 16.4
1 Type or import the data into two adjacent columns.	Open file **Xm16-04**.
2 Click **Tools, Data Analysis Plus,** and **Sign Test**.	
3 Specify the **Input Range:**.	A1:B101
4 Click **Labels**, if applicable.	
5 Specify the value of α (**Alpha**), and click **OK**.	.05

INTERPRET

There is relatively strong evidence to indicate that people perceive the European car to provide a more comfortable ride than the North American car. There are, however, two aspects of the experiment that may detract from the conclusion that European cars provide a more comfortable ride. First, did the respondents know in which car they were riding? If so, they may have answered on their preconceived bias that European cars are more expensive and therefore better. If the subjects were blindfolded, we would be more secure in our conclusion. Second, was the order in which each subject rode the two cars varied? If all of the subjects rode the North American car first and the European car second, that may have influenced their ratings. The experiment should have been conducted so that the car each subject rode in first was randomly determined.

Figure 16.9 describes the sampling distribution of the test statistic.

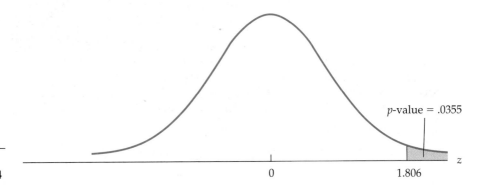

Figure 16.9

Sampling distribution of the test statistic for Example 16.4

CHECKING THE REQUIRED CONDITIONS

As was the case with the Wilcoxon rank sum test, the sign test requires that the populations be identical in shape and spread. The histograms (Figures 16.10 and 16.11) confirm that these conditions are satisfied in this example. The other condition is that the sample size exceeds 10.

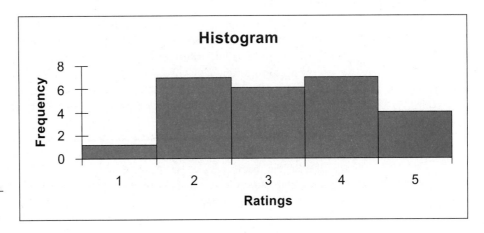

Figure 16.10

Histogram of ratings of European car in Example 16.4

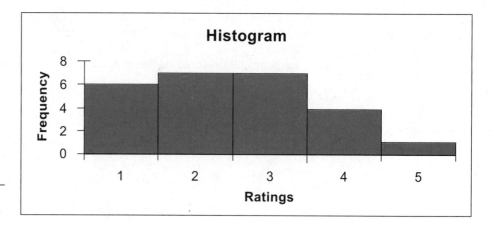

Figure 16.11

Histogram of ratings of North American car in Example 16.4

WILCOXON SIGNED RANK SUM TEST

The Wilcoxon signed rank sum test is used under the following circumstances:

1. The problem objective is to compare two populations.
2. The data are interval and extremely nonnormal.
3. The samples are matched pairs.

This test is the nonparametric counterpart of the t-test of μ_D.

Because the data are interval, we can refer to the Wilcoxon signed rank sum test as a test of μ_D. However, to be consistent with the other nonparametric techniques and to avoid confusion, we will express the hypotheses to be tested in the same way as in Section 16.2.

TEST STATISTIC

We begin by computing the paired differences $D = x_1 - x_2$. As we did in the sign test, we eliminate all differences where $D = 0$. Next we rank the absolute values of D where 1 = smallest value of $|D|$ and n = largest value of $|D|$ and where n = number of nonzero differences. (We average the ranks of tied observations.) The sum of the ranks of the positive differences (denoted T^+) and the sum of the ranks of the negative differences (denoted T^-) are then calculated. We arbitrarily select T^+, which we label T, as our test statistic.

If the null hypothesis (H_0 : The two population locations are the same) is true, we expect that the number of positive and negative differences would be similar. Moreover, the magnitudes of the positive and negative differences are also expected to be similar. However, if either of these circumstances is untrue, then there are indications that the null hypothesis is false. Consider the following small-scale illustrations.

Sample 1

X1	X2	D	|D|	Rank+	Rank−
7	16	−9	9		6
9	29	−20	20		10
14	6	8	8	5	
21	7	14	14	9	
11	4	7	7	4	
17	7	10	10	7	
8	11	−3	3		1
8	14	−6	6		3
3	15	−12	12		8
11	6	5	5	2	
			Totals	27	28

In sample 1, there are five positive and five negative differences. Moreover, the magnitudes of the positive and negative differences are similar. As a result, the rank sums T^+ and T^- are approximately the same. This statistical result is consistent with a true null hypothesis.

Sample 2

X1	X2	D	|D|	Rank+	Rank−
10	7	3	3	3	
9	14	−5	5		5
14	6	8	8	8	
21	12	9	9	9	
11	4	7	7	7	
17	7	10	10	10	
9	8	1	1	1	
14	8	6	6	6	
5	3	2	2	2	
8	4	4	4	4	
			Totals	50	5

In the second sample, there are nine positive differences and only one negative difference. The rank sums are sufficiently different to allow us to conclude that the null hypothesis is false. If we were to apply the sign test here we would draw the same conclusion.

Sample 3

X1	X2	D	|D|	Rank+	Rank−
7	10	−3	3		1
9	14	−5	5		3
24	6	18	18	7	
63	12	51	51	10	
33	4	29	29	8	
7	15	−8	8		4
47	8	39	39	9	
24	14	10	10	5	
20	5	15	15	6	
4	8	−4	4		2
			Totals	45	10

Sample 3 features matched pairs where there are six positive differences and four negative differences. However, the four smallest absolute differences came from matched pairs where the differences were negative. The six largest absolute differences came from matched pairs where the differences were positive. The resulting rank sums are quite different, providing some evidence that the null hypothesis is false.

Applying the sign test to sample 3 would not produce a test statistic sufficiently large or small to reject the null hypothesis. As you can see, the magnitude of the differences is critical in this test.

For sample sizes greater than 30, T is approximately normally distributed with mean

$$E(T) = \frac{n(n+1)}{4}$$

and standard deviation

$$\sigma_T = \sqrt{\frac{n(n+1)(2n+1)}{24}}$$

Thus, the standardized test statistic is

$$z = \frac{T - E(T)}{\sigma_T}$$

▼ EXAMPLE 16.5

Traffic congestion on roads and highways costs industry billions of dollars annually as workers struggle to get to and from work. Several suggestions have been made about how to improve this situation, one of which is called flextime, which involves allowing workers to determine their own schedules (provided they work a full shift). Such workers will likely choose an arrival and departure time to avoid rush-hour traffic. In a preliminary experiment designed to investigate such a program, the general manager of a large company wanted to compare the times it took workers to travel from their homes to work at 8:00 AM with travel time under the flextime program. A random sample of 32 workers was selected. The employees recorded the time it took to arrive at work at 8:00 AM on Wednesday of one week. The following week, the same employees arrived at work at times of their own choosing. The travel time on Wednesday of that week was recorded. These results are stored in columns A (arrive at 8:00 AM) and B (flextime) in file Xm16-05 with some of the numbers listed below. Can we conclude that travel times under the flextime program are different from travel times to arrive at work at 8:00 AM?

Time: 8:00 Arrival	Time: Flextime
34	31
35	31
43	44
.	
.	
20	19
19	21
42	38

Solution

IDENTIFY

The objective is to compare two populations; the data are interval and were produced from a matched pairs experiment. If matched pairs differences are normally distributed, we should apply the *t*-test of μ_D. To judge whether the data are normal, we instructed Excel to compute the paired differences and draw the histogram. Figure 16.12 depicts this histogram. Apparently, the distribution of differences is extremely nonnormal, indicating that we should employ the Wilcoxon signed rank sum test.

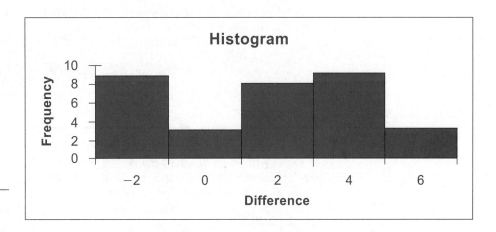

Figure 16.12

Histogram of matched pairs difference for Example 16.5

Because we want to know whether the two populations of times differ, we perform a two-tail test:

H_0 : The two population locations are the same.

H_1 : The location of population 1 (8:00 arrival) is different from the location of population 2 (flextime program).

COMPUTE

Microsoft Excel Output for Example 16.5

Wilcoxon Signed Rank Sum Test	
Difference	8:00-Arr - Flextime
T+	367.5
T−	160.5
Observations (for test)	32
z Stat	1.935
P(Z<=z) one-tail	0.0265
z Critical one-tail	1.6449
P(Z<=z) two-tail	0.053
z Critical two-tail	1.96

The output includes the rank sums $T^+ = 367.5$ and $T^- = 160.5$. The test statistic is $z = 1.935$ with a *p*-value of .053.

GENERAL COMMANDS	EXAMPLE 16.5
1 Type or import the data into two adjacent columns.	Open file **Xm16-05**.
2 Click **Tools, Data Analysis Plus,** and **Wilcoxon Signed Rank Sum Test**.	
3 Specify the **Input Range:**.	A1:B33
4 Click **Labels,** if applicable.	
5 Specify the value of α **(Alpha),** and click **OK**.	.05

INTERPRET

There is not enough evidence to infer that flextime commutes are different from the commuting times under the current schedule. This conclusion may be due primarily to the way in which this experiment was performed. All of the drivers recorded their travel time with 8:00 AM arrival on the first Wednesday and their flextime travel time on the second Wednesday. If the second day's traffic was heavier than usual, that may account for the conclusion reached. As we pointed out in Example 16.4, the order of schedules should have been randomly determined for each employee. In this way, the effect of varying traffic conditions could have been minimized.

Figure 16.13 depicts the sampling distribution in this example.

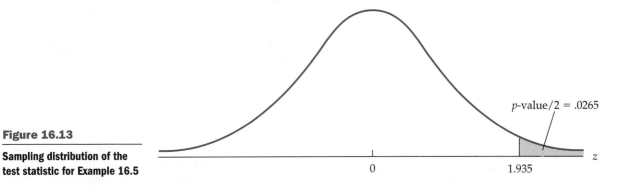

Figure 16.13

Sampling distribution of the test statistic for Example 16.5

Here is how we recognize when to use the two techniques introduced in this section.

> ### *Factors That Identify the Sign Test*
> 1. Problem objective: compare two populations.
> 2. Data type: ordinal.
> 3. Experimental design: matched pairs.

> ### *Factors That Identify the Wilcoxon Signed Rank Sum Test*
> 1. Problem objective: compare two populations.
> 2. Data type: interval.
> 3. Distribution of differences: extremely nonnormal.
> 4. Experimental design: matched pairs.

EXERCISES

16.16 A random sample of 50 people was asked to rate two brands of ice cream using the following responses:

Delicious
OK
Not bad
Terrible

The responses were converted to codes 4, 3, 2, and 1, respectively, and stored in columns A (Brand A) and B (Brand B) in file Xr16-16. Can we infer that Brand A is preferred?

16.17 Refer to Exercise 16.16. The responses were recoded 28, 25, 16, and 3, respectively, and stored in columns A and B in file Xr16-17.
  a Can we infer that brand A is preferred?
  b Compare your answer in part (a) with that of Exercise 16.16.

16.18 A random sample of 80 statistics students was asked to rate both the course and the instructor. The ratings are

Excellent
Very good
Good
Fair
Poor

The responses were coded 5, 4, 3, 2, and 1, respectively, and stored in columns A (course) and B (instructor) in file Xr16-18. Can we infer that students rate the instructor more highly than the course?

16.19 Refer to Exercise 16.18. The responses were recoded 85, 83, 78, 70, and 25, respectively, and stored in columns A (course) and B (instructor) in file Xr16-19.

a Can we infer that students rate the instructor more highly than the course?
b Compare your answer in part (a) with that of Exercise 16.18.

16.20 Refer to Example 16.4. Suppose that the responses have been recoded in the following way.

6 = Ride is very uncomfortable.
24 = Ride is quite uncomfortable.
28 = Ride is neither uncomfortable nor comfortable.
53 = Ride is quite comfortable.
95 = Ride is very comfortable.

The results are stored in columns A (European car ratings) and B (North American car ratings) in file Xr16-20.

a Do these data allow us to conclude that the European car is perceived to be more comfortable than the North American car?
b Compare your answer in part (a) with that of Example 16.4.

16.21 Data from a matched pairs experiment were stored in columns A and B in file Xr16-21.

a Use the sign test to determine whether the location of population A is different from the location of population B.
b Repeat part (a) using the Wilcoxon signed rank sum test.
c Compare the results of the tests in parts (a) and (b).

16.22 Data from a matched pairs experiment were stored in columns A and B in file Xr16-22.

a Use the sign test to determine whether the location of population A is different from the location of population B.
b Repeat part (a) using the Wilcoxon signed rank sum test.
c Compare the results of the tests in parts (a) and (b).

16.23 Research scientists at a pharmaceutical company have recently developed a new nonprescription sleeping pill. They decide to test its effectiveness by measuring the time it takes for people to fall asleep after taking the pill. Preliminary analysis indicates that the time to fall asleep varies considerably from one person to another. Consequently, they organize the experiment in the following way. A random sample of 100 volunteers who regularly suffer from insomnia is chosen. Each person is given one pill containing the newly developed drug and one placebo. (A placebo is a pill that contains absolutely no medication.) Participants are told to take one pill one night and the second pill one night a week later. (They do not know whether the pill they are taking is the placebo or the real thing, and the order of use is random.) Each participant is fitted with a device that measures the time until sleep occurs. The data are stored in columns A and B of file Xr16-23. Can we conclude that the new drug is effective? (This exercise is identical to Exercise 12.46, except for the data.)

16.24 Suppose the housework study referred to in Exercise 16.8 was repeated with some changes. In the revised experiment, 60 women were asked last year and again this year how many hours of housework they perform weekly. The results are stored in file Xr16-24. (Column A = hours of housework this year and column B = hours of housework last year.) Can we conclude that women as a group are doing less housework now than last year?

16.25 At the height of the energy shortage during the 1970s, governments were actively seeking ways to persuade consumers to reduce their energy consumption. Among other efforts undertaken, several advertising campaigns were launched. To provide input on how to design effective advertising messages, a poll was taken in which people were asked how concerned they were about shortages of gasoline and electricity. There were four possible responses to the questions.

Not concerned at all
Not too concerned
Somewhat concerned
Very concerned

A poll of 150 individuals produced the results stored in file Xr16-25, where column A = concern about gasoline shortage and column B = concern about electricity shortage. (The responses were coded 1, 2, 3, and 4, respectively.) Do these data provide enough evidence to allow us to infer that concern about a gasoline shortage exceeded concern about an electricity shortage?

16.26 A locksmith is in the process of selecting a new key-cutting machine. If there is a difference in key-cutting speed between the two machines under consideration, he will purchase the faster one. If there is no difference, he will purchase the cheaper machine. The times (in seconds) required to cut each of the 24 most common types of keys are stored in columns A (machine A) and B (machine B, the cheaper one) of file Xr16-26. What should he do?

16.27 A large sporting-goods store located in Florida is planning a renovation that will result in an increase in the floor space for one department. The manager of the store has narrowed her choice about which department's floor space to increase to two possibilities: the tennis-equipment department or the swimming-accessories department. The manager would like to enlarge the tennis-equipment department, because she believes that this department improves the overall image of the store. She decides, however, that if the swimming-accessories department can be shown to have higher gross sales, she will choose that department. She has collected each of the two departments' weekly gross sales data for the past six months; these are stored in columns A (tennis-equipment sales) and B (swimming-accessories sales) of file Xr16-27. Which department should be enlarged?

16.28 Does the brand name of an ice cream affect consumers' perceptions of it? The marketing manager of a major dairy pondered this question. She decided to ask 60 randomly selected people to taste the same flavor of ice cream in two different dishes. The dishes contained exactly the same ice cream but were labeled differently. One was given a name that suggested that its maker was European and sophisticated; the other was given a name that implied that the product was domestic and inexpensive. The tasters were asked to rate each ice cream using one of the following responses:

Excellent
Very good
Good
Fair
Poor

The responses were coded using a 5-4-3-2-1 system and stored in file Xr16-28 (column A = ratings for "European" ice cream, and column B = ratings for "domestic" ice cream). Do the results allow the manager to conclude that the "European" brand is preferred?

16.29 Do children feel less pain than adults? That question was addressed by nursing professors at the Universities of Alberta and Saskatchewan (reported in the *Toronto Star*, 14 June 1991). Suppose that, in a preliminary study, 50 eight-year-old children and their mothers were subjected to moderately painful pressure on their hands. Each was asked to rate the level of pain as very severe (4), severe (3), moderate (2), or weak (1). The data were stored in file Xr16-29 using the codes in parentheses (column A = responses from children and column B = responses from their mothers). Can we conclude that children feel less pain than adults?

16.30 In a study to determine whether sex affects salary offers for graduating MBA students, 25 pairs of students were selected. Each pair consisted of a male and a female student who had almost identical grade-point averages, courses taken, ages, and previous work experience. The highest salary offered to each student upon graduation was recorded and stored in columns A (female offers) and B (male offers) of file Xr16-30. Is there sufficient evidence to allow us to conclude that salary offers differ between men and women?

16.31 Admissions officers at universities and colleges face the problem of comparing grades achieved at different high schools. As a step toward developing a more informed interpretation of such grades, an admissions officer at a large state university conducts the following experiment. The records of 100 students from the same local high school who just completed their first year at the university were selected. Each of these students was paired (according to average grade in the last year of high school) with a student from another local high school who also just completed the first year at the university. For each matched pair, the average letter grades (4 = A, 3 = B, 2 = C, 1 = D, or 0 = F) in the first year of university study were recorded. The results are stored in file Xr16-31 (column A = grades of students from high school A and column B = grades of students from high school B). Do these results allow us to conclude that, in comparing two students with the same high-school average (one from high school A and the other from high school B), preference in admissions should be given to the student from high school A?

16.32 Some movie studios believe that by adding sexually explicit scenes to the home video version of movies they can increase the movies' appeal and profitability (*Wall Street Journal*, 14 October 1988). A studio executive decided to test this belief. She organized a study that involved 40 movies that were rated PG-13. Versions of each movie were created by adding scenes that changed the rating to R. The two versions of the movies were then made available to rental shops. For each of the 40 pairs of movies, the total number of rentals in one major city during a one-week period was recorded and stored in file Xr16-32 (column A = rentals of PG-13 version and column B = rentals of R version). Do these data provide enough evidence to support the belief?

16.33 Do waiters or waitresses earn larger tips? To answer this question, a restaurant consultant undertook a preliminary study. The study involved measuring the percentage of the total bill left as a tip for one randomly selected waiter and one randomly selected waitress in each of 50 restaurants during a one-week period. The data are stored in columns A and B of file Xr16-33. What conclusions can be drawn from these data? (This exercise is identical to Exercise 12.44, except for the data.)

16.4 KRUSKAL-WALLIS TEST

In this section we introduce the first of two statistical procedures designed to compare two or more populations. The **Kruskal-Wallis test** is applied to problems with the following characteristics:

1. The problem objective is to compare two or more populations.
2. The data are either ordinal or interval and extremely nonnormal.
3. The samples are independent.

When the data are interval and normal, we apply the analysis of variance F-test presented in Chapter 14 to determine whether differences exist. When the data are extremely nonnormal, we will treat the data as if they were ordinal and employ the Kruskal-Wallis test.

The null and alternative hypotheses for this test are similar to those we specified in the analysis of variance. Because the data are ordinal or are treated as ordinal, however, we test population locations instead of population means. In all applications of the Kruskal-Wallis test, the null and alternative hypotheses are (where k is the number of populations):

H_0: The locations of all k populations are the same.

H_1: At least two population locations differ.

TEST STATISTIC

The test statistic is calculated in a way that closely resembles the way in which the Wilcoxon rank sum test was calculated. The first step is to rank all the observations. As before, 1 = smallest observation and n = largest observation, where $n = n_1 + n_2 + \cdots + n_k$. In case of ties, average the ranks.

If the null hypothesis is true, the ranks should be evenly distributed among the k samples. The degree to which this is true is judged by calculating the rank sums (labeled $T_1, T_2, \ldots, T_k$). The last step is to calculate the test statistic, which is denoted H.

Test Statistic for Kruskal-Wallis Test

$$H = \left[\frac{12}{n(n+1)} \sum_{j=1}^{k} \frac{T_j^2}{n_j} \right] - 3(n+1)$$

Although it is impossible to see from this formula, if the rank sums are similar, the test statistic will be small. As a result, a small value of H supports the null hypothesis. Conversely, if considerable differences exist among the rank sums, the test statistic will be large. To judge the value of H, we need to know its sampling distribution.

SAMPLING DISTRIBUTION OF THE TEST STATISTIC

The distribution of the test statistic can be derived in the same way we derived the sampling distribution of the test statistic in the Wilcoxon rank sum test. That is, we can list all possible combinations of ranks and their probabilities to yield the sampling distribution. A table of critical values can then be determined. However, this is only necessary for small sample sizes. For sample sizes greater than or equal to 5, the test statistic H is approximately chi-squared distributed with $\nu = k - 1$ degrees of freedom. As we noted above, large values of H are associated with different population locations. Consequently, we want to reject the null hypothesis if H is sufficiently large. Thus, the p-value of the test is

$$p\text{-value} = P(\chi^2 > H)$$

Figure 16.14 describes this sampling distribution and the p-value.

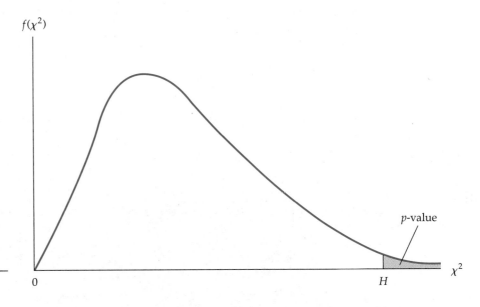

Figure 16.14

Sampling distribution of H

▼ EXAMPLE 16.6

The managers of fast-food restaurants are extremely interested in knowing how their customers rate the quality of food and service and the cleanliness of the restaurants. Customers are given the opportunity to fill out customer comment cards. Suppose that one franchise wanted to compare how customers rate the three shifts (4:00 PM to midnight, midnight to 8:00 AM, and 8:00 AM to 4:00 PM). In a preliminary study, 50 customer cards were randomly selected from each shift. The responses to the question concerning speed of service were stored in columns A, B, and C, respectively, in file Xm16-06 (4 = excellent, 3 = good, 2 = fair, and 1 = poor). A partial list of the data are provided. Do these data provide sufficient evidence to indicate whether customers perceive the speed of service to be different among the three shifts?

COMMENTS: 4:00–MIDNIGHT	COMMENTS: MIDNIGHT–8:00	COMMENTS: 8:00–4:00
2	1	3
4	4	4
1	2	4
.	.	.
.	.	.
.	.	.
3	2	1
4	2	1
4	4	3

Solution

IDENTIFY

The problem objective is to compare three populations (the ratings of the three shifts); the data are ordinal and the samples are independent. These three factors are sufficient to determine the use of the Kruskal-Wallis test. The complete test follows.

H_0 : The locations of all three populations are the same.

H_1 : At least two population locations differ.

Test statistic: $H = \left[\dfrac{12}{n(n+1)} \sum_{j=1}^{k} \dfrac{T_j^2}{n_j} \right] - 3(n+1)$

COMPUTE

Microsoft Excel Output for Example 16.6

Kruskal-Wallis Test

Group	Rank Sum	Observations
4:00-mid	4106	50
Mid-8:00	3519	50
8:00-4:00	3700	50
H Stat		1.915
df		2
p-value		0.3839
chi-squared Critical		5.9915

The rank sums are $T_1 = 4106$, $T_2 = 3519$, and $T_3 = 3700$. The value of the test statistic is $H = 1.915$, and the p-value .3839.

GENERAL COMMANDS	EXAMPLE 16.6
1 Type or import the data into k adjacent columns.	Open file **Xm16-06**.
2 Click **Tools, Data Analysis Plus**, and **Kruskal Wallis Test**.	
3 Specify the **Input Range:**.	A1:C51
4 Click **Labels**, if applicable.	
5 Specify the value of α **(Alpha)**, and click **OK**.	.05

INTERPRET

There is not enough evidence to infer that a difference in speed of service exists among the three shifts. Management should assume that all three of the shifts are equally rated, and any action to improve service should be applied to all three shifts. Management should also bear in mind that the data were generated from a self-selecting sample. (See the story of the *Literary Digest* poll presented in Chapter 7.)

Figure 16.15 exhibits the chi-squared sampling distribution.

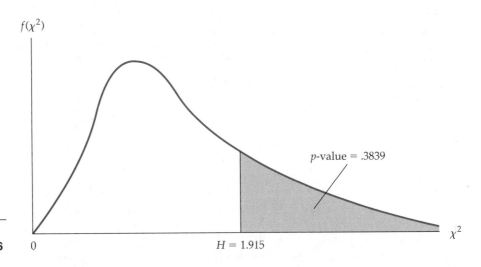

Figure 16.15

Sampling distribution of the test statistic for Example 16.6

KRUSKAL-WALLIS TEST AND THE WILCOXON RANK SUM TEST

The Kruskal-Wallis test can be used to test for a difference between two populations. It will produce the same outcome as the two-tail Wilcoxon rank sum test. However, the Kruskal-Wallis test can only determine whether a *difference* exists. To determine whether one population is left of another or one population is right of another, we must apply the Wilcoxon rank sum test.

We complete this section with a review of how to recognize the use of the Kruskal-Wallis test.

> **Factors That Identify the Kruskal-Wallis Test**
>
> 1 Problem objective: compare two or more populations.
> 2 Data type: ordinal or interval and extremely nonnormal.
> 3 Experimental design: independent samples.

EXERCISES

16.34 Random samples of 50 people each were asked to rate four different computer printers in terms of their ease of use. The responses are

Very easy to use
Easy to use
Difficult to use
Very difficult to use

The responses were coded using a 4-3-2-1 system and stored in columns A, B, and C in file Xr16-34. Do these data yield enough evidence to infer that differences in ratings exist among the four printers?

16.35 Refer to Exercise 16.34. The responses were recoded using a 25-22-5-2 system.
  a Do these data yield enough evidence to infer that differences in ratings exist among the four printers?
  b Compare your answer in part (a) with Exercise 16.34.

16.36 Refer to Example 16.6. The responses were recoded using a 100-90-55-25 system and stored in file Xr16-36.

  a Do these data yield enough evidence to infer that differences in ratings exist among the three shifts?
  b Compare your answer in part (a) with Example 16.6.

16.37 In an effort to determine whether differences exist among three methods of teaching statistics, a professor of business taught his course differently in each of three large sections. In the first section, he taught by lecturing; in the second, he taught by the case method; and in the third, he used a computer software package extensively. At the end of the semester, each student was asked to evaluate the course on a 7-point scale, where 1 = atrocious, 2 = poor, 3 = fair, 4 = average, 5 = good, 6 = very good, and 7 = excellent. From each section, the professor chose 25 evaluations at random. The data are in columns A, B, and C of file Xr16-37. Is there evidence that differences in student satisfaction exist with respect to at least two of the three teaching methods?

16.38 Applicants to law schools must write the Law School Admission Test (LSAT). There are several companies that offer assistance in preparing for the test. To determine whether they work, and, if so, which one is best, an experiment was conducted. Several hundred law school applicants were surveyed and asked to report their LSAT score and which if any LSAT preparation course they took. The scores are stored in file Xr16-38 using the following format:

Column A: LSAT score for applicants with preparatory course A
Column B: LSAT score for applicants with preparatory course B
Column C: LSAT score for applicants with preparatory course C
Column D: LSAT score for applicants with no preparatory course

Do these data allow us to infer that there are differences among the four groups of LSAT scores? (This exercise is identical to Exercise 14.12, except for the data.)

16.39 A consumer testing service is comparing the effectiveness of four different brands of drain cleaners. The experiment consists of using each product on 50 different clogged sinks and measuring the amount of time that elapses until each drain became unclogged. The recorded times, measured in minutes, are stored in columns A through D in file Xr16-39.
  a Which techniques should be considered as possible procedures to apply to determine if differences exist? What are the required conditions? How do you decide?
  b If a statistical analysis has shown that the times are not normally distributed, can the service conclude that differences exist among the speeds at which the four brands perform?

16.40 During the last presidential campaign, the Gallup organization surveyed a random sample of 30 registered Democrats in January, another 30 in February, and yet another 30 in March. All 90 Democrats were asked to "rate the chances of the Democrats winning the presidential race in your state." The responses and their numerical codes were excellent (4), good (3), fair (2), and poor (1). The results of the surveys are stored in columns A to C, respectively, in file Xr16-40. Do these data allow us to infer that Democrats' ratings of their chances of winning the presidency changed over the three-month period?

16.41 Because there are no national or regional standards, it is difficult for university admission committees to compare graduates of different high schools. University administrators have noted that an 80% average at a high school with low standards may be equivalent to a 70% average at another school with higher standards of grading. In an effort to more equitably compare applications, a pilot study was initiated. Random samples of students who were admitted the previous year were drawn. All of the students entered the business program with averages between 74% and 76% from a random sample of four local high schools. Their average grades in the first year at the university were computed and stored in columns A through D of file Xr16-41. Can the university admissions officer conclude that there are differences in grading standards among the four high schools? (This exercise is identical to Exercise 14.5, except for the data.)

16.42 It is common practice in the advertising business to create several different advertisements and then ask a random sample of potential customers to rate the ads on several different dimensions. Suppose that an advertising firm developed four different ads for a new breakfast cereal and asked a sample of 400 shoppers to rate the believability of the advertisements. One hundred people viewed ad 1, another 100 viewed ad 2, another 100 saw ad 3, and another 100 saw ad 4. The ratings were very believable (4), quite believable (3), somewhat believable (2), and not believable at all (1). The responses are stored in columns A to D, respectively, in file Xr16-42. Can the firm's management conclude that differences exist in believability among the four ads?

16.43 Do university students become more supportive of their varsity teams as they progress through their four-year stint? To help answer this question, a sample of students was drawn. Each was asked whether she or he was a freshman, sophomore, junior, or senior and to what extent she or he supported the university's football team, the Hawks. The responses to this question are:

Wildly fanatic
Support the Hawks wholeheartedly
Support the Hawks, but not that enthusiastically
Who are the Hawks?

The responses were coded using a 4-3-2-1 numbering system. The data were stored in columns A through D for freshman, sophomore, junior, and senior, respectively, in file Xr16-43. Can we conclude that the four levels of students differ in their support for the Hawks?

16.44 In anticipation of buying a new scanner, a student turned to a Web site that reported the results of surveys of users of the different scanners. A sample of 133 responses was listed showing the ease of use of five different brands. The survey responses were:

Very easy
Easy
Not easy
Difficult
Very difficult

The responses were assigned numbers from 1 to 5 and stored in columns A through E for Brands A through E, respectively, in file Xr16-44. Can we infer that there are differences in perceived ease of use among the five brands of scanners?

16.5 FRIEDMAN TEST

This section introduces another statistical technique whose objective is to compare two or more populations of ordinal or interval data. In Section 14.4, we presented the randomized block experiment of the analysis of variance. In this section, we present its nonparametric counterpart. The **Friedman test** is applied to problems with the following characteristics:

1 The problem objective is to compare two or more populations.

2 The data are either ordinal or interval and extremely nonnormal.

3 The data are generated from a randomized block experiment.

The null and alternative hypotheses are identical to the ones tested in the Kruskal-Wallis test. That is,

H_0 : The locations of all k populations are the same.

H_1 : At least two population locations differ.

TEST STATISTIC

To calculate the test statistic, we first rank each observation within each block, where 1 = smallest observation and k = largest observation, averaging the ranks of ties. Then we compute the rank sums, which we label $T_1, T_2, \ldots, T_k$. The test statistic is defined as follows. (Recall that b = number of blocks.)

> *Test Statistic for the Friedman Test*
>
> $$F_r = \left[\frac{12}{b(k)(k+1)} \sum_{j=1}^{k} T_j^2 \right] - 3b(k+1)$$

SAMPLING DISTRIBUTION OF THE TEST STATISTIC

The test statistic is approximately chi-squared distributed with $\nu = k - 1$ degrees of freedom, provided that either k or b is greater than or equal to 5. As was the case with the Kruskal-Wallis test, we reject the null hypothesis when the test statistic is large. Hence, the p-value of the test is

$P(\chi^2 > F_r)$

(See Figure 16.16.) This test, like all the other nonparametric tests, requires that the populations being compared are identical in shape and spread.

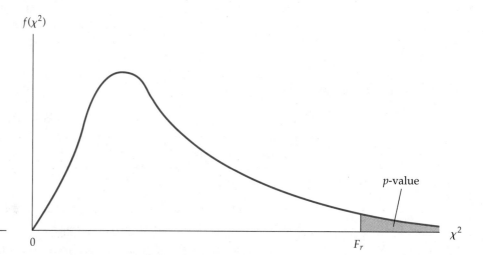

Figure 16.16

Sampling distribution of F_r

EXAMPLE 16.7

At a large university, applicants to the teachers' college must submit their academic records together with a letter of reference from one of their professors. A committee of four faculty members then rates each applicant relative to other applicants that year. The ratings are:

1 The candidate is in the top 5% of applicants.
2 The candidate is in the top 10% of applicants, but not in the top 5%.
3 The candidate is in the top 25% of applicants, but not in the top 10%.
4 The candidate is in the top 50% of applicants, but not in the top 25%.
5 The candidate is in the bottom 50% of applicants.

The dean is concerned about inconsistencies in the ratings. Some applicants are rated highly by one or two members, while other members of the committee may rate the same applicant considerably lower. To test for inconsistencies in the rating system, she takes a random sample of the evaluations of 20 applicants. The results are stored in columns A through D in file Xm16-07 and listed below. What conclusions can the dean draw from these data?

PROFESSOR A'S RATINGS	PROFESSOR B'S RATINGS	PROFESSOR C'S RATINGS	PROFESSOR D'S RATINGS
3	2	3	4
1	1	2	2
2	3	2	3
2	4	3	4
4	5	5	5
3	5	3	2
1	2	1	3
1	3	2	4
1	2	2	2
3	2	4	3
2	4	5	4
2	4	4	4
3	5	5	4
4	4	2	3
3	1	3	3
4	3	4	5
2	1	2	2
1	4	2	3
3	2	3	2
1	4	3	3

Solution

IDENTIFY

The problem objective is to compare the four populations of professors' evaluations, which we can see are ordinal data. This experiment is identified as a randomized block design because all 20 applicants were evaluated by all four professors on the committee. (The treatments are the professors, and the blocks are the applicants.) The appropriate statistical technique is the Friedman test. The null and alternative hypotheses are as follows:

H_0: The locations of all four populations are the same.

H_1: At least two population locations differ.

The test statistic: $F_r = \left[\dfrac{12}{b(k)(k+1)} \sum_{j=1}^{k} T_j^2\right] - 3b(k+1)$

which is chi-squared distributed with $k - 1$ degrees of freedom.

COMPUTE

Microsoft Excel Output for Example 16.7

Friedman Test

Group	Rank Sum
Professor A	36.5
Professor B	51.5
Professor C	53.5
Professor D	58.5
Fr Stat	8.07
df	3
p-value	0.0446
chi-squared Critical	7.8147

The rank sums are $T_1 = 36.5$, $T_2 = 51.5$, $T_3 = 53.5$, and $T_4 = 58.5$. The value of the test statistic is $F_r = 8.07$. The *p*-value = .0446.

GENERAL COMMANDS	EXAMPLE 16.7
1 Type or import the data into *k* adjacent columns.	Open file **Xm16-07**.
2 Click **Tools, Data Analysis Plus,** and **Friedman Test.**	
3 Specify the **Input Range:**.	A1:D21
4 Click **Labels,** if applicable.	
5 Specify the value of α **(Alpha),** and click **OK**.	.05

INTERPRET

There appears to be sufficient evidence to indicate that the professors' evaluations differ. The dean should attempt to determine why the evaluations differ. Is the problem the way in which the assessment is conducted, or is it that some professors are using different criteria? Figure 16.17 describes the sampling distribution and *p*-value.

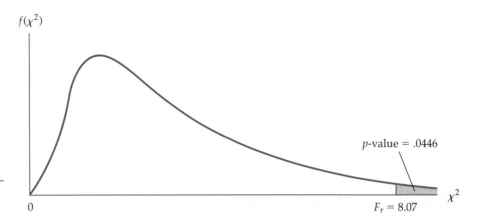

Figure 16.17

Sampling distribution of the test statistic for Example 16.7

THE FRIEDMAN TEST AND THE SIGN TEST

The relationship between the Friedman and sign tests is the same as the relationship between the Kruskal-Wallis and Wilcoxon rank sum tests. That is, we can use the Friedman test to determine whether two populations differ. The conclusion will be the same as that produced from the sign test. However, we can only use the Friedman test to determine whether a difference exists. If we want to determine whether one population is to the left or to the right of another, we must use the sign test.

Here is a list of the factors that tell us when to use the Friedman test.

Factors That Identify the Friedman Test

1 Problem objective: compare two or more populations.

2 Data type: ordinal or interval and extremely nonnormal.

3 Experimental design: blocked samples.

EXERCISES

16.45 A random sample of 30 people was asked to rate each of four different premium brands of coffee. The ratings are:

Poor
Fair
Good
Excellent

The responses were assigned numbers 1 through 4, respectively, and stored in columns A through D in file Xr16-45. Can we infer that differences exist among the ratings of the four brands of coffee?

16.46 Refer to Exercise 16.45. Suppose that the codes were 12, 31, 66, and 72, respectively. The data are stored in file Xr16-46.

a Can we infer that differences exist among the ratings of the four brands of coffee?

b Compare your answer in part (a) to that in Exercise 16.45.

16.47 Refer to Example 16.7. Suppose that the responses were recoded so that the numbers equaled the midpoint of the range of percentiles. That is,

97.5 = The candidate is in the top 5% of applicants.
92.5 = The candidate is in the top 10% of applicants but not in the top 5%.
82.5 = The candidate is in the top 25% of applicants but not in the top 10%.
62.5 = The candidate is in the top 50% of applicants but not in the top 25%.
25 = The candidate is in the bottom 50% of applicants.

The data are stored in columns A through D in file Xr16-47.

a Can we conclude that differences exist among the ratings assigned by the four professors?

b Compare your answer in part (a) to that in Example 16.7.

16.48 Ten judges were asked to test the sensory quality of four different brands of orange juice. The judges assigned scores using a 5-point scale where 1 = bad, 2 = poor, 3 = average, 4 = good, and 5 = excellent. The results are stored in file Xr16-48. Can we conclude that there are differences in sensory quality among the four brands of orange juice?

16.49 The manager of a personnel company is in the process of examining her company's advertising programs. Currently, the company advertises in each of the three local newspapers for a wide variety of positions, including computer programmers, secretaries, and receptionists. The manager has decided that only one newspaper will be used if it can be determined that there are differences in the number of inquiries generated among the newspapers. The following experiment was performed. For one week (six days), six different jobs were advertised in each of the three newspapers. The number of inquiries was counted, and the results appear in the accompanying table. (They are also stored in columns A, B, and C in file Xr16-49.)

Job Advertised	Newspaper		
	1	2	3
Receptionist	14	17	12
Systems analyst	8	9	6
Junior secretary	25	20	23
Computer programmer	12	15	10
Legal secretary	7	10	5
Office manager	5	9	4

a What techniques should be considered to apply in reaching a decision? What are the required conditions? How do we determine whether the conditions are satisfied?

b Assuming that the data are extremely nonnormally distributed, can we conclude that differences exist among the newspapers' abilities to attract potential employees?

16.50 In recent years, lack of confidence in the Postal Service has led many companies to send all of their correspondence by private courier. A large company is in the process of selecting one of three possible couriers to act as its sole delivery method. To help in making the decision, an experiment was performed whereby letters were sent using each of the three couriers at 12 different times of the day to a delivery point across town. The number of minutes required for delivery was recorded and stored in columns A, B, and C in file Xr16-50. Can we conclude that there are differences in delivery times among the three couriers? (This exercise is identical to Exercise 14.19, except for the data.)

16.51 A well-known soft-drink manufacturer has used the same secret recipe for its product since its introduction over 100 years ago. In response to a decreasing market share, however, the president of the company is contemplating changing the recipe. He has developed two alternative recipes. In a preliminary study, he asked 20 people to taste the original recipe and the two new recipes. He asked each to evaluate the taste of the product on a 5-point scale, where 1 = awful, 2 = poor, 3 = fair, 4 = good, and 5 = wonderful. These data are stored in columns A (original recipe), B (new recipe 1), and C (new recipe 2) on file Xr16-51. The president decides that unless significant differences exist among evaluations of the products, he will not make any changes. Can we conclude that there are differences in the ratings of the three recipes?

16.52 The manager of a chain of electronic-products retailers is trying to decide on a location for its newest store. After a thorough analysis, the choice has been narrowed to three possibilities. An important factor in the decision is the number of people passing each location. The number of people passing each location per day was counted during 30 days. The data are stored in columns A (site 1), B (site 2), and C (site 3) in file Xr16-52.

a Which techniques should be considered to determine whether the locations differ? What are the required conditions? How do you select a technique?

b Can management conclude that there are differences in the numbers of people passing the three locations if the number of people passing each location is extremely nonnormally distributed?

16.53 Many North Americans suffer from high levels of cholesterol, which can lead to heart attacks. For those with very high levels (over 280), doctors prescribe drugs to reduce cholesterol levels. A pharmaceutical company has recently developed four such drugs. To determine if any differences exist in their benefits, an experiment was organized. The company selected 25 groups of four men, each of whom had levels in excess of 280. In each group, the men were matched according to age and weight. The drugs were administered over a two-month period, and the reduction in cholesterol was recorded. The data are stored in columns A, B, C, and D in file Xr16-53. Do these results allow the company to conclude that differences exist among the three new drugs? (This exercise is identical to Example 14.2, except for the data.)

16.6 SUMMARY

Nonparametric statistical tests are applied to problems where the data are either ordinal or interval and extremely nonnormal. The Wilcoxon rank sum test is used to compare two populations of ordinal or interval data when the data are generated from independent samples. The sign test is used to compare two populations of ordinal data drawn from a matched pairs experiment. The Wilcoxon signed rank sum test is employed to compare two populations of nonnormal interval data taken from a matched pairs experiment. When the objective is to compare two or more populations of independently sampled ordinal or interval and extremely nonnormal data, the Kruskal-Wallis test is employed. The Friedman test is used instead of the Kruskal-Wallis test when the samples are blocked.

IMPORTANT TERMS

Nonparametric statistical techniques
Population locations
Distribution-free statistics
Wilcoxon rank sum test
Sign test
Wilcoxon signed rank sum test
Kruskal-Wallis test
Friedman test

SYMBOLS

Symbol	Pronounced	Represents
T_i	T-sub-i or T-i	Rank sum of sample i ($i = 1, 2, \ldots, k$)
T^+	T-plus	Rank sum of positive differences
T^-	T-minus	Rank sum of negative differences
σ_T	sigma-sub-T or sigma-T	Standard deviation of the sampling distribution of T

FORMULAS

Wilcoxon rank sum test statistic

$$T = T_1$$

$$E(T) = \frac{n_1(n_1 + n_2 + 1)}{2}$$

$$\sigma_T = \sqrt{\frac{n_1 n_2(n_1 + n_2 + 1)}{12}}$$

$$z = \frac{T - E(T)}{\sigma_T}$$

Sign test statistic

$$x = \text{number of positive differences}$$

$$z = \frac{x - .5n}{.5\sqrt{n}}$$

Wilcoxon signed rank sum test statistic

$$T = T^+$$

$$E(T) = \frac{n(n + 1)}{4}$$

$$\sigma_T = \sqrt{\frac{n(n + 1)(2n + 1)}{24}}$$

$$z = \frac{T - E(T)}{\sigma_T}$$

Kruskal-Wallis Test

$$H = \left[\frac{12}{n(n + 1)} \sum_{j=1}^{k} \frac{T_j^2}{n_j}\right] - 3(n + 1)$$

Friedman Test

$$F_r = \left[\frac{12}{b(k)(k + 1)} \sum_{j=1}^{k} T_j^2\right] - 3b(k + 1)$$

MICROSOFT EXCEL OUTPUT AND INSTRUCTIONS

Technique	Page
Wilcoxon rank sum test	504
Sign test	514
Wilcoxon signed rank sum test	520
Kruskal-Wallis test	526–527
Friedman test	532

SUPPLEMENTARY EXERCISES

16.54 In a study to determine which of two teaching methods is perceived to be better, two sections of an introductory psychology course were taught in different ways by the same professor. At the course's completion, each student rated the course on a boring/stimulating spectrum. In this process, 1 = very boring, 2 = somewhat boring, 3 = a little boring, 4 = neither boring nor stimulating, 5 = a little stimulating, 6 = somewhat stimulating, and 7 = very stimulating. The results are and stored in columns A (section 1) and B (section 2) of file Xr16-54. Can we conclude that the ratings of the two teaching methods differ?

16.55 The researchers at a large carpet manufacturer have been experimenting with a new dyeing process in hopes of reducing the streakiness that frequently occurs with the current process. As an experiment, 15 carpets are dyed using the new process, and another 15 are dyed using the existing method. Each carpet is rated on a 5-point scale of streakiness, where 5 is extremely streaky, 4 is quite streaky, 3 is somewhat streaky, 2 is a little streaky, 1 is not streaky at all. The results are stored in columns A (new process) and B (existing process) of file Xr16-55. Is there enough evidence to infer that the new method is better?

16.56 Some of the researchers described in Exercise 16.55 felt that the color of the dye might be a factor. The experiment was rerun using the new and current processes to dye carpets 15 different colors. For each color, both types of carpets were rated on the 5-point scale. These data are stored in columns A (new process) and B (existing process) of file Xr16-56. Can we conclude that the new process is superior?

16.57 The editor of the student newspaper was in the process of making some major changes in the newspaper's layout. He was also contemplating changing the typeface of the print used. To help make a decision, he set up an experiment in which 20 individuals were asked to read four newspaper pages, with each page printed in a different typeface. If the reading speed differed, the typeface that was read fastest would be used. However, if there were not enough evidence to allow the editor to conclude that such differences exist, the current typeface would be continued. The times (in seconds) to completely read one page were recorded and stored in columns A through D in file Xr16-57. Determine the course of action the editor should follow, assuming that the times are extremely nonnormal. (This exercise is identical to Exercise 14.48, except the data.)

16.58 Large potential profits for pharmaceutical companies exist in the area of hair growth drugs. The head chemist for a large pharmaceutical company is conducting experiments to determine which of two new drugs is more effective in growing hair among balding men. One experiment was conducted as follows. A total of 30 pairs of men—each pair matched according to their degree of baldness—were selected. One man used drug A, and the other used drug B. After 10 weeks, the men's new hair growth was examined, and the new growth was judged using the following ratings.

0 = no growth
1 = some growth
2 = moderate growth

The results are stored in columns A and B, respectively, in file Xr16-58. Do these data provide sufficient evidence that drug B is more effective?

16.59 Suppose that a precise measuring device for new hair growth has been developed and is used in the experiment described in Exercise 16.58. The percentages of new hair growth for the 30 pairs of men involved in the experiment are stored in the first two columns of file Xr16-59. Column A stores the results from drug A, and column B stores the results from drug B. Do these data allow the chemist to conclude that drug B is more effective?

16.60 The printing department of a publishing company wants to determine whether there are differences in durability among three types of book bindings. Twenty-five books with each type of binding was selected and placed in machines that continually open and close them. The number of openings and closings until the pages separate from the binding was recorded and stored in file Xr16-60.

a What techniques should be considered to determine whether differences exist among the types of bindings? What are the required conditions? How do you decide which technique to use?

b If we know that the number of openings and closings is extremely nonnormally distributed, test to determine whether differences exist among the types of bindings.

16.61 In recent years, consumers have become more safety conscious, particularly about children's products. A manufacturer of children's pajamas is looking for material that is as nonflammable as possible. In an experiment to compare a new fabric with the kind now being used, 50 pieces of each kind were exposed to an open flame, and the number of seconds until the fabric burst into flames was recorded. The data are stored in columns A (new material) and B (old material) of file Xr16-61. Because the new material is much more expensive than the current material, the manufacturer will switch only if the new material can be shown to be better. On the basis of these data, what should the manufacturer do?

16.62 Samuel's is a chain of family restaurants. Like many other service companies, Samuel's surveys its customers on a regular basis to monitor their opinions. Two questions (among others) asked in the survey are as follows.

  A While you were at Samuel's, did you find the restaurant
     1 slow   2 moderate   3 fast
  B What day was your visit to Samuel's?

The responses of a random sample of 269 customers are stored in file Xr16-62. (Column A stores the responses on Sunday, column B contains the responses for Monday, etc.) Can the manager infer that there are differences in customer perceptions of the speed of service among the days of the week?

16.63 An advertising firm wants to determine the relative effectiveness of two recently produced commercials for a car dealership. An important attribute of such commercials is their believability. In order to judge this aspect of the commercials, 60 people were randomly selected. Each watched both commercials and then rated them on a 5-point scale where 1 = not believable, 2 = somewhat believable, 3 = moderately believable, 4 = quite believable, and 5 = very believable. The ratings are stored in columns A (commercial 1) and B (commercial 2) of file Xr16-63. Do these data provide sufficient evidence to indicate that there are differences in believability between the two commercials?

16.64 Researchers at the U.S. National Institute of Aging in Bethesda, Maryland, have been studying hearing loss. They have hypothesized that as men age they will lose their hearing faster than comparatively aged women because many more men than women have worked at jobs where noise levels have been excessive. To test their beliefs, the researchers randomly selected one man and one woman aged 45, 46, 47, ..., 78, 79, 80 and measured the percentage hearing loss for each person. The data are stored by ascending ages in columns A (men's hearing loss) and B (women's hearing loss) in file Xr16-64. What conclusions can be drawn from these data?

16.65 In a Gallup poll this year, 200 people were asked, "Do you feel that the newspaper you read most does a good job of presenting the news?" The same question was asked of another 200 people 10 years ago. The results of both surveys are stored in columns A (this year) and B (10 years ago) in file Xr16-65. The possible responses were as follows:

3 = good job
2 = fair job
1 = not a good job

Do these data provide enough evidence to infer that people perceive newspapers as doing a better job 10 years ago than today?

16.66 The increasing number of traveling businesswomen represents a large potential clientele for the hotel industry. Many hotel chains have made changes designed to attract more women. To help direct these changes, a hotel chain commissioned a study to determine whether major differences exist between male and female business travelers. A total of 100 male and 100 female executives were questioned on a variety of topics, one of which was the number of trips they had taken in the previous 12 months. The data are stored in columns A (male) and B (female) in file Xr16-66. We would like to know whether these data provide enough evidence to allow us to conclude that businesswomen and businessmen differ in the number of business trips taken per year.

16.67 To examine the effect that a tough midterm test has on student evaluations of professors, a statistics professor had her class evaluate her teaching effectiveness before the midterm test. The questionnaire asked for opinions on a number of dimensions, but the last question is considered the most important. It is, "How would you rate the overall performance of the instructor?" The possible responses are: 1 = poor, 2 = fair, 3 = good, 4 = excellent. After a difficult test, the evaluation was redone. The evaluation scores before and after the test for each of the 40 students in the class are stored in file Xr16-67 as follows.

Column A: evaluation score before the test
Column B: evaluation score after the test

Do the data allow the professor to conclude that the results of the midterm negatively influence student opinion?

16.68 The Wilfrid Laurier University bookstore conducts annual surveys of its customers. One question asks respondents to rate the prices of textbooks. The wording is, "The bookstore's prices of textbooks are reasonable." The responses are:

1 Strongly disagree
2 Disagree
3 Neither agree nor disagree
4 Agree
5 Strongly agree

The respondents were also categorized by year. The responses for first-, second-, third-, and fourth-year students are stored in columns A through D in file Xr16-68. Can we infer that the perceptions of the bookstore's prices vary by what year the student is in?

16.69 The town of Stratford, Ontario, is very much dependent upon the Shakespearean Festival it holds every summer for its financial well-being. Thousands of people visit Stratford to attend one or more Shakespearean plays and spend money in hotels, restaurants, and gift shops. As a consequence, any sign that the number of visitors will decrease in the future is cause for concern. Two years ago, a survey of 100 visitors asked how likely it was that they would return within the next two years. This year the survey was repeated with another 100 visitors. The likelihood of returning within two years was measured as

4 = very likely
3 = somewhat likely
2 = somewhat unlikely
1 = very unlikely

The data are stored in column A (survey results from two years ago) and column B (survey results from this year) in file Xr16-69. Conduct whichever statistical procedures you deem necessary to determine whether the citizens of Stratford should be concerned about the results of the two surveys.

16.70 The Nickels restaurant chain regularly conducts surveys of its customers. Respondents are asked to assess food quality, service, and price. The responses are:

1 Excellent
2 Good
3 Fair

They are also asked whether they would come back. Responses are:

1 Yes
2 No
3 Undecided

The responses to the question about quality of food for a sample of customers are stored in columns A (responded that they would return), B (responded that they would not return), and C (undecided) in file Xr16-70. Can we infer that the ratings differ according to whether the customer responds yes, no, or undecided to the question about return?

16.71 Scientists have been studying the effects of lead in children's blood, bones, and tissue for a number of years. It is known that lead reduces intelligence and can cause a variety of other problems. A study directed by Dr. Herman Needleman, a psychiatrist at the University of Pittsburgh Medical Center, examined some of these problems. Two hundred boys attending public schools in Pittsburgh were recruited. Each boy was categorized as having low or high levels of lead in his bones. Each boy was then assessed by his teachers on a four-point scale (1 = low, 2 = moderate, 3 = high, and 4 = extreme) on degrees of aggression. These data are stored so that column A = degree of aggression for boys with low levels of lead and column B = degree of aggression for boys with high levels of lead in file Xr16-71. Is there evidence to infer that boys with high levels of lead are more aggressive than boys with low levels of lead?

16.72 How does gender affect teaching evaluations? Several researchers have addressed this question over the past decade. (See "The Use and Abuse of Student Ratings of Professors" by Peter Seldin, published in the *Chronicle of Higher Education*, July 2, 1993.) In one study, several female and male professors in the same department with similar backgrounds were selected. A random sample of 100 female students was drawn. Each student evaluated a female professor and a male professor. A sample of 100 male students was drawn, and each also evaluated a female professor and a male professor. The ratings were based on a four-point scale where 1 = poor, 2 = fair, 3 = good, and 4 = excellent. The evaluations are stored in file Xr16-72 in the following way:

Column A = female student
Column B = female professor ratings
Column C = male professor ratings
Column D = male student
Column E = female professor ratings
Column F = male professor ratings

a Can we infer that female students rate female professors higher than they rate male professors?
b Can we infer that male students rate male professors higher than they rate female professors?

16.73 Clublink is a corporation that owns golf courses across North America. Guests of members are asked to complete a questionnaire that asks a variety of questions including the guest's evaluation of the overall course condition. The responses to this question are

1 Eagle
2 Birdie
3 Par
4 Bogey

Other questions deal with details about the guests. A random sample of 200 questionnaires was stored in columns A (female guests) and B (male guests) in file Xr16-73. Can we infer than men and women differ in their assessment of the overall condition of the course?

16.74 It is an unfortunate fact of life that the characteristics that one is born with play a critical role in later life. For example, race is a critical factor in almost all aspects of North American life. Height and weight also determine how friends, teachers, employers, and customers will treat you. And now we may add physical attractiveness to this list. A recent study conducted by economists Jeff Biddle of Michigan State University and Daniel Hamermesh of the University of Texas followed the careers of students from a prestigious U.S. law school. A panel of independent raters examined the graduation yearbook photos of the students and rated their appearance as unattractive, neither attractive nor unattractive, or attractive. The annual incomes in thousands of dollars five years after graduation were recorded and stored in columns A to C, respectively, in file Xr16-74. Assuming that incomes are extremely nonnormally distributed, can we infer that incomes of lawyers are affected by physical attractiveness?

16.75 According to a CNN news report broadcast on November 26, 1995, 9% of full-time workers telecommute. This means that they do not work in their employer's offices but instead perform their work at home using a computer and modem. To ascertain whether such workers are more satisfied than their nontelecommuting counterparts, a study was undertaken. A random sample of telecommuters and regular office workers was taken. Each was asked how satisfied they were with their current employment. The responses (1 = very unsatisfied; 2 = somewhat unsatisfied; 3 = somewhat satisfied; 4 = very satisfied) are stored in columns A (telecommuters) and B (regular office workers). The data are stored in file Xr16-75. What conclusions can we draw from these data?

16.76 Can you become addicted to exercise? In a study conducted at the University of Wisconsin at Madison, a random sample of dedicated exercisers who usually work out every day was drawn. Each completed a questionnaire that gauged his or her mood on a 5-point scale, where 5 = very relaxed and happy, 4 = somewhat relaxed and happy, 3 = neutral feeling, 2 = tense and anxious, and 1 = very tense and anxious. The group was then instructed to abstain from all workouts for the next three days. Moreover they were told to be as physically inactive as possible. Each day their mood was measured using the same questionnaire. The data are stored in file Xr16-76. Column A stores the code identifying the respondent and columns B through E store the measures of mood for the day before the experiment began and for the three days of the experiment, respectively.

a Can we infer that for each day the exercisers abstained from physical activity they were less happy than when they were exercising?

b Do the data indicate that by the third day moods were improving?

16.77 Casino Windsor has been in operation for several years. Recently, casino gambling was allowed across the river in Detroit. Concerned about the competition, the casino conducted a survey to determine the opinions of its customers. Among other questions, respondents were asked to give their opinion about "Your overall impression of Casino Windsor." The responses are

Excellent
Good
Average
Poor
Unacceptable

The responses were recorded as 5, 4, 3, 2, 1, respectively. Additionally, the gender of the respondent was noted. The responses about overall impression were stored in columns A (female) and B (male) in file Xr16-77. Can we infer that female and male gamblers differ in their overall impression of Casino Windsor?

16.78 How does alcohol affect judgment? To provide some insight, an experiment was conducted. A random sample of customers of an Ohio bar was selected. Each respondent was asked to assess the attractiveness of members of the opposite sex who were in the bar at the time. The assessment was to be made on a five-point scale (1 = very unattractive, 2 = unattractive,

3 = neither attractive nor unattractive, 4 = attractive, and 5 = very attractive). The survey was conducted three hours before closing and again just before closing using another group of respondents. The data have been stored in file Xr16-78. Column A contains the assessments three hours before closing, and column B stores the assessments made just before closing.

Can we conclude that the assessments made just before closing are higher than those made three hours earlier? If so, what does this imply about the effects of alcohol on judgments? (The survey results were reported in the September 1997 edition of the *Report on Business*.)

Chapter 17

Simple Linear Regression and Correlation

17.1 Introduction

17.2 Model

17.3 Estimating the Coefficients

17.4 Error Variable: Required Conditions

17.5 Assessing the Model

17.6 Using the Regression Equation

17.7 Coefficients of Correlation

17.8 Regression Diagnostics I

17.9 Summary

17.1 INTRODUCTION

Chapters 17 and 18 address problems whose objective is to analyze the relationship between interval variables. **Regression analysis** is used to predict the value of one variable on the basis of other variables. This technique may be the most commonly used statistical procedure because, as you can easily appreciate, almost all companies and government institutions forecast variables such as product demand, interest rates, inflation rates, prices of raw materials, and labor costs.

The technique involves developing a mathematical equation that describes the relationship between the variable to be forecast, which is called the **dependent variable,** and variables that the statistics practitioner believes are related to the dependent variable. The dependent variable is denoted y, while the related variables are called **independent variables** and are denoted $x_1, x_2, \ldots, x_k$ (where k is the number of independent variables).

If we are interested only in determining whether a relationship exists, we employ correlation analysis. We have already introduced this technique. In Chapter 2, we presented the graphical method to describe the relationship between two interval variables—the scatter diagram. We introduced covariance and the coefficient of correlation in Chapter 3.

Because regression analysis involves a number of new techniques and concepts, we divided the presentation into two chapters. In this chapter, we present techniques that allow us to determine the relationship between only two variables. In Chapter 18, we expand our discussion to more than two variables.

Here are three illustrations of regression analysis.

Illustration 1

The admissions officer at a university would like to develop a better method of deciding which high school graduates to admit to the university. She would like a method that predicts the university grade point average. As a first step she lists the variables that she believes affect a student's performance at university:

High school graduating average

Scholastic Assessment Test (SAT) score

Number of hours of extracurricular activity per week

Number of hours of work at a part-time job

Ratings of letters of reference

Illustration 2

An actuary would like to predict a person's longevity. After some thought, he created the following list of variables:

Longevity of parents

Longevity of grandparents

Number of cigarettes smoked per week

Amount of exercise per week

Amount overweight

Illustration 3

A real estate agent wants to more accurately predict the selling price of houses. She believes that the following variables affect the price of a house:

Size of the house (in square feet)

Number of bedrooms

Frontage of the lot

Condition

Location

In each of these illustrations, the primary motive for using regression analysis is forecasting. Nonetheless, analyzing the relationship among variables can also be quite useful to the statistics practitioner. For instance, in Illustration 2 the effect of smoking can be determined, providing more incentive to quit.

Regardless of why regression analysis is performed, the next step in the technique is to develop a mathematical equation or model that accurately describes the nature of the relationship that exists between the dependent variable and the independent variables. This stage—which is only a small part of the total process—is described in the next section. In the ensuing sections of this chapter (and in Chapter 18), we will spend considerable time assessing and testing how well the model fits the actual data. Only when we're satisfied with the model do we use it to estimate and forecast.

17.2 MODEL

The job of developing a mathematical equation can be quite complex, because we need to have some idea about the nature of the relationship between each of the independent variables and the dependent variable. For example, the admissions officer in illustration 1 needs to know how a part-time job affects grade point average. If she proposes a linear relationship, that may imply that as the amount of work per week rises (or falls), the university grade point average will rise or fall. A quadratic relationship may suggest that the grade point average will increase over a certain range of hours of part-time work but will decrease over a different range. Perhaps certain combinations of values of part-time work and other independent variables influence the grade point average in one way, while other combinations change in other ways. The number of different mathematical models that could be proposed is virtually infinite.

You might have encountered various models in previous courses. For instance, the following represent relationships in the natural sciences.

$E = mc^2$, where E = energy, m = mass, and c = speed of light

$F = ma$, where F = force, m = mass, and a = acceleration

$S = at^2/2$, where S = distance, t = time, and a = gravitational acceleration

These are all examples of **deterministic models,** so named because—except for small measurement errors—such equations allow us to determine the value of the dependent variable (on the left side of the equation) from the value of the independent variables. In many practical applications of interest to us, deterministic models are unrealistic. In Illustration 3, is it reasonable to believe that we can determine the selling price of a house solely on the basis of those independent variables. Unquestionably, they affect the price, but many other variables (some of which may not be measurable) also influence price. What must be included in most practical models is a method to represent the randomness that is part of a real-life process. Such a model is called a **probabilistic model.**

To create a probabilistic model, we start with a deterministic model that approximates the relationship we want to model. We then add a random term that measures the error of the deterministic component. Suppose that, in Illustration 3, the real estate agent knows that the cost of building a new house is about $75 per square foot and that most lots sell for about $25,000. The approximate selling price would be

$$y = 25{,}000 + 75x$$

where y = selling price and x = size of the house in square feet. A house of 2,000 square feet would therefore be estimated to sell for

$$y = 25{,}000 + 75(2{,}000) = 175{,}000$$

We know, however, that the selling price is not likely to be exactly $175,000. Prices may actually range from $100,000 to $250,000. In other words, the deterministic model is not really suitable. To represent this situation properly, we should use the probabilistic model

$$y = 25{,}000 + 75x + \varepsilon$$

where ε (the Greek letter *epsilon*) represents the random term (also called the error variable)—the difference between the actual selling price and the estimated price based on the size of the house. The random term thus accounts for all the variables, measurable and immeasurable, that are not part of the model. The value of ε will vary from one sale to the next, even if x remains constant. That is, houses of exactly the same size will sell for different prices because of differences in location and number of bedrooms and bathrooms, as well as other variables.

We will present only probabilistic models. Additionally, to simplify the presentation, all models will be linear. In this chapter, we restrict the number of independent variables to one. The model to be used in this chapter is called the **first-order linear model**—sometimes called the **simple linear regression model.** Figure 17.1 depicts the deterministic component of the model.

First-Order Linear Model

$$y = \beta_0 + \beta_1 x + \varepsilon$$

where

y = dependent variable

x = independent variable

β_0 = y-intercept

β_1 = slope of the line (defined as rise/run)

ε = error variable

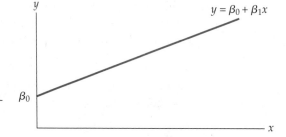

Figure 17.1

First-order linear model: deterministic component

The problem objective addressed by the model is to analyze the relationship between two variables, x and y, both of which must be interval. To define the relationship between x and y, we need to know the value of the coefficients of the linear model β_0 and β_1. However, these coefficients are population parameters, which are almost always unknown. In the next section, we discuss how these parameters are estimated.

17.3 ESTIMATING THE COEFFICIENTS

We estimate the parameters β_0 and β_1 in a way similar to the methods used to estimate all the other parameters discussed in this book. We draw a random sample from the population of interest and calculate the sample statistics we need. Because β_0 and β_1 represent the coefficients of a straight line, their estimators are based on drawing a straight line through the sample data.

In Chapters 2 and 3 we introduced the descriptive methods employed to describe the relationship between two interval variables. We pointed out that we are particularly interested in determining whether a linear relationship exists. In Section 2.6 we introduced the **scatter diagram** from which we attempted to fit a straight line. We used the **least squares method** (presented in Section 3.6) to draw the straight line that minimizes the sum of squared differences between the points and the line. The formulas, which were derived using calculus, are summarized below.

Calculation of b_0 and b_1

$$b_1 = \frac{\text{cov}(x,y)}{s_x^2}$$

$$b_0 = \bar{y} - b_1 \bar{x}$$

Statisticians have shown that b_0 and b_1 are unbiased estimators of β_0 and β_1, respectively. We will use the least squares method to produce the sample regression line

$$\hat{y} = b_0 + b_1 x$$

The following example illustrates how these coefficients are determined and interpreted.

▼ EXAMPLE 17.1

Car dealers across North America use the *Red Book* to help them determine the value of used cars that their customers trade in when purchasing new cars. The book, which is published monthly, lists the trade-in values for all basic models of cars. It provides alternative values for each car model according to its condition and optional features. The values are determined on the basis of the average paid at recent used-car auctions. (These auctions are the source of supply for many used-car dealers.) However, the *Red Book* does not indicate the value determined by the odometer reading, despite the fact that a critical factor for used-car buyers is how far the car has been driven. To examine this issue, a used-car dealer randomly selected 100 three-year-old Ford Tauruses that were sold at auction during the past month. Each car was in top condition and equipped with automatic transmission, AM/FM cassette tape player, and air conditioning. The dealer recorded the price and the number of miles on the odometer. These data are stored in file Xm17-01; some of the data are listed below. The dealer wants to find the regression line.

Car	Odometer Reading	Auction Selling Price
1	37,388	$5,318
2	44,758	5,061
3	45,833	5,008
.	.	.
.	.	.
.	.	.
100	36,392	5,133

Solution

IDENTIFY

Notice that the problem objective is to analyze the relationship between two interval variables. Because we want to know how the odometer rating affects the selling price, we identify the former as the independent variable, which we label x, and the latter as the dependent variable, which we label y.

COMPUTE

The complete printouts are shown below. The printouts include more statistics than we need right now. However, we will be discussing the rest of the printouts later. We have also included the scatter diagrams, which is often a first step in the regression analysis. Notice that there does appear to be a straight-line relationship between the two variables.

Microsoft Excel Output for Example 17.1

Summary Output

Regression Statistics

Multiple R	0.8063
R Square	0.6501
Adjusted R Square	0.6466
Standard Error	151.6
Observations	100

ANOVA

	df	SS	MS	F	Significance F
Regression	1	4183528	4183528	182.1	0.0000
Residual	98	2251362	22973		
Total	99	6434890			

	Coefficients	Standard Error	t Stat	P-value
Intercept	6533	84.51	77.31	0.0000
Odometer	20.0312	0.00231	213.49	0.0000

GENERAL COMMANDS	EXAMPLE 17.1
1 Type or import the data into two columns.	Open file **Xm17-01**.
2 Click **Tools, Data Analysis ...**, and **Regression**.	
3 Specify the **Input Y Range**.	B1:B101
4 Specify the **Input X Range**. Click **Labels**, if applicable. Click **OK**.	A1:A101
5 To draw the scatter diagram click **Line Fit Plots** before clicking **OK**.	

6 You can also draw the scatter diagram (shown below) using the commands described in Chapter 2.

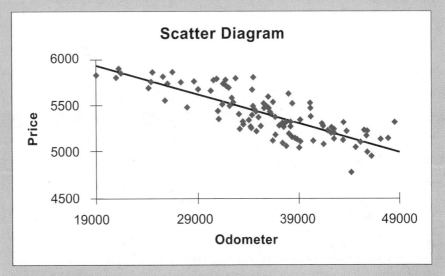

The scatter diagram will be drawn with the two axes starting at zero. This may cause some of the points to be bunched together, leaving large blank spaces in the diagram. To modify the chart, proceed as follows.

7 Activate the chart by double-clicking anywhere within the boundaries of the box.

8 Double-click the Y-axis, click **Scale**, and change the **Minimum, Maximum,** and/or **Major** and **Minor Units.** Click **OK**.

4500 (Minimum)
6000 (Maximum)

9 Repeat for the X-axis. Click **OK**.

19000 49000

10 To draw the least squares line, click **Add Trendline . . . ,** and specify **Linear.**

INTERPRET

The sample regression line is

$$\hat{y} = 6{,}533 - 0.0312x$$

The slope coefficient b_1 is -0.0312, which means that for each additional mile on the odometer, the price decreases by an average of $0.0312 (3.12 cents).

The intercept is $b_0 = 6{,}533$. Technically, the intercept is the point at which the regression line and the y-axis intersect. This means that when $x = 0$ (i.e., the car was not driven at all) the selling price is $6,533. We might be tempted to interpret this number as the price of cars that have not been driven. However, in this case, the intercept is probably meaningless. Because our sample did not include any cars with zero miles on the odometer, we have no basis for interpreting b_0. As a general rule, we cannot determine the value of y for a value of x that is far outside the range of the sample values of x. In this example, the smallest and largest values of x are 19,057 and 49,223, respectively. Because $x = 0$ is not in this interval, we cannot safely interpret the value of y when $x = 0$.

In the sections that follow, we will return to this problem and the computer output to introduce other statistics associated with regression analysis.

EXERCISES

17.1 The term "regression" was originally used in 1885 by Sir Francis Galton in his analysis of the relationship between the heights of children and parents. He formulated the "law of universal regression," which specifies that "each peculiarity in a man is shared by his kinsmen, but on average in a less degree." (Evidently, people spoke this way in 1885.) In 1903, two statisticians, K. Pearson and A. Lee, took a random sample of 1,078 father–son pairs to examine Galton's law ("On the Laws of Inheritance in Man, I. Inheritance of Physical Characteristics," *Biometrika* 2:457–462). Their sample regression line was

son's height = 33.73 + .516 × father's height

a Interpret the coefficients.
b What does the regression line tell you about the heights of sons of tall fathers?
c What does the regression line tell you about the heights of sons of short fathers?

17.2 Suppose that a statistics practitioner wanted to update the study described in Exercise 17.1. She collected data on 400 father–son pairs and stored the data in columns 1 (fathers' heights in inches) and 2 (sons' heights in inches) in file Xr17-02.
a Determine the sample regression line.
b What does the value of b_0 tell you?
c What does the value of b_1 tell you?

17.3 Refer to Exercise 2.45. Determine the sample regression line that describes how one's mark in calculus affects the mark in statistics. What do the coefficients indicate about this relationship?

17.4 Refer to Exercise 2.52.

a Determine the sample regression line that depicts how the metabolism rate is affected by the amount of exercise.
b What does the value of b_0 tell you?
c What does the value of b_1 tell you?

17.5 Refer to Exercise 2.53. Find the sample regression line that describes how the duration of flu is affected by age. Discuss what the coefficients tell you about the relationship.

17.6 In television's early years, most commercials were 60 seconds long. Now, however, commercials can be any length. The objective of commercials remains the same—to have as many viewers as possible remember the product in a favorable way and eventually buy it. In an experiment to determine how the length of a commercial affects people's memory of it, 60 randomly selected people were asked to watch a one-hour television program. In the middle of the show, a commercial advertising a brand of toothpaste appeared. Some viewers watched a commercial that lasted for 20 seconds, others watched one that lasted for 24 seconds, 28 seconds, ..., 60 seconds. The essential content of the commercials was the same. After the show, each person was given a test to measure how much he or she remembered about the product. The commercial times and test scores (on a 30-point test) are stored in file Xr17-06.

a Draw a scatter diagram of the data to determine whether a linear model appears to be appropriate.
b Determine the least squares line.
c Interpret the coefficients.

17.7 After several semesters without much success, Pat Statsdud (a student in the lower quarter of a statistics course) decided to try to improve. Pat needed to know the secret of success for university and college students. After many hours of discussion with other, more successful, students, Pat postulated a rather radical theory: the longer one studied, the better one's grade. To test the theory, Pat took a random sample of 100 students in an economics course and asked each to report the average amount of time he or she studied economics and the final mark received. These data are stored in columns A (study time in hours) and B (final mark out of 100) in file Xr17-07.

  a Determine the sample regression line.
  b Interpret the coefficients.
  c Is the sign of the slope logical? If the slope had had the opposite sign, what would that tell you?

17.8 The growing interest and use of the Internet has forced many companies into considering ways to sell their products on the Web. Therefore, it is of interest to these companies to determine who is using the Web. A statistics practitioner undertook a study to determine how education and Internet use are connected. She took a random sample of 200 adults (20 years of age and older) asked each to report the years of education they had completed and the number of hours of Internet use in the previous week. These data are stored in columns A and B (education and Internet use, respectively) in file Xr17-08.

  a Perform a regression analysis to describe how the two variables are related.
  b Interpret the coefficients.

17.9 The human resource manager of a telemarketing firm is concerned about the rapid turnover of the firm's telemarketers. It appears that many telemarketers do not work very long before quitting. There may be a number of reasons, including relatively low pay, personal unsuitability for the work, and the low probability of advancement. Because of the high cost of hiring and training new workers the manager decided to examine the factors that influence workers to quit. He reviewed the work history of a random sample of workers who had quit in the last year and recorded the number of weeks on the job before quitting and the age of each worker when originally hired. These data are stored in file Xr17-09 (column A contains the age and column B contains the length of employment in weeks).

  a Use regression analysis to describe how the length of employment and age are related.
  b Briefly discuss what the coefficients tell you.

17.10 The Trans-Alaska Pipeline System carries crude oil from Prudhoe Bay on Alaska's North Slope 800 miles to the port of Valdez, on the south coast of Alaska. The pipeline carries a mixture of different qualities of oil. Quality of oil is measured in API Gravity degrees—the higher the degrees API, the higher the quality. Because the pipeline mixes oils of different degrees, shippers in Valdez receive oil of different quality than they purchased. To compensate shippers, a "quality bank" was established. The owners of the pipeline proposed compensating shippers 15 cents per barrel for every degree below the level to which the shippers agreed. However, a refinery near Fairbanks, which receives 26-degree oil and mixes it with 20-degree oil, objected to the proposal. It suggested a 3.09- to 5.35-cent differential. Because oil carriers are required to establish "just and reasonable" rates, a hearing before an administrative law judge was held. At the hearing, an expert hired by the shippers produced the table below to show the relationship between quality and price per barrel of Mideast oil. (The data are stored in file Xr17-10.) Use regression analysis to determine the appropriate compensation.

Mideast oil degrees API	27.0	28.5	30.8	31.3	31.9	34.5	34.0
Price per barrel	12.02	12.04	12.32	12.27	12.49	12.70	12.80
Mideast oil degrees API	34.7	37.0	41.1	41.0	38.8	39.3	
Price per barrel	13.00	13.00	13.17	13.19	13.22	13.27	

Source: M. O. Finkelstein and B. Levin, *Statistics for Lawyers* (Springer-Verlag, 1990), 338–39.

17.11 All Canadians have government-funded health insurance, which pays for any medical care they require. However, when traveling out of the country, Canadians usually acquire supplementary health insurance to cover the difference between the costs incurred for emergency treatment and what the government program pays. In the United States, this cost differential can be prohibitive. Until recently, private insurance companies (such as Blue Cross) charged everyone the same weekly rate, regardless of age. However, because of rising costs and the realization that older people frequently incur greater medical emergency expenses, insurers had to change their premium plans. They decided to offer rates that depend on the age of the customer. To help determine the new rates, one insurance company gathered data concerning the age and mean daily medical expenses of a random sample of 1,348 Canadians during the previous 12-month period. The data are stored in file Xr17-11 (column A = age; column B = mean daily medical expense).

a Determine the sample regression line.
b Interpret the coefficients.
c What rate plan would you suggest?

17.12 The four C's, carats, cut, clarity, and color, determine the price of diamonds. Carats refer to the weight of the diamond. One carat equals .2 grams. An advertisement in a Singapore newspaper (*Straits Times,* February 29, 1992) featured 48 ladies' diamond rings in which the stones varied in weight from .12 carats to .35 carats. The ad listed the weights of the stones together with the nonnegotiable price in Singapore dollars. These data are stored in columns A (weights) and B (price) in file Xr17-12.

a Use regression analysis to determine how weight and price are related.
b What do the coefficients tell you?

17.4 ERROR VARIABLE: REQUIRED CONDITIONS

In the previous section, we used the least squares method to estimate the coefficients of the linear regression model. A critical part of this model is the error variable ε. In the next section, we will present an inferential method that determines whether there is a linear relationship. Later we will show how we use the regression equation to estimate and predict. For these methods to be valid, however, four requirements involving the probability distribution of the **error variable** must be satisfied.

Required Conditions for the Error Variable

1 The probability distribution of ε is normal.
2 The mean of the distribution is zero; that is, $E(\varepsilon) = 0$.
3 The standard deviation of ε is σ_ε, which is a constant no matter what the value of x is.
4 The value of ε associated with any particular value of y is independent of ε associated with any other value of y.

Requirements 1, 2, and 3 can be interpreted in another way: For each value of x, y is a normally distributed random variable whose mean is

$$E(y) = \beta_0 + \beta_1 x$$

and whose standard deviation is σ_ε. Notice that the mean depends on x. The standard deviation, however, is not influenced by x, because it is a constant over all values of x. Figure 17.2 depicts this interpretation. Notice that for each value of x, $E(y)$ changes, but the shape of the distribution of y remains the same. That is, for each x, y is normally distributed with the same standard deviation.

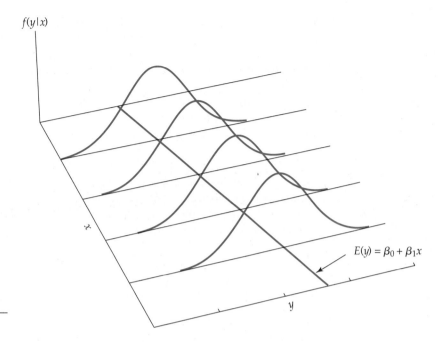

Figure 17.2

Distribution of y given x

In Section 17.8, we will discuss how departures from these required conditions affect the regression analysis and how they are identified.

OBSERVATIONAL AND EXPERIMENTAL DATA

In Chapter 12 we described the difference between observational and experimental data. We pointed out that statistics practitioners often design controlled experiments to enable them to interpret the results of their analyses more clearly than would be the case after conducting an observational study. Example 17.1 is an illustration of observational data. In that example we merely observed the odometer reading and auction selling price of 100 randomly selected cars.

If you examine Exercise 17.6, you will see experimental data gathered through a controlled experiment. To determine the effect of the length of a television commercial on its viewers' memories of the product advertised, the statistics practitioner arranged for 60 television viewers to watch a commercial of differing lengths and then tested their memories of that commercial. Each viewer was randomly assigned a commercial length. The values of x ranged from 20 to 60 and were set by the statistics practitioner as part of the experiment. For each value of x, the distribution of the memory test scores is assumed to be normally distributed with a constant variance.

We can summarize the difference between the experiment described in Example 17.1 and the one described in Exercise 17.6. In Example 17.1, both the odometer reading and the auction selling price are random variables. We hypothesize that for each possible odometer reading, there is a theoretical population of auction selling prices that are normally distributed with a mean that is a linear function of the odometer reading and a standard deviation that is constant. In Exercise 17.6 the length of commercial is not a random variable but a series of values selected by the statistics practitioner. For each commercial length, the memory test scores are required to be normally distributed with a constant standard deviation.

Regression analysis can be applied to data generated from either observational or controlled experiments. In both cases our objective is to determine how the independent variable affects the dependent variable. However, observational data can be analyzed in another way. When the data are observational both variables are random variables. We need not specify that one variable is independent and the other is dependent. We can simply determine *whether* the two variables are related. The equivalent of the required conditions described above is that the two variables are bivariate normally distributed. A bivariate normal distribution is described in Figure 17.3. As you can see, it is a three-dimensional bell curve. The dimensions are the variables x, y, and the joint density function $f(x,y)$.

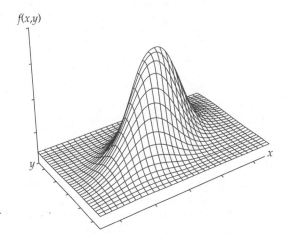

Figure 17.3

Bivariate normal distribution

In Section 17.7 we will discuss the statistical technique that is used when both x and y are random variables and they are bivariate normally distributed. We will also introduce a procedure applied when the normality requirement is not satisfied.

EXERCISES

17.13 Describe what the required conditions mean in Exercise 17.2. Do these requirements seem reasonable?

17.14 If the required conditions are satisfied in Exercise 17.10, what can you say about the distribution of the price per barrel?

17.15 Assuming that the required conditions are satisfied in Exercise 17.11, what does this tell you about the distribution of mean daily expenses?

17.5 ASSESSING THE MODEL

The least squares method produces the best straight line. However, there may in fact be no relationship or perhaps a nonlinear (e.g., quadratic) relationship between the two variables. If so, a linear model is likely to be impractical. Consequently, it is important for us to assess how well the linear model fits the data. If the fit is poor, we should discard the linear model and seek another one.

Several methods are used to evaluate the model. In this section, we present two statistics and one test procedure to determine whether a linear model should be employed. They are the standard error of estimate, the *t*-test of the slope, and the coefficient of determination. All of these methods are based on the sum of squares for error.

SUM OF SQUARES FOR ERROR

The least squares method is based on finding the coefficients that minimize the sum of squared differences between the points and the line defined by the coefficients. We can measure how well the straight line fits the data by calculating the value of the sum of squared differences. The difference between the points and the line are called **residuals.** That is,

residual for point $i = y_i - \hat{y}_i$

Residuals are observed values of the error variable. Consequently, the minimized sum of squared differences is called the **sum of squares for error,** denoted SSE.

Sum of Squares for Error

$$SSE = \sum_{i=1}^{n}(y_i - \hat{y}_i)^2$$

The direct method of calculating SSE can be long and tedious. It requires that for each value of x we compute the value of $\hat{y}$. That is, for $i = 1$ to n,

$\hat{y}_i = b_0 + b_1 x_i$

For each point we then compute the difference between the actual value of y and the value calculated at the line, which is the residual. We square each residual and sum the squared values. Table 17.1 shows these calculations for the first three and last three cars in Example 17.1.

The sum of squares for error is 2,251,362.47. This statistic plays a role in every statistical technique that follows.

Table 17.1 Calculation of SSE for Example 17.1

Car i	Odometer Reading x_i	Auction Selling Price y_i	Regression* Equation $\hat{y}_i = 6533 - .0312x_i$	Residual $(y_i - \hat{y}_i)$	Residual Squared $(y_i - \hat{y}_i)^2$
1	37388	5318	5366.49	−48.49	2351.71
2	44758	5061	5132.07	−71.07	5051.60
3	45833	5008	5098.43	−90.43	8177.06
.	.	.	.	.	.
.	.	.	.	.	.
.	.	.	.	.	.
98	33190	5259	5494.15	−235.15	55296.93
99	39196	5356	5306.17	49.83	2483.51
100	36392	5133	5393.93	−260.93	68084.67
					SSE = 2251362.47

*We used the actual (not the rounded) coefficients.

STANDARD ERROR OF ESTIMATE

In Section 17.4, we pointed out that the error variable ε is normally distributed with mean zero and standard deviation σ_ε. If σ_ε is large, some of the errors will be large, which implies that the model's fit is poor. If σ_ε is small, the errors tend to be close to the mean (which is zero), and, as a result, the model fits well. Hence, we could use σ_ε to measure the suitability of using a linear model. Unfortunately, σ_ε is a population parameter and, like most parameters, is unknown. We can, however, estimate σ_ε from the data. The estimate is based on SSE. The unbiased estimator of the variance of the error variable σ_ε^2 is

$$s_\varepsilon^2 = \frac{SSE}{n-2}$$

The square root of s_ε^2 is called the **standard error of estimate.**

> **Standard Error of Estimate**
>
> $$s_\varepsilon = \sqrt{\frac{SSE}{n-2}}$$

▼ EXAMPLE 17.2

Find the standard error of estimate for Example 17.1, and describe what it tells you about the model's fit.

Solution The standard error of estimate can be found in the Excel printout on page 548.

COMPUTE

Standard Error 151.6

INTERPRET

The smallest value that s_ε can assume is zero, which occurs when SSE = 0, that is, when all the points fall on the regression line. Thus, when s_ε is small, the fit is excellent, and the linear model is likely to be an effective analytical and forecasting tool. If s_ε is large, the model is a poor one, and the statistics practitioner should improve it or discard it. We judge the value of s_ε by comparing it to the values of the dependent variable y or more specifically to the sample mean $\bar{y}$. In this example, because $s_\varepsilon = 151.6$ and $\bar{y} = 5,411.4$, we would have to admit that the standard error of estimate is not very small. On the other hand, it is not a large number. Because there is no predefined upper limit on s_ε, it is difficult to assess the model in this way (except in cases where s_ε is obviously a small number). In general, the standard error of estimate cannot be used as an absolute measure of the model's utility.

Nonetheless, s_ε is useful in comparing models. If the statistics practitioner has several models from which to choose, the one with the smallest value of s_ε should generally be the one used. As you'll see, s_ε is also an important statistic in other procedures associated with regression analysis.

▲

TESTING THE SLOPE

To understand this method of assessing the linear model, consider the consequences of applying the regression technique to two variables that are not at all linearly related. If we could observe the entire population and draw the regression line, we would observe the graph shown in Figure 17.4. The line is horizontal, which means that the value of y is unaffected by the value of x. Recall that a horizontal straight line has a slope of zero, that is, $\beta_1 = 0$.

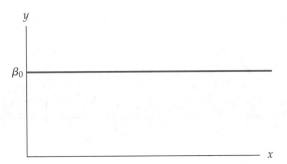

Figure 17.4

$\beta_1 = 0$

Because we rarely examine complete populations, the parameters are unknown. However, we can draw inferences about the population slope β_1 from the sample slope b_1.

The process of testing hypotheses about β_1 is identical to the process of testing any other parameter. We begin with the hypotheses. The null hypothesis specifies that there is no linear relationship, which means that the slope is zero. We can conduct one- or two-tail tests of β_1. In most cases, we perform a two-tail test to determine whether there is sufficient evidence to infer that a linear relationship exists. Consequently, we test

$$H_0: \beta_1 = 0$$
$$H_1: \beta_1 \neq 0$$

The test statistic is

$$t = \frac{b_1 - \beta_1}{s_{b_1}}$$

where s_{b_1} is the standard error of b_1 (which is the standard deviation of the sampling distribution of b_1). It is defined as

$$s_{b_1} = \frac{s_\varepsilon}{\sqrt{(n-1)s_x^2}}$$

If the error variable is normally distributed, the test statistic is Student t distributed with $\nu = n - 2$ degrees of freedom.

▼ EXAMPLE 17.3

Test to determine whether there is enough evidence in Example 17.1 to infer that there is a linear relationship between the auction price and the odometer reading.

Solution

IDENTIFY

We test the hypotheses

$$H_0: \beta_1 = 0$$
$$H_1: \beta_1 \neq 0$$

If the null hypothesis is true, no linear relationship exists. If the alternative hypothesis is true, some linear relationship exists between the two variables.

COMPUTE

The following is part of the original output that appeared on page 548.

	Coefficients	Standard Error	t Stat	P-value
Intercept	6533	84.51	77.31	0.0000
Odometer	20.0312	0.00231	213.49	0.0000

INTERPRET

The value of the test statistic is $t = -13.49$, with a p-value of 0. There is overwhelming evidence to infer that a linear relationship exists. This means is that the odometer reading does affect the auction selling price of the cars. However, as was the case when we interpreted the y-intercept, the conclusion we draw here is valid only over the range of the values of the independent variable. That is, we can infer that there is a linear relationship between odometer reading and auction price for the three-year-old Ford Tauruses whose odometer reading lies between 19,057 and 49,223 miles (the minimum and maximum values of x in the sample). Because we have no observations outside this range, we do not know how, or even whether, the two variables are related. This issue is particularly important to remember when we use the regression equation to estimate or forecast. (See Section 17.6.)

TESTING THE y-INTERCEPT

The output also includes a test of β_0. The null and alternative hypotheses are similar to those specified in the test of the slope. That is we test

$H_0 : \beta_0 = 0$

$H_1 : \beta_0 \neq 0$

As you can see, the test results are $t = 77.31$ with a p-value of 0. This tells us that there is overwhelming evidence to infer that the population y-intercept is not equal to zero. However, as we have already discussed, in virtually all cases the y-intercept has little practical meaning. Consequently, we are seldom interested in its value and whether or not it is equal to zero. In most applications, this test is ignored.

COEFFICIENT OF DETERMINATION

The test of β_1 addresses only the question of whether there is enough evidence to infer that a linear relationship exists. In many cases, however, it is also useful to measure the strength of that linear relationship, particularly when we want to compare several different models. The statistic that performs this function is the coefficient of determination.

The **coefficient of determination,** denoted R^2, is computed in the following way.

Coefficient of Determination

$$R^2 = \frac{[\text{cov}(x,y)]^2}{s_x^2 s_y^2}$$

With a little algebra, statisticians can show that

$$R^2 = 1 - \frac{\text{SSE}}{\sum(y_i - \bar{y})^2}$$

The significance of the second formula is based on the analysis of variance technique. In Chapter 14, we partitioned the total sum of squares into two sources of variation. Here, we begin the discussion by observing that the deviation between y_i and $\bar{y}$ can be decomposed into two parts. That is,

$$(y_i - \bar{y}) = (y_i - \hat{y}_i) + (\hat{y}_i - \bar{y})$$

This equation is represented graphically (for $i = 1$) in Figure 17.5.

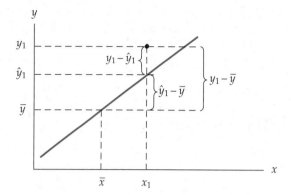

Figure 17.5

Analysis of the deviation

Now we ask, why are the values of y different from one another? In Example 17.1, we observe that the auction selling prices of the cars vary, and we'd like to explain why. From Figure 17.5, we see that part of the difference between y_i and $\bar{y}$ is the difference between $\hat{y}_i$ and $\bar{y}$, which is accounted for by the difference between x_i and $\bar{x}$. That is, some of the price variation is explained by the odometer reading. The other part of the difference between y_i and $\bar{y}$, however, is accounted for by the difference between y_i and $\hat{y}_i$. This difference is the residual, which to some degree represents variables not otherwise represented by the model. (These variables likely include the local supply and demand for this type of used car, the color of the car, and other relatively small details.) As a result, we say that this part of the difference is *unexplained* by the odometer variation.

If we now square both sides of the equation, sum over all sample points, and perform some algebra, we produce

$$\sum (y_i - \bar{y})^2 = \sum (y_i - \hat{y}_i)^2 + \sum (\hat{y}_i - \bar{y})^2$$

The quantity on the left side of this equation is a measure of the variation in the dependent variable (selling price). The first quantity on the right side of the equation is SSE, and the second term is denoted SSR, for **sum of squares for regression.** We can rewrite the equation as

Variation in y = SSE + SSR

As we did in the analysis of variance, we've partitioned the variation of y into two parts: SSE, which measures the amount of variation in y that remains unexplained; and SSR, which measures the amount of variation in y that is explained by the variation in the independent variable (odometer reading). With a little more algebra we produce another formula:

$$R^2 = 1 - \frac{SSE}{\sum (y_i - \bar{y})^2} = \frac{\sum (y_i - \bar{y})^2 - SSE}{\sum (y_i - \bar{y})^2} = \frac{SSR}{\sum (y_i - \bar{y})^2}$$

It follows that R^2 measures the proportion of the variation in y that is explained by the variation in x. Incidentally, the notation R^2 is derived from the fact that the coefficient of determination is the coefficient of correlation squared. Recall that we introduced the sample coefficient of correlation in Chapter 3 and labeled it r. (To be consistent with computer output, we capitalize r in the definition of the coefficient of determination.) We discuss the coefficient of correlation in Section 17.7.

▼ EXAMPLE 17.4

Find the coefficient of determination for Example 17.1, and describe what this statistic tells you about the regression model.

Solution Refer to page 548.

COMPUTE

R Square .6501

Excel prints a second R^2 statistic called the *coefficient of determination adjusted for degrees of freedom*. We will define and describe this statistic in Chapter 18.

INTERPRET

We found that R^2 is equal to .6501. This statistic tells us that 65.01% of the variation in the auction selling prices is explained by the variation in the odometer readings. The remaining 34.99% is unexplained. Unlike the value of a test statistic, the coefficient of determination does not have a critical value that enables us to draw conclusions. We know that the higher the value of R^2, the better the model fits the data. From the t-test of β_1 we know already that there is evidence of a linear relationship. The coefficient of determination merely supplies us with a measure of the strength of that relationship. As you will discover in the next chapter, when we improve the model, the value of R^2 increases.

▲

OTHER PARTS OF THE EXCEL PRINTOUT

The other part of the printout shown on page 548 relates to our discussion of the interpretation of the value of R^2, when its meaning is derived from the partitioning of the variation in y. The values of SSR and SSE are shown in an analysis of variance table similar to the tables introduced in Chapter 14. The general form of the table is shown below. The F-test performed in the ANOVA table will be explained in Chapter 18.

General Form of the ANOVA Table in the Simple Linear Regression Model

Source	d.f.	Sums of Squares	Mean Squares	F-Value
Regression	1	SSR	MSR = SSR/1	F = MSR/MSE
Residual	$n - 2$	SSE	MSE = SSE/$(n - 2)$	
Total	$n - 1$	Variation in y		

DEVELOPING AN UNDERSTANDING OF STATISTICAL CONCEPTS

Once again, we encounter the concept of explained variation. We first discussed the concept in Section 12.4 when we introduced the matched pairs experiment, where the experiment was designed to reduce the variation among experimental units. This concept was extended in the analysis of variance, where we partitioned the total variation into two or more sources (depending on the experimental design). And now, in regression analysis, we use the concept to measure how the dependent variable is affected by the independent variable. We partition the variation of the dependent variable into two sources: the variation explained by the variation in the independent variable and the unexplained variation. The greater the explained variation, the better the model is. We often refer to the coefficient of determination as a measure of the *explanatory power* of the model.

CAUSE-AND-EFFECT RELATIONSHIP

A common mistake is made by many students when they attempt to interpret the results of a regression analysis when there is evidence of a linear relationship. They imply that changes in the independent variable cause changes in the dependent variable. It must be emphasized that we cannot infer a causal relationship from statistics alone. Any inference about the cause of the changes in the dependent variable must be justified by a reasonable theoretical relationship. For example, statistical tests established that the more one smoked, the greater the probability of developing lung cancer. However, this analysis did not prove that smoking causes lung cancer. It only demonstrated that smoking and lung cancer were somehow related. Only when medical investigations established the connection were scientists able to confidently declare that smoking causes lung cancer.

As another illustration, consider Example 17.1, where we showed that the odometer reading is linearly related to the auction price. While it seems reasonable to conclude that decreasing the odometer reading would cause the auction price to rise, the conclusion may not be entirely true. It is theoretically possible that the price is determined by the overall condition of the car and that the condition generally worsens when the car is driven longer. Another analysis would be needed to establish the veracity of this conclusion.

Be cautious about the use of the terms "explained variation" and "explanatory power of the model." Do not interpret the word "explained" to mean "caused." We say that the coefficient of determination measures the amount of variation in y that is explained (not caused) by the variation in x. Thus, regression analysis can only show that a statistical relationship exists. We cannot infer that one variable causes another.

EXERCISES

17.16 Refer to Exercise 17.2.

   a What is the standard error of estimate? Interpret its value.

   b Describe how well the heights of the fathers and sons are linearly related.

   c Are the heights of fathers and sons linearly related?

17.17 Refer to Exercise 17.3. Apply the three methods of assessing the model to determine how well the linear model fits.

17.18 Refer to Exercise 17.4. Are the two variables linearly related?

17.19 Refer to Exercise 17.5. Use two statistics to measure the strength of the linear association. What do these statistics tell you?

17.20 Refer to Exercise 17.6.
  a Determine the standard error of estimate, and describe what this statistic tells you about the regression model.
  b Determine the coefficient of determination. What does this statistic tell you about how well the linear regression model fits?
  c Can we infer that the length of commercial and memory test score are linearly related?

17.21 Refer to Exercise 17.7.
  a Test to determine whether there is evidence of a linear relationship between study time and the final mark.
  b Determine the coefficient of determination. What does this statistic tell you about the regression line?

17.22 Refer to Exercise 17.8.
  a Determine the standard error of estimate, and describe what this statistic tells you about the regression line.
  b Can we conclude that education and Internet use are linearly related?
  c Determine the coefficient of determination, and discuss what its value tells you about the two variables.

17.23 Refer to Exercise 17.9. Are length of employment and age linearly related? (Conduct a statistical test to decide.) If so, provide a statistic that measures the strength of the association.

17.24 Refer to Exercise 17.10. Use whatever statistics you think useful to describe the reliability of your suggested compensation plan.

17.25 Refer to Exercise 17.11. Use whatever statistics you think useful to describe the reliability of your insurance premium plan.

17.26 Refer to Exercise 17.12.
  a Determine the standard error of estimate and describe what this statistic tells you about the regression model.
  b Determine the coefficient of determination. What does this statistic tell you about how well the linear regression model fits?
  c Can we infer that the weight and price of the diamonds are linearly related?

17.27 An economist wanted to investigate the relationship between office rents and vacancy rates. Accordingly, he took a random sample of monthly office rents and the percentage of vacant office space in 30 different cities. The results were stored in file Xr17-27 (column A = vacancy rates in percent and column B = monthly rents in dollars per square foot).
  a Determine the regression line.
  b Interpret the coefficients.
  c Can we conclude from these data that higher vacancy rates result in lower rents?
  d Measure how well the linear model fits the data. Discuss what this (these) measure(s) tell you.

17.28 Physicians have been recommending more exercise for their patients, particularly those who are overweight. One benefit of regular exercise appears to be a reduction in cholesterol, a substance associated with heart disease. In order to study the relationship more carefully, a physician took a random sample of 50 patients who do not exercise. He measured their cholesterol levels. He then started them on regular exercise programs. After four months, he asked each patient how many minutes per week (on average) he or she exercised and also measured their cholesterol levels. The results are stored in file Xr17-28 (column A = weekly exercise in minutes; column B = cholesterol level before exercise program; and column C = cholesterol level after exercise program).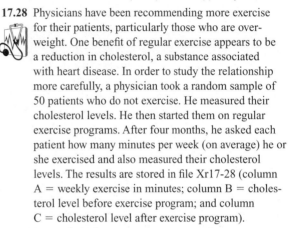
  a Determine the regression line that relates exercise time with cholesterol reduction.
  b Interpret the coefficients.
  c Can we conclude that the amount of exercise is linearly related to cholesterol reduction?
  d Measure how well the linear model fits.

17.29 Although a large number of tasks in the computer industry are robotic, a number of operations require human workers. Some jobs require a great deal of dexterity to properly position components into place. A large North American computer maker routinely tests applicants for these jobs by giving a dexterity test that involves a number of intricate finger and hand movements. The tests are scored on a 100-point scale. Only those who have scored above 70 are hired. To determine whether the tests are valid predictors of job performance, the personnel manager drew a random sample of 45 workers who were hired two months ago. He recorded their test scores and the percentage of nondefective computers they produced in the last week. These data are stored in columns A and B, respectively, in file Xr17-29. Can the manager infer that the test is a valid predictor? That is, can he infer that higher test results are associated with higher percentages of nondefective units?

17.6 USING THE REGRESSION EQUATION

Using the techniques in Section 17.5, we can assess how well the linear model fits the data. If the model fits satisfactory, we can use it to forecast and estimate values of the dependent variable. To illustrate, suppose that, in Example 17.1, the used-car dealer wanted to predict the selling price of a three-year-old Ford Taurus with 40,000 miles on the odometer. Using the regression equation, with $x = 40,000$, we get

$$\hat{y} = 6,533 - 0.0312x = 6,533 - 0.0312(40,000) = 5,285$$

Thus, the dealer would predict that the car would sell for \$5,285.

By itself, however, this value does not provide any information about how closely the value will match the true selling price. To discover that information, we must use an interval. In fact, we can use one of two intervals: the prediction interval of a particular value of y or the interval estimate of the expected value of y.

PREDICTING THE PARTICULAR VALUE OF y FOR A GIVEN x

The first interval we present is used whenever we want to predict one particular value of the dependent variable, given a specific value of the independent variable. This interval, often called the **prediction interval,** is calculated as follows.

Prediction Interval

$$\hat{y} \pm t_{\alpha/2} s_\varepsilon \sqrt{1 + \frac{1}{n} + \frac{(x_g - \bar{x})^2}{(n-1)s_x^2}}$$

where x_g is the given value of x and

$$\hat{y} = b_0 + b_1 x_g$$

ESTIMATING THE EXPECTED VALUE OF y FOR A GIVEN x

The conditions described in Section 17.4 imply that, for a given value of x, there is a population of values of y whose mean is

$$E(y) = \beta_0 + \beta_1 x$$

To estimate the mean of y, given x, we would use the following interval.

Interval Estimator of the Expected Value of y

$$\hat{y} \pm t_{\alpha/2} s_\varepsilon \sqrt{\frac{1}{n} + \frac{(x_g - \bar{x})^2}{(n-1)s_x^2}}$$

Unlike the formula for the prediction interval described above, this formula does not include the 1 under the square-root sign. As a result, the interval estimate of the expected value of y will be narrower than the prediction interval for the same given value of x and confidence level. This is because there is less error in estimating a mean value as opposed to predicting an individual value.

It must be understood that the widths of prediction interval and the interval estimator of the expected value depend entirely on how well the regression equation fits the data. That is, if the equation provides a poor fit, the intervals will be quite wide, and an equation that fits quite well will yield narrow intervals. Moreover, the prediction is based on the assumption that the relationship between the independent and dependent variables is stable. If the relationship changes, the use of the regression equation for analysis as well as forecasting is likely to produce a poor outcome.

▼ EXAMPLE 17.5

a A used-car dealer is about to bid on a three-year-old Ford Taurus equipped with automatic transmission, air conditioner, and AM/FM cassette tape player, and with 40,000 miles on the odometer. To help him decide how much to bid, he needs to predict the selling price.

b The used-car dealer mentioned in part (a) has an opportunity to bid on a lot of cars offered by a rental company. The rental company has 250 Ford Tauruses, all equipped with automatic transmission, air conditioning, and AM/FM cassette tape players. All of the cars in this lot have about 40,000 miles on the odometer. The dealer would like an estimate of the selling price of all the cars in the lot.

Solution

IDENTIFY

a The dealer would like to predict the selling price of a single car. Thus, he needs to employ the prediction interval

$$\hat{y} \pm t_{\alpha/2}\, s_\varepsilon \sqrt{1 + \frac{1}{n} + \frac{(x_g - \bar{x})^2}{(n-1)s_x^2}}$$

b The dealer wants to determine the mean price of a large lot of cars, so he needs to calculate the interval estimate of the expected value.

$$\hat{y} \pm t_{\alpha/2}\, s_\varepsilon \sqrt{\frac{1}{n} + \frac{(x_g - \bar{x})^2}{(n-1)s_x^2}}$$

Technically, this formula is used for infinitely large populations. However, we can interpret our problem as attempting to determine the average selling price of all Ford Tauruses equipped as described above, all with 40,000 miles on the odometer. The critical factor in part (b) is the need to estimate the mean price of a number of cars. We arbitrarily select a 95% confidence level.

COMPUTE

Microsoft Excel Output for Example 17.5

Prediction Interval

Price
Predicted value   5287.07

Prediction Interval
Lower limit   4984.24
Upper limit   5589.91

Interval Estimate of Expected Value
Lower limit   5251.87
Upper limit   5322.27

GENERAL COMMANDS	EXAMPLE 17.5
1 Type or import the data into two columns.	Open file **Xm17-01.**
2 Type the given value of *x* into any cell. We suggest the next available row in the column containing the independent variable.	**40000** (in cell A102)
3 Click **Tools, Data Analysis Plus,** and **Prediction Interval.**	
4 Specify the **Input Y Range:** (range of the dependent variable).	**B1:B101**
5 Specify the **Input X Range:** (range of the independent variable).	**A1:A101**
6 Click **Labels,** if applicable.	
7 Specify the **Given X Range:** (range of the given value of x).	**A102**
8 Specify the **Confidence Level (1-Alpha),** and click **OK.**	**.95**

INTERPRET

We predict that one car will sell for between $4,984.24 and $5,589.91. The average selling price of the population of three-year-old Ford Tauruses is estimated to lie between $5,251.87 and $5,322.27. Because predicting the selling price of one car is more difficult than estimating the mean selling price of all similar cars, the prediction interval is wider than the interval estimate of the expected value.

We repeat our caution concerning the use of the prediction interval. If the relationship between the odometer reading and the auction selling price remains unchanged, it is likely that the actual price of the three-year-old Taurus will be between $4,984.24 and $5,589.91. If some events occur to alter the relationship, the prediction interval may be useless.

THE EFFECT OF THE GIVEN VALUE OF x ON THE INTERVALS

If the two intervals were calculated for various values of x and graphed, Figure 17.6 would be produced. Notice that both intervals are represented by curved lines. This is due to the fact that the farther the given value of x is from $\bar{x}$, the greater the estimated error becomes. This factor is measured by

$$\frac{(x_g - \bar{x})^2}{(n-1)s_x^2}$$

which appears in both the prediction interval and the interval estimate of the expected value.

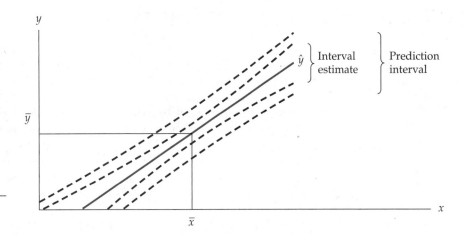

Figure 17.6

Interval estimates and prediction intervals

EXERCISES

17.30 Refer to Exercise 17.2. A statistics practitioner wants to produce a prediction interval of the height of a man whose father is 72 inches tall. What formula should be used? Produce such an interval using a confidence level of 99%.

17.31 Refer to Exercise 17.3. Predict with 95% confidence the mark in statistics when the calculus mark is 70.

17.32 Refer to Exercise 17.6.
   a Predict with 95% confidence the memory test score of a viewer who watches a 36-second commercial.
   b Estimate with 95% confidence the mean memory test score of people who watch 36-second commercials.

17.33 Refer to Exercise 17.7.
   a Predict with 90% confidence the final mark of a student who studies for 25 hours.
   b Estimate with 90% confidence the average mark of all students who study for 25 hours.

17.34 Refer to Exercise 17.8. Estimate with 90% confidence the mean amount of time spent on the Internet by people with 15 years of education.

17.35 Refer to Exercise 17.9. The company has just hired a 25-year-old telemarketer. Predict with 95% confidence how long he will stay with the company.

17.36 Refer to Exercise 17.10. Predict with 90% confidence the price of a barrel of oil if the API is 42.0.

17.37 Refer to Exercise 17.11.
  a Predict with 95% confidence the daily emergency medical expense of a 65-year-old Canadian.
  b Estimate with 95% confidence the mean daily emergency medical expense of all 65-year-old Canadians.

17.38 Refer to Exercise 17.12. Predict with 90% confidence the price of a diamond that weighs .35 carats.

17.39 Refer to Exercise 17.27. Predict with 95% confidence the monthly office rent in a city when the vacancy rate is 10%.

17.40 Refer to Exercise 17.28. Predict with 95% confidence the reduction in cholesterol level of an individual who plans to exercise for 300 minutes per week for a total of four months.

17.41 Refer to Exercise 17.27. Suppose that an individual whose cholesterol level is 250 is planning to exercise for 250 minutes per week. Predict with 95% confidence his cholesterol level after four months.

17.7 COEFFICIENTS OF CORRELATION

In Section 17.5, we noted that the coefficient of determination is the coefficient of correlation squared. When we introduced the coefficient of correlation (also called the **Pearson coefficient of correlation**) in Chapter 3, we observed that it is used to measure the strength of association between two variables. Why then do we use the coefficient of determination as our measure of the regression model's fit? The answer: The coefficient of determination is a better measure than the coefficient of correlation, because the values of R^2 can be interpreted more precisely. That is, R^2 is defined as the proportion of the variation in y that is explained by the variation in x. Except for $r = -1$, 0, and 1, the coefficient of correlation cannot be interpreted quantitatively. (When $r = -1$ or 1, every point falls on the regression line, and when $r = 0$, there is no linear pattern.) However, the coefficient of correlation can be useful in another way. We can use it to test for a linear relationship between two variables.

In cases where we're interested in determining *how* the independent variable affects the dependent variable, we estimate and test the linear regression model. The *t*-test of the slope presented in Section 17.5 allows us to determine whether a linear relationship actually exists. As we pointed out in Section 17.4, the statistical test requires that for each value of x, there exists a population of values of y that are normally distributed with a constant variance. This condition is required whether the data are experimental or observational.

In many circumstances we're interested in determining only *whether* a linear relationship exists and not the form of the relationship. When the data are observational and the two variables are bivariate normally distributed (see Section 17.4), we can calculate the coefficient of correlation and use it to test for linear association.

As we noted in Chapter 3, the population coefficient of correlation is denoted ρ (the Greek letter *rho*). Because ρ is a population parameter (which is almost always unknown), we must estimate its value from the sample data. Recall that the sample coefficient of correlation is defined as follows.

Sample Coefficient of Correlation

$$r = \frac{\text{cov}(x,y)}{s_x s_y}$$

TESTING THE COEFFICIENT OF CORRELATION

When there is no linear relationship between the two variables, $\rho = 0$. To determine whether we can infer that ρ is zero, we test the following hypotheses.

$H_0: \rho = 0$

$H_1: \rho \neq 0$

The test statistic is defined in the following way.*

Test Statistic for Testing $\rho = 0$

$$t = r\sqrt{\frac{n-2}{1-r^2}}$$

which is Student t distributed with $v = n - 2$ degrees of freedom, provided that the variables are bivariate normally distributed.

EXAMPLE 17.6

Using the data in Example 17.1, test to determine whether we can infer that a linear relationship exists by testing the population correlation coefficient.

Solution

IDENTIFY

We test the hypotheses

$H_0: \rho = 0$

$H_1: \rho \neq 0$

---

*This test statistic is used only when testing $\rho = 0$. To test other values, another test statistic must be employed.

> **COMPUTE**

Microsoft Excel Output for Example 17.6

Correlation	
Odometer and Price	
Pearson Coefficient of Correlation	−0.806
t Stat	−13.49
df	98
P(T<=t) one-tail	0
t Critical one-tail	1.6606
P(T<=t) two-tail	0
t Critical two-tail	1.9845

We have omitted part of the actual printout. That part will be discussed below. The value of the test statistic is $t = -13.49$, which has a p-value of 0.

GENERAL COMMANDS	EXAMPLE 17.6
1 Type or import the data into two adjacent columns.	Open file **Xm17-01**.
2 Click **Tools, Data Analysis Plus,** and **Correlation.**	
4 Specify the **Variable 1 Range:**.	A1:A101
5 Specify the **Variable 2 Range:**.	B1:B101
6 Click **Labels,** if applicable.	
7 Specify the value of α (**Alpha**), and click **OK**.	.05

> **INTERPRET**

As was the case in Example 17.3, where we tested β_1, there is overwhelming evidence of a linear relationship.

▲

If you review Example 17.3, you will find the same value of the test statistic and p-value. This is not a coincidence; the two tests are identical. This should be no surprise, because the data are the same, and the objective is the same: to determine whether two variables are linearly related. Hence, it is necessary to perform only one test, either the t-test of β_1 or the t-test of ρ. (We performed both tests for you to show you that they are identical and to motivate the next technique.)

SPEARMAN RANK CORRELATION COEFFICIENT

In the previous sections of this chapter, we have dealt only with interval variables and have assumed that all of the conditions for the validity of the hypothesis tests and interval estimates have been met. In many situations, however, one or both variables may be ordinal, or if both variables are interval, the bivariate distribution may be extremely nonnormal. In such cases, we measure and test to determine if a relationship exists by employing a nonparametric technique, the **Spearman rank correlation coefficient.**

The Spearman rank correlation coefficient is calculated like all of the previously introduced nonparametric methods by first ranking the data. We then calculate the Pearson correlation coefficient of the ranks.

The population Spearman correlation coefficient is labeled ρ_s, and the sample statistic used to estimate its value is labeled r_s.

> **Sample Spearman Rank Correlation Coefficient**
>
> $$r_s = \frac{\text{cov}(a,b)}{s_a s_b}$$
>
> where a and b are the ranks of x and y, respectively.

We can test to determine if a relationship exists between the two variables. The hypotheses to be tested are

$H_0: \rho_s = 0$

$H_1: \rho_s \neq 0$

(We also can conduct one-tail tests.)

When n is greater than 30, r_s is approximately normally distributed with mean 0 and standard deviation $1/\sqrt{n-1}$. Thus, for $n > 30$, the test statistic is

> **Test Statistic for Testing $\rho_s = 0$ when $n > 30$**
>
> $$z = \frac{r_s - 0}{1/\sqrt{n-1}} = r_s \sqrt{n-1}$$
>
> which is standard normally distributed.

▼ EXAMPLE 17.7

In the teaching evaluations that are completed by students in the last week of a course, students are asked (among other questions) to assess the quality of the course and their expected grade. The responses to the first question are

    1 Very poor

    2 Poor

    3 Fair

    4 Average

    5 Good

6 Very good

7 Excellent

The responses to the second question are the letter grades A, B, C, D, and F. A random sample of 100 evaluations from a statistics course are stored in columns A (course quality using the codes 1, 2, 3, 4, 5, 6, and 7) and B (expected grade using codes 4, 3, 2, 1, and 0, respectively) in file Xm17-07. Can we infer that a student's assessment of the quality of the course increases with his or her expected grade?

Solution

IDENTIFY

The problem objective is to analyze the relationship between two variables. Both variables are ordinal, which makes the appropriate technique the Spearman rank correlation coefficient test.

To answer the question, we specify the hypotheses as

$H_0: \rho_s = 0$

$H_1: \rho_s > 0$

The test statistic is

$z = r_s \sqrt{n-1}$

COMPUTE

Microsoft Excel Output for Example 17.7

Correlation

Course and Grade

Spearman Rank Correlation	0.5165
z Stat	5.1388
P(Z<=z) one-tail	0
z Critical one-tail	1.6449
P(Z<=z) two-tail	0
z Critical two-tail	1.96

As we did in Example 17.6, we edited the output. In this case we've omitted the first half of the printout, which shows the Pearson correlation coefficient and test statistic, which are not relevant to this problem.

The Spearman rank correlation coefficient is .5165. The value of the test statistic is $z = 5.1388$, whose *p*-value is 0.

INTERPRET

There is enough evidence to infer that as the expected grade increases, so does the assessment of the course. This suggests that instructors of courses that students find difficult must be judged somewhat differently from instructors of easy or popular courses.

EXERCISES

17.42 The weekly returns of two stocks are recorded for a 52-week period. These data are stored in file Xr17-42.
  a Assuming that the returns are normally distributed, can we infer that the stocks are correlated?
  b Assuming that the returns are extremely nonnormally distributed, can we infer that the stocks are correlated?

17.43 The general manager of an engineering firm wants to know if a draftsman's experience influences the quality of his work. She selects 34 draftsmen at random and records their years of work experience and their quality rating (as assessed by their supervisors). These data are stored in file Xr17-43 (column 1 = work experience in years; and column 2 = quality rating, where 5 = excellent, 4 = very good, 3 = average, 2 = fair, and 1 = poor). Can we infer from these data that years of work experience is a factor in determining the quality of work performed?

17.44 Refer to Exercise 17.6.
  a Determine the coefficient of correlation.
  b Test the coefficient of correlation to determine whether a linear relationship exists between the length of commercial and memory test score.
  c Assume that the conditions for the test conducted in Exercise 17.6 are not met. Do the data allow us to conclude that the longer the commercial, the higher the memory test score will be?

17.45 Assume that the normality requirement in Exercise 17.7 is not met. Test to determine whether grade and study time are positively related.

17.46 Refer to Exercise 17.8.
  a Determine the coefficient of correlation.
  b Test the coefficient of correlation to determine whether a linear relationship exists between Internet use and education.
  c Assume that the conditions for the test conducted in Exercise 17.8 are not met. Do the data allow us to conclude that people with more education use the Internet more?

17.47 If the normality requirement in Exercise 17.9 has been violated, can we infer that age and length of employment are related?

17.48 Assume that the price and quality of the oil in Exercise 17.10 are not bivariate normally distributed. Conduct a test to determine whether the higher prices are related to higher quality.

17.49 Refer to Exercise 17.11.
  a Determine the coefficient of correlation.
  b Test the coefficient of correlation to determine whether a linear relationship exists between age and medical expense.
  c Assume that the normality requirement for the test conducted in Exercise 17.11 is not met. Do the data allow us to conclude that older Canadians incur higher medical expenses?

17.50 Refer to Exercise 17.12. If we assume that price and weight are not bivariate normally distributed, can we infer that the two variables are related?

17.51 Refer to Exercise 17.27. If office rents and vacancy rates are not bivariate normally distributed, can we infer that higher vacancy rates result in lower rents?

17.52 The production manager of a firm wants to examine the relationship between aptitude test scores given prior to hiring of production-line workers and performance ratings received by the employees three months after starting work. The results of the study would allow the firm to decide how much weight to give to these aptitude tests relative to other work-history information obtained, including references. The aptitude test results range from 0 to 100. The performance ratings are as follows:

  1 = Employee has performed well below average.

  2 = Employee has performed somewhat below average.

  3 = Employee has performed at the average level.

  4 = Employee has performed somewhat above average.

  5 = Employee has performed well above average.

A random sample of 50 production workers yielded the results stored in columns A and B of file Xr17-52. Can the firm's manager infer that aptitude test scores are correlated with performance rating?

17.8 REGRESSION DIAGNOSTICS I

In Section 17.4, we described the required conditions for the validity of regression analysis. Simply put, the error variable must be normally distributed with a constant variance, and the errors must be independent of each other. In this section, we show how to diagnose violations. Additionally, we discuss how to deal with observations that are unusually large or small. Such observations must be investigated to determine if an error was made in recording them.

RESIDUAL ANALYSIS

Most departures from required conditions can be diagnosed by examining the residuals, which we discussed in Section 17.5. Most computer packages allow you to output the values of the residuals and apply various graphical and statistical techniques to this variable.

We can also compute the standardized residuals. We standardize residuals in the same way we standardize all variables, by subtracting the mean and dividing by the standard deviation. The mean of the residuals is zero, and because the standard deviation σ_ε is unknown, we must estimate its value. The simplest estimate is the standard error of estimate s_ε. Thus

$$\text{Standardized residual for point } i = \frac{r_i}{s_\varepsilon}$$

Excel calculates the standardized residuals by dividing the residuals by the standard deviation of the residuals. (The difference between the standard error of estimate and the standard deviation of the residuals is that in the formula of the former the denominator is $n - 2$, whereas in the formula for the latter the denominator is $n - 1$.) Part of the printout for Example 17.1 follows.

Excel Printout of Predicted Values, Residuals and Standardized Residuals for Example 17.1

Residual Output

Observation	Predicted Price	Residuals	Standard Residuals
1	5368.4575	−50.4575	−0.3346
2	5138.8250	−77.8250	−0.5161
3	5105.3304	−97.3304	−0.6454
98	5499.2577	−240.2577	−1.5932
99	5312.1243	43.8757	0.2910
100	5399.4906	−266.4906	−1.7672

Commands

Proceed with the first four steps of regression analysis described on page 548. Before clicking **OK**, select **Residuals** and **Standardized Residuals**. The predicted values, residuals, and standardized residuals will be printed.

An analysis of the residuals will allow us to determine if the error variable is nonnormal, whether the error variance is constant, and whether the errors are independent. We begin with nonnormality.

NONNORMALITY

As we've done throughout this book, we check for normality by drawing the histogram of the residuals. Excel's version is shown below. As you can see, the histogram is bell shaped, leading us to believe that the error is normally distributed.

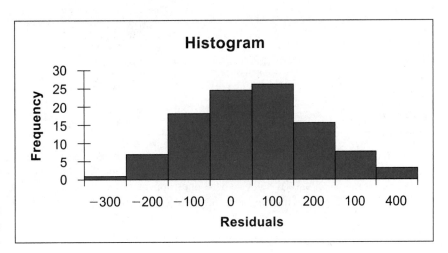

Microsoft Excel Histogram of Residuals for Example 17.1

HETEROSCEDASTICITY

The variance of the error variable σ_ε^2 is required to be constant. When this requirement is violated, the condition is called **heteroscedasticity.** (You can impress friends and relatives by using this term. If you can't pronounce it, try **homoscedasticity,** which refers to the condition where the requirement is satisfied.) One method of diagnosing heteroscedasticity is to plot the residuals against the predicted values of y. We then look for a change in the spread of the plotted points. Figure 17.7 describes such a situation. Notice that, in this illustration, σ_ε^2 appears to be small when $\hat{y}$ is small and large when $\hat{y}$ is large. Of course, many other patterns could be used to depict this problem.

Figure 17.8 illustrates a case in which σ_ε^2 is constant. As a result, there is no apparent change in the variation of the residuals.

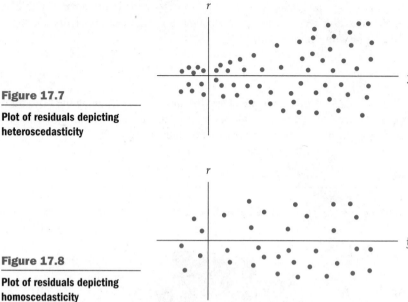

Figure 17.7

Plot of residuals depicting heteroscedasticity

Figure 17.8

Plot of residuals depicting homoscedasticity

Excel's plot of the residuals versus the predicted values of y for Example 17.1 is shown below. There does appear to be a decrease in the variance for large values of $\hat{y}$. However, it is far from clear that there is a problem here.

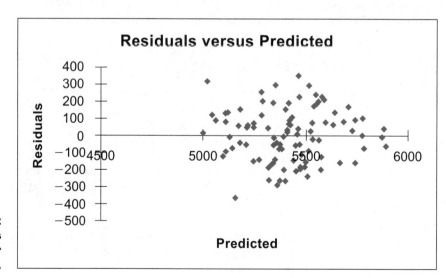

Microsoft Excel Plot of Predicted Values versus Residuals for Example 17.1

NONINDEPENDENCE OF THE ERROR VARIABLE

In Chapter 2, we briefly described the difference between cross-sectional and time-series data. Cross-sectional data are observations made at approximately the same time, whereas a time series is a set of observations taken at successive points of time.

The data in Example 17.1 are cross sectional because all of the prices and odometer readings were taken at about the same time. If we were to observe the auction price of cars every week for (say) a year, that would constitute a time series.

Condition 4 states that the values of the error variable are independent. When the data are time series, the errors often are correlated. Error terms that are correlated over time are said to be autocorrelated or serially correlated. For example, suppose that, in an analysis of the relationship between annual gross profits and some independent variable, we observe the gross profits for the years 1980 to 1999. The observed values of y are denoted $y_1, y_2, \ldots, y_{20}$, where y_1 is the gross profit for 1980, y_2 is the gross profit for 1981, and so on. If we label the residuals $r_1, r_2, \ldots, r_{20}$, then—if the independence requirement is satisfied—there should be no relationship among the residuals. However, if the residuals are related, it is likely that **autocorrelation** exists.

We can often detect autocorrelation by graphing the residuals against the time periods. If a pattern emerges, it is likely that the independence requirement is violated. Figures 17.9 (alternating positive and negative residuals) and 17.10 (increasing residuals) exhibit patterns indicating autocorrelation. (Notice that we joined the points to make it easier to see the patterns.) Figure 17.11 shows no pattern (the residuals appear to be randomly distributed over the time periods), and thus likely represents the occurrence of independent errors.

In Chapter 18, we introduce the Durbin-Watson test, which is another statistical test to determine if one form of this problem is present. We also describe a number of remedies to violations of the required condition.

Figure 17.9

Plot of residuals versus time indicating autocorrelation (alternating)

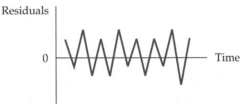

Figure 17.10

Plot of residuals versus time indicating autocorrelation (increasing)

Figure 17.11

Plot of residuals versus time indicating independence

OUTLIERS

An outlier is an observation that is unusually small or unusually large. To illustrate, consider Example 17.1, where the range of odometer readings was 19,057 to 49,223 miles. If we had observed a value of 5,000 miles, we would identify that point as an outlier. There are several possibilities that we need to investigate.

1 *There was an error in recording the value.* To detect an error we would check the point or points in question. In Example 17.1, we could check the car's odometer to determine if a mistake was made. If so, we would correct it before proceeding with the regression analysis.

2 *The point should not have been included in the sample.* Occasionally, measurements are taken from experimental units that do not belong with the sample. We can check to ensure that the car with the 5,000-mile odometer reading was actually three years old. We should also investigate the possibility that the odometer was rolled back. In either case, the outlier should be discarded.

3 *The observation was simply an unusually large or small value that belongs to the sample and that was recorded properly.* In this case, we would do nothing to the outlier. It would be judged to be valid.

Outliers can be identified from the scatter diagram. Figure 17.12 depicts a scatter diagram with one outlier. The statistics practitioner should check to determine if the measurement was recorded accurately and whether the experimental unit should be included in the sample.

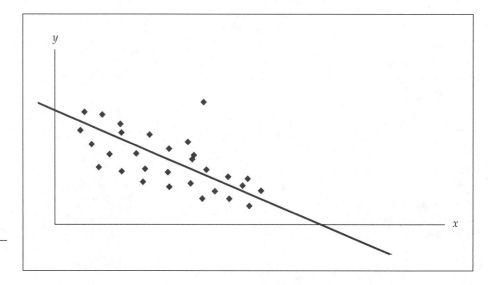

Figure 17.12

Scatter diagram with one outlier

The standardized residuals also can be helpful in identifying outliers. We suggest that if the absolute value of a standardized residual exceeds 2.0, the values of the independent and dependent variables should be investigated.

PROCEDURE FOR REGRESSION DIAGNOSTICS

The order of the material presented in this chapter is dictated by pedagogical requirements. Consequently, we presented the least squares method, methods of assessing the model's fit, predicting and estimating using the regression equation, coefficients of correlation, and finally, the regression diagnostics. In a practical application, the regression diagnostics would be conducted earlier in the process. It is appropriate to investigate violations of the required conditions when the model is assessed and before using the regression equation to predict and estimate. The following steps describe the entire process.

1. *Develop a model that has a theoretical basis.* That is, for the dependent variable in question, find an independent variable that you believe is linearly related to it.
2. *Gather data for the two variables.* Ideally, conduct a controlled experiment. If that is not possible, collect observational data.
3. *Draw the scatter diagram to determine whether a linear model appears to be appropriate.* Identify possible outliers.
4. *Determine the regression equation.*
5. *Calculate the residuals, and check the required conditions.*

    Is the error variable nonnormal?

    Is the variance constant?

    Are the errors independent?

    Check the outliers.

6. *Assess the model's fit.*

    Compute the standard error of estimate,

    Test to determine whether there is a linear relationship. (Test β_1 or ρ.)

    Compute the coefficient of determination.

7. *If the model fits the data, use the regression equation to predict a particular value of the dependent variable and/or estimate its mean.*

EXERCISES

17.53 Refer to Exercise 17.5.

   a Determine the residuals and the standardized residuals.

   b Draw the histogram of the residuals. Does it appear that the errors are normally distributed? Explain.

   c Identify possible outliers.

   d Plot the residuals versus the predicted values of y. Does it appear that heteroscedasticity is a problem? Explain.

17.54 Refer to Exercise 17.6.

   a Determine the residuals and the standardized residuals.

   b Draw the histogram of the residuals. Does it appear that the errors are normally distributed? Explain.

   c Identify possible outliers.

   d Plot the residuals versus the predicted values of y. Does it appear that heteroscedasticity is a problem? Explain.

17.55 Refer to Exercise 17.7.
  a Does it appear that the errors are normally distributed? Explain.
  b Does it appear that heteroscedasticity is a problem? Explain.

17.56 Are the required conditions satisfied in Exercise 17.8?

17.57 Refer to Exercise 17.9.
  a Determine the residuals and the standardized residuals.
  b Draw the histogram of the residuals. Does it appear that the errors are normally distributed? Explain.
  c Identify possible outliers.
  d Plot the residuals versus the predicted values of y. Does it appear that heteroscedasticity is a problem? Explain.

17.58 Refer to Exercise 17.10. Are the required conditions satisfied?

17.59 Are the required conditions satisfied in Exercise 17.11?

17.60 Refer to Exercise 17.12. Does it appear that the required conditions are satisfied?

17.9 SUMMARY

Simple linear regression and correlation are techniques for analyzing the relationship between two interval variables. Regression analysis assumes that the two variables are linearly related. The least squares method produces estimates of the intercept and the slope of the regression line. Considerable effort is expended in assessing how well the linear model fits the data. We calculate the standard error of estimate, which is an estimate of the standard deviation of the error variable. We test the slope to determine whether there is sufficient evidence of a linear relationship. The strength of the linear association is measured by the coefficient of determination. When the model provides a good fit, we can use it to predict the particular value and to estimate the expected value of the dependent variable. We can also use the Pearson correlation coefficient to measure and test the relationship between two normally distributed variables. The Spearman rank correlation coefficient analyzes the relationship between two variables, at least one of which is ordinal. It can also be used when the variables are extremely nonnormal. We completed this chapter with a discussion of how to diagnose violations of the required conditions.

IMPORTANT TERMS

Regression analysis
Dependent variable
Independent variable
Deterministic model
Probabilistic model
First-order linear model
Simple linear regression model
Scatter diagram
Least squares method
Error variable
Residuals
Sum of squares for error
Standard error of estimate
Coefficient of determination
Sum of squares for regression
Prediction interval
Pearson coefficient of correlation
Spearman rank correlation coefficient
Heteroscedasticity
Homoscedasticity
Autocorrelation

SYMBOLS

Symbol	Pronounced	Represents
β_0	*Beta-sub-zero* or *beta-zero*	y-intercept coefficient
β_1	*Beta-sub-one* or *beta-one*	Slope coefficient
ε	*Epsilon*	Error variable
$\hat{y}$	*y-hat*	Fitted or calculated value of y
b_0	*b-sub-zero* or *b-zero*	Sample y-intercept coefficient
b_1	*b-sub-one* or *b-one*	Sample slope coefficient
σ_ε	*Sigma-sub-epsilon* or *sigma-epsilon*	Standard deviation of error variable
s_ε	*s-sub-epsilon* or *s-epsilon*	Standard error of estimate
s_{b_1}	*s-sub-b-sub-one* or *s-b-one*	Standard error of b_1
R^2	*R-squared*	Coefficient of determination
x_g	*x-sub-g* or *x-g*	Given value of x
ρ	*Rho*	Pearson coefficient of correlation
r		Sample coefficient of correlation
ρ_s	*Rho-sub-s* or *rho-s*	Spearman rank correlation
r_s	*r-sub-s* or *r-s*	Sample Spearman rank correlation
r_i	*r-sub-i* or *r-i*	Residual of ith point

FORMULAS

Sample slope

$$b_1 = \frac{\text{cov}(x,y)}{s_x^2}$$

Sample y-intercept

$$b_0 = \bar{y} - b_1 \bar{x}$$

Sum of squares for error

$$\text{SSE} = \sum_{i=1}^{n}(y_i - \hat{y}_i)^2$$

Standard error of estimate

$$s_\varepsilon = \sqrt{\frac{\text{SSE}}{n-2}}$$

Test statistic for the slope

$$t = \frac{b_1 - \beta_1}{s_{b_1}}$$

Standard error of b_1

$$s_{b_1} = \frac{s_\varepsilon}{\sqrt{(n-1)s_x^2}}$$

Coefficient of determination

$$R^2 = \frac{[\text{cov}(x,y)]^2}{s_x^2 s_y^2} = 1 - \frac{\text{SSE}}{\sum (y_i - \bar{y})^2}$$

Prediction interval

$$\hat{y} \pm t_{\alpha/2} s_\varepsilon \sqrt{1 + \frac{1}{n} + \frac{(x_g - \bar{x})^2}{(n-1)s_x^2}}$$

Interval estimator of the expected value of y

$$\hat{y} \pm t_{\alpha/2} s_\varepsilon \sqrt{\frac{1}{n} + \frac{(x_g - \bar{x})^2}{(n-1)s_x^2}}$$

Sample coefficient of correlation

$$r = \frac{\text{cov}(x,y)}{s_x s_y}$$

Test statistic for testing $\rho = 0$

$$t = r\sqrt{\frac{n-2}{1-r^2}}$$

Sample Spearman rank correlation coefficient

$$r_s = \frac{\text{cov}(a,b)}{s_a s_b}$$

Test statistic for testing $\rho_s = 0$ when $n > 30$

$$z = \frac{r_s - 0}{1/\sqrt{n-1}} = r_s \sqrt{n-1}$$

MICROSOFT EXCEL OUTPUT AND INSTRUCTIONS

Technique	Page
Regression	548–549
Prediction interval	566
Pearson coefficient of correlation and test	570
Spearman rank correlation and test	572

SUPPLEMENTARY EXERCISES

17.61 The manager of Colonial Furniture has been reviewing weekly advertising expenditures. During the past six months, all advertisements for the store have appeared in the local newspaper. The number of ads per week has varied from one to seven. The store's sales staff has been tracking the number of customers who enter the store each week. The number of ads and the number of customers per week for the past 26 weeks have been stored in the file Xr17-61.

  a Determine the sample regression line.
  b Interpret the coefficients.
  c Can the manager infer that the larger the number of ads, the larger the number of customers?
  d Find and interpret the coefficient of determination.
  e In your opinion, is it a worthwhile exercise to use the regression equation to predict the number of customers who will enter the store, given that Colonial intends to advertise five times in the newspaper? If so, find a 95% prediction interval. If not, explain why not.

17.62 The president of a company that manufactures car seats has been concerned about the number and cost of machine breakdowns. The problem is that the machines are old and becoming quite unreliable. However, the cost of replacing them is quite high, and the president is not certain that the cost can be made up in today's slow economy. To help make a decision about replacement, he gathered data about last month's costs for repairs and the ages (in months) of the plant's 20 welding machines. These data are stored in file Xr17-62.

  a Find the sample regression line.
  b Interpret the coefficients.
  c Determine the coefficient of determination, and discuss what this statistic tells you.
  d Conduct a test to determine whether the age of a machine and its monthly cost of repair are linearly related.
  e Is the fit of the simple linear model good enough to allow the president to predict the monthly repair cost of a welding machine that is 120 months old? If so, find a 95% prediction interval. If not, explain why not.

17.63 Several years ago, Coca-Cola attempted to change its 100-year-old recipe. One reason why the company's management felt this was necessary was competition from Pepsi-Cola. Respondents of surveys of Pepsi drinkers indicated that they preferred Pepsi because it was sweeter than Coke. As part of the analysis that led to Coke's ill-fated move, the management of Coca-Cola performed extensive surveys wherein consumers tasted various versions of the new Coke. Suppose that a random sample of 200 cola drinkers was given versions of Coke with different amounts of sugar. After tasting the product, each drinker was asked to rate the taste quality. The possible responses were as follows.

  5 = excellent

  4 = good

  3 = average

  2 = fair

  1 = poor

The responses and sugar content (percent by volume) of the version tasted were recorded in columns A and B, respectively, of file Xr17-63. Can management infer that sugar content affects drinkers' ratings of the cola?

17.64 An agronomist wanted to investigate the factors that determine crop yield. Accordingly, she undertook an experiment wherein a farm was divided into 30 one-acre plots. The amount of fertilizer applied to each plot was varied. Corn was then planted, and the amount of corn harvested at the end of the season was recorded. These data were stored in file Xr17-64.

  a Find the sample regression line, and interpret the coefficients.
  b Can the agronomist conclude that there is a linear relationship between the amount of fertilizer and the crop yield?
  c Find the coefficient of determination, and interpret its value.
  d Does the simple linear model appear to be a useful tool in predicting crop yield from the amount of fertilizer applied? If so, produce a 95% prediction interval of the crop yield when 300 pounds of fertilizer are applied. If not, explain why not.

17.65 Auto manufacturers are required to test their vehicles for a variety of pollutants in the exhaust. The amount of pollutant varies even among identical vehicles, so that several vehicles must be tested. The engineer in charge of testing has collected data (in grams per kilometer driven) on the amounts of two pollutants, carbon monoxide and nitrous oxide, for 50 identical vehicles. These data are stored in columns A and B in file Xr17-65. The engineer believes the company can save money by testing for only one of the pollutants because the two pollutants are closely linked. That is, if a car is emitting a large amount of carbon monoxide, it will also emit a large amount of nitrous oxide. Do the data support the engineer's belief?

17.66 It is doubtful that any sport collects more statistics than baseball. This surfeit of statistics allows fans to conduct a great variety of statistical analyses. For example, fans are always interested in determining which factors lead to successful teams. A statistics practitioner determined the team batting average and the team winning percentage for the 14 American League teams at the end of a recent season. We will assume that these data represent a random sample of the relationship between batting average and winning percentage for all time. These data are stored in file Xr17-66.

    a Find the sample regression line, and interpret the coefficients.

    b Find the standard error of estimate, and describe what this statistic tells you.

    c Do these data provide sufficient evidence to conclude that higher team batting averages lead to higher winning percentages?

    d Find the coefficient of determination, and interpret its value.

    e Predict with 90% confidence the winning percentage of a team whose batting average is .275.

17.67 In an effort to further analyze a baseball team's winning percentage, the statistics practitioner determined each team's earned run average (ERA). (An earned run average is the number of earned runs a baseball team gives up in an average nine-inning game.) These data, together with the team's winning percentage, are stored in file Xr17-67.

    a Find the sample regression line, and interpret the coefficients.

    b Find the standard error of estimate, and describe what this statistic tells you.

    c Do these data provide sufficient evidence to conclude that lower earned run averages lead to higher winning percentages?

    d Find the coefficient of determination, and interpret its value.

    e Predict with 90% confidence the winning percentage of a team whose ERA is 4.00.

17.68 In the last decade, society in general and the judicial system in particular have altered their opinions on the seriousness of drunken driving. In most jurisdictions, driving an automobile with a blood-alcohol level in excess of .08 is a felony. Because of a number of factors, it is difficult to provide guidelines on when it is safe for someone who has consumed alcohol to drive a car. In an experiment to examine the relationship between blood-alcohol level and the weight of a drinker, 50 men of varying weights were each given three beers to drink, and one hour later their blood-alcohol level was measured. These data are stored in file Xr17-68.

    a If we assume that the two variables are normally distributed, can we conclude that blood-alcohol level and weight are related?

    b After examining the data, the statistics practitioner in charge of the experiment concluded that a regression analysis was invalid (because he determined that the error term was nonnormal). What conclusions can you draw from these data about the relationship between blood-alcohol level and weight?

17.69 One general belief held by observers of the business world is that taller men earn more money than shorter men. In a University of Pittsburgh study (reported in the *Wall Street Journal,* 30 December 1986), 250 MBA graduates, all about 30 years old, were polled and asked to report their height (in inches) and their annual income (to the nearest $1,000). These data are stored in columns A and B, respectively, in file Xr17-69.

    a Determine the sample regression line, and interpret the coefficients.

    b Do these data provide sufficient statistical evidence to infer that taller MBAs earn more money than shorter ones?

    c Provide a measure of the strength of the linear relationship between income and height.

    d Do you think that this model is good enough to be used to estimate and predict income on the basis of height? If not, explain why not. If so, estimate with 95% confidence the mean income of all six-foot men with MBAs and predict with 95% confidence the income of a man 5 feet 10 inches tall with an MBA.

17.70 Every year the United States Federal Trade Commission rates cigarette brands according to their levels of tar and nicotine, substances that are hazardous to smokers' health. Additionally, the commission includes the amount of carbon monoxide, a by-product of burning tobacco, which seriously affects the heart. A random sample of 25 brands was taken. The data are stored in file Xr17-70. Column A stores the brand name, column B stores the tar content in milligrams, column C holds the nicotine content in milligrams, and column D contains the carbon monoxide in milligrams.

a Are the levels of tar and nicotine linearly related?
b Are the levels of nicotine and carbon monoxide linearly related?

17.71 The analysis the human resources manager performed in Exercise 17.29 indicated that the dexterity test is not a predictor of job performance. However, before discontinuing the test, he decided that the problem is that the statistical analysis was flawed in that it only examined the relationship between test score and job performance for those who scored well in the test. Recall that only those who scored above 70 were hired. Applicants who achieved scores below 70 were not hired. The manager decided to perform another statistical analysis. A sample of 50 job applicants who scored above 50 were hired, and as before the workers' performance was measured. The test scores and percentages of nondefective computers produced are stored in columns A and B, respectively, in file Xr17-71. On the basis of these data, should the manager discontinue the dexterity tests?

17.72 Some critics of television complain that the amount of violence shown on television contributes to violence in our society. Others point out that television also contributes to the high level of obesity among children. We may have to add financial problems to the list. A sociologist theorized that people who watch television frequently are exposed to many commercials, which in turn leads them to buy more, finally resulting in increasing debt. To test this belief, a sample of 430 families was drawn. For each, the total debt and the number of hours the television is turned on per week were recorded. These data are stored in columns A and B, respectively in file Xr17-72. Perform a statistical procedure to help test the theory.

17.73 The New York Marathon is run in May when temperatures vary considerably. Because of the enormous strain running more than 26 miles has on the human body, higher temperatures sap the strength of runners and generally result in slower times. To examine the effect of temperature on male runners, the winning times were measured together with the temperature for the years 1978 to 1998. These data are stored in columns A (winning times), and B (temperature in degrees Fahrenheit) in file Xr17-73 (and Xr02-78).

a Draw a scatter diagram of temperature versus winning times.
b Does it appear that there is a linear relationship?
c Conduct a regression analysis, and interpret the coefficients.
d Is there evidence to infer that there is a linear relationship?
e Determine the coefficient of determination, and interpret its value.

17.74 Refer to Exercise 17.73. The winning times for female runners (and temperatures) are stored in file Xr17-74 (and Xr02-79). Perform a similar analysis to the one above.

Case 17.1 Predicting University Grades from High School Grades*

Ontario high school students must complete a minimum of six Ontario Academic Credits (OACs) to gain admission to a university in the province. Most students take more than six OACs because universities take the average of the best six in deciding which students to admit. Most programs at universities require high school students to select certain courses. For example, science programs require two of chemistry, biology, and physics. Students applying to engineering must complete at least two mathematics OACs as well as physics. In recent years, one business program began an examination of all aspects of their program including the criteria used to admit students. Students are required to take English and calculus OACs, and the minimum high school average is about 85%. Strangely enough, even though students are required to complete English and calculus, the marks in these subjects are not included in the average unless they

*The author is grateful to Leslie Grauer for her help in gathering the data for this case.

are in the top six courses in a student's transcript. To examine the issue, the registrar took a random sample of students who recently graduated with the BBA (Bachelor of Business Administration) degree. He recorded the university GPA (range 0 to 12), the high school average based on the best six courses, and the high school average using English and calculus and the four next best marks. These data are stored in columns A to C, respectively, in file C17-01.

a Is there a relationship between university grades and high school average using the best six OACs?

b Is there a relationship between university grades and high school average using the best four OACs plus calculus and English?

c What should the university do about the information provided by this case?

Case 17.2 Insurance Compensation for Lost Revenues[*]

In July 1990, a rock-and-roll museum opened in Atlanta, Georgia. The museum was located in a large city block containing a variety of stores. In late July 1992, a fire that started in one of these stores burned the entire block, including the museum. Fortunately, the museum had taken out insurance to cover the cost of rebuilding as well as lost revenue. As a general rule, insurance companies base their payment on how well the company performed in the past. However, the owners of the museum argued that the revenues were increasing, and hence they are entitled to more money under their insurance plan. The argument was based on the revenues and attendance figures of an amusement park that was opened nearby, featuring rides and other similar attractions. The amusement park opened in December 1991. The two entertainment facilities were operating jointly during the last 4 weeks of 1991 and the first 28 weeks of 1992 (the point at which the fire destroyed the museum). In April 1995, the museum reopened with considerably more features than the original one.

The attendance for both facilities for December 1991 to October 1995 are listed in columns A (museum) and B (amusement park) in file C17-02. During the period when the museum was closed, the data show zero attendance.

The owners of the museum argue that the weekly attendance from the 29th week of 1992 to the 16th week of 1995 should be estimated using the most current data (17th to 42nd week of 1995). The insurance company argues that the estimates should be based on the 4 weeks of 1991 and the 28 weeks of 1992, when both facilities were operating and before the museum reopened with more features than the original museum.

a Estimate the coefficients of the simple regression model based on the insurance company's argument. That is, use the attendance figures for the last 4 weeks in 1991 and the next 28 weeks in 1992 to estimate the coeffi-

[*]The case and the data are real. The names have been changed to preserve anonymity. The author wishes to thank Dr. Kevin Leonard for supplying the problem and the data.

cients. Then, use the model to calculate point predictions for the museum's weekly attendance figures when the museum was closed. Calculate the predicted total attendance.

b Repeat part (a) using the museum's argument. That is, use the attendance figures after the reopening in 1995 to estimate the regression coefficients, and use the equation to predict the weekly attendance when the museum was closed. Calculate the total attendance that was lost because of the fire.

c In your opinion, which figure should be used to calculate how much the insurance company should award the museum? How should that compensation be determined?

Chapter 18

Multiple Regression

- **18.1** Introduction
- **18.2** Model and Required Conditions
- **18.3** Estimating the Coefficients and Assessing the Model
- **18.4** Regression Diagnostics II
- **18.5** Regression Diagnostics III (Time Series)
- **18.6** Nominal Independent Variables
- **18.7** Summary

18.1 INTRODUCTION

In the previous chapter, we employed the simple linear regression model to analyze how one interval variable (the dependent variable y) is affected by another interval variable (the independent variable x). The restriction of using only one independent variable was motivated by the need to simplify the introduction to regression analysis. Although there are a number of applications where we purposely develop a model with only one independent variable, in general we prefer to include as many independent variables as can be shown to significantly affect the dependent variable. Arbitrarily limiting the number of independent variables also limits the usefulness of the model.

In this chapter, we allow for any number of independent variables. In so doing, we expect to develop models that fit the data better than would a simple linear regression model. We proceed in a manner similar to that in Chapter 17. We begin by describing the multiple regression model and listing the required conditions. We let the computer produce the required statistics and use them to assess the model's fit and diagnose violations of the required conditions. We employ the model by predicting the particular value of the dependent variable and estimating its expected value.

18.2 MODEL AND REQUIRED CONDITIONS

We now assume that k independent variables are potentially related to the dependent variable. Thus, the model is represented by the following equation.

$$y = \beta_0 + \beta_1 x_1 + \beta_2 x_2 + \cdots + \beta_k x_k + \varepsilon$$

where y is the dependent variable, $x_1, x_2, \ldots, x_k$ are the independent variables, $\beta_0, \beta_1, \ldots, \beta_k$ are the coefficients, and ε is the error variable. The independent variables may actually be functions of other variables. For example, we might define some of the independent variables as follows:

$$x_2 = x_1^2$$

$$x_5 = x_3 x_4$$

$$x_7 = \log(x_6)$$

The error variable is retained because, even though we have included additional independent variables, deviations between predicted values of y and actual values of y will still occur. Incidentally, when there is more than one independent variable in the regression model, we refer to the graphical depiction of the equation as a **response surface** rather than as a straight line. Figure 18.1 depicts a scatter diagram of a response surface with $k = 2$. (When $k = 2$, the regression equation creates a **plane**.) Of course, whenever k is greater than 2, we can only imagine the response surface; we cannot draw it.

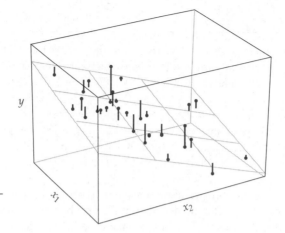

Figure 18.1

Scatter diagram and response surface with $k = 2$

An important part of the regression analysis comprises several statistical techniques that evaluate how well the model fits the data. These techniques require the following conditions, which we introduced in the previous chapter.

> *Required Conditions for Error Variable*
>
> **1** The probability distribution of the error variable ε is normal.
> **2** The mean of the error variable is zero.
> **3** The standard deviation of ε is σ_ε, which is a constant.
> **4** The errors are independent.

In Section 17.7, we discussed how to recognize when the requirements are unsatisfied. Those same procedures can be used to detect violations of required condition in the multiple regression model.

We now proceed as we did in Chapter 17; we discuss how the model's coefficients are estimated and how we assess the model's fit.

18.3 ESTIMATING THE COEFFICIENTS AND ASSESSING THE MODEL

The general form of the sample regression equation is

$$\hat{y} = b_0 + b_1 x_1 + b_2 x_2 + \cdots + b_k x_k$$

The procedures introduced in Chapter 17 are extended to the multiple regression model. However, in Chapter 17, we discussed how to interpret the coefficients first, followed by a discussion of how to assess the model's fit. In practice, we reverse the process. That is, the first step is to determine how well the model fits. If the model's fit is poor, there is no point in a further analysis of the coefficients of that model. A much higher priority is assigned to the task of improving the model. In this chapter, we show how a regression analysis is performed. The steps we use are as follows.

1. Use a computer and software to generate the coefficients and the statistics used to assess the model.

2. Diagnose violations of the required conditions. If there are problems, we attempt to remedy them.

3. Assess the model's fit. Three statistics perform this function: the standard error of estimate, the coefficient of determination, and the F-test of the analysis of variance. The first two were introduced in Chapter 17; the third will be introduced here.

4. If we are satisfied with the model's fit and that the required conditions are met, we can attempt to interpret the coefficients and test them as we did in Chapter 17. We use the model to predict or estimate the expected value of the dependent variable.

We illustrate these techniques with the following example.

▼ EXAMPLE 18.1[*]

La Quinta Motor Inns is a moderately priced chain of motor inns located across the United States. Its market is the frequent business traveler. The chain recently launched a campaign to increase market share by building new inns. The management of the chain is aware of the difficulty in choosing locations for new motels. Moreover, making decisions without adequate information often results in poor decisions. Consequently, they acquired data on 100 randomly selected inns belonging to La Quinta. The objective was to predict which sites are likely to be profitable.

To measure profitability La Quinta used operating margin, which is the ratio of the sum of profit, depreciation, and interest expenses divided by total revenue. (Although occupancy is often used as measure of a motel's success, the statistics practitioner concluded that occupancy was too unstable, especially during economic turbulence.) The higher the operating margin, the greater the success of the inn. La Quinta defines profitable inns as those with an operating margin in excess of 50% and unprofitable ones as those with margins of less than 30%. After a discussion with a number of experienced managers La Quinta decided to select one or two independent variables from each of these categories: competition, market awareness, demand generators, demographics, and physical qualities. To measure the degree of competition they determined the total number of motel and hotel rooms within three miles of each La Quinta inn. Market awareness was measured by the number of miles to the nearest La Quinta inn. Two variables that represent sources of customers were chosen. The amount of office space and college and university enrollment in the surrounding community are demand generators. Both of these are measures of economic activity. A demographic variable that describes the community is the median household income. Finally, as a measure of the physical qualities of the location, La Quinta

---

[*]Adapted from Sheryl E. Kimes and James A. Fitzsimmons, "Selecting Profitable Hotel Sites at La Quinta Motor Inns," *INTERFACES 20,* March–April 1990: 12–20.

chose the distance to the downtown core. These data are stored in file Xm18-01, where column A stores the inn selected (INN), column B contains the operating margin (MARGIN in percent), and columns C through H (ROOMS, NEAREST in miles, OFFICE in thousands of square feet, COLLEGE enrollment in thousands, INCOME in thousands of dollars, and DISTTWN in miles, respectively) store the six independent variables. Some of these data are shown below. Conduct a regression analysis and analyze the results.

INN	MARGIN	ROOMS	NEAREST	OFFICE	COLLEGE	INCOME	DISTTWN
1	55.5	3,203	0.1	549	8.0	37	12.1
2	33.8	2,810	1.5	496	17.5	39	0.4
3	49.0	2,890	1.9	254	20.0	39	12.2
...	...	...	...	...	...	...	...
98	40.0	3,397	2.7	855	19.5	42	11.7
99	39.8	3,823	0.7	202	17.0	36	10.0
100	35.2	3,251	2.6	275	13.0	39	10.5

Solution Excel is called upon to perform the calculations, the results of which are seen below.

COMPUTE

Microsoft Excel Output for Example 18.1

Summary Output

Regression Statistics

Multiple R	0.7246
R Square	0.5251
Adjusted R Square	0.4944
Standard Error	5.51
Observations	100

ANOVA

	df	SS	MS	F	Significance F
Regression	6	3123.8	520.6	17.14	0.0000
Residual	93	2825.6	30.4		
Total	99	5949.5			

	Coefficients	Standard Error	t Stat	P-value
Intercept	72.45	7.89	9.18	0.0000
ROOMS	20.0076	0.0013	26.07	0.0000
NEAREST	21.65	0.63	22.60	0.0108
OFFICE	0.020	0.0034	5.80	0.0000
COLLEGE	0.21	0.13	1.59	0.1159
INCOME	20.41	0.14	22.96	0.0039
DISTTWN	0.23	0.18	1.26	0.2107

GENERAL COMMANDS	EXAMPLE 18.1
1 Type or import the data into adjacent columns. 2 Click **Tools, Data Analysis...**, and **Regression**. 3 Specify the **Input Y Range:**. 4 Specify the **Input X Range:**. 5 Click **Labels**, if applicable, and click **OK**.	Open file **Xm18-01**. B1:B101 C1:H101

The regression equation is

MARGIN = 72.45 − 0.0076 ROOMS − 1.65 NEAREST + 0.020 OFFICE
     + 0.21 COLLEGE − 0.41 INCOME + 0.23 DISTTWN

We assess the model in three ways: the standard error of estimate, the coefficient of determination (both introduced in Chapter 17), and the F-test of the analysis of variance (presented below).

STANDARD ERROR OF ESTIMATE

Recall that σ_ε is the standard deviation of the error variable ε and that, because σ_ε is a population parameter, it is necessary to estimate its value by using s_ε. In multiple regression, the standard error of estimate is defined as follows.

Standard Error of Estimate

$$s_\varepsilon = \sqrt{\frac{SSE}{n - k - 1}}$$

Excel outputs s_ε for Example 18.1 as:

Standard Error 5.51

Recall that we judge the magnitude of the standard error of estimate relative to the values of the dependent variable and particularly to the mean of y. In this example, $\bar{y} = 45.739$ (not shown in printouts). It appears that the standard error of estimate is not particularly small.

COEFFICIENT OF DETERMINATION

Recall from Chapter 17 that the coefficient of determination is defined as

$$R^2 = 1 - \frac{SSE}{\sum(y_i - \bar{y})^2}$$

For Example 18.1, Excel prints

R-Square = .5251

This means that 52.51% of the variation in operating margin is explained by the six independent variables, while 47.49% remains unexplained.

Notice that Excel prints a second R^2 statistic, called the **coefficient of determination adjusted for degrees of freedom,** which has been adjusted to take into account the sample size and the number of independent variables. The rationale for this statistic is that, if the number of independent variables k is large relative to the sample size n, the unadjusted R^2 value may be unrealistically high. To understand this point, consider what would happen if the sample size is 2 in a simple linear regression model. The line will fit the data perfectly, resulting in $R^2 = 1$ when in fact there may be no linear relationship. To avoid creating a false impression, the adjusted R^2 is often calculated. Its formula follows.

Coefficient of Determination Adjusted for Degrees of Freedom

$$\text{Adjusted } R^2 = 1 - \frac{\text{SSE}/(n-k-1)}{\sum(y_i - \bar{y})^2/(n-1)}$$

If n is considerably larger than k, the actual and adjusted R^2 values will be similar. But if SSE is quite different from zero and k is large compared to n, the actual and adjusted values of R^2 will differ substantially. If such differences exist, the analyst should be alerted to a potential problem in interpreting the coefficient of determination. In Example 18.1, the adjusted coefficient of determination is .4944, indicating that, no matter how we measure the coefficient of determination, the model's fit is moderately good.

TESTING THE VALIDITY OF THE MODEL

In the simple linear regression model, we tested the slope coefficient to determine whether sufficient evidence existed to allow us to conclude that there was a linear relationship between the independent variable and the dependent variable. However, because there is only one independent variable in that model, the t-test also tested to determine whether that model is valid. When there is more than one independent variable, we need another method to test the overall validity of the model. The technique is a version of the analysis of variance, which we introduced in Chapter 14.

To test the validity of the regression model, we specify the following hypotheses:

$H_0: \beta_1 = \beta_2 = \beta_3 = \beta_4 = \beta_5 = \beta_6 = 0$

$H_1:$ At least one β_i is not equal to zero

If the null hypothesis is true, none of the independent variables $x_1, x_2, \ldots, x_6$ is linearly related to y, and therefore the model is invalid. If at least one β_i is not equal to zero, the model does have some validity.

When we introduced the coefficient of determination in Chapter 17, we noted that the variation in the dependent variable [measured by $\sum(y_i - \bar{y})^2$] can be decomposed into two parts: the explained variation (measured by SSR) and the unexplained variation (measured by SSE). That is,

Variation in y = SSR + SSE

Furthermore, we established that, if SSR is large relative to SSE, the coefficient of determination will be high—signifying a good model. On the other hand, if SSE is large, most of the variation will be unexplained, which indicates that the model provides a poor fit and consequently has little validity.

The test statistic is the same one we encountered in Section 14.2, where we tested for the equality of k population means. To judge whether SSR is large enough relative to SSE to allow us to infer that at least one coefficient is not equal to zero, we compute the ratio of the two mean squares. (Recall that the mean square is the sum of squares divided by its degrees of freedom; recall, too, that the ratio of two mean squares is F distributed as long as the underlying population is normal—a required condition for this application.) The calculation of the test statistic is summarized in an analysis of variance (ANOVA) table, which in general appears as follows. The Excel ANOVA table is shown below.

Analysis of Variance Table for Regression Analysis

Source of Variation	Degrees of Freedom	Sums of Squares	Mean Squares	F-Statistic
Regression	k	SSR	MSR = SSR/k	F = MSR/MSE
Residual	$n - k - 1$	SSE	MSE = SSE/$(n - k - 1)$	
Total	$n - 1$	$\sum(y_i - \bar{y})^2$		

Below we reprint Excel's ANOVA table. (See page 592.)

Microsoft Excel Analysis of Variance for Example 18.1

ANOVA

	df	SS	MS	F	Significance F
Regression	6	3123.8	520.6	17.14	0.0000
Residual	93	2825.6	30.4		
Total	99	5949.5			

A large value of F indicates that most of the variation in y is explained by the regression equation and that the model is valid. A small value of F indicates that most of the variation in y is unexplained. The rejection region allows us to determine whether F is large enough to justify rejecting the null hypothesis. The rejection region is

$$F > F_{\alpha, k, n-k-1}$$

In Example 18.1, the rejection region (assuming $\alpha = .05$) is

$$F > F_{\alpha, k, n-k-1} = F_{.05, 6, 93} \approx 2.17$$

As you can see from the printout, $F = 17.14$. The printout also includes the p-value (**Significance F**) of the test, which is

$$p\text{-value} = P(F > 17.14) = 0$$

Obviously, there is a great deal of evidence to infer that the model is valid.

Although each assessment measurement offers a different perspective, all agree in their assessment of how well the model fits the data, because all are based on the sum of squares for error, SSE. The standard error of estimate is

$$s_\varepsilon = \sqrt{\frac{SSE}{n-k-1}}$$

and the coefficient of determination is

$$R^2 = 1 - \frac{SSE}{\sum(y_i - \bar{y})^2}$$

When the response surface hits every single point, SSE = 0. Hence $s_\varepsilon = 0$, and $R^2 = 1$.

If the model provides a poor fit, we know that SSE will be large [its maximum value is $\sum(y_i - \bar{y})^2$], s_ε will be large, and [because SSE is close to $\sum(y_i - \bar{y})^2$] R^2 will be close to zero.

The F-statistic also depends on SSE. Specifically,

$$F = \frac{(\sum(y_i - \bar{y})^2 - SSE)/k}{SSE/(n-k-1)}$$

When SSE = 0,

$$F = \frac{\sum(y_i - \bar{y})^2/k}{0/(n-k-1)}$$

which is infinitely large. When SSE is large, SSE is close to $\sum(y_i - \bar{y})^2$ and, as a result, F is quite small.

The relationship among s_ε, R^2, and F is summarized in Table 18.1.

If we're satisfied that the model fits the data as well as possible, and that the required conditions are satisfied (see Section 18.4), we can interpret and test the individual coefficients and use the model to predict and estimate.

INTERPRETING THE COEFFICIENTS

The intercept is $b_0 = 72.45$. This is the average operating margin when all of the independent variables are zero. As we observed in Chapter 17, it is often misleading to try to interpret this value, particularly if zero is outside the range of the values of the independent variables (as is the case here).

The relationship between MARGIN and ROOMS is described by $b_1 = -.0076$. From this number we learn that in this model, for each additional room within three miles of the La Quinta inn, the operating margin decreases on average by .0076% (assuming that the other independent variables in this model are held constant). Changing the units, we can interpret b_1 to say that for each additional 1,000 rooms the margin decreases by 7.6%.

The coefficient $b_2 = -1.65$ specifies that for each additional mile that the nearest competitor is to a La Quinta inn, the average operating margin decreases by 1.65% (assuming the constancy of the other independent variables).

The nature of the relationship between ROOMS and MARGIN and between NEAREST and MARGIN was expected. Obviously, more competitors closer to the inn will decrease the profitability of the inn.

Table 18.1 **Relationship among s_ε, R^2, and F**

SSE	s_ε	R^2	F	Assessment of Model
0	0	1	∞	Perfect
Small	Small	Close to 1	Large	Good
Large	Large	Close to 0	Small	Poor
$\sum(y_i - \bar{y})^2$	$\sqrt{\dfrac{\sum(y_i - \bar{y})^2}{n - k - 1}}$*	0	0	Useless

*When n is large and k is small, this quantity is approximately equal to the standard deviation of y.

The relationship between OFFICE and MARGIN is expressed by $b_3 = .020$. Because OFFICE is measured in thousands of square feet (of office space in the nearby community), we interpret this number as the average increase in operating margin for each additional thousand square feet of office space (keeping the other independent variables fixed). So, for every extra 100,000 square feet of office the operating margin increases on average by 2.0%.

The relationship between COLLEGE and MARGIN is specified by $b_4 = .21$, which we interpret to mean that for each additional thousand students the average operating margin increases by .21% (when the other variables are constant).

Both OFFICE and COLLEGE produced positive coefficients indicating that these measures of economic activity are positively related to margin.

The relationship between INCOME and MARGIN is described by $b_5 = -.41$. For each additional thousand dollar increase in median household income, the average operating margin *decreases* by .41% (holding all other variables constant). This result was unexpected. However, one interpretation of this result is that in more industrial areas the incomes tend to be lower. And, operating margins are higher in such communities.

The last variable in the model is DISTTWN. Its relationship with MARGIN is described by $b_6 = .23$. This tells us that for each additional mile to the downtown center, the operating margin increases on average by .23% (keeping the other independent variables constant).

TESTING THE COEFFICIENTS

In Chapter 17, we described how to test to determine whether there is sufficient evidence to infer that in the simple linear regression model x and y are linearly related. The null and alternative hypotheses were

$H_0: \beta_1 = 0$

$H_1: \beta_1 \neq 0$

The test statistic was

$$t = \frac{b_1 - \beta_1}{s_{b_1}}$$

which is Student t distributed with $v = n - 2$ degrees of freedom.

In the multiple regression model, we have more than one independent variable. For each such variable, we can test to determine if there is enough evidence of a linear relationship between it and the dependent variable.

> **Testing the Coefficients**
>
> $H_0: \beta_i = 0$
>
> $H_1: \beta_i \neq 0$
>
> (for $i = 1, 2, \ldots, k$); the test statistic is
>
> $$t = \frac{b_i - \beta_i}{s_{b_i}}$$
>
> which is Student t distributed with $v = n - k - 1$

To illustrate, we test each of the coefficients in the multiple regression model in Example 18.1. The tests that follow are performed just as all other tests in this book have been performed. We set up the null and alternative hypotheses, identify the test statistic, and use the computer to calculate the value of the test statistic and its p-value. For each independent variable, we test ($i = 1, 2, 3, 4, 5, 6$).

$H_0: \beta_i = 0$

$H_1: \beta_i \neq 0$

Refer to page 592 and examine the computer output for Example 18.1. The output includes the t-tests of β_i.

Test of β_1 Value of the test statistic: $t = -6.07$; p-value $= 0$

There is overwhelming evidence to infer that the number of motel and hotel rooms within three miles of the La Quinta inn and the operating margin are linearly related.

Test of β_2 Value of the test statistic: $t = -2.60$; p-value $= .0108$

There is evidence to conclude that the distance to the nearest motel and the operating margin of the La Quinta inn are linearly related.

Test of β_3 Value of the test statistic: $t = 5.80$; p-value $= 0$

This test allows us to infer that there is a linear relationship between the operating margin and the amount of office space around the inn.

Test of β_4 Value of the test statistic: $t = 1.59$; p-value $= .1159$

From this statistical test we discover that there is no evidence of a linear relationship between college enrollment in the community around the inn and the operating margin.

Test of β_5 Value of the test statistic: $t = -2.96$; p-value $= .0039$

There is overwhelming statistical evidence to indicate that the operating margin and the median household income are linearly related.

Test of β_6 Value of the test statistic: $t = 1.26$; p-value $= .2107$

There is not enough evidence to infer the existence of a linear relationship between the distance to the downtown center and the operating margin of the La Quinta inn.

INTERPRET

We have discovered that in this model the number of hotel and motel rooms, distance to the nearest motel, amount of office space, and median household income are linearly related to the operating margin. Moreover, in this model we found no evidence to infer that college enrollment and distance to downtown center are linearly related to operating margin. The *t*-tests tell La Quinta's management that, in choosing the site of a new motel, they should look for locations where no or few other motels are present, where there is a great deal of office space, and where the surrounding households are relatively affluent.

A CAUTIONARY NOTE ABOUT INTERPRETING THE RESULTS

Care should be taken when interpreting the results of this and other regression analyses. We might find that in one model there is enough evidence to conclude that a particular independent variable is linearly related to the dependent variable but that in another model no such evidence exists. Consequently, whenever a particular *t*-test is *not* significant, we state that there is not enough evidence to infer that the independent and dependent variable are linearly related *in this model*. The implication is that another model may yield different conclusions.

Furthermore, if one or more of the required conditions are violated, the results may be invalid. In Section 17.8 we introduced the procedures that allow the statistics practitioner to examine the model's requirements. We will add to this discussion in Section 18.4. We also remind you that it is dangerous to extrapolate far outside the range of the observed values of the independent variables.

▲

t-TESTS AND THE ANALYSIS OF VARIANCE

The *t*-tests of the individual coefficients allow us to determine whether $\beta_i \neq 0$ (for $i = 1, 2, \ldots, k$), which tells us whether a linear relationship exists between x_i and y. There is a *t*-test for each independent variable. Consequently, the computer automatically performs k *t*-tests. (It actually conducts $k + 1$ *t*-tests, including the one for β_0, which we usually ignore.) The **F-test** in the analysis of variance combines these *t*-tests into a single test. That is, we test all the β_i at one time to determine if at least one of them is not equal to zero. The question naturally arises: Why do we need the F-test if it is nothing more than the combination of the previously performed *t*-tests? Recall that we addressed this issue before. In Chapter 14, we pointed out that we could replace the analysis of variance by a series of *t*-tests of the difference between two means. However, by doing so we increase the probability of making a Type I error. That means that even when there is no linear relationship between each of the independent variables and the dependent variable, multiple *t*-tests will likely show some are significant. As a result, we will conclude erroneously that, since at least one β_i is not equal to zero, the model is valid. The F-test, on the other hand, is performed only once. Because the probability that a Type I error will occur in a single trial is equal to α, the chance of erroneously concluding that the model is valid is substantially less with the F-test than with multiple *t*-tests.

There is another reason why the F-test is superior to multiple *t*-tests. Because of a commonly occurring problem called *multicollinearity*, the *t*-tests may indicate that some independent variables are not linearly related to the dependent variable, when in fact they are. The problem of multicollinearity does not affect the F-test, nor does it inhibit us from developing a model that fits the data well. Multicollinearity is discussed in Section 18.4.

THE *F*-TEST AND THE *t*-TEST IN THE SIMPLE LINEAR REGRESSION MODEL

It is useful for you to know that we can use the *F*-test to test the validity of the simple linear regression model. However, this test is identical to the *t*-test of β_1, which in the simple linear regression model tells us whether that independent variable is linearly related to the dependent variable. However, because there is only one independent variable, the *t*-test of β_1 also tells us whether the model is valid, which is the purpose of the *F*-test.

The relationship between the *t*-test of β_1 and the *F*-test can be explained mathematically. Statisticians can show that if we square a *t*-statistic with v degrees of freedom we produce an *F*-statistic with 1 and v degrees of freedom. (We briefly discussed this relationship in Chapter 14.) To illustrate, consider Example 17.1 on page 547. We found the *t*-test of β_1 to be -13.49, with degrees of freedom equal to 98 ($v = n - 2 = 100 - 2 = 98$). The *p*-value was 0. The output included the analysis of variance table where $F = 182.11$ and *p*-value = 0. The *t*-statistic squared is $t^2 = (-13.49)^2 = 181.98$, which is approximately 182.11. (The difference is due to rounding errors.) Notice that the degrees of freedom of the *F*-statistic are 1 and 98. Thus, we can use either test to test the validity of the simple linear regression model.

USING THE REGRESSION EQUATION

As was the case with simple linear regression, we can use the multiple regression equation in two ways: We can produce the prediction interval for a particular value of *y*, and we can produce the interval estimate of the expected value of *y*. Like the other calculations associated with multiple regression, we call on the computer to do the work.

Suppose that in Example 18.1 a manager investigated a potential site for a La Quinta inn and found the following characteristics:

> There are 3,815 rooms within three miles of the site, and the closest other hotel or motel is 3.4 miles away. The amount of office space is 476,000 square feet. There is one college and one university nearby with a total enrollment of 24,500 students. From the census, the manager learns that the median household income in the area (rounded to the nearest thousand) is $39,000. Finally, the distance to the downtown center has been measured at 3.6 miles.

The manager wants to predict the operating margin if and when the inn is built.

In Chapter 17 we introduced the prediction interval and interval estimate of the expected value of *y*. We can use these procedures to produce a prediction interval for one inn and an interval estimate of the expected (average) operating margin for all sites with the given variables.

Excel Prediction and Interval Estimate

Prediction Interval
MARGIN
Predicted value   37.09
Prediction Interval
Lower limit   25.40
Upper limit   48.79
Interval Estimate of Expected Value
Lower limit   32.97
Upper limit   41.21

COMMANDS

See the commands on page 566.
In cells C102 to H102 we input the values **3815, 3.4, 476, 24.5, 39, 3.6**, respectively. We specified 95% confidence.

INTERPRET

We predict that the operating margin will fall between 25.40% and 48.79%. This interval is quite wide, confirming the need to have extremely well-fitting models to make accurate predictions. However, management defines a profitable inn as one with an operating margin greater than 50% and an unprofitable inn as one with an operating margin below 30%. As you can see, the entire prediction interval is below 50%, and part of it is below 30%. The management of La Quinta will pass on this site.

The expected operating margin of all sites that fit this category is estimated to be between 32.97% and 41.21%. We interpret this to mean that if we built inns on an infinite number of sites that fit the category described above, the average inn would not be profitable.

EXERCISES

18.1 A developer who specializes in summer cottage properties is considering purchasing a large tract of land adjoining a lake. The current owner of the tract has already subdivided the land into separate building lots and has prepared the lots by removing some of the trees. The developer wants to forecast the value of each lot. From previous experience, she knows that the most important factors affecting the price of the lot are size, number of mature trees, and distance to the lake. From a nearby area, she gathers the relevant data for 60 recently sold lots. These data are stored in file Xr18-01. (Column A = price in thousands of dollars, column B = lot size in thousands of square feet, column C = number of mature trees, and column D = distance to the lake in feet.) Conduct a multiple regression analysis.

  a What is the standard error of estimate? Interpret its value.
  b What is the coefficient of determination? What does this statistic tell you?
  c What is the coefficient of determination, adjusted for degrees of freedom? Why does this value differ from the coefficient of determination? What does this tell you about the model?
  d Test the validity of the model. What does the p-value of the test statistic tell you?
  e Interpret each of the coefficients.
  f Test to determine whether each of the independent variables is linearly related to the price of the lot.

18.2 After analyzing the results of Exercise 17.7, Pat decided that a certain amount of studying could actually improve final grades. However, too much studying would not be warranted, because Pat's ambition (if that's what one could call it) was ultimately to graduate with the absolute minimum level of work. Pat was registered in a statistics course, which had only three weeks to go before the final exam, and where the final grade was determined in the following way.

$$\text{Total mark} = 20\% \text{ (Assignment)} \\ + 30\% \text{ (Midterm test)} \\ + 50\% \text{ (Final exam)}$$

In order to determine how much work to do for the remaining three weeks, Pat needed to be able to predict the final exam mark on the basis of the assignment mark and the midterm mark. Pat's marks on these were 12/20 and 14/30, respectively. Accordingly, Pat undertook the following analysis. The final exam mark, assignment mark, and midterm test mark for 30 students who took the statistics course last year were collected. These data are stored in columns 1 to 3, respectively, in file Xr18-02. Perform a multiple regression analysis.

  a What is the standard error of estimate? Briefly describe how you interpret this statistic.
  b What is the coefficient of determination? What does this statistic tell you?
  c What is the coefficient of determination, adjusted for degrees of freedom? What do this statistic and the one alluded to in part (b) tell you about the model?
  d Test the validity of the model. What does the p-value of the test statistic tell you?
  e Interpret each of the coefficients.
  f Can Pat infer from these results that the assignment mark is linearly related to the final grade?
  g Can Pat infer from these results that the midterm mark is linearly related to the final grade?

18.3 The president of a company that manufactures drywall wants to analyze the variables that affect demand for his product. Drywall is used to construct walls in houses and offices. Consequently, the president decides to develop a regression model in which the dependent variable is monthly sales of drywall (in hundreds of 4×8 sheets) and the independent variables are

Number of building permits issued in the county
Five-year mortgage rates (in percentage points)
Vacancy rate in apartments (in percentage points)
Vacancy rate in office buildings (in percentage points)

To estimate a multiple regression model, he took the monthly observations from the past two years. The data are stored in columns A to E, respectively, in file Xr18-03. Use Excel to calculate the statistics in the multiple regression model

  a What is the standard error of estimate? Can you use this statistic to assess the model's fit? If so, how?
  b What is the coefficient of determination, and what does it tell you about the regression model?
  c What is the coefficient of determination, adjusted for degrees of freedom? What do this statistic and the statistic referred to in part (b) tell you about how well this model fits the data?
  d Test the validity of the model. What does the p-value of the test statistic tell you?
  e Interpret each of the coefficients.
  f Test to determine whether each of the independent variables is linearly related to drywall demand.

18.4 Suppose that the statistics practitioner who did the analysis described in Exercise 17.2 wanted to investigate other variables that determine heights. As part of the same study, she also recorded the heights of the mothers. These values are stored in column C of file Xr18-04. (Columns A and B contain the data from Exercise 17.2.) Analyze the data using multiple regression.

  a What is the standard error of estimate, and what does this statistic tell you?
  b What is the coefficient of determination? What does this statistic tell you?
  c What is the coefficient of determination, adjusted for degrees of freedom? What do this statistic and the one referred to in part (b) tell you about how well the model fits the data?
  d Test the validity of the model. What does the test result tell you?
  e Interpret each of the coefficients.
  f Do these data allow the statistics practitioner to infer that the heights of the sons and the fathers are linearly related?
  g Do these data allow the statistics practitioner to infer that the heights of the sons and the mothers are linearly related?

18.5 When one company buys another company, it is not unusual that some workers are terminated. The severance benefits offered to the laid-off workers are often the subject of dispute. Suppose that the Laurier Company recently bought the Western Company and subsequently terminated 20 of Western's employees. As part of the buyout agreement, it was promised that the severance packages offered to the former Western employees would be equivalent to those offered to Laurier employees who had been terminated in the past year.

Thirty-six-year-old Bill Smith, a Western employee for the past 10 years, earning $32,000 per year, was one of those let go. His severance package included an offer of five weeks' severance pay. Bill complained that this offer was less than that offered to Laurier's employees when they were laid off, in contravention of the buyout agreement. A statistician was called in to settle the dispute. The statistician was told that severance is determined by three factors: age, length of service with the company, and pay. To determine how generous the severance package had been, a random sample of 50 Laurier ex-employees was taken. For each, the following variables were recorded. (The data are stored in columns A to D, respectively, of file Xr18-05.)

Number of weeks of severance pay
Age of employee
Number of years with the company
Annual pay (in thousands of dollars)

a Determine the regression equation. Interpret the coefficients.
b Comment on how well the model fits the data.
c Do all the independent variables belong in the equation? Explain.
d Are the required conditions satisfied? Explain.
e Perform an analysis to determine if Bill is correct in his assessment of the severance package.

18.6 The admissions officer of a university is trying to develop a formal system of deciding which students to admit to the university. She believes that determinants of success include the standard variables—high school grades and SAT scores. However, she also believes that students who have participated in extracurricular activities are more likely to succeed than those who have not. To investigate the issue, she randomly sampled 100 fourth-year students and recorded the following variables:

GPA for the first three years at the university (range: 0 to 12)
GPA from high school (range: 0 to 12)
SAT score (range: 200 to 800)
Number of hours on average spent per week in organized extracurricular activities in the last year of high school

The data are stored in columns A to D, respectively, of file Xr18-06.

a Develop a model that helps the admissions officer decide which students to admit, and use the computer to generate the usual statistics.
b What is the standard error of estimate? What does this statistic tell you?
c What is the coefficient of determination? Interpret its value.
d What is the coefficient of determination, adjusted for degrees of freedom? Interpret its value.
e Test the validity of the model. What does the p-value of the test statistic tell you?
f Interpret each of the coefficients.
g Test to determine whether each of the independent variables is linearly related to the dependent variable.
h Predict with 95% confidence the GPA for the first three years of university for a student whose high school GPA is 10, whose SAT score is 600, and who worked an average of two hours per week on organized extracurricular activities in the last year of high school.
i Estimate with 90% confidence the mean GPA for the first three years of university for all students whose high school GPA is 8, whose SAT score is 550, and who worked an average of 10 hours per week on organized extracurricular activities in the last year of high school.

18.7 The marketing manager for a chain of hardware stores needed more information about the effectiveness of the three types of advertising that the chain used. These are localized direct mailing (in which flyers describing sales and featured products are distributed to homes in the area surrounding a store), newspaper advertising, and local television advertisements. To determine which type is most effective, the manager collected one week's data from 25 randomly selected stores. For each store, the following variables were recorded.

Weekly gross sales
Weekly expenditures on direct mailing
Weekly expenditures on newspaper advertising
Weekly expenditures on television commercials

All variables were recorded in thousands of dollars and stored in columns A to D, respectively, in file Xr18-07.

a Find the regression equation.
b What are the coefficient of determination and the coefficient of determination adjusted for degrees of freedom? What do these statistics tell you about the regression equation?
c What does the standard error of estimate tell you about the regression model?
d Test the validity of the model. What does the p-value of the test statistic tell you?
e Which independent variables are linearly related to weekly gross sales? Explain.

f Predict with 95% confidence next week's gross sales if a local store spent $800 on direct mailing, $1,200 on newspaper advertisements, and $2,000 on television commercials.

g Estimate with 95% confidence the mean weekly gross sales for all stores that spend $800 on direct mailing, $1,200 on newspaper advertising, and $2,000 on television commercials.

h Discuss the difference between the two intervals found in parts (f) and (g).

18.8 In an effort to explain to customers why their electricity bills have been so high lately, and how, specifically, they could save money by reducing the thermostat settings on both space heaters and water heaters, an electric utility company has collected total kilowatt consumption figures for last year's winter months, as well as thermostat settings on space and water heaters, for 100 homes. The data are stored in columns A (consumption), B (space heater thermostat setting), and C (water heater thermostat setting) of file Xr18-08.

a Determine the regression equation.
b Determine the standard error of estimate, and comment about what it tells you.
c Determine the coefficient of determination, and comment about what it tells you.
d Test the validity of the model, and describe what this test tells you.
e Predict with 95% confidence the electricity consumption of a house whose space heater thermostat is set at 70 and whose water healer thermostat is set at 130.
f Estimate with 95% confidence the average electricity consumption for houses whose space heater thermostat is set at 70 and whose water heater thermostat is set at 130.

18.9 In Exercise 17.27, an economist examined the relationship between office rents and the city's office vacancy rate. The model appears to be quite poor. It was decided to add another variable that measures the state of the economy. The city's unemployment rate was chosen for this purpose. The data are stored in file Xr18-09. Column A contains the rents, column B stores the vacancy rates (these data are identical to file Xr17-27), and column C contains the unemployment rate in percent.

a Determine the regression equation.
b Determine the coefficient of determination, and describe what this value means.
c Test the model's validity.
d Determine which of the two independent variables is linearly related to rents.

e Predict with 95% confidence the office rent in a city whose vacancy rate is 10% and whose unemployment rate is 7%.

18.10 Exercise 17.8 analyzed the relationship between Internet use and education. In an effort to determine whether other variables affected Internet use, another survey was performed. A random sample of two hundred adult Internet users was interviewed. Each person was asked to report his or her age and income. These data are stored in columns A (weekly internet use in hours), B (age), and C (annual income in thousands of dollars.) in file Xr18-10.

a Determine the regression equation.
b Determine the coefficient of determination, and describe what this value means.
c Test the model's validity.
d Predict with 90% confidence the Internet use for an individual who is 40 years old earning $50,000.
e Estimate with 95% confidence the mean Internet use of all individuals who are 30 years old and who earn $35,000.

18.11 An MBA program that was started two decades ago wanted to analyze the factors that affect student performance. The dean of the School of Business decided to build a multiple regression model where the dependent variable is the MBA grade point average (GPA) for each of 100 randomly selected MBA students who graduated in the past three years. The independent variables are the undergraduate GPA, the Graduate Management Admissions Test score (GMAT), and the number of years of work experience prior to entering the program. These data are stored in columns A to D, respectively, in file Xr18-11.

a Conduct a multiple regression analysis.
b Briefly describe what the coefficients tell you.
c Test to determine which independent variables affect the dependent variable.
d Use whatever statistics you wish to assess the model's fit.
e Can we infer that the regression model is valid?
f Predict with 95% confidence the MBA GPA for an applicant whose undergraduate GPA is 8, GMAT score = 630, and who has worked for five years.

18.12 Life insurance companies are keenly interested in predicting how long their customers will live, because their premiums and profitability depend on such numbers. An actuary for one insurance company gathered data from 100 recently deceased male customers. He recorded the age at death of the customer plus the ages at death of his mother and father, the mean ages at death of his grandmothers, and the mean ages at death of his grandfathers. These data are recorded in columns A to E, respectively, of file Xr18-12.

a Perform a multiple regression analysis on these data.
b Is the model valid?
c Are the required conditions satisfied?
d Is multicollinearity a problem here?
e Interpret and test the coefficients.
f Predict with 95% confidence the longevity of a man whose parents lived to the age of 70, whose grandmothers averaged 80 years, and whose grandfathers averaged 75.
g Estimate with 95% confidence the mean longevity of men whose mothers lived to 75, whose fathers lived to 65, whose grandmothers averaged 85 years, and whose grandfathers averaged 75.

18.13 One of the critical factors that determine the success of a catalog store chain is the availability of products that consumers want to buy. If a store is sold out, future sales to that customer are less likely. Accordingly, delivery trucks operating from a central warehouse regularly resupply stores. In an analysis of a chain's operations, the general manager wanted to determine the variables that affect how long it takes to unload delivery trucks. A random sample of 50 deliveries to one store was observed. The times (in minutes) to unload the truck, the total number of boxes, and the total weight (in hundreds of pounds) of the boxes were recorded and stored in file Xr18-13.

a Determine the multiple regression equation.
b How well does the model fit the data? Explain.
c Are the required conditions satisfied?
d Is multicollinearity a problem?
e Interpret and test the coefficients. What does this analysis tell you?
f Produce a prediction interval for the amount of time needed to unload a truck with 100 boxes weighing 5,000 pounds.
g Produce an interval estimate of the average amount of time needed to unload trucks with 100 boxes weighing 5,000 pounds.

18.14 Refer to Exercise 18.1.

a Predict with 90% confidence the selling price of a 40,000-square-foot lot that has 50 mature trees and is 25 feet from the lake.
b Estimate with 90% confidence the average selling price of 50,000-square-foot lots that have 10 mature trees and are 75 feet from the lake.

18.15 Refer to Exercise 18.2.

a Predict Pat's final exam mark with 95% confidence.
b Predict Pat's total mark with 95% confidence.

18.16 Refer to Exercise 18.3. Predict next month's drywall sales with 95% confidence if the number of building permits is 50, the five-year mortgage rate is 9.0%, and the vacancy rates are 3.6% in apartments and 14.3% in office buildings.

18.4 REGRESSION DIAGNOSTICS II

In Section 17.8, we discussed how to determine whether the required conditions are unsatisfied. The same procedures can be used to diagnose problems in the multiple regression model. Here is a brief summary of the diagnostic procedure we described in Chapter 17. Calculate the residuals and check the following.

1 *Is the error variable extremely nonnormal?* Draw the histogram of the residuals. How closely does it resemble a bell shape?

2 *Is the error variance constant?* Plot the residuals versus the predicted values of y. Is the spread constant?

3 *Are the errors independent (time-series data)?* Plot the residuals versus the time periods. Are the points random?

4 *Are there observations that are inaccurate or do not belong to the target population?* Double-check the accuracy of outliers.

If the error is extremely nonnormal and/or the variance is not a constant, several remedies can be attempted. These are described at the end of this section.

Outliers are checked by examining the data in question to ensure accuracy.

Nonindependence of a time series can sometimes be detected by graphing the residuals and the time periods and looking for evidence of autocorrelation. In Section 18.5, we introduce the Durbin-Watson test, which tests for one form of autocorrelation. We will offer a corrective measure for nonindependence.

There is another problem that is applicable to multiple regression models only. It is called *multicollinearity* and its consequences are described below.

MULTICOLLINEARITY

Multicollinearity (also called *collinearity* and *intercorrelation*) is a condition that exists when the independent variables are correlated with one another. The adverse effect of multicollinearity is that some or all of the estimated regression coefficients ($b_1, b_2, \ldots, b_k$) tend to have large sampling variability. That is, some of the standard errors ($s_{b_1}, s_{b_2}, \ldots, s_{b_k}$) will be large, which means that in repeated sampling there will be a great deal of variation in the coefficients. This may result in estimated coefficients that are far from their true values. In some cases we will observe negative coefficients when in fact there is a positive relationship between that independent variable and the dependent variable. Moreover, because the standard error is the denominator in the test statistic, a large standard error will yield small t-statistics, which means that even when the independent and dependent variables are linearly related, the test statistic may indicate that there is no evidence of a linear relationship. Fortunately, multicollinearity does not affect the F-test or the other statistics that assess the model.

▼ EXAMPLE 18.2

A real estate agent wanted to develop a model to predict the selling price of a home. The agent believed that the most important variables in determining the price of a house are its size, number of bedrooms, and lot size. Accordingly, he took a random sample of 100 homes that recently sold and recorded the selling price, the number of bedrooms, the size (in square feet), and the lot size (in square feet). These data are stored in columns A through D of file Xm18-02. Some of the data follow. Analyze the relationship between the selling price and the other three variables.

PRICE	BEDROOMS	HOUSE SIZE	LOT SIZE
$124,100	3	1,290	3,900
218,300	4	2,080	6,600
117,800	3	1,250	3,750
.	.	.	.
.	.	.	.
117,500	3	1,570	4,950
157,400	3	1,560	5,100
155,900	4	1,620	4,800

Solution The proposed multiple regression model is

$$\text{PRICE} = \beta_0 + \beta_1 \text{ BEDROOMS} + \beta_2 \text{ HOUSE SIZE} + \beta_3 \text{ LOT SIZE} + \varepsilon$$

COMPUTE

Microsoft Excel Output for Example 18.2

Summary Output

Regression Statistics

Multiple R	0.7483
R Square	0.5600
Adjusted R Square	0.5462
Standard Error	25023
Observations	100

ANOVA

	df	SS	MS	F	Significance F
Regression	3	76501718347	25500572782	40.73	0.0000
Residual	96	60109046053	626135896		
Total	99	136610764400			

	Coefficients	Standard Error	t Stat	P-value
Intercept	37717.59	14176.74	2.66	0.0091
BEDROOMS	2306.08	6994.19	0.33	0.7423
HOUSE SIZE	74.30	52.98	1.40	0.1640
LOT SIZE	24.36	17.02	20.26	0.7982

INTERPRET

The statistics that assess the overall model are as follows:

$F = 40.73$, p-value $= 0$

which tells us that the model is valid.

The coefficient of determination is

$R^2 = .5600$

This statistic says that 56.00% of the variation in prices is explained by the three independent variables.

The standard error of estimate is

$s_\varepsilon = 25{,}023$

We computed the mean price to be $154,066. From these statistics we conclude that the model fits moderately well.

Now examine the coefficients and their standard errors, *t*-statistics, and *p*-values. They are listed below.

Independent Variable	Coefficient	Standard Error	t-Statistic	p-Value
BEDROOMS	2306.08	6994.19	.33	.7423
HOUSE SIZE	74.30	52.98	1.40	.1640
LOT SIZE	−4.36	17.02	−.26	.7982

The relatively large standard errors produced small *t*-statistics with large *p*-values. This tells us that there is no evidence to infer that any of the independent variables are linearly related to the selling price.

In total, the statistics tell us that we have a valid, moderately well-fitting model, all of whose independent variables are apparently not linearly related to the selling price. This contradictory conclusion is the result of multicollinearity.

If we run three simple regression models where the independent variable is (1) the number of bedrooms, (2) the house size, and (3) the lot size, the printouts below are produced. This result tells us that each of the independent variables is strongly related to selling price.

Microsoft Excel Output: Regression of PRICE versus BEDROOMS in Example 18.2

Summary Output

Regression Statistics

Multiple R	0.6454
R Square	0.4166
Adjusted R Square	0.4106
Standard Error	28519
Observations	100

ANOVA

	df	SS	MS	F	Significance F
Regression	1	56905922988	56905922988	69.97	0.0000
Residual	98	79704841412	813314708		
Total	99	136610764400			

	Coefficients	Standard Error	t Stat	P-value
Intercept	25422	15642	1.63	0.1073
BEDROOMS	35439	4237	8.36	0.0000

The *t*-statistic and its *p*-value are 8.36 and 0, respectively. In the absence of the other two independent variables, we conclude that there is overwhelming evidence of a linear relationship between the number of bedrooms and the selling price.

Microsoft Excel Output: Regression of PRICE versus HOUSE SIZE in Example 18.2

Summary Output

Regression Statistics

Multiple R	0.7478
R Square	0.5591
Adjusted R Square	0.5547
Standard Error	24790
Observations	100

ANOVA

	df	SS	MS	F	Significance F
Regression	1	76385713057	76385713057	124.30	0.0000
Residual	98	60225051343	614541340		
Total	99	136610764400			

	Coefficients	Standard Error	t Stat	P-value
Intercept	40066	10521	3.81	0.0002
HOUSE SIZE	64.20	5.76	11.15	0.0000

As was the case with the regression of price on bedrooms, we see that there is evidence ($t = 11.15$, p-value $= 0$) of a linear relationship between house size and price.

Microsoft Excel Output: Regression of PRICE versus LOT SIZE in Example 18.2

Summary Output

Regression Statistics

Multiple R	0.7409
R Square	0.5489
Adjusted R Square	0.5443
Standard Error	25077
Observations	100

ANOVA

	df	SS	MS	F	Significance F
Regression	1	74984894543	74984894543	119.24	0.0000
Residual	98	61625869857	628835407		
Total	99	136610764400			

	Coefficients	Standard Error	t Stat	P-value
Intercept	38940	10837	3.59	0.0005
LOT SIZE	20.98	1.92	10.92	0.0000

Once again, in the absence of the other independent variables, this regression analysis leads to the conclusion that lot size and price are linearly related ($t = 10.92$, p-value $= 0$). Notice as well that in the multiple regression analysis the coefficient of lot size was *negative*. But as you can see above, the relationship between lot size and price is more likely to be positive. As we pointed out in the introduction to multicollinearity, the large variation in the sampling distribution of the coefficients caused by multicollinearity may produce wildly differing values far from the true value of the parameter.

INTERPRET

The t-tests in the multiple regression model infer that no independent variable is linearly related to the selling price. The three simple linear regression models contradict this conclusion. They tell us that the number of bedrooms, the house size, and the lot size are all linearly related to the price. How do we account for this contradiction? The answer is that the three independent variables are correlated with each other. It is reasonable to believe that larger houses have more bedrooms and are situated on larger lots, and that smaller houses have fewer bedrooms and are located on smaller lots. To confirm this belief, we computed the correlation among the three independent variables.

Microsoft Excel Output: Correlation among Independent Variables in Example 18.2

	BEDROOMS	HOUSE SIZE	LOT SIZE
BEDROOMS	1		
HOUSE SIZE	0.8465	1	
LOT SIZE	0.8374	0.9936	1

The coefficient of correlation between number of bedrooms and house size is .8465; the correlation between number of bedrooms and lot size is .8374; the correlation between house size and lot size is .9936. In the multiple regression model, multicollinearity affected the t-tests so that they inferred that none of the independent variables is linearly related to price when, in fact, all are.

▲

Another problem caused by multicollinearity is the interpretation of the coefficients. We interpret the coefficients as measuring the change in the dependent variable when the corresponding independent variable increases by one unit while all the other independent variables are held constant. This interpretation may be impossible when the independent variables are highly correlated, because when the independent variable increases by one unit, some or all of the other independent variables will change. In the multiple regression model in this example, the coefficient of BEDROOMS is 2,306. Without multicollinearity, we would interpret this coefficient to mean that for each additional bedroom the average price increases by $2,306, provided that the other variables are held constant. However, because BEDROOMS is correlated with HOUSE SIZE and LOT SIZE, it is impossible to increase BEDROOMS by 1 and hold the other variables constant.

This raises two important questions for the statistics practitioner. First, how do we recognize the problem when it occurs; and second, how do we avoid or correct it?

Multicollinearity exists in virtually all multiple regression models. In fact, finding two completely uncorrelated variables is rare. The problem becomes serious, however, only when two or more independent variables are highly correlated. Unfortunately, we do not have a critical value that indicates when the correlation between two independent variables is large enough to cause problems. To complicate the issue, multicollinearity also occurs when a combination of several independent variables is correlated with another independent variable or with a combination of other independent variables. Consequently, even with access to all of the correlation coefficients, determining when the multicollinearity problem has reached the serious stage may be extremely difficult. A good indicator of the problem is a large F-statistic but small t-statistics.

Minimizing the effect of multicollinearity is often easier than correcting it. The statistics practitioner must try to include independent variables that are independent of each other. For example, the real estate agent wanted to include house size, the number of bedrooms, and the lot size, three variables that are clearly related. Rather than developing a model that uses all such variables, the statistics practitioner may choose to include only house size, plus several other variables that measure other aspects of a house's value.

REMEDYING VIOLATIONS OF REQUIRED CONDITIONS

The most commonly used method to remedy nonnormality or heteroscedasticity is to transform the dependent variable. There are several points to note about this procedure. First, the actual form of the transformation depends on which condition is unsatisfied and on the specific nature of the violation. Because there are many different ways to violate the required conditions of the statistical techniques, the list of transformations given here is unavoidably incomplete. Second, these transformations can be useful in improving the model. That is, if the linear model appears to be quite poor, we often can improve the model's fit by transforming y. Third, many computer software systems allow us to make transformations quite easily. You might want to experiment to see the effect these transformations have on your statistical results.

Here is a brief list of the most commonly used transformations.

1 *Log Transformation:* $y' = \log y$ (provided $y \geq 0$). The log transformation is used when (a) the variance of the error variable increases as y increases or (b) the distribution of the error variable is positively skewed.

2 *Square Transformation:* $y' = y^2$. Use this transformation when (a) the variance is proportional to the expected value of y or (b) the distribution of the error variable is negatively skewed.

3 *Square-Root Transformation:* $y' = \sqrt{y}$ (provided that $y \geq 0$). The square-root transformation is helpful when the variance is proportional to the expected value of y.

4 *Reciprocal Transformation:* $y' = 1/y$. When the variance appears to significantly increase when y increases beyond some critical value, the reciprocal transformation is recommended.

EXERCISES

18.17 Refer to Exercise 18.1. Calculate the residuals. Does it appear that the residuals are extremely nonnormal? Is the variance of the error variable constant? Explain.

18.18 Refer to Exercise 18.1. Determine the correlation among the independent variables. What do these statistics tell you about the *t*-tests of the coefficients in the multiple regression model?

18.19 Refer to Exercise 18.2. Calculate the residuals. Does it appear that the residuals are extremely nonnormal? Is the variance of the error variable constant? Explain.

18.20 In Exercise 18.2 the correlation between ASSIGNMENT AND MIDTERM is .104. What does this statistic tell you about the *t*-tests of the coefficients in the multiple regression model?

18.21 Refer to Exercise 18.4. Calculate the residuals. Does it appear that the residuals are extremely nonnormal? Is the variance of the error variable constant? Explain.

18.22 Refer to Exercise 18.4. The correlation between FATHER and MOTHER is .3978.
  a What does this correlation tell you about the independent variables?
  b What does it say about the *t*-tests of β_1 and β_2 in the multiple regression model?

18.23 Refer to Exercise 18.5.
  a Is multicollinearity a problem? Explain.
  b Determine the residuals and predicted values using the regression equation.
  c Draw a histogram of the residuals. Does it appear that the error variable is normally distributed?
  d Plot the residuals (on the vertical axis) and the predicted values (on the horizontal axis). Is the variance of the error variable constant?
  e Identify observations that should be checked for accuracy.

18.24 Refer to Exercise 18.6. Conduct an analysis of the residuals to determine whether any of the required conditions are violated. Identify any observations that should be checked for accuracy.

18.25 Determine whether there are violations of the required conditions in the regression model used in Exercise 18.11. Which, if any, observations should be checked to ensure that they were correctly recorded?

18.26 Determine whether the required conditions are satisfied in Exercise 17.61.

18.27 Refer to Exercise 18.7.

  a Use whatever techniques you deem necessary to check the normality requirement and for heteroscedasticity.
  b Is multicollinearity a problem? Explain.

18.28 Refer to Exercise 18.8.

  a Determine whether the required conditions are satisfied.
  b Is multicollinearity a problem?

18.29 Refer to Exercise 18.9. Determine whether the required conditions are satisfied.

18.30 Determine whether the required conditions for the regression analysis conducted in Exercise 18.10 are satisfied.

18.5 REGRESSION DIAGNOSTICS III (TIME SERIES)

In Chapter 17, we pointed out that we check to see if the errors are independent when the data constitute a times series—data gathered sequentially over a series of time periods. In Section 17.8, we described the graphical procedure for determining whether the required condition that the errors are independent is violated. We plot the residuals versus the time periods and look for patterns. In this section, we augment that procedure with the **Durbin-Watson test**.

DURBIN-WATSON TEST

The Durbin-Watson test allows the statistics practitioner to determine whether there is evidence of **first-order autocorrelation**—a condition in which a relationship exists between consecutive residuals r_i and r_{i-1}, where i is the time period. The Durbin-Watson statistic is defined below.

Durbin-Watson Statistic

$$d = \frac{\sum_{i=2}^{n}(r_i - r_{i-1})^2}{\sum_{i=1}^{n} r_i^2}$$

The range of the values of d is

$$0 \leq d \leq 4$$

where small values of d ($d < 2$) indicate a positive first-order autocorrelation and large values of d ($d > 2$) imply a negative first-order autocorrelation. Positive first-order autocorrelation is a common occurrence in time series. It occurs when consecutive residuals tend to be similar. In that case, $(r_i - r_{i-1})^2$ will be small, producing a small value for d. Negative first-order autocorrelation occurs when consecutive residuals differ widely. For example, if positive and negative residuals generally alternate, $(r_i - r_{i-1})^2$ will be large, and as a result, d will be greater than 2. Figures 18.2 and 18.3 depict positive first-order autocorrelation, whereas Figure 18.4 illustrates negative autocorrelation. Notice that in Figure 18.2 the first residual is a small number; the second residual, also a small number, is somewhat larger, and that trend continues. In Figure 18.3, the first residual is large, and in general, succeeding residuals decrease. In both figures, consecutive residuals are similar. In Figure 18.4, the first residual is a positive number, which is followed by a negative residual. The remaining residuals follow this pattern (with some exceptions). Consecutive residuals are quite different.

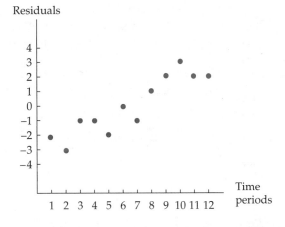

Figure 18.2

Positive first-order autocorrelation

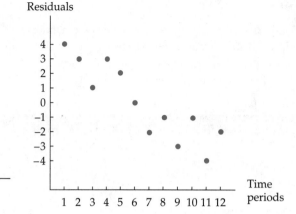

Figure 18.3

Positive first-order autocorrelation

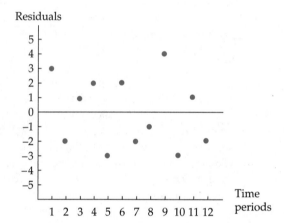

Figure 18.4

Negative first-order autocorrelation

Table 7 in Appendix B is designed to test for positive first-order autocorrelation by providing values of d_L and d_U for a variety of values of n and k and for $\alpha = .01$ and .05.

The decision is made in the following way. If $d < d_L$, we conclude that there is enough evidence to show that positive first-order autocorrelation exists. If $d > d_U$, we conclude that there is not enough evidence to show that positive first-order autocorrelation exists. And if $d_L \leq d \leq d_U$, the test is inconclusive. The recommended course of action when the test is inconclusive is to continue testing with more data until a conclusive decision can be made.

For example, to test for positive first-order autocorrelation with $n = 20$, $k = 3$, and $\alpha = .05$, we test the following hypotheses:

H_0 : There is no first-order autocorrelation

H_1 : There is positive first-order autocorrelation

The decision is made as follows.

If $d < d_L = 1.00$, reject the null hypothesis in favor of the alternative hypothesis

If $d > d_U = 1.68$, do not reject the null hypothesis

If $1.00 \leq d \leq 1.68$, the test is inconclusive

To test for negative first-order autocorrelation, we change the critical values. If $d > 4 - d_L$, we conclude that negative first-order autocorrelation exists. If $d < 4 - d_U$, we conclude that there is not enough evidence to show that negative first-order autocorrelation exists. If $4 - d_U \leq d \leq 4 - d_L$, the test is inconclusive.

We can also test simply for first-order autocorrelation by combining the two one-tail tests. If $d < d_L$ or $d > 4 - d_L$, we conclude that autocorrelation exists. If $d_U \leq d \leq 4 - d_U$, we conclude that there is no evidence of autocorrelation. If $d_L \leq d \leq d_U$ or $4 - d_U \leq d \leq 4 - d_L$, the test is inconclusive. The significance level will be 2α (where α is the one-tail significance level). Figure 18.5 describes the range of values of d and the conclusion for each interval.

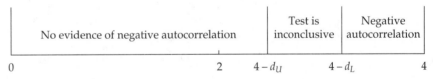

Figure 18.5

Durbin-Waston test

For time-series data, we add the Durbin-Watson test to our list of regression diagnostics. That is, we determine whether the error variable is normally distributed with constant variance (as we did in Section 17.8), we identify outliers that should be verified, and we conduct the Durbin-Watson test.

EXAMPLE 18.3

Christmas week is a critical period for most ski resorts. Because many students and adults are free from other obligations, they are able to spend several days indulging in their favorite pastime, skiing. A large proportion of gross revenue is earned during this period. A ski resort in Vermont wanted to determine the effect that weather had on their sales of lift tickets. The manager of the resort collected the number of lift tickets sold during Christmas week, the total snowfall (in inches), and the average temperature (in degrees Fahrenheit) for the past 20 years. These data are listed below and stored in columns A to C, respectively, of file Xm18-03. Develop the multiple regression model, and diagnose any violations of the required conditions.

Tickets	Snowfall	Temperature
6,835	19	11
7,870	15	−19
6,173	7	36
7,979	11	22
7,639	19	14
7,167	2	−20
8,094	21	39
9,903	19	27
9,788	18	26
9,557	20	16
9,784	19	−1
12,075	25	−9
9,128	3	37
9,047	17	−15
10,631	0	22
12,563	24	2
11,012	22	32
10,041	7	18
9,929	21	32
11,091	11	−15

Solution We estimated the model

$$\text{TICKETS} = \beta_0 + \beta_1 \text{ SNOWFALL} + \beta_2 \text{ TEMPERATURE} + \varepsilon$$

Microsoft Excel Output for Example 18.3

SUMMARY OUTPUT

Regression Statistics

Multiple R	0.3465
R Square	0.1200
Adjusted R Square	0.0165
Standard Error	1712
Observations	20

ANOVA

	df	SS	MS	F	Significance F
Regression	2	6793798	3396899	1.16	0.3373
Residual	17	49807214	2929836		
Total	19	56601012			

	Coefficients	Standard Error	t Stat	P-value
Intercept	8308	904	9.19	0.0000
SNOWFALL	74.59	51.57	1.45	0.1663
TEMPERATURE	28.75	19.70	20.44	0.6625

As you can see, the coefficient of determination is low ($R^2 = 12.00\%$ and adjusted $R^2 = 1.65\%$), and the p-value of the F-test is .3373, which indicates that the model is a poor one. We determined the residuals and the predicted values. We drew the histogram, plotted the residuals and the predicted values, and plotted the residuals and the time periods. (The observations constitute a time series because we observed the results for each of the past 20 years.)

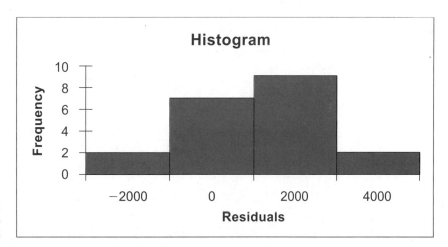

Microsoft Excel Histogram of Residuals for Example 18.3

The histogram of the residuals is approximately bell shaped, precluding the possibility that the error variable is extremely nonnormal.

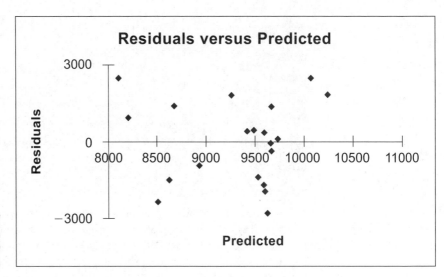

Microsoft Excel Graph of Residuals versus Predicted for Example 18.3

The graph of the residuals versus the predicted values indicates that the variance of the error variable is constant.

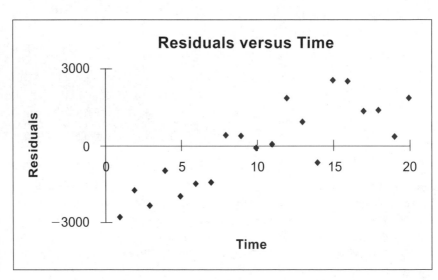

Microsoft Excel Graph of Residuals versus Time Periods for Example 18.3

This graph reveals a serious problem. There appears to be a strong relationship between consecutive values of the residuals, which indicates that the requirement that the errors are independent has been violated. To confirm this diagnosis, we instructed Excel to calculate the Durbin-Watson statistic.

Microsoft Excel Output of the Durbin-Watson Statistic for Example 18.3

Durbin-Watson Statistic

d = 0.5931

> **COMMANDS**
>
> Proceed through the usual steps to conduct a regression analysis, and print the residuals (see page 574). Click **Tools, Data Analysis Plus,** and **Durbin-Watson Statistic.** Specify the **Input Range:** of the residuals, click **Labels,** if appropriate, and click **OK**.

The critical values are determined by noting that $n = 20$ and $k = 2$ (there are two independent variables in the model). If we wish to test for positive first-order autocorrelation with $\alpha = .05$, we find in Table 7(a) in Appendix B

$d_L = 1.10$ and $d_U = 1.54$

The null and alternative hypotheses are

H_0 : There is no first-order autocorrelation

H_1 : There is positive first-order autocorrelation

The rejection region is $d < d_L = 1.10$. Because $d = .5931$, we reject the null hypothesis and conclude that there is enough evidence to infer that positive first-order autocorrelation exists.

Autocorrelation usually indicates that the model is incomplete, which means that we need to include one or more independent variables that have time-ordered effects on the dependent variable. The simplest such independent variable represents the time periods. To illustrate, we included a third independent variable that records the year since the data were gathered. Thus, YEARS = 1, 2, ..., 20. The new model is

$$\text{TICKETS} = \beta_0 + \beta_1 \text{ SNOWFALL} + \beta_2 \text{ TEMPERATURE} + \beta_3 \text{ YEARS} + \varepsilon$$

Microsoft Excel Output for Example 18.3 with YEARS Included in Model

SUMMARY OUTPUT

Regression Statistics

Multiple R	0.8608
R Square	0.7410
Adjusted R Square	0.6924
Standard Error	957
Observations	20

ANOVA

	df	SS	MS	F	Significance F
Regression	3	41940217	13980072	15.26	0.0001
Residual	16	14660795	916300		
Total	19	56601012			

	Coefficients	Standard Error	t Stat	P-value
Intercept	5966	631	9.45	0.0000
SNOWFALL	70.18	28.85	2.43	0.0271
TEMPERATURE	29.23	11.02	20.84	0.4145
YEARS	229.97	37.13	6.19	0.0000

As we did before, we calculate the residuals and conduct regression diagnostics using Excel. The results follow.

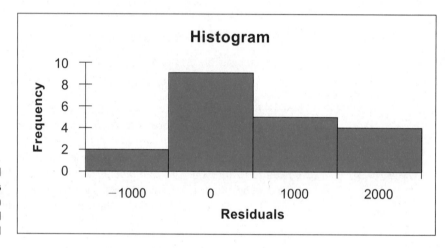

Microsoft Excel Histogram of Residuals for Example 18.3 with YEARS Included in Model

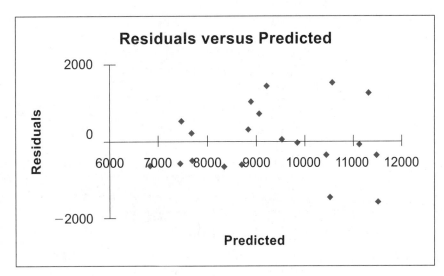

Microsoft Excel Graph of Residuals versus Predicted for Example 18.3 with YEARS Included in Model

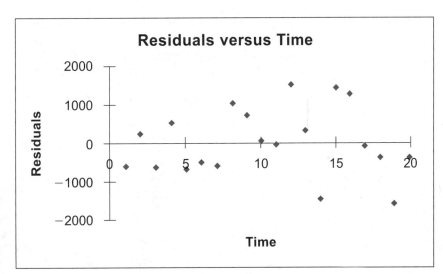

Microsoft Excel Graph of Residuals versus Time Periods for Example 18.3 with YEARS Included in the Model

Microsoft Excel Output of the Durbin-Watson Statistic for Example 18.3 with YEARS Included in the Model

Durbin-Watson Statistic

d = 1.885

The graphs show no signs of a violation of the required conditions. The Durbin-Watson statistic is $d = 1.885$. From Table 7(a) in Appendix B, we find the critical values of the Durbin-Watson test. With $k = 3$ and $n = 20$, we find

$$d_L = 1.00 \quad \text{and} \quad d_U = 1.68$$

Because $d > 1.68$, we conclude that there is not enough evidence to infer the presence of positive first-order autocorrelation.

▲

As you can see, the model has improved dramatically. The F-test tells us that the model is valid. The t-tests tell us that both the amount of SNOWFALL and YEARS are significantly linearly related to the number of lift tickets. This information could prove useful in advertising for the resort. For example, if there has been a recent snowfall, they could emphasize that in their advertising. If no new snow has fallen, they may emphasize their snow-making facilities.

DEVELOPING AN UNDERSTANDING OF STATISTICAL CONCEPTS

Notice that the addition of the variable YEARS explained a large proportion of the variation in the number of lift tickets sold. That is, the resort experienced a relatively steady increase in sales over the past 20 years. Once this variable was included in the model, the variable SNOWFALL became significant, because it was able to explain some of the remaining variation in lift ticket sales. Without YEARS, SNOWFALL and TEMPERATURE were unable to explain a significant proportion of the variation in ticket sales. The graph of the residuals versus the time periods and the Durbin-Watson test enabled us to identify the problem and correct it. In overcoming the autocorrelation problem, we improved the model so that we identified SNOWFALL as an important variable in determining ticket sales. This result is quite common. Correcting a violation of a required condition will frequently improve the model.

EXERCISES

18.31 Given the following information. perform the Durbin-Watson test to determine whether first-order autocorrelation exists.

$n = 25 \quad k = 5 \quad \alpha = .10 \quad d = .90$

18.32 Test the following hypotheses with $\alpha = .05$:

H_0 : There is no first-order autocorrelation

H_1 : There is positive first-order autocorrelation

$n = 50 \quad k = 2 \quad d = 1.38$

18.33 Test the following hypotheses with $\alpha = .02$:

H_0 : There is no first-order autocorrelation

H_1 : There is first-order autocorrelation

$n = 90 \quad k = 5 \quad d = 1.60$

18.34 Test the following hypotheses with $\alpha = .05$:

H_0 : There is no first-order autocorrelation

H_1 : There is negative first-order autocorrelation

$n = 33 \quad k = 4 \quad d = 2.25$

18.35 One hundred observations of variables y, x_1, and x_2 were taken over 100 consecutive time periods. The data are stored in the first three columns, respectively, in file Xr18-35.

a Conduct a regression analysis of these data.
b Calculate the residuals and standardized residuals.
c Identify observations that should be checked.
d Draw the histogram of the residuals. Does it appear that the normality requirement is satisfied?
e Plot the residuals versus the predicted values of y. Is the error variance constant?
f Plot the residuals versus the time periods. Perform the Durbin-Watson test. Is there evidence of autocorrelation? Use $\alpha = .10$.
g If autocorrelation was detected in part (f), propose an alternative regression model to remedy the problem. Use the computer to generate the statistics associated with this model.
h Redo parts (a) through (f). Compare the two models.

18.36 Weekly sales of a company's product (y) and those of its main competitor (x) were recorded for one year. These data are stored in chronological order in columns A (company's sales) and B (competitor's sales) in file Xr18-36.
  a Conduct a regression analysis of these data.
  b Calculate the residuals and standardized residuals.
  c Identify observations that should be checked.
  d Draw the histogram of the residuals. Does it appear that the normality requirement is satisfied?
  e Plot the residuals versus the predicted values of y. Is the error variance constant?
  f Plot the residuals versus the time periods. Perform the Durbin-Watson test. Is there evidence of autocorrelation? Use $\alpha = .10$.
  g If autocorrelation was detected in part (f), propose an alternative regression model to remedy the problem. Use the computer to generate the statistics associated with this model.
  h Redo parts (a) through (f). Compare the two models.

18.37 Observations of variables y, x_1, x_2, and x_3 were taken over 80 consecutive time periods. The data are stored in the first four columns, respectively, in file Xr18-37.
  a Conduct a regression analysis of these data.
  b Calculate the residuals and standardized residuals.
  c Identify observations that should be checked.
  d Draw the histogram of the residuals. Does it appear that the normality requirement is satisfied?
  e Plot the residuals versus the predicted values of y.
  f Plot the residuals versus the time periods. Perform the Durbin-Watson test. Is there evidence at the 10% significance level of autocorrelation?
  g If autocorrelation was detected in part (f), propose an alternative regression model to remedy the problem. Use the computer to generate the statistics associated with this model.
  h Redo parts (a) through (f). Compare the two models.

18.38 Refer to Exercise 18.3.
  a Draw the histogram of the residuals. Does it appear that the normality requirement is violated? Explain.
  b Graph the residuals and the predicted values y. Is the error variance constant? Explain.

18.39 Refer to Exercise 18.3. Calculate the coefficients of correlation among the independent variables.
  a What do these correlations tell you about the independent variables?
  b Is it likely that the t-tests of the coefficients are meaningful? Explain.

18.40 Refer to Exercise 18.3. Calculate the Durbin-Watson statistic, and test to determine whether there is evidence of positive first-order autocorrelation. Use $\alpha = .05$.

18.41 The manager of a tire store in Minneapolis has been concerned with the high cost of inventory. The current policy is to stock all the snow tires that are predicted to sell over the entire winter at the beginning of the season (end of October). The manager can reduce inventory costs by having suppliers deliver snow tires regularly from October to February. However, he needs to be able to predict weekly sales to avoid stockouts that will ultimately lose sales. To help develop a forecasting model, he records the number of snow tires sold weekly during the last winter and the amount of snowfall (in inches) in each week. These data are stored in columns A and B, respectively, in file Xr18-41.
  a Develop a regression model, and use Excel to produce the statistics.
  b Perform a complete diagnostic analysis to determine whether the required conditions are satisfied.
  c If one or more conditions are unsatisfied, attempt to remedy the problem.
  d Use whatever procedures you wish to assess how well the new model fits the data.
  e Interpret and test each of the coefficients.

18.6 NOMINAL INDEPENDENT VARIABLES

When we introduced regression analysis, we pointed out that all the variables must be interval. But in many real-life cases, one or more independent variables are nominal. For example, suppose that the used-car dealer in Example 17.1 believed that the color of a car is a factor in determining its auction price. Color is clearly a nominal variable. If we assign numbers to each possible color, these numbers will be completely arbitrary, and using them in a regression model will usually be pointless. For example, if the dealer believes the colors that are most popular, white and silver, are likely to lead to higher prices than other colors, he may assign a code of 1 to white cars, a code of 2 to silver cars, and a code of 3 to all other colors. Columns A and B of file Xm17-01A contain the auction price and the odometer reading (identical to file Xm17-01). Column C includes codes identifying the color of the Ford Tauruses referred to in the original problem. If we now conduct a multiple regression analysis, the results below would be obtained.

Microsoft Excel Output: Regression of PRICE Versus BEDROOMS in Example 18.2

Summary Output

Regression Statistics

Multiple R	0.8095
R Square	0.6552
Adjusted R Square	0.6481
Standard Error	151.2
Observations	100

ANOVA

	df	SS	MS	F	Significance F
Regression	2	4216263	2108132	92.17	0.0000
Residual	97	2218627	22872		
Total	99	6434890			

	Coefficients	Standard Error	t Stat	P-value
Intercept	6580	92.96	70.79	0.0000
Odometer	20.0313	0.0023	213.56	0.0000
Color	221.67	18.11	21.20	0.2345

The regression equation is

PRICE = 6,580 − .0313 ODOMETER − 21.67 COLOR

Aside from the inclusion of the variable COLOR, this equation is very similar to the one we produced in the simple regression model (PRICE = 6,533 − 0.0312 ODOMETER). An examination of the output above reveals that the variable representing color is not linearly related to price (t-statistic = −1.20, and p-value = .2345). There are two possible explanations for this result. First, there is no relationship between color and price. Second, color is a factor in determining the car's price, but the way in which the dealer assigned the codes to the colors made detection of that fact impossible. That is, the dealer treated the nominal variable, color, as an interval variable. To further understand why we cannot use nominal data in regression analysis, try to interpret the coefficient of COLOR. Such an effort is similar to attempting to interpret the mean of a sample of nominal data. It is futile. Even though this effort failed, it is possible to include nominal variables in the regression model. This is accomplished through the use of indicator variables.

An **indicator variable** (also called a **dummy variable**) is a variable that can assume either of only two values (usually 0 and 1), where one value represents the existence of a certain condition and the other value indicates that the condition does not hold. In this illustration we would create two indicator variables to represent the color of the car.

$I_1 = 1$ (if color is white)
$ = 0$ (if color is not white)

and

$I_2 = 1$ (if color is silver)
$ = 0$ (if color is not silver)

Notice that we need only two indicator variables to represent the three colors. A white car is represented by $I_1 = 1$ and $I_2 = 0$. A silver car is represented by $I_1 = 0$ and $I_2 = 1$. Because cars that are painted some other color are neither white nor silver, they are represented by $I_1 = 0$ and $I_2 = 0$. It should be apparent that we cannot have $I_1 = 1$ and $I_2 = 1$, as long as we assume that no Ford Taurus is two-toned.

The effect of using these two indicator variables is to create three equations, one for each of the three colors. As you're about to discover, we can use the equations to determine how the car's color affects its auction selling price.

In general, to represent a nominal variable with m categories, we must create $m - 1$ indicator variables.

INTERPRETING AND TESTING THE COEFFICIENTS OF INDICATOR VARIABLES

In columns 4 and 5 of file Xm17-01A, we stored the values of I_1 and I_2. We then performed a multiple regression analysis using variables ODOMETER, I_1, and I_2. The Excel printout follows.

Microsoft Excel Output for Example 17.1 with Indicator Variables in the Model

Summary Output

Regression Statistics

Multiple R	0.8355
R Square	0.6980
Adjusted R Square	0.6886
Standard Error	142.3
Observations	100

ANOVA

	df	SS	MS	F	Significance F
Regression	3	4491749	1497250	73.97	0.0000
Residual	96	1943141	20241		
Total	99	6434890			

	Coefficients	Standard Error	t Stat	P-value
Intercept	6350	92.17	68.90	0.0000
Odometer	20.0278	0.00237	211.72	0.0000
I-1	45.24	34.08	1.33	0.1876
I-2	147.74	38.18	3.87	0.0002

The regression equation is

PRICE = $6{,}350 - .0278$ ODOMETER $+ 45.24 I_1 + 147.74 I_2$

The intercept (b_0) and the coefficient of ODOMETER (b_1) are interpreted in the usual manner. When ODOMETER = $I_1 = I_2 = 0$, the dependent variable PRICE equals 6,350. For each additional mile on the odometer, the auction price decreases, on average, by 2.78 cents. Now examine the remaining two coefficients.

$b_2 = 45.24$

$b_3 = 147.74$

These tell us that, in this sample, on average, a white car sells for $45.24 more than other colors, and a silver car sells for $147.74 more than other colors. The reason both comparisons are made with other colors is that such cars are represented by $I_1 = I_2 = 0$. Thus, for a nonwhite and nonsilver car, the equation becomes

PRICE = $b_0 + b_1$ ODOMETER $+ b_2(0) + b_3(0)$

which is

PRICE = $6{,}350 - .0278$ ODOMETER

For a white car ($I_1 = 1$ and $I_2 = 0$), the regression equation is

PRICE = $b_0 + b_1$ ODOMETER $+ b_2(1) + b_3(0)$

which is

PRICE = $6{,}350 - .0278$ ODOMETER $+ 45.24$
$= 6{,}395.24 - .0278$ ODOMETER

Finally, for a silver car ($I_1 = 0$ and $I_2 = 1$), the regression equation is

PRICE = $b_0 + b_1$ ODOMETER $+ b_2(0) + b_3(1)$

which simplifies to

PRICE = $6{,}350 - .0278$ ODOMETER $+ 147.74$
$= 6{,}497.74 - .0278$ ODOMETER

Figure 18.6 depicts the graph of PRICE versus ODOMETER for the three different color categories. Notice that the three lines are parallel (with slope = $b_1 = -.0278$) while the intercepts differ.

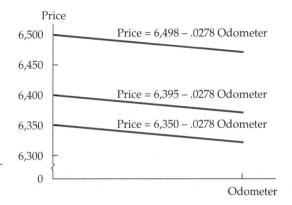

Figure 18.6

Price versus odometer for three colors

We can also perform the usual t-tests of β_2 and β_3; however, because the variables I_1 and I_2 represent different groups (the three color categories), these t-tests allow us to draw inferences about the differences in auction prices between the groups. The test of β_2 is conducted as follows.

$H_0 : \beta_2 = 0$

$H_1 : \beta_2 \neq 0$

Test statistic: $t = 1.33$ (p-value $= .1876$)

There is insufficient evidence to infer that white Tauruses sell for more or less than do nonwhite, nonsilver Tauruses. To determine if silver-colored Tauruses sell for a different price than other colors, we test

$H_0 : \beta_3 = 0$

$H_1 : \beta_3 \neq 0$

Test statistic: $t = 3.87$ (p-value $= .0002$)

We can conclude that there are differences in price between silver-colored Tauruses and the other category.

EXERCISES

18.42 Create and identify indicator variables to represent the following nominal variables.

  a Religious affiliation (Catholic, Protestant, and others)

  b Working shift (8:00 A.M. to 4:00 P.M., 4:00 P.M. to 12:00 midnight, and 12:00 midnight to 8:00 A.M.)

  c Supervisor (Jack Jones, Mary Brown, George Fosse, and Elaine Smith)

18.43 In a study of computer applications, a survey asked which microcomputer a number of companies used. The following indicator variables were created:

$I_1 = 1$ if IBM
$ = 0$ if not

$I_2 = 1$ if Macintosh
$ = 0$ if not

What computer is being referred to by each of the following pairs of values?

a $I_1 = 0$; $I_2 = 1$
b $I_1 = 1$; $I_2 = 0$
c $I_1 = 0$; $I_2 = 0$

18.44 Suppose that, in Exercise 18.11, the dean believed that the type of undergraduate degree also influenced the student's GPA as a graduate student. The most common undergraduate degrees of students attending the graduate school of business are BBA (Business Administration), BEng, BSc, and BA. Because the type of degree is a nominal variable, the following three indicator variables were created.

$I_1 = 1$ if degree is BBA
$\quad = 0$ if not

$I_2 = 1$ if degree is BEng
$\quad = 0$ if not

$I_3 = $ if degree is BSc
$\quad = 0$ if not

The data for the 100 students are stored in columns A to F of file Xr18-44. (Columns A to D are identical to columns A to D of file Xr18-11.) Conduct a multiple regression analysis.

a Can we conclude that, on average, the BBA graduate performs better than the BA graduate?
b Can we conclude that, on average, the BEng graduate performs better than the BA graduate?
c Can we conclude that, on average, the BSc graduate performs better than the BA graduate?
d Find the point prediction for graduate school GPA of a BEng whose undergraduate GPA was 9.0, whose GMAT score was 700, and who has had 10 years of work experience.
e Repeat part (c) for a BA student.

18.45 Refer to Exercise 18.12, where a multiple regression analysis was performed to predict men's longevity based on the parents' and grandparents' longevity. Suppose that in addition to these data the actuary also recorded whether the man was a smoker (1 = yes and 0 = no). These data are stored in column F of file Xr18-45. (Columns A to E are identical to columns A to E of file Xr18-12.)

a Perform a multiple regression analysis, and compare the results with those produced in Exercise 18.12. Describe the differences.
b Does smoking affect length of life? Explain.

18.46 The manager of an amusement park would like to be able to predict daily attendance, in order to develop more accurate plans about how much food to order and how many ride operators to hire. After some consideration, he decided that the following three factors are critical.

Yesterday's attendance
Weekday or weekend
Predicted weather

He then took a random sample of 40 days. For each day, he recorded the attendance, the previous day's attendance, day of the week, and weather forecast. The first independent variable is interval, but the other two are nominal. Accordingly, he created the following sets of indicator variables.

$I_1 = 1$ if weekend
$\quad = 0$ if not

$I_2 = 1$ if mostly sunny is predicted
$\quad = 0$ if not

$I_3 = 1$ if rain is predicted
$\quad = 0$ if not

These data are stored in file Xr18-46. Conduct a multiple regression analysis on these data.

a Is this model valid? Explain.
b Can we conclude that weather is a factor in determining attendance?
c Do these results provide sufficient evidence that weekend attendance is, on average, larger than weekday attendance?

18.47 The real estate agent described in Example 18.2 has become so fascinated by the multiple regression technique that he decided to improve the model. Recall that the agent believed that the most important variables in determining the price of a house are its size, number of bedrooms, and lot size. He took a random sample of 100 houses that were recently sold and recorded the price of the house plus the other three variables. After some consideration, he decided that structure of the house was also a factor. There are four structures. They are two story, side split, back split, and ranch. Each house was classified. Three indicator variables were created. They are

$I_1 = 1$ (if two story)
$\quad = 0$ (if not)

$I_2 = 1$ (if side split)
$\quad = 0$ (if not)

$I_3 = 1$ (if back split)
$\quad = 0$ (if not)

Variables I_1, I_2, and I_3 are stored in columns E, F, and G, respectively, in file Xr18-47 (columns A to D are identical to columns A to D of file Xm18-02).

a Perform a multiple regression analysis, and compare your results with Example 18.2.
b Interpret and test the coefficients of the indicator variables.

18.48 After reviewing the results of Exercise 17.9, the human resources manager tried one last time to improve the model. He recorded the gender of the worker where 1 = female and 0 = male. These codes were stored in column 3 of file Xr18-48. Columns A and B are identical to those in file Xr17-09. Can we infer that female telemarketers stay at their jobs longer than male telemarketers?

18.49 Recall Exercise 17.6, where a statistics practitioner analyzed the relationship between the length of a commercial and viewers' memory of the commercial's product. The experiment was repeated, measuring the same two variables. However, in this experiment not only was the length varied but also the type of commercial. There were three types: humorous (1), musical (2), and serious (3). The memory test scores, lengths, and type of commercial (using the codes in parentheses) were recorded in columns A to C of file Xr18-49.

a Perform a regression analysis using the codes provided in the data file.
b Can we infer that the memory test score is affected by the type of commercial?
c Create indicator variables to describe the type of commercial, and perform another regression analysis.
d Repeat part (b) using the second model.
e Discuss the reasons for the differences between parts (b) and (d).

18.50 Refer to Exercise 18.13, where the amount of time to unload a truck was analyzed. The manager realized that another variable may affect unloading time, the time of day. He recorded the following codes: 1 = morning; 2 = early afternoon; 3 = late afternoon. These codes are stored in column D of file Xr18-50. (Columns A to C are identical to columns A to C of file Xr18-13.)

a Run a regression using the codes for time of day.
b Create indicator variables to represent time of day. Perform a regression analysis with these new variables.
c Which model fits better? Explain.
d Does time of day affect the time to unload?

18.51 Profitable banks are ones that make good decisions on loan applications. Credit scoring is the statistical technique that helps banks make that decision. However, many branches overturn credit scoring recommendations, while other banks do not use the technique. In an attempt to determine the factors that affect loan decisions, a statistics practitioner surveyed 103 banks and recorded the percentage of bad loans (any loan that is not completely repaid); the average size of the loan; and whether a scorecard is used and, if so, whether scorecard recommendations are overturned more than 10% of the time. These results are stored in columns A (percentage bad loans); B (average loan); and C (code 1 = no scorecard; 2 = scorecard overturned more than 10% of the time; and 3 = scorecard overturned less than 10% of the time) in file Xr18-51.

a Create indicator variables to represent the codes.
b Perform a regression analysis.
c How well does the model fit the data?
d Is multicollinearity a problem?
e Interpret and test the coefficients. What does this tell you?
f Predict with 95% confidence the percentage of bad loans for a bank whose average loan is $10,000 and that does not use a scorecard.

18.52 Refer to Exercise 17.62, where a simple linear regression model was used to analyze the relationship between welding machine breakdowns and the age of the machine. The analysis proved to be so useful to company management that they decided to expand the model to include other machines. Data were gathered for two other machines. These data as well as the original data are stored in file Xr18-52 in the following way.

Column A: cost of repairs
Column B: age of machine
Column C: machine (1 = welding machine; 2 = lathe; 3 = stamping machine)

a Develop a multiple regression model.
b Interpret the coefficients.
c Can we conclude that welding machines cost more to repair than other machines?

18.53 Absenteeism is a serious employment problem in most countries. It is estimated that absenteeism reduces potential output by more than 10%. Two economists launched a research project to learn more about the problem. They randomly selected 100 organizations to participate in a one-year study. For each organization, they recorded the average number of days absent per employee and several variables thought to affect absenteeism. File Xr18-53 contains the following information:

Column A: average employee wage
Column B: percentage of part-time employees

Column C: percentage of unionized employees
Column D: availability of shiftwork (1 = yes; 0 = no)
Column E: union–management relationship
 (1 = good; 0 = not good)
Column F: average number of days absent per employee

a Conduct a regression analysis.
b Can we infer that the availability of shiftwork affects absenteeism?
c Is there enough evidence to infer that in organizations where the union–management relationship is good, absenteeism is lower?

18.7 SUMMARY

The multiple regression model extends the model introduced in Chapter 17. The statistical concepts and techniques are similar to those presented in simple linear regression. We assess the model in three ways: standard error of estimate, the coefficient of determination (and the coefficient of determination adjusted for degrees of freedom), and the F-test of the analysis of variance. We can use the t-tests of the coefficients to determine whether each of the independent variables is linearly related to the dependent variable. As we did in Chapter 17, we showed how to diagnose violations of the required conditions and to identify other problems. We introduced multicollinearity and demonstrated its effect and its remedy. We presented the Durbin-Watson test to detect first-order autocorrelation and we discussed how indicator variables allow us to use nominal variables.

IMPORTANT TERMS

Response surface
Plane
Coefficient of determination adjusted for degrees of freedom
Significance F
F-test
Multicollinearity
Durbin-Watson test
First-order autocorrelation
Indicator variable
Dummy variable

SYMBOLS

Symbol	Pronounced	Represents
β_i	Beta-sub-i or beta-i	Coefficient of ith independent variable
b_i	b-sub-i or b-i	Sample coefficient
I_1	I-sub-1 or I-1	Indicator variable

FORMULAS

Standard error of estimate

$$s_\varepsilon = \sqrt{\frac{\text{SSE}}{n-k-1}}$$

Test statistic for β_i

$$t = \frac{b_i - \beta_i}{s_{b_i}}$$

Coefficient of determination

$$R^2 = \frac{[\text{cov}(x,y)]^2}{s_x^2 s_y^2} = 1 - \frac{\text{SSE}}{\sum(y_i - \bar{y})^2}$$

Adjusted coefficient of determination

$$\text{Adjusted } R^2 = 1 - \frac{\text{SSE}/(n - k - 1)}{\sum(y_i - \bar{y})^2/(n - 1)}$$

Mean square for error

$$\text{MSE} = \text{SSE}/k$$

Mean square for regression

$$\text{MSR} = \text{SSR}/(n - k - 1)$$

F-statistic

$$F = \text{MSR}/\text{MSE}$$

MICROSOFT EXCEL OUTPUT AND INSTRUCTIONS

Technique	Page
Regression	592–593
Prediction interval	601
Durbin-Watson statistic	619

SUPPLEMENTARY EXERCISES

18.54 After analyzing whether the number of ads affected the number of customers, the manager in Exercise 17.61 decided to determine whether where he advertised made any difference. As a result, he reorganized the experiment. Each week he advertised several times per week, but in only one of the advertising media. He again recorded the weekly number of customers, the number of ads, and the location of that week's advertisement (1 = newspaper, 2 = radio, and 3 = television). These data are stored in columns A to C, respectively, in file Xr18-54.

  a Create indicator variables to describe the advertising medium.
  b Conduct a regression analysis. Test to determine whether the model is valid.
  c Does the advertising medium make a difference? Explain.

18.55 Supermarkets frequently price products such as bread and milk to attract customers to the store. A manager of a dairy that supplies milk to a supermarket wanted to know how sales of milk are affected by different prices. Consequently, she recorded the weekly sales of milk at one supermarket, the price of a quart of her company's brand (price A), and the price of a quart of her competitor's brand (price B). The data for the past 52 weeks are stored in columns A through C, respectively, in file Xr18-55.

  a Develop a regression model, and use a software package to produce the statistics.
  b Use whatever procedures you wish to assess how well the model fits the data.
  c Interpret each of the coefficients.
  d Can we infer that each of the independent variables is linearly related to the weekly sales of milk?
  e Test to determine whether the model is valid.
  f Predict with 90% confidence the sales of milk when the company's price is 65 cents and the competitor's price is 45 cents.

18.56 Refer to Exercise 18.55.

  a Analyze the results, and determine whether the required conditions are satisfied.
  b Is multicollinearity a problem that affects your answer in part (d)? Explain.

18.57 The general manager of the Cleveland Indians baseball team is in the process of determining which minor-league players to draft. He is aware that his team needs home-run hitters and would like to find a way to predict the number of home runs a player will hit. Being an astute statistics practitioner, he gathers a random sample of players and records the number of home runs each player hit in his first two full years as a major-league player, the number of home runs he hit in his last full year in the minor leagues, his age, and the number of years of professional baseball. These data are stored in columns A through D, respectively, in file Xr18-57.

  a Develop a regression model, and use a software package to produce the statistics.
  b Interpret each of the coefficients.
  c How well does the model fit?
  d Test the model's validity.
  e Do each of the independent variables belong in the model?
  f Predict with 95% confidence the number of home runs in the first two years of a player who is 25 years old, has played professional baseball for 7 years, and hit 22 home runs in his last year in the minor leagues.
  g Estimate with 95% confidence the expected number of home runs in the first two years of players who are 27 years old, have played professional baseball for 5 years, and hit 18 home runs in their last year in the minors.

18.58 Refer to Exercise 18.57.
  a Determine whether the required conditions are satisfied.
  b Is multicollinearity a problem? Could we have known about the multicollinearity before the model was created? Explain.

18.59 The agronomist referred to in Exercise 17.64 believed that the amount of rainfall as well as the amount of fertilizer used would affect the crop yield. She redid the experiment in the following way. Thirty greenhouses were rented. In each, the amount of fertilizer and the amount of water were varied. At the end of the growing season, the amount of corn was recorded with the data stored in file Xr18-59 (column A = crop yield in kilograms; column B = amount of fertilizer applied in kilograms; column C = amount of water in liters per week).

  a Determine the sample regression line, and interpret the coefficients.
  b Do these data allow us to infer that there is a linear relationship between the amount of fertilizer and the crop yield?
  c Do these data allow us to infer that there is a linear relationship between the amount of water and the crop yield?
  d What can you say about the multiple regression model's fit?
  e Predict the crop yield when 100 kilograms of fertilizer and 1,000 liters of water are applied. Use a confidence level of 95%.

18.60 Refer to Exercise 18.59. Perform a complete diagnostic analysis to determine whether the required conditions are satisfied. Which conditions, if any, are unsatisfied? Suggest a way to remedy the problem.

18.61 A baseball fan has been collecting data from a newspaper on the various American League teams. She wants to explain each team's winning percentage as a function of its batting average and its earned run average plus an indicator variable for whether or not the team fired its manager within the last 12 months (code = 1 if it did and 0 if it did not). The data for 50 randomly selected teams over the last five seasons are stored in file Xr18-61.
  a Perform a regression analysis.
  b Do these data provide sufficient evidence that a team that fired its manager within the last 12 months wins less frequently than a team that did not fire its manager?

18.62 Regression analysis is often used in medical research to examine the variables that affect various biological processes. A study performed by medical scientists investigated nutritional effects on preweaning mouse pups. In the experiment, the amount of nutrients was varied by rearing the pups in different litter sizes. After 32 days, the body weight and brain weight (both measured in grams) were recorded. These data are stored in file Xr18-62 (column A = brain weight; column B = litter size; column C = body weight).*
  a Conduct a multiple regression analysis where the dependent variable is the brain weight. Interpret the coefficients.
  b Can we infer that there is a linear relationship between litter size and brain weight?

---

*Source: Matthews, D. E., and Farewell, V. T., *Using and Understanding Medical Statistics* (Karger, 1988).

c Can we infer that there is a linear relationship between body weight and brain weight?
d What is the coefficient of determination, and what does it tell you about this model?
e Test the validity of the model.
f Predict with 95% confidence the brain weight of a mouse pup that came from a litter of 10 pups and whose body weight is 8 grams.
g Estimate with 95% confidence the mean weight of all mouse pups that came from litters of 6 pups and whose body weight is 7 grams.

18.63 Refer to Exercise 18.62. Suppose that the experiment did not record the body weights of the mice.

a Conduct a simple linear regression analysis where the dependent variable is brain weight and the independent variable is litter size. Interpret the coefficients.
b Can we infer that there is a linear relationship between brain weight and litter size?
c What is the coefficient of determination, and what does it tell you about this model?
d Test the validity of the model. Compare the results of this test with the test performed in part (b).
e Predict with 95% confidence the brain weight of a mouse pup that came from a litter of 10 pups.
f Estimate with 95% confidence the mean weight of all mouse pups that came from litters of 6 pups.
g Compare the results of this analysis with the analysis undertaken in Exercise 18.62.

18.64 The administrator of a school board in a large county was analyzing the average mathematics test scores in the schools under her control. She noticed that there were dramatic differences in scores among the schools. In an attempt to improve the scores of all the schools, she attempted to determine the factors that account for the differences. Accordingly, she took a random sample of 40 schools across the county and, for each, determined the mean test score last year, the percentage of teachers in each school who have at least one university degree in mathematics, the mean age, and the mean annual income of the mathematics teachers. These data are stored in columns A to D, respectively, of file Xr18-64.

a Conduct a regression analysis to develop the equation.
b Is the model valid? Explain.
c Are the required conditions satisfied? Explain.
d Is multicollinearity a problem? Explain.
e Interpret and test the coefficients.
f Predict with 95% confidence the test score at a school where 50% of the mathematics teachers have mathematics degrees, the mean age is 43, and the mean annual income is $48,300.

18.65 University students often complain that universities reward professors for research but not for teaching, and they argue that professors react to this situation by devoting more time and energy to the publication of their findings and less time and energy to classroom activities. Professors counter that research and teaching go hand in hand; more research makes better teachers. A student organization at one university decided to investigate the issue. They randomly selected 50 economics professors who are employed by a multicampus university. The students recorded the salaries of the professors, their average teaching evaluations (on a 10-point scale), and the total number of journal articles published in their careers. These data are stored in columns A to C, respectively, in file Xr18-65. Perform a complete analysis (produce the regression equation, assess it, and diagnose it), and report your findings.

18.66 Lotteries have become important sources of revenue for governments. Many people have criticized lotteries, however, referring to them as a tax on the poor and uneducated. In an examination of the issue, a random sample of 100 adults was asked how much they spend on lottery tickets and was interviewed about various socioeconomic variables. The purpose of this study is to test the following beliefs.

1 Relatively uneducated people spend more on lotteries than do relatively educated people.
2 Older people buy more lottery tickets than younger people.
3 People with more children spend more on lotteries than people with fewer children.
4 Relatively poor people spend a greater proportion of their income on lotteries than relatively rich people.

The following data were stored in columns A to E, respectively, of file Xr18-66.

Amount spent on lottery tickets as a percentage of total household income
Number of years of education
Age
Number of children
Personal income (in thousands of dollars)

a Develop the multiple regression equation.
b Is the complete model valid?
c Are the required conditions satisfied?
d Is multicollinearity a problem?
e Test each of the beliefs. What conclusions can you draw?

18.67 Refer to Exercise 17.73. After analyzing the scatter diagram and regression output, the statistics practitioner proposed a *quadratic* model that is expressed as follows.

$$y = y = \beta_0 + \beta_1 x_1 + \beta_2 x_2 + \varepsilon$$

where

y = winning time
x_1 = temperature
x_2 = temperature^2

a Conduct a multiple regression.
b Compare your results with those of Exercise 17.73. Which model appears to be better?

18.68 Repeat Exercise 18.67, using the female runner data. (See Exercise 17.74.)

Case 18.1 Quebec Referendum Vote: Was There Electoral Fraud?*

As we described in Case 13.1 (page 403), Quebecers have been debating whether to separate from Canada and form an independent nation. A referendum was held on October 30, 1995, in which the people of Quebec voted not to separate. The vote was extremely close, with the "No" side winning by only 52,448 votes. A large number of "No" votes was cast by the non-Francophone (non-French-speaking) people of Quebec, who make up about 20% of the population and who very much want to remain Canadians. The remaining 80% are Francophones, a majority of whom voted "Yes."

After the votes were counted, it became clear that the tallied vote was much closer than it should have been. Supporters of the "No" side charged that poll scrutineers, all of whom were appointed by the pro-separatist provincial government, rejected a disproportionate number of ballots in ridings where the percentage of "Yes" votes was low and where there are large numbers of Allophone (people whose first language is neither English nor French) and Anglophone (English-speaking) residents. (Electoral laws require that ballots that do not appear to be properly marked are supposed to be rejected.) They were outraged that in a strong democracy like Canada votes would be rigged much like in many nondemocratic countries around the world.

If, in ridings where there was a low percentage of "Yes" votes, there was a high percentage of rejected ballots, this would be evidence of electoral fraud. Moreover, if, in ridings where there were large percentages of Allophone and/or Anglophone voters, there were high percentages of rejected ballots, this too would constitute evidence of fraud on the part of the scrutineers and possibly the government.

In order to determine the veracity of the charges the following variables were recorded for each riding.

Riding number

Percentage of rejected ballots in referendum

Percentage of "Yes" votes

*This case is based on "Voting Irregularities in the 1995 Referendum on Quebec Sovereignty," Jason Cawley and Paul Sommers, *Chance*, Vol. 9, No. 4, Fall 1996. I am grateful to Dr. Paul Sommers, Middlebury College, for his assistance in writing this case.

Percentage of Allophones

Percentage of Anglophones

These data are stored in columns A through E, respectively, in file C18-01.

- **a** Perform an analysis to determine how the percentage of "Yes" votes, Allophones, and Anglophones affects the percentage of rejected ballots.
- **b** Can we infer that electoral fraud took place? If so, how did it manifest itself?

Case 18.2 Quebec Referendum Vote: The Rebuttal

Refer to Case 18.1. Government supporters acknowledged that the highest percentage of rejected ballots occurred in ridings where large numbers of Allophones live. Because the ballots were printed in English and French only, it is reasonable to believe that a greater number of voters would not be able to understand instructions and, thus, inadvertently spoil their ballots. If they are right, there should be a relationship between the percentages of rejected ballots in this referendum and in the previous provincial election held in 1994. Both variables are stored in columns B and C, respectively, in file C18-02 (column A stores the riding number). What do these data tell you about the government's rebuttal?

Chapter 19

Statistical Inference: Conclusion

19.1 Introduction

19.2 Identifying the Correct Technique: Summary of Statistical Inference

19.3 The Last Word

19.1 INTRODUCTION

You now have been introduced to about 35 statistical techniques. If you are like most students, you probably understand statistical inference and are capable of interpreting computer output. However, at this point you may not be confident that you can apply statistical techniques in real life. The main problem is that it is difficult to ascertain which statistical procedure to apply. In this chapter, we attempt to calm your fears. We begin by displaying the flowchart that allows statistics practitioners to determine the appropriate technique to apply. As we did in Chapter 13, we provide a guide detailing the test statistics, interval estimators, and required conditions. The flowchart is augmented by an Excel macro that performs the same function. It is described below.

Use the flowchart and guide to determine how each of the exercises and cases is to be addressed. Because these exercises and cases were drawn from a wide variety of applications and require the use of many of the methods introduced in this book, they provide the same kind of challenge faced by real statistics practitioners. By attempting to solve these problems, you will be getting a realistic exposure to statistical applications. Incidentally, this also provides practice in the approach required to succeed in a statistics course final examination.

19.2 IDENTIFYING THE CORRECT TECHNIQUE: SUMMARY OF STATISTICAL INFERENCE

A list of the inferential techniques that are applied in describing populations, comparing populations, and analyzing relationships among variables follows on page 638. We have not included the chi-squared test of normality and Bartlett's test. Both are investigative statistical procedures employed to help identify which of the following statistical methods to use to solve a problem or to confirm that a required condition has been satisfied.

LIST OF STATISTICAL METHODS INTRODUCED IN CHAPTERS 11–18

t-test and estimator of μ

Chi-squared test and estimator of σ^2

z-test and estimator of p

t-test and estimator of $\mu_1 - \mu_2$ (equal-variances formulas)

t-test and estimator of $\mu_1 - \mu_2$ (unequal-variances formulas)

t-test and estimator of μ_D

F-test and estimator of σ_1^2/σ_1^2

z-test (Cases 1 and 2) and estimator of $p_1 - p_2$

F-test of the analysis of variance: independent samples

F-tests of the analysis of variance: randomized blocks

F-tests of the analysis of variance: independent samples—two factors

LSD multiple comparison method (and Bonferroni adjustment)

Tukey's multiple comparison method

Chi-squared goodness-of-fit test

Chi-squared test of a contingency table

Wilcoxon rank sum test

Sign test

Wilcoxon signed rank sum test

Kruskal-Wallis test

Friedman test

Simple linear regression and correlation

Spearman rank correlation

Multiple regression

Figure 19.1 on pages 642–643 depicts the flowchart used to identify which technique to employ to solve any problem (except for investigative procedures). We have also created an Excel macro that asks the questions that appear in the flowchart. When you supply the answers, Excel identifies the technique. Simply click **Tools, Data Analysis Plus,** and **Technique Identification,** and follow instructions.

We will illustrate the use of the guide and flowchart with a series of cases beginning on page 644.

Table 19.1   Summary of Statistical Inference Techniques: Chapters 11 through 18

Problem objective: Describe a single population.
  Data type: Interval
    Descriptive measurement: Central location
      Parameter: μ
      Test statistic: $t = \dfrac{\bar{x} - \mu}{s/\sqrt{n}}$
      Interval Estimator: $\bar{x} \pm t_{\alpha/2} \dfrac{s}{\sqrt{n}}$
      Required condition: Population is normal.
    Descriptive measurement: Variability
      Parameter: σ^2
      Test statistic: $\chi^2 = \dfrac{(n-1)s^2}{\sigma^2}$
      Interval estimator: LCL $= \dfrac{(n-1)s^2}{\chi^2_{\alpha/2}}$ UCL $= \dfrac{(n-1)s^2}{\chi^2_{1-\alpha/2}}$
      Required condition: Population is normal.

Data type: Nominal
  Number of categories: Two
    Parameter: p
    Test statistic: $z = \dfrac{\hat{p} - p}{\sqrt{p(1-p)/n}}$
    Interval estimator: $\hat{p} \pm z_{\alpha/2}\sqrt{\dfrac{\hat{p}(1-\hat{p})}{n}}$
    Required condition: $np \geq 5$ and $n(1-p) \geq 5$ (for test)
    $\quad\quad\quad\quad\quad\quad\quad\; n\hat{p} \geq 5$ and $n(1-\hat{p}) \geq 5$ (for estimate)
  Number of categories: Two or more
    Parameters: $p_1, p_2, \ldots, p_k$
    Test statistic: $\chi^2 = \sum\limits_{i=1}^{k} \dfrac{(f_i - e_i)^2}{e_i}$
    Required condition: $e_i \geq 5$
Problem objective: Compare two populations.
  Data type: Interval
    Descriptive measurement: Central location
      Experimental design: Independent samples
        Population variances: $\sigma_1^2 = \sigma_2^2$
        Parameter: $\mu_1 - \mu_2$
        Test statistic $t = \dfrac{(\bar{x}_1 - \bar{x}_2) - (\mu_1 - \mu_2)}{\sqrt{s_p^2\left(\dfrac{1}{n_1} + \dfrac{1}{n_2}\right)}}$
        Interval estimator: $(\bar{x}_1 - \bar{x}_2) \pm t_{\alpha/2}\sqrt{s_p^2\left(\dfrac{1}{n_1} + \dfrac{1}{n_2}\right)}$
        Required condition: Populations are normal.
        If populations are nonnormal, apply Wilcoxon rank sum test.
        Population variances: $\sigma_1^2 \neq \sigma_2^2$
        Parameter: $\mu_1 - \mu_2$
        Test statistic: $t = \dfrac{(\bar{x}_1 - \bar{x}_2) - (\mu_1 - \mu_2)}{\sqrt{\dfrac{s_1^2}{n_1} + \dfrac{s_2^2}{n_1}}}$

        $\nu = \dfrac{(s_1^2/n_1 + s_2^2/n_2)^2}{\left(\dfrac{(s_1^2/n_1)^2}{n_1 - 1} + \dfrac{(s_2^2/n_2)^2}{n_2 - 1}\right)}$

        Interval estimator: $(\bar{x}_1 - \bar{x}_2) \pm t_{\alpha/2}\sqrt{\dfrac{s_1^2}{n_1} + \dfrac{s_2^2}{n_2}}$
        Required condition: Populations are normal.
      Experimental design: Matched pairs
        Parameter: μ_D
        Test statistic: $t = \dfrac{\bar{x}_D - \mu_D}{s_D/\sqrt{n_D}}$
        Interval estimator: $\bar{x}_D \pm t_{\alpha/2}\dfrac{s_D}{\sqrt{n_D}}$
        Required condition: Differences are normal.
        If differences are nonnormal, apply Wilcoxon signed rank sum test.
        Nonparametric technique: Wilcoxon signed rank sum test.
        Test statistic: $z = \dfrac{T - E(T)}{\sigma_T}$
        Required condition: Populations are identical in shape and spread.

Descriptive measurement: Variability
  Parameter: σ_1^2/σ_2^2
  Test statistic: $F = s_1^2/s_2^2$
  Interval estimator: $\text{LCL} = \left(\dfrac{s_1^2}{s_2^2}\right)\dfrac{1}{F_{\alpha/2,v_1,v_2}}$ $\text{UCL} = \left(\dfrac{s_1^2}{s_2^2}\right)F_{\alpha/2,v_2,v_1}$
  Required condition: Populations are normal.
Data type: Ordinal
  Experimental design: Independent samples
    Nonparametric technique: Wilcoxon rank sum test
    Test statistic: $z = \dfrac{T - E(T)}{\sigma_T}$
    Required condition: Populations are identical in shape and spread.
  Experimental design: Matched pairs
    Nonparametric technique: Sign test
    Test statistic: $z = \dfrac{x - .5n}{.5\sqrt{n}}$
    Required condition: Populations are identical in shape and spread.
Data type: Nominal
  Number of categories: Two
    Parameter: $p_1 - p_2$
    Test statistic:
    Case 1: $H_0: (p_1 - p_2) = 0$
    $$z = \dfrac{(\hat{p}_1 - \hat{p}_2)}{\sqrt{\hat{p}(1 - \hat{p})\left(\dfrac{1}{n_1} + \dfrac{1}{n_2}\right)}}$$
    Case 2: $H_0: (p_1 - p_2) = D$ $(D \neq 0)$
    $$z = \dfrac{(\hat{p}_1 - \hat{p}_2) - (p_1 - p_2)}{\sqrt{\dfrac{\hat{p}_1(1 - \hat{p}_1)}{n_1} + \dfrac{\hat{p}_2(1 - \hat{p}_2)}{n_2}}}$$
    Interval estimator: $(\hat{p}_1 - \hat{p}_1) \pm z_{\alpha/2}\sqrt{\dfrac{\hat{p}_1(1 - \hat{p}_1)}{n_1} + \dfrac{\hat{p}_2(1 - \hat{p}_2)}{n_2}}$
    Required conditions: $n_1\hat{p}_1$, $n_1(1 - \hat{p}_1)$, $n_2\hat{p}_2$, and $n_2(1 - \hat{p}_2) \geq 5$
  Number of categories: Two or more
    Statistical technique: Chi-squared test of a contingency table
    Test statistic: $\chi^2 = \sum\limits_{i=1}^{k}\dfrac{(f_i - e_i)^2}{e_i}$
    Required condition: $e_i \geq 5$
Problem objective: Compare two or more populations.
  Data type: Interval
    Experimental design: Independent samples (1 and 2 factors)
      Parameters: $\mu_1, \mu_2, \ldots, \mu_k$
      Test statistic: $F = \dfrac{\text{MST}}{\text{MSE}}$
      Required conditions: Populations are normal with equal variances. If populations are nonnormal, apply Kruskal–Wallis test.
    Experimental design: Randomized blocks
      Parameters: $\mu_1, \mu_2, \ldots, \mu_k$
      Test statistic: $F = \dfrac{\text{MST}}{\text{MSE}}$
      Required conditions: Populations are normal with equal variances. If populations are nonnormal, apply Friedman test.

Data type: Ordinal
  Experimental design: Independent samples
  Nonparametric technique: Kruskal–Wallis test
  Test statistic: $H = \left[\dfrac{12}{n(n+1)} \sum_{j=1}^{k} \dfrac{T_j^2}{n_j}\right] - 3(n+1)$
  Required condition: Populations are Identical in shape and spread and $n_j \geq 5$.
  Experimental design: Randomized blocks
  Nonparametric technique: Friedman Test
  Test statistic: $F_r = \left[\dfrac{12}{b(k)(k+1)} \sum_{j=1}^{k} T_j^2\right] - 3b(k+1)$
  Required condition: Populations are identical in shape and spread and b or $k \geq 5$.
Data type: Nominal
  Statistical technique: Chi-squared test of a contingency table
  Test statistic: $\chi^2 = \sum_{i=1}^{k} \dfrac{(f_i - e_i)^2}{e_i}$
  Required condition: $e_j \geq 5$

Problem objective: Analyze the relationship between two variables.
Data type: Interval
  Parameters: β_0, β_1 (simple linear regression)
  Test statistic: $t = \dfrac{b_1 - \beta_1}{s_{b_1}}$
  Interval estimator: $\hat{y} \pm t_{\alpha/2} s_\varepsilon \sqrt{\dfrac{1}{n} + \dfrac{(x_g - \bar{x})^2}{(n-1)s_x^2}}$
  Prediction Interval: $\hat{y} \pm t_{\alpha/2} s_\varepsilon \sqrt{1 + \dfrac{1}{n} + \dfrac{(x_g - \bar{x})^2}{(n-1)s_x^2}}$
  Required conditions: ε is normally distributed with mean zero and standard deviation σ_ε; ε values are independent.
  To test whether two bivariate normally distributed variables are linearly related:
  Parameter: ρ
  Test statistic: $t = r\sqrt{\dfrac{(n-2)}{1-r^2}}$
  If x and y are not bivariate normally distributed, apply Spearman rank correlation coefficient test.
Data type: Ordinal
  Statistical technique: Spearman rank correlation coefficient test
  Parameter: ρ_s
  Test statistic: $z = r_s \sqrt{n-1}$
  Required condition: none
Data type: Nominal
  Statistical technique: Chi-squared test of a contingency table
  Test statistic: $\chi^2 = \sum_{i=1}^{k} \dfrac{(f_i - e_i)^2}{e_i}$
  Required condition: $e_j \geq 5$

Problem objective: Analyze the relationship among two or more variables.
Data type: Interval
  $\beta_0, \beta_1, \beta_2, \ldots, \beta_k$ (multiple regression)
  Test statistics: $t = \dfrac{b_i - \beta_i}{s_{b_i}}$ $\quad (i = 1, 2, \ldots, k)$
  $F = \dfrac{\text{MSR}}{\text{MSE}}$
  Required conditions: ε is normally distributed with mean zero and standard deviation σ_ε; ε values are independent.

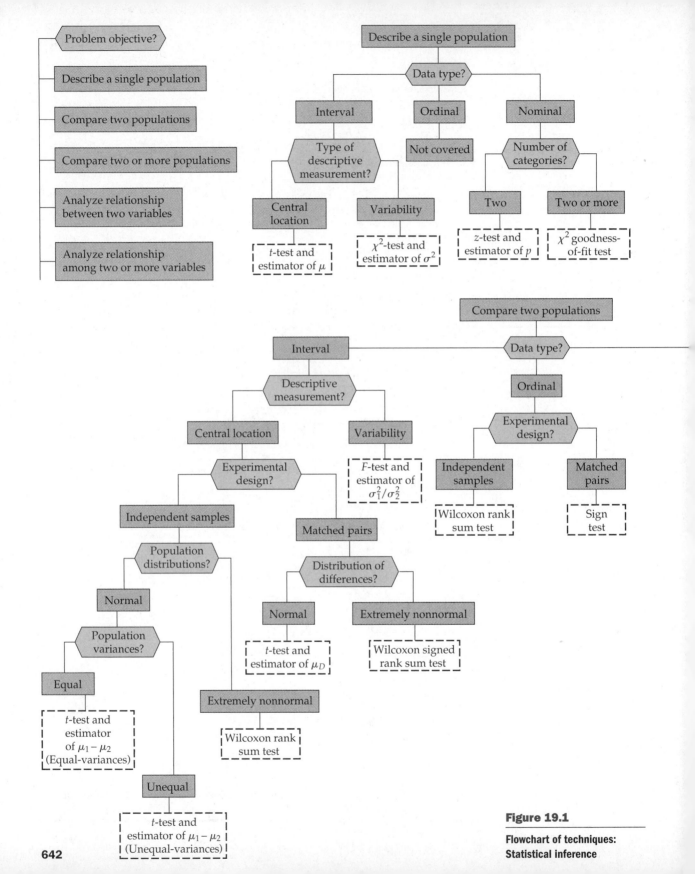

Figure 19.1

Flowchart of techniques: Statistical inference

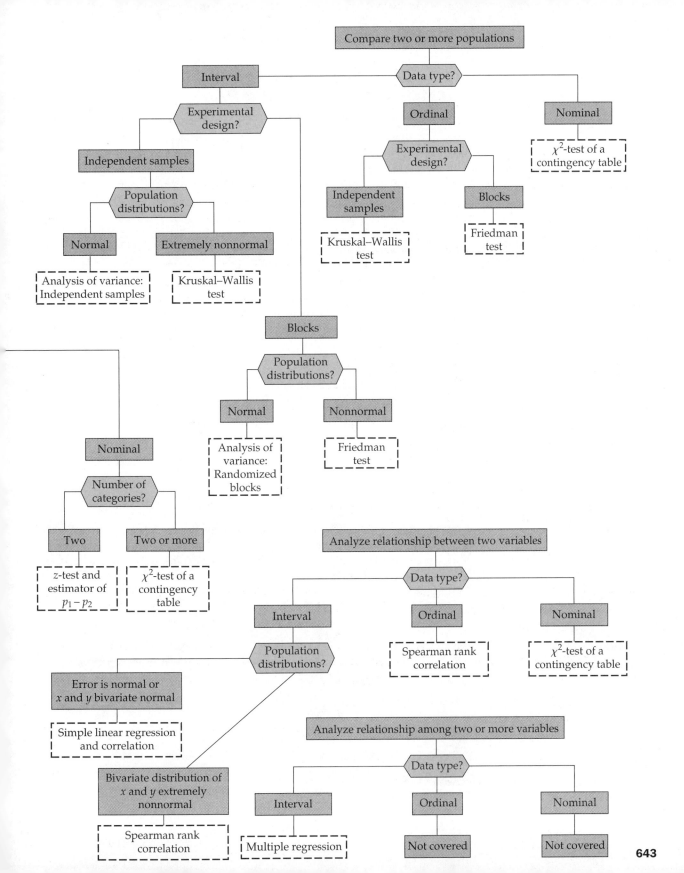

CASE 19.1 DO BANKS DISCRIMINATE AGAINST WOMEN BUSINESS OWNERS? I*

Increasingly, more women are becoming owners of small businesses. However, questions about how they are treated by banks and other financial institutions have been raised by women's groups. Banks are particularly important to small businesses, because studies show that bank financing represents about one-quarter of total debt and that for medium-sized businesses the proportion rises to approximately one-half. If women's requests for loans are rejected more frequently than are men's requests, or if women must pay higher interest charges than men do, women have cause for complaint. Banks might then be subject to criminal as well as civil suits. To examine this issue, a research project was launched.

The researchers surveyed a total of 1,165 business owners, of whom 115 were women. The percentage of women in the sample, 9.9%, compares favorably with other sources that indicate that women own about 10% of established small businesses. The survey asked a series of questions to men and women business owners who applied for loans during the previous month. It also determined the nature of the business, its size, and its age. Additionally, the owners were asked about their experiences in dealing with banks. The questions asked in the survey included the following:

1 What is the gender of the owner?

  1 female

  2 male

2 Was the loan approved?

  1 no

  2 yes

3 If it was approved, what interest rate (points above prime) did you get?

Of the 115 women who asked for a loan, 14 were turned down. A total of 98 men who asked for a loan were rejected. The rates above prime for all loans that were granted were recorded. These data are stored in columns A (rates paid by women) and B (rates paid by men) in file C19-01.

What do these data disclose about possible gender bias by the banks?

Solution

IDENTIFY

The problem objective is to compare two populations: small businesses owned by women and those owned by men. We can compare them in two ways: whether their loan applications are denied; and for loans granted, how much above prime they pay

---

*Adapted from A. L. Riding and C. S. Swift, "Giving Credit Where It's Due: Women Business Owners and Canadian Financial Institutions," Carleton University Working Paper Series WPS 89-07, 1989.

in interest. Whether the loans are approved is a nominal variable. There are two possible "values" of this variable: "approve the loan" and "don't approve the loan." The appropriate technique is the z-test of $p_1 - p_2$.

To test whether gender bias exists, we test to determine if the proportions of loans denied is greater for women (p_1) than for men (p_2).

$H_0: p_1 - p_2 = 0$

$H_1: p_1 - p_2 > 0$

This is an application of the Case 1 test procedure. Thus, the test statistic is

$$z = \frac{(\hat{p}_1 - \hat{p}_2)}{\sqrt{\hat{p}(1-\hat{p})\left(\frac{1}{n} + \frac{1}{n_2}\right)}}$$

COMPUTE

Because we know the values of the sample proportions, we use the **Z-Test of 2 Proportions (Case 1)** worksheet in **Stats-summary.xls**.

Microsoft Excel Output of the z-Test of Two Proportions of Loans Granted: Case 19.1

z-Test of the Difference Between Two Proportions (Case 1)

Sample 1	
Sample proportion	0.1217
Sample size	115
Sample 2	
Sample proportion	0.0933
Sample size	1050
z Stat	0.9810
P(Z<=z) one-tail	0.1633
P(Z<=z) two-tail	0.3266

The value of the test statistic is $z = .9810$, and its p-value $= .1633$.

IDENTIFY

The interest rate is an interval variable, the descriptive measurement is central location, and the samples are independent. The histograms (Figures 19.2 and 19.3) are bell shaped, indicating that the variables are approximately normal. The F-test of σ_1^2/σ_2^2 (shown below) indicates that there is not enough evidence to infer that the population variances are unequal. Putting all these factors together gives us the equal-variances t-test of $\mu_1 - \mu_2$.

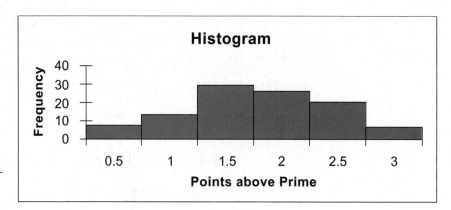

Figure 19.2

Histogram of points above prime: Women's loans

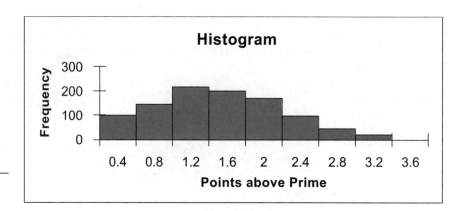

Figure 19.3

Histogram of points above prime: Men's loans

Microsoft Excel Output of the *F*-Test of the Ratio of Two Variances (Points above Prime): Case 19.1

F-Test Two-Sample for Variances

	W Rate	M Rate
Mean	1.545	1.278
Variance	0.405	0.446
Observations	101	952
df	100	951
F	0.91	
P(F<=f) one-tail	0.2784	
F Critical one-tail	0.7713	

The test statistic is $F = .91$. The two-tail *p*-value is $2 \times .2784 = .5568$.

For loans granted, we need to know whether there is evidence that the mean loan rate granted to women (μ_1) is greater than the mean loan rate granted to men (μ_2). Thus, we test

$$H_0: \mu_1 - \mu_2 = 0$$

$$H_1: \mu_1 - \mu_2 > 0$$

The test statistic is

$$t = \frac{(\bar{x}_1 - \bar{x}_2) - (\mu_1 - \mu_2)}{\sqrt{s_p^2\left(\dfrac{1}{n_1} + \dfrac{1}{n_2}\right)}}$$

Microsoft Excel Output of the *t*-Test of the Difference between Two Means of Points above Prime: Case 19.1

t-test: Two-Sample Assuming Equal Variances

	W Rate	M Rate
Mean	1.55	1.28
Variance	0.405	0.446
Observations	101	952
Pooled Variance	0.442	
Hypothesized Mean Difference	0	
df	1051	
t Stat	3.85	
P(T<=t) one-tail	0.0001	
t Critical one-tail	1.646	
P(T<=t) two-tail	0.0001	
t Critical two-tail	1.962	

INTERPRET

The value of the test statistic is $t = 3.85$, which has a *p*-value of .0001 **[P(T<=t) one-tail]**. There is overwhelming evidence to infer that the mean loan rate for women is greater than the mean loan rate for men. Earlier we discovered that in the *z*-test of $p_1 - p_2$, the *p*-value was .1633. Thus, although there is no evidence to infer that women are denied loans more frequently than men, we can conclude that there is evidence of gender bias by banks.

CASE 19.2 DO BANKS DISCRIMINATE AGAINST WOMEN BUSINESS OWNERS? II*

To help explain the apparent discrimination against women documented in Case 19.1, researchers performed further analyses. In the original study, the following pieces of information were gathered for each company:

1 Form of business
  a proprietorship
  b partnership
  c corporation
2 Annual gross sales
3 Age of the firm

---

*Adapted from A. L. Riding and C. S. Swift, "Giving Credit Where It's Due: Women Business Owners and Canadian Financial Institutions," Carleton University Working Paper Series WPS 89-07, 1989.

These data, together with the data from Case 19.1, are stored in file C19-02 in the following way:*

Column A: Rates above prime paid by women

Column B: Type of women's business (1 = proprietorship; 2 = partnership; 3 = corporation)

Column C: Annual gross sales of women's businesses (in thousands of dollars)

Column D: Age of women's businesses

Column E: Rates above prime paid by men

Column F: Type of men's business (1 = proprietorship; 2 = partnership; 3 = corporation)

Column G: Annual gross sales of men's businesses (in thousands of dollars)

Column H: Age of men's businesses

What do these data tell you about the alleged discrimination against women by banks?

Solution

IDENTIFY

We are looking for additional reasons to account for the results obtained from Case 19.1. Specifically, we would like to know whether there is a relationship between annual gross sales and interest rates and between age of firm and interest rates. If there are such relationships, we need to know whether men's businesses are older and have higher gross sales. If so, the apparent gender bias shown in Case 19.1 may be explained by banks' legitimate desire to grant favorable terms to more established businesses with higher sales. We can use the data about the type of business to determine whether banks view certain types of businesses more favorably than others, and if men's businesses tend to be of the types favored by banks. Before we can discuss the consequences of such findings, we need to conduct several tests. We will identify the technique to apply by asking specific questions:

1 Do banks grant lower interest rates to businesses with higher sales?

2 Do banks grant lower interest rates to older businesses?

---

*It was necessary to manipulate the data to apply some of the techniques described below.

For each question, we wish to analyze the relationship between two interval variables. If the two variables are bivariate normally distributed, we test the Pearson correlation coefficient. If they are extremely nonnormal the appropriate technique is the Spearman rank correlation. Because it is difficult to determine the distribution, it is probably best to perform both procedures. For each pair of variables, we want to know whether the relationship is negative. We test as follows:

Pearson Correlation Coefficient:

$H_0: \rho = 0$

$H_1: \rho < 0$

The test statistic is

$$t = r\sqrt{\frac{(n-2)}{1-r^2}}$$

Spearman Rank Correlation Coefficient:

$H_0: \rho_s = 0$

$H_1: \rho_s < 0$

The test statistic is

$$z = r_s\sqrt{n-1}$$

COMPUTE

Microsoft Excel Output of the Tests of Correlation between Rates and Sales: Case 19.2

Correlation

Rates and Sales

Pearson Coefficient of Correlation	−0.27
t Stat	−9.0896
df	1051
P(T<=t) one-tail	0
t Critical one-tail	1.6463
P(T<=t) two-tail	0
t Critical two-tail	1.9622
Spearman Rank Correlation	−0.2629
z Stat	−8.5265
P(Z<=z) one-tail	0
z Critical one-tail	1.6449
P(Z<=z) two-tail	0
z Critical two-tail	1.96

The *p*-values of the tests of the Pearson coefficient of correlation and the Spearman rank correlation both equal 0.

Microsoft Excel Output of the Tests of Correlation between Rates and Age: Case 19.2

Correlation	
Rates and Age	
Pearson Coefficient of Correlation	−0.1915
t Stat	−6.3244
df	1051
P(T<=t) one-tail	0
t Critical one-tail	1.6463
P(T<=t) two-tail	0
t Critical two-tail	1.9622
Spearman Rank Correlation	−0.1853
z Stat	−6.0106
P(Z<=z) one-tail	0
z Critical one-tail	1.6449
P(Z<=z) two-tail	0
z Critical two-tail	1.96

Here too, both test statistics produce *p*-values of 0.

INTERPRET

There is overwhelming statistical evidence to infer that banks grant lower rates to older businesses and business with higher sales.

This result presents two additional questions:

3 Do women's businesses have lower gross sales than men's businesses?

4 Are women's businesses younger than men's businesses?

IDENTIFY

The problem objective is to compare two populations of interval data. Using a similar analysis to that employed in Case 19.1, we discover that the appropriate technique for each question is the *t*-test of $\mu_1 - \mu_2$. For both questions, we test the following hypotheses (where μ_1 equals the mean [annual gross sales and age] of women's businesses and μ_2 equals the mean [annual gross sales and age] of men's businesses).

$H_0: \mu_1 - \mu_2 = 0$

$H_1: \mu_1 - \mu_2 < 0$

The test statistic is

$$t = \frac{(\bar{x}_1 - \bar{x}_2) - (\mu_1 - \mu_2)}{\sqrt{s_p^2 \left(\frac{1}{n_1} + \frac{1}{n_2}\right)}}$$

COMPUTE

Microsoft Excel Output of the *t*-Test of the Difference between Two Means (Sales): Case 19.2

t-Test: Two-Sample Assuming Equal Variances

	W Sales	M Sales
Mean	552	1183
Variance	60133	128618
Observations	101	952
Pooled Variance	122102	
Hypothesized Mean Difference	0	
df	1051	
t Stat	−17.27	
P(T<=t) one-tail	0.0000	
t Critical one-tail	1.646	
P(T<=t) two-tail	0.0000	
t Critical two-tail	1.962	

Microsoft Excel Output of the *t*-Test of the Difference between Two Means (Age): Case 19.2

t-Test: Two-Sample Assuming Equal Variances

	W Age	M Age
Mean	9.24	12.58
Variance	15.98	27.05
Observations	101	952
Pooled Variance	26.00	
Hypothesized Mean Difference	0	
df	1051	
t Stat	−6.261	
P(T<=t) one-tail	0.0000	
t Critical one-tail	1.646	
P(T<=t) two-tail	0.0000	
t Critical two-tail	1.962	

INTERPRET

There is very strong statistical evidence to infer that men own businesses that are older than women's businesses and that businesses owned by men have larger gross sales than businesses owned by women. This result is not surprising, because only in the last 25 years have women become involved in full-time work in large numbers. Up until about 1975 ownership of businesses had been a male domain. Moreover, more established businesses tend to have higher gross sales.

There is one more factor that may explain the outcome of Case 19.1. It is the type of business:

5 Do the data allow us to infer that certain types of businesses have different rates of interest?

IDENTIFY

There are three populations of interval data to compare: interest rates for proprietorships, for partnerships, and for corporations. The appropriate technique is the single-factor independent samples analysis of variance. (We can show that the required conditions are satisfied.)

$H_0: \mu_1 = \mu_2 = \mu_3$

$H_1:$ At least two means differ

The test statistic is the F-statistic ($F = $ MST/MSE) of the analysis of variance.

COMPUTE

Microsoft Excel Output of the Analysis of Variance (Points above Prime for Business Types): Case 19.2

Anova: Single Factor

Summary

Groups	Count	Sum	Average	Variance
Proprietorship	193	273.4	1.417	0.411
Partnership	86	104.2	1.211	0.467
Corporation	774	994.9	1.285	0.451

ANOVA

Source of Variation	SS	df	MS	F	P-value	F crit
Between Groups	3.455	2	1.728	3.88	0.0209	3.00
Within Groups	467.4	1050	0.445			
Total	470.9	1052				

INTERPRET

The statistical analysis reveals that there is enough evidence to infer that there are differences in rates among the three types of businesses.

Finally, we need to know whether men and women own different types of businesses:

6 Is there a relationship between gender and business type?

IDENTIFY

The problem objective is to analyze the relationship between two nominal variables, gender and business type. The appropriate method is the chi-squared test of a contingency table:

H_0: The two variables are independent

H_1: The two variables are dependent

The test statistic is $\chi^2 = \sum_{i=1}^{k} \dfrac{(f_i - e_i)^2}{e_i}$

COMPUTE

We created the variable gender where 1 = female business owner and 2 = male business owner.

Microsoft Excel Output of the Contingency Table (Gender and Business Type): Case 19.2

Contingency Table

Gender	Business 1	Business 2	TOTAL
1	31	162	193
2	10	76	86
3	60	714	774
TOTAL	101	952	1053

chi-squared Stat	12.7504
df	2
p-value	0.0017
chi-squared Critical	5.9915

INTERPRET

There is evidence of a relationship between gender and type of business. The proportion of women's businesses in this study is 101/1053 = 9.6%. From the contingency table, you can see that the proportion of proprietorships owned by women is 31/193 = 16.1%. From the analysis of variance printout, we discover that, on average, proprietorships pay the highest interest rates. Apparently, women own businesses that banks view as riskier, resulting in higher rates of interest.

INTERPRETING THE RESULTS OF CASES 19.1 AND 19.2

Cases 19.1 and 19.2 present statistics practitioners with real challenges. The results of the cases are contradictory. One case presents evidence of bias, and the other provides evidence to suggest that the apparent bias is actually attributable to other factors. Results such as these have led less-than-competent observers to state that one can prove anything with statistics. Although there are often several different interpretations of data based on different analyses, we can perform further analyses to arrive at the truth. Consider how we can resolve the apparent contradictions in Cases 19.1 and 19.2. It is necessary to consider carefully which data and techniques we should use. You will encounter our answer eventually.

19.3 THE LAST WORD

We have come to the end of the journey that began with the words, "Statistics is a way to get information from data." You will shortly write the final examination in your statistics course. (We assume that readers of this book are taking a statistics course and not just reading it for fun.) If you believe that this event will be the point where you and statistics part company, you could not be more wrong. In the world into which you are about to graduate, the potential applications of statistical techniques are virtually limitless. No matter the profession you ultimately pursue, you will have thousands of opportunities to learn more about your world. And, many of those opportunities will require you to convert data into information or interpret someone else's statistical analysis. However, if you are unable or unwilling to employ statistics, you cannot consider yourself to be competent.

In our experience, we have come across far too many people who display an astonishing ignorance of statistics. This raises the question, "What statistical concepts and techniques will you need for your life after the final exam?" We don't expect students to remember the formulas (or computer commands) that calculate the interval estimates or test statistics. (Statistics reference books are available for that purpose.) However, you must know what you can and cannot do with statistical techniques. You must remember a number of important principles that were covered in this book. To assist you, we have selected the 12 most important concepts and listed them. They are drawn from the "Developing an Understanding of Statistical Concepts" subsections that are scattered throughout the book. We hope that they prove useful to you.

TWELVE STATISTICAL CONCEPTS YOU NEED FOR LIFE AFTER THE STATISTICS FINAL EXAM

1 Statistical techniques are methods that convert data into information. Descriptive techniques describe and summarize; inferential techniques allow us to make estimates and draw conclusions about populations from samples.

2 We need a large number of techniques because there are numerous objectives and types of data. There are three types of data: interval (real numbers), nominal (categories), and ordinal (ratings). Each combination of data type and objective requires specific techniques.

3 We gather data by various sampling plans. However, the validity of any statistical outcome is dependent on the validity of the sampling. "Garbage in/garbage out" very much applies in statistics.

4 The sampling distribution is the source of statistical inference. The interval estimator and the test statistic are derived directly from the sampling distribution. All inferences are actually probability statements based on the sampling distribution.

5 All tests of hypotheses are conducted similarly. We assume that the null hypothesis is true. We then compute the value of the test statistic. If the difference between what we have observed (and calculated) and what we expect to observe is too large, we reject the null hypothesis. The standard that decides what is "too large" is determined by the probability of a Type I error.

6 In any test of hypothesis (and in most decisions) there are two possible errors, Type I and Type II errors. The relationship between the probability of these errors helps us decide where to set the standard. If we set the standard so high that the probability of a Type I error is very small, we increase the probability of a Type II error. A procedure designed to decrease the probability of a Type II error will have a relatively large probability of a Type I error.

7 We can improve the precision of an interval estimate or decrease the probability of a Type II error by increasing the sample size. More data means more information, which results in narrower intervals or lower probabilities of making mistakes, which produces better decisions.

8 The sampling distributions that are used for interval data are the Student t and the F. These distributions are related so that the various techniques for interval data are themselves related. For example, we can use the analysis of variance in place of a two-tail t-test of two means.

9 In analyzing interval data, we attempt to explain as much of the variation as possible. By doing so, we can learn a great deal about whether populations differ and what variables affect the response (dependent) variable.

10 The techniques used on nominal data require that we count the number of times each category occurs. The counts are then used to compute statistics. The sampling distributions we use for nominal data are the z (standard normal) and the chi-squared distributions. These distributions are related, which also describes the relationship among the techniques.

11 The techniques used on ordinal data are based on a ranking procedure. Statisticians call these techniques nonparametric. Because the requirements for the use of nonparametric techniques are less than those for parametric procedures, we often use nonparametric techniques in place of parametric ones when the required conditions for the parametric test are not satisfied. To ensure the validity of a statistical technique, we must check the required conditions.

12 We can obtain data through experimentation or by observation. Observational data lend themselves to several conflicting interpretations. Data gathered by an experiment are more likely to lead to a definitive interpretation. In addition to designing experiments, statistics practitioners can also select particular sample sizes to produce the accuracy and confidence they desire.

We believe that these concepts will serve you well in reading and understanding reports that use statistics. They will also allow you to exploit fully the sets of data that you will encounter in your professional life and apply the appropriate statistical technique that converts the data into information. In so doing, you will be a better decision-maker.

EXERCISES

19.1 Mutual funds minimize risks by diversifying the investments they make. There are mutual funds that specialize in particular types of investments. For example, the TD Precious Metal Mutual Fund buys shares in gold mining companies. The value of this mutual fund depends on a number of factors related to the companies in which the fund invests as well as on the price of gold. To investigate the relationship between the value of the fund and the price of gold, an economics student gathered the daily fund price and the daily price of gold for a 28-day period. These data are stored in columns A and B, respectively, of file Xr19-01. Can we infer from these data that there is a positive linear relationship between the value of the fund and the price of gold? (The authors are grateful to Jim Wheat for writing this exercise.)

19.2 Scientists at the Genetics and IVF Institute, a private fertility center in Virginia, announced the development of a new technique that allows parents to choose the sex of their baby before conception. A group of women who wanted their babies to be female were impregnated using the new technology. Of the 17 resulting babies, 15 were girls. Does this provide sufficient evidence to infer that the new technology does affect the sex of the baby?

19.3 The widespread use of salt on roads in Canada and the northern United States during the winter and acid precipitation throughout the year combine to cause rust on cars. Car manufacturers and other companies offer rustproofing services to help purchasers preserve the value of their cars. A consumer protection agency decides to determine whether there are any differences between the rust protection provided by automobile manufacturers and that provided by two competing types of rustproofing services. As an experiment, 60 identical new cars are selected. Of these, 20 are rustproofed by the manufacturer. Another 20 are rustproofed using a method that applies a liquid to critical areas of the car. The liquid hardens, forming a (supposedly) lifetime bond with the metal. The last 20 are treated with oil and are re-treated every 12 months. The cars are then driven under similar conditions in a Minnesota city. The number of months until the first rust appears is recorded and stored in columns A, B, and C, respectively, in file Xr19-03. Is there sufficient evidence to conclude that at least one rustproofing method is different from the others?

19.4 In the door-to-door selling of vacuum cleaners, various factors influence sales. The Birk Vacuum Cleaner Company considers its sales pitch and overall package to be extremely important. As a result, it often thinks of new ways to sell its product. Because the company's management dreams up so many new sales pitches each year, there is a two-stage testing process. In stage 1, a new plan is tested with a relatively small sample. If there is sufficient evidence that the plan increases sales, a second, considerably larger, test is undertaken. The statistical test is performed so that there is only a 1% chance of concluding that the new pitch is successful in increasing sales when it actually does not increase sales. In a stage 1 test to determine if the inclusion of a "free" ten-year service contract increases sales, 100 sales representatives were selected at random from the company's list of several thousand. The monthly sales of these representatives were recorded for one month prior to use of the new sales pitch and for one month after its introduction. The results are stored in file Xr19-04. Should the company proceed to stage 2?

19.5 Two drugs are used to treat heart attack victims. Streptokinase, which has been available since 1959, costs about $500. The second drug is t-PA, a genetically engineered product that sells for about $3,000. Both streptokinase and t-PA work by opening the arteries and dissolving blood clots, which are the cause of heart attacks. Several previous studies have failed to reveal any differences between the effects of the two drugs. Consequently, in many countries where health care is funded by governments, physicians are required to use the less expensive streptokinase. However, t-PA's maker, Genentech, Inc., contended that in the earlier studies showing no difference between the two drugs, their drug was not used in the right way. Genentech decided to sponsor a more thorough experiment. The experiment was organized in 15 countries, including the United States and Canada, and involved a total of 41,000 patients. In this study, t-PA was given to patients in 90 minutes instead of 3 hours as in previous trials. Half of the sample of 41,000 patients were treated by a rapid injection of t-PA with intravenous heparin, while the other half received streptokinase along with heparin. The number of deaths in each sample was recorded. A total of 1,497 patients treated with streptokinase died, while 1,292 patients who received t-PA died.

a Can we infer that t-PA is better than streptokinase in preventing deaths?
b Estimate with 95% confidence the cost per life saved by using t-PA.

19.6 A small but important part of a university library's budget is the amount collected in fines on overdue books. Last year, a library collected $75,652.75 in fine payments; however, the head librarian suspects that some employees are not bothering to collect the fines on overdue books. In an effort to learn more about the situation, she asked a sample of 400 students (out of a total student population of 50,000) how many books they had returned late to the library in the previous 12 months. They were also asked how many days overdue the books had been. The results indicated that the total number of days overdue ranged from 0 to 55 days. The number of days overdue was stored in file Xr19-06.

a Estimate with 95% confidence the average number of days overdue for all 50,000 students at the university.
b If the fine is 25 cents per day, estimate the amount that should be collected annually. Should the librarian conclude that not all the fines were collected?

19.7 The practice of therapeutic touch is used in hospitals all over the world and is taught in some medical and nursing schools. In this therapy, trained practitioners manipulate something that they call the "human energy field." The manipulation is carried out without actually touching the patient's body. Practitioners claim that anyone can be trained to feel this energy field. Some researchers say that no reliable evidence exists showing that this technique actually heals patients. James Randi, a professional magician who is also known as a skeptic of some types of alternative medicine, has tried for years to test the practice of therapeutic touch. So far, only one practitioner has agreed to submit to his test, and she did no better than chance in detecting the energy field. Emily Rosa, an 11-year-old Colorado girl working on a science fair project, was more successful in recruiting trained therapists. For her experiment, Emily placed a screen between a practitioner's eyes and hands, and then held her own hand over one of the practitioner's hands. If the human energy field can be felt, then practitioners should be able to identify which of their hands Emily held hers over. Emily conducted 280 tests with the 21 subjects. The results of these tests are stored in file Xr19-07 where 2 = correct location and 1 = incorrect location. What can we infer about touch therapy from these data?

19.8 Despite the increase in the number of working women, it appears that women continue to perform most of the household functions such as cooking and cleaning. Two researchers from the University of Arizona undertook a study to examine how much housework teenagers are performing. The researchers were particularly interested in the differences between teenage boys and girls. A random sample of teenagers was selected. Each teenager reported the number of hours of housework he or she performs in an average week. The data are stored in column A (boys) and column B (girls) in file Xr19-08. Can we conclude that teenage girls do more housework than teenage boys?

19.9 Refer to Exercise 19.8. The survey also asked whether the mother worked outside of the home. The amount of housework performed by each teenager is stored in file Xr19-09. In this file, column A contains the amount of housework performed during an average week for teenagers whose mothers work at a job outside the home and column B stores the same variable for teenagers whose mothers work inside the home. Can we infer that children whose mothers work outside the home do more housework than children whose mothers work inside the home?

19.10 One of the ways in which advertisers measure the value of television commercials is by telephone surveys conducted shortly after commercials are aired. Respondents who watched a certain television station at a given time period, during which the commercial appeared, are asked if they can recall the name of the product in the commercial. Suppose an advertiser wants to compare the recall proportions of two commercials. The first commercial is relatively inexpensive. A second commercial shown a week later is quite expensive to produce. The advertiser decides that the second commercial is viable only if its recall proportion is more than 15% higher than the recall proportion of the first commercial. Two surveys of 500 television viewers each were conducted after each commercial was aired. Each person was asked whether he or she remembered the product name. The results are stored in columns A (commercial 1) and B (commercial 2) (2 = remembered the product name and 1 = did not remember the product name) in file Xr19-10. Can we infer that the second commercial is viable?

19.11 A professor has noted that the exam results of students who write their exams at 1:00 appear to be better than the marks of students who write their exams at 4:00. He forms the belief that the difference is caused by energy levels that are affected by the timing of the most recent meal. In other words, eating a meal before the exam will produce higher grades. To test this belief he randomly selects 100 students to write a test at 12:30. Half the class is provided lunch, and the other half is deprived of lunch. The test scores are recorded in columns A (test score of those who had lunch) and B (test scores of those with no lunch) in file Xr19-11. Do the data support the professor's belief?

19.12 Refer to Exercise 19.11. The professor would like to know whether the type of meal affects test scores. He randomly selects 200 students; 50 are fed a lunch of pizza, another 50 are fed tuna sandwiches, another 50 are fed hamburgers, and the remaining 50 eat a salad for lunch. The test scores are stored on columns A through D, respectively, in file Xr19-12. Do these data indicate that the type of meal affects test scores?

19.13 How consistent are professional athletes? Do they perform at about the same level year in and year out, or do they have great seasons interspersed with bad ones? To answer these questions for hockey players, an MBA student randomly selected 50 National Hockey League players who played in the 1992–1993 and 1993–1994 seasons. For each player, he recorded the average points per game and their plus/minus scores. (Plus/minus scores measure the number of goals a player's team scores minus the number of goals the opposing team scores while that player is on the ice.) These data are stored in file Xr19-13 in the following way:

Column A: Player
Column B: 1993–1994 points per game

Column C: 1992–1993 points per game
Column D: 1993–1994 plus/minus
Column E: 1992–1993 plus/minus

Can we conclude from these data that players' are inconsistent from one year to the next in terms of points per game and their plus/minus ratio? (The authors are grateful to Gordon Barnett for writing this exercise.)

19.14 Professional athletes in North America are paid very well for their ability to play games that amateurs play for fun. To determine the factors that influence a team to pay a hockey player's salary, an MBA student randomly selected 50 hockey players who played in the 1992–1993 and 1993–1994 seasons. He recorded their salaries at the end of the 1993–1994 season as well as a number of performance measures in the previous two seasons. The following data were recorded in file Xr19-14:

Columns A and B: games played in 1992–1993 and 1993–1994

Columns C and D: goals scored in 1992–1993 and 1993–1994

Columns E and F: assists recorded in 1992–1993 and 1993–1994

Columns G and H: plus/minus score in 1992–1993 and 1993–1994

Columns H and I: penalty minutes served in 1992–1993 and 1993–1994

Column J: salary in U.S. dollars

(Plus/minus is the number of goals scored by his team minus the number of goals scored by the opposing team while the player is on the ice.) Develop a model that analyzes the relationship between salary and the performance measures. Describe your findings. (The authors wish to thank Gordon Barnett for writing this exercise.)

19.15 Winter is the influenza season in North America. Each winter thousands of elderly and sick people die from the flu and its attendant complications. Consequently, many elderly people receive flu shots in the fall. It has generally been accepted that young healthy North Americans need not receive flu shots because, although many contract the disease, few die from it. However, there are economic consequences. Sick days cost both the employee and employer. A study published in the *New England Journal of Medicine* reported the results of an experiment to determine whether it is useful for young healthy people to take flu shots. A random sample of working adults was selected. Half received a flu shot in November; the other half received a placebo. The number of sick days over the next six-month period were recorded in columns A (flu shot) and B (placebo). Columns C (flu shot) and D (placebo) contain the number of visits to the doctor. All the data are in file Xr19-15.

a Can we conclude that the number of sick days is less for those who take flu shots?

b Can we conclude that those who take flu shots visit their doctors less frequently?

19.16 Sales of a product may depend on its placement in a store. Cigarette manufacturers frequently offer discounts to retailers who display their products more prominently than competing brands. To examine this phenomenon more carefully, a cigarette manufacturer (with the assistance of a national chain of restaurants) planned the following experiment. In 20 restaurants, the manufacturer's brand was displayed behind the cashier's counter with all the other brands (this was called position 1). In another 20 restaurants, the brand was placed separately but close to the other brands (position 2). In a third group of 20 restaurants, the cigarettes were placed in a special display next to the cash register (position 3). The number of cartons sold during one week at each restaurant was recorded and is stored in columns A, B, and C, respectively, in file Xr19-16. Is there sufficient evidence to infer that sales of cigarettes differ according to placement?

19.17 After a recent study, researchers reported on the effects of folic acid on the occurrence of spina bifida—a birth defect in which there is incomplete formation of the spine. A sample of 2,000 women who gave birth to children with spina bifida was recruited. Prior to attempting to get pregnant again, half the sample were given regular doses of folic acid, and the other half were given a placebo. After 18 months, there were 1,209 births. The number of births and the number of children born with spina bifida are shown. Determine whether we can infer that folic acid reduces the incidence of spina bifida?

	Group Taking Folic Acid	Group Taking Placebos
Number of births	597	612
Number of children born with spina bifida	6	21

19.18 The image of the U.S. Postal Service has suffered in recent years. One reason may be the perception that postal workers are rude in their dealings with the public. In an effort to improve its image, the Postal Service is contemplating the introduction of public relations seminars for all of its inside workers.

Because of the substantial costs involved, the Postal Service decided to institute the seminars only if there is more than a 25% reduction in the number of written customer complaints about personnel who took the seminars. As a trial, the employees of 30 large postal centers attended the seminar. The monthly average number of complaints per center (for all centers) was 640 before the trial. The number of complaints in the 30 centers is stored in file Xr19-18. Can we conclude that the seminars should be instituted?

19.19 Discrimination in hiring has been illegal for a number of years. It is illegal to discriminate against any person on the basis of race, sex, or religion. It is also illegal to discriminate because of a person's handicap if it in no way prevents that person from performing that job. In recent years, the definition of "handicap" has widened. Several people have successfully sued companies because they were denied employment for no other reason than that the applicant was overweight. A study was conducted to examine attitudes toward overweight people. The experiment involved showing a number of subjects videotape of an applicant being interviewed for a job. Prior to the interview, the subject was given a description of the job. Following the interview, the subject was asked to score the applicant in terms of how well the applicant was suited for the job. The score was out of 100, where higher scores described greater suitability. The same procedure was repeated for each subject. However, the sex and weight (average and overweight) of the applicants varied. The results are stored in file Xr19-19 using the following format:

Column A: score for average-weight males
Column B: score for overweight males
Column C: score for average-weight females
Column D: score for overweight females

a Can we infer that the scores of the four groups of applicants differ?
b Are the differences detected in part (a) due to weight, sex, or some interaction? (You must alter the format of the data to apply the technique to answer this question.)

19.20 The Scholastic Assessment Test (SAT), which is organized by the Educational Testing Service (ETS), is important to high school students seeking admission to colleges and universities throughout the United States. A number of companies offer courses to prepare students for the SAT. The Stanley H. Kaplan Educational Center claims that its students gain, on average, more than 110 points by taking its course. ETS, however, insists that preparatory courses can improve a score by no more than 40 points. (The minimum and maximum scores of the SAT are 400 and 1,600, respectively.) Suppose a random sample of 40 students wrote the exam, then took the Kaplan preparatory course, and then wrote the exam again. The results are stored in columns A (after Kaplan preparatory course) and B (before) of file Xr19-20.

a Do these data provide sufficient evidence to refute the ETS claim?
b Do these data provide sufficient evidence to refute Kaplan's claim?

19.21 It is generally believed that salespeople who are paid on a commission basis outperform salespeople who are paid a fixed salary. Some management consultants argue, however, that in certain industries the fixed-salary salesperson may sell more because the consumer will feel less sales pressure and respond to the salesperson less as an antagonist. In an experiment to study this, a random sample of 180 salespeople from a retail clothing chain was selected. Of these, 90 salespeople were paid a fixed salary, and the remaining 90 were paid a commission on each sale. The total dollar amount of one month's sales for each was recorded and stored in columns A (fixed salary) and B (commission) in file Xr19-21. Can we conclude that the commission salesperson outperforms the fixed-salary salesperson?

19.22 Obesity among children in North America is said to be at near-epidemic proportions. Some experts blame television for the problem, citing the statistic that children watch on average about 26 hours per week. During this time, children are not engaged in any physical activity, which results in weight gains. However, the problem may be compounded by a reduction in metabolic rate. In an experiment to address this issue (the study results were published in the February 1993 issue of the medical journal *Pediatrics*), scientists from Memphis State University and the University of Tennessee at Memphis took a random sample of 223 children aged 8 to 12. Each child's metabolic rate (the amount of calories burned per hour) was measured while at rest and also measured while the child watched a television program (*The Wonder Years*). These data are stored in columns A and B, respectively, in file Xr19-22. Do these data allow us to conclude that there is a decrease in metabolism when children watch television?

19.23 The game of Scrabble is one of the oldest and most popular board games. It is played all over the world, and there is even an annual world championship competition. The game is played by forming words and placing them on the board to obtain the maximum number of points. It is generally believed that a large vocabulary is the only skill required to be successful. However, there is a strategic element to the game that suggests that mathematical skills are just as necessary. To determine which skills are most in demand, a statistics practitioner recruited a random sample of fourth-year university English and mathematics majors and asked them to play the game. A total of 500 games were played by different pairs of English and mathematics majors. The scores in each game are stored in file Xr19-23 (column A: English major scores; column B: mathematics major scores).

a Can we conclude that mathematics majors win more frequently than do English majors?
b Do these data allow us to infer that the average score obtained by English majors is greater than that for mathematics majors?
c Why are the results of parts (a) and (b) not the same?

(The authors would like to thank Scott Bergen for his assistance in writing this exercise.)

19.24 The cost of workplace injuries is high for the individual worker, the company, and for society. It is in everyone's interests to rehabilitate the injured worker as quickly as possible. A statistics practitioner working for an insurance company has investigated the problem. He believes that a major determinant in how quickly a worker returns to his or her job after sustaining an injury is physical condition. To help determine whether he is on the right track, he organized an experiment. He took a random sample of male and female workers who were injured in the last year. He recorded their gender, their physical condition, and the number of working days until they returned to their job. These data are stored in file Xr19-24 in the following way. Columns B and C store the number of working days until return to work for men and women, respectively. In each column the first 25 observations relate to those who are physically fit, the next 25 rows relate to individuals who are moderately fit, and the last 25 observations are for those who are in poor physical shape. Can we infer that the six groups differ? If differences exist, determine whether the differences are due to gender, physical fitness, or some combination of gender and physical fitness.

19.25 Some psychologists believe that there are three different personality types: type A is the aggressive workaholic; type B is the relaxed underachiever; type C displays various characteristics of types A and B. The personnel manager of a large insurance company believes that, among life insurance salespersons, there are equal numbers of all three personality types and that their degree of satisfaction and sales ability are not influenced by personality type. In a survey of 150 randomly selected salespersons, he determined the type of personality, measured their job satisfaction on a seven-point scale (where 1 = very dissatisfied and 7 = very satisfied), and determined the total amount of life insurance sold by each during the previous year. The results are stored in file Xr19-25 using the following format:

Column A: personality type (A = 1; B = 2; C = 3)
Column B: job satisfaction
Column C: total amount of life insurance sold (in hundreds of thousands of dollars)

Test all three of the personnel manager's beliefs. That is, test to determine whether there is enough evidence to justify the following conclusions:

i The proportions of each personality type are different.
ii The job satisfaction measures for each personality type are different.
iii The life insurance sales for each personality type are different.

(You must alter the format of the data to apply the techniques in parts [ii] and [iii].)

19.26 Simco, Inc., is a manufacturer that purchased a new piece of equipment designed to reduce costs. After several months of operation, the results were quite unsatisfactory. The operations manager believes that the problem lies with the machine's operators, who were unable to master the required skills. It was decided to establish a training program to upgrade the skills of those workers with the greatest likelihood of success. To do so, the company needed to know which skills are most needed to run the machine. Experts identified six such skills. They are dexterity, attention to detail, teamwork skills, mathematical ability, problem-solving skills, and technical knowledge. To examine the issue, a random sample of workers was drawn. Workers were measured on each of the six skills through a series of paper-and-pencil tests and through supervisor ratings. Additionally, each worker received a score on the quality of his or her actual work on the machine. These data are stored in columns A through G of file Xr19-26. (Columns A to F are the scores on the skill tests and column G stores the quality-of-work scores.) Identify the skills that affect the quality of work. (We are grateful to Scott Bergen for writing this exercise.)

19.27 The high cost of medical care makes it imperative that hospitals operate efficiently and effectively. As part of a larger study, patients leaving a hospital were surveyed. They were asked how satisfied they were with the treatment they received. The responses were recorded with a measure of the degree of severity of their illness (as determined by the admitting physician) and the length of stay. These data are recorded in file Xr19-27 in the following way:

Column A: severity of illness (1 = least severe and 10 = most severe)
Column B: satisfaction level (1 = very unsatisfied; 2 = somewhat unsatisfied; 3 = neither satisfied nor dissatisfied; 4 = somewhat satisfied; 5 = very satisfied)
Column C: number of days in hospital

a Is the satisfaction level related to the severity of illness?

b Is the satisfaction level higher for patients who stay for shorter periods of time?

19.28 Since germs have been discovered, parents have been telling their children to wash their hands. Common sense tells us that this should help minimize the spread of infectious diseases and lead to better health. A study reported in the *University of California at Berkeley Wellness Letter* (Volume 13, Issue 6, March 1997) may confirm the advice our parents gave us. A study in Michigan tracked a random sample of children, some of whom washed their hands four or more times during the school day. The number of sick days due to colds and flu were recorded for the past year and stored in columns A (number of sick days for children who washed their hands four or more times per day) and B (number of sick days of children who washed their hands fewer than four times per day) in file Xr19-28. Do these data allow us to infer that a child who washed his or her hands four or more times during the school day will have fewer sick days due to cold and flu than other children?

19.29 The experiment to determine the effect of taking a preparatory course to improve SAT scores in Exercise 19.20 was criticized by other statistics practitioners. They argued that the first test would provide a valuable learning experience that would produce a higher test score on the second exam even without the preparatory course. Consequently, another experiment was performed. Forty students wrote the SAT without taking any preparatory course. At the next scheduled exam (three months later), these same students took the exam again (again with no preparatory course). The scores for both exams are stored in columns A (first test scores) and B (second test scores) in file Xr19-29. Can we infer that repeating the SAT produces higher exam scores even without the preparatory course?

19.30 In recent years, North Americans have experienced flooding in various parts of the continent. In an effort to develop flood-forecasting tools, scientists wanted to determine the relationship between river flows and precipitation and evaporation. In one study, scientists gathered annual data for the total discharge of water on the Grand River at Galt, Ontario. They also recorded precipitation (snow and rain) and a measure of potential evaporation (which is a function of temperature, humidity, and wind) for each year between 1914 and 1980 around Galt. These data are stored in columns A (potential evaporation); B (precipitation in millimeters); and C (river flow in cubic decimeters) in file Xr19-30. It is generally believed that flow and precipitation should be positively related, and flow and potential evaporation should be negatively related. Do the data confirm the scientists' beliefs? (This exercise was prepared by Lynette Snelgrove. The data are from *The Impact of Climate Change on the Water in the Grand River Basin, Ontario,* Department of Geography Publication Series No. 40, University of Waterloo.)

19.31 Betting on the results of National Football League games is a popular North American activity. In many states and provinces, it is legal to do so provided that wagers are made through the government-authorized betting organization. In the province of Ontario, Pro-Line serves that function. Bettors can choose any team on which to wager, and Pro-Line sets the odds, which determine the winning payoffs. It is also possible to bet that in any game a tie will be the result. (A tie is defined as a game in which the winning margin is three or fewer points. A win occurs when the winning margin is greater than three.) To assist bettors, Pro-Line lists the favorite for each game and predicts the point spread between the two teams. To judge how well Pro-Line predicts outcomes, the Creative Statistics Company tracked the results of the 1993 season. It recorded whether a team was favored by three or fewer points (1); 3.5 to 7 points (2); 7.5 to 11 points (3); or 11.5 or more points (4). It also recorded whether the favored team won (1); lost (2); or tied (3). These data are recorded in columns A (Pro-Line's predictions) and B (game results) in file Xr19-31. Can we conclude that Pro-Line's forecasts are useful for bettors?

Note: You must alter the format of the data to apply the techniques to answer the questions in the next three exercises.

19.32 At the completion of repair work at one of a chain of automotive centers, customers are asked to fill out the following form.

Tell us what you think.

Are you satisfied?	Good	Fair	Poor
1 Quality of work performed			
2 Fairness of price			
3 Explanation of work and guarantee			
4 Checkout process			
5 Will return in future			
1 no			
2 yes			
Comments?			

A random sample of 134 responses was drawn. The results were recorded in file Xr19-32. The responses to questions 1 through 4 (1 = poor; 2 = fair; 3 = good) are stored in columns A through D, respectively. Responses to question 5 are stored in column E. If a positive comment was made, 1 is recorded in column F, 2 if a negative comment was made, and 3 if no comment was made. Can we infer that those who say they will return assess each category higher than those who will not return?

19.33 Refer to Exercise 19.32. Is there sufficient evidence to infer that those who make positive comments, negative comments, and no comments differ in their assessment of each category?

19.34 Refer to Exercise 19.32. Suppose that 100 responses from each of three stores in the chain are recorded in file Xr19-34 using the format below.

Columns A to F: Responses to questions 1 to 5 plus comments code
Column G: Code representing stores 1, 2, and 3

Can we conclude that differences exist among the three stores?

19.35 Several years ago we heard about the "Mommy Track," the phenomenon of women being underpaid in the corporate world because of what is seen as their divided loyalties between home and office. There may also be a "Daddy Differential." The "Daddy Differential" refers to the situation where men whose wives stay at home earn more than men whose wives work. It is argued that the differential occurs because bosses reward their male employees if they come from "traditional families." Linda Stroh of Loyola University of Chicago studied a random sample of 348 male managers employed by 20 Fortune 500 companies. Each manager reported his annual income and whether his wife stayed at home to care for their children or worked outside the home. The incomes (in thousands of dollars) are stored in file Xr19-35. The incomes of the managers whose wives stay at home are stored in column A. Column B contains the incomes of managers whose wives work outside the home.

a Can we conclude that men whose wives stay at home earn more than men whose wives work outside the home?

b If your answer in part (a) is affirmative, does this establish a case for discrimination? Can you think of another cause-and-effect scenario? Explain.

19.36 A popular game of chance is craps, which uses two dice. The "shooter" rolls both dice, winning on the first roll if he throws a 7 or 11, and losing if he rolls 2, 3, or 12. If he rolls any other number, he continues to roll the dice until he repeats the number and wins or throws a 7 and loses. Elementary rules of probability allow even students of probability and statistics to calculate the probability of winning and losing. These probabilities are based on the probability distribution of the total of two fair dice. They are as follows.

Total	Probability
2	1/36
3	2/36
4	3/36
5	4/36
6	5/36
7	6/36
8	5/36
9	4/36
10	3/36
11	2/36
12	1/36

A professor of statistics suspects that the dice are not fairly balanced and records each of 1,000 throws. These data are stored in file Xr19-36.

a Do these data allow us to infer that the dice are not fair?

b Can we conclude that the dice are set up so that the probability of 7 is less than 6/36?

19.37 A programmer for a small software company has almost completed the company's newest network program. He just finished the user interface and was about to perform a routine backup. As a rule, a backup is performed once a day to guard against power, hardware, or software failure. Once the backup has started, the information is unavailable until the backup is completed.

The amount of time that the program takes to complete the backup is critical. For systems that are in constant use, a portion of time must be set aside to back up, during which work on that system ceases. The company has developed three different systems of backing up files that are basically equivalent in memory usage. However, because time is critical, the company wanted to determine whether differences existed in the amount of time taken for each to work. A programmer randomly selected 26 different programs and backed up each of them using the three backing systems. The amount of time taken for each was recorded (in minutes) and stored in columns A to C of file Xr19-37. (The rows represent the 26 different programs.) Can we infer that the backup times differ among the three systems? (The author is grateful to James Pong, Diana Mansour, and the Ortech Company for their assistance in this exercise.)

19.38 There are enormous differences between health care systems in the United States and Canada. In a study to examine one dimension of these differences, 300 heart attack victims in each country were randomly selected. (Results of the study conducted by Dr. Daniel Mark of Duke University Medical Center, Dr. David Naylor of Sunnybrook Hospital in Toronto, and Dr. Paul Armstrong of the University of Alberta were published in the *Toronto Sun*, 27 October 1994.) Each patient was asked the following questions regarding the effect of his or her treatment:

1 How many days did it take you to return to work?
2 Do you still have chest pain? (This question was asked one month, six months, and twelve months after the patients' heart attacks.)

The responses are stored in file Xr19-38 in the following way:

Column A: code representing nationality: 1 = U.S.; 2 = Canada
Column B: responses to question 1
Column C: responses to question 2—one month after heart attack: 2 = yes; 1 = no
Column D: responses to question 2—six months after heart attack: 2 = yes; 1 = no
Column E: responses to question 2—twelve months after heart attack: 2 = yes; 1 = no

Can we conclude that recovery is faster in the United States? (You must alter the format of the data to apply the techniques to answer this question.)

19.39 As all baseball fans know, first base is the only base that the base runner may overrun. At second and third base the runner may be tagged out if he runs past them. Consequently, on close plays at second and third base, the runner will slide, enabling him to stop at the base. In recent years, however, several players have chosen to slide headfirst when approaching first base, claiming that this is faster than simply running over the base. In an experiment to test this claim, the 25 players on one National League team were recruited. Each player ran to first base with and without sliding, and the times to reach the base were recorded. These data are stored in columns A (no slide), and B (slide) in file Xr19-39. Can we conclude that sliding is slower than not sliding?

19.40 Du Pont is always looking for new ways to improve their products and reduce costs. Researchers in the Teflon coatings laboratory recently uncovered a compound that degrades into two potentially saleable products, which were labeled C and Y. The research team carried out several experiments to determine the factors that affect the production of C and Y. In particular, they examined how the temperature and flow rate determined the percentage yield of each component. The higher the temperature and flow rate, the more expensive the production becomes. However, this may be offset by greater yields. The experiment used three different flow rates (5, 35, and 80 milliliters per minute) and two temperatures (360°C and 400°C). The experiment was repeated several times for each combination of flow rate and temperature. The percentage yields were recorded and stored in file Xr19-40 using the following format:

Column A: percent yield of C
Column B: percent yield of Y
Column C: flow rate
Column D: temperature

Do the data allow us to conclude that for each of products Y and C

a flow rate affects yield?
b temperature affects yield?
c particular combinations of flow rate and temperature affect yields?

(You must alter the format of the data to apply the techniques to answer these questions.) (The authors are grateful to James Fong and Diana Mansour for creating this exercise.)

19.41 How does mental outlook affect a person's health? The answer to this question may allow physicians to care more effectively for their patients. In an experiment to examine the relationship between attitude and physical health, Dr. Daniel Mark, a heart specialist at Duke University, studied 1,719 men and women who recently had undergone a heart catheterization, a procedure that checks for clogged arteries. Patients undergo this procedure when heart disease results in chest pain. All of the patients in the experiment were in about the same condition. In interviews, 14% of the patients doubted that they would recover sufficiently to resume their daily routines. Dr. Mark identified these individuals as pessimists; the others were (by default) optimists. After one year, Dr. Mark recorded how many patients were still alive. The data are stored in columns A (1 = optimist, 2 = pessimist) and B (2 = alive, 1 = dead) in file Xr19-41. Do these data allow us to infer that pessimists are less likely to survive than optimists with similar physical ailments? (You must alter the format of the data to apply the techniques to answer this question.)

19.42 Can music make you smarter? And if so, which kind of music works best? Two University of California at Irvine professors addressed these questions (as reported on *Dateline* in September 1994). A random sample of 135 students was given tests that measured the ability to reason. One-third of the students were then put in a room where rock and roll music was played. A second group of 45 students was placed in a room and listened to music composed by Mozart. The last group was placed in a room where no music was played. The students then took another test. The differences (second test score minus first test score) were recorded and stored in columns A to C, respectively, in file Xr19-42. Can we infer that the type of music affects test results?

19.43 As a follow-up to the experiment described in Exercise 19.42, the students again were placed in their chosen rooms. After leaving their rooms, the students waited ten minutes before being tested. The differences (third test score minus first test score) were recorded and stored in file Xr19-43 (using the same format as in file Xr19-42). Does it appear that the effects of the music wear off after ten minutes?

19.44 A study in the journal *Neurology* reported that ibuprofen (sold commercially as Motrin, Advil, and Nuprin) may slow or even prevent the progression of Alzheimer's disease. The study was based on a sample of 1,500 volunteers of varying ages who kept detailed records of what medications they took. Researchers at the Johns Hopkins School of Public Health noticed that those who had taken ibuprofen for at least two years showed a 60 percent lower risk of getting Alzheimer's. Aspirin had a smaller effect, and acetaminophen (the main ingredient in Tylenol) had no effect. Further analysis revealed that almost all of those who had taken ibuprofen for at least two years did so to relieve the symptoms of arthritis. Assuming that the 60 percent reduction in the risk of Alzheimer's seen in the sample is statistically significant, can we infer that taking ibuprofen regularly will reduce the incidence of Alzheimer's? Suggest another way to interpret the results that does not lead to the conclusion about the relationship between ibuprofen and Alzheimer's.

19.45 Researchers in both the business world and the academic world often treat college students as representative of the adult population. This practice reduces sampling costs enormously, but its effectiveness is open to question. An experiment was performed to determine the suitability of using student surrogates in research. The study used two groups of people.

1 The first consisted of 60 adults (18 years of age or older) chosen so that they represented by age and occupation the adult population of a Midwestern state.

2 The second consisted of 60 students enrolled in an introductory marketing course at a public university.

The experiment involved showing each group a 60-second television commercial advertising a financial institution. Each respondent was asked to assess the commercial's believability. The responses were recorded as follows:

4 very believable
3 somewhat believable
2 not very believable
1 not at all believable

These data are stored in columns A and B, respectively, in file Xr19-45.

a Can we infer that there are differences among the two groups of respondents?
b What conclusions can you draw regarding the suitability of using students as surrogates in marketing research?

(Adapted from R. Kesevan, D. G. Anderson, and O. Mascarenhas, "Students as Surrogates in Advertising Research," *Developments in Marketing Science* 7, 1984: 438–41.)

19.46 The battle between customers and car dealerships is often intense. Customers want the lowest price, and dealers want to extract as much money as possible. One source of conflict is the trade-in car. Most dealers will offer a relatively low trade-in in anticipation of negotiating the final package. In an effort to determine how dealers operate, a consumer organization undertook an experiment. Seventy-two individuals were recruited. Each solicited an offer on "their" five-year-old Ford Taurus. The exact same car was used throughout the experiment. The only variables were the age and gender of the "owner." The ages were categorized as (1) young, (2) middle, and (3) senior. The cash offers are stored in columns B and C in file Xr19-46. Column B stores the data for female owners, and column C contains the offers made to male owners. The first twelve rows in both columns represent the offers made to young people, the next twelve rows represent the middle group, and the last twelve rows represent the elderly owners.

a Can we infer that differences exist among the six groups?

b If differences exist, determine whether the differences are due to gender, age, or some interaction.

19.47 After studying the results of the analyses performed in Cases 18.1 and 18.2, a statistics practitioner observed that one needed to determine the amount of increase or decrease in the percentage of rejected ballots between the referendum and the 1994 provincial election and how the change relates to whether the majority voted "Yes" or "No." Accordingly, the statistics practitioner recorded the percentage of rejected ballots in the referendum, the percentage of rejected ballots in the 1994 provincial election, and whether the riding majority voted "Yes" (1) or "No" (2). These data are stored in columns B, C, and D, respectively, in file Xr19-47. (Column A contains the riding number.)

a Can we infer from these results that, in ridings with a majority of "No" votes, there was an increase in the percentage of rejected ballots?

b Can we infer that, in ridings with a majority "Yes" votes, there was a decrease in the percentage of rejected ballots?

19.48 Repeat Cases 19.1 and 19.2, assuming that interest rates, annual gross sales, and ages of businesses are extremely nonnormally distributed.

Case 19.3   Ambulance and Fire Department Response Interval Study*

Every year, thousands of people die of heart attacks, partly because of delays in waiting for emergency medical care to arrive. One form of heart attack is ventricular fibrillation rhythm, which is treated by a defibrillator. However, immediate medical attention is critical. In general, if a patient receives treatment within eight minutes, he or she is very likely to survive. It is estimated that the probability of survival is reduced by 7 to 10% for each minute thereafter that defibrillation is delayed.

The region in the Ambulance and Fire Department Response Interval Study is composed of the three cities of Cambridge, Waterloo, and Kitchener. Each city has a fire department, and the region has a 911 emergency telephone system. When a medical-related call is received by the Police Dispatch Center, it is relayed to the Central Ambulance Communication Center (CACC). The CACC dispatches both the ambulance and the fire department to certain calls that match one of several criteria indicating the need for fire department personnel. There are two ambulance services that cover the region: the Cambridge Memorial Hospital Ambulance Service and the Kitchener-Waterloo Regional Ambulance Service.

Currently, ambulance personnel sent to the patient after a 911 call perform defibrillation. A city counselor recently suggested that, because the fire department has more centers, it is likely that fire department personnel could arrive at the scene more quickly than ambulance personnel. A study was undertaken to determine whether fire department personnel should be trained in the use of defibrillators and sent to treat ventricular fibrillation rhythm.

*The author is grateful to Bruce Jermyn for supplying this case. The data are real. However, the sample size was reduced to ease disk-storage problems.

Between March 1, 1994, and August 31, 1994, all calls that involved both ambulance and fire department personnel were monitored. The times for each service to arrive at the scene were recorded and stored in file C19-03, using the format below.

Column A: call number for Cambridge calls

Column B: time in minutes for the ambulance to arrive

Column C: time for fire truck to arrive

Column D: call number for Kitchener calls

Column E: time in minutes for the ambulance to arrive

Column F: time for fire truck to arrive

Column G: call number for Waterloo calls

Column H: time in minutes for the ambulance to arrive

Column I: time for fire truck to arrive

It has been decided that the training of fire department personnel is warranted only if it can be shown that a fire truck arrives at the scene on average more than one minute sooner than an ambulance and that the frequency of arrival within eight minutes is greater for the fire department.

What conclusions can be drawn from the data?

Case 19.4 *PC Magazine* Survey

In general, consumers of personal computers are not particularly brand conscious. Most people decide which brand of computer to buy on the basis of price and features. However, brand-name manufacturers are attempting to win back customers by providing a number of services and high-quality products. To examine whether personal computer users perceive differences among manufacturers, a marketing research consultant undertook a survey of PC users. He asked a random sample of Apple, Compaq, Dell, IBM, and Packard-Bell owners several questions. The questions and possible responses are shown below.

Did you ever call the computer manufacturer to help with a problem?

The following questions were asked only of those who answered "yes."

1 How long approximately (in seconds) did it take to reach a technician?

2 How long approximately (in minutes) did it take to solve the problem?

3 Rate the quality of the technical support.

  1 = poor; 2 = fair; 3 = average; 4 = good; 5 = excellent

4 Would you recommend this brand of computer to a friend?

  2 = yes; 1 = no

The results are stored in file C19-04. Columns A through D store the responses to questions 1 to 4, respectively. Column E identifies the computer brand (1 = Apple; 2 = Compaq; 3 = Dell; 4 = IBM; 5 = Packard-Bell). For each question, can we infer that differences among the brands exist?

(The author is grateful to Alex Eliadis and David Jones for their help in this case. The case is based on an actual survey conducted by Willard Shullman for *PC Magazine*.)

Case 19.5 WLU Graduate Survey*

Every year, the graduates of Wilfrid Laurier University are surveyed to determine their employment status and annual income. The university offers undergraduate degrees in arts and music, business administration, and science, as well as several master's degrees. The survey asked a random sample of 1994 graduates the following questions:

1 With which degree did you graduate?
  1 arts and music
  2 business administration—nonaccounting
  3 business administration—accounting
  4 science
  5 master's

2 What is your current employment status?
  1 completing additional education
  2 employed
  3 other
  4 unemployed

3 If you are employed, what is your annual income?

The data are stored in columns A to G in file C19-05 in the following way:

Column A: degree

Column B: employment status

Columns C–G: income for those employed with degrees in arts and music, business (nonaccounting), business (accounting), science, and master's, respectively.

High school students who are about to choose a university program would like to have the following information.

*Source: "Wilfrid Laurier 1994 Graduate Survey Report" as printed in the *Atrium*, 8 November 1995.

Are there differences in the employment rates among the five groups of graduates? (Employment rate is defined as the percentage of graduates who are employed.)

Are there differences in income among the five groups of graduates?

Among business administration graduates, is there a difference in the employment rates and income between accounting and nonaccounting graduates?

Case 19.6 Evaluation of a New Antidepressant Drug

Clinical depression afflicts many people and costs businesses billions of dollars. Fortunately, there are several drugs that effectively treat this disease. Pharmaceutical companies are constantly experimenting to develop new and better drugs. Statistical techniques are used to determine whether and to what degree the new drug is better and whether there are certain people in whom the drug appears to be more effective. Suppose that a random sample of men and women who suffer from moderate depression are given a new antidepressant. Each person is asked to rate its effectiveness by filling out the form below. The form asks people to assess the drug after 7 and 14 days, because antidepressants usually take several days to take effect. Similar drugs often have different effects on men and women. Consequently, the gender of the patient is recorded. Also recorded is whether the patient suffered from the most common side effect, headaches.

To evaluate your experience with this medication, place a check mark to show how you have felt since you began taking it.

After Therapy (Day)	7	14
1 very much worse		
2 much worse		
3 a little worse		
4 no change		
5 a little better		
6 much better		
7 very much better		

The responses are stored in file C19-06 in the following way.

Columns A and B: evaluations for days 7 and 14 for women

Columns C and D: evaluations for days 7 and 14 for men

Column E: headaches among women (2 = yes; 1 = no)

Column F: headaches among men (2 = yes; 1 = no)

The company wants answers to the following questions.

a Among female patients, is there improvement between days 7 and 14?

b Among male patients, is there improvement between days 7 and 14?

c Does the frequency of headaches differ between men and women?

Case 19.7   Nutrition Education Programs*

Nutrition education programs, which teach their clients how to lose weight or reduce cholesterol levels through better eating patterns, have been growing in popularity. The nurse in charge of one such program at a local hospital wanted to know whether the programs actually work. A random sample of 33 clients who attended a nutrition education program for those with elevated cholesterol levels was drawn. The study recorded the weight, cholesterol levels, total dietary fat intake per average day, total dietary cholesterol intake per average day, and percent of daily calories from fat. These data were gathered both before and three months after the program. The researchers also determined the gender, age, and height of the clients.

The data are stored in file C19-07 in the following way.

Column A: gender (1 = female; 2 = male)

Column B: age

Column C: height (in meters)

Columns D and E: weight, before and after (in kilograms)

Columns F and G: cholesterol level, before and after

Columns H and I: total dietary fat intake per average day, before and after (in grams).

Columns J and K: dietary cholesterol intake per average day, before and after (in milligrams)

Columns L and M: percent daily calories from fat, before and after

The nurse would like the following information.

a In terms of each of weight, cholesterol level, fat intake, cholesterol intake, and calories from fat, is the program a success?

b Does gender affect the amount of reduction in each of weight, cholesterol level, fat intake, cholesterol intake, and calories from fat?

c Does age affect the amount of reduction in weight, cholesterol level, fat intake, cholesterol intake, and calories from fat?

*The author would like to thank Karen Cavrag for writing this case.

Case 19.8 Do Banks Discriminate against Women Business Owners? III

A statistics practitioner made a final effort to determine whether banks discriminate against women business owners because of their gender. For each of the women business owners who had received a loan, he attempted to find a male business owner whose characteristics closely matched. The matching was done on the basis of type of business (proprietorship, partnership, or corporation), gross sales, and age of company. A match was made when the type of business was the same, the gross sales were within $10,000 of each other, and the ages were within one year of each other. The interest rates (points above prime) for each pair are recorded and stored in columns B (women's rates) and C (men's rates)—column A stores the pair number—in file C19-08. What do these data tell you? Is this analysis definitive? Explain.

Appendix A

Sample Statistics from Data Files in Chapters 9 and 10

Chapter 9

9.36 $\bar{x} = 9.85$
9.37 $\bar{x} = 69.88$
9.38 $\bar{x} = 22.54$
9.39 $\bar{x} = 252.38$
9.40 $\bar{x} = 1810.16$
9.41 $\bar{x} = 12.10$
9.42 $\bar{x} = 10.21$
9.43 $\bar{x} = .5096$
9.44 $\bar{x} = 26.81$

Chapter 10

10.37 $\bar{x} = 5064.96$
10.38 $\bar{x} = 29{,}119.52$
10.39 $\bar{x} = 569.00$
10.40 $\bar{x} = 10.50$
10.41 $\bar{x} = 19.13$
10.42 $\bar{x} = 4.84$ $n = 63$
10.43 $\bar{x} = 20.06$
10.44 $\bar{x} = 5.64$
10.45 $\bar{x} = -1.20$

10.46 $\bar{x} = 55.80$
10.47 $\bar{x} = 5.04$
10.48 $\bar{x} = 19.39$
10.49 $\bar{x} = 105.70$
10.50 $\bar{x} = 17.55$
10.51 $\bar{x} = 29.92$ $n = 110$
10.52 $\bar{x} = 231.56$

Appendix. B

Tables

Table 1

Binomial Probabilities

Tabulated values are $P(X \leq k) = \sum_{x=0}^{k} p(x)$. (Values are rounded to three decimal places.)

$n = 5$

k								p							
	.01	.05	.10	.20	.25	.30	.40	.50	.60	.70	.75	.80	.90	.95	.99
0	.951	.774	.590	.328	.237	.168	.078	.031	.010	.002	.001	.000	.000	.000	.000
1	.999	.977	.919	.737	.633	.528	.337	.187	.087	.031	.016	.007	.000	.000	.000
2	1.000	.999	.991	.942	.896	.837	.683	.500	.317	.163	.104	.058	.009	.001	.000
3	1.000	1.000	1.000	.993	.984	.969	.913	.812	.663	.472	.367	.263	.081	.023	.001
4	1.000	1.000	1.000	1.000	.999	.998	.990	.969	.922	.832	.763	.672	.410	.226	.049

$n = 6$

k								p							
	.01	.05	.10	.20	.25	.30	.40	.50	.60	.70	.75	.80	.90	.95	.99
0	.941	.735	.531	.262	.178	.118	.047	.016	.004	.001	.000	.000	.000	.000	.000
1	.999	.967	.886	.655	.534	.420	.233	.109	.041	.011	.005	.002	.000	.000	.000
2	1.000	.998	.984	.901	.831	.744	.544	.344	.179	.070	.038	.017	.001	.000	.000
3	1.000	1.000	.999	.983	.962	.930	.821	.656	.456	.256	.169	.099	.016	.002	.000
4	1.000	1.000	1.000	.998	.995	.989	.959	.891	.767	.580	.466	.345	.114	.033	.001
5	1.000	1.000	1.000	1.000	1.000	.999	.996	.984	.953	.882	.822	.738	.469	.265	.059

$n = 7$

k								p							
	.01	.05	.10	.20	.25	.30	.40	.50	.60	.70	.75	.80	.90	.95	.99
0	.932	.698	.478	.210	.133	.082	.028	.008	.002	.000	.000	.000	.000	.000	.000
1	.998	.956	.850	.577	.445	.329	.159	.063	.019	.004	.001	.000	.000	.000	.000
2	1.000	.996	.974	.852	.756	.647	.420	.227	.096	.029	.013	.005	.000	.000	.000
3	1.000	1.000	.997	.967	.929	.874	.710	.500	.290	.126	.071	.033	.003	.000	.000
4	1.000	1.000	1.000	.995	.987	.971	.904	.773	.580	.353	.244	.148	.026	.004	.000
5	1.000	1.000	1.000	1.000	.999	.996	.981	.937	.841	.671	.555	.423	.150	.044	.002
6	1.000	1.000	1.000	1.000	1.000	1.000	.998	.992	.972	.918	.867	.790	.522	.302	.068

Table 1

continued

$n = 8$

k	.01	.05	.10	.20	.25	.30	.40	.50	.60	.70	.75	.80	.90	.95	.99
0	.923	.663	.430	.168	.100	.058	.017	.004	.001	.000	.000	.000	.000	.000	.000
1	.997	.943	.813	.503	.367	.255	.106	.035	.009	.001	.000	.000	.000	.000	.000
2	1.000	.994	.962	.797	.679	.552	.315	.145	.050	.011	.004	.001	.000	.000	.000
3	1.000	1.000	.995	.944	.886	.806	.594	.363	.174	.058	.027	.010	.000	.000	.000
4	1.000	1.000	1.000	.990	.973	.942	.826	.637	.406	.194	.114	.056	.005	.000	.000
5	1.000	1.000	1.000	.999	.996	.989	.950	.855	.685	.448	.321	.203	.038	.006	.000
6	1.000	1.000	1.000	1.000	1.000	.999	.991	.965	.894	.745	.633	.497	.187	.057	.003
7	1.000	1.000	1.000	1.000	1.000	1.000	.999	.996	.983	.942	.900	.832	.570	.337	.077

$n = 9$

k	.01	.05	.10	.20	.25	.30	.40	.50	.60	.70	.75	.80	.90	.95	.99
0	.914	.630	.387	.134	.075	.040	.010	.002	.000	.000	.000	.000	.000	.000	.000
1	.997	.929	.775	.436	.300	.196	.071	.020	.004	.000	.000	.000	.000	.000	.000
2	1.000	.992	.947	.738	.601	.463	.232	.090	.025	.004	.001	.000	.000	.000	.000
3	1.000	.999	.992	.914	.834	.730	.483	.254	.099	.025	.010	.003	.000	.000	.000
4	1.000	1.000	.999	.980	.951	.901	.733	.500	.267	.099	.049	.020	.001	.000	.000
5	1.000	1.000	1.000	.997	.990	.975	.901	.746	.517	.270	.166	.086	.008	.001	.000
6	1.000	1.000	1.000	1.000	.999	.996	.975	.910	.768	.537	.399	.262	.053	.008	.000
7	1.000	1.000	1.000	1.000	1.000	1.000	.996	.980	.929	.804	.700	.564	.225	.071	.003
8	1.000	1.000	1.000	1.000	1.000	1.000	1.000	.998	.990	.960	.925	.866	.613	.370	.086

Table 1

continued

n = 10

k	p														
	.01	.05	.10	.20	.25	.30	.40	.50	.60	.70	.75	.80	.90	.95	.99
0	.904	.599	.349	.107	.056	.028	.006	.001	.000	.000	.000	.000	.000	.000	.000
1	.996	.914	.736	.376	.244	.149	.046	.011	.002	.000	.000	.000	.000	.000	.000
2	1.000	.988	.930	.678	.526	.383	.167	.055	.012	.002	.000	.000	.000	.000	.000
3	1.000	.999	.987	.879	.776	.650	.382	.172	.055	.011	.004	.001	.000	.000	.000
4	1.000	1.000	.998	.967	.922	.850	.633	.377	.166	.047	.020	.006	.000	.000	.000
5	1.000	1.000	1.000	.994	.980	.953	.834	.623	.367	.150	.078	.033	.002	.000	.000
6	1.000	1.000	1.000	.999	.996	.989	.945	.828	.618	.350	.224	.121	.013	.001	.000
7	1.000	1.000	1.000	1.000	1.000	.998	.988	.945	.833	.617	.474	.322	.070	.012	.000
8	1.000	1.000	1.000	1.000	1.000	1.000	.998	.989	.954	.851	.756	.624	.264	.086	.004
9	1.000	1.000	1.000	1.000	1.000	1.000	1.000	.999	.994	.972	.944	.893	.651	.401	.096

n = 15

k	p														
	.01	.05	.10	.20	.25	.30	.40	.50	.60	.70	.75	.80	.90	.95	.99
0	.860	.463	.206	.035	.013	.005	.000	.000	.000	.000	.000	.000	.000	.000	.000
1	.990	.829	.549	.167	.080	.035	.005	.000	.000	.000	.000	.000	.000	.000	.000
2	1.000	.964	.816	.398	.236	.127	.027	.004	.000	.000	.000	.000	.000	.000	.000
3	1.000	.995	.944	.648	.461	.297	.091	.018	.002	.000	.000	.000	.000	.000	.000
4	1.000	.999	.987	.836	.686	.515	.217	.059	.009	.001	.000	.000	.000	.000	.000
5	1.000	1.000	.998	.939	.852	.722	.403	.151	.034	.004	.001	.000	.000	.000	.000
6	1.000	1.000	1.000	.982	.943	.869	.610	.304	.095	.015	.004	.001	.000	.000	.000
7	1.000	1.000	1.000	.996	.983	.950	.787	.500	.213	.050	.017	.004	.000	.000	.000
8	1.000	1.000	1.000	.999	.996	.985	.905	.696	.390	.131	.057	.018	.000	.000	.000
9	1.000	1.000	1.000	1.000	.999	.996	.966	.849	.597	.278	.148	.061	.002	.000	.000
10	1.000	1.000	1.000	1.000	1.000	.999	.991	.941	.783	.485	.314	.164	.013	.001	.000
11	1.000	1.000	1.000	1.000	1.000	1.000	.998	.982	.909	.703	.539	.352	.056	.005	.000
12	1.000	1.000	1.000	1.000	1.000	1.000	1.000	.996	.973	.873	.764	.602	.184	.036	.000
13	1.000	1.000	1.000	1.000	1.000	1.000	1.000	1.000	.995	.965	.920	.833	.451	.171	.010
14	1.000	1.000	1.000	1.000	1.000	1.000	1.000	1.000	1.000	.995	.987	.965	.794	.537	.140

Table 1

continued

$n = 20$

k	p=.01	.05	.10	.20	.25	.30	.40	.50	.60	.70	.75	.80	.90	.95	.99
0	.818	.358	.122	.012	.003	.001	.000	.000	.000	.000	.000	.000	.000	.000	.000
1	.983	.736	.392	.069	.024	.008	.001	.000	.000	.000	.000	.000	.000	.000	.000
2	.999	.925	.677	.206	.091	.035	.004	.000	.000	.000	.000	.000	.000	.000	.000
3	1.000	.984	.867	.411	.225	.107	.016	.001	.000	.000	.000	.000	.000	.000	.000
4	1.000	.997	.957	.630	.415	.238	.051	.006	.000	.000	.000	.000	.000	.000	.000
5	1.000	1.000	.989	.804	.617	.416	.126	.021	.002	.000	.000	.000	.000	.000	.000
6	1.000	1.000	.998	.913	.786	.608	.250	.058	.006	.000	.000	.000	.000	.000	.000
7	1.000	1.000	1.000	.968	.898	.772	.416	.132	.021	.001	.000	.000	.000	.000	.000
8	1.000	1.000	1.000	.990	.959	.887	.596	.252	.057	.005	.001	.000	.000	.000	.000
9	1.000	1.000	1.000	.997	.986	.952	.755	.412	.128	.017	.004	.001	.000	.000	.000
10	1.000	1.000	1.000	.999	.996	.983	.872	.588	.245	.048	.014	.003	.000	.000	.000
11	1.000	1.000	1.000	1.000	.999	.995	.943	.748	.404	.113	.041	.010	.000	.000	.000
12	1.000	1.000	1.000	1.000	1.000	.999	.979	.868	.584	.228	.102	.032	.000	.000	.000
13	1.000	1.000	1.000	1.000	1.000	1.000	.994	.942	.750	.392	.214	.087	.002	.000	.000
14	1.000	1.000	1.000	1.000	1.000	1.000	.998	.979	.874	.584	.383	.196	.011	.000	.000
15	1.000	1.000	1.000	1.000	1.000	1.000	1.000	.994	.949	.762	.585	.370	.043	.003	.000
16	1.000	1.000	1.000	1.000	1.000	1.000	1.000	.999	.984	.893	.775	.589	.133	.016	.000
17	1.000	1.000	1.000	1.000	1.000	1.000	1.000	1.000	.996	.965	.909	.794	.323	.075	.001
18	1.000	1.000	1.000	1.000	1.000	1.000	1.000	1.000	.999	.992	.976	.931	.608	.264	.017
19	1.000	1.000	1.000	1.000	1.000	1.000	1.000	1.000	1.000	.999	.997	.988	.878	.642	.182

Table 1

continued

$n = 25$

k	.01	.05	.10	.20	.25	.30	.40	.50	.60	.70	.75	.80	.90	.95	.99
0	.778	.277	.072	.004	.001	.000	.000	.000	.000	.000	.000	.000	.000	.000	.000
1	.974	.642	.271	.027	.007	.002	.000	.000	.000	.000	.000	.000	.000	.000	.000
2	.998	.873	.537	.098	.032	.009	.000	.000	.000	.000	.000	.000	.000	.000	.000
3	1.000	.966	.764	.234	.096	.033	.002	.000	.000	.000	.000	.000	.000	.000	.000
4	1.000	.993	.902	.421	.214	.090	.009	.000	.000	.000	.000	.000	.000	.000	.000
5	1.000	.999	.967	.617	.378	.193	.029	.002	.000	.000	.000	.000	.000	.000	.000
6	1.000	1.000	.991	.780	.561	.341	.074	.007	.000	.000	.000	.000	.000	.000	.000
7	1.000	1.000	.998	.891	.727	.512	.154	.022	.001	.000	.000	.000	.000	.000	.000
8	1.000	1.000	1.000	.953	.851	.677	.274	.054	.004	.000	.000	.000	.000	.000	.000
9	1.000	1.000	1.000	.983	.929	.811	.425	.115	.013	.000	.000	.000	.000	.000	.000
10	1.000	1.000	1.000	.994	.970	.902	.586	.212	.034	.002	.000	.000	.000	.000	.000
11	1.000	1.000	1.000	.998	.989	.956	.732	.345	.078	.006	.001	.000	.000	.000	.000
12	1.000	1.000	1.000	1.000	.997	.983	.846	.500	.154	.017	.003	.000	.000	.000	.000
13	1.000	1.000	1.000	1.000	.999	.994	.922	.655	.268	.044	.011	.002	.000	.000	.000
14	1.000	1.000	1.000	1.000	1.000	.998	.966	.788	.414	.098	.030	.006	.000	.000	.000
15	1.000	1.000	1.000	1.000	1.000	1.000	.987	.885	.575	.189	.071	.017	.000	.000	.000
16	1.000	1.000	1.000	1.000	1.000	1.000	.996	.946	.726	.323	.149	.047	.000	.000	.000
17	1.000	1.000	1.000	1.000	1.000	1.000	.999	.978	.846	.488	.273	.109	.002	.000	.000
18	1.000	1.000	1.000	1.000	1.000	1.000	1.000	.993	.926	.659	.439	.220	.009	.000	.000
19	1.000	1.000	1.000	1.000	1.000	1.000	1.000	.998	.971	.807	.622	.383	.033	.001	.000
20	1.000	1.000	1.000	1.000	1.000	1.000	1.000	1.000	.991	.910	.786	.579	.098	.007	.000
21	1.000	1.000	1.000	1.000	1.000	1.000	1.000	1.000	.998	.967	.904	.766	.236	.034	.000
22	1.000	1.000	1.000	1.000	1.000	1.000	1.000	1.000	1.000	.991	.968	.902	.463	.127	.002
23	1.000	1.000	1.000	1.000	1.000	1.000	1.000	1.000	1.000	.998	.993	.973	.729	.358	.026
24	1.000	1.000	1.000	1.000	1.000	1.000	1.000	1.000	1.000	1.000	.999	.996	.928	.723	.222

Table 2

Poisson Probabilities

Tabulated values are $P(X \leq k) = \sum_{x=0}^{k} p(x)$. (Values are rounded to three decimal places.)

k	.10	.20	.30	.40	.50	1.0	1.5	2.0	2.5	3.0	3.5	4.0	4.5	5.0	5.5	6.0
0	.905	.819	.741	.670	.607	.368	.223	.135	.082	.050	.030	.018	.011	.007	.004	.002
1	.995	.982	.963	.938	.910	.736	.558	.406	.287	.199	.136	.092	.061	.040	.027	.017
2	1.000	.999	.996	.992	.986	.920	.809	.677	.544	.423	.321	.238	.174	.125	.088	.062
3	1.000	1.000	1.000	.999	.998	.981	.934	.857	.758	.647	.537	.433	.342	.265	.202	.151
4	1.000	1.000	1.000	1.000	1.000	.996	.981	.947	.891	.815	.725	.629	.532	.440	.358	.285
5						.999	.996	.983	.958	.916	.858	.785	.703	.616	.529	.446
6						1.000	.999	.995	.986	.966	.935	.889	.831	.762	.686	.606
7							1.000	.999	.996	.988	.973	.949	.913	.867	.809	.744
8								1.000	.999	.996	.990	.979	.960	.932	.894	.847
9									1.000	.999	.997	.992	.983	.968	.946	.916
10										1.000	.999	.997	.993	.986	.975	.957
11											1.000	.999	.998	.995	.989	.980
12												1.000	.999	.998	.996	.991
13													1.000	.999	.998	.996
14														1.000	.999	.999
15															1.000	.999
16																1.000
17																
18																
19																
20																

Table 2

continued

k	6.5	7.0	7.5	8.0	8.5	9.0	μ 9.5	10	11	12	13	14	15
0	.002	.001	.001	.000	.000	.000	.000	.000	.000	.000	.000	.000	.000
1	.011	.007	.005	.003	.002	.001	.001	.000	.000	.000	.000	.000	.000
2	.043	.030	.020	.014	.009	.006	.004	.003	.001	.001	.000	.000	.000
3	.112	.082	.059	.042	.030	.021	.015	.010	.005	.002	.001	.000	.000
4	.224	.173	.132	.100	.074	.055	.040	.029	.015	.008	.004	.002	.001
5	.369	.301	.241	.191	.150	.116	.089	.067	.038	.020	.011	.006	.003
6	.527	.450	.378	.313	.256	.207	.165	.130	.079	.046	.026	.014	.008
7	.673	.599	.525	.453	.386	.324	.269	.220	.143	.090	.054	.032	.018
8	.792	.729	.662	.593	.523	.456	.392	.333	.232	.155	.100	.062	.037
9	.877	.830	.776	.717	.653	.587	.522	.458	.341	.242	.166	.109	.070
10	.933	.901	.862	.816	.763	.706	.645	.583	.460	.347	.252	.176	.118
11	.966	.947	.921	.888	.849	.803	.752	.697	.579	.462	.353	.260	.185
12	.984	.973	.957	.936	.909	.876	.836	.792	.689	.576	.463	.358	.268
13	.993	.987	.978	.966	.949	.926	.898	.864	.781	.682	.573	.464	.363
14	.997	.994	.990	.983	.973	.959	.940	.917	.854	.772	.675	.570	.466
15	.999	.998	.995	.992	.986	.978	.967	.951	.907	.844	.764	.669	.568
16	1.000	.999	.998	.996	.993	.989	.982	.973	.944	.899	.835	.756	.664
17		1.000	.999	.998	.997	.995	.991	.986	.968	.937	.890	.827	.749
18			1.000	.999	.999	.998	.996	.993	.982	.963	.930	.883	.819
19				1.000	.999	.999	.998	.997	.991	.979	.957	.923	.875
20					1.000	1.000	.999	.998	.995	.988	.975	.952	.917
21							1.000	.999	.998	.994	.986	.971	.947
22								1.000	.999	.997	.992	.983	.967
23									1.000	.999	.996	.991	.981
24										.999	.998	.995	.989
25										1.000	.999	.997	.994
26											1.000	.999	.997
27												.999	.998
28												1.000	.999
29													1.000

Table 3

Normal Curve Areas

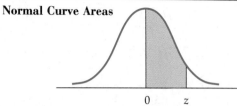

z	.00	.01	.02	.03	.04	.05	.06	.07	.08	.09
0.0	.0000	.0040	.0080	.0120	.0160	.0199	.0239	.0279	.0319	.0359
0.1	.0398	.0438	.0478	.0517	.0557	.0596	.0636	.0675	.0714	.0753
0.2	.0793	.0832	.0871	.0910	.0948	.0987	.1026	.1064	.1103	.1141
0.3	.1179	.1217	.1255	.1293	.1331	.1368	.1406	.1443	.1480	.1517
0.4	.1554	.1591	.1628	.1664	.1700	.1736	.1772	.1808	.1844	.1879
0.5	.1915	.1950	.1985	.2019	.2054	.2088	.2123	.2157	.2190	.2224
0.6	.2257	.2291	.2324	.2357	.2389	.2422	.2454	.2486	.2517	.2549
0.7	.2580	.2611	.2642	.2673	.2704	.2734	.2764	.2794	.2823	.2852
0.8	.2881	.2910	.2939	.2967	.2995	.3023	.3051	.3078	.3106	.3133
0.9	.3159	.3186	.3212	.3238	.3264	.3289	.3315	.3340	.3365	.3389
1.0	.3413	.3438	.3461	.3485	.3508	.3531	.3554	.3577	.3599	.3621
1.1	.3643	.3665	.3686	.3708	.3729	.3749	.3770	.3790	.3810	.3830
1.2	.3849	.3869	.3888	.3907	.3925	.3944	.3962	.3980	.3997	.4015
1.3	.4032	.4049	.4066	.4082	.4099	.4115	.4131	.4147	.4162	.4177
1.4	.4192	.4207	.4222	.4236	.4251	.4265	.4279	.4292	.4306	.4319
1.5	.4332	.4345	.4357	.4370	.4382	.4394	.4406	.4418	.4429	.4441
1.6	.4452	.4463	.4474	.4484	.4495	.4505	.4515	.4525	.4535	.4545
1.7	.4554	.4564	.4573	.4582	.4591	.4599	.4608	.4616	.4625	.4633
1.8	.4641	.4649	.4656	.4664	.4671	.4678	.4686	.4693	.4699	.4706
1.9	.4713	.4719	.4726	.4732	.4738	.4744	.4750	.4756	.4761	.4767
2.0	.4772	.4778	.4783	.4788	.4793	.4798	.4803	.4808	.4812	.4817
2.1	.4821	.4826	.4830	.4834	.4838	.4842	.4846	.4850	.4854	.4857
2.2	.4861	.4864	.4868	.4871	.4875	.4878	.4881	.4884	.4887	.4890
2.3	.4893	.4896	.4898	.4901	.4904	.4906	.4909	.4911	.4913	.4916
2.4	.4918	.4920	.4922	.4925	.4927	.4929	.4931	.4932	.4934	.4936
2.5	.4938	.4940	.4941	.4943	.4945	.4946	.4948	.4949	.4951	.4952
2.6	.4953	.4955	.4956	.4957	.4959	.4960	.4961	.4962	.4963	.4964
2.7	.4965	.4966	.4967	.4968	.4969	.4970	.4971	.4972	.4973	.4974
2.8	.4974	.4975	.4976	.4977	.4977	.4978	.4979	.4979	.4980	.4981
2.9	.4981	.4982	.4982	.4983	.4984	.4984	.4985	.4985	.4986	.4986
3.0	.4987	.4987	.4987	.4988	.4988	.4989	.4989	.4989	.4990	.4990

SOURCE: Abridged from Table 1 of A. Hald, *Statistical Tables and Formulas* (New York: Wiley & Sons, Inc.), 1952. Reproduced by permission of A. Hald and the publisher, John Wiley & Sons, Inc.

Table 4
Critical Values of t

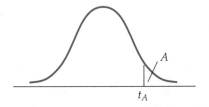

DEGREES OF FREEDOM	$t_{.100}$	$t_{.050}$	$t_{.025}$	$t_{.010}$	$t_{.005}$	DEGREES OF FREEDOM	$t_{.100}$	$t_{.050}$	$t_{.025}$	$t_{.010}$	$t_{.005}$
1	3.078	6.314	12.706	31.821	63.657	24	1.318	1.711	2.064	2.492	2.797
2	1.886	2.920	4.303	6.965	9.925	25	1.316	1.708	2.060	2.485	2.787
3	1.638	2.353	3.182	4.541	5.841	26	1.315	1.706	2.056	2.479	2.779
4	1.533	2.132	2.776	3.747	4.604	27	1.314	1.703	2.052	2.473	2.771
5	1.476	2.015	2.571	3.365	4.032	28	1.313	1.701	2.048	2.467	2.763
6	1.440	1.943	2.447	3.143	3.707	29	1.311	1.699	2.045	2.462	2.756
7	1.415	1.895	2.365	2.998	3.499	30	1.310	1.697	2.042	2.457	2.750
8	1.397	1.860	2.306	2.896	3.355	35	1.306	1.690	2.030	2.438	2.724
9	1.383	1.833	2.262	2.821	3.250	40	1.303	1.684	2.021	2.423	2.705
10	1.372	1.812	2.228	2.764	3.169	45	1.301	1.679	2.014	2.412	2.690
11	1.363	1.796	2.201	2.718	3.106	50	1.299	1.676	2.009	2.403	2.678
12	1.356	1.782	2.179	2.681	3.055	60	1.296	1.671	2.000	2.390	2.660
13	1.350	1.771	2.160	2.650	3.012	70	1.294	1.667	1.994	2.381	2.648
14	1.345	1.761	2.145	2.624	2.977	80	1.292	1.664	1.990	2.374	2.639
15	1.341	1.753	2.131	2.602	2.947	90	1.291	1.662	1.987	2.369	2.632
16	1.337	1.746	2.120	2.583	2.921	100	1.290	1.660	1.984	2.364	2.626
17	1.333	1.740	2.110	2.567	2.898	120	1.289	1.658	1.980	2.358	2.617
18	1.330	1.734	2.101	2.552	2.878	140	1.288	1.656	1.977	2.353	2.611
19	1.328	1.729	2.093	2.539	2.861	160	1.287	1.654	1.975	2.350	2.607
20	1.325	1.725	2.086	2.528	2.845	180	1.286	1.653	1.973	2.347	2.603
21	1.323	1.721	2.080	2.518	2.831	200	1.286	1.653	1.972	2.345	2.601
22	1.321	1.717	2.074	2.508	2.819	∞	1.282	1.645	1.960	2.326	2.576
23	1.319	1.714	2.069	2.500	2.807						

SOURCE: From M. Merrington, "Table of Percentage Points of the t-Distribution," *Biometrika* 32 (1941): 300. Reproduced by permission of the Biometrika Trustees.

Table 5

Critical Values of χ^2

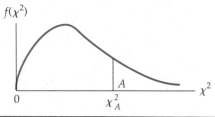

DEGREES OF FREEDOM	$\chi^2_{.995}$	$\chi^2_{.990}$	$\chi^2_{.975}$	$\chi^2_{.950}$	$\chi^2_{.900}$	$\chi^2_{.100}$	$\chi^2_{.050}$	$\chi^2_{.025}$	$\chi^2_{.010}$	$\chi^2_{.005}$
1	0.0000393	0.0001571	0.0009821	0.0039321	0.0157908	2.70554	3.84146	5.02389	6.63490	7.87944
2	0.0100251	0.0201007	0.0506356	0.102587	0.210720	4.60517	5.99147	7.37776	9.21034	10.5966
3	0.0717212	0.114832	0.215795	0.351846	0.584375	6.25139	7.81473	9.34840	11.3449	12.8381
4	0.206990	0.297110	0.484419	0.710721	1.063623	7.77944	9.48773	11.1433	13.2767	14.8602
5	0.411740	0.554300	0.831211	1.145476	1.61031	9.23635	11.0705	12.8325	15.0863	16.7496
6	0.675727	0.872085	1.237347	1.63539	2.20413	10.6446	12.5916	14.4494	16.8119	18.5476
7	0.989265	1.239043	1.68987	2.16735	2.83311	12.0170	14.0671	16.0128	18.4753	20.2777
8	1.344419	1.646482	2.17973	2.73264	3.48954	13.3616	15.5073	17.5346	20.0902	21.9550
9	1.734926	2.087912	2.70039	3.32511	4.16816	14.6837	16.9190	19.0228	21.6660	23.5893
10	2.15585	2.55821	3.24697	3.94030	4.86518	15.9871	18.3070	20.4831	23.2093	25.1882
11	2.60321	3.05347	3.81575	4.57481	5.57779	17.2750	19.6751	21.9200	24.7250	26.7569
12	3.07382	3.57056	4.40379	5.22603	6.30380	18.5494	21.0261	23.3367	26.2170	28.2995
13	3.56503	4.10691	5.00874	5.89186	7.04150	19.8119	22.3621	24.7356	27.6883	29.8194
14	4.07468	4.66043	5.62872	6.57063	7.78953	21.0642	23.6848	26.1190	29.1413	31.3193
15	4.60094	5.22935	6.26214	7.26094	8.54675	22.3072	24.9958	27.4884	30.5779	32.8013
16	5.14224	5.81221	6.90766	7.96164	9.31223	23.5418	26.2962	28.8454	31.9999	34.2672
17	5.69724	6.40776	7.56418	8.67176	10.0852	24.7690	27.5871	30.1910	33.4087	35.7185
18	6.26481	7.01491	8.23075	9.39046	10.8649	25.9894	28.8693	31.5264	34.8053	37.1564
19	6.84398	7.63273	8.90655	10.1170	11.6509	27.2036	30.1435	32.8523	36.1908	38.5822
20	7.43386	8.26040	9.59083	10.8508	12.4426	28.4120	31.4104	34.1696	37.5662	39.9968
21	8.03366	8.89720	10.28293	11.5913	13.2396	29.6151	32.6705	35.4789	38.9321	41.4010
22	8.64272	9.54249	10.9823	12.3380	14.0415	30.8133	33.9244	36.7807	40.2894	42.7956
23	9.26042	10.19567	11.6885	13.0905	14.8479	32.0069	35.1725	38.0757	41.6384	44.1813
24	9.88623	10.8564	12.4011	13.8484	15.6587	33.1963	36.4151	39.3641	42.9798	45.5585
25	10.5197	11.5240	13.1197	14.6114	16.4734	34.3816	37.6525	40.6465	44.3141	46.9278
26	11.1603	12.1981	13.8439	15.3791	17.2919	35.5631	38.8852	41.9232	45.6417	48.2899
27	11.8076	12.8786	14.5733	16.1513	18.1138	36.7412	40.1133	43.1944	46.9630	49.6449
28	12.4613	13.5648	15.3079	16.9279	18.9392	37.9159	41.3372	44.4607	48.2782	50.9933
29	13.1211	14.2565	16.0471	17.7083	19.7677	39.0875	42.5569	45.7222	49.5879	52.3356
30	13.7867	14.9535	16.7908	18.4926	20.5992	40.2560	43.7729	46.9792	50.8922	53.6720
40	20.7065	22.1643	24.4331	26.5093	29.0505	51.8050	55.7585	59.3417	63.6907	66.7659
50	27.9907	29.7067	32.3574	34.7642	37.6886	63.1671	67.5048	71.4202	76.1539	79.4900
60	35.5346	37.4848	40.4817	43.1879	46.4589	74.3970	79.0819	83.2976	88.3794	91.9517
70	43.2752	45.4418	48.7576	51.7393	55.3290	85.5271	90.5312	95.0231	100.425	104.215
80	51.1720	53.5400	57.1532	60.3915	64.2778	96.5782	101.879	106.629	112.329	116.321
90	59.1963	61.7541	65.6466	69.1260	73.2912	107.565	113.145	118.136	124.116	128.299
100	67.3276	70.0648	74.2219	77.9295	82.3581	118.498	124.342	129.561	135.807	140.169

SOURCE: From C. M. Thompson, "Tables of the Percentage Points of the χ^2-Distribution," *Biometrika* 32 (1941): 188–89. Reproduced by permission of the Biometrika Trustees.

Table 6(a)

Percentage Points of the F Distribution, A = .05

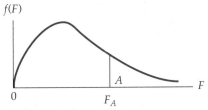

ν_2 \ ν_1	1	2	3	4	5	6	7	8	9
1	161.4	199.5	215.7	224.6	230.2	234.0	236.8	238.9	240.5
2	18.51	19.00	19.16	19.25	19.30	19.33	19.35	19.37	19.38
3	10.13	9.55	9.28	9.12	9.01	8.94	8.89	8.85	8.81
4	7.71	6.94	6.59	6.39	6.26	6.16	6.09	6.04	6.00
5	6.61	5.79	5.41	5.19	5.05	4.95	4.88	4.82	4.77
6	5.99	5.14	4.76	4.53	4.39	4.28	4.21	4.15	4.10
7	5.59	4.74	4.35	4.12	3.97	3.87	3.79	3.73	3.68
8	5.32	4.46	4.07	3.84	3.69	3.58	3.50	3.44	3.39
9	5.12	4.26	3.86	3.63	3.48	3.37	3.29	3.23	3.18
10	4.96	4.10	3.71	3.48	3.33	3.22	3.14	3.07	3.02
11	4.84	3.98	3.59	3.36	3.20	3.09	3.01	2.95	2.90
12	4.75	3.89	3.49	3.26	3.11	3.00	2.91	2.85	2.80
13	4.67	3.81	3.41	3.18	3.03	2.92	2.83	2.77	2.71
14	4.60	3.74	3.34	3.11	2.96	2.85	2.76	2.70	2.65
15	4.54	3.68	3.29	3.06	2.90	2.79	2.71	2.64	2.59
16	4.49	3.63	3.24	3.01	2.85	2.74	2.66	2.59	2.54
17	4.45	3.59	3.20	2.96	2.81	2.70	2.61	2.55	2.49
18	4.41	3.55	3.16	2.93	2.77	2.66	2.58	2.51	2.46
19	4.38	3.52	3.13	2.90	2.74	2.63	2.54	2.48	2.42
20	4.35	3.49	3.10	2.87	2.71	2.60	2.51	2.45	2.39
21	4.32	3.47	3.07	2.84	2.68	2.57	2.49	2.42	2.37
22	4.30	3.44	3.05	2.82	2.66	2.55	2.46	2.40	2.34
23	4.28	3.42	3.03	2.80	2.64	2.53	2.44	2.37	2.32
24	4.26	3.40	3.01	2.78	2.62	2.51	2.42	2.36	2.30
25	4.24	3.39	2.99	2.76	2.60	2.49	2.40	2.34	2.28
26	4.23	3.37	2.98	2.74	2.59	2.47	2.39	2.32	2.27
27	4.21	3.35	2.96	2.73	2.57	2.46	2.37	2.31	2.25
28	4.20	3.34	2.95	2.71	2.56	2.45	2.36	2.29	2.24
29	4.18	3.33	2.93	2.70	2.55	2.43	2.35	2.28	2.22
30	4.17	3.32	2.92	2.69	2.53	2.42	2.33	2.27	2.21
40	4.08	3.23	2.84	2.61	2.45	2.34	2.25	2.18	2.12
60	4.00	3.15	2.76	2.53	2.37	2.25	2.17	2.10	2.04
120	3.92	3.07	2.68	2.45	2.29	2.17	2.09	2.02	1.96
∞	3.84	3.00	2.60	2.37	2.21	2.10	2.01	1.94	1.88

SOURCE: From M. Merrington and C. M. Thompson, "Tables of Percentage Points of the Inverted Beta (F)-Distribution," *Biometrika* 33 (1943): 73–88. Reproduced by permission of the Biometrika Trustees.

Table 6(a)

continued

ν_2 \ ν_1	10	12	15	20	24	30	40	60	120	∞
1	241.9	243.9	245.9	248.0	249.1	250.1	251.1	252.2	253.3	254.3
2	19.40	19.41	19.43	19.45	19.45	19.46	19.47	19.48	19.49	19.50
3	8.79	8.74	8.70	8.66	8.64	8.62	8.59	8.57	8.55	8.53
4	5.96	5.91	5.86	5.80	5.77	5.75	5.72	5.69	5.66	5.63
5	4.74	4.68	4.62	4.56	4.53	4.50	4.46	4.43	4.40	4.36
6	4.06	4.00	3.94	3.87	3.84	3.81	3.77	3.74	3.70	3.67
7	3.64	3.57	3.51	3.44	3.41	3.38	3.34	3.30	3.27	3.23
8	3.35	3.28	3.22	3.15	3.12	3.08	3.04	3.01	2.97	2.93
9	3.14	3.07	3.01	2.94	2.90	2.86	2.83	2.79	2.75	2.71
10	2.98	2.91	2.85	2.77	2.74	2.70	2.66	2.62	2.58	2.54
11	2.85	2.79	2.72	2.65	2.61	2.57	2.53	2.49	2.45	2.40
12	2.75	2.69	2.62	2.54	2.51	2.47	2.43	2.38	2.34	2.30
13	2.67	2.60	2.53	2.46	2.42	2.38	2.34	2.30	2.25	2.21
14	2.60	2.53	2.46	2.39	2.35	2.31	2.27	2.22	2.18	2.13
15	2.54	2.48	2.40	2.33	2.29	2.25	2.20	2.16	2.11	2.07
16	2.49	2.42	2.35	2.28	2.24	2.19	2.15	2.11	2.06	2.01
17	2.45	2.38	2.31	2.23	2.19	2.15	2.10	2.06	2.01	1.96
18	2.41	2.34	2.27	2.19	2.15	2.11	2.06	2.02	1.97	1.92
19	2.38	2.31	2.23	2.16	2.11	2.07	2.03	1.98	1.93	1.88
20	2.35	2.28	2.20	2.12	2.08	2.04	1.99	1.95	1.90	1.84
21	2.32	2.25	2.18	2.10	2.05	2.01	1.96	1.92	1.87	1.81
22	2.30	2.23	2.15	2.07	2.03	1.98	1.94	1.89	1.84	1.78
23	2.27	2.20	2.13	2.05	2.01	1.96	1.91	1.86	1.81	1.76
24	2.25	2.18	2.11	2.03	1.98	1.94	1.89	1.84	1.79	1.73
25	2.24	2.16	2.09	2.01	1.96	1.92	1.87	1.82	1.77	1.71
26	2.22	2.15	2.07	1.99	1.95	1.90	1.85	1.80	1.75	1.69
27	2.20	2.13	2.06	1.97	1.93	1.88	1.84	1.79	1.73	1.67
28	2.19	2.12	2.04	1.96	1.91	1.87	1.82	1.77	1.71	1.65
29	2.18	2.10	2.03	1.94	1.90	1.85	1.81	1.75	1.70	1.64
30	2.16	2.09	2.01	1.93	1.89	1.84	1.79	1.74	1.68	1.62
40	2.08	2.00	1.92	1.84	1.79	1.74	1.69	1.64	1.58	1.51
60	1.99	1.92	1.84	1.75	1.70	1.65	1.59	1.53	1.47	1.39
120	1.91	1.83	1.75	1.66	1.61	1.55	1.50	1.43	1.35	1.25
∞	1.83	1.75	1.67	1.57	1.52	1.46	1.39	1.32	1.22	1.00

Numerator degrees of freedom: column headers (ν_1). Denominator degrees of freedom: row labels (ν_2).

Table 6(b)

Percentage Points of the F Distribution, $A = .025$

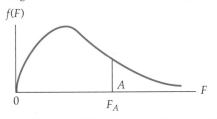

v_2 \ v_1	1	2	3	4	5	6	7	8	9
1	647.8	799.5	864.2	899.6	921.8	937.1	948.2	956.7	963.3
2	38.51	39.00	39.17	39.25	39.30	39.33	39.36	39.37	39.39
3	17.44	16.04	15.44	15.10	14.88	14.73	14.62	14.54	14.47
4	12.22	10.65	9.98	9.60	9.36	9.20	9.07	8.98	8.90
5	10.01	8.43	7.76	7.39	7.15	6.98	6.85	6.76	6.68
6	8.81	7.26	6.60	6.23	5.99	5.82	5.70	5.60	5.52
7	8.07	6.54	5.89	5.52	5.29	5.12	4.99	4.90	4.82
8	7.57	6.06	5.42	5.05	4.82	4.65	4.53	4.43	4.36
9	7.21	5.71	5.08	4.72	4.48	4.32	4.20	4.10	4.03
10	6.94	5.46	4.83	4.47	4.24	4.07	3.95	3.85	3.78
11	6.72	5.26	4.63	4.28	4.04	3.88	3.76	3.66	3.59
12	6.55	5.10	4.47	4.12	3.89	3.73	3.61	3.51	3.44
13	6.41	4.97	4.35	4.00	3.77	3.60	3.48	3.39	3.31
14	6.30	4.86	4.24	3.89	3.66	3.50	3.38	3.29	3.21
15	6.20	4.77	4.15	3.80	3.58	3.41	3.29	3.20	3.12
16	6.12	4.69	4.08	3.73	3.50	3.34	3.22	3.12	3.05
17	6.04	4.62	4.01	3.66	3.44	3.28	3.16	3.06	2.98
18	5.98	4.56	3.95	3.61	3.38	3.22	3.10	3.01	2.93
19	5.92	4.51	3.90	3.56	3.33	3.17	3.05	2.96	2.88
20	5.87	4.46	3.86	3.51	3.29	3.13	3.01	2.91	2.84
21	5.83	4.42	3.82	3.48	3.25	3.09	2.97	2.87	2.80
22	5.79	4.38	3.78	3.44	3.22	3.05	2.93	2.84	2.76
23	5.75	4.35	3.75	3.41	3.18	3.02	2.90	2.81	2.73
24	5.72	4.32	3.72	3.38	3.15	2.99	2.87	2.78	2.70
25	5.69	4.29	3.69	3.35	3.13	2.97	2.85	2.75	2.68
26	5.66	4.27	3.67	3.33	3.10	2.94	2.82	2.73	2.65
27	5.63	4.24	3.65	3.31	3.08	2.92	2.80	2.71	2.63
28	5.61	4.22	3.63	3.29	3.06	2.90	2.78	2.69	2.61
29	5.59	4.20	3.61	3.27	3.04	2.88	2.76	2.67	2.59
30	5.57	4.18	3.59	3.25	3.03	2.87	2.75	2.65	2.57
40	5.42	4.05	3.46	3.13	2.90	2.74	2.62	2.53	2.45
60	5.29	3.93	3.34	3.01	2.79	2.63	2.51	2.41	2.33
120	5.15	3.80	3.23	2.89	2.67	2.52	2.39	2.30	2.22
∞	5.02	3.69	3.12	2.79	2.57	2.41	2.29	2.19	2.11

Numerator degrees of freedom across top; Denominator degrees of freedom down left.

SOURCE: From M. Merrington and C. M. Thompson, "Tables of Percentage Points of the Inverted Beta (F)-Distribution," *Biometrika* 33 (1943): 73–88. Reproduced by permission of the Biometrika Trustees.

Table 6(b)

continued

ν_2 \ ν_1	10	12	15	20	24	30	40	60	120	∞
1	968.6	976.7	984.9	993.1	997.2	1,001	1,006	1,010	1,014	1,018
2	39.40	39.41	39.43	39.45	39.46	39.46	39.47	39.48	39.49	39.50
3	14.42	14.34	14.25	14.17	14.12	14.08	14.04	13.99	13.95	13.90
4	8.84	8.75	8.66	8.56	8.51	8.46	8.41	8.36	8.31	8.26
5	6.62	6.52	6.43	6.33	6.28	6.23	6.18	6.12	6.07	6.02
6	5.46	5.37	5.27	5.17	5.12	5.07	5.01	4.96	4.90	4.85
7	4.76	4.67	4.57	4.47	4.42	4.36	4.31	4.25	4.20	4.14
8	4.30	4.20	4.10	4.00	3.95	3.89	3.84	3.78	3.73	3.67
9	3.96	3.87	3.77	3.67	3.61	3.56	3.51	3.45	3.39	3.33
10	3.72	3.62	3.52	3.42	3.37	3.31	3.26	3.20	3.14	3.08
11	3.53	3.43	3.33	3.23	3.17	3.12	3.06	3.00	2.94	2.88
12	3.37	3.28	3.18	3.07	3.02	2.96	2.91	2.85	2.79	2.72
13	3.25	3.15	3.05	2.95	2.89	2.84	2.78	2.72	2.66	2.60
14	3.15	3.05	2.95	2.84	2.79	2.73	2.67	2.61	2.55	2.49
15	3.06	2.96	2.86	2.76	2.70	2.64	2.59	2.52	2.46	2.40
16	2.99	2.89	2.79	2.68	2.63	2.57	2.51	2.45	2.38	2.32
17	2.92	2.82	2.72	2.62	2.56	2.50	2.44	2.38	2.32	2.25
18	2.87	2.77	2.67	2.56	2.50	2.44	2.38	2.32	2.26	2.19
19	2.82	2.72	2.62	2.51	2.45	2.39	2.33	2.27	2.20	2.13
20	2.77	2.68	2.57	2.46	2.41	2.35	2.29	2.22	2.16	2.09
21	2.73	2.64	2.53	2.42	2.37	2.31	2.25	2.18	2.11	2.04
22	2.70	2.60	2.50	2.39	2.33	2.27	2.21	2.14	2.08	2.00
23	2.67	2.57	2.47	2.36	2.30	2.24	2.18	2.11	2.04	1.97
24	2.64	2.54	2.44	2.33	2.27	2.21	2.15	2.08	2.01	1.94
25	2.61	2.51	2.41	2.30	2.24	2.18	2.12	2.05	1.98	1.91
26	2.59	2.49	2.39	2.28	2.22	2.16	2.09	2.03	1.95	1.88
27	2.57	2.47	2.36	2.25	2.19	2.13	2.07	2.00	1.93	1.85
28	2.55	2.45	2.34	2.23	2.17	2.11	2.05	1.98	1.91	1.83
29	2.53	2.43	2.32	2.21	2.15	2.09	2.03	1.96	1.89	1.81
30	2.51	2.41	2.31	2.20	2.14	2.07	2.01	1.94	1.87	1.79
40	2.39	2.29	2.18	2.07	2.01	1.94	1.88	1.80	1.72	1.64
60	2.27	2.17	2.06	1.94	1.88	1.82	1.74	1.67	1.58	1.48
120	2.16	2.05	1.94	1.82	1.76	1.69	1.61	1.53	1.43	1.31
∞	2.05	1.94	1.83	1.71	1.64	1.57	1.48	1.39	1.27	1.00

NUMERATOR DEGREES OF FREEDOM (columns); DENOMINATOR DEGREES OF FREEDOM (rows)

Table 6(c)

Percentage Points of the F Distribution, $A = .01$

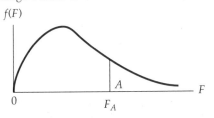

ν_2 \ ν_1	1	2	3	4	5	6	7	8	9
1	4,052	4,999.5	5,403	5,625	5,764	5,859	5,928	5,982	6,022
2	98.50	99.00	99.17	99.25	99.30	99.33	99.36	99.37	99.39
3	34.12	30.82	29.46	28.71	28.24	27.91	27.67	27.49	27.35
4	21.20	18.00	16.69	15.98	15.52	15.21	14.98	14.80	14.66
5	16.26	13.27	12.06	11.39	10.97	10.67	10.46	10.29	10.16
6	13.75	10.92	9.78	9.15	8.75	8.47	8.26	8.10	7.98
7	12.25	9.55	8.45	7.85	7.46	7.19	6.99	6.84	6.72
8	11.26	8.65	7.59	7.01	6.63	6.37	6.18	6.03	5.91
9	10.56	8.02	6.99	6.42	6.06	5.80	5.61	5.47	5.35
10	10.04	7.56	6.55	5.99	5.64	5.39	5.20	5.06	4.94
11	9.65	7.21	6.22	5.67	5.32	5.07	4.89	4.74	4.63
12	9.33	6.93	5.95	5.41	5.06	4.82	4.64	4.50	4.39
13	9.07	6.70	5.74	5.21	4.86	4.62	4.44	4.30	4.19
14	8.86	6.51	5.56	5.04	4.69	4.46	4.28	4.14	4.03
15	8.68	6.36	5.42	4.89	4.56	4.32	4.14	4.00	3.89
16	8.53	6.23	5.29	4.77	4.44	4.20	4.03	3.89	3.78
17	8.40	6.11	5.18	4.67	4.34	4.10	3.93	3.79	3.68
18	8.29	6.01	5.09	4.58	4.25	4.01	3.84	3.71	3.60
19	8.18	5.93	5.01	4.50	4.17	3.94	3.77	3.63	3.52
20	8.10	5.85	4.94	4.43	4.10	3.87	3.70	3.56	3.46
21	8.02	5.78	4.87	4.37	4.04	3.81	3.64	3.51	3.40
22	7.95	5.72	4.82	4.31	3.99	3.76	3.59	3.45	3.35
23	7.88	5.66	4.76	4.26	3.94	3.71	3.54	3.41	3.30
24	7.82	5.61	4.72	4.22	3.90	3.67	3.50	3.36	3.26
25	7.77	5.57	4.68	4.18	3.85	3.63	3.46	3.32	3.22
26	7.72	5.53	4.64	4.14	3.82	3.59	3.42	3.29	3.18
27	7.68	5.49	4.60	4.11	3.78	3.56	3.39	3.26	3.15
28	7.64	5.45	4.57	4.07	3.75	3.53	3.36	3.23	3.12
29	7.60	5.42	4.54	4.04	3.73	3.50	3.33	3.20	3.09
30	7.56	5.39	4.51	4.02	3.70	3.47	3.30	3.17	3.07
40	7.31	5.18	4.31	3.83	3.51	3.29	3.12	2.99	2.89
60	7.08	4.98	4.13	3.65	3.34	3.12	2.95	2.82	2.72
120	6.85	4.79	3.95	3.48	3.17	2.96	2.79	2.66	2.56
∞	6.63	4.61	3.78	3.32	3.02	2.80	2.64	2.51	2.41

Denominator degrees of freedom (ν_2); Numerator degrees of freedom (ν_1).

SOURCE: From M. Merrington and C. M. Thompson, "Tables of Percentage Points of the Inverted Beta (F)-Distribution," *Biometrika* 33 (1943): 73–88. Reproduced by permission of the Biometrika Trustees.

Table 6(c)

continued

ν_2 \ ν_1	10	12	15	20	24	30	40	60	120	∞
				NUMERATOR DEGREES OF FREEDOM						
1	6,056	6,106	6,157	6,209	6,235	6,261	6,287	6,313	6,339	6,366
2	99.40	99.42	99.43	99.45	99.46	99.47	99.47	99.48	99.49	99.50
3	27.23	27.05	26.87	26.69	26.60	26.50	26.41	26.32	26.22	26.13
4	14.55	14.37	14.20	14.02	13.93	13.84	13.75	13.65	13.56	13.46
5	10.05	9.89	9.72	9.55	9.47	9.38	9.29	9.20	9.11	9.02
6	7.87	7.72	7.56	7.40	7.31	7.23	7.14	7.06	6.97	6.88
7	6.62	6.47	6.31	6.16	6.07	5.99	5.91	5.82	5.74	5.65
8	5.81	5.67	5.52	5.36	5.28	5.20	5.12	5.03	4.95	4.86
9	5.26	5.11	4.96	4.81	4.73	4.65	4.57	4.48	4.40	4.31
10	4.85	4.71	4.56	4.41	4.33	4.25	4.17	4.08	4.00	3.91
11	4.54	4.40	4.25	4.10	4.02	3.94	3.86	3.78	3.69	3.60
12	4.30	4.16	4.01	3.86	3.78	3.70	3.62	3.54	3.45	3.36
13	4.10	3.96	3.82	3.66	3.59	3.51	3.43	3.34	3.25	3.17
14	3.94	3.80	3.66	3.51	3.43	3.35	3.27	3.18	3.09	3.00
15	3.80	3.67	3.52	3.37	3.29	3.21	3.13	3.05	2.96	2.87
16	3.69	3.55	3.41	3.26	3.18	3.10	3.02	2.93	2.84	2.75
17	3.59	3.46	3.31	3.16	3.08	3.00	2.92	2.83	2.75	2.65
18	3.51	3.37	3.23	3.08	3.00	2.92	2.84	2.75	2.66	2.57
19	3.43	3.30	3.15	3.00	2.92	2.84	2.76	2.67	2.58	2.49
20	3.37	3.23	3.09	2.94	2.86	2.78	2.69	2.61	2.52	2.42
21	3.31	3.17	3.03	2.88	2.80	2.72	2.64	2.55	2.46	2.36
22	3.26	3.12	2.98	2.83	2.75	2.67	2.58	2.50	2.40	2.31
23	3.21	3.07	2.93	2.78	2.70	2.62	2.54	2.45	2.35	2.26
24	3.17	3.03	2.89	2.74	2.66	2.58	2.49	2.40	2.31	2.21
25	3.13	2.99	2.85	2.70	2.62	2.54	2.45	2.36	2.27	2.17
26	3.09	2.96	2.81	2.66	2.58	2.50	2.42	2.33	2.23	2.13
27	3.06	2.93	2.78	2.63	2.55	2.47	2.38	2.29	2.20	2.10
28	3.03	2.90	2.75	2.60	2.52	2.44	2.35	2.26	2.17	2.06
29	3.00	2.87	2.73	2.57	2.49	2.41	2.33	2.23	2.14	2.03
30	2.98	2.84	2.70	2.55	2.47	2.39	2.30	2.21	2.11	2.01
40	2.80	2.66	2.52	2.37	2.29	2.20	2.11	2.02	1.92	1.80
60	2.63	2.50	2.35	2.20	2.12	2.03	1.94	1.84	1.73	1.60
120	2.47	2.34	2.19	2.03	1.95	1.86	1.76	1.66	1.53	1.38
∞	2.32	2.18	2.04	1.88	1.79	1.70	1.59	1.47	1.32	1.00

DENOMINATOR DEGREES OF FREEDOM

Table 7(a)

Critical Values for the Durbin–Watson d Statistic, $\alpha = .05$

	$k = 1$		$k = 2$		$k = 3$		$k = 4$		$k = 5$	
n	d_L	d_U	d_L	d_U	d_L	d_U	d_L	d_U	d_L	d_U
15	1.08	1.36	.95	1.54	.82	1.75	.69	1.97	.56	2.21
16	1.10	1.37	.98	1.54	.86	1.73	.74	1.93	.62	2.15
17	1.13	1.38	1.02	1.54	.90	1.71	.78	1.90	.67	2.10
18	1.16	1.39	1.05	1.53	.93	1.69	.82	1.87	.71	2.06
19	1.18	1.40	1.08	1.53	.97	1.68	.86	1.85	.75	2.02
20	1.20	1.41	1.10	1.54	1.00	1.68	.90	1.83	.79	1.99
21	1.22	1.42	1.13	1.54	1.03	1.67	.93	1.81	.83	1.96
22	1.24	1.43	1.15	1.54	1.05	1.66	.96	1.80	.86	1.94
23	1.26	1.44	1.17	1.54	1.08	1.66	.99	1.79	.90	1.92
24	1.27	1.45	1.19	1.55	1.10	1.66	1.01	1.78	.93	1.90
25	1.29	1.45	1.21	1.55	1.12	1.66	1.04	1.77	.95	1.89
26	1.30	1.46	1.22	1.55	1.14	1.65	1.06	1.76	.98	1.88
27	1.32	1.47	1.24	1.56	1.16	1.65	1.08	1.76	1.01	1.86
28	1.33	1.48	1.26	1.56	1.18	1.65	1.10	1.75	1.03	1.85
29	1.34	1.48	1.27	1.56	1.20	1.65	1.12	1.74	1.05	1.84
30	1.35	1.49	1.28	1.57	1.21	1.65	1.14	1.74	1.07	1.83
31	1.36	1.50	1.30	1.57	1.23	1.65	1.16	1.74	1.09	1.83
32	1.37	1.50	1.31	1.57	1.24	1.65	1.18	1.73	1.11	1.82
33	1.38	1.51	1.32	1.58	1.26	1.65	1.19	1.73	1.13	1.81
34	1.39	1.51	1.33	1.58	1.27	1.65	1.21	1.73	1.15	1.81
35	1.40	1.52	1.34	1.58	1.28	1.65	1.22	1.73	1.16	1.80
36	1.41	1.52	1.35	1.59	1.29	1.65	1.24	1.73	1.18	1.80
37	1.42	1.53	1.36	1.59	1.31	1.66	1.25	1.72	1.19	1.80
38	1.43	1.54	1.37	1.59	1.32	1.66	1.26	1.72	1.21	1.79
39	1.43	1.54	1.38	1.60	1.33	1.66	1.27	1.72	1.22	1.79
40	1.44	1.54	1.39	1.60	1.34	1.66	1.29	1.72	1.23	1.79
45	1.48	1.57	1.43	1.62	1.38	1.67	1.34	1.72	1.29	1.78
50	1.50	1.59	1.46	1.63	1.42	1.67	1.38	1.72	1.34	1.77
55	1.53	1.60	1.49	1.64	1.45	1.68	1.41	1.72	1.38	1.77
60	1.55	1.62	1.51	1.65	1.48	1.69	1.44	1.73	1.41	1.77
65	1.57	1.63	1.54	1.66	1.50	1.70	1.47	1.73	1.44	1.77
70	1.58	1.64	1.55	1.67	1.52	1.70	1.49	1.74	1.46	1.77
75	1.60	1.65	1.57	1.68	1.54	1.71	1.51	1.74	1.49	1.77
80	1.61	1.66	1.59	1.69	1.56	1.72	1.53	1.74	1.51	1.77
85	1.62	1.67	1.60	1.70	1.57	1.72	1.55	1.75	1.52	1.77
90	1.63	1.68	1.61	1.70	1.59	1.73	1.57	1.75	1.54	1.78
95	1.64	1.69	1.62	1.71	1.60	1.73	1.58	1.75	1.56	1.78
100	1.65	1.69	1.63	1.72	1.61	1.74	1.59	1.76	1.57	1.78

SOURCE: From J. Durbin and G. S. Watson, "Testing for Serial Correlation in Least Squares Regression, II," *Biometrika* 30 (1951): 159–78. Reproduced by permission of the Biometrika Trustees.

Table 7(b)

Critical Values for the Durbin–Watson d Statistic, $\alpha = .01$

	$k = 1$		$k = 2$		$k = 3$		$k = 4$		$k = 5$	
n	d_L	d_U	d_L	d_U	d_L	d_U	d_L	d_U	d_L	d_U
15	.81	1.07	.70	1.25	.59	1.46	.49	1.70	.39	1.96
16	.84	1.09	.74	1.25	.63	1.44	.53	1.66	.44	1.90
17	.87	1.10	.77	1.25	.67	1.43	.57	1.63	.48	1.85
18	.90	1.12	.80	1.26	.71	1.42	.61	1.60	.52	1.80
19	.93	1.13	.83	1.26	.74	1.41	.65	1.58	.56	1.77
20	.95	1.15	.86	1.27	.77	1.41	.68	1.57	.60	1.74
21	.97	1.16	.89	1.27	.80	1.41	.72	1.55	.63	1.71
22	1.00	1.17	.91	1.28	.83	1.40	.75	1.54	.66	1.69
23	1.02	1.19	.94	1.29	.86	1.40	.77	1.53	.70	1.67
24	1.04	1.20	.96	1.30	.88	1.41	.80	1.53	.72	1.66
25	1.05	1.21	.98	1.30	.90	1.41	.83	1.52	.75	1.65
26	1.07	1.22	1.00	1.31	.93	1.41	.85	1.52	.78	1.64
27	1.09	1.23	1.02	1.32	.95	1.41	.88	1.51	.81	1.63
28	1.10	1.24	1.04	1.32	.97	1.41	.90	1.51	.83	1.62
29	1.12	1.25	1.05	1.33	.99	1.42	.92	1.51	.85	1.61
30	1.13	1.26	1.07	1.34	1.01	1.42	.94	1.51	.88	1.61
31	1.15	1.27	1.08	1.34	1.02	1.42	.96	1.51	.90	1.60
32	1.16	1.28	1.10	1.35	1.04	1.43	.98	1.51	.92	1.60
33	1.17	1.29	1.11	1.36	1.05	1.43	1.00	1.51	.94	1.59
34	1.18	1.30	1.13	1.36	1.07	1.43	1.01	1.51	.95	1.59
35	1.19	1.31	1.14	1.37	1.08	1.44	1.03	1.51	.97	1.59
36	1.21	1.32	1.15	1.38	1.10	1.44	1.04	1.51	.99	1.59
37	1.22	1.32	1.16	1.38	1.11	1.45	1.06	1.51	1.00	1.59
38	1.23	1.33	1.18	1.39	1.12	1.45	1.07	1.52	1.02	1.58
39	1.24	1.34	1.19	1.39	1.14	1.45	1.09	1.52	1.03	1.58
40	1.25	1.34	1.20	1.40	1.15	1.46	1.10	1.52	1.05	1.58
45	1.29	1.38	1.24	1.42	1.20	1.48	1.16	1.53	1.11	1.58
50	1.32	1.40	1.28	1.45	1.24	1.49	1.20	1.54	1.16	1.59
55	1.36	1.43	1.32	1.47	1.28	1.51	1.25	1.55	1.21	1.59
60	1.38	1.45	1.35	1.48	1.32	1.52	1.28	1.56	1.25	1.60
65	1.41	1.47	1.38	1.50	1.35	1.53	1.31	1.57	1.28	1.61
70	1.43	1.49	1.40	1.52	1.37	1.55	1.34	1.58	1.31	1.61
75	1.45	1.50	1.42	1.53	1.39	1.56	1.37	1.59	1.34	1.62
80	1.47	1.52	1.44	1.54	1.42	1.57	1.39	1.60	1.36	1.62
85	1.48	1.53	1.46	1.55	1.43	1.58	1.41	1.60	1.39	1.63
90	1.50	1.54	1.47	1.56	1.45	1.59	1.43	1.61	1.41	1.64
95	1.51	1.55	1.49	1.57	1.47	1.60	1.45	1.62	1.42	1.64
100	1.52	1.56	1.50	1.58	1.48	1.60	1.46	1.63	1.44	1.65

SOURCE: From J. Durbin and G. S. Watson, "Testing for Serial Correlation in Least Squares Regression, II," *Biometrika* 30 (1951): 159–78. Reproduced by permission of the Biometrika Trustees.

Appendix C

Answers to Selected Even-Numbered Exercises

All answers have been double-checked for accuracy. However, we cannot be absolutely certain that there are no errors. Students should not automatically assume that answers that don't match ours are wrong. When and if we discover mistakes we will post corrected answers on our Web page. (See page 8 for the address.) If you find any errors, please e-mail the author (address on Web page). We will be happy to acknowledge you with the discovery.

Chapter 1

1.4 a The population is the set of all the manufacturer's chips. **b** The sample is the set of 1,000 chips. **c** The parameter is the proportion of defectives in all of the manufacturer's chips. **d** The statistic is the 7.5% defective rate in the sample. **e** Parameter **f** Statistic **g** We estimate that the proportion of defectives in all the manufacturer's chips is 7.5%. Because this value is less than 10%, we reject the claim.
1.6 a Flip the coin (say) 100 times and record the numbers of heads and tails. **b** The population is the set of heads and tails, assuming that the coin is flipped an infinite number of times. **c** The sample is the set of heads and tails recorded in part (a). **d** The parameter is the proportion of heads (or tails). **e** The statistic is the sample proportion of heads (or tails) calculated from part (a). **f** Compute the sample proportion and compare it to 50%. If the difference is large conclude that the coin is not fair.
1.8 a The population of interest is the set of fuel consumption figures for the entire fleet. **b** The parameter is the average of the fleet's fuel consumption. **c** The sample is the set of 50 observations stored on the disk. **d** The statistic is the average of the 50 observations. **e** The owner can use the sample average to estimate the population average.

Chapter 2

2.2 a Interval **b** Interval **c** Nominal **d** Ordinal
2.4 a Nominal **b** Interval **c** Nominal **d** Interval **e** Ordinal
2.6 a Interval **b** Interval **c** Nominal **d** Ordinal **e** Interval
2.8 a 9 or 10 **c** The histogram is positively skewed. **d** There is more than one modal class. **e** The histogram is not bell shaped.
2.10 a 9 or 10 **c** The histogram appears to be slightly positively skewed, unimodal, and bell shaped. **d** Most salaries lie between $40,000 and $80,000 with a small fraction of salaries over $90,000.
2.12 The histogram is unimodal, bell shaped, and roughly symmetric. Most of the heights lie between 18 and 23 inches.
2.14 The histogram is unimodal, symmetric, and bell shaped. Most tomatoes weigh between 2 and 7 ounces with a small fraction weighing less than 2 ounces or more than 7 ounces.
2.16 The histogram of the number of books shipped daily is negatively skewed, It appears that there is a maximum number that the company can ship.
2.18 b The histogram is positively skewed, bimodal, and far from bell shaped. **c** About a quarter of employees spends no time on the Internet. Of those who do, most spend between 2 and 12 hours per week.
2.20 Approximately 2/3 of drivers do not use the ETR. Of the remainder, most bills range from $20 to $80.
2.22 Expenses are slightly positively skewed and range from under $400 to over $2,000.
2.24 Democrats have the support of 41% of voters, Republicans, 39%; and Independents, 20%.
2.26 c Most applicants are BAs. The remainder is more or less equally divided among the other three categories.
2.28 59% of students are single, 27% are married, 9% are divorced, and the rest are widowed.
2.30 Dell is most popular with a 40% proportion, followed by other, 26%; IBM, 21%; and Compaq, 13%.
2.32 50% said the economy would worsen, 17% said it would get better, and the rest stated that the economy would remain the same.

2.34 Most accidents (22%) occur on landing. The second most frequent type of accident is descent (19%).
2.36 Most male-owned businesses are in the service sector and the retail/wholesale trade.
2.38 Over the 50 days there was a small upward trend.
2.40 Although there was a great deal of day-to-day variation, the last price was the same as the first.
2.42 Over the first year, the president's popularity was stable. For the first eight months of the second year, his popularity decreased, and over the past four months it has recovered somewhat.
2.44 c There is a positive relationship between the two variables. **d** The relationship is not linear.
2.46 b There is a moderately strong positive linear relationship between mark and study time.
2.48 b There is a weak negative linear relationship.
2.50 b The relationship is positive. **c** There is a strong positive linear relationship between amount overweight and amount of television.
2.52 b There is a strong positive linear relationship between metabolism rate and exercise.
2.54 b There is a moderately strong linear relationship between the two variables.
2.56 There is a very weak linear relationship between the amounts purchased on credit and debit cards.
2.58 There is a relationship between the two variables, but it isn't linear.
2.60 There is a moderately strong positive linear relationship between temperature and the number of lift tickets sold.
2.62 The histogram is bell shaped. Most bills are less than $16.
2.64 There is a strong positive linear relationship between mean temperature and the number of beers sold.
2.66 Almost half (48%) the students play bridge, 25% play hearts, and 15% play poker.
2.68 There has been a steady upward increase in faculty salaries.
2.70 The histogram is positively skewed. Almost all those interviewed skied for less than nine days.
2.72 More than half (53%) of those surveyed sought whiter teeth, 29% opted for straighter teeth, and the rest (18%) preferred fuller lips.
2.74 More than half (54%) rated Red Lobster as very good or excellent. However, 16% responded fair or poor. The rest (30%) rated the restaurant good.
2.76 Almost two-thirds (61%) of the respondents rated Casino Windsor excellent or good, but 15% rated it poor or unacceptable. The remaining 24% rated the casino average.
2.78 There is a weak positive linear relationship between the winning time for male runners and temperature.
2.80 The histogram is bell shaped with a modal class of 20 to 25 pounds.
2.82 More than a third (37%) of televisions were turned off. 22% of those surveyed reported that they watched some other channel. The remaining 41% were approximately equally divided.

Chapter 3

3.2 a 6, 5, 5 **b** The mean number of sick days is 6, the median is 5, and the most frequent number of sick days is 5.
3.4 a 39.3, 38, all **b** The mean amount of time is 39.3 minutes. Half the group took less than 38 minutes.
3.6 a 24,329, 24,461 **b** The mean starting salary is $24,329. Half the sample earn less than $24,461.
3.8 a 53.25, 52.50 **b** The mean service time is 53.25 minutes. Half the sample was served in less than 52.50 minutes. **c** Both statistics are useful. The mean provides us with a statistic that measures overall service, and the median provides a statistic we can use to make comparisons. **d** The information is more precise.
3.10 a 175.73, 160 **b** The mean withdrawal was $175.73. Half the sample withdrew less than $160.
3.12 a 20.02, 20.11 **b** The mean length is 20.02 inches. Half the sample was less than 20.11 inches. **c** Information is more precise.
3.14 median = 59,651; half the sample earn less than $59,651.
3.16 6.57
3.18 54.19
3.24 a 64, 167.09, 12.93 **b** standard deviation **c** Information is more precise.
3.26 Punter 1: 40.22, 6.34; punter 2: 14.81, 3.85; punter 3: 3.63, 1.91 **b** Punter 3 is most consistent.
3.28 a 1.13 **b** There is relatively little variation.
3.30 a 16% **b** 97.5% **c** 16%
3.32 40.73, 6.38
3.34 a 17.93 **b** About 68% of the marks lie within 17.93 of the mean.
3.36 3, 5, 7
3.38 39.4, 86.5
3.40 6.6, 17.6
3.42 b 44, 52.5, 60 **c** Yes, the six numbers 86 and larger
3.44 b 50,889, 59,651, 70,131 **c** 98,999, 100,716, 102,341
3.46 b 1.529, 1.549, 1.569 **c** No
3.48 $-7.1429, -.9109$
3.50 a 18,964.38 **b** .9037 **c** There is a strong positive linear relationship.
3.52 There is a negative linear relationship.
3.54 .6850; there is a moderately strong positive linear relationship.

3.56 a .8748 **b** There is a strong positive linear relationship.
3.58 .6868; there is a moderately strong positive linear relationship.
3.60 a 29,913, 30,660 **b** 148,213,791, 12,174 **d** The number of coffees sold varies considerably.
3.62 a mean, median, standard deviation **b** 93.9, 7.717 **c** Chris is not a good golfer.
3.64 a 26.32, 26 **b** 88.57, 9.41
3.66 $\bar{x} = 150.77$, median = 150.5, and $s = 19.76$
3.68 a 335.62, .3936 **b** yield = $120.37 + .18$ fertilizer **c** There is a relatively weak positive linear relationship.

Chapter 4

4.2 a Subjective approach **b** In the long run, this Yankee team would win 25% of World Series.
4.4 a Subjective approach **b** In the long run, if conditions remain unchanged, the Dow Jones Industrial Index will increase on 60% of days.
4.6 {Adams wins, Brown wins, Collins wins, Dalton wins}
4.8 a {0, 1, 2, 3, 4, 5} **b** {4, 5} **c** .10 **d** .65 **e** 0
4.10 P(Contractor 1 wins) = 2/6, P(Contractor 2 wins) = 3/6, P(Contractor 3 wins) = 1/6
4.12 a 40% **b** 90%
4.14 a P(Single) = 15%, P(Married) = 50%, P(Divorced) = 25%, P(Widowed) = 10% **b** Relative frequency approach
4.16 $P(A_1) = .3$, $P(A_2) = .4$, $P(A_3) = .3$, $P(B_1) = .6$, $P(B_2) = .4$
4.18 a $P(A_1 | B_1) = 4/7$ **b** $P(A_2 | B_1) = 3/7$ **c** Yes. It is not a coincidence.
4.20 No, because $P(A_1 | B_1) \neq P(A_1)$
4.22 Yes, because $P(A_1 | B_1) = .25$ and $P(A_1) = .25$
4.24 $P(A_1) = .40$, $P(A_2) = .45$, $P(A_3) = .15$, $P(B_1) = .45$, $P(B_2) = .55$
4.26 a .85 **b** .75 **c** .50
4.28 a .36 **b** .489 **c** .83
4.30 a .40 **b** .24 **c** .275 **d** Yes, the events are dependent.
4.32 a .35 **b** .538 **c** .714 **d** (a) is the joint probability and (b) and (c) are conditional probabilities.
4.34 a .24 **b** .39 **c** No, because P(Consult map | Male) $\neq P$(Consult map)
4.36 a .563 **b** No. P(Above average testosterone | Murderer) = .563 and P(Above average testosterone) = .51
4.38 $P(A \text{ and } B) = .32$, $P(A^C \text{ and } B) = .14$
4.40 $P(A \text{ and } B) = .24$, $P(A^C \text{ and } B) = .06$
4.42 .538
4.44 a .81 **b** .01 **c** .18 **d** .99
4.46 a .809 **b** .00909 **c** .1818 **d** .9908
4.48 a .122 **b** .484
4.50 .327
4.52 .526
4.54 .0344
4.56 a .1296 **b** .3889
4.58 a .3548 **b** .0354
4.60 a .29 **b** .2840 **c** No, the probabilities in (a) and (b) differ.
4.62 a .19 **b** .517 **c** No, P(Procedure is successful) = .81 and P(Procedure is successful | factor less than −8) = .517.
4.64 .2950
4.66 .825
4.68 a .698 **b** .954
4.70 .203

Chapter 5

5.2 a .48 **b** .52
5.4 a 1/4 **b** 1/2 **c** 1/4 **d** 3/4
5.6 a 3/8 **b** 3/8 **c** 7/8 **d** 1/2
5.8 a .06 **b** 0 **c** .35 **d** .65
5.10 a .21 **b** .31 **c** .26
5.12 $\mu = 1.7$, $\sigma = .9$
5.14 y 2 5 8 11
$p(y)$ .4 .3 .2 .1
5.16 $\mu = 5.0$, $\sigma = 3.0$
5.18 Amount .25 .50 .75 1.00 1.25 1.50 1.75
Probability .05 .15 .15 .25 .20 .10 .10
5.20 $\mu = 1.03$, $\sigma = .41$
5.22 $\mu = 2.76$, $\sigma = 1.52$
5.24 $\mu = 1.47$, $\sigma = 1.59$
5.26 $\mu = 9.65$, $\sigma = 6.51$
5.28 a .267 **b** .103 **c** 0
5.30 a .07776 **b** .3456 **c** .91296 **d** .66304
5.32 a .07776 **b** .3456 **c** .91296 **d** .66304
5.34 a .171194 **b** .091636 **c** .909528 **d** .810564
5.36 .972901
5.38 a .3369 **b** .6157
5.40 a .2990 **b** .8415
5.42 a .0646 **b** .9666 **c** .9282 **d** 22.5
5.44 .5987
5.46 .005921
5.48 a .07179 **b** .9020 **c** .4629
5.50 .05095
5.52 .1244
5.54 a .0887 **b** .8162 **c** .9400
5.56 a .1734 **b** .2388
5.58 a .0137 **b** .00159
5.60 a .1087 **b** .3906
5.62 a .01370 **b** .4471
5.64 a .82 **b** 1.04
5.66 a 15 **b** 3.57 **c** .1065
5.68 a 7.27 **b** 1.09
5.70 a .00793 **b** 56 **c** 4.10
5.72 a .1616 **b** .00947 **c** .0132
5.74 a .1346 **b** .3828 **c** .0800
5.76 .08755

Chapter 6

6.2 b .5 **c** .25
6.4 a $f(x) = .10 - .005x$ $0 \leq x \leq 20$ **b** .25 **c** .33
6.6 .4345
6.8 .4441
6.10 .4893

6.12 .9251
6.14 .0475
6.16 .1196
6.18 .0010
6.20 0
6.22 1.70
6.24 .0122
6.26 .4435
6.28 a .6759 **b** .3745 **c** .1469
6.30 .6915
6.32 a .2023 **b** .3372
6.34 56.36 months
6.36 a .0301 **b** .0025
6.38 a .0038 **b** .9082 **c** .6568
6.40 7.65 hours
6.42 4.05 hours
6.44 9,636 pages
6.46 a .3336 **b** .0314 **c** .0436
6.48 a .0018 **b** .6440
6.50 $12.88
6.52 a 1.341 **b** 1.319 **c** 1.990 **d** 1.653
6.54 a 1.3406 **b** 1.3195 **c** 1.9890 **d** 1.6527
6.56 a .0189 **b** .0341 **c** .0927 **d** .0324
6.58 a 9.23635 **b** 135.807 **c** 9.39046 **d** 37.4848
6.60 a 73.3441 **b** 102.946 **c** 16.3382 **d** 24.7690
6.62 a .2688 **b** 1.0 **c** .9903 **d** 0
6.64 a 4.35 **b** 8.89 **c** 3.29 **d** 2.50
6.66 a 1.48 **b** 1.76 **c** 1.82 **d** 1.16
6.68 a .0510 **b** .1634 **c** .0222 **d** .2133

Chapter 8

8.4 a .1056 **b** .1587 **c** .0062
8.6 a .4435 **b** .7333 **c** .8185
8.8 a .1191 **b** .2347 **c** .2902
8.10 We can answer part (c) and possibly part (b), depending on how nonnormal the population is.
8.12 a .0918 **b** .0104 **c** .00077
8.14 a .3085 **b** 0
8.16 a .0038 **b** It appears to be false.
8.20 a .1337, .1378 **b** .9095, .9049 **c** .9558, .9599
8.22 a .7635, .7611 **b** .8578, .8577 **c** .8582, .8577 **d** .2160, .2177 **e** .0239, .0241
8.24 .0256
8.26 .0035
8.28 .1151; the claim may be true
8.30 0; the claim appears to be false
8.32 a .1587 **b** The claim may be true.
8.34 .0071
8.36 .0033
8.38 .8413
8.40 .8413
8.42 a .3974 **b** .3050
8.44 1

Chapter 9

9.10 125 ± 3.920
9.12 125 ± 1.645
9.14 125 ± 6.579
9.16 a increases **b** decreases **c** decreases
9.18 25 ± 1.663
9.20 25 ± 2.326
9.22 25 ± .987
9.24 a decreases **b** increases **c** increases
9.26 52,000 ± 3,675
9.28 52,000 ± 9,659
9.30 52,000 ± 9,659
9.32 a decreases **b** increases **c** increases
9.34 43.75 ± 6.93; we estimate that the mean age of men who frequent the bar lies between 36.82 and 50.68.
9.36 9.85 ± 2.94 (LCL = 6.91, UCL = 12.79)
9.38 22.54 ± 1.29 (LCL = 21.25, UCL = 23.83)
9.40 1810.16 ± 98.00 (LCL = 1712.16, UCL = 1908.16)
9.42 10.21 ± .57 (LCL = 9.64, UCL = 10.78)
9.44 26.81 ± .36; the mean assembly time is estimated to lie between 26.45 and 27.17.
9.46 384
9.48 a 100 ± 5 **b** The sample was designed to produce the interval width in (a).
9.50 100 ± 2
9.52 2,148
9.54 1,082
9.56 15 ± .59; the mean pulse-recovery time for all 40- to 50-year-old women is estimated to lie between 14.41 and 15.59 minutes.
9.58 216
9.60 14.98 ± .31 (LCL = 14.67, UCL = 15.29)
9.62 411.3 ± 4.9 (LCL = 406.4, UCL = 416.2)

For all remaining chapters, our answers to the questions requiring a test of hypothesis where no significance level is specified are based on a 5% significance level.

Chapter 10

10.6 O. J. Simpson
10.8 $z = .60$; $z > 1.88$; p-value = .2743; there is no evidence to infer that $\mu > 50$.
10.10 $z = 0$; $z < -1.96$ or $z > -1.96$; p-value = 1.0; there is no evidence to infer that $\mu \neq 100$.
10.12 $z = -1.33$; $z < -1.645$; p-value = .0912; there is not enough evidence to infer that $\mu < 50$.
10.14 $z = 1.00$; p-value = .1587; there is no evidence to infer that the mean is greater than 60 minutes.
10.16 $z = 2.83$; p-value = .0023; there is evidence to infer that the mean is greater than 60 minutes.
10.18 $z = .71$; p-value = .2397; there is no evidence to infer that the mean is greater than 60 minutes.
10.20 a z increases **b** z increases **c** z increases
10.22 p-value = .0304; there is enough evidence to infer that the

average student spends less than the recommended amount of time.
10.24 p-value = 0; there is evidence to infer that the average student spends less than the recommended amount of time.
10.26 p-value = 0; there is evidence to infer that the average student spends less than the recommended amount of time.
10.28 a p-value decreases **b** p-value decreases **c** p-value decreases
10.30 a $z = -2.50$; p-value = .0062; p-value decreases **b** $z = -.50$; p-value = .3085; p-value increases **c** $z = -2.50$; p-value = .0062; p-value decreases
10.32 $z = -1.41$; p-value = .0786; no
10.34 $z = 1.87$; p-value = .0620; there is not enough evidence to infer that the population mean differs from 25.
10.36 $z = 2.00$; p-value = .0455; yes
10.38 $z = -2.06$; p-value = .0197; yes
10.40 $z = 1.91$; p-value = .0279; there is enough evidence to infer that the hypothesis is correct.
10.42 $z = 3.34$; p-value = .0008; yes
10.44 $z = 1.60$; p-value = .0584; there is enough evidence to support the claim.
10.46 $z = 2.26$; p-value = .0118; no
10.48 $z = -1.22$; p-value = .1108; no
10.50 $z = -.78$; p-value = .4380; no
10.52 $z = 1.56$; p-value = .0594; there is not enough evidence to infer that Nike is correct.
10.54 .0176
10.58 .1492
10.60 .6480
10.66 .1054
10.68 .0187

Chapter 11
11.2 500 ± 24.80
11.4 500 ± 56.84
11.6 500 ± 29.64
11.8 200 ± 35.52
11.10 a Interval widens. **b** Interval widens. **c** Interval widens. **d** Interval width is unchanged.
11.12 33 ± 1.71
11.14 33 ± .33
11.16 33 ± .99
11.18 133 ± .83
11.20 a Interval widens. **b** Interval widens. **c** Interval widens. **d** Interval width is unchanged.
11.22 $t = 1.05$; p-value = .1597; no
11.24 $t = 2.68$; p-value = .0074; yes
11.26 $t = .50$; p-value = .3125; no
11.28 a p-value increases **b** p-value increases **c** p-value decreases
11.30 50 ± 9.30
11.32 1000 ± 104.94
11.34 The Student t distribution is more widely dispersed than the standard normal.
11.36 15,500 ± 517.55
11.38 150 ± 2.92
11.40 The sample size is large.
11.42 p-value = .0786
11.44 p-value = .1631
11.46 There is a relatively large difference between the z and t critical values.
11.48 p-value = .0038
11.50 p-value = .0576
11.52 With $n = 250$, the Student t and standard normal distributions are very similar.
11.54 29.33 ± 5.94 (LCL = 23.39, UCL = 35.27)
11.56 a 22.60 ± 1.41 (LCL = 21.19, UCL = 24.01) **b** $t = 3.81$; p-value = .0004; yes **c** Variable is required to be normal. The histogram is not bell shaped.
11.58 7.15 ± .23 (LCL = 6.92, UCL = 7.38)
11.60 $t = 3.12$; p-value = .0012; yes
11.62 a 62.90 ± 1.92 (LCL = 60.98, UCL = 64.82) **b** Prices are required to be normal. The histogram is approximately bell shaped.
11.64 14.75 ± 2.32 (LCL = 12.43, UCL = 17.08)
11.66 $t = -.48$; p-value = .3142; no
11.68 $\chi^2 = 35.93$; p-value = .1643; no
11.70 The interval narrows.
11.72 p-value = .0714; no
11.74 The p-value decreases
11.76 $\chi^2 = 6.79$; p-value = .2544; there is not enough evidence to infer that the variance is less than 500.
11.78 LCL = .2186; UCL = 2.0712
11.80 $\chi^2 = 25.98$; p-value = .7088; no
11.82 $\chi^2 = 305.85$; p-value = .0044; yes
11.84 .48 ± .0438
11.86 .48 ± .0310
11.88 .10 ± .0263
11.90 .10 ± .0698
11.92 .10 ± .0247
11.94 .90 ± .0698
11.96 .84 ± .0246
11.98 $z = 1.80$; p-value = .0359
11.100 p-value = .0694
11.102 a .5 ± .03 **b** The sample size was designed to produce this interval.
11.104 564
11.106 a .92 ± .0188 **b** We expected a wider interval. **c** Yes, because the interval was no wider than .03.
11.108 $z = 1.13$; p-value = .1303; no
11.110 Proportion: LCL = .0846, UCL = .1104; Total: LCL = 8.46 million, UCL = 11.04 million

11.112 $z = 1.20$; p-value $= .1151$; no
11.114 $z = 1.80$; p-value $= .0362$; yes
11.116 a $.4581 \pm .0374$ (LCL $= .4207$, UCL $= .4956$) **b** $z = .75$; p-value $= .2275$; no
11.118 $\chi^2 = 30.71$; p-value $= .0435$; problems appear to be likely.
11.120 a 71.88 ± 2.85 (LCL $= 69.03$, UCL $= 74.73$) **b** $t = 2.74$; p-value $= .0043$; yes
11.122 $.632 \pm .0502$ (LCL $= .5818$, UCL $= .6822$)
11.124 $6.352 \pm .301$ (LCL $= 6.051$, UCL $= 6.653$)
11.126 $.667 \pm .109$ (LCL $= .558$, UCL $= .776$)
11.128 $z = -1.33$; p-value $= .0912$; no
11.130 a $t = -2.97$; p-value $= .0018$; yes **b** $\chi^2 = 101.58$; p-value $= .0011$; yes
11.132 Proportion: LCL $= .1245$, UCL $= .1821$
Total: LCL $= 49,800$, UCL $= 72,840$
11.134 Number of cars: $t = .46$; p-value $= .3351$. Amount of time: $t = 7.00$; p-value $= 0$. The employee is stealing by lying about the amount of time.

For all exercises in Chapters 12, 13, and 19 we employed the F-test of two variances at the 5% significance level to decide which one of the equal-variances or unequal-variances t-test and estimator of the difference between two means to use to solve the problem. Additionally, for exercises that compare two populations and are accompanied by data files, our answers were derived by defining the sample from population 1 as the data stored in the first column (often column A). The data stored in the second column represent the sample from population 2. Paired differences were defined as the difference of the variable in the first column minus the variable in the second column.

Chapter 12

12.2 a $t = .689$, p-value $= .4941$ **b** 55.00 ± 160.49
12.4 a $t = 2.78$, p-value $= .0060$ **b** 55.00 ± 39.01
12.6 a $t = 1.94$, p-value $= .0280$ **b** 3.00 ± 2.575
12.8 The p-value increases and the interval widens.
12.10 The p-value decreases and the interval narrows.
12.12 a $t = -.809$, p-value $= .2117$ **b** -79.0 ± 197.41
12.14 a $t = -2.31$, p-value $= .0118$ **b** -79.0 ± 68.09
12.16 a $t = -2.26$, p-value $= .0147$ **b** -113.0 ± 101.06
12.18 The p-value decreases, and the width of the interval does not change.
12.20 a $t = 2.36$, p-value $= .0101$; yes **b** 7.14 ± 5.03 (LCL $= 2.11$, UCL $= 12.17$) **c** Both variables must be normally distributed with unequal variances. **d** The requirements appear to be satisfied.
12.22 $t = -2.20$, p-value $= .0156$; yes
12.24 $t = -2.40$, p-value $= .0092$; yes
12.26 $t = 1.07$, p-value $= .1440$; switch to supplier B
12.28 a $t = 1.79$, p-value $= .0737$; no **b** 6.03 ± 6.66 (LCL $= -.57$, UCL $= 12.63$) **c** Both histograms are somewhat bell shaped.
12.30 a $t = -11.61$, p-value $= 0$; yes **b** $-2.780 \pm .472$ (LCL $= -3.252$, UCL $= -2.308$)
12.36 $t = -1.90$, p-value $= .0898$; no
12.38 a $t = 1.16$, p-value $= .2581$; there is no evidence to infer that the means differ. **b** 9.58 ± 17.12 **c** $t = 7.25$, p-value $= 0$; there is evidence to infer that the means differ. **d** 9.42 ± 2.86
12.40 a $t = 2.10$, p-value $= .0474$; there is evidence to infer that the means differ. **b** 8.92 ± 8.81 **c** $t = 1.68$, p-value $= .1208$; there is no evidence to infer that the means differ. **d** 9.08 ± 11.89
12.42 a $t = -2.02$, p-value $= .0344$; yes **b** -2.08 ± 2.27 (LCL $= -4.35$, UCL $= .19$ **c** The paired differences are required to be normally distributed. **d** The histogram is somewhat bell shaped.
12.44 $t = -3.70$, p-value $= .0005$; there is evidence to infer that waiters and waitresses earn different amounts in tips.
12.46 $t = -6.09$, p-value $= 0$; yes
12.48 LCL $= .238$, UCL $= 1.050$
12.50 $F = .262$, p-value $= .0914$; there is not enough evidence to infer that the population variances differ.
12.52 $F = .593$, p-value $= .0350$; yes
12.54 $F = 1.59$, p-value $= .3050$; yes
12.56 $F = .203$, p-value $= 0$; yes
12.58 $F = .306$, p-value $= 0$; yes
12.60 $.060 \pm .1361$
12.62 The width of the interval decreases.
12.64 $.040 \pm .220$
12.66 The width of the interval increases.
12.68 p-value $= .1085$
12.70 The p-value decreases.
12.72 p-value $= .0986$
12.74 The p-value increases.
12.76 a $z = 4.31$, p-value $= 0$; yes **b** $z = 2.16$, p-value $.0153$; yes **c** $.100 \pm .045$ (LCL $= .055$, UCL $= .145$)
12.78 a $z = 2.54$, p-value $= .0055$;

yes **b** .1657 ± .1163 (LCL = .0494, UCL = .2820)
12.80 $z = 1.62$, p-value = .0525; no
12.82 $z = -2.51$, p-value .0061; yes
12.84 $t = 1.56$, p-value = .1204; no
12.86 a $t = 2.65$, p-value = .0059; yes **b** $t = .877$, p-value = .1931; no **c** The gross sales are required to be normally distributed with equal variances.
12.88 $z = -2.30$, p-value = .0106; seatbelt usage increased.
12.90 $t = -1.06$, p-value = .2980; no
12.92 $t = 2.26$, p-value = .0119; yes
12.94 $t = -4.28$, p-value = 0; yes
12.96 a $t = 2.62$, p-value = .0059; there is enough evidence to infer that 40-year-old men have more iron in their bodies than do 20-year-old men. **b** $t = 4.08$, p-value = .0001; there is enough evidence to infer that 40-year-old women have more iron in their bodies than do 20-year-old women.
12.98 $t = -11.21$, p-value = 0; yes
12.100 $z = 1.26$, p-value = .1037; no
12.102 $t = -1.30$, p-value = .0979; no
12.104 $t = -11.79$, p-value = 0; yes
12.106 $t = 1.26$, p-value = .1057; no
12.108 $t = 2.22$, p-value = .0141; yes
12.110 $t = 6.57$, p-value = 0; yes
12.112 $t = -3.97$, p-value = .0002; yes
12.114 $t = 1.38$, p-value = .0855; no
12.116 $t = 19.08$, p-value = 0; the mean vocabulary of children whose parents read to them regularly is larger than the mean vocabulary of children whose parents do not read to them regularly.

Chapter 13

13.2 a $z = 2.83$, p-value = .0024; yes **b** $t = .90$, p-value = .1853; no **c** .4336 ± .0914 (LCL = .3423, UCL = .5250) **d** 97.38 ± 7.16 (LCL = 90.22, UCL = 104.55)
13.4 a $z = 1.54$, p-value = .0619; no **b** $z = 3.02$, p-value = .0013; yes
13.6 Speeds: $t = 1.07$, p-value = .1424, proper stops: $t = -.84$, p-value = .2021; there is no evidence to infer that the speed bumps are effective.
13.8 $t = 3.73$, p-value = .0002; yes
13.10 $t = .96$, p-value = .1711; no
13.12 $z = 1.02$, p-value = .1548; no
13.14 $t = 14.06$, p-value = 0; yes
13.16 Memory: $t = 3.27$, p-value = .0008; reasoning: $t = 4.11$, p-value = 0; reaction time: $t = -.58$, p-value = .5637; vocabulary: $t = 1.25$, p-value = .2163; there is enough evidence to infer that bridge players have better memory and reasoning than do nonplayers. There is not enough evidence to infer that bridge players have different reaction time and vocabularies.
13.18 Mean: LCL = .2042, UCL = .4358. Number: LCL = 204.2, UCL = 435.8
13.20 $t = -3.27$, p-value = .0011; yes
13.22 a $z = 1.70$, p-value = .0450; yes **b** $z = 1.07$, p-value = .1417; no
13.24 a $t = -6.00$, p-value = 0; yes **b** $z = -.29$, p-value = .3867; no
13.26 $z = 5.33$, p-value = 0; yes
13.28 $t = -.79$, p-value = .2151; no
13.30 $t = 1.40$, p-value = .0813; no

Chapter 14

14.4 a $F = 4.75$, p-value = .0017; yes **b** The variables are normally distributed with the same variance. **c** The histograms are bell shaped, and the sample variances are similar.
14.6 a $F = 2.94$, p-value = .0363; there is evidence to infer that there are differences in completion times between the forms. **b** The times for each form are normally distributed with the same variance. **c** The histograms are bell shaped, and the sample variances are similar.
14.8 $F = 1.73$, p-value = .1783; no
14.10 a $F = 13.95$, p-value = 0; there is enough evidence to infer that the leaf sizes differ between the three groups. **b** $F = 101.47$, p-value = 0; there is enough evidence to infer that the amounts of nicotine differ between the three groups. **c** The executives were wrong.
14.12 $F = 4.06$, p-value = .0074; yes
14.14 a $F = 9.73$, p-value = .0101; yes **b** $F = 6.82$, p-value = .0285; yes
14.16 a $F = 7.68$, p-value = .0002; yes **b** $F = 15.94$, p-value = 0; yes **c** The response is normally distributed with the same variance. **d** The histograms are bell shaped, and the sample variances are similar.
14.18 a $F = 123.36$, p-value = 0; yes **b** $F = 323.16$, p-value = 0; there is evidence to infer a difference between plots.
14.20 a $F = 21.16$, p-value = 0; yes **b** $F = 66.02$, p-value = 0; the randomized block experiment was appropriate.
14.22 $F = 11.91$, p-value = 0; there is enough evidence to infer a difference in listening time between the days of the week.
14.24 a $F = .52$, p-value = .6025 **b** $F = 9.99$, p-value = .0012 **c** $F = .04$, p-value = .8394
14.26 a There are two factors—class configuration and time. **b** The

response is the number of times students ask and answer questions. **c** There are two levels of class configuration and three levels of time. **d** Test for interaction: $F = 12.28$, p-value $= .0002$; there is evidence of interaction.

14.28 a Factor A (columns) is the form, and factor B (samples) is the income group. **b** The response is the completion time. **c** $a = 4$, $b = 3$ **d** $F = 2.56$, p-value $= .0586$ **e** $F = 4.11$, p-value $= .0190$ **f** test for interaction **g** $F = 1.04$, p-value $= .4030$; there is no evidence of interaction. There is enough evidence to infer that there are differences in completion times between the income brackets.

14.30 a The p-values for devices, alloys, and interaction are .0775, .5798, and .7584, respectively. There are no sources of variation.

14.32 High schools A and C differ.
14.34 Forms 1 and 4 differ.
14.36 Lacquers 2 and 3 differ.
14.38 a $F = 3.56$, p-value $= .0236$; yes **b** Flares C and D differ. **c** Flares A and C differ from flare D.
14.40 $B = .38$, p-value $= .8258$; there is not enough evidence to infer that the population variances differ.
14.42 $B = 1.56$, p-value $= .6692$; there is not enough evidence to infer that the population variances differ.
14.44 $B = 1.89$, p-value $= .7561$; there is not enough evidence to infer that the population variances differ.
14.46 $B = .15$, p-value $= .9281$; there is not enough evidence to infer that the population variances differ.
14.48 $F = 13.79$, p-value $= 0$; switch to the typeface that is read the fastest.
14.50 a $F = 7.67$, p-value $= .0001$; yes **b** μ_1 differs from μ_2, μ_3, and μ_4.

14.52 a There are 16 levels representing one of four ranks within four departments. **b** $F = 2.94$, p-value $= .0019$; there is evidence of a difference between the 16 groups. **c** Factor A (columns) is the faculty. The levels are business, engineering, arts, and science. Factor B (samples) is the rank. The levels are professor, associate professor, assistant professor, and lecturer. **d** $F = .61$, p-value $= .6109$ **e** $F = 4.49$, p-value $= .0064$ **f** test for interaction **g** $F = 3.04$, p-value $= .0044$; there is evidence of interaction.
14.54 $F = 7.72$, p-value $= .0070$; yes
14.56 Test for interaction: $F = .44$, p-value $= .6418$. Test for age category differences: $F = 58.78$, p-value $= 0$; there is evidence of differences between age categories. Test for gender differences: $F = 3.66$, p-value $= .0576$; there is not enough evidence of differences between genders.
14.58 Test for interaction: $F = 12.36$, p-value $= 0$; yes
14.60 a $F = 9.54$, p-value $= .0002$; yes **b** μ_1 differs from μ_2 and μ_3.
14.62 $F = 14.47$, p-value $= 0$; yes
14.64 $F = 13.84$, p-value $= 0$; yes

Chapter 15

15.2 $\chi^2 = 2.27$, p-value $= .6868$
15.4 The p-value increases.
15.6 $\chi^2 = 9.96$, p-value $= .0189$
15.8 $\chi^2 = 6.85$, p-value $= .0769$
15.10 $\chi^2 = 14.07$, p-value $= .0071$; yes
15.12 $\chi^2 = 33.85$, p-value $= 0$; there is evidence to infer that the ageing schedule has changed.
15.14 $\chi^2 = 31.53$, p-value $= 0$; yes
15.16 $\chi^2 = 19.10$, p-value $= 0$
15.18 $\chi^2 = 4.78$, p-value $= .0289$
15.20 $\chi^2 = 4.40$, p-value $= .1110$
15.22 $\chi^2 = 2.35$, p-value $= .3087$; no

15.24 $\chi^2 = 46.37$, p-value $= 0$; yes
15.26 $\chi^2 = 31.48$, p-value $= 0$; yes
15.28 $\chi^2 = 41.76$, p-value $= 0$; yes
15.30 $\chi^2 = 20.89$, p-value $= .0019$; yes. The editor can more efficiently direct the sales representatives.
15.32 $\chi^2 = 31.78$, p-value $= 0$; there is evidence of nonnormality.
15.34 $\chi^2 = 16.62$, p-value $= .0002$; there is evidence to infer that time spent working at part-time jobs is not normally distributed.
15.36 $\chi^2 = 8.43$, p-value $= .0037$; there is evidence to infer that returns are not normally distributed.
15.38 Supplements: $\chi^2 = 2.27$, p-value $= .1317$. Placebo: $\chi^2 = 1.39$, p-value $= .2386$; no
15.40 $\chi^2 = .71$, p-value $= .7004$; no
15.42 $\chi^2 = 6.43$, p-value $= .0401$; yes
15.44 Course A: $\chi^2 = .52$, p-value $= .4691$; course B: $\chi^2 = 2.27$, p-value $= .3219$; course C: $\chi^2 = .69$, p-value $= .4058$; no course: $\chi^2 = .29$, p-value $= .8646$; there is no evidence to infer nonnormality.
15.46 $\chi^2 = 19.68$, p-value $= .0001$; the return rate can be influenced.
15.48 $\chi^2 = 5.42$, p-value $= .2465$; no
15.50 $\chi^2 = 23.81$, p-value $= .0025$; yes **b** $z = 2.09$, p-value $= .0364$; yes
15.52 $\chi^2 = 86.62$, p-value $= 0$; yes
15.54 $\chi^2 = 15.88$, p-value $= .0032$; yes
15.56 $\chi^2 = 13.55$, p-value $= .0011$; yes
15.58 $\chi^2 = 4.23$, p-value $= .1206$; no
15.60 $\chi^2 = 13.71$, p-value $= .0011$; yes

Chapter 16

16.2 a $z = -.27$, p-value $= .3929$; no **b** identical results
16.4 a $z = 3.61$, p-value $= .0004$; yes **b** identical results

16.6 $z = 1.16$, p-value $= .1238$; no
16.8 $z = -1.91$, p-value $= .0279$; yes
16.10 $z = 1.96$, p-value $= .0494$; yes
16.12 $z = .80$, p-value $= .2124$; no
16.14 $z = 3.14$, p-value $= .0016$; yes
16.16 $z = 1.00$, p-value $= .1587$; no
16.18 $z = -.80$, p-value $= .2117$; no
16.20 a $z = 1.81$, p-value $= .0355$; yes **b** identical results
16.22 a $z = -1.44$, p-value $= .1490$ **b** $z = -2.91$, p-value $= .0036$ **c** Wilcoxon signed rank sum test produced a smaller p-value.
16.24 $z = -5.67$, p-value $= 0$; yes
16.26 $z = -2.97$, p-value $= .0030$; purchase faster machine
16.28 $z = 3.57$, p-value $= .0002$; yes
16.30 $z = .74$, p-value $= .4594$; no
16.32 $z = -4.77$, p-value $= 0$; yes
16.34 $H = .90$, p-value $= .8257$; no
16.36 a $H = 1.91$, p-value $= .3839$; no **b** Identical results
16.38 $H = 6.65$, p-value $= .0838$; no
16.40 $H = 3.78$, p-value $= .1507$; no
16.42 $H = 4.34$, p-value $= .2269$; no
16.44 $H = 18.73$, p-value $= .0009$; yes
16.46 a $F_r = 8.00$, p-value $= .0460$; yes **b** identical results
16.48 $F_r = 9.42$, p-value $= .0242$; yes
16.50 $F_r = 2.63$, p-value $= .2691$; no
16.52 a Analysis of variance (randomized block) and Friedman test. The former requires that the data are normally distributed. Choose Kruskal-Wallis test if the data are extremely nonnormal. **b** $F_r = 11.47$, p-value $= .0032$; yes

16.54 $z = -2.27$, p-value $= .0230$; yes
16.56 $z = 1.60$, p-value $= .0544$; no
16.58 $z = -3.58$, p-value $= .0002$; yes
16.60 a Analysis of variance (single factor, independent samples) or Kruskal-Wallis test. The former requires that the data are normally distributed. Choose Kruskal-Wallis test if the data are extremely nonnormal. **b** $H = 3.55$, p-value $= .1699$; there is no evidence of differences.
16.62 $H = 14.87$, p-value $= .0213$; yes
16.64 $z = 1.12$, p-value $= .1309$; there is no evidence of differences between men's and women's hearing loss.
16.66 $z = .70$, p-value $= .4840$; no
16.68 $H = 3.04$, p-value $= .3859$; no
16.70 $H = 7.85$, p-value $= .0197$; yes
16.72 a $z = 5.27$, p-value $= 0$; yes **b** $z = -1.39$, p-value $= .0828$; no
16.74 $H = 42.59$, p-value $= 0$; yes
16.76 Day 1 vs. Before: $z = 8.19$, p-value $= 0$; day 2 vs. Before: $z = 7.68$, p-value $= 0$; day 3 vs. Before: $z = 5.92$, p-value $= 0$; yes **b** Day 3 vs. day 2: $z = 5.38$, p-value $= 0$; yes
16.78 $z = -10.06$, p-value $= 0$; yes

Chapter 17

17.2 a $\hat{y} = 36.54 + .479x$ **b** nothing **c** For each additional inch of father's height, the son's height increases on average by .479 inch.
17.4 a $\hat{y} = 33.67 + 2.37x$ **b** $b_0 = 33.67$ is the value of the y-intercept. $b_1 = 2.37$; for each additional hour of exercise, the metabolism rate increases on average by 2.37.
17.6 b $\hat{y} = 3.64 + .267x$ **c** $b_1 = .267$; for each additional second of commercial, the memory test score increases on average by .267. $b_0 = 3.64$ is the y-intercept.
17.8 a $\hat{y} = -2.03 + .788x$ **b** $b_1 = .788$; for each additional year of education, Internet use increases by .788 hour. $b_0 = -2.03$ is the y-intercept.
17.10 $\hat{y} = 9.44 + .0949x$; the appropriate compensation is 9.49 cents per degree API.
17.12 a $\hat{y} = -259.6 + 3721x$ **b** $b_1 = 3721$; for each additional carat of weight, the price increases on average by \$3721. $b_0 = -259.6$ cannot be interpreted.
17.14 It is normally distributed with constant variance and a mean that is a linear function of the quality of oil.
17.16 a $s_\varepsilon = 3.22$ **b** $R^2 = .2665$ **c** $t = 12.03$. There is enough evidence to infer that the heights of sons and fathers are linearly related.
17.18 $t = 12.79$, p-value $= 0$. There is enough evidence to infer that the two variables are linearly related.
17.20 a $s_\varepsilon = 5.89$ **b** $R^2 = .2893$ **c** $t = 4.86$. There is enough evidence to infer that the length of commercial and the memory test score are linearly related.
17.22 a $s_\varepsilon = 4.45$ **b** $t = 4.93$. There is enough evidence to infer that educational level and Internet use are linearly related.
17.24 $s_\varepsilon = .1325$, $t = 11.48$. There is enough evidence to infer that oil quality and price are linearly related. $R^2 = .9229$
17.26 a $s_\varepsilon = 31.84$ **b** $R^2 = .9783$ **c** $t = 45.50$. There is enough evidence to infer that the price and weight of diamonds are linearly related.
17.28 a $\hat{y} = 2.05 + .0909x$ **b** $b_1 = .0909$; for each additional minute of exercise, cholesterol is reduced on average by .0909. $b_0 = 2.05$ cannot be interpreted.

c $t = 7.06$; there is enough evidence to infer that exercise and cholesterol reduction are linearly related. **d** $R^2 = .5095$; the model fits moderately well.

17.30 lower prediction limit = 62.65, upper prediction limit = 79.40

17.32 a lower prediction limit = 1.38, upper prediction limit = 25.15 **b** lower confidence limit = 11.73, upper confidence limit = 14.80

17.34 lower confidence limit = 8.62, upper confidence limit = 10.96

17.36 lower prediction limit = 13.16, upper prediction limit = 13.70

17.38 lower prediction limit = 985.14, upper prediction limit = 1100.33

17.40 lower prediction limit = 7.95, upper prediction limit = 50.72

17.42 a $t = 2.66$, p-value = .0104; there is evidence to infer that the returns on the two stocks are linearly related. **b** $z = 2.30$, p-value = .0214; there is evidence to infer that the returns on the two stocks are linearly related.

17.44 a $r = .5378$ **b** $t = 4.86$, p-value = 0. There is sufficient evidence to infer that commercial length and memory test score are linearly related. **c** $r_s = .546$, $z = 4.19$, p-value = 0. There is sufficient evidence to infer that commercial length and memory test score are linearly related.

17.46 a $r = .3308$ **b** $t = 4.93$, p-value = 0. There is sufficient evidence to infer that education and Internet use are linearly related. **c** $r_s = .3657$, $z = 5.16$, p-value = 0. There is sufficient evidence to infer that education and Internet use are linearly related.

17.48 $r_s = .9381$. There is sufficient evidence to infer that higher quality oil is associated with higher prices.

17.50 $r_s = .9656$, $z = 6.62$, p-value = 0. There is sufficient evidence to infer that the price and weight of diamonds are linearly related.

17.52 $z = 4.82$, p-value = 0; there is enough evidence to infer that aptitude test scores are correlated with performance ratings.

17.54 b It appears that the errors are normally distributed. **c** Observations 18, 25, 53, and 59 have standardized residuals whose absolute values exceed 2.0. **d** heteroscedasticity does not appear to be a problem.

17.56 The histogram is bimodal, suggesting that the errors are not normally distributed. The plot of the residuals versus the predicted values indicates some degree of heteroscedasticity.

17.58 The histogram of the residuals is bell shaped. The graph of the residuals versus the predicted is too small to determine heteroscedasticity.

17.60 The histogram of the residuals is bell shaped, and there is no sign of heteroscedasticity.

17.62 a $\hat{y} = 114.85 + 2.47x$ **b** $b_1 = 2.47$; for each additional month of age, repair costs increase on average by $2.47. $b_0 = 114.85$ is the y-intercept. **c** $R^2 = .5659$; 56.59% of the variation in repair costs is explained by the variation in age. **d** $t = 4.84$, p-value = .0001. There is enough evidence to infer that repair costs and age are linearly related. **e** lower prediction limit = 318.12, upper prediction limit = 505.18

17.64 a $\hat{y} = 60.50 + .0606x$. The slope is .0606, which tells us that for each additional unit of fertilizer, corn yield increases on average by .0606. The y-intercept is 60.50, which has no real meaning. **b** $t = 1.04$, p-value = .3072. There is no evidence of a linear relationship between amount of fertilizer and corn yield. **c** $R^2 = .0372$; 3.72% of the variation in corn yield is explained by the variation in amount of fertilizer. **d** The model is too poor to be used to predict.

17.66 a $\hat{y} = -.227 + 2.794x$. The slope is $-.227$; for each additional point increase in batting average, the team's winning percentage increases on average by 2.794 points. $s_\varepsilon = .0567$. This statistic is large relative to the average winning percentage, .500. The model is poor. **c** $t = 1.69$, p-value = .0580. There is not enough evidence to infer a positive linear relationship between team batting average and winning percentage. **d** $R^2 = .1931$; 19.31% of the variation in winning percentage is explained by the variation in team batting average. **e** lower prediction limit = .4283, upper prediction limit = .6549

17.68 a $r = .4177$, $t = 3.19$, p-value = .0026. There is sufficient evidence to infer that weight and blood-alcohol level are linearly related. **b** $r_s = .3702$, $z = 2.59$, p-value = .0096. There is sufficient evidence to infer that weight and blood-alcohol level are linearly related.

17.70 a $r = .9766$, $t = 21.78$, p-value = 0. There is sufficient evidence to infer that levels of tar and nicotine are linearly related. **b** $r = .9259$, $t = 11.76$, p-value = 0. There is sufficient evidence to infer that levels of nicotine and carbon monoxide are linearly related.

17.72 $\hat{y} = 48,040 + 2,582x$; $t = 13.77$, p-value = 0; $R^2 = .3069$; there is evidence of a linear relationship. The coefficient of determination is 30.69%.

17.74 b The scatter diagram indicates a linear relationship.

c $\hat{y} = 140.2 + .119x$; $b_0 = 140.2$ is the y-intercept; $b_1 = .119$; for each additional degree the winning time increases on average by .119 minute. **d** $t = 3.26$, p-value = .0042; there is evidence of a linear relationship. **e** $R^2 = .3581$; 35.81% of the variation in winning times is explained by the variation in temperature.

Chapter 18

18.2 a $s_\varepsilon = 3.75$; this statistic is an estimate of the standard deviation of the error variable. **b** $R^2 = .7629$; 76.29% of the variation in final marks is explained by the variation in the independent variables in the model. **c** R^2 (adjusted) = .7453; it differs from R^2 because it includes an adjustment for the number of independent variables. **d** $F = 43.43$, p-value = 0; there is sufficient evidence to infer that the model is valid. **e** $t = .97$, p-value = .3417; there is no evidence to infer that the final mark and assignment mark are linearly related. **f** $t = 9.12$, p-value = 0; there is enough evidence to infer that the final mark and midterm mark are linearly related.

18.4 a $s_\varepsilon = 3.22$; this statistic is an estimate of the standard deviation of the error variable. **b** $R^2 = .2712$; 27.12% of the variation in heights is explained by the variation in the independent variables in the model. **c** R^2 (adjusted) = .2676; it differs from R^2 because it includes an adjustment for the number of independent variables. **d** $F = 73.88$, p-value = 0; there is sufficient evidence to infer that the model is valid. **f** $t = 11.69$, p-value = 0; there is enough evidence to infer that the heights of fathers and sons are linearly related. **g** $t = -1.60$, p-value = .1105; there is no evidence to infer that the heights of mothers and sons are linearly related.

18.6 a $\hat{y} = .72 + .611$ HS GPA + .0027 SAT + .0463 Activities **b** $s_\varepsilon = 2.03$; this statistic is an estimate of the standard deviation of the error variable. **c** $R^2 = .2882$; 28.82% of the variation in three-year GPA is explained by the variation in the independent variables in the model. **d** R^2 (adjusted) = .2660; it differs from R^2 because it includes an adjustment for the number of independent variables. **e** $F = 12.96$, p-value = 0; there is sufficient evidence to infer that the model is valid. **g** HS GPA: $t = 6.06$, p-value = 0. SAT: $t = .94$, p-value = .3482. Activities: $t = .72$, p-value = .4720. Only HS GPA is linearly related to three-year GPA. **h** 4.45, 12 (maximum) **i** 6.90, 8.22

18.8 a $\hat{y} = 577 + 90.6$ Space + 9.66 Water **b** $s_\varepsilon = 213.7$; this statistic is an estimate of the standard deviation of the error variable. **c** $R^2 = .7081$; 70.81% of the variation in electricity consumption is explained by the variation in the independent variables in the model. **d** $F = 117.64$, p-value = 0; there is sufficient evidence to infer that the model is valid. **e** 7748, 8601 **f** 8127, 8222

18.10 a $\hat{y} = 13.03 - .279$ Age + .0938 Income **b** $R^2 = .1985$; 19.85% of the variation in Internet use is explained by the variation in the independent variables in the model. **c** $F = 24.39$, p-value = 0; there is sufficient evidence to infer that the model is valid. **d** $-.40$ (change to 0, which is the minimum value), 13.50 **e** 6.27, 9.61

18.12 b $F = 67.97$, p-value = 0. There is evidence to infer that the model is valid. **c** The errors appear to be normally distributed and the variance appears to be constant. **d** The correlations between independent variables indicate the presence of multicollinearity. **e** $b_1 = .451$; for each one-year increase in the mother's age of death, the customer's age at death increases on average by .451 year provided that the other variables are held constant (which may not be possible because of multicollinearity). $b_2 = .411$; for each one-year increase in the father's age at death, the customer's age at death increases on average by .411 year provided that the other variables are held constant (which may not be possible because of multicollinearity). $b_3 = .0166$; for each one-year increase in the grandmothers' mean age of death, the customer's age at death increases on average by .0166 year provided that the other variables are held constant (which may not be possible because of multicollinearity). $b_4 = .0869$; for each one-year increase in the grandfathers' mean age of death, the customer's age at death increases on average by .0869 year provided that the other variables are held constant (which may not be possible because of multicollinearity). Mothers: $t = 8.27$, p-value = 0; Fathers: $t = 8.26$, p-value = 0; Grandmothers: $t = .25$, p-value = .8028; Grandfathers: $t = 1.32$, p-value = .1890. The ages at death of the mothers and fathers are linearly related to the ages at death of their children. **f** 65.54, 77.31 **g** 68.75, 74.66

18.14 a lower prediction limit = $35,498, upper prediction limit = $172,240 **b** lower confidence limit = $39,291, upper confidence limit = $90,304

18.16 lower prediction limit = 16,712, upper prediction limit = 35,293

18.18 The independent variables are only weakly correlated. The conclusions from the t-tests are valid.

18.20 The two variables are weakly correlated. The results of the two t-tests are valid.

18.22 a The heights of the mothers and fathers are correlated. **b** The two t-tests may be invalid.

18.24 The histogram is slightly skewed; the errors are not extremely nonnormal. The error variance appears to be constant. The correlations between the independent variables is small; multicollinearity is not a problem. Observation 91's standardized residual exceeds 2.0. The variables should be checked at this point.

18.26 The histogram of the residuals is bell shaped. There is no sign of heteroscedasticity. There is no discernable pattern in the plot of time versus residuals; there is no sign of autocorrelation.

18.28 The variance of the error variable appears to be constant. The histogram of the residuals is bell shaped. Multicollinearity does not appear to be a problem.

18.30 The variance of the error variable is not constant, and the histogram of the residuals is negatively skewed.

18.32 There is evidence of positive first-order autocorrelation.

18.34 There is no evidence to infer negative first-order autocorrelation.

18.36 a $\hat{y} = 2260 + .423x$ **c** Observation 4 **d** The histogram is bell shaped. **e** The error variable variance appears to be constant. **f** $d = .7859$; there is evidence of autocorrelation. **g** $y = \beta_0 + \beta_1 x + \beta_3 t + \varepsilon$; $\hat{y} = 446.2 + 1.10x + 38.92t$ **h** The second model fits better.

18.38 a The histogram of the residuals is bell shaped. **b** The variance of the error variable appears to be constant.

18.40 The test is inconclusive.

18.42 a Let $I_1 = 1$, if Catholic, and 0, if not. Let $I_2 = 1$, if Protestant, and 0, if not. **b** Let $I_1 = 1$, if 8:00 AM to 4:00 PM, and 0, if not. Let $I_2 = 1$, if 4:00 PM to midnight, and 0, if not. **c** Let $I_1 = 1$, if Jack Jones, and 0, if not. Let $I_2 = 1$, if Mary Brown, and 0, if not. $I_3 = 1$, if George Fosse, and 0, if not

18.44 a $t = 3.04$, p-value $= .0016$; there is enough evidence to infer that BBA graduates outperform BA graduates. **b** $t = 3.06$, p-value $= .0015$; there is enough evidence to infer that BEng graduates outperform BA graduates. **c** $t = .51$, p-value $= .3067$; there is not enough evidence to infer that BSc graduates outperform BA graduates. **d** 10.37 **e** 9.69

18.46 a $F = 20.43$, p-value $= 0$; there is enough evidence to support the model's validity. **b** I_2: $t = 1.86$, p-value $= .0713$; I_2: $t = -1.58$, p-value $= .1232$; there is not enough evidence to infer that weather is a factor. **c** $t = 3.30$, p-value $= .0012$; there is enough evidence to infer that on average weekend attendance is greater than weekday attendance.

18.48 $t = 2.00$, p-value $= .0246$; there is enough evidence to infer that female telemarketers stay at their jobs longer than male telemarketers.

18.50 a $\hat{y} = -41.42 + .644$ Boxes $+ .349$ Weight $+ 4.54$ Codes **b** Let $I_1 = 1$, if morning, and 0, if not. Let $I_2 = 1$, if early afternoon, and 0, if not. **c** Model 1: $s_\varepsilon = 6.25$ and $R^2 = .8525$. Model 2: $s_\varepsilon = 3.82$ and $R^2 = .9461$. Model 2 fits better. **d** I_1: $t = -4.51$, p-value $= 0$; I_2: $t = 4.47$, p-value $= 0$; there is enough evidence to infer that time of day is a factor.

18.52 a Let $I_1 = 1$, if welding machine, and 0, if not. Let $I_2 = 1$, if lathe, and 0, if not. $\hat{y} = 119.3 + 2.54$ Age $-11.76\ I_1 - 199.4\ I_2$ **b** $b_1 = 2.54$; for each additional month, repair costs increase on average by \$2.54. $b_2 = -11.76$; in this sample repair costs for welding machines cost on average \$11.76 less than the repair costs for stamping machines. $b_3 = -199.4$; in this sample repair costs for lathes cost on average \$199.40 less than the repair costs for stamping machines. **c** $t = -.60$, p-value $= .2766$; there is no evidence to infer that welding machines costs less to repair than do stamping machines.

18.54 a Let $I_1 = 1$, if newspaper ad, and 0, if not. Let $I_2 = 1$, if radio ad, and 0, if not. **b** $\hat{y} = 282.6 + 25.23$ Ads $- 23.36\ I_1 - 46.59\ I_2$. $F = 14.91$, p-value $= 0$; there is enough evidence to infer that the model is valid. **c** I_1: $t = -1.48$, p-value $= .1467$. $I_2 = -2.83$, p-value $= .0067$; there is enough evidence to infer that the number of customers generated from radio ads is different from the number of customers generated from television ads.

18.56 a The histogram is bell shaped. The variance of the error variable appears to be constant. The Durbin-Watson statistic is .5971 (with $\alpha = .05$, $k = 2$, and $n = 50$ [closest to 52], $d_L = 1.46$ and $d_U = 1.63$); there is evidence of autocorrelation. **b** The correlation between the two prices is .0485; multicollinearity is not a problem.

18.58 a The histogram is bell shaped. The variance of the error variable is not constant. **b** The correlation between age and years as a professional is .7355; multicollinearity is a problem.

18.60 The histogram is bell shaped.

The variance of the error variable is not constant. A transformation may fix the problem. Because of the way in which the experiment was conducted the correlation between the two independent variables is zero.
18.62 a $\hat{y} = .178 + .0067$ Lsize $+ .0243$ Bdywght **b** $t = 2.14$, p-value $= .0475$; there is enough evidence to infer that litter size and brain weight are linearly related. **c** $t = 3.59$, p-value $= .0023$; there is enough evidence to infer that body weight and brain weight are linearly related. **d** $R^2 = .6505$; 65.05% of the variation in brain weight is explained by the variation in the independent variables in the model. **e** $F = 15.82$, p-value $= .0001$; there is sufficient evidence to infer that the model is valid. **f** .407, .472 **g** .367, .410
18.64 a $\hat{y} = 35.7 + .247$ MathDgr $+ .245$ Age $+ .133$ Income **b** $F = 6.66$, p-value $= .0011$; there is sufficient evidence to infer that the model is valid. **c** The error variable appears to be normal with constant variance. **d** Age and Income are correlated, which likely affects the t-tests. **f** 49.01, 81.02
18.66 a $\hat{y} = 11.91 - .430$ Education $+ .0292$ Age $+ .0934$ Children $- .0745$ Income **b** $F = 18.17$, p-value $= 0$; there is sufficient evidence to infer that the model is valid. **c** The histogram of the residuals is approximately bell shaped, but the variance may not be constant. **d** The correlation between Education and Income may distort the t-tests. **e** 1. $t = -3.26$, p-value $= .0008$. 2. $t = 1.16$, p-value $= .1251$. 3. $t = .42$, p-value $= .3390$. 4. $t = -2.69$, p-value $= .0043$. Beliefs 1 and 4 appear to be true.
18.68 a $\hat{y} = 128.6 + .49$ Temperature $- .0029$ Temperature^2 **b** $R^2 = .3727$, R^2-adjusted $= .3030$, $s_\varepsilon = 1.628$; comparing the adjusted coefficients of determination and the standard errors of estimate, we conclude that the linear model fits better.

Chapter 19

19.2 z-test of p: $z = 3.15$, p-value $= .0016$; there is evidence to infer the new technology affects the sex of the baby.
19.4 t-test of μ_D: $t = -2.29$, p-value $= .0121$; there is not enough evidence at the 1% significance level to infer that the company should proceed to stage 2.
19.6 a t-estimator of μ: LCL $= 6.40$, UCL $= 7.77$ **b** LCL $= \$80,000$, UCL $= \$97,125$
19.8 Equal-variances t-test of $\mu_1 - \mu_2$: $t = -6.27$, p-value $= 0$; there is evidence to infer that teenage girls do more housework than do teenage boys.
19.10 z-test of $p_1 - p_2$ (case 2): $z = -2.28$, p-value $= .0114$; there is evidence to infer that the second commercial is viable.
19.12 Analysis of variance, single-factor, independent samples experimental design: $F = 1.37$, p-value $= .2543$; there is no evidence to infer that the type of meal affects test scores.
19.14 Multiple regression, t-tests of β_i: High degree of multicollinearity produced t-tests that indicate that none of the independent variables is linearly related to salary. After deleting all the 1992–93 season variables the following results were obtained. Games played: $t = -1.83$, p-value $= .0740$. Goals: $t = 3.74$, p-value $= .0005$. Assists: $t = 3.57$, p-value $= .0009$. Plus/minus: $t = -.47$, p-value $= .6389$. Penalty minutes: $t = .55$, p-value $= .5859$. Players are paid on the basis of goals and assists.
19.16 Analysis of variance, single-factor, independent samples: $F = 1.09$, p-value $= .3441$; there is no evidence to infer that sales of cigarettes differ according to placement.
19.18 t-test of μ: $t = -1.31$, p-value $= .0994$; there is not enough evidence to infer that the seminars should be instituted.
19.20 t-tests of μ_D: **a** $t = 2.98$, p-value $= .0024$; there is enough evidence to infer that the ETS claim is false. **b** $t = -3.39$, p-value $= .0008$; there is enough evidence to infer that the Kaplan claim is false.
19.22 t-test of μ_D: $t = 8.13$, p-value $= 0$; there is enough evidence to infer that there is a decrease in the metabolism rate when children watch television.
19.24 Analysis of variance, two-factor design: test for interaction: $F = 1.66$, p-value $= .1935$; there is no evidence of interaction. Test for differences between genders: $F = 3.77$, p-value $= .0541$; there is not enough evidence of a difference between men and women. Test for difference between fitness levels: $F = 39.97$, p-value $= 0$; there is evidence of a difference between fitness levels.
19.26 Multiple regression, t-tests of β_i: Only problem-solving skill ($t = 6.27$, p-value $= 0$) and technical knowledge ($t = 11.24$, p-value $= 0$) are linearly related to quality.
19.28 Unequal-variances t-test of $\mu_1 - \mu_2$: $t = -3.71$, p-value $= .0001$; there is evidence to infer that children who wash their hands at least four times per day have fewer sick days due to cold and flu than children who don't wash at least four times per day.
19.30 Multiple regression, t-tests of β_i: Potential evaporation: $t = 1.36$,

p-value = .1772. Precipitation: t = .98, p-value = .3295; neither variable is linearly related to flow.

19.32 Wilcoxon rank sum test: Question 1: $z = -4.95$, p-value = 0; there is evidence to infer that customers who say they will return assess the quality of work performed higher than customers who do not plan to return. Question 2: $z = -1.56$, p-value = .0589; there is not enough evidence to infer that customers who say they will return assess fairness of price higher than customers who do not plan to return. Question 3: $z = -.56$, p-value = .2888; there is not enough evidence to infer that customers who say they will return assess explanation of work and guarantee higher than customers who do not plan to return. Question 4: $z = -1.65$, p-value = .0490; there is enough evidence to infer that customers who say they will return assess the checkout process higher than customers who do not plan to return.

19.34 Questions 1 to 4: Kruskal-Wallis tests: Question 1: $H = 2.63$, p-value = .2678; there is no evidence to infer that there are differences in the assessment of quality of work performed between the three stores. Question 2: $H = 5.25$, p-value = .0723; there is not enough evidence to infer that there are differences in the assessment of fairness of price between the three stores. Question 3: $H = .37$, p-value = .8304; there is no evidence to infer that there are differences in the assessment of explanation of work and guarantee between the three stores. Question 4: $H = 1.25$, p-value = .5341; there is no evidence to infer that there are differences in the assessment of the checkout process between the three stores. Question 5 and comments: χ^2-test of a contingency table. Question 5: $\chi^2 = 2.48$, p-value = .2897; there is no evidence to infer that there is a difference between the three stores with respect to whether the customer will return in future. Comments: $\chi^2 = 30.98$, p-value = 0; there is evidence to infer that there is a difference between the three stores with respect to customer comments.

19.36 a χ^2 goodness-of-fit test: $\chi^2 = 9.40$, p-value = .4943; there is no evidence to indicate that the dice are not fairly balanced. **b** z-test of p: $z = -.48$, p-value = .3143; there is no evidence to infer that the dice are set up so that the probability of 7 is less than 6/36.

19.38 Question 1: Equal-variances t-test of $\mu_1 - \mu_2$: $t = -4.00$, p-value = 0; there is evidence to infer that recovery is faster in the United States. Question 2: z-tests of $p_1 - p_2$: One month after heart attack: $z = -1.55$, p-value = .0609; there is not enough evidence to infer that recovery is faster in the United States. Six months after heart attack: $z = .43$, p-value = .6646; there is not enough evidence to infer that recovery is faster in the United States. Twelve months after heart attack: $z = .26$, p-value = .6016; there is not enough evidence to infer that recovery is faster in the United States.

19.40 Analysis of variance, two-factor design: Product Y: Test for interaction: $F = 14.30$, p-value = 0; there is enough evidence to infer that temperature and flow interact to affect product yield. Product C: Test for interaction: $F = 1.03$, p-value = .3628; there is no evidence of interaction. Test for flow: $F = 3.69$, p-value = .0315; there is evidence to infer that there are differences in product yield between the three levels of flow. Test for temperature: $F = 14.06$, p-value = .0004; there is evidence to infer that there are differences in product yield between the two levels of temperature.

19.42 Analysis of variance, single-factor, independent samples design: $F = 34.35$, p-value = 0; there is enough evidence to infer that the type of music affects test results.

19.44 The data are observational. There may be a link between arthritis and Alzheimer's disease that explains the statistical result.

19.46 Analysis of variance, two-factor design: Test for interaction: $F = .72$, p-value = .4927; there is no evidence of interaction. Test for age: $F = 6.37$, p-value = .0030; there is enough evidence to infer that age affects offers. Test for gender: $F = 7.98$, p-value = .0062; there is enough evidence to infer that gender affects offers.

19.48 Case 19.1: test of difference of interest rates between male-owned and female-owned businesses: Wilcoxon rank sum test: $z = 3.83$, p-value = .0001; there is enough evidence to infer that female business owners pay higher rates of interest than male business owners. Case 19.2: Test of relationship between interest rates and sales: Spearman rank correlation: $r_s = -.2629$, $z = -8.53$, p-value = 0; there is evidence to infer that interest rates and sales are linearly related. Test of relationship between interest rates and ages: Spearman rank correlation: $r_s = -.1853$, $z = -6.01$, p-value = 0; there is evidence to infer that interest rates and ages are linearly related. Differences in sales between male-owned and female-owned businesses: Wilcoxon rank sum test: $z = -14.09$, p-value $= 0$; there is evidence to infer that female-owned businesses have lower sales than male-owned businesses. Differences in ages between male-owned and female-owned businesses: Wilcoxon rank sum test: $z = -6.26$, p-value $= 0$; there is evidence to infer that male-owned businesses are older than female-owned businesses. Difference in interest rates between the three types of business: Kruskal-Wallis test: $H = 7.22$, p-value $= .0270$; there is enough evidence to infer that there are differences in interest rates between the three types of businesses.

Index

A

ABS (example), 392–397
Accounting course exemptions (case), 387
Addition rule, 105–106
   formula, 105, 114
Alternative hypothesis, 255–257
   setting up to define Type I and Type II errors, 286–287
Ambulance and fire department responses (case), 665–666
Analysis of variance (ANOVA), 357–358, 405–463
   ANOVA table, 414–415
   Bartlett's test, 455–456
   Bonferroni adjustment, 451–452, 454
   compared with t-tests of the difference between two means, 418–419
   for complete factorial experiment, 444–445
   computation via Excel, 429
   defined, 406
   experimental designs, 423–425
   Fisher's least significant difference method, 449–451, 454
   multifactor experiment, 423
   multiple comparisons, 449–454
   randomized block design. *See* Randomized block design
   relationship between F-statistic and t-statistic, 420
   single-factor ANOVA. *See* Single-factor analysis of variance
   three examples, 406
   Tukey's multiple comparison method, 452–454
   two-factor, 434–445
   two-factor experiment, 423
   See also Randomized blocks; Single-factor analysis of variance, Two-factor analysis of variance
Announcer commercials (case), 4, 403–404
ANOVA. *See* Analysis of variance (ANOVA)
ANOVA table, 414–415
   for randomized block design, 427
Antidepressant drug evaluation (case), 668–669
Apple juice concentrate (example), 407–416, 434–436
Art exhibit (example with histogram), 21–25
Autocorrelation, 577
Average, 54

B

Balanced (design), 437
Bank discrimination against women (cases), 644–653, 670
Bar chart, 33
   creating via Excel, 34
   use in magazines, newspapers, 35–36
   when to use, 36
Barnes Exhibit (example with histogram), 21–25, 147–149
Bartlett's test, 455–456
   Excel computation, 455–456
   formula, 455–456, 460
Bayes' Law, 109–110
Bell-shaped histogram, 28, 29
Beta, determining via Excel for test hypothesis, 285–286
Between-treatments variation, 409
Bimodal histogram, 28
Binomial distribution, 128–136
   binomial table, 132–133
   cumulative probability, 132
   formula for mean, 134
   formula for standard deviation, 134
   formula for variance, 134
   testing hypotheses about proportions by using, 320
Binomial experiment, 128
Binomial probability distribution, formula, 130, 142
Binomial random variable, 128–130
   defined, 128, 129
Binomial table, 132–133
Bivariate normal distribution, 554
Bivariate techniques, 42
Block, defined, 424
Bonanza International, case, 386–387
Bonferroni adjustment, 451–452
   Excel computation, 453–454
Bottleneck in production line (example), 294–295
Bound, 236
Box plot, 73–75
   defined, 73
Bunting (cases), 116–117, 145
Business school graduates (example of bar and pie charts), 33–34
Business school graduates' income (example), 208

C

Calculations, Excel v. manual, 295
Cases
   accounting course exemptions, 387

ambulance and fire department responses, 665–666
antidepressant drug evaluation, 668–669
bank discrimination against women, 644–653, 670
Bonanza International, 386–387
to bunt or not to bunt (part 1), 116–117
to bunt or not to bunt (part 2), 145
exposure to a code of ethics, 494–495
host selling and announcer commercials, 4–5, 403–404
insurance company compensation for lost revenues, 586–587
Let's Make a Deal, 116
nutrition education programs, 669
PC Magazine survey, 666–667
Pepsi's exclusivity agreement with a university, 2–4, 327–328
Pepsi's exclusivity agreement with a university, Coke side, 328
predicting outcomes of basketball and baseball games, 493
predicting university grades from high school grades, 585–586
Quebec: electoral fraud?, 634–635
Quebec separation, 4, 403
uninsured motorists, 328–329
Wilfred Laurier University graduate survey, 667–668
Casino example (histogram), 31
Catsup labeling (example), 268–271
Central limit theorem, 205
Chebysheff's Theorem, 65–66
Chi-squared density function, formula, 177
Chi-squared distribution, 177–182
   computation via Excel, 181–182
   determining chi-squared values, 177–182
   mathematical derivation, 180–181
Chi-squared statistic, 305
   Excel computation, 476–478
Chi-squared test and estimate: variance, via Excel, 308
Chi-squared tests, 464–495
   computing chi-squared statistic (Excel), 476–478
   of a contingency table, 472–480
   factors that identify, 310
   goodness-of-fit test, 465–472, 490
   for normality, 484–489
   situations in which they apply (examples), 465

I-1

Cholesterol drug (example), 428–429
Choosing correct inferential technique, 389–397, 637–643
  flowchart, 390, 642–643
Classes, and frequency distribution, 22
Class intervals, selecting the number of, 24–25
Cluster sample, defined, 195
Cluster sampling, 195–196
Coefficient of correlation, 77–78, 568–569
  compared with scatter diagram and covariance, 78
  Excel calculation, 80, 570
  population, 78
  sample, 78
  test statistic for testing coefficient of correlation = 0, 569, 582
Coefficient of determination, formula, 559, 582
Coefficient of determination adjusted for degrees of freedom, 594, 631
Comfortability of ride (example), 513–515
Complement rule, 103
  formula, 103, 114
Complete factorial experiment
  conducting analysis of variance, 444–445
  defined, 436
Completely randomized design, 416
Computer keyboard production (example), 271–273
Computers, and statistics, 7
Concepts, twelve statistical concepts you need for life, 654–655
Conditional probability, 97–98
  formula, 98, 114
Confidence interval estimator of μ, 234, 235
Confidence level, 7, 233, 234
Consistency, 230
Container-filling machines (example), 307–309
Contingency table
  chi-squared test of, 472–480
  defined, 473
  Excel computation, 653
  expected frequencies for a, 475
Continuity correction factor, 216
Continuous probability distributions, 146–187
  chi-squared distribution, 177–182
  F distribution, 182–185
  finding values of Z, 161–165
  normal distribution, 153–170
  probability density functions, 147–153
  Student t distribution, 170–176
  uniform probability distribution, 149–150
  z_A and percentiles, 165–168
Continuous random variable, 120, 147
  using to approximate a discrete distribution, 151–153

Covariance, 76–77
  calculating via Excel, 80
  compared with scatter diagram and coefficient of correlation, 78
  magnitude, 77
  population covariance formula, 76
  sample covariance formula, 76
  sign, 77
Critical number omega, 453
Cross-classification table, defined, 473
Cross-sectional data, defined, 38
Cumulative normal probability, computation via Excel, 167–168
Cumulative probability, 132
D
Data
  calculations for types of, 18–19
  categorical, 17
  cross-sectional, 38
  defined, 16
  experimental, 348–349
  general guidelines for exploring, 84
  hierarchy of, 19
  interval. See Interval data
  missing data, 317
  nominal. See Nominal data
  numerical, 16
  observational, 348–349
  ordinal. See Ordinal data
  qualitative, 17–18
  quantitative, 16
  ratio, 16
  recoding, 317–318
  time-series, 38
  types, 16–20
Degrees of freedom, 171, 296
Denominator degrees of freedom, 362
Dependent variable, defined, 543
Derivations of statistical formulas, 290–291
Descriptive statistics, 3
Descriptive techniques, 654
Deterministic model, 545
Difference between two sample means, defined, 220
Discrete random variable, 120
Distribution-free statistics, 497
  See also Nonparametric statistical techniques
"Double-blind" experiment, 254
Drop-down menus, 11
Dummy variable, 625
Durbin-Watson statistic, 612
Durbin-Watson test, 612–615
E
Education costs (example), 42–44
Election
  inferring winner (example), 294
  Senate seat exit poll (example), 313–315
Election forecast (example), 218–219
Empirical Rule, 64–65, 66
Employment duration (example), 505–507

Equal-variances interval estimator, formula, 334, 379
Equal-variances test statistic, formula, 334, 379
Equal-variances t-test and estimator of difference between two means, factors that identify, 344
Error of estimation, 236
  bound on, 236
Error variable
  defined, 552
  required conditions, 552–554
Estimate, defined, 189
Estimation, 227–252
  computing z-estimate of a mean manually, 238
  computing z-estimate of a mean via Excel, 238–241
  consistency, 230
  correctness/incorrectness of interval estimates, 249–250
  error of estimation, 236
  how to apply estimation techniques, 237
  interpreting interval estimate, 240–243
  interval estimator, 228–231
  interval estimator of μ, 233
  objective of, 228
  point estimator, 228
  of population mean when population standard deviation is known, 232–245
  relative efficiency, 231
  selecting sample size, 245–247
  unbiased estimator, 229–230
Estimator of population variance, factors that identify, 310
Ethics, exposure to a code of (case), 494–495
Event, defined, 93
Events
  independent, 98–99
  union of, 99–100
Excel, 7
  ANOVA: two-factor with replication, 442–443
  ANOVA, 429, 652
  arithmetic mean, 55–56
  bar chart creation, 34
  Bartlett's test, 455–456
  Bonferroni adjustment, 453–454
  cell, 11
  chi-squared distribution, 181–182
  chi-squared goodness-of-fit test, 469
  chi-squared statistic, 476–478
  chi-squared test and estimate: variance, 308
  chi-squared test of normality, 487–488
  coefficient of correlation, 570, 649–650
  contingency table, 653
  Control key, 11
  correlation, 80
  covariance, 80

INDEX I-3

and creating simple random sample, 192–193
cumulative normal probability, 167
cumulative probability, 133
determining Beta, 285–286
Durbin-Watson statistic, 619, 621
F-estimate of the ratio of two variables, 365–366
F-test two-sample for variances, 363–364, 395, 646–647
F value, 184–185
File (box), 11
Fisher's least significant difference method, 453–454
Formula bar, 11
Friedman test, 532
histogram creation, 23–24
individual probability of binomial random variable, 133
installing, 10
introduction to, 10–11
Kruskal-Wallis test, 526–527
least squares line, 81
line chart creation, 39
macros, 7
median, 57
Menu bar, 11
mode via, 58
multiple regression, 592–593, 607–610, 616–618, 619–621
multiple regression with indicator variables, 625–626
multiple regression with nominal independent variables, 624
Open command, 11
percentiles, 72–73
pie chart creation, 34
Poisson random variables, 140
prediction and interval estimate, 601
prediction interval, 566
regression line, 548–550
residual analysis, 574
sample distribution creation by computer simulation, 212–215
sample variance, 63
scatter diagram creation, 43
screen, 11
sign test, 514
single-factor analysis of variance, 416
Student t distribution, 176
t-estimate of a mean computing, 300–301
t-estimate of the difference between two means (equal-variances), 341–342, 396
t-estimate of the difference between two means (unequal-variances), 339
t-test: paired two sample for means, 353–354
t-test: two-sample assuming equal variances, 340–341, 395, 651
t-test: two-sample assuming unequal variances, 337–338
t-test and estimate: mean, 298–299, 354–355
Title bar, 11
Toolbars, 11
Tukey's multiple comparison method, 453–454
v. manual calculations, 295
Wilcoxon rank sum test, 504
workbook, 11
worksheet, 11
z-estimate of a mean, 238–241
z-test and estimate: proportion, 314, 316
z-test and estimate: two proportions, 371–375, 394, 645
z-test of a mean, 267–268
Exhaustive outcome, defined, 91
Exit poll, 294
Expected frequency, 467
Expected value (of X), 125
 formula, 124, 125, 142
Expected value, laws of, 126
Expected value of the difference between two means, formula, 226
Expected value of the sample mean (formula), 225
Expected value of the sample proportion, formula, 225
Experimental data, 348–349
Experimental unit, defined, 408
Experiments, design of, 358
Explained variation, 561–562
Explanatory power, 562

F

Fabric softener market share (example), 466–469
Factor, 406
Factorial experiments, 434
Factor level, 406
F Crit, 416
F density function, formula, 182
F distribution, 182–185
 determining values of F, 182–185
 determining values of F with Excel, 184–185
F-estimate of the ratio of two variables, Excel computation, 365–366
F-statistic, 631
F-test two-sample for variances, Excel computation, 363–364, 395
First-order autocorrelation, 613
First-order linear model, 545
 formula, 546
Fisher's least significant difference method, 449–451
 Excel computation, 453–454
Fixed-effects experiment, 424
Flextime (example), 518–521
Flowchart for identifying correct technique, 390, 642–643
Forecasting, 254–255
Formula population coefficient of correlation, 78, 86
Formulas
 addition rule, 105, 114
Bartlett's test, 455–456, 460
binomial probability distribution, 130, 142
chi-squared density function, 177
coefficient of determination, 559, 582
coefficient of determination adjusted for degrees of freedom, 594, 631
complement rule, 103, 114
conditional probability, 98, 114
equal-variances interval estimator, 334, 379
equal-variances test statistic, 334, 379
expected value (of X), 124, 125, 142
expected value of the difference between two means, 226
expected value of the sample proportion, 225
F density function, 182
F-estimator of ratio of two variances, 380
F-test of ratio of two variances, 379
first-order linear model, 546
Friedman test, 530, 536
independent samples single-factor experiment, 409–410, 413, 458
interval estimator of μ, 233, 251, 296, 323
interval estimator of mean of the population of differences, 354, 379
interval estimator of population variance, 306, 324
interval estimator of the expected value of y, 564, 582
laws of expected value, 126, 142
laws of variance, 126, 142
least significant difference method, 450, 459
mean square for error, 413, 458, 631
mean square for regression, 631
mean square for treatments, 413, 458
multiplication rule, 103, 114
normal density function, 153
Poisson probability distribution, 137, 142
population mean, 54, 86
population standard deviation, 64, 86
population variance, 61, 86, 125, 142
prediction interval, 564, 582
randomized block experiment, 426–427, 458–459
range, 60–86
sample coefficient of correlation, 87, 569, 582
sample mean, 54, 86
sample size, 245, 251
sample slope, 547, 581
sample Spearman rank correlation coefficient, 571, 582
sample standard deviation, 64, 86
sample variance, 61, 86
sample y–intercept, 547, 581
slope coefficient, 81, 87
standard deviation, 142
standard deviation of the difference between two means, 226

Formulas *continued*
  standard deviation of the sample mean, 225
  standard deviation of the sample proportion, 225
  standard error of estimate, 556, 581, 593, 630
  standard error of sample slope coefficient, 558, 582
  standardizing the difference between two sample means, 226
  standardizing the sample mean, 225
  standardizing the sample proportion, 225
  Student t density function, 171
  sum of squares for error, 555, 581
  sums of squares in two-factor experiment, 440, 459
  test statistic for β_i, 630
  test statistic for chi-squared goodness-of-fit test, 467, 490
  test statistic for μ, 296, 323
  test statistic for mean of the population of differences, 353, 379
  test statistic for population variance, 305, 324
  test statistic for single-factor analysis of variance, 413, 458
  test statistic for slope, 558, 581
  Tukey's multiple comparison method, 453, 459
  two-factor experiment, 440, 441, 459
  unequal-variances interval estimator, 335, 379
  unequal-variances test statistic, 335, 379
  uniform probability density function, 150
  variance of the difference between two means, 226
  variance of the sample mean, 225
  variance of the sample proportion, 225
  Wilcoxon rank sum test statistic, 502, 536
  y-intercept, 81, 87
  z-interval estimator of difference of two proportions, 380
  z-test and estimator of difference of two proportions, 380
Frequencies, 23
Frequency distribution, 22
Friedman test, 529–533
  Excel computation, 532
  factors that identify, 533
  formula, 530, 536
  and sign test, 533

G
Gambling example (histogram), 31
Gasoline sales (example), 150–151
Goodness-of-fit test (chi-squared), 465–472
  test statistic for, 467
  when to use, 470–471
Gosset, William S., and t-statistic, 296

Grades and course assessments (example), 571–572
Graphical descriptive techniques, 3, 15–51
  bar chart, 33–36
  compared with numerical techniques, 68
  frequency distribution, 22–23
  histogram, 22–32
  least squares line, 46–47
  line chart, 38–41
  pie chart, 33–36
  scatter diagrams, 42–47
Guide to statistical inference (where each technique is introduced), 290

H
Heteroscedasticity, 575–576
High-fiber cereals (example), 336–339
Highly significant, 266, 267
Histogram, 22
  bell-shaped, 28, 29
  bimodal, 28
  creating a, 22–25
  information looked for, 53
  kurtosis, 70
  shapes of, 25–27
  skewness, 25, 27, 70
  symmetric, 25, 26
  unimodal, 27
  when to use, 32
Holmes, Oliver Wendell, 256
Home selling price (example), 606–610
Homoscedasticity, 575
Host selling commercials (case), 4, 403–404
Hypothesis testing, 253–292
  alternative hypothesis, 255–257
  calculating probability of Type II error, 279–287
  conclusions of a test of hypothesis, 262
  effect on β of changing μ, 281–282
  effect on β of changing probability of Type I error, 282–283
  guidelines, 266–268
  illustrations of, 254–255
  interpreting results of a test, 261–262
  judging the test, 283–284
  null hypothesis, 255–257
  one-tail tests, 271
  population mean/population standard deviation known, 257–279
  power of a test, 285
  p-value method, 262–266
  reducing probability of error by increasing sample size, 284
  rejection region method, 258–260
  research hypothesis, 255–257
  setting up alternative hypothesis to define Type I and Type II errors, 286–287
  standardized test statistic, 260
  testing hypotheses and interval estimators, 274, 275
  test statistic, 257
  two-tail tests, 271
  Type I error, 255–256
  Type II error, 255–256

I
Independence, defined, 98–99
Independent samples
  inference about difference between two means, 332–345
  matched pairs compared with, 355–356
  single-factor ANOVA, 409–410, 413, 421, 458
  two-factor ANOVA, 332–347
  Wilcoxon rank sum test, 499–510
Independent variables, defined, 543
Indicator variable, 625
Inference, 6–7, 90, 223–224, 228, 293–404
  choosing correct inferential technique, 389–397, 637–643
  flowchart of techniques, 390, 642–643
  guide showing where each technique is introduced, 290
  inferential techniques, 654
  single population. *See* Inference about a single population
  summary of, 391–392, 637–643
  two populations. *See* Inference about two populations
Inference about a single population, 293–329
  about mean when standard deviation is unknown, 295–305
  factors that identify z-test and interval estimator of p, 321
  missing data, 317
  population proportion, 311–323
  population variance, 305–311
  recoding data, 317–318
  selecting sample size to estimate proportion, 318–320
  statistic and sampling distribution, 312
  summarized, 391
  testing hypotheses about proportions by using binomial distribution, 320
  three examples, 294–295
  See also Inference; Inference about two populations
Inference about two populations, 330–387
  decision rule: equal-variances or unequal-variances t-tests and estimators, 335–336
  difference between two population proportions, 367–378
  four examples, 331–332
  independent samples, 332–347
  independent samples compared with matched pairs, 355–356
  matched pairs experiment, 349–361
  ratio of two variables, 361–367
  sampling distribution of difference between two means, 332–333

summarized, 391–392
See also Inference; Inference about a single population
Inferential statistics, 3–4
Insurance company compensation for lost revenues (case), 586–587
Interquartile range, 73
Intersection, defined, 96
Interval data, 16, 19, 289, 654
 analyzing, 655
 calculations for, 18
 graphically describing, 20–33
 numerical descriptive techniques for, 52–88
Interval estimate
 correctness/incorrectness of, 249–250
 interpreting, 240–243
 width of, 242, 245
Interval estimator, 228–231
 as probability statement about sample mean, 242–243
Interval estimator for mean of the population of differences, 354, 358
 factors that identify, 359
 formula, 354, 379
Interval estimator for population variance, 306
Interval estimator of difference between two population proportions, 370, 375, 380
Interval estimator of μ, 233
 formula, 233, 251, 296
Interval estimator of p, 312–313, 324
Interval estimator of population variance, formula, 306, 324
Interval estimator of ratio of two population variances, 365, 380
Interval estimator of the expected value of y, formula, 564, 582
Interval variables, describing relationship between two (scatter diagrams), 42–48

J
Joint probability, 96
 given v. required, 111

K
Kruskal-Wallis test, 524–528
 Excel computation, 526–527
 factors that identify, 528
 formula, 524, 536
 Wilcoxon rank sum test and, 527
Kurtosis, 70

L
La Quinta Inns (example), 591–599
Laws of expected value, 126, 142
Laws of variance, 126, 142
Least significant difference method, 449–451
 formula, 450, 459
Least squares line, 46–47, 80–81
 calculating coefficients, 81
 calculating via Excel, 81
Least squares method, 80–84, 547
Let's Make a Deal (case), 116
Level, 406
Likelihood probability, 110
Linear regression. *See* Simple linear regression and correlation
Linear relationship, 44
 positive and negative, 45
Line chart, 38–42
Literary Digest poll, 190–191, 198
Lower confidence level, 233

M
Marginal probability, 96–97
Marks (in courses with and without computer—example), 28–30
Marks (in last year's final exam—example), 20–21
Matched pairs, defined, 352
Matched pairs experiment, 349–361
 compared with independent samples, 355–356
 estimating the mean difference, 354–355
 observational and experimental data, 357
 required condition, 357
Mean
 arithmetic, 54–56
 computing via Excel, 55–56
 population, 54, 86, 124, 232–245
 sample, 54, 86
 v. mode and median, 58
Mean absolute deviation, 63
Mean of the population of differences, 352–353
Mean square for error, formula, 413, 458
Mean square for treatments, formula, 413, 458
Mean squares, defined, 413
Measures of central location, 54–59
 arithmetic mean, 54–56
 mean, median, mode compared, 58
 median, 56–57
 mode, 57–58
 for ordinal and nominal data, 59
Median, 19, 56–57
 computing via Excel, 57
 v. mean and mode, 58
Mensa membership (example), 165
Missing data, 317
Modal class, 27
Mode, 27, 57–58
 computing via Excel, 58
 v. mean and median, 58
Models, 544–546
Money earned by college students (example), 159–161
Multicollinearity, 606–611
Multinomial experiment, 466
Multiple regression, 588–635
 Durbin-Watson test, 612–615
 estimating coefficients and assessing the model, 590–605
 Excel calculation, 592–593
 F-test and the t-test in simple linear regression model, 600
 model and required conditions, 589–590
 multicollinearity, 606–611
 nominal independent variables, 623–627
 regression diagnostics, 605–612
 remedying violations of required conditions, 611
 required conditions for error variable, 590
 t-tests and analysis of variance, 599
 using regression equation, 600
Multiplication rule, 103–104
 formula, 103, 114
Mutually exclusive outcome, defined, 91

N
Negative linear relationship, 45
Nielsen ratings, 190, 295, 315
Nightline phone-in show, 190–191
Nominal data, 16–17, 19, 289, 654
 calculations for, 18–19
 measures for central location, 59
 statistical techniques for, 483
 summary of tests on, 482–484
 techniques, 655
Nonparametric, defined, 508
Nonparametric statistical techniques, 496–541
 defined, 497
 Friedman test, 529–533
 Kruskal-Wallis test, 524–528
 sign test, 511–515
 Wilcoxon rank sum test, 499–510
 Wilcoxon signed rank sum test, 516–521
Nonresponse error, 197
Nonsampling error
 errors in data acquisition, 197
 nonresponse error, 197
 selection bias, 198
Normal approximation to the binomial, 215–218
Normal density function, formula, 153
Normal distribution, 153–170
Normal random variable, 153
Null hypothesis, 255–257
Numerator degrees of freedom, 362
Numerical descriptive techniques for interval data, 3, 52–88
 arithmetic mean, 54–56, 58
 box plot, 73–75
 Chebysheff's Theorem, 65–66
 coefficient of correlation, 77–78
 compared with graphical techniques, 68
 covariance, 76–77
 covariance compared with scatter diagram and coefficient of correlation, 78
 Empirical Rule, 64–65, 66
 general guidelines for exploring data, 84
 interquartile range, 73
 least squares method, 80–84
 measures of central location, 54–60

Numerical descriptive techniques for interval data *continued*
  measures of linear relationship, 76–84
  measures of variability, 60–69
  percentiles, 71–73
  range, 60–61
  sample kurtosis, 70
  sample skewness, 70
  standard deviation, 64–67
  variance, 61–64
Nutrition education programs (case), 669

O
Objective (type of information produced by histogram), 32
Observational data, 348–349
Odometer reading (example), 558–559
Oil production, table and bar chart compared, 35–36
One-sided interval estimators, 274
One-tail tests, defined, 271
One-way analysis of variance, 407–423
Ordinal data, 17, 19, 289, 654
  calculations for, 19
  graphically describing, 35
  measures for central location, 59
  measures of relative standing and variability for, 75
  and nonparametric techniques, 497
  techniques, 655
  treatment in a time series, 41
Outliers, 74, 578
Oy vey, 191

P
p
  interval estimator of, 312–313, 324
  test statistic for, 312, 324
p-value
  defined, 263
  describing, 164–166
  interpreting, 263–264
p-value method, 262–266
Parameter, 6
Parameters, 189
PC Magazine survey (case), 666–667
Pearson coefficient of correlation, 568
Pepsi
  exclusivity agreement with university (case), 327–328
  exclusivity agreement with university, Coke side (case), 328
Percentiles, 71–73
  defined, 71
  formulas, 71
  locating, 71–72
  and z_A, 165–166
Pie chart, 33
  creating via Excel, 34
  use in magazines, newspapers, 35–36
  when to use, 36
Plane, 589
Point estimator, 228–231, 305
Point estimator of ratio of two variances, 361
Poisson distribution, 136–141

Poisson experiment, 136
Poisson probability distribution, 137
  formula, 137, 142
Poisson random variable
  defined, 136–137
  Excel computations, 140
Poisson table, 139–140
Pooled proportion estimator, 369
Pooled variance estimator, 334
Population, 6
Population coefficient of correlation, formula, 78, 86
Population locations, 497
Population mean, 54
  estimating when standard deviation is known, 232–245
  formula, 54, 86, 124
Population parameters, 189
  using sampling distribution to infer, 209–210
Population proportion, inference about a, 311–323
Population standard deviation, formula, 64, 86, 125
Population variance
  formula, 125, 142
  inference about, 305–311
  testing and estimating, 306
Positive linear relationship, 45
Posterior probability, 110
Predicting the outcomes of basketball and baseball games (case), 493
Predicting university grades from high school grades (case), 585–586
Prediction interval
  Excel computation, 566
  formula, 564, 582
Prior probability, 110
Probabilistic model, defined, 545
Probability, 89–187
  addition rule, 105–106
  applicability illustrated, 90
  assigning to events, 90–95
  classical approach, 92
  complement rule, 103
  conditional, 97–98
  continuous probability distributions. *See* Continuous probability distributions
  cumulative, 132
  distribution. *See* Probability distribution
  of an event, 93–94
  identifying method to use, 111
  interpreting, 94
  intersection of two events, 96
  joint, 96, 111
  likelihood, 110
  marginal, 96–97
  multiplication rule, 103–104
  posterior, 110
  prior, 110
  probability density function, defined, 149

  random experiment, 90–91
  relative frequency approach, 92
  requirements of, 92
  subjective approach, 92–93
Probability density function, defined, 149
Probability distribution
  binomial, 128–136
  binomial experiment, 128
  binomial random variable, 128, 129–130
  defined, 120
  discrete, 121
  formula for discrete, 121
  Poisson distribution, 136–141
  relationship to a population, 122
  and statistical inference, 223–224
Probability trees, 107–110
Problem objectives, 289–290
Proportion, 311
  sampling distribution of, 215–220

Q
Quartiles, defined, 71
Quebec separation (cases), 4, 403

R
Random-effects experiment, 424
Random experiment, 90–91
Randomized block design, 425–432
  defined, 424
  factors that identify, 432
  formulas, 426–427, 458–459
  mean squares, 426
  sums of squares, 426
  test statistic for blocks, 427
  test statistic for treatments, 427
Random variable
  continuous, 120, 147
  defined, 119–120
  discrete, 120
  standardizing, 154
Range, 60–61
  formula, 60, 86
Rate of return, 21
Recoding data, 317–318
Recycling newspapers (example), 297–299
Red Book, 547
Regression analysis
  defined, 543
  examples, 543–544
Regression diagnostics, 574–579, 605–623
Regression line, Excel calculation, 548–550
Rejection region, defined, 258
Rejection region method, 258–260
Relative efficiency, 231
  simulation experiment, 248–249
Repeated measures, 424
Replicate, defined, 437
Research hypothesis, 255–257
Residual analysis, Excel computation, 574
Residuals, 555
Responses, 408

Response surface, 589
Response variable, defined, 408
Return on investment, example, 30–31
Robust (techniques), defined, 299
Rule of five, 470, 479

S

Salaries of MBAs at Wilfred Laurier University (example), 221–222
Sample, 6
Sample coefficient of correlation, formula, 87, 569, 582
Sampled population, 189
Sample kurtosis, formula, 70
Sample mean, 54
 formula, 54, 86
Sample size, 4, 196
 to estimate a mean, 245
 formula, 245, 251
 reducing probability of error by increasing, 284
 selecting, 245–247
 selecting to estimate proportion, 318–320
Sample skewness, formula, 70
Sample slope, formula, 547, 581
Samples of size 2 and their means (table), 201
Sample space, 91–92
Sample Spearman rank correlation coefficient, formula, 571, 582
Sample statistic, difference from hypothesized value of parameter, 275
Sample y-intercept, formula, 547, 581
Sampling, 188–198, 654
 cluster sampling, 195–196
 estimate, 189
 Literary Digest poll, 190–191, 198
 Nielsen ratings, 190
 Nightline phone-in show, 190–191
 nonsampling error, 197–198
 sample size, 196
 sampling error, 196–197
 sampling plans, 191–196
 simple random, 191–193
 stratified random sampling, 193–195
Sampling distribution, 119, 654, 199–226
 approximate sampling distribution of a sample proportion, 218–219
 central limit theorem, 205
 creating by computer simulation, 212–215
 creating empirically, 206
 defined, 200
 difference between two means, 220–222, 332–333
 of the mean, 200–215
 normal approximation to the binomial, 215–218
 of a proportion, 215–220
 of the sample mean (formula), 206
 and statistical inference, 223–224
 of s^2, 305
 use for inference, 209–210

Sampling distribution of a proportion, 215–220
Sampling distribution of difference between two means, 220–222, 332–333
Sampling distribution of difference between two proportions, 368
Sampling distribution of the mean, 200–215
Sampling error, 196–197
 defined, 196
Scatter diagrams, 42–48, 546
 compared with coefficient of correlation and covariance, 78
 creating via Excel, 43
 defined, 42
 patterns of, 44–45
 when to use, 47
Selection bias, 198
Self-selected samples, defined, 190
Significance level, 7, 255
Significant, 266, 267
Sign test, 511–515
 defined, 511
 Excel computation, 514
 factors that identify, 521
 formula, 512, 536
 and Friedman test, 533
Simple event, defined, 93
Simple linear regression and correlation, 542–587
 assessing the model, 555–562
 cause-and-effect relationship, 562
 coefficients of correlation, 568–569
 error variable: required conditions, 552–554
 estimating coefficients, 546–550
 F-test and t-test in, 600
 heteroscedasticity, 575–576
 models, 544–546
 nonnormality, 575
 outliers, 578
 regression diagnostics, 574–579
 residual analysis, 574–575
 Spearman rank correlation coefficient, 570–571
 using regression equation, 564–567
 See also Multiple regression
Simple linear regression model, 545
Simple random sample
 defined, 191
 use of Excel, 192–193
Single-factor analysis of variance, 407–423
 Excel computation, 416
 factors that identify the independent samples, 421
 notation for, 409
 randomized blocks, 425–432
 sampling distribution of test statistic, 413–416
 single factor experiment, 423
 test statistic, 409–413
 See also Analysis of variance (ANOVA); Two-factor analysis of variance

Single population
 inference about, 293–329
 inference about mean when standard deviation is unknown, 295–305
Skewness, 70
Ski resort ticket sales (example), 616–622
SLOP (self-selected opinion poll), 191
Slope, testing, 557–558
Slope coefficient, formula, 81, 87
Soda pop bottle contents (example), 206–208
Solving problems, basic approach, 237
Spearman rank correlation coefficient, 570–571
SS(Total), 415
SSB, 425
SSE, 410, 458, 555, 581
SST, 409–458
Standard deviation, 64–67
 defined, 64
 interpreting, 64–67
Standard deviation of the difference between two means, formula, 226
Standard deviation of the sample mean, formula, 225
Standard deviation of the sample proportion, formula, 225
Standard error of estimate, 556–557
 formula, 556, 581, 593, 630
Standard error of sample slope coefficient, formula, 558, 582
Standard error of the mean, 206
Standardized test statistic
 defined, 260
 formula, 260, 292
Standardizing the difference between two sample means, formula, 226
Standardizing the sample mean, formula, 225
Standardizing the sample proportion, formula, 225
Standard normal random variable, defined, 155
Standard standard deviation, formula, 64, 86, 125, 142
Statistic, 6
Statistical inference, 6–7, 90, 223–224, 228
 See also Inference
Statistically significant, defined, 260
Statistical methods, 654
 specific objectives of, 289–290
Statistician, 2
Statistics, 2
Statistics practitioners, 2
Stratified random sample, defined, 194
Stratified random sampling, 193–195
 criteria for separating a population, 194
Student t distribution, 170–176, 296
 computing via Excel, 176
Sum of squares for blocks (SSB), 425
Sum of squares for error (SSE), formula, 410, 458, 555, 581

Sum of squares for Treatments (SST), formula, 409, 458
Sums of squares, in two-factor analysis of variance, 440
Symbols, defined, 49, 85–86, 141, 187, 224–225, 251, 291, 323, 490, 535

T
Target population, 189
Tax audit (example), 192–193
Television
  determining ratings (example), 295
  Tonight Show statistics (example), 315–316
  viewing effects (example), 237–240, 256
Test, power of, 285
t-estimate of a mean, computing via Excel, 300–301
t-estimate of the difference between two means (equal-variances), Excel computation, 341–342, 396
t-estimate of the difference between two means (unequal-variances), Excel computation, 339
t-statistic, defined, 296, 300
t-test: paired two-sample for means, Excel computation, 353–354
t-test: two-sample assuming equal variances, Excel computation, 340–341, 350–351, 395
t-test: two-sample assuming unequal variances, Excel computation, 337–338
t-test and estimate: mean, Excel computation, 354–355
t-test and estimator of μ, 301
t-tests of the difference between two means, analysis of variance (ANOVA) compared with, 418–419
Testing the slope, 557–558
Test statistic, 257
Test statistic for μ, formula, 296, 323
Test statistic for mean of the population of differences, 353, 358, 379
  factors that identify, 359
Test statistic for p, 312, 324
Test statistic for population variance, 306
Test statistic for ratio of two variances, 362, 379
Test statistic for single-factor analysis of variance, formula, 413, 458
Test statistic for slope, formula, 558, 581
Test statistic for σ^2, formula, 305, 324
Test statistic for the difference between two proportions, 369, 370, 375, 380
Time-series data
  defined, 38
  describing, 38–42
Time series regression diagnostics, 612–623
Tire longevity (example), 349–355
Trial (example of hypothesis testing), 255–256
Tukey's multiple comparison method
  Excel computation, 453–454
  formula, 453, 459
Twelve statistical concepts you need for life, 654–655
Two-factor analysis of variance, 434–445
  Excel computation, 442–443
  factors that identify, 445
  F-tests in, 440–441
  sums of squares in, 440
  See also Analysis of variance (ANOVA); Single-factor analysis of variance
Two-tail tests, 271
  computing, 272–273
Type I error, 255–256, 654
  costs of, 266–267
  setting up alternative hypothesis to define, 286–287
Type II error, 255–256, 654
  calculating probability of, 279–287
  costs of, 266–267
  setting up alternative hypothesis to define, 286–287

U
Unbiased estimator, 229–230
  simulation experiment, 247–248
Unequal-variances interval estimator, formula, 335, 379
Unequal-variances test statistic, formula, 335, 379
Unequal-variances t-test and estimator of difference between two means, factors that identify, 345
Uniform probability density function, formula, 150
Uniform probability distribution, 149–150
Unimodal histogram, 27
Uninsured motorists (case), 328–329
Union of events, 99–100
Univariate techniques, 42
Upper confidence level, 233
Used car prices (example), 547–550, 565–566

V
Variable, 16
Variance, 61–64
  computing via Excel, 63
  interpreting, 64
  laws of, 126
  population, 61
  population variance formula, 61, 86, 125, 142
  sample, 61
  sample variance formula, 61, 86
  See also Analysis of variance (ANOVA)
Variance of the difference between two means, formula, 226
Variance of the sample mean, formula, 225
Variance of the sample proportion, formula, 225

W
Web page for book, 7–8
Whiskers, 74
Wilcoxon rank sum test, 499–510
  Excel computation, 504
  factors that identify, 508
  and Kruskal-Wallis test, 527
  required conditions, 508
Wilcoxon rank sum test statistic, formula, 502, 536
Wilcoxon signed rank sum test, 516–521
  Excel computation, 520
  factors that identify, 521
  formula, 518, 536
Wilfred Laurier University graduate survey (case), 667–668
Within-treatment variation, 410

X
$\bar{X}$
  computing cumulative normal probabilities with Excel, 167–168
  defined, 119, 121
  expected value, 125

Y
y-intercept, formula, 81, 87

Z
Z
  computing cumulative normal probabilities with Excel, 167
  defined, 155
  finding values of, 161–165
z_A
  defined, 161
  and percentiles, 165–166
z-estimate of a mean, computing manually, 238
z-test and estimate: proportion
  factors that identify, 321
  via Excel, 314, 316
z-test and estimate: two proportions, Excel computation, 371–375, 394
z-test of a mean, via Excel, 267–268

LICENSING AND WARRANTY AGREEMENT

Notice to Users: Do not install or use the CD-ROM until you have read and agreed to this agreement. You will be bound by the terms of this agreement if you install or use the CD-ROM or otherwise signify acceptance of this agreement. If you do not agree to the terms contained in this agreement, do not install or use any portion of this CD-ROM.

License: The material in the CD-ROM (the "Software") is copyrighted and is protected by United States copyright laws and international treaty provisions. All rights are reserved to the respective copyright holders. No part of the Software may be reproduced, stored in a retrieval system, distributed (including but not limited to over the www/Internet), decompiled, reverse engineered, reconfigured, transmitted, or transcribed, in any form or by any means — electronic, mechanical, photocopying, recording, or otherwise — without the prior written permission of Brooks/Cole (the "Publisher"). Adopters of Keller's *Applied Statistics with Microsoft Excel* may place the Software on the adopting school's network during the specific period of adoption for classroom purposes only in support of that text. The Software may not under any circumstances, be reproduced and/or downloaded for sale. For further permission and information, contact Brooks/Cole, 511 Forest Lodge Road, Pacific Grove, California 93950.

U.S. Government Restricted Rights: The enclosed Software and associated documentation are provided with RESTRICTED RIGHTS. Use, duplication, or disclosure by the Government is subject to restrictions as set forth in subdivision(c)(1)(ii) of the Rights in Technical Data and Computer Software clause at DFARS 252.277.7013 for DoD contracts, paragraphs(c)(1) and (2) of the Commercial Computer Software-Restricted Rights clause in the FAR (48 CFR 52.227-19) for civilian agencies, or in other comparable agency clauses. The proprietor of the enclosed software and associated documentation is Brooks/Cole, 511 Forest Lodge, Pacific Grove, CA 93950.

Limited Warranty: The warranty for the media on which the Software is provided is for ninety (90) days from the original purchase and valid only if the packaging for the Software was purchased unopened. If, during that time, you find defects in the workmanship or material, the Publisher will replace the defective media. The Publisher provides no other warranties, expressed or implied, including the implied warranties of merchantability or fitness for a particular purpose, and shall not be liable for any damages, including direct, special, indirect, incidental, consequential, or otherwise.

For Technical Support:
Voice: 1-800-423-0563
Fax: 1-859-647-5045
E-mail: support@kdc.com